PROSTAGLANDIN ABSTRACTS

A Guide to the Literature
Volume 2: 1971-1973

PROSTAGLANDIN ABSTRACTS

A Guide to the Literature
Volume 2: 1971-1973

Ronald A. Shalita

Population Information Program
Department of Medical and Public Affairs
The George Washington University Medical Center
Washington, D. C.

SPRINGER SCIENCE+BUSINESS MEDIA, LLC

Library of Congress Cataloging in Publication Data

Sparks, Richard M
  Prostaglandin abstracts.

  Abstracts prepared by authors of articles or members of the Science Communication Division staff, George Washington University Medical Center.
  Vol. 2 edited by Ronald A. Shalita.
  CONTENTS: v. 1 1906-1970–v. 2. 1971-1973.
  1. Prostaglandin–Abstracts. 2. Prostaglandin–Bibliography. I. Shalita, Ronald A. II. George Washington University, Washington, D.C. Medical Center. Science Communication Division. III. Title. [DNLM: 1. Prostaglandin–Abstracts. ZQU90 S736p]
QP801.P68S65   574.1'9247   73-21780

ISBN 978-1-4684-6158-9   ISBN 978-1-4684-6156-5 (eBook)
DOI 10.1007/978-1-4684-6156-5

The preparation of this volume was supported by the United States Agency for International Development through a contract with the George Washington University.

CONTENTS

ACKNOWLEDGEMENTS

The editor expresses his gratitude to those who have contributed to the completion and success of this project, including the Upjohn Company of Kalamazoo, Michigan, and particularly Dr. Udo Axen; Informatics Inc. of Rockville, Maryland; Mary Janet Normandy for technical advice; and the staff of the Population Information Program, especially Dr. Phyllis Piotrow, Project Director, and Helen Kolbe, Senior Information Specialist, for their guidance and support, abstractors Leah Jeanne Gail and Harvey Chernov, and editorial assistant Rosanne Sadosky.

FOREWORD

Prostaglandin Abstracts Volume II contains abstracts of the significant scientific literature published in 1971, 1972, and 1973 on the role of prostaglandins in human reproduction. Of the more than 4000 articles which were published between 1971 and 1973 and deal with prostaglandins, endogenous and exogenous, in males and females, nearly 1100 involve human reproduction. *Prostaglandin Abstracts*, Volume I (Plenum Publishers, 1973) contains abstracts of all articles on prostaglandins published between 1906, when the action of these compounds was first noted, and 1970, when clinical research with these compounds was well underway.

Clinical use of prostaglandins during the 1960s proved safe but provided evidence of troublesome side effects. During the 1970s, the development of new analogues, various protocols for international testing, and improved routes of administration indicates that prostaglandins still hold promise as a useful drug in the management of fertility. Research into potential contraceptive and other clinical uses of prostaglandins continues to be an exciting and expanding field of fundamental and clinical research.

The citations and abstracts which are included in this volume have been selected and reproduced from POPINFORM, a computerized on-line information retrieval system which includes in its data base more than 25,000 documents on regulation of fertility. Also included as an Appendix is a bibliography (without abstracts) of 1971-1973 material on aspects of prostaglandin research that do not involve reproduction. This bibliography was prepared by the Upjohn Company of Kalamazoo, Michigan and is printed here with permission.

It is hoped that these two volumes will serve as a comprehensive source for interested researchers throughout the world. Certainly the critical role of prostaglandins in human reproduction and the increasing world concern over population problems warrant continued research and attention. The Office of Population, United States Agency for International Development, has supported the preparation of these two volumes by the Population Information Program of George Washington University so that scientists throughout the world and particularly those in developing countries who may not have ready access to scientific journals and reports can be fully informed of scientific and medical progress in this area.

March 1, 1975

R. T. Ravenholt, M.D., M.P.H.
Director

J. Joseph Speidel, M.D., M.P.H.
Chief, Research Division

Office of Population
United States Agency for International Development

EXPLANATORY NOTE

Every citation in this bibliography was verified with a copy on file in the library of the Population Information Program of the George Washington University Medical Center. If a satisfactory author's or prepared abstract accompanied the article, and was used, credit was given. Otherwise, abstracts were prepared by members of the Population Information Program staff. Those individuals, identified by their initials at the end of the abstracts are: Harvey Chernov, Ingrid Cruz, Leah Jeanne Gail, James R. Heath III, Margaret L. Hume, John S. Lee, Mary Janet Normandy, Ronald A. Shalita, Richard M. Sparks, Nancy Stillerman, Michael Towers, Arthur Turner, and George Wolfhard. If a prepared abstract was used with some alteration, it has been identified as "Author modified."

Since this volume is a guide to all aspects of prostaglandin research in reproductive physiology, abstracts have been prepared for all the types of literature—journal articles, reviews, books, book chapters, newspaper articles, abstracts, symposia—in which prostaglandins have been discussed. In the preparation of abstracts primary consideration has been given to the research significance of the article, that is, to the amount of new information contained in it. A comprehensive bibliography of prostaglandin citations in fields other than reproductive physiology follows the abstracts.

Citations are grouped chronologically by year and alphabetically by the surname of the senior author. Editorials, notes, and materials with no author indicated are listed under Anonymous, alphabetically by title.

0001
ANDERSON, G.G.; CORDERO, L.; HOBBINS, J.C.; SPEROFF, L.
> Clinical use of prostaglandins as oxytocin substances.
> In: Ramwell, P. and Shaw, J., eds. Prostaglandins. Annals of the New York Academy of Sciences 180: 499-512. 1971.

A study has been made of induced labor at term in 42 patients out of an anticipated group of 200 to 300. The purpose of the study was to assess the effect of $PGF_2\alpha$ and PGE_2 on the pregnant human uterus at term in relation to their possible roles in inducing labor and delivery of a normal fetus. Therefore, a double blind study was formulated, using $PGF_2\alpha$, PGE_2, and 'Syntocinon' in patients between 36 and 43 weeks gestation. The Bishop inducibility scoring system was used before starting intravenous infusions of any of the oxytocic substances; patients with a score of 6 or less were considered difficult inductions, whereas those scoring 7 to 13 were considered inducible. Primigravidas were deleted from the protocol. Apgar scores were recorded for the infants at 1 minute and 5 minutes after birth. All patients with a Bishop score of 7 or greater delivered no matter what drug was given, while only 4 out of 17 patients with a score of 6 or less delivered. No maternal or fetal complications were noted. Higher dosage rates were tried and results are described. An equal distribution among the 3 drugs was found for the development of moderate diuresis, while the form, frequency, and intensity of contractions were indistinguishable. The length of labor in the successful inductions cannot be meaningfully compared due to the small number of patients receiving each drug at various Bishop score levels. (ART)   002718

0002
ANONYMOUS
> How safe is abortion?
> Lancet 2: 1239-1244. 1971.

This editorial briefly mentions that prostaglandins have proven to be safer than abdominal hysterotomy in terminating midtrimester pregnancies. (JRH)   003147

0003
ANONYMOUS
> Prostaglandins and abortion.
> Lancet 2: 536. 1971.

This is a brief report of a March 1971 meeting of 39 workers with clinical experience in using prostaglandins, which was held in Stockholm under the auspices of the WHO Research and Training Center on Human Reproduction. It discussed the planning of multicenter trials and the monitoring of side effects of fertility-regulating agents and described clinical experiences in the use of prostaglandins in the induction of abortion. The paucity of side effects when PGE_2 or $PGF_2\alpha$ were administered by the intravaginal or intrauterine route was contrasted with their occurrence after intravenous administration. Confusion about how prostaglandins work in the human subject and the need for caution in the clinical application of prostaglandins were areas of general agreement at the conference. (ART)   002784

0004
ANONYMOUS
> Prostaglandins and spontaneous abortion.
> Clinical Medicine 78(8): 41. 1971.

It is mentioned that PGE_2 and $PGF_2\alpha$ are found in abnormally high amounts in human amniotic fluid and in venous blood of patients undergoing spontaneous abortion. The possible role of these prostaglandins in causing spontaneous abortion is mentioned. It is stated that final confirmation could come when prostaglandin antagonists become available. (JRH)   003327

0005
ANONYMOUS

Prostaglandins.
Reports on Population/Family Planning (1): 13-14. 1971.

PGE_2 and $PGF_2\alpha$ influence contractions of uterine and tubal musculature, but do not cause luteolysis in women. Therefore they are being tested as antigestational agents for postcoital use to interrupt normal zygote transport, for missed menstruation, and for abortion. Side effects and unsuitability of the oral route limit PGs to clinical or hospital settings for abortion where intravenous, intraamniotic, or transcervical intrauterine routes can be managed. Vaginal routes are being tested, but high frequency of failures and incomplete abortions will limit use of natural PGs. (LJG)   003379

0006
ANONYMOUS

Prostaglandins: a possible view of birth control of the future.
In: Contraceptive Technology 1971. Atlanta, Emory University Family Planning Program, 1971. p. 26.

The known PGs are 17 unsaturated hydroxy-acids having the properties of lipids. They were first discovered in human semen and occur there in high concentrations. They are active in the physiology of sperm transport, labor, luteolysis, and oviduct motility. PGs may have a future as once-a-month contraceptives and for inducing labor, treating dysmenorrhea, and aborting pregnancies up to 22 weeks of gestation. (LJG)   003397

0007
ANONYMOUS

The 'penicillin' of contraception may be here.
Medical Opinion 7: 66-71. 1971.

2 medical doctors discuss new contraceptive techniques for application to various countries. Although IUD and contraceptive pills were discussed, great importance was given to the prostaglandins for the use of postfertility and postcoital fertility control. Consideration is also given to the use of these compounds as menstrual regulators and as inducers of labor at will toward the end of gestation. (GW)   002610

0008
ANONYMOUS

The use of prostaglandins in obstetrics.
London Clinic Medical Journal 12: 9-11. January 1971.

In the nonpregnant woman PGEs inhibit contractions of the uterus, whereas by early pregnancy both PGEs and PGFs have been shown to stimulate contractions, their effectiveness varying at different stages of pregnancy. Injected intravenously, PGs have been used successfully to induce labor, $PGF_2\alpha$ being required in amounts 5-10 times as high as PGE_1 or PGE_2. Increasing the dose produces shorter latent intervals but also uterine hypertonus. In the doses necessary to induce labor no adverse side effects have been observed. Intravenous infusions of $PGF_2\alpha$ during early pregnancy are thought to prevent ovum implantation though the mechanism is not understood; the success rate for early therapeutic abortion is inversely related to the gestational interval. PGEs and PGFs are needed in amounts 10 times as great for abortion as for induction of labor. Side effects of nausea, vomiting, and diarrhea are reported, and occur in greater frequency with $PGF_2\alpha$ than with PGE_2. Intermittent injections between the fetal sac and uterine wall have been shown to induce abortion with doses 1/10 those required for intravenous injection; side effects are negligible. It has been reported also that PGs can be administered vaginally in the first and second trimesters to induce abortion. (MLH)   003507

0009
ARCHAMBAULT, G.F.
    Prostaglandins.
    Hospital Formulary Management 6: 22. January 1971.

    A brief history of prostaglandins and their various potential uses are discussed. It was suggested that prostaglandins for the induction of labor might be available by late 1971. (RAS)   003432

0010
BARRETT, S.; BLOCKEY, M.A. de B.; BROWN, J.M.; CUMMING, I.A.; GODING, J.R.; MOLE, B.J.; OBST, J.M.
    Initiation of the oestrous cycle in the ewe by infusions of $PGF_2\alpha$ to the autotransplanted ovary.
    Journal of Reproduction and Fertility 24: 136-137. 1971.

    Authors report on experiments with infusion of $PGF_2\alpha$ into the ovarian arterial circulation of sheep with ovarian autotransplants. In 1 experiment, secretion rates of progesterone fell abruptly in all animals when given PMSG on withdrawal of implants of progesterone and the commencement of $PGF_2\alpha$ infusion 7 days later. In another experiment, $PGF_2\alpha$ was infused into a ewe with a persistent corpora lutea with no pretreatment of progesterone or PMSG. $PGF_2\alpha$ again caused a sustained fall in the secretion of progesterone; this was followed by a rise in $PGF_2\alpha$ with a second peak 16 hours later. These experiments show that $PGF_2\alpha$ is capable of initiating a new cycle in sheep with persisting corpora lutea. (GW)   002763

0011
BEAZLEY, J.M.; GILLESPIE, A.
    Double-blind trial of prostaglandin E_2 and oxytocin in induction of labour.
    Lancet 1: 152-155. 1971.

    A double-blind trial of oxytocin (Syntocinon) and prostaglandin E_2 in the induction of labor in 300 patients resulted in equally high success rates for both drugs. 73% of patients in each group were delivered or achieved 6 cm. dilatation of the cervix within 12 hours of the start of the infusion. The patients were not matched for factors known to affect labor, and so no statistical analysis of the results was attempted. In the successful cases there were no maternal complications of treatment. (Authors)   002952

4

0012
BEAZLEY, J.M.
  The induction of labour with prostaglandins.
  Research in Prostaglandins 1(2): 1-2. 1971.

This review article is focused on the use of the prostaglandins in the induction of labor at or near term as reported in 15 papers by research scientists and their coworkers. It is noted that the first report of the use of a prostaglandin ($PGF_2\alpha$) to induce labor in 10 women at or near term was published in 1968. The relative activity or effects on the human myometrium of prostaglandins of the E, F, A, and B series are discussed. Threshold doses, routes of administration, stimulation of contractions, and uterine hypotonicity are discussed. Side effects of the E and F prostaglandins are briefly mentioned. (ART)   002828

0013
BEDWANI, J.R.; MARLEY, P.B.
  Effect of ovariectomy and of various hormones on the synthesis of progesterone by guinea-pig
    placenta in vitro.
  Journal of Reproduction and Fertility 26: 343-349. 1971.

Placental slices from intact and ovariectomized guinea pigs, incubated in Krebs-Ringer solution containing nicotinamide and NAD, synthesized no progesterone de novo but were able to convert pregnenolone to progesterone. Placental slices from animals ovariectomized 6 days before the incubation synthesized twice as much progesterone from pregnenolone ($16 \mu g/g/3$ hr) as slices from intact animals ($7 \mu g/g/3$ hr, $p<.01$). This suggests that the 3-beta-hydroxysteroid dehydrogenase activity of the placenta increases after ovariectomy. Cyclic AMP significantly reduced the conversion of pregnenolone to progesterone ($p<.02$), whereas gonadotrophic hormones (LH, or a combination of HCG and PMSG) and prostaglandins E_1 and $F_2\alpha$ had no effect in vitro. (Authors' modified)   003071

0014
BEDWANI, J.R.; HORTON, E.W.
  Interaction between prostaglandins and gonadotrophins in the rabbit ovary.
  British Journal of Pharmacology 43: 794-803. 1971.

In vitro studies of the interaction between prostaglandins and gonadotropins were performed using minced ovaries from pseudopregnant rabbits incubated with or without LH or cAMP and various PG concentrations. In addition, plasma samples were obtained from the ovarian veins of anesthetized pseudopregnant rabbits before and after the injection of LH. Progesterone and PGs were extracted from ovarian and plasma samples and isolated by TLC. Progesterone was quantitated spectrophotometrically; PGs were assayed biologically. PGE_1 $1-100 \mu g/ml$, but not PGE_2, inhibited the production of progesterone stimulated by LH. PGE_1 had no effect on progesterone production stimulated by cAMP. LH did not affect the concentration of PGE_1, PGE_2, or $PGF_2\alpha$ found in rabbit ovaries. Similarly, in the in vivo experiments LH stimulated the release of progesterone, but there was no evidence of PG in the effluent. The authors concluded that it is unlikely that PGE has a physiological role in the regulation of steroidogenesis. Further, they do not support the hypothesis that PG functions as modulators of hormonal actions mediated by cAMP. (HC)   003446

0015
BEHRMAN, H.; YOSHINAGA, K.; GREEP, R.
   Extraluteal effects of prostaglandins.
   In: Ramwell, P. and Shaw, J., eds. Prostaglandins. Annals of the New York Academy of Sciences 180:
      426-435. 1971.

This report concerns experiments designed to test the very early effects of $PGF_2\alpha$ on secretion of progesterone and 20-alpha-dihydroprogesterone (20-alpha-ol) and to compare these effects with luteinizing hormone (LH) similarly administered (via the femoral vein) in normal and hypophysectomized, pseudopregnant rats. Ovarian venous blood samples were collected immediately after administration of 20 mg of LH and 10 mg of $PGF_2\alpha$, alone or in combination. Progesterone and 20-alpha-ol were separated from a lipid extract from the blood samples by thin layer chromatography, identified, isolated, then quantitated by gas-liquid chromatogrpahy. The secretion rate of progesterone and 20-alpha-ol in intact animals is shown in a table. LH was found to have no stimulatory effect on either steroid in intact animals. $PGF_2\alpha$ produced a marked decrease in progesterone secretion but little change in 20-alpha-ol secretion. When LH was administered simultaneously with $PGF_2\alpha$, progesterone secretion was normal but 20-alpha-ol secretion was increased significantly. Hypophysectomy of itself produced a significant decline in progesterone secretion compared with the intact control rate, but not in 20-alpha-ol secretion. LH administration to hypophysectomized animals prevented the decrease in progesterone secretion but increased 20-alpha-ol secretion over that observed after hypophysectomy alone. $PGF_2\alpha$ produced a small but significant increase in progesterone secretion in hypophysectomized animals, but a more dramatic increase in 20-alpha-ol secretion was observed. Simultaneous administration of LH and $PGF_2\alpha$ to hypophysectomized animals produced an effect that was lower than LH alone. Ovarian venous flow rates are shown in a table. The significance of the observed data is discussed. (ART)   002712

0016
BEHRMAN, H.R.; YOSHINAGA, K.; WYMAN, H.; GREEP, R.O.
   Effects of prostaglandin on ovarian steroid secretion and biosynthesis during pregnancy.
   American Journal of Physiology 221: 189-193. 1971.

The corpus luteum of the rat synthesizes 2 major progestins--progesterone and 20alpha-hydroxy-pregn-4-en-3-one (20alpha-ol)--but the latter compound has little progestational activity. The present study was conducted 1) to measure ovarian venous progestin secretion at various times after a single injection of $PGF_2\alpha$ (as a means of determining the sequence of events induced by $PGF_2\alpha$ prior to the abortive response) and 2) to learn something of the mechanism of action of this compound. The progestin output was also measured from incubated ovarian slices in order to directly compare the effects of prostaglandin with an in vivo and an in vitro system. After injection of $PGF_2\alpha$ in vivo, progesterone secretion decreased more than 50% within 6 hours and more than 80% after 12 hours. 20alpha-ol secretion increased approximately 2.8-fold after 6 hours and about 3.7-fold 12 hours after a single administration of $PGF_2\alpha$. However, progestin and estrogen secretion did not change significantly, although there was an increase in the latter at 6 hours. Prostaglandins are cleared very rapidly from the circulatory system, and the continued effect produced by a single injection indicates that $PGF_2\alpha$ may be bound in some manner to (or in) the luteal cell and/or initiates a response that is self-generating. (ART)   002798

0017

BEHRMAN, H.R.; MACDONALD, G.J.; GREEP, R.O.

Regulation of ovarian cholesterol esters: evidence for the enzymatic sites of prostaglandin-induced loss of corpus luteum function.

Lipids 6(11): 791-796. 1971.

The effect of $PGF_2\alpha$ on corpus luteum function was investigated. Female rats received Human Chorionic Gonadotrophin (HCG) to induce highly luteinized ovaries and 3 days later hypophysectomy was carried out; treatments followed immediately. $PGF_2\alpha$ was administered subcutaneously at a dosage of .5 mg/kg b.i.d., prolactin at 2 IU b.i.d. and luteinizing hormone (LH) at 10 ug b.i.d., all in 15% gelatin. Ovarian tissue was pooled from animals within each treatment group, one portion from each for incubation of the tissue and another portion for assay of cholesterol esterase and cholesterol synthetase activity. $PGF_2\alpha$ was shown to produce a 75% decrease in the ovarian content of cholesterol esters in animals with an intact pituitary, an effect equivalent to that produced by hypophysectomy alone. Prolactin treatment maintained cholesterol esters at twice the level seen in the hypophysectomized control group. When $PGF_2\alpha$ was administered with prolactin, a loss in the trophic action of prolactin to maintain ovarian cholesterol esters occurred. Simultaneous administration of LH with prolactin did not prevent this action of $PGF_2\alpha$ in hypophysectomized animals. The specific activity of cholesterol ester synthetase and cholesterol esterase was assayed by identical treatments. $PGF_2\alpha$ depressed synthetase activity by 75% when administered to intact animals; hypophysectomy reduced activity to a similar extent but the addition of prolactin maintained activity. Prolactin alone was more effective than were LH and prolactin administered together. Effects on esterase activity were analagous but not as severe as those observed on synthetase activity. The effect of $PGF_2\alpha$ on progesterone output can be directly correlated with the effect on enzyme activities. Following these experiments, it was suggested that $PGF_2\alpha$ acted by producing a lesion in the luteal cell which prevented adequate storage of cholesterol esters and impaired the ability of the corpus luteum to produce progesterone. (MLH)　003488

0018

BEHRMAN, H.R.; GREEP, R.O.

Sites of prostaglandin-induced lesions in the corpus luteum.

Biology of Reproduction 5: 86. 1971.

The present study was conducted to provide information on the mechanism of luteolysis induced by prostaglandin. Immature, 30 day-old rats (22) were treated with 50 IU PMS followed 60-65 hours later with 25 IU HCG. Hypophysectomy (APX) was performed 3 days later, and these animals were treated b.i.d. for 2 days with either prolactin (NIH-S9; 2 IU), $PGF_2\alpha$ (.5 mg/kg), or prolactin + $PGF_2\alpha$. Intact animals were treated with $PGF_2\alpha$ and control groups (intact and APX) received no treatments. In incubated tissues progesterone levels decreased from 50.2 ± 5.9 to 28.3 ± 5.2 $\mu g/g$ with $PGF_2\alpha$ in intact animals, and $12.6 \pm .7$ $\mu g/g$ with APX alone. In the APX + prolactin group, progesterone was maintained at 37.2 ± 4.2 $\mu g/g$, but $PGF_2\alpha$ prevented this action of prolactin ($13.2 \pm .9$ $\mu g/g$). Cholesterol ester synthetase and cholesterol esterase activities were reduced 67% and 40%, respectively, by $PGF_2\alpha$ in intact animals, and 81% and 77% by APX, but prolactin maintained normal enzyme activites after APX. $PGF_2\alpha$ prevented maintenance of synthetase but not esterase activity with prolactin. Tissue levels of cholesterol ester were reduced 62% by $PGF_2\alpha$ in intact animals and 80% by APX. Prolactin increased cholesterol ester levels (threefold) in APX animals, but $PGF_2\alpha$ prevented the luteotrophic expression of prolactin, causing a lesion in enzymes regulating cholesterol ester turnover and steroidogenesis. (Authors' modified)　003052

0019

BERGSTROM, S.; SPEROFF, L.

Further clinical studies with prostaglandins in reproductive physiology.

In: Ramwell, P. and Shaw, J., eds. Prostaglandins. Annals of the New York Academy of Sciences 180: 553-568. 1971.

This is an account of the general discussion by 19 participants, including questions, answers, comments, and brief presentations of new data. The general unifying theme of the discussion is clinical experience with the use of prostaglandins, particularly PGE_2 and $PGF_2\alpha$. Side effects are discussed at some length, including nausea, vomiting, diarrhea, venous erythema, pyrexia, headache, blurring of vision, and tachycardia. Criteria, or lack of them, are discussed by which success or failure can be determined in utilizing prostaglandin to terminate pregnancy. The expulsion rates for intrauterine devices (IUDs) in different parts of the world are discussed. Need for further research is emphasized. (ART)   002724

0020

BERGSTROM, S.; BYGDEMAN, M.; SAMUELSSON, B.; WIQVIST, N.

The prostaglandins and human reproduction.

Hospital Practice 6(2): 51-57. 1971.

The history of prostaglandin research, structure, occurrence, metabolic effects, relationship to cAMP, use in pregnancy termination, and side effects are briefly reviewed. (RMS)   003296

0021

BLAND, K.P.; HORTON, E.W.; POYSER, N.L.

Levels of prostaglandin $F_2\alpha$ in the uterine venous blood of sheep during the oestrous cycle.

Life Sciences (I)10: 509-517. 1971.

Concentration of prostaglandin $F_2\alpha$ in sheep uterine venous blood was monitored by silicic acid column chromatography of blood samples drawn by laparotomy on selected days of the estrous cycle. Results show that $PGF_2\alpha$ is present toward the end of the cycle on Days 14, 15, and 16 at levels of 3.3 to 8.0 ng/ml. No prostaglandin $F_2\alpha$ was detected in blood from the jugular vein of 1 sheep on Day 15. These results support the hypothesis that the uterine substance which suppresses the corpus luteum via local circulation on Day 15 of the estrous cycle is prostaglandin $F_2\alpha$. (RMS)   002727

0022

BLATCHLEY, F.R.; DONOVAN, B.T.; POYSER, N.L.; HORTON, E.W.; THOMPSON, C.J.; LOS, M.

Identification of prostaglandin $F_2\alpha$ in the utero-ovarian blood of guinea-pig after treatment with oestrogen.

Nature New Biology 230: 243-244. 1971.

Guinea pigs were injected subcutaneously with 10 μg of estradiol benzoate per day from Days 4-6 of the cycle. On Day 7, each was anesthetized with pentobarbitone sodium and injected with 5,000 IU of heparin intravenously. The utero-ovarian vein on one side was exposed. Blood was collected and centrifuged. Prostaglandins were extracted by solvent partition and partially purified by silicic acid column chromatography. Bioassay of the eluted fractions on the rat fundal strip detected PGF-like activity in the plasma extracts from intact guinea pigs treated with estrogens but not in the controls, nor in an estrogen treated hysterectomized group. None of the samples contained any PGE-like activity. The methylester-trimethylsilyl ether derivative of the extracted material was prepared. Gas chromatography gave a peak retention time corresponding to this derivative of authentic $PGF_2\alpha$. Mass spectra showed all the principal m/e peaks of $PGF_2\alpha$. Plasma from the other groups was also investigated by these methods and did not contain $PGF_2\alpha$. In 6 out of 8 experiments, bioassay showed more $PGF_2\alpha$ in the intact animals pretreated with estrogen than in controls. That this secretion of $PGF_2\alpha$ seems to be favored by steroid satisfactorily accounts for the luteolytic action of estrogen in the intact guinea pig. (MT)   002987

0023
BODIN, M.
Prostaglandins and habitual abortion.
British Medical Journal 2(5761): 587. 1971.

The author hypothesizes that the habitual aborter (excluding those in whom other possible causes have been suspected or identified) may be the woman who is hypersensitive to any form of mechanical stimuli which are capable of causing spontaneous release of prostaglandins in pregnancy, especially early pregnancy. For example, mild trauma, such as a blow or fall of so slight a degree that it might not be recognized, remembered, or associated as possibly significant, or the abdominal and/or vaginal/cervical stimuli of intercourse in early pregnancy, perhaps even before pregnancy had been suspected or diagnosed are possible causes. Such women might be identified by prostaglandin sensitivity tests and appropriate clinical management can then be devised. (ART)   002740

0024
BRAAKSMA, J.T.; BRENNER, W.E.; FISHBURNE, J.I. Jr.; STAUROVSKY, L.
Intrauterine extra-amniotic administration of prostaglandin $F_2\alpha$ for therapeutic abortion: early myometrial effects.
American Journal of Obstetrics and Gynecology 114(4): 511-515. October 15, 1971.

Prostaglandin ($PGF_2\alpha$) was administered extraamniotically to abort 7 women, 8-20 weeks pregnant, and the intraamniotic pressure was recorded at various dosages, dose sequences, and at 3 injection sites. $PGF_2\alpha$ was given through a calibrated polyethlene catheter that had been inserted intravaginally within a ureteral catheter, subsequently removed. Catheters were left in place in the fundus (only 1 site for 5 patients), extraamniotically halfway between the fundus and cervix, and 2 cm above the internal cervical os (all 3 sites for 2 patients). $PGF_2\alpha$ was administered at doses of 5 μg/ml to 2 mg/ml in 1 cc volumes. Intraamniotic pressure and maternal radial artery pressure and pulse were recorded directly. 6 of 7 women aborted in a mean of 9.4 hours, after receiving a total dose of 1.2 to 38.5 mg $PGF_2\alpha$. Other than nausea and vomiting, and severe vomiting and pain in 2 women who were given large initial doses, no side effects and no changes in blood pressure or pulse were observed. In the 2 patients who received doses of 5, 10, 20, 40, 100, and 200 μg PG at all 3 sites, the uterus responded with the greatest basal tone from injection at the deep fundal site. In those given 5, 10, 20, 50, 100, and 200, μg PG at the deep fundal site only, patients in earlier pregnancy responded with a greater increase in basal tone than did those in later pregnancy. Patients 13 weeks pregnant or less responded to 50 μg; a patient 17 weeks pregnant responded to 200 μg; but a patient 20 weeks pregnant failed to respond to 200 μg PG with any increase in uterine tone. (LJG)   004259

0025
BRUMMER, H.C.
Interaction of E prostaglandins and Syntocinon on the pregnant human myometrium.
Journal of Obstetrics and Gynaecology of the British Commonwealth 78: 305-309. 1971.

The possibility that ketonic prostaglandins (PGEs) and oxytocin (Syntocinon, Sandoz) may be used to supplement each other on the pregnant human uterus was investigated in vitro. The enhancement effect of E prostaglandins on the myometrial response to oxytocin is described both for midtrimester and term myometrial strips. An hypothesis regarding the role of prostaglandins in labor is put forward and discussed. The application of this sensitizing effect to the induction of midtrimester abortion is also considered. (Author's modified)   002675

0026
BRUNTON, W.J.
    Ion transport by isolated rhesus monkey oviduct and stimulation by human seminal plasma and
        prostaglandin E_1.
    Biology of Reproduction 5: 105. 1971.

In order to obtain information on the secretory process of the rhesus monkey oviduct, the behavior of an in vitro preparation of this tissue has been studied. The ampulla of the oviduct, mounted as a sheet in a double plastic chamber, maintains steady transmembrane potential (4.8 $\pm$ .9 mv, lumen negative) and short-circuit current (47 $\pm$ 4 μa/sq cm for many hours. Bidirectional ion flux across the short-circulated oviduct was measured using radiosotopes. With identical media on both sides of the membrane and the transmembrane potential reduced to zero, there was a net flux of chloride into the lumen and a net flux of of sodium in the opposite direction. Human seminal plasma (1 μl/ml) or prostaglandin E_1(1 μg/ml) increased PD and SCC. The results indicate that the isolated oviduct of the rhesus monkey actively transports chloride and sodium and that human seminal plasma or prostaglandin E_1has a marked effect on the electrical parameters of this preparation. (Author's modified)   003054

0027
BYGDEMAN, M.; WIQVIST, N.
    Early abortion in the human.
    In: Ramwell, P. and Shaw, J., eds. Prostaglandins, Annals of the New York Academy of Sciences 180:
        473-482. 1971.

A brief history is given of the events concerning discovery, identification, and synthesis of the prostaglandins, and of some of the research on their effects in animals and human subjects. Clinical usage of $PGF_2\alpha$, PGE_1, and PGE_2as abortifacients is reported. Side effects in terms of nausea and diarrhea have been reported after systematic administration of $PGF_2\alpha$ by many investigators. Studies of clinical cases reported here indicate that the dose range of $PGF_2\alpha$ (given intravenously) which stimulates effective contractions (indicated by dysmenorrheic pain), but does not cause side effects (nausea and diarrhea), is rather limited. Intrauterine administration of PGE_2or $PGF_2\alpha$ between the fetal membranes and the uterine wall resulted in sustained, intensive contractions generally lasting for several hours. Abortion was accomplished in all 12 cases reported, and the dose required was around 1/10 of that required for iv administration. General side effects were eliminated by this method. The efficiency of the prostaglandins as abortifacients is related to the stage of gestation, being more effective when given during the first 8 weeks of pregnancy. (ART)   002716

0028
BYGDEMAN, M.; TOPPOZADA, M.; WIQVIST, N.
    Induction of midtrimester abortion by intra-amniotic administration of prostaglandin $F_2\alpha$. A
        preliminary report.
    Acta Physiologica Scandinavica 82: 415-416. 1971.

The aim of this study was to investigate a local route of administration of prostaglandins by intraamniotic injection of PGF$_2\alpha$ through a polyethylene catheter introduced into the amniotic cavity by abdominal puncture. This route of administration was used in 9 women in Weeks 14 to 20 of gestation and avoided the risk of intravenous leakage and local erythema at the site of venepuncture. The uterus responded within 60 minutes to separate doses of PGF$_2\alpha$, ranging between 5 and 15 mg, at intervals of 3 to 14 hours, depending on the contractile response. Abortion was induced in all 9 women; 4 were complete and 5 retained the placenta after expulsion of the fetus. No generalized side effects were observed except for 2 patients who vomited. It is believed that PGF$_2\alpha$ acts locally on the myometrium after diffusion through the fetal membranes. (ART)   002779

0029
BYGDEMAN, M.; WIQVIST, N.
Prostaglandins -- a new group of abortifacient substances.
In: Sobrero, A.J. and Harvey, R.M., eds. Advances in planned parenthood -- Vol. 7. [Proceedings of the ninth annual meeting of the American Association of Planned Parenthood Physicians, Kansas City, Missouri, April 5-6, 1971.] Princeton, Excerpta Medica, 1972. p. 139-145.

The brief history from 1966-1970 of PGs as abortifacients is told, including comparative potency, results, and side effects of intravenous and intrauterine application of PGs for early and midtrimester abortion. By repeated single iv injections with uterine tone as the endpoint, the potency of PGF$_2\alpha$ was 8 times less and of PGF$_1\alpha$ 40 times less than PGE$_1$ and PGE$_2$. In the first trials 91% of pregnancies up to 8 weeks were terminated (negative pregnancy test), but only 11%-32% of later pregnancies were aborted (expulsion of fetus) by infusing PGF$_2\alpha$ iv, increased stepwise, with a mean dose of 76.9-89.5 μg/minute. Another trial compared a group of 34 women given 25 μg/minute PGF$_2\alpha$ increased up to 100 μg, with a group of 23 given 50 μg increased by 25 μg/hour up to 150 μg/minute. The mean doses were 54.1 and 108.1 μg/minute, and success rates were 18% and 83%. Side effects of vomiting and diarrhea were increased in proportion to about 1 episode per hour in the high-dose group. In comparison, the success rate in Uganda was 88% with 50 μg/minute, and 53% overall among 13 trials collected by the WHO early in 1971. In contrast to the iv route, intrauterine injections produced a more gradual onset but sustained and intense contractions, except in cases where systemic leakage had apparently occurred. Using .2-1 mg PGF$_2\alpha$ or 20-100 μg PGE$_2$ (about 1/10 the iv dose), success rates were 78%-80% in 19 women pregnant 6-12 weeks and 57% in 4 women pregnant 13-17 weeks. Only 23% experienced vomiting and none had diarrhea. (LJG)   004495

0030
CALDWELL, B.V.; MOOR, R.M.
Further studies on the role of the uterus in the regulation of corpus luteum function in sheep.
Journal of Reproduction and Fertility 26: 133-135. 1971.

The concept that uterine plasma acquires selective lytic properties around the critical 14th day of the estrous cycle in ewes is supported by the observation that the corpus luteum is rapidly destroyed by the infusion of 14-day uterine plasma but is not affected by similar infusions of either 8-day uterine plasma or 14-day plasma obtained from the jugular vein. It has been suggested that the lytic substance may be carried to the ovaries in sheep by means of a 'countercurrent' exchange between the uterine vein and ovarian artery which run along together in this species. It has been shown that merely separating the ovarian artery from the uterine vein and placing a piece of omentum between the 2 vessels results in prolonged luteal function. The recently reported effects of prostaglandins and their possible association with the mechanisms for regulating luteal function make these compounds of considerable interest in this connection. (ART)   002789

0031
CARPENTER, M.P.
The lipid composition of maturing rat testis. The effect of alphatocopherol.
Biochimica et Biophysica Acta 231: 52-79. 1971.

Reference to prostaglandins is very short. The author notes on page 76 that known pathways for the metabolism of arachidonic acid include conversion to PGE_2 and $PGF_2\alpha$. Literature is cited on synthesis of PGs in the rat testis. (RMS)   002732

0032
CASILLAS, E.R.; HOSKINS, D.D.
Adenyl cyclase activity and cyclic 3', 5'-AMP content of ejaculated monkey spermatozoa.
Archives of Biochemistry and Biophysics 147: 148-155. 1971.

The effects of hormones, PGs, and manganese ion were assessed on cAMP release by ejaculated spermatozoa of rhesus monkeys, and intracellular cAMP content was measured after exposure to some of these hormones. Accumulation of cAMP, measured with carbon-14-ATP, was increased 1.54-fold by thyroxine 1 μg/ml and 1.90-fold by triiodothyronine 1 μg/ml to a range of 42-72 pM per billion sperm per 20 minutes. cAMP release was not changed substantially by gonadotropins, androgens, estrogens, catecholamines, PGE_1, or PGE_2. Manganese ion .5-20 mM stimulated cAMP release from 2- to 50-fold and increased anaerobic fructolysis and motility. Mean intracellular cAMP, measured enzymatically, was 103 pM per billion sperm; level of intracellular cAMP was not altered by hormones, but was inhibited 40% by manganese ion without further effect by added hormones. (LJG)   003439

0033
CHANNING, C.P.
Prostaglandin stimulation of luteinization of rhesus monkey granulosa cell cultures.
In: Abstracts, 4th Annual Meeting of the Society for the Study of Reproduction, Boston, Massachusetts, June 29-July 1, 1971. (Abstract 15)

Various prostaglandins (PG) obtained from the Upjohn Co. were added at each medium change to cultures of granulosa cells (GC) harvested from 3-5 mm follicles of human menopausal gonadotrophin (HMG) treated and normal monkeys. Cell morphology was monitored by phase contrast microscopy and progestin secretion estimated by protein binding assay. GC harvested from HMG-treated monkeys did not luteinize in culture and resembled GC harvested from 3-5 mm medium-sized follicles of untreated monkeys. In 6 experiments .1 or 1.0 μg/ml PGE_1 and PGE_2 brought about morphological luteinization and a 10-1000-fold stimulation in progestin secretion in a manner similar to dibutryl cyclic-3',5'-AMP or human LH. PGA_1 was about 10 times less effective compared to PGE_2, and $PGF_2\alpha$ had little or no effect at a dose of l μg/ml. Under no condition did any of the prostaglandins exert a lytic effect upon the GC cultures. This direct stimulatory action of PG upon GC contrasts to a luteolytic action of PG observed by others when the PG are administered in vivo. Most likely the in vivo luteolytic effect is mediated by an indirect action. (Author)   003006

0034
CHANNING, C.P.
Prostaglandin stimulation of luteinization of rhesus monkey granulosa cell cultures.
Biology of Reproduction 5: 87. 1971.

Various prostaglandins obtained from the Upjohn Company were added at each medium change to cultures of granulosa cell (GC) harvested from 3-5 mm follicles of human menopausal gonadotrophin (HMG)-treated and normal monkeys. Cell morphology was monitered by phase-contrast microscopy and progestin secretion estimated by protein-binding assay. GC harvested from HMG-treated monkeys did not luteinize in culture and resembled GC harvested from 3-5 mm medium-sized follicles of untreated monkeys. In 6 experiments .1 or 1.0 μg/ml PGE_1 and PGE_2 brought about morphological luteinization and a 10-1000-fold stimulation in progestin secretion in a manner similar to dibutryl cyclic-3',5'-AMP or human LH. PGA_1 was about 10 times less effective than PGE_2, and $PGF_2\alpha$ had little or no effect at a dose of 1 μg/ml. Under no condition did any of the prostaglandins exert a lytic effect upon the GC cultures. This direct stimulatory action of PG upon GC contrasts to a luteolytic action of PG observed by others when the PGs are administered in vivo. Most likely the in vivo luteolytic effect is mediated by an indirect action. (Author's modified)   003055

0035
CHAUDHURI, G.
   Intrauterine device: possible role of prostaglandins.
   Lancet 1: 480. 1971.

   A hypothesis is presented to explain the mechanism of action of the IUD. It is suggested that the IUD, by a mild traumatic action, releases a prostaglandin and that it is this substance which prevents conception. (Author)   002618

0036
COLLIER, J.G.; FLOWER, R.J.
   Effect of aspirin on human seminal prostaglandins.
   Lancet 2: 852-853. 1971.

   The concentration of prostaglandins was measured in the semen of healthy subjects before, during, and in 1 subject after a 7-day course of oral aspirin. Prostaglandins were extracted by the method of Bygdeman and Samuelsson (1966). In each subject the volumes of semen were unaffected by aspirin. When aspirin was taken there was a fall in concentration of both E and F type prostaglandins in all trials. For PGE, the concentration in the first sample after aspirin was significantly lower than that of the mean of the PGE controls, but the reduction was not completely maintained. With PGF, the fall was maintained: The means of the PGF concentrations during aspirin were significantly lower than the means of the controls. In the 1 subject in whom the procedure was repeated, the PGF concentration had returned to normal by 14 days after stopping aspirin, and the reduction in prostaglandin concentration induced by aspirin was similar on both occasions. (MT)   003073

0037
COUTINHO, E.M.
   Physiologic and pharmacologic studies of the human oviduct.
   Fertility and Sterility 22(12): 807-815. December 1971.

   The author developed a technique of recording tubal motility in women with an indwelling catheter and pressure transducer. The human oviduct exhibits bursts of contractions approximately every hour, more often in menstruation, more intense and lasting in proliferative phase but less so in luteal phase. The contractions are asynchronous with the other tube, other parts of the same tube, and the uterus, and are independent of pituitary control. Contractions are stimulated by catecholamines, oxytocin, alkaloids such as ergonovine, and PGFs, but inhibited by alpha-adrenergic antagonists and PGEs. (LJG)   003442

0038

COUTINHO, E.M.; MAIA, H.S.

The contractile response of the human uterus, fallopian tubes, and ovary to prostaglandins in vivo. Fertility and Sterility 22: 539-543. 1971.

Both prostaglandins $F_2\alpha$ and E_2 stimulate the nonpregnant human uterus in vivo. On a weight basis, PGE_2 appears more potent than $PGF_2\alpha$ on the myometrium. The 2 prostaglandins have opposing effects on the fallopian tube. $PGF_2\alpha$ stimulates, whereas PGE_2 inhibits, tubal motility in vivo. A contractile response of the ovary is recorded following intravenous administration of $PGF_2\alpha$. It is proposed that relaxation of the tubal isthmus induced by seminal PGE_2 is a prerequisite for sperm penetration into the tube. (Authors)   002864

0039

COUTINHO, E.M.

Tubal and uterine motility.

In: Diczfalusy, E., and Borell, U., eds. Nobel Symposium 15, Control of human fertility. Stockholm, Almqvist and Wiksell, 1971. p.97-115.

The author discusses the effects of various hormones and pharmacological agents on uterine and tubal motility in human patients in vivo. The effects of prostaglandins are mentioned briefly. PGE compounds were found to be inhibitory and PGF compounds stimulatory. Prostaglandins are most effective during the luteal phase of the menstrual cycle. (JRH)   003221

0040

CRAFT, I.L.; CULLUM, A.R.; MAY, D.T.L.; NOBLE, A.D.; THOMAS, D.J.

Prostaglandin E_2 compared with oxytocin for the induction of labour.

British Medical Journal 3: 276-279. 1971.

A comparison has been made between the effectiveness of infusing prostaglandin E_2 with Syntocinon for the induction of labor in the presence of intact membranes. Rapid titration schedules were used to induce an early uterine response. All 15 subjects receiving prostaglandin E_2 achieved cervical dilatation, whereas this occurred in only 9 out of 15 patients receiving Syntocinon. (Authors' modified)   002974

0041

CSAPO, A.I.; SAUVAGE, J.P.; WIEST, W.G.

The efficacy and acceptability of intravenously administered prostaglandin $F_2\alpha$ as an abortifacient. American Journal of Obstetrics and Gynecology 111: 1059-1063. 1971.

The efficacy and acceptability of prostaglandin $F_2\alpha$ ($PGF_2\alpha$) as an abortifacient was examined during intravenous infusions in 10 obstetrically normal midtrimester patients. Infusions of up to 200 μg per minute of $PGF_2\alpha$ induced complete abortions in 3 and incomplete abortions in an additional 3 patients. 4 patients failed to abort. All patients experienced undesirable side effects, i.e., nausea, vomiting, and diarrhea, of various degrees. This form of therapy cannot be recommended for routine therapeutic use; however, current clinical studies complemented by model experiments in animals suggest that the efficacy and acceptability of $PGF_2\alpha$ can be increased by topical intrauterine administration. Preliminary deductions pertinent to the mechanism of the abortifacient effect of $PGF_2\alpha$ have been made, based on measurements of intraamniotic pressure, circulating estradiol-17β and progesterone levels. (Authors' modified)   003313

0042

DAVID, H.P.

Abortion: public health concerns and needed psychosocial research.
American Journal of Public Health 61: 510-516. 1971.

Prostaglandins are mentioned as having promise of being the ideal abortifacient which can be self-administered as a suppository without toxic side effects. (JRH)   003358

0043

DE CUMINSKY, B.S.; MERCURI, O.

The identification of prostaglandin E_1 in rat seminal vesicle gland.
Lipids 6: 278-280. 1971.

The isolation of PGE_1 by solvent partition and thin-layer chromatography procedures from rat vesicle gland is outlined. Biological assay and spectrophotometric quantitative determinations are described. (RMS)   002734

0044

DEIS, R.P.

Induction of lactogenesis and abortion by prostaglandin $F_2\alpha$ in pregnant rats.
Nature 229: 568. 1971.

The regression of the corpus lutuem in rats as a possible response to $PGF_2\alpha$ was ivestigated in 8 pregnant rats. Each received 4 doses of .3 mg PG every 4 hours on Day 17 or 18; the day following the last dose, an oxytocin test was performed to determine onset of lactogenesis. No control rat (normal saline) was positive. All treated rats showed visible milk. Abortion was not induced by oxytocin 30 minutes after the final dose of prostaglandin. On Day 20 of the normal 22-day pregnancy, all treated rats delivered small fetuses. 7 of these rats showed normal maternal behavior with foster litters. (RMS)   002621

0045

DEMERS, L.M.

The morning-after pill.
New England Journal of Medicine 284: 1034-1036. 1971.

In this editorial on postcoital antifertility agents the ideal compound is seen as a nonsteroid which could affect the estrogen-progesterone ratio during the period before egg implantation and yet have localized specificity. Prostaglandins are mentioned as compounds which can possibly satisfy these requirements. (JRH)   003356

0046
DONOVAN, B.T.
　　The control of ovarian function.
　　Acta Endocrinologica 66: 1-15. 1971.

This Laqueur Memorial Lecture deals with several aspects of the control of ovarian function. Some early work involving hypophysectomy, pituitary stalk section, and transplantation of the hypophysis, which led to the definition of neurohumoral agents affecting pituitary hormone discharge, is described. The shortcomings in the use of the term 'releasing factor' are emphasized and its replacement by 'neurohumor' urged. In discussing the induction of ovulation, the integration within the brain of neural and hormonal factors is touched upon with particular reference to progesterone. Trophic and lytic mechanisms affecting corpus luteum function are dealt with, and the role of the uterus in producing a luteolytic agent outlined. Prostaglandin $F_2\alpha$ possesses luteolytic properties and may be of physiological importance in this regard since it has been detected in uterine vein blood. (Author's modified)　　002920

0047
DOYLE, L.L.; BARCLAY, D.L.; DUNCAN, G.W.; KIRTON, K.T.
　　Human luteal function following hysterectomy as assessed by plasma progestin.
　　American Journal of Obstetrics and Gynecology 110: 92-97. 1971.

The effect of the uterus on the corpus luteum was tested utilizing 5 patients undergoing elective hysterectomy with the finding that normal ovarian cyclicity is maintained in the absence of the uterus. Several papers are cited, which report that prostaglandins fail to induce luteolysis in primates but are luteolytic in species responding to hysterectomy, as evidence of nonuterine involvement in human luteolysis. (RMS)　　002645

0048
DUNCAN, G.W.; KIRTON, K.T.
　　Present status of prostaglandins in fertility control.
　　Journal of Reproductive Medicine 7: 18-22. 1971.

The basic structure of prostanoic acid and alterations in the cyclopentane ring that characterize the 4 main categories of prostaglandins (PGE, PGF, PGA, and PGB) are shown in figures. Only PGE_2 and $PGF_2\alpha$ are discussed. Their action as abortifacients raises the question of how they interrupt pregnancy. Are they luteolytic and do they inhibit luteal progesterone synthesis? Answers to these questions are indicated by the work of Pharriss; Gutknecht, Wyngarden, and Pharriss; and Johnston and Hunter. The effects of prostaglandin on ovarian blood flow was studied by monitoring hydrogen desaturation, which permits measurement of tissue blood flow without surgical alteration of the immediate vascular system or interference with the vascular pool. Acute intravenous injection of $PGF_2\alpha$ to pseudopregnant rabbits (Day 9) reduced blood flow up to 50% for approximately 20 minutes. The same dose continued for 4 days was sufficient to induce luteolysis. Cardiac output, heart rate, and systemic blood pressure were reduced. The data are not inconsistent with the hypothesis that a marked reduction in ovarian perfusion, even for a limited period of time and repeated over 2 days, perhaps associated with limitation of metabolism, precursors, or accumulation of end products, could result in luteal demise. Intravaginal administration of either PGE_2 (2 mg) or $PGF_2\alpha$ (20 mg) to monkeys during the 1st or 3rd trimester of pregnancy initiated increased uterine activity within 1 to 8 minutes and resulted in abortion. Clinical trials of the use of prostaglandins as abortifacients, reported by Karim and Filshie and by Bygdeman et al. are discussed. Side effects noted by these researchers are reviewed. It is concluded that we are not yet able to adequately comprehend the effects of the prostaglandins on their target tissues. (ART)   002770

0049
DUNCAN, G.W.
Prostaglandins and fertility control.
In: Cento workshop series on clinical and applied research in family planning. Ankara, Turkey, AID, Communications and Medic Branch, 1971. p. 225-234.

The general and reproductive physiology of PGs are reviewed. A few remarks are included about PGs in smooth muscle, eye, nervous system, cardiovascular system, kidney, metabolism, and blood clotting. PGEs contract the proximal human fallopian tube and relax the distal portion. Since the PGE content of human seminal plasma is often subnormal in infertile men, PGE from semen may affect sperm transport or capacitation in the oviduct. PGs are known to inhibit neuromuscular transmission in vas deferens, and to increase glucose oxidation by sperm, but PGs do not alter sperm motility or morphology. PGEs and $PGF_2\alpha$ levels are high in amniotic fluid just before labor contractions, after spontaneous abortion, in menstrual fluid (but not in anovulatory cycles). PGEs inhibit, but $PGF_2\alpha$ stimulates, the nonpregnant myometrium in vitro, particularly the upper segment. PGs have been used to induce labor successfully in 50 women by intravenous infusion. Labor lasted 10 hours, was normal without maternal or fetal complications. Intravenous PGE_2 and $PGF_2\alpha$, 5 and 50 μg per minute, induced abortion in 77% and 90% of 2 study groups; 40 and 100 mg intravaginally brought on uterine bleeding within 1-6 hours in 11 of 12 women 2-7 days after expected menses; intravaginal PGs aborted 45 women 7-23 weeks pregnant. (LJG)   003398

0050
EINHORN, V.; TAYLOR, G.S.
The effect of prostaglandins on junction potentials in the mouse vas deferens.
In: Abstracts, Australian Physiology and Pharmacology Society Meeting, Canberra, Australia, May 19-21, 1971. p. 38.

In response to transmural sympathetic nerve stimulation mouse vas deferens smooth muscle cell excitatory junction potentials (EJP) were not altered by 10^{-11}g per ml PGE_1, or PGE_2; contractions were inhibited. $PGF_2\alpha$ produced a small decrease in resting membrane potential, an initial increase in frequency, followed by a decrease in amplitude of the EJP and contractions. Higher concentrations of PGE_1, or PGE_2 produced effects similar to $PGF_2\alpha$, but with concentrations as high as 10^{-5}g per ml the EJP was not abolished. (HC)   003500

0051
EMBREY, M.P.
Induction of abortion by prostaglandins E (PGE_1 and PGE_2).
Journal of Reproductive Medicine 6: 15-18. 1971.

In 30 patients recommended for termination of pregnancy at 9 to 28 weeks of gestation, PGE_1 was used in 2 patients and PGE_2 in the remaining 28. Induction of abortion was successful in 28 of the 30 patients, while in the other 2, 1 patient received an inadequate amount of PGE_2 and 1 patient given PGE_1 aborted only after introduction of a Foley catheter into the cervix. No comparison of PGE_1 and PGE_2 was possible in this small series. Unlike oxytocin, the potency of which increases many times as pregnancy advances, longer infusion times and the total dose of PGE needed was approximately 5 to 10 times that which was found to be necessary for the successful induction of labor at term. The oxytocic effects of the E prostaglandins on uterine myometrium appear to be much more pronounced than those of $PGF_2\alpha$, which requires a much higher dose. (ART)   002771

0052

EMBREY, M.P.

PGE compounds for induction of labour and abortion.

In: Ramwell, P. and Shaw, J., eds. Prostaglandins. Annals of the New York Academy of Sciences 180: 518-523. 1971.

Experience in the induction of labor and abortion using E prostaglandins are outlined. Figure 1 (p. 519) is an almost continuous record of the induction of labor using an infusion of 2 μg/min over 4.5 hours. Figure 2 p. 521 is a continuous record of induction of abortion using 4 μg/min over 6 hours. Tables are presented giving data on the induction of labor in 28 out of 30 women and the induction of abortion in 13 of 15 women. Early experiments on the synergistic notion of prostaglandins and oxytocin with consequent reduction in the amount of prostaglandin needed to induce abortion are discussed briefly. (RMS)   002725

0053

EMBREY, M.P.; HILLIER, K.

Therapeutic abortion by intrauterine instillation of prostaglandins.

British Medical Journal 1: 588-590. 1971.

In a preliminary study of 15 patients the clinical effectiveness of prostaglandins (PGE_2 and $PGF_2\alpha$) as abortifacients when administered by intrauterine instillation compared favorably with previous studies using the intravenous route. Abortion was successfully induced in 14 patients. The average total dose of prostaglandins required was about one third of the amount needed intravenously, and side effects were minimal. (Authors' modified)   002637

0054

FILSHIE, G.M.

The use of prostaglandin E_2 in the management of intrauterine death, missed abortion, and hydatidiform mole.

Journal of Obstetrics and Gynaecology of the British Commonwealth 78: 87-90. 1971.

Successful evacuation of the uterus is reported in a series of 7 cases of intrauterine death, 5 cases of missed abortion, and 1 case of hydatidiform mole using prostaglandin E_2 infused at rates of .5 to 5.0 μg/min depending on gestational ages, which ranged from 12 to 40 weeks. 2 cases were monitored for plasma fibrinogen and showed no significant alteration during infusion. (RMS)   002625

0055

FILSHIE, M.

The use of prostaglandins in obstetrics and gynaecology.

Midwife and Health Visitor 7: 99-102. March 1971.

The folklore, history, structure, nomenclature and occurrence of PGs are mentioned briefly before a slightly more detailed elaboration of their use in inducing labor and midtrimester abortion. PGE_2, .5 μg per minute, or $PGF_2\alpha$, .05 μg per kg per minute intravenously will induce term labor without the prior rupture of the membranes and will not cause fluid retention as does oxytocin. Thus PGs are good for women with toxemia, risk of prolapsed cord, and intrauterine death when there is risk of hypofibrinogenaemia. At dosages 10 times higher PGs can be used for midtrimester abortion, with shorter abortion interval than hypertonic saline and no need for abdominal hysterotomy. PGE_2 offers fewer side effects, thus less chance of infection from diarrhea. Research is being conducted on both the vaginal route and extraovular routes for abortion and PG as a luteolytic. PG antagonists may be developed for habitual abortion, premature labor, and dysmenorrhea. (LJG)   003448

0056
FLYNN, V.T.
    Coitus and premature labour.
    Medical Journal of Australia 1: 1350-1351. 1971.

The author suggested that coitus rather than orgasm was the precipitating factor in causing premature labor in some women. It has been suggested be Bygdeman et al. that absorption of prostaglandins through the vaginal wall may cause premature labor in susceptible people. The vasodilation that is known to occur after orgasm may increase the absorption of prostaglandins from the vagina and so enhance the stimulatory effect of prostaglandins. The absence of coitus may explain the successful treatment of premature labor by bed rest in hospital, as opposed to bed rest at home. It is possible that some of the women who have uterine contractions after coitus may have increased absorption or increased sensitivity to prostaglandins, or that their husbands may have a higher concentration than normal of prostaglandins. (MT)   002776

0057
FORT, A.T.
    Prenatal intrusion into the amnion. A review of the diagnostic, therapeutic and research results.
    American Journal of Obstetrics and Gynecology 110: 432-455. 1971.

In this extensive review article on injection and measurement of substances in the amniotic fluid a brief mention is made of prostaglandins. The work of Karim and Devlin (1967) is cited, in which PGF was found in increased amounts in the amniotic fluid during labor. (JRH)   003326

0058
FUCHS, A.-R.
    Uterine activating hormones.
    In: Fuchs, F. and Klopper, A. eds., Endocrinology of pregnancy. New York, Harbor and Row, 1971. p.
        286-305.

The responses of the uterus and oviduct to prostaglandin administration in both in vivo and in vitro systems is discussed. The author hypothesizes that prostaglandins in the semen may facilitate ejaculation and then stimulate a contractile response in the female genital tract in such a way as to facilitate fertilization. (RAS)   003032

0059
FUCHS, F.; PRIETO, M.; MARCUS, R.
    Effect of prostaglandins of uterine activity in pregnant rhesus monkeys.
    In: Ramwell, P. and Shaw, J., eds. Prostaglandins. Annals of the New York Academy of Sciences 180:
        531-532. 1971.

The effects of PGE_2 and $PGF_2\alpha$ were tested in 6 pregnant rhesus monkeys, 4 close to term (150 days) and 2 in midpregnancy. None of the animals delivered during the experiment, but 3 delivered within 48 hours following termination of the experiment. 1 fetus was stillborn. Discussion following the presentation of this paper embraces such subjects as dosage levels, vaginal pessary administration of prostaglandins, side effects, local action of prostaglandins administered intravaginally, ocular pressure problems, amount of prostaglandins in human semen, degree of fertility, nature of the diarrhea that has been observed in some cases, mode of actions of cholera toxin, action of prostaglandins on blood platelets and the clotting mechanism, and percentage of failures with oxytocin. (ART)   002722

0060
GILLESPIE, A.; DEWHURST, C.J.; BEAZLEY, J.M.
Prostaglandin-induced labour.
British Medical Journal 2: 222. 1971.

In a letter to the editor, the authors refer to recently reported results in the induction of labor by the use of PGE_2 (Roberts and Turnbull, Brit Med J, 27 March, 1971), noting that prior amniotomy and high initial dosages cause this series to differ from those previously reported and possibly explain the observed uterine hypertonus. (RMS)   002624

0061
GILLESPIE, A.; BEAZLEY, J.M.; VAN DORP, D.A.
The use of an 'unnatural' prostaglandin in the termination of pregnancy.
Journal of Obstetrics and Gynaecology of the British Commonwealth 78: 301-304. 1971.

3 midtrimester abortions were successfully carried out using an 'unnatural' prostaglandin (omega-homo-PGE_1). The effects of this prostaglandin are described and its possible advantages discussed. (Authors)   002676

0062
GILLESPIE, A.
Use of prostaglandins for induction of abortion and labor.
In: Ramwell, P. and Shaw, J., eds. Prostaglandins. Annals of the New York Academy of Sciences 180: 524-527. 1971.

This paper reports on the use of omega-homo-PGE_1, a synthesized unnatural prostaglandin having 21 carbon atoms, in the production of abortion in 3 women at 14, 15 and 17 weeks of pregnancy, and the use of PGE_2 or oxytocin in a double blind study for induction of labor at term in 250 patients. Success was attained by using an increased infusion rate short of producing significant side effects. (ART) 002720

0063
GODING, J.R.; BAIRD, D.T.; CUMMING, I.A.; McCRACKEN, J.A.
Functional assessment of autotransplanted endocrine organs.
Acta Endocrinologica. Suppl. 158: 169-199. 1971.

This paper critically reviews autotransplants of the adrenal gland and of the ovary. Impairment of cyclicity was shown to result from separation of ovary from the uterus. The uterus was shown to affect luteal function by secreting $PGF_2\alpha$. $PGF_2\alpha$ gained access to the ovarian circulation by a countercurrent mechanism which was demonstrated to be quantitatively competent to cause luteolysis in the normal animals. The minimum infusion rate of $PGF_2\alpha$ into the uterine vein of sheep on Day 7 of the cycle which would cause luteolysis was about 20 μg/hour. This is similar to the observed rate of $PGF_2\alpha$ secretion in the uterine vein of ewes at the time of luteolysis. Hence $PGF_2\alpha$ can be regarded as 'the' luteolytic hormone in the sheep, and the ovine uterus as an endocrine organ. (Authors' modified) 003344

0064

GOODLIN, R.C.; KELLER, D.W.; RAFFIN, M.
Orgasm during late pregnancy. Possible deleterious effects.
Obstetrics and Gynecology 38: 916-920. 1971.

The authors briefly mention that intravaginal application of prostaglandins during late pregnancy can induce uterine contractions and therefore the prostaglandins in seminal fluid could induce labor. (JRH)   003264

0065

GRANSTROM, E.; SAMUELSSON, B.
On the metabolism of prostaglandin $F_2\alpha$ in female subjects.
Journal of Biological Chemistry 246: 5254-5263. 1971.

Metabolites of prostaglandin $F_2\alpha$ administered intravenously to female subjects were isolated. The structures of the 2 main compounds were 5alpha,7alpha-dihydroxy- 11-keto-tetranorprosta-1,16-dioic acid and its delta-lactone. The elucidation of their structures was based on gas chromatography and mass spectrometry of several derivatives and on infrared and nuclear magnetic resonance spectrometry. The structure of the delta-lactone was further established by its conversion into 5alpha,7alpha-dihydroxy-11-ketotetranor-prosta-1,16-dioic acid by alkaline hydrolysis and by mass spectrometry of a derivative obtained by borodeuteride reduction. A pathway for the formation of 5alpha,7alpha-dihydroxy-11- ketotetranor-prosta-1,16-dioic acid from prostaglandin $F_2\alpha$ is proposed. (Authors)   003128

0066

GUSTAVII, B.; BRUNK, U.
Decidual-cell necrosis after injection of hypertonic saline for therapeutic abortion.
Lancet 2: 826. 1971.

Prostaglandins are briefly mentioned as a possible mechanism by which hypertonic saline provides abortion, involving a release of prostaglandin from damaged decidual cells. A similar mechanism may explain the induction of labor after artificial rupture of the membranes, or after sweeping the membranes off the lower uterine segment. (MT)   003212

0067

GUTKNECHT, G.D.; DUNCAN, G.W.; WYNGARDEN, L.J.
Inhibition of prostaglandin $F_2\alpha$ or LH-induced luteolysis by 17β-estradiol.
Biology of Reproduction 5: 87. 1971.

Administration of prostaglandin $F_2\alpha$ ($PGF_2\alpha$) induces luteolysis and terminates pregnancy either by reducing ovarian and plasma progesterone levels or by stimulating uterine smooth muscle contractions. The present study was designed to compare the luteolytic action of $PGF_2\alpha$ and luteinizing hormone (LH) by administering these substances alone or in combination with estrogen to 48 pseudopregnant Dutch Belted rabbits. $PGF_2\alpha$ (500 μg administered subcutaneously b.i.d. on Days 9-12 of pseudopregnancy) reduced ($p < .001$) corpora lutea weights by Day 14 and lowered peripheral plasma progestin levels on Days 11 ($p < .011$) and 14 ($p < .05$). LH (50 μg intravenously on Day 9) caused a similar reduction in corpora lutea weights. In animals receiving 17β-estradiol (20mcg/day sc on Days 9-12) plus LH, corpora lutea weights were similar to controls. Corpora lutea weights in rabbits receiving 17β-estradiol plus $PGF_2\alpha$ were greater ($p < .05$) than weights in animals receiving $PGF_2\alpha$ alone but less ($p < .05$) than controls. In rabbits receiving estrogen plus $PGF_2\alpha$ plasma progestin concentrations on Days 11 and 14 were not significantly less than controls and were higher ($p < .05$) than concentrations in animals receiving $PGF_2\alpha$ alone. A five-fold increase in the dose of $PGF_2\alpha$ overcame tha antagonistic effect of 17β-estradiol. It is suggested that in the rabbit the mechanisms by which LH and $PGF_2\alpha$ induce luteolysis are not the same. (Authors)   003056

0068

GUTKNECHT, G.D.; WYNGARDEN, L.J.; PHARRISS, B.B.
The effect of prostaglandin $F_2\alpha$ on ovarian and plasma progesterone levels in the pregnant hamster.
Proceedings of the Society for Experimental Biology and Medicine 136: 1151-1157. 1971.

The effect of prostaglandin $F_2\alpha$ ($PGF_2\alpha$) on peripheral plasma and ovarian progesterone concentrations has been studied in pregnant hamsters. $PGF_2\alpha$ administered subcutaneously at a dose of .1 mg/day on Days 5 through 7 postcoitus lowered both plasma and ovarian progesterone levels and terminated pregnancy in all animals. Ovaries from $PGF_2\alpha$ treated hamsters showed histological evidence of luteal disorganization on Days 6 and 7 postcoitus. A 10-fold increase in the dose of $PGF_2\alpha$ (1 mg/day) caused an indication of luteal breakdown within 5 hours of the initial treatment. Ovarian weights, plasma progesterone levels, and luteal morphology of females after treatment and termination were similar to progesterone maintained pregnancy in $PGF_2\alpha$ treated females. (Authors) 002650

0069

HAMBERG, M.; SAMUELSSON, B.
On the metabolism of prostaglandins E_1 and E_2 in man.
Journal of Biological Chemistry 246(22): 6713-6721. 1971.

7-alpha-hydroxy-5,11-diketotetranor-prosta-1,16-dioic acid was identified as the major urinary metabolite after intravenous administration of tritium-labeled prostaglandin E_2, prostaglandin E_1, 11-alpha-hydroxy-9,15-diketoprosta-5,13-dienoic acid, 11-alpha- hydroxy-9,15-diketoprost-5-enoic acid, and 11-alpha-hydroxy-9,15-diketoprostanoic acid to male and female subjects. Injected prostaglandin E_2 was rapidly converted initially into 11-alpha-hydroxy-9,15-diketoprost-5-enoic acid. Kinetic parameters for these 2 compounds are given. A method for quantitative determination of the major urinary metabolite was developed. In 3 subjects, 7, 16, and 27 μg/24 hours were excreted. (Authors) 003131

0070

HARBON, S.; CLAUSER, H.

Cyclic adenosine 3',5' monophosphate levels in rat myometrium under the influence of epinephrine, prostaglandins and oxytocin. Correlations with uterus motility.

Biochemical and Biophysical Research Communications 44(6): 1496-1503. 1971.

L-epinephrine and the prostaglandins of the E series (PGE_1 and PGE_2) greatly stimulate the activity of adenylcyclase in rat myometrium. It is demonstrated that the same adenylcyclase, but different hormone receptors, are involved in these stimulations. Oxytocin and prostaglandin $F_2\alpha$ ($PGF_2\alpha$) have no effect nor do they inhibit the stimulation induced by the former agonists. The latter agents strongly contract the rat uterus, an action which is antagonized by epinephrine and N^6-2'-0-dibutyryl cyclic AMP, but not by PGE_1 nor PGE_2. It is concluded that the intracellular level of cyclic AMP is not the sole parameter regulating uterine motility. (Authors' modified)   003219

0071

HARGROVE, J.L.; JOHNSON, J.M.; ELLIS, L.C.

Prostaglandin E_1 induced inhibition of rabbit testicular contractions in vitro.

Proceedings of the Society for Experimental Biology and Medicine 136: 958-961. 1971.

Intact rabbit testes contracted autorhythmically in vitro with a frequency of 2-3 beats/min. When PGE_1 was added to the bathing media, there were alterations in amplitude of contractions, rate, overall tone, and mode of contraction. At relatively high levels of PGE_1 (greater than 36 nM) the contractions were completely effaced; with lower concentrations of PGE_1, recovery was evident after 5 to 6 minutes. The effects of PGE_1 on amplitude, rate, and tonus were dose-dependent. The above observations, plus those of other workers 1) that prostaglandins are present in mammalian testes, 2) that a mechanism for their synthesis appears to be present in this organ, and 3) that these compounds modulate androgen synthesis and inhibit testicular capsular contraction, suggest a possible physiological role for these unsaturated fatty acid derivatives in normal testicular function. (Authors)   002926

0072

HENDRICKS, C.

Effect of PGE_2 and $PGF_2\alpha$ on uterine contractility.

In: Ramwell, P. and Shaw, J., eds. Prostaglandins. Annals of the New York Academy of Sciences 180: 528-530. 1971.

It is reported that, compared with the use of oxytocin to induce abortion, prostaglandins were notably more effective as abortifacients, although both drugs produce the same type of increases in uterine contractility pattern. It is postulated that there must be a crucial difference in the actions of the 2 drugs. Presumably, this involves some direct effect of the prostaglandins upon the cervix and lower uterine segment which acts very efficiently to permit the cervix to yield; as a result, the products of conception are easily expelled by the increased uterine activity. (ART)   002721

0073

HUSSON, J.F.

Les prostaglandines et leurs roles dans la physiologie de la reproduction. [Prostaglandins and their roles in the physiology of reproduction.]

Gynecologie et Obstetrique 70: 361-368. 1971.

After a brief historical and structural introduction to the prostaglandins, the author surveys their different roles in the physiology of reproduction. He discusses the research that has been done on the effects of PGs on the gravid and nongravid uterus, as well as reviewing the possibilities of their action in labor and in male and female fertility. (NES)   003388

0074

INGELMAN-SUNDBERG, A.; SANDBERG, F.; RYDEN, G.

In vitro studies of the motility of the human fallopian tube.

In: Abstracts, Seventh World Congress on Fertility and Sterility. Amsterdam, Excerpta Medica, 1971. (International Congress Series 234b) p. 119.

Acetylcholine, epinephrine, and norephinephrine had a strong and oxytocin had a weak stimulatory effect in vitro on the longitudinal musculature of the human fallopian tube. PGE, and PGE_2 stimulated the isthmic part and relaxed the rest of the tube. $PGF_1\alpha$, $PGF_1\beta$, and $PGF_2\alpha$ had a stimulatory effect while PGE_3 and $PGF_2\beta$ had a relaxing action. PGE relaxed the circular musculature. (HC)   003496

0075

INGELMAN-SUNDBERG, A.; SANDBERG, F.; RYDEN, G.; MOLFESE, A.

The effect of prostaglandin E_1 on the circular and longitudinal musculature of the human oviduct in vitro.

Acta Obstetricia et Gynecologica Scandinavica 50: 51. 1971.

In previous experiments in vitro with Magnus-Kehrer technique, we were able to demonstrate a specific action of prostaglandin E_1 and E_2 on the longitudinal musculature of the human oviduct. It consisted of a strong contraction of the proximal segment and relaxation of the rest of the tube. With the same technique we have now studied the influence of prostaglandin E_1 on the spontaneous motility of muscle rings from the isthmus and the ampulla as well as on the longitudinal musculature from the same parts of the tube. It is evident that prostaglandin E_1 causes strong increase in tonus of the longitudinal musculature in the isthmus region, whereas it acts slightly relaxing on the circular musculature. In the ampulla the effect is relaxing on both muscle systems. In order to study the effect of this finding on the passage through the tube, perfusion experiments were carried out. (Authors' modified)   003015

0076

INGELMAN-SUNDBERG, A.; SANDBERG, F.; RYDEN, G.

The effect of prostaglandins on the uterus and fallopian tube.

In: Sherman, H.I., ed. Pathways to conception. The cervix and the oviduct in reproduction. Springfield, C.C. Thomas, 1971. p. 18-29.

The action of several PGs on human pregnant and nonpregnant uterus and oviduct in vivo and in vitro are summarized, following a general introduction to PG structure and physiology. In the nonpregnant uterus in vivo, 10-25% of a labeled dose of PGE_1 previously deposited in the posterior fornix, was excreted 24-32 hours later. The authors suggest that probably too little of the 1 mg mixed PGs in a normal ejaculate would be absorbed for biologic action. In vitro PGEs decrease, but PGFs increase motility of uterine muscle strips. Both groups of PGs stimulate the pregnant uterus in vivo, with decreasing sensitivity as pregnancy progresses. In vitro small PGE doses stimulate and higher doses inhibit, whereas PGF only stimulates pregnant uterine motility. In vivo studies on oviduct by intravaginal instillation of PGs showed random increases in resistance. In vitro, longtudinal strips from the proximal tube were stimulated by most PGs, but the distal portions were relaxed. Circular muscle rings from all regions of the oviduct were relaxed by PGE_1. Understanding the role of PGs in conception is lacking because the absorption of PGs from semen, endogenous production, and threshold for activity of the uterus and oviduct are not known. (LJG)   003424

0077
ITO, H.; KATAYAMA, T.
Male infertility and prostaglandins.
In: Abstracts, Seventh World Congress on Fertility and Sterility, Amsterdam, Excerpta Medica, 1971.
  (International Congress Series 234a) Abstract 30. p. 9-10

PGs were measured by bioassay and absorbance at 278 nm and no significant differences were found between normal and oligospermic men or men with normal or subnormal sperm motility. The men were grouped into 12 with sperm count of 0, 40 with 0-9 million, 48 with 10-49 million, 35 with 50 million or more sperm and 7 controls, and into motility groups of less than or more than 50%. None of the fertile oligozoospermic men had low PG levels. Therefore abnormal semen PG level may be one of the causes of some types of male infertility. (LJG)   003499

0078
JANSZEN, F.H.A.; NUGTEREN, D.H.
Histochemical localisation of prostaglandin synthetase.
Histochemie 27: 159-164. 1971.

A method has been developed for the detection of the prostaglandin synthesizing enzyme system in tissue sections. The localization of the enzyme is indicated by a brown staining when cryotome sections are incubated with the prostaglandin precursor 8,11,14- eicosatrienoic acid in the presence of 3,3'-diaminobenzidine and KCN. The method is shown to be specific for prostaglandin synthetase. Results obtained sheep vesicular glands and rabbit kidney are presented. (Authors)   003136

0079
JOHNSON, J.M.; HARGROVE, J.L.; ELLIS, L.C.
Prostaglandin $F_1\alpha$ induced stimulation of rabbit testicular contractions in vitro.
Proceedings of the Society for Experimental Biology and Medicine 138: 378-381. 1971.

Excised rabbit testes were observed to contract autorhythmically in vitro at a frequency of 2-3 times/min. When $PGF_1\alpha$ was added to the bathing media, an increase in rate and overall tone with a change in amplitude was observed. At lower concentrations of prostaglandin, the amplitude increased; but at higher levels (greater than 71 nM), amplitude was decreased. The response of the testicular capsule was dose dependent from .4 to 213 nM concentration of $PGF_1\alpha$. When $PGF_1\alpha$ was added to a testicular preparation during the latent period, before contractions were initiated spontaneously, an overall increase in tone was observed at 14 nM concentration, and contractions appeared approximately 7 minutes after treatment. At 57 nM concentration, $PGF_1\alpha$ initiated a series of modulated contractions within 1-3 minutes that increased in amplitude with time. The available data now suggest that prostaglandins may modulate testicular contractions in vitro: one group inhibiting and a second group stimulating this activity. (Authors)   003148

0080
KARIM, S.M.M.
Action of prostaglandin in the pregnant woman.
In: Ramwell, P. and Shaw, J., eds. Annals of the New York Academy of Sciences 180: 483-498. 1971.

PGE_2 and $PGF_2\alpha$ can induce abortion or labor, depending upon when given. Approximately 1,000 women were treated with these prostaglandins in the Mulago Hospital, Kampala, Uganda. PGE_2 was used successfully to induce labor in 398 out of 400 women, and $PGF_2\alpha$ was effective in 93 of 100 women at term. When used to produce therapeutic abortions, there were 5 failures in 150 cases given PGE_2, and 6 failures out of 50 cases given $PGF_2\alpha$. In a recent study it was shown that intravenous infusion of ethyl alcohol antagonized uterine activity induced in pregnant women by PGE_2 and $PGF_2\alpha$ but not that induced by infusion of oxytocin. In a double blind clinical trial of PGE_2, $PGF_2\alpha$, and oxytocin for induction of labor, there were 4 failures in 100 cases treated with PGE_2, 33 failures in another 100 cases given $PGF_2\alpha$, and 44 failures out of 100 patients given oxytocin. Different routes of prostaglandin administration are discussed, including single intravenous injection, intramuscular and subcutaneous injection, oral administration, and intravaginal administration. By the oral route, 10 times higher concentrations of these prostaglandins are required to stimulate the uterus during early pregnancy to produce abortion as at term to induce labor, and at such dose levels they produce severe diarrhea and vomiting. The use of prostaglandins administered once monthly to induce menstruation and serve as fertility control agents is discussed. (ART)   002717

0081
KARIM, S.M.M.
Effects of oral administration of prostaglandins E_2 and $F_2\alpha$ on the human uterus.
Journal of Obstetrics and Gynaecology of the British Commonwealth 78: 289-293. 1971.

The effects of oral administration of prostaglandins E_2 and $F_2\alpha$ have been studied in 42 male and nonpregnant female volunteers and in pregnant women. Oral doses of .5 mg prostaglandin E_2 or 5 mg prostaglandin $F_2\alpha$ stimulated the pregnant uterus in vivo at term to contract. Single oral doses of up to 3 mg of prostaglandin E_2 or 30 mg prostaglandin $F_2\alpha$ did not produce any noticeable effect on the gastrointestinal tract or the cardiovascular system. Doses of 4.5 mg prostaglandin E_2 and 40.50 mg prostaglandin $F_2\alpha$ produced contractions of the uterus in early pregnancy and in addition produced diarrhea and vomiting. The implications of these findings with respect to the clinical use of prostaglandins as oral abortifacients and oxytocics are discussed. (Author)   002678

0082
KARIM, S.M.M.
Induction of abortion with prostaglandins.
Research in Prostaglandins 1(3): 1-3. December 1971.

A short review of the induction of abortion discusses intravenous infusion, intrauterine administration, and the vaginal route for administering PGE_2 and $PGF_2\alpha$. The role of PGs in spontaneous abortion, some general conclusions on technique, and side effects are also mentioned. (RMS)   003065

0083
KARIM, S.M.M.
Once-a-month vaginal administration of prostaglandins E_2 and $F_2\alpha$ for fertility control.
Contraception 3: 173-183. 1971.

In 12 women, either 40 mg PGE$_2$or 100 mg PGF$_2\alpha$ was inserted into the posterior fornix of the vagina in 2 divided doses at 4-hour intervals. At the time of the intravaginal administration, these women had passed their expected day of menstruation by 2 to 7 days. In 10 of the 12 women, menstrual-like uterine bleeding started within 1 to 6 hours after vaginal insertion of the prostaglandin. A third prostaglandin tablet inserted the next day induced menstrual bleeding in 1 patient, but was ineffective in the remaining case, the 12th patient, for whom pregnancy was terminated at 10 weeks' gestation by an intravenous infusion of PGE$_2$. The possible mechanism of action of the prostaglandin is discussed. (ART)   002890

0084
KARIM, S.M.M.; SHARMA, S.D.
Oral administration of prostaglandins for the induction of labour.
British Medical Journal 1: 260-262. 1971.

Prostaglandins E$_2$and F$_2\alpha$ administered by mouth were used to induce labor in 100 patients between 35 and 44 weeks of gestation. The usual effective dose of prostaglandin E$_2$was .5 and of F$_2\alpha$ 5 mg. These were repeated every 2 hours until labor was established. Induction was successful in 79 out of 80 women treated with oral prostaglandin E$_2$and in 16 out of 20 women treated with F$_2\alpha$. (Authors' modified)   002623

0085
KARIM, S.M.M.
Prostaglandins and reproduction.
In: MacDonald, R.R., ed. Scientific basis of obstetrics and gynaecology. London, Churchill, 1971. p. 315-346.

Many aspects of the role of prostaglandins in reproduction are reviewed. Topics discussed are structure and nomenclature, occurrence and release, synthesis, metabolism, pharmacological action on the uterus, and prostaglandins in relation to parturition and spontaneous abortion. (RAS) 003415

0086
KARIM, S.M.M.
Prostaglandins as abortifacients.
New England Journal of Medicine 285: 1534-1535. 1971.

The author discusses and compares the relative effectiveness of PGE$_2$and PGF$_2\alpha$ and various routes of administration in inducing abortions in the first and second trimesters of pregnancy. He also compares the use of prostaglandins with other methods of inducing abortions. Because of the side effects, intravenous infusion of prostaglandins is not as useful as intravaginal, intrauterine, and intraamniotic routes, which permit the use of much smaller doses. In the first trimester the author feels that existing methods of abortion are more effective than prostaglandins, but in the second trimester prostaglandins given by either the intravaginal, intrauterine or intraamniotic route offer clear advantages over other methods. (JRH)   003354

0087
KARIM, S.M.M.; SHARMA, S.D.
Second trimester abortion with single intra-amniotic injection of prostaglandin E_2or $F_2\alpha$.
Lancet 2: 47-48. 1971.

The use of intravenous infusion of prostaglandins for terminating first and second trimester pregnancies has aroused a great deal of interest. The discomfort of the long duration of infusion and the side effects, such as nausea, vomiting, diarrhea, and erythema at the site of injection, are some of the disadvantages. By the use of intravaginal and intrauterine administration of prostaglandins, some of these side effects have been eliminated. Based on the observations that prostaglandins $E_2$and $F_2\alpha$ are present in samples of amniotic fluid obtained during spontaneous abortion, we have studied the effects of intraamniotic administration of these prostaglandins on the second trimester pregnant uterus. In 10 women from 13 to 22 weeks pregnant, single intraamniotic administration of 2.5-5.0 mg PGE_2or 25 mg $PGF_2\alpha$ stimulated the uterus to contract and resulted in abortion in every case. Abortion was complete in 8 women and 2 required manual removal of the placenta. The injection-abortion interval ranged from 4 1/2 hours to 18 hours, with a mean of 11.4 hours. The prostaglandins contained in .1-.5 ml of solution were administered through an epidural catheter into the amniotic cavity, which was connected to a pen recorder and used to monitor uterine activity. The procedure does not require the withdrawal of any amniotic fluid. 4 women had nausea and 3 vomited. Otherwise the procedure was free from side effects. It seems that prostaglandins administered by this route produce a local action on the uterus. (Authors)   002925

0088
KARIM, S.M.M.; SHARMA, S.D.
The effect of ethyl alcohol on prostaglandins $E_2$and $F_2\alpha$ induced uterine activity in pregnant women.
Journal of Obstetrics and Gynaecology of the British Commonwealth 78: 251-254. 1971.

Intravenous infusion of 500 ml of 10% ethanol per hour for 1 hour inhibited uterine activity initiated and maintained by an infusion of either prostaglandin E_2or prostaglandin $F_2\alpha$. This study was carried out in 7 women with intrauterine death of the fetus in the third trimester of pregnancy and in 4 women in whom therapeutic termination of pregnancy in the second trimester was planned. The significance of these findings in relation to the physiological role of prostaglandins in parturition is discussed. (Authors)   002627

0089
KARIM, S.M.M.; HILLIER, K.; SOMERS, K.; TRUSSELL, R.R.
The effects of prostaglandins $E_2$and $F_2\alpha$ administered by different routes on uterine activity and the cardiovascular system in pregnant and non-pregnant women.
Journal of Obstetrics and Gynaecology of the British Commonwealth 78: 172-179. 1971.

Prostaglandin $E_2$and prostaglandin $F_2\alpha$ both stimulate the intact nonpregnant and pregnant human uterus. Doses (up to 40 μg/min) required to provide long-lasting stimulation when given by intravenous, subcutaneous, and intramuscular routes had no effect on the cardiovascular system. Both the nonpregnant and the pregnant uterus during early pregnancy showed a similar sensitivity to prostaglandin $E_2$and $F_2\alpha$. The nonpregnant uterus was not inhibited by prostaglandin E_2at any dose level. (Authors)   002630

0090
KARIM, S.M.M.
  The role of prostaglandins in human parturition.
  Proceedings of the Royal Society of Medicine 64: 10-12. 1971.

  The relationship of prostaglandin content in aminiotic fluid and in blood to labor is discussed. A table giving details of the induction of labor at or near term using prostaglandin E_2 in 10 patients is presented. The author also mentions use of prostaglandins in successful termination of pregnancy and reviews the case for a normal physiological role of prostaglandins in human parturition. (RMS) 002664

0091
KARIM, S.M.M.; TRUSSELL, R.R.
  The use of prostaglandins in obstetrics.
  East African Medical Journal 48(1): 1-12. January 1971.

  The literature on the use of prostaglandins in the induction of labor and abortion, and in missed labor and abortion and hydatidiform mole is reviewed in detail. (RMS)   003162

0092
KARIM, S.M.M.; SHARMA, S.D.
  Therapeutic abortion and induction of labour by the intravaginal administration of prostaglandins E_2 and $F_2\alpha$.
  Journal of Obstetrics and Gynaecology of the British Commonwealth 78: 294-300. 1971.

  In 45 women pregnancy was terminated in the first and second trimesters, and in 10, labor was induced at term by the intravaginal administration of prostaglandins E_2 and $F_2\alpha$. With prostaglandin E_2 the average induction-abortion interval in 30 women was 12 1/2 hours, and in the 15 women in whom abortion was induced with prostaglandin $F_2\alpha$ the average induction-abortion interval was 14 hours 50 minutes. (Authors)   002677

0093
KARIM, S.M.M.
  Use of prostaglandins for fertility control.
  In: Abstracts, Seventh World Congress on Fertility and Sterility, Amsterdam, Excerpta Medica, 1971.
    (International Congress Series 234a) Abstract 35. p. 11.

  The author's preliminary reports of pregnancy termination will be extended by over 500 cases in 3 categories: 1) abortions of 6-week or more gestation by iv, intravaginal, intrauterine, and oral PGs; 2) intravaginal PG for missed menstruation of 10-day duration; and 3) PG self-administered once monthly as a postcoital contraceptive. Data will show patient descriptions, procedures, results, complications, and follow-up details. (LJG)   003407

0094
KAWAI, Y.
  Studies of the effects of prostaglandins on human myometrium in vivo.
  In: Abstracts, 7th World Congress on Fertility and Sterility, Amsterdam. Excerpta Medica, 1971.
    (International Congress Series, 234a) Abstract 31. p. 10.

The authors examined amniotic fluid obtained during labor and different stages of pregnancy and analyzed cervical mucus from women with normal cycles and from dysmennorrheic patients. The effects of the PGs found in these materials on pregnant and nonpregnant human myometrium in vivo were determined. The authors suggest that the mechanism of action of PG differs in various stages of pregnancy. (HC)   003497

0095

KINOSHITA, K.; WAGATSUMA, T.; HOGAKI, M.; SAKAMOTO, S.
The induction of abortion by prostaglandin $F_2\alpha$.
American Journal of Obstetrics and Gynecology 111: 855-857. 1971.

$PGF_2\alpha$ was used for induction of therapeutic abortion in women 6 to 29 weeks pregnant. Intravenous infusion of $PGF_2\alpha$ (11-33 μg/min) produced uterine contraction and in most cases dilation of cervix followed by abortion. It was found that less $PGF_2\alpha$ was required to produce satisfactory uterine contractions in the more advanced pregnancies. The side effects consisted of diarrhea in all patients, nausea, vomiting, hypogastric pain, headache, cough, rubedo, and phlebitis in some patients. Most of these side effects were transient and responded to symptomatic treatment. The authors conclude that intravenous infusion of $PGF_2\alpha$ should be the method of choice in terminating middle pregnancies. (JRH)   003105

0096

KINOSHITA, K.; WAGATSUMA, T.; HOGAKI, M.; SAKAMOTO, S.
The induction of labor with prostaglandin $F_2\alpha$.
Acta Obstetricia et Gynaecologica Japonica 18(2): 87-94. 1971.

The effect of the intravenous infusion of $PGF_2\alpha$ on the activity of the pregnant uterus in vivo was studied in 30 women at or near term. Vaginal delivery occurred in all 30 patients. No fetal complications attributable to $PGF_2\alpha$ were observed, nor any maternal cardiovascular or pulmonary effects. As for the infusion rate, it is suggested that a slow rate of $PGF_2\alpha$ infusion should be used initially then increased steadily until an effective infusion rate is obtained. No uterine hypertonus was demonstrated within the administered dose. The pattern of uterine activity produced by $PGF_2\alpha$ was different from that of oxytocin, in that the former showed at the beginning the characteristic irregular uncoordinated contractions; the latter, however, displayed the regular coordinated contractions. This suggests that the mechanism of action of the 2 differs from each other; the possible physiological role of $PGF_2\alpha$ in the process of parturition was discussed. (Authors' modified)   003325

0097

KIRTON, K.; DUNCAN, G.; OESTERLING, T.; FORBES, A.
Prostaglandins and reproduction in the rhesus monkey.
In: Ramwell, P. and Shaw, J., eds. Prostaglandins. Annals of the New York Academy of Sciences 180: 445-455. 1971.

It has been postulated that prostaglandins such as $PGF_2\alpha$, which have unique venoconstrictive activity, have a luteolytic property which accounts for their pregnancy-terminating effects. Studies made on laboratory primates as a forerunner of clinical studies have shown that $PGF_2\alpha$ effectively terminated pregnancy when injected early (10 to 12 days postovulation), a time when active luteal steroidogenesis is needed to maintain pregnancy. It was found that $PGF_2\alpha$ was equally effective when injected subcutaneously or given by intravenous infusion, if administered before Day 30 or after Day 100. The essentiality of the unique venoconstrictive properties of $PGF_2\alpha$ to the abortifacient properties was questioned by the finding that PGE_2 also terminates pregnancy effectively. Uterine contractility was monitored in order to quantify the effects of prostaglandins. Quantitative comparison of uterine stimulation was correlated with the effectiveness of prostaglandins in interrupting pregnancy. The uterus was more sensitive either very early in pregnancy or during later stages. PGE_2 was about 10 times as potent as $PGF_2\alpha$ in its ability to stimulate uterine contractility. This study demonstrated that PGE_2 and $PGF_2\alpha$ initiated rhythmic uterine contractions throughout pregnancy in rhesus monkeys when administered by subcutaneous injection or by iv infusion. This increased uterine motility will nearly always terminate pregnancy if continued for a sufficient length of time. The amplitude and frequency of contractions induced during the second trimester of pregnancy were less than those induced in the first or third trimester. The increased potency of PGE_2 over $PGF_2\alpha$ measured in terms of uterine stimulation, was corroborated in human subjects when administered during the first half of pregnancy or near term to induce labor. Experimental results are described in which prostaglandins were administered intravaginally. (ART)   002714

0098

KIRTON, K.T.; FORBES, A.D.
Abortifacient efficacy of $PGF_2\alpha$ administered intra-amniotically to rhesus-monkeys.
Contraception 4: 31-35. 1971.

The study was undertaken to determine if prostaglandins would stimulate uterine contractility when administered into amniotic fluid. $PGF_2\alpha$ was used on 4 mature rhesus monkeys. Uterine contractions were not initiated in 1 animal by 100 μg of $PGF_2\alpha$ but were by an additional 200 μg given approximately 1 hour later. 2 animals given 1.0 mg of $PGF_2\alpha$ each aborted between 36 and 48 hours later. One monkey received a 5.25 mg dose and aborted 12 to 18 hours later. No detectable prostaglandin dehydrogenase activity was noted in samples of the amniotic fluid. The results indicated that the amniotic fluid acted as a reservoir from which the $PGF_2\alpha$ was released to the uterus through exchange of the fluid with the placental-maternal circulation, or by diffusion directly across the placental membranes. Simplicity of administration and duration of uterine activity after a single application are obvious advantages of this method over other currently used delivery systems for prostaglandin-induced abortions. (ART)   002894

0099

KIRTON, K.T.; WYNGARDEN, L.J.; FORBES, A.D.
Effects of prostaglandins on luteal cell function in vivo.
In: Hormonal Steroids. [Proceedings of the Third International Congress, Hamburg, September 7-12, 1970.] Amsterdam, Excerpta Medica 1971. (International Congress Series No. 219) p. 685-690.

Previous studies indicate that prostaglandins are luteolytic in vivo in various laboratory animals and more recent investigations indicate a similar activity of $PGF_2\alpha$ in sheep. Although intravenous PGE_2 initially stimulates progestin levels in rhesus monkey, no permanent change in corpus luteum function has been observed. PGE_2 and $PGF_2\alpha$ appear to stimulate steroidogenesis directly in primates. (RAS) 002998

0100

LABHSETWAR, A.P.
Luteolysis and ovulation induced by prostaglandin $F_2\alpha$ in the hamster.
Nature 230: 528-529. 1971.

Female hamsters in the first day of pregnancy were used to determine the effect of racemic $PGF_2\alpha$ on luteolysis and ovulation. Some of the animals were bilateraly hysterectomized (uterus removed) on Day 4 of pregnancy. Experimental animals were injected subcutaneously with 50-100 μg/day dl $PGF_2\alpha$ on Days 4-6 inclusive. All animals were killed on Day 8 of pregnancy and examined for number and condition of corpora lutea and number of implantation sites. No embryos were found in normal animals that received $PGF_2\alpha$ and their ovaries had 2 sets of corpora lutea, 1 set degenerating and a newly formed set. There were also new ova in the fallopian tubes of these animals while untreated controls had none, suggesting that the treated hamsters had recently ovulated. It was also found that progesterone given concurrently with the $PGF_2\alpha$ maintained the pregnancy, even though the corpora lutea still degenerated. Progesterone alone did not cause luteal degeneration. The corpora lutea of hysterectomized animals given $PGF_2\alpha$ also degenerated, but hysterectomy alone did not cause this effect. The effects of $PGF_2\alpha$ could be due either to its luteolytic effect causing a decrease in progesterone levels thus stimulating LH secretion which in turn stimulates ovulation, or $PGF_2\alpha$ could directly stimulate LH secretion and the LH causes the luteolysis and subsequent ovulation. (JRH) 002641

0101

LABHSETWAR, A.P.

Prostaglandin $F_2\alpha$ and the implantation process in the rat.

Journal of Endocrinology 50: 353-354. 1971

Experiments were performed to determine if the antifertility effects of $PGF_2\alpha$ are due to its effect on implantation as seen by the Pontamine blue reaction in rats and hamsters. It was found that there was no significant difference in the number of implantation sites judged by the number of blue patches in the uterine horns of mated females that had received 500 μg of $PGF_2\alpha$ subcutaneously twice daily on Days 4-7 of pregnancy and mated control females. However, autopsy on Day 9 of pregnancy showed implantation sites in only one of the $PGF_2\alpha$ treated females and they were degenerating. It was concluded that the antifertility property of $PGF_2\alpha$ is not mediated through inhibition of implantation and that if $PGF_2\alpha$ reduced the blood flow to the uterus, its effect is not enough to affect implantation. (JRH)   003182

0102

LAMPRECHT, S.A.; ZOR, U.; TSAFRIRI, A.; LINDNER, H.R.

Action of prostaglandin E_2 and luteinizing hormone on cyclic adenosine 3',5'-monophosphate production in fetal, early postnatal and adult rat ovaries.

Israel Journal of Medical Sciences 7: 704-705. 1971.

Whole ovaries of adult rats incubated in Krebs-Ringer bicarbonate buffer responded to addition of PGE_2 to the medium within 1 minute with a twofold rise in production of cAMP from prelabeled endogenous tritiated ATP. After 1 minute the rise was fivefold. A similar response was obtained when fully luteinized ovaries or isolated graafian follicles collected on the morning of proestrus were incubated with PGE_2. These effects of PGE_2 mimic those produced by luteinizing hormone (LH) on rat ovarian tissue in vitro. In contrast to ovaries of adult rats (100 days) and prepubertal rats (12 days), which responded to both LH and PGE_2, prenatal and early postnatal rat ovaries (1 day before birth to 2 days postpartum) were unresponsive to stimulation by LH; incubation of these ovaries with PGE_2, however, resulted in a significant rise in cAMP production. A current theory postulates that the mechanism of action of cAMP involves the stimulation of protein kinase (PK). PK activity was determined in the 27,000 x g supernatant of ovaries derived from rats aged 28, 12 and 7 days. In the 28- and 12-day-old ovaries, exogenous cAMP brought about a four- to fivefold increase in PK activity. However, in 7-day-old ovaries, stimulation of PK activity was only of the order of 30%. Whole ovaries from 28-day-old rats incubated for 20 minutes with either PGE_2 or LH had four- to fivefold higher PK activity than controls, and no further stimulation was obtained by exogenous cAMP. It appears that the competence of the ovarian adenyl cyclase-cAMP system to respond to LH, and of the PK system to respond to cAMP, is acquired gradually during early postnatal development. (Authors)   003192

0103
LEE, C.M.
Prostaglandins and the reproductive system.
Fertility Control 6: 21-26. 1971.

This brief review of prostaglandins and the reproductive system discusses the male and female reproductive system, induction of labor, therapeutic abortion, missed abortion, missed labor, hydatidiform mole, and postconception contraceptives. (MT)   002752

0104
LEVY, B.; LINDNER, H.R.
Selective blockade of the vasodepressor response to prostaglandin $F_2\alpha$ in the anaesthetized rabbit.
British Journal of Pharmacology 43: 236-241. 1971.

The prostaglandin-blocking activity of meclofenamic acid, N-(2,6-dichloro-m-toly) anthranilic acid (CI-583), was analyzed in the anesthetized rabbit. $PGF_2\alpha$, PGE_1, and isoprenaline were injected before and after meclofenamic acid infusion. Isoprenaline produced a fall in blood pressure, a reduction in oviduct motility, and a reduction in uterine motility if the uterus showed marked spontaneous motility. $PGF_2\alpha$ uniformly produced a fall in blood pressure and an increase in both uterine and oviduct contractility. PGE_1 produced a fall in blood pressure, a reduction in oviduct motility, and no consistent effect on uterine motility. Meclofenamic acid selectively blocked the vasodepressor response to $PGF_2\alpha$. The vasodepressor responses to PGE_1 and isoprenaline as well as the effects of all 3 agonists on uterine and oviduct contractility were not reduced by treatment with meclofenamic acid. Polyphloretin phosphate (PPP) administered to 2 rabbits in a cumulative dose of 94 mg/kg showed no significant blocking action on the vasodepressor or uterine and oviduct contractor responses to PGE_1 or $PGF_2\alpha$ though this compound, like meclofenamic acid, has been reported to antagonize the actions of PGE_1 and $PGF_2\alpha$ on isolated smooth muscle preparation. (Authors' modified)   003309

0105
LIGGINS, G.C.
Hormonal steroid contraceptives. II: clinical considerations.
Drugs 1: 461-483. 1971.

The bulk of this comprehensive review covers the worldwide acceptability, use, efficacy, comparative minor side effects, and clinical guidelines for selecting among the 4 main types of steroid contraceptives: the combined, sequential, continuous oral, and injectible progestagen-only contraceptives. Listed as new hormonal contraceptives undergoing trial are reverse sequential, progestagen implants, postcoital estrogen, and PGE_2 as a menstrual inducer. (LJG)   003440

0106
LIGGINS, G.C.; GRIEVES, S.
Possible role for prostaglandin $F_2\alpha$ in parturition in sheep.
Nature 232: 629-631. 1971.

In sheep, the mechanism of onset of labor seems to be controlled chiefly by corticosteroids secreted by the adrenals of the fetal lamb. The role of $PGF_2\alpha$ was investigated in premature parturition induced by infusion of a glucocorticoid into the fetal lamb. In one experimental group, fetuses were given an intracarotid infusion of dexamethasone either for 24 hours or until labor contractions were observed (at 46 to 52 hours). In another experimental group the fetuses were infused for 52 to 57 hours, but the onset of labor was prevented by giving the ewe progesterone in oil 100 mg/12 hours. Before and during corticosteroid induced labor, 100 ml samples of plasma were obtained by plasmaphoresis from chronically implanted catheters in the maternal jugular and uterine veins. $PGF_2\alpha$ was not found in any of the samples taken before labor but was present in the uterine vein plasma during labor in 9 or 10 ewes. Samples of maternal cotyledons, fetal cotyledons, and myometrium were obtained by rapid excision of these tissues from sacrificed ewes. Control animals were normal pregnant ewes of known gestation length. The concentration of $PGF_2\alpha$ in maternal cotyledons was raised above the levels found in 2 control animals after dexamethasone had been infused into the fetus for 24 hours, but the concentration in myometrium was unchanged. In labor, a furthur increase in concentration in maternal cotyledons was found and, in addition, the concentration in the myometrium was increased. Similarly increased concentrations of $PGF_2\alpha$ were also found in maternal cotyledons and myometrium of ewes that would have been in labor had they not received progesterone, indicating that the appearance of $PGF_2\alpha$ is not a consequence of labor. The concentration of $PGF_2\alpha$ in fetal cotyledons was relatively low in all samples. (ART)   002751

0107

LIGGINS, G.C.; GRIEVES, S.

The role of prostaglandin $F_2\alpha$ in parturition of the ewe.

Asia and Oceania Congress of Endocrinology, 4th, University of Aukland, New Zealand, 1971. (Abstract 169)

Using gas-liquid chromatography for assay, $PGF_2\alpha$ was detected in uterine vein plasma of ewes during labor. Concentration on Days 126 and 140 of gestation was 82 and 164 ng per g in maternal cotyledons and 70 and 80 ng per g in the myometrium, respectively. After 24 hour intracarotid-infusion of dexamethasone to the fetus, concentration rose to 315 ng per g in the cotyledon and 97 ng per g in the myometrium. During labor, values increased again to 630 ng per g and 383 ng per g in the cotyledon and myometrium, respectively. The authors suggest that the rise in $PGF_2\alpha$ may play a role in the mechanism by which fetal adrenal activity initiates parturition in the ewe. (HC)   003417

0108

LIMA, F.; GIMENO, M.F.; GOLDRAIJ, A.; BEDNERS, A.S.; GIMENO, A.L.

Prostaglandin $F_2\alpha$ and motility of the myometrium isolated from ovariectomized rats. Effects of 17-beta estradiol.

Life Sciences 10: 999-1008. 1971.

The presence of prostaglandin $F_2\alpha$ ($PGF_2\alpha$) in extracts of uterine horns removed from ovariectomized rats, from nonovariectomized rats, and from ovariectomized rats injected with 17-beta estradiol, was studied. Thin layer chromatography determinations demonstrate the presence of $PGF_2\alpha$ in extracts of uterine horns from ovariectomized animals and ovariectomized animals injected with a very small dose of 17-beta estradiol (.1 μg/day for 3 days) but not in extracts from nonovariectomized or from ovariectomized rats injected with with a higher dose of the hormone (60 μg/day for 3 days). The relative isometric developed tension (IDT), expressed on dry-tissue weight basis, as well as the stability of the IDT and of the frequency of contractions of myometrial strips isolated from castrated rats, were greater than those exhibited by preparations from noncastrated animals and from castrated animals injected with 17-beta estradiol (60 μg/day for 3 days). The influence of ovarian hormones on uterine $PGF_2\alpha$ and the possibility of some interrelation between uterine $PGF_2\alpha$ and the functional activity of isolated rat myometrium is discussed. (Authors)   003134

0109
LOUNG, K.C.; BUCKLE, A.E.R.; ANDERSON, M.M.
Results in 1,000 cases of therapeutic abortion managed by vacuum aspiration.
British Medical Journal 4: 477-479. 1971.

Intravenous PGE_2, intrauterine PGE_2 and $PGF_2\alpha$, and vaginal tablets of $PGF_2\alpha$ and PGE_2 used for therapeutic abortion are briefly mentioned and are still undergoing clinical trials. Authors state that most gynecologists will have to rely, at present, on more conventional methods for inducing abortion. (MT)   003048

0110
McCRACKEN, J.
Prostaglandin $F_2\alpha$ and corpus luteum regression.
In: Ramwell, P. and Shaw, J., eds. Prostaglandins. Annals of the New York Academy of Sciences 180: 456-472. 1971.

In the sheep, it has been shown that the luteolytic effect of the uterus (which is bicornuate) on the ovary is of a unilateral nature, i.e., each ovary is under the control of the adjacent contiguous horn of the uterus. Ligation of vascular connections between the uterus and ovary in the sheep resulted in luteal retention, whereas ligation of the uterine artery alone had no effect. This and other evidence suggested that a lytic factor (or factors) was present in the uterine vein blood. To test the suggestion that $PGF_2\alpha$ might be the factor, a series of experiments was performed to study the effects of $PGF_2\alpha$ on ovarian venous blood flow and progesterone secretion rate from the sheep ovary autotransplanted with vascular anastomoses to the vessels of the neck. The technique for continuous intraarterial infusion of the ovary is shown in a diagram. The biological potency of the $PGF_2\alpha$ used in the experiments was tested in a bioassay system using gerbil colon and was found to be potent. The effective dose levels of $PGF_2\alpha$ was established. There was an initial rise in progesterone secretion rate which, however, fell within 1 hour to about 50% of control values, and thereafter fell slowly during the remainder of the infusion and continued to fall after the infusion was stopped. After 24 hours the progesterone was almost undetectable, and after 48 hours the animal showed a typical estrus response with a distinct peak in plasma LH during estrus. Higher concentrations of $PGF_2\alpha$ resulted in a marked vasoconstrictive effect, and estrus was observed 18 hours later. Progesterone secretion rate began to rise by the fourth day after estrus, suggesting that ovulation and corpus luteum formation had taken place. The results suggest that $PGF_2\alpha$ may be the luteolysin in the sheep. The possibility of a countercurrent mechanism allowing substances to pass from the utero-ovarian vein into the very closely adherent and very tortuous ovarian artery is discussed. (ART)   002715

0111
McCRACKEN, J.A.; BAIRD, D.T.; GODING, J.R.
Factors affecting the secretion of steroids from the transplanted ovary in sheep.
Recent Progress in Hormone Research 27: 537-582. 1971.

The authors review the techniques for transplanting the ovary alone or the ovary and uterus together to a more accessible site in the neck of the sheep. They describe the types of experiments that have been performed with this preparation. In 1 section, the authors present the evidence that $PGF_2\alpha$ is the uterine luteolytic factor in sheep. (JRH)   003376

0112
McNEILLY, A.S.; FOX, C.A.
  The effect of prostaglandins on the guinea-pig mammary gland.
  Journal of Endocrinology 51: 603-604. 1971.

The effect of close intraarterial injection of PGE_1, PGE_2, $PGF_1\alpha$ and $PGF_2\alpha$ on milk ejection from lactating guinea pig mammary glands was measured. All 4 prostaglandins stimulated milk ejection and the order of activity was PGE_1, PGE_2, $PGF_2\alpha$ and $PGF_1\alpha$. The prostaglandins (up to 30 μg) did not inhibit the mild ejection activity of oxytocin. It was concluded that the PGs did not produce their effect by stimulating the release of endogenous oxytocin, since the latent period for PG and oxytocin was about the same. (JRH)   003279

0113
MANSEL-JONES, D.
  Clinical trails of prostaglandins.
  British Medical Journal 2(5752): 50. 1971.

The author, a medical assessor for the Committee on Safety of Drugs, warns researchers to use only those preparations of prostaglandins which have clearance for specific clinical trials. (RAS)   002606

0114
MARCUS, S.L.
  Effect of intravaginal prostaglandin E_2 and of semen on uterine contractility in the rhesus monkey.
  In: Abstracts, 7th World Congress on Fertility and Sterility, Amsterdam, Excerpta Medica, 1971.
    (International Congress Series 234a) Abstract 33. p. 10-11.

Only during the ovulatory phase was an inhibition of uterine contractility noted within 20 minutes after the intravaginal deposition of PGE_2 in rhesus monkeys. (HC)   003498

0115
MARSH, J.
  The effect of prostaglandins on the adenyl cyclase of the bovine corpus luteum.
  In: Ramwell, P. and Shaw, J., eds. Prostaglandins. Annals of the New York Academy of Sciences 180:
    416-425. 1971.

Studies of the steroidogenesis of the corpus luteum and its control by gonadotropins have been extended to an investigation of the effects of prostaglandins on this tissue and to determine if they also play a role in the control mechanism. Luteinizing hormone (LH) was found to produce a marked increase in steroid synthesis. A study was made of the mechanism by which LH produced this effect and the possible role of cyclic adenosine monophosphate (AMP) in this hormonal action. It was found that exogenous cyclic AMP could produce a stimulation of progesterone synthesis about equal to that of LH. The effect of LH was also found not to be additive to a saturation amount of cyclic AMP, indicating that this cyclic nucleotide could be a mediator of the action of LH. A study was then made of the adenyl cyclase and cyclic nucleotide phosphodiesterase enzyme systems in homogenates of bovine corpora lutea in order to determine if LH elicited an increase in endogenous cyclic AMP by activating the former system or by inhibiting the latter. It was found that LH increased the adenyl cyclase activity and had no effect on the phosphodiesterase system. It has now been reported that $PGF_2\alpha$ given in vivo was found to exert a luteolytic action on the corpus luteum of rhesus monkeys, rats, guinea pigs, and rabbits. In contrast to their effects in vivo, studies made in vitro on the effects of

$PGF_2\alpha$ and other prostaglandins (PGE_2, PGE_1, PGA_1) on ovarian tissue of rats, rabbits, and cows have shown that these substances stimulate steroidogenesis rather than inhibit it. Thus, the prostaglandins appeared to be possible candidates for the mediatory role between LH and its action on the adenyl cyclase system of the corpus luteum. An investigation was undertaken to determine if the prostaglandins are in fact mediators of the action of LH. It was found that several prostaglandins significantly stimulate adenyl cyclase activity in homogenates of bovine corpora lutea, and confirm the previous report that these substances stimulate steroidogenesis in tissue slices. It is considered probable that prostaglandins bring about this steroidogenic effect by the mediation of cyclic AMP. However, conflicting evidence has rendered uncertain the mediator role of the prostaglandins for LH effects. (ART)   002711

0116

MASON, N.R.; TOOMEY, R.E.

Stimulation of cyclic AMP in rat ovaries.

In: Abstracts, 53rd Meeting Endocrine Society, San Francisco, June 24-26, 1971. (Abstract 118)

The effect of gonadotrophins and prostaglandins on 3',5'-cyclic AMP levels in incubating rat ovary slices was measured by the protein binding assay of Gilman (Proc. Nat'l. Acad. Sci., 67: 305 [1970]). LH (NIH-LH-S7) added at 5 μg/ml to ovary slices from PMS-HCG treated rats caused a 4-8-fold increase in cyclic AMP levels within 15 minutes (control level = .3-.4 pmoles/mg tissue). A significant increase was observed as early as 1 minute after the addition of LH, and maximum levels of the cyclic nucleotide were obtained between 15 and 30 minutes. Prolactin (NIH-P-S5) at 10 μg/ml had no effect, but FSH (NIH-FSH-S2) at 10 μg/ml gave a 3-fold stimulation. Both LH and FSH were found to have a stimulatory effect on cyclic AMP levels in slices from ovaries taken from immature rats. LH at .1, 1, and 5 μg/ml gave increases of 1.6, 3.7, and 6.3 times the control values, respectively, while the stimulation by FSH at .5, l, and 10 μg/ml was 1.2, 1.4, and 4.7 times. A combination of .1 μg LH and 1 μg FSH gave no effect over that of .1 μg LH alone. Prolactin had no effect on cyclic AMP levels in the immature ovaries. PGE_1 or PGE_2 at 5 μg/ml stimulated cyclic AMP levels at least as much as did LH using slices from immature ovaries. These data indicate that cyclic AMP levels are stimulated by both LH and FSH in tissue from immature as well as luteinized ovaries. Prostaglandins also will affect the immature ovary tissue. (Authors)   003005

0117

MILLER, N.N.

Uganda and the wonder drug: a new approach to population control.

American Universities Field Staff Reports, East Africa Series 10(4): 1-8. 1971.

The testing of PG as contraceptive, abortifacient, and inducer of labor by S.M.M. Karim at Makerere University, Kampala, Uganda, is described and embellished with biographical details. Karim initiated trials of PGs after his career at London University. The author predicts that largely as a result of his work with PGs manufactured in Uganda, PGs may become the safe, inexpensive, nontoxic, reliable, noncoital contraceptive for use after ovulation, before menstruation, or after missed menstruation, within 4 years if Karim's trials are successful. PGs may also be used for therapeutic abortion, male infertility, labor induction, and for abortion in animals. It is especially fitting that the drug that may 'bridge the technology gap in contraceptives' has been developed in Uganda with its 3.3% annual growth rate and no national family planning program. (LJG)   003334

0118

NAKAJIMA, A.

The effects of prostaglandins on the electrical activity of the fallopian tube.

In: Abstracts, Seventh World Congress on Fertility and Sterility, Amsterdam, Excerpta Medica, 1971. (International Congress Series 234a) Abstract 32. p. 10.

The basic pattern of electrical activity of rabbit fallopian tube as well as the effects of $PGF_2\alpha$ and PGE_1 on this pattern were recorded. The ampulla exhibited a monophasic action potential resembling regular peristalsis. The isthmus showed bursts of spike discharges from descending peristalsis merging with tetanic contractions. PGE_1 inhibited the whole oviduct by hyperpolarizing it; $PGF_2\alpha$ excited the tube by depolarization. Catecholamine antagonists had no effect. (LJG)   003491

0119

NAKANO, J.; PRANCAN, A.V.

Metabolic degradation of prostaglandin E_1 in the rat plasma and in rat brain, heart, lung, kidney and testicle homogenates.

Journal of Pharmacy and Pharmacology 23: 231-232. 1971.

Prostaglandins are metabolized by oxidation of the secondary alcohol group at C-15 in swine lungs by the catalytic action of NAD-dependent 15-hydroxy-prostaglandin dehydrogenase. Recently, PGE_1 was converted to less polar metabolites in dog isolated kidneys. This study compares the rate of metabolism of PGE_1 in rat plasma, brain, heart, kidney, and testicle with that in rat lung. The pooled tissues of lungs, kidney, heart, brain, and testicles of male Holtzman rats were homogenized at 4 in 4 volumes of Bucher medium by a tissue grinder. The homogenates were centrifuged at 10000 g for 20 minutes and the protein in the supernatant determined. The supernatant was shaken with .1 Ci/ml of tritiated PGE_1 (28 Ci/millimole), 50 ng/ml of PGE_1, and 2 millimole of NAD at 37.5. Before and at 2, 5, 10, 20, 40, and 60 minutes after incubation was initiated, a 4-ml aliquot of the sample was pipetted into tubes containing .5 ml of N HCL solution to acidify the aliquot to pH 3.0. Tritiated PGE_1 and its metabolites were extracted twice with ethyl acetate. The extract was separated by chromatography. PGE_1 was eluted with 70% ethyl acetate in toluene while a less polar metabolite, 15-keto PGE_1 was eluted with 40% ethyl acetate in toluene. 15 ml of the scintillation fluid was added to 4-ml aliquots of each chromatography fraction and the radioactivity counted. Before incubation the silicic acid chromatography of the extracted rat plasma showed a single peak of tritiated PGE_1; which was eluted in fractions 11-14 of an ethyl acetate-toluene 70:30 solvent system. An aliquot of rat plasma incubated for 1 hour at 37 gave similar results, indicating no metabolic degradation of PGE_1. In contrast the chromatography of the rat lung homogenate fractions 11-14 showed no tritiated PGE_1 peak after 20 minutes of incubation; however fractions 6-9 gave prominent peak, apparently due to a less polar PGE_1 metabolite identical with 15-keto-PGE_1. Very slow metabolic degradation of tritiated PGE was observed in rat plasma and in the brain and heart homogenates. In contrast tritiated PGE_1 metabolism did occur in rat testicle, kidney, and lung homogenates. The kidney and lung homogenates metabolized 95% of tritiated PGE_1 within 20 minutes, while the rat testicle homogenates, converted 80% of tritiated PGE_1 in the same time period. In summary, rat homogenized lung, liver, and testicle inactivate PGE_1. Rat plasma and homogenized brain and heart have little ability to metabolize PGE_1. (FDB)   002608

0120

NAKANO, J.; MONTAGUE, B.; DARROW, B.

Metabolism of prostaglandin E_1 in human plasma, uterus and placenta, in swine ovary and in rat testicle.

Biochemical Pharmacology 20: 2512-2514. 1971.

The metabolism of tritiated PGE_1, by homogenates of human plasma, uterus, and by swine ovary and rat testicle and kidney was measured. It was found that the PGE_1 was rapidly metabolized in rat kidney, testicle, and human placenta, but little metabolism occurred in human plasma, uterus, and swine ovary. The metabolite was most likely 15 keto- PGE_1 produced by PGDH. The significance of these results are discussed. (JRH)   003173

0121
PEARSE, W.H.; McCLURG, J.
The mode of action of cyclic AMP as a contraceptive.
American Journal of Obstetrics and Gynecology 109: 724-731. 1971.

Prostalgandins and their relationship to cAMP are mentioned in the text. (MT)   003121

0122
PERKLEV, T.; AHREN, K.
Effect of prostaglandins, LH and polyphloretin phosphate on the lactic acid production of the prepubertal rat ovary.
Life Sciences 10: 1387-1393. 1971.

A dose-response relation was established between the in vitro effect of prostaglandins (PG) and lactic acid production by the prepubertal rat ovary. The ovary was sensitive to a concentration as low as .025 μg PGE_1/ml incubation medium. When human luteinizing hormone (LH) and PGE_1 were added in combination to isolated ovaries, no additive effect on lactic acid production was seen. Polyphloretin phosphate, an antagonist to the smooth muscle contracting action of prostaglandins, antagonized the stimulatory effect of both LH and PGE_1. The results open the possibility that the formation of prostaglandins might be an intermediate step in the action of LH on ovarian glycolysis. (Authors) 003272

0123
PHARRISS, B.B.
Interactions of prostaglandins and hormones.
In: Foa, P.P., ed. The action of hormones. Genes to population. Springfield Ill., Charles C Thomas, 1971. p. 262-279.

Insulinlike effects and interactions with anterior and posterior pituitary hormones are discussed. Some recent findings on the possible role of $PGF_2\alpha$ in luteolysis are briefly mentioned. (RMS) 003380

0124
PHARRISS, B.B.; HUNTER, K.K.
Interrelationships of prostaglandin $F_2\alpha$ and gonadotropins in the immature female rat.
Proceedings of the Society of Experimental Biology and Medicine 136: 503-506. 1971.

Prostaglandin $F_2\alpha$ was tested in immature female rats to measure its effect on gonadotropin activity. Neither ovarian weight gain nor ovulation was stimulated by $PGF_2\alpha$ administration. Prostaglandin $F_2\alpha$ did have an inhibiting effect on the results of PMS (pregnant mare serum) and HCG (human chorionic gonadotropin) treatment, decreasing ovarian weight gain, and ovulation. PMS stimulation of uterine growth was enhanced by $PGF_2\alpha$. This enhancement is probably due to increased secretion of estrogen

by the ovary or by altering the ratio of estrogens to progestogens in the circulation. (Authors)
002728

0125
PHARRISS, B.B.
  Prostaglandin induction of luteolysis.
  In: Ramwell, P. and Shaw, J., eds. Prostaglandins. Annals of the New York Academy of Sciences 180:
    436-444. 1971.

Discussed are possible mechanisms whereby prostaglandins are luteolytic, the role of luteolysis in pregnancy, and the functionality of this information for design of a monthly antifertility preparation for human fertility control. The mechanism of luteolysis is hypothesized. The dichotomy is noted of the effects of $PGF_2\alpha$ in vivo and in vitro. The discovery of the effectiveness of $PGF_2\alpha$ as a luteolytic agent in certain animal species suggested that the pituitary was not important for prostaglandin-induced luteolysis. It is concluded that the mechanism whereby prostaglandins are luteolytic is still unanswered. However, the concept that luteolysis is the mechanism of pregnancy termination by prostaglandins, at least in laboratory animals, is clear. This mechanism is discussed. In primates, one of the possible mechanisms of pregnancy termination is the induction of menses by removing ovarian hormonal support (luteolysis), but at present the luteolytic phenomena have not been described in human subjects. Due to the similarity of the hormonal patterns in rhesus monkeys and human females, it is predicted that the prostaglandin will be effective in regulating pregnancy by terminating luteal progesterone synthesis and thereby inducing a 'natural menstruation.' The key to luteolysis in the human subject probably lies in correct timing. (ART)   002713

0126
PHARRISS, B.B.
  Prostaglandins in fertility
  Proceeding of the Royal Society of Medicine 64: 10. 1971.

The hypothesis that a venoconstrictor substance, liberated by the uterus, inhibits the corpus luteum by reducing the perfusion at a critical period was tested by continuously infusing or injecting $PGF_2\alpha$ into pseudopregnant rats and rabbits, cycling guinea pigs, and pregnant rats, rabbits, hamsters, and rhesus monkeys. A total resorption of fetal tissue and immediate menstruation resulted with evidence that $PGF_2\alpha$ operates directly on the ovary or its vascular supply. (RMS)   002663

0127
PICKLES, V.R.
  Bibliography on effects of prostaglandins on the reproductive system.
  Bibliography of Reproduction 18(1-6): 155-160. 1971.

The author states that this bibliography is not complete, even on this limited topic, nor is it intended as a kind of honors list. It has been compiled with the intention of most readily introducing the English-speaking newcomer to an already complex field. Consequently, the title has been followed strictly, and many preliminary or repetitive accounts or others that have been 'overtaken by events' are omitted. 152 citations are included. (MT)   002972

0128
PICKLES, V.R.
Prostaglandins and the I.U.D.
Lancet 1: 756. 1971.

A letter to the editor comments on a hypothesis (Chaudhuri, March 6, p. 480) that IUDs owe their effect to the release of prostaglandins from the endometrium, and gives supporting evidence. (RMS)
002617

0129
PINTO, R.M.; GIMENEZ, H.G.; DUNAIEVSKY, J.
The actions of prostaglandin PGE_1 and estradiol upon the contractile activity of the three separate layers of the pregnant human uterus.
International Journal of Gynaecology and Obstetrics 9: 8-11. 1971.

A study was made of the action of PGE_1 upon the 3 layers of the pregnant human uterus in vitro and its behavior in the presence of estrogens and oxytocin. Low doses of PGE_1 were found to stimulate the activity of the 3 layers of the pregnant human uterus, whereas high doses of PGE_1 brought about decrease and blockage of such activity. Addition to the bath of high doses of estradiol blocked the action of PGE_1. No differences in the action of PGE_1 upon the 3 layers of either the uterine body or the lower segment were observed. (Authors)   003126

0130
PION, R.J.; WABREK, A.J.; WILSON, W.B., Jr.
Innovative methods in prevention of the need for abortion.
Clinical Obstetrics and Gynecology 14: 1313-1316. 1971.

The authors mentioned the possible development of a prostaglandin tampon to be used as a menstrual inducer. An early abortion (1-2 weeks following a missed menstrual period) would lower the morbidity and mortality rate, and a vaginally administered prostaglandin may serve this purpose. (RAS)   003441

0131
PORTER, D.G.; BEHRMAN, H.R.
Prostaglandin-induced myometrial activity inhibited by progesterone.
Nature 232: 627-628. 1971.

$PGF_2\alpha$ has been shown to stimulate the myometrium in vitro, and in vivo it has an abortifacient effect. Little is known of the influence of other myometrial regulatory agents, such as the ovarian steroids, on the uterine response to prostaglandins. This paper reports a study of the influence of progesterone on myometrial responses to $PGF_2\alpha$ in the rabbit, a species in which progesterone is known to have a blocking action on the uterine muscle. $PGF_2\alpha$ was administered before and at the height of maximal inhibitory effect, and after recovery from the effects of progesterone given subcutaneously. Results demonstrate that progesterone inhibits the myometrial response to $PGF_2\alpha$ in the rabbit. Whatever the underlying cause, it seems clear that $PGF_2\alpha$ is a much less effective oxytocic agent in the presence of the progesterone block. If progesterone is a major uterine regulatory agent in the human female, the results of these experiments suggest that alternatives to a direct action on uterine muscle, such as an interference with steroidogenesis, should not be overlooked in seeking the mechanism by which prostaglandins induce abortion. (ART)   002746

0132

POYSER, N.L.; HORTON, E.W.; THOMPSON, C.J.; LOS, M.

Identification of prostaglandin $F_2\alpha$ released by distension of guinea-pig uterus in vitro.

Nature 230: 526-528. 1971.

The possibility that $PGF_2\alpha$ accounts for the luteolytic effects observed after distension of the guinea pig uterus was investigated. Both uterine horns were removed from 1 animal on Day 3 of the cycle and distended by insertion of a 3 x 30 mm piece of polyethylene tubing, which was then immediately removed from the control horn. Both horns were incubated for 3 hours at 37 degrees C in Tyrode solution gassed with 5% carbon dioxide in oxygen. The experimental horn sample was found by rat bioassay to contain 65 ng PGE_1 equivalent/ml activity, which resembled pharmacologically prostaglandin acitivity. 8 of 9 experimental horn preparations showed more smooth muscle contracting activity than untouched controls in a second experiment. Pooled samples were extracted for prostaglandinlike material. The experimental sample showed 40 ng of PGE_1 equivalent, while the control sample showed no detectable activity. Further tests eliminated the simple presence of tubing as a factor. Identification of the active factor evidently released in vastly different quantities from experimental and control pooled samples of 35 animals was $PGF_2\alpha$ based on column chromatography, bioassay, gas chromatography and mass spectrometry. (RMS)   002604

0133

RAMWELL, P.W.; SHAW, J.E.

The biological and clinical implications of the prostaglandins.

In: Abstracts, 162nd National Meeting of the American Chemical Society, September 12-17, 1971, Washington, D.C. (Abstract Medi 5)

PGs are a unique class of naturally occurring substances having diverse biological effects but strict structure-activity relationships. Cyclic AMP is implicated in their mechanism of action but it is not obligatory. Their clinical potential is great but side effects cause limitations. Manipulation of endogenous PGs, molecular modifications and greater knowledge of their cell biology may lead to greater utility for these substances. (HC)   003506

0134

RANGARAJAN, N.S.; LaCROIX, G.E.; MOGHISSI, K.S.

Induction of labor with prostaglandin.

Obstetrics and Gynecology 38: 546-550. 1971.

Labor was induced at term with oxytocin and prostaglandin $F_2\alpha$ in 40 patients who were distributed into 2 groups according to the state of cervical dilatation prior to induction. Average parity was similar in the 2 groups. 95% of the patients induced with oxytocin and 80% of those who received $PGF_2\alpha$ delivered. These results reveal that $PGF_2\alpha$ is not as effective as oxytocin in inducing labor. 40% of the patients who received $PGF_2\alpha$ exhibited 1 or more prolonged contractions during the course of induction. 2 instances of fetal bradycardia occurred, both associated with hypertonic contractions. More studies are needed before the safety of $PGF_2\alpha$ at term can be confirmed. (Authors)   002955

0135

RAVENHOLT, R.; SPEIDEL, J.

Prostaglandins in family planning strategy.

In: Ramwell, P. and Shaw, J., eds. Prostaglandins. Annals of the New York Academy of Sciences 180: 537-552. 1971.

World population statistics are briefly reviewed and the efforts of the Population Program Assistance element of the United States Agency for International Development (AID) in sponsoring research on means for improved control of fertility (especially for application in less developed countries), as well as the development of family planning programs, are discussed in some detail. The possible role of prostaglandins in achieving desired population or fertility control in the world is discussed briefly. (ART)   002723

0136
ROBERTS, G.; CASSIE, R.; TURNBULL, A.C.
Therapeutic abortion by intrauterine instillation of prostaglandin E_2.
Journal of Obstetrics and Gynaecology of the British Commonwealth 78: 834-37. 1971.

Intrauterine instillation of prostaglandin E_2 was used in an attempt to induce abortion in 20 pregnancies of less than 12 weeks gestation. The technique failed altogether in 7 patients, and achieved only incomplete abortion in the remaining 13. Evidence of uterine infection developed in 4 patients. The results obtained have not encouraged the continuing use of the method for terminating early pregnancy in this center. (Authors)   002958

0137
ROBERTS, G.; TURNBULL, A.C.
Uterine hypertonus during labour induced by prostaglandins.
British Medical Journal 1: 702-705. 1971.

Labor was induced successfully at or near term in 34 out of 35 cases by combined amniotomy and intravenous infusion of either prostaglandins $F_2\alpha$, E_2, or E_1. Of particular importance is the finding of hypertonus in 4 of the 18 cases induced with prostaglandin E_2. (Authors' modified)   002659

0138
ROTH-BRANDEL, U.; WIQVIST, N.; BYGDEMAN, M.
Effect of prostaglandins on the contractility of the non-pregnant human uterus in vivo.
Acta Obstetricia et Gynecologica Scandinavica 50: 35. 1971.

The effect on uterine contractility of intravenous administration of prostaglandin E_1 (PGE_1) and $F_2\alpha$ ($PGF_2\alpha$) was studied in nonpregnant women at various stages of the menstrual cycle. Uterine contractility was continously recorded by a micro balloon introduced into the uterine cavity. At 19 experiments on 15 women the substances were given as single intravenous injections in increasing doses (PGE_1 5-100 μg and $PGF_2\alpha$ 20-200 μg). The results showed that contrary to the usual inhibitory effect of PGE_1 on human myometrium in vitro, doses above threshold levels of both substances invariably induced stimulation of the uterus in terms of elevation of tone and, following the higher doses, increased motility. Quantitative evaluation assessed as tone elevation in mm Hg did not prove any strict dose-response relationship. The threshold doses defined as the lowest dose of each substance inducing a minimum tone elevation of 5 mm Hg were found to be around 20 μg for PGE_1 and 50 μg for $PGF_2\alpha$ in the majority of cases. That means that the nonpregnant uterus seems to be more sensitive to $PGF_2\alpha$ than the uterus at midpregnancy investigated in earlier studies, whereas the sensitivity to PGE_1 is about the same at both stages. There were too few experiments to permit any evaluation of the relation between uterine sensitivity and the stages of the menstrual cycle. The influence of continous intravenous infusion of $PGF_2\alpha$ on uterine contractility was illustrated in one case. This pressure curve showed that except an initial and lasting elevation of uterine tone reaching nearly 50 mm Hg, also a gradual increment of the motility was induced. Following an infusion period of 2 to 3 hours, a rather irregular uterine activity exhibiting contractions with a mean intensity of 75-100 mm Hg and a frequency of 15-20 contractions per 10 minutes was demonstrated. Effects of intravenously administered pure prostaglandin compounds studied by the methods described do not permit any conclusions concerning the role these substances may have in the process of human fertility. However, knowledge of the isolated effects of these highly active stimulatory substances might prove them suitable for therapeutic purposes. This has already been tried with some degree of success as far as the pregnant human uterus is concerned, e.g. for induction of labor at term and for initiation of therapeutic abortion at early stages of pregnancy. (Authors' modified)   003016

0139
ROTH-BRANDEL, U.
Prostaglandinernas effekt pa uterus kontraktilitet under in vivo forhallanden: en kliniak experimentell studie. [The effect of prostaglandins on uterine contractility in in vivo experiments: a clinical experimental study.]
Dissertation, Karolinska Institute, Stockholm. 1971. 35 p.

The effect of prostaglandins on the contractility of the pregnant and nonpregnant human was studied under in vivo conditions. Uterine contractility was recorded by measuring the amniotic pressure or by using the microballoon method and the tracings were analyzed by qualitative and quantitative methods. Prostaglandin $E_1(PGE_1)$, $E_2(PGE_2)$, $F_1\alpha$ ($PGF_1\alpha$) and $F_2\alpha$ ($PGF_2\alpha$) were administered as single intravenous injections or by continuous intravenous infusion. The intramuscular, intraamniotic, and vaginal routes of administration were applied in a limited number of cases. Both the PGE and PGF compunds had a stimulatory effect on uterine contractility. Single intravenous injections given during the first and second trimester of pregnancy elicited a rapid elevation of uterine tonus. The magnitude of this response as measured in mm Hg turned out to be dose dependent. A rough determination of the relative potency of the 4 compounds could be made by compiling the results obtained in the second trimester of gestation. Single intravenous injections given to nonpregnant women resulted as well in elevation of uterine tonus. However, any significant dose-response relationship could not be established in nonpregnant women, probably due to difficulties related to the method of recording. Uterine sensitivity was therefore investigated by determining the dose of prostaglandin that elicited a threshold response. This dose was approximately the same for the pregnant and nonpregnant uterus. The results did not indicate any major differences in sensitivity during the various phases of the menstrual cycle. Constant intravenous infusion of prostaglandin induced a gradual increment of uterine activity both in the second trimester of gestation and at term. Low infusion rates of PGE_1, PGE_2, and $PGF_2\alpha$ administered at or near term induced reasonably coordinated activity similar to that seen during spontaneous labor at term. The same low infusion rate of PGE_2 given at midpregnancy induced incoordinated activity with low intensity of the contractions. Administration of high doses of PGE_1 during the second trimester initially caused elevation of tonus superimposed on small frequent contractions. This state of uterine incoordination was followed by the development of more coordinated contractions of high intensity. The stimulatory response of single intravenous injections of PGE_1 and PGE_2 was compared in the first and second trimester of pregnancy with the corresponding effect of oxytocin and ergometrin. PGE_1 induced an elevation of uterine tonus of similar magnitude as that of oxytocin, although the duration of the prostaglandin response was approximately 3 times longer. Injection of ergometrin under the same experimental conditions resulted in a less pronounced elevation of tonus, but the response had considerably longer duration. Intravenous infusion of oxytocin during the second trimester resulted in more coordinated contractions than those obtained by PGE_1 and PGE_2. Intraamniotic and vaginal administration was not accompanied by increased uterine contractility. However, the doses applied were very low. Possible mechanisms of action of prostaglandin are discussed on the basis of these results. (Author's modified)   003159

0140
ROTH-BRANDEL, U.
Response of pregnant human uterus to low and high doses of prostaglandin E_1 and E_2.
Acta Obstetrica et Gynecologica Scandinavica 50: 159-166. 1971.

Uterine sensitivity to low doses of prostaglandin E_2 was compared at midpregnancy and at term. An analysis of the amniotic pressure recordings revealed that the uterus in midpregnancy responded by increased contractility to an infusion rate of the same order as that commonly used for induction of labor at term. However, the absolute intensity of the contractions was significantly higher at term than at midpregnancy. Intravenous infusion of high doses of prostaglandin E_1 (maximum dose without consistent subjective side effects) at midpregnancy stimulated the uterus to frequent contractions (6-8 contractions per 10 minutes) of gradually increasing intensity, which reached an average value of 35 mm Hg following 6 hours of infusion. The individual uterine response to prostaglandin varied considerably from one case to another. The 'maximum tolerable infusion rate' which could be administered without major subjective side effects also varied within a wide range. (Author's modified)
  002917

0141
RUSSELL, P.T.
Prostaglandin biosynthesis in the human placenta.
Journal of the American Oil Chemists' Society 48: 94A. 1971.

The biosynthesis of prostaglandins (PG) in human placental tissue has been investigated. Acetone powder preparations of placentas were incubated in buffered solution with tritiated arachidonic acid for 0, 5, and 30 minute intervals. Ethyl acetate extracts of the incubates acidified to pH 2 were separated into crude fractions on silicic acid columns using increasing concentrations of ethyl acetate in benzene. The radioactivity of peaks eluting with standard prostaglandin increased with the length of incubation time. The PG-like radioactivity was, further fractionated by 2 thin-layer chromatographic systems using Silica Gel G, with and without silver nitrate, and by columns of Amberlyst 15 ion exchange resin impregnated with silver nitrate. Only radioactivity cochromatographing with standard prostaglandins in each of the 4 systems was considered to be enzymatically formed prostaglandin. Results from these incubations indicate that the acetone powder preparation of the human placenta is capable of prostaglandin formation. The predominate prostaglandinlike radioactivity cochromatographed with PGA: standard on the argentation Silica Gel G TLC system. Less radioactivity was associated with PGE_2 standard. There was no indication of $PGF_2\alpha$ formation. The observed formation of PGA_2 and PGE_2 diene prostaglandins form arachidonic acid is consistent with the known pathways of prostaglandin biosynthesis. (Author)   003022

0142
SEGAL, S.J.; TIETZE, C.
Contraceptive technology: current and prospective methods.
Reports on Population/Family Planning 1: 1-24. July 1971.

This review article describes methods of contraception and contains a section on prostaglandins. It is suggested that prostaglandins, because of failures and side effects, will probably not replace present surgical methods in abortion, but they may be of value if some of these difficulties can be overcome. (JRH)   003288

0143
SELLNER, R.G.
Effects of prostaglandins on steroidogenesis by bovine luteal and rabbit ovarian tissue in vitro.
Ph.d. Dissertation, Pennsylvania State University, University Park. 1971. 47 p.

Following reports that prostaglandins of the E and F series exert luteolytic activity in the mammalian corpus luteum in vivo but not in vitro, an investigation was undertaken to determine the effects of prostaglandins E_1, E_2 and $F_2\alpha$ on progesterone production by bovine luteal slices and rabbit ovarian tissue during in vitro incubation. 3 bovine corpora lutea of pregnancy were incubated with each of the prostaglandins (E_1, E_2, and $F_2\alpha$) at concentration of .001, .100, and 10.0 μg/ml of KRB, both alone and in combination with LH (NIH-LH-B5, .05 μg/ml medium). Following 2 hour incubations, progesterone concentration of each sample was determined by a procedure involving ethanolic extraction followed by silica gel column and paper chromatography. Final quantitation was by ultraviolet spectrophotometry. All 3 of the prostaglandins tested increased progesterone production at the 10.0 μg/ml concentration ($p < .01$) when compared to progesterone values obtained from incubated control samples. Increases were noted at the .100 level for PGE_1 ($p < .01$) and for PGE_2 ($p < .05$). None of the 3 prostaglandins tested produced significant increases when used at the .001 μg/ml concentration. Sliced ovarian tissue from 9 sexually mature New Zealand white rabbits was incubated with each of the 3 prostaglandins at concentration of 1.0 μg/ml KRB. Each of the 3 prostaglandins was used in each of the 9 incubations. After 2 hour incubations, tissue samples were analyzed for progesterone content by a method involving homogenization with NaOH, extraction with ethyl acetate, purification by thin-layer chromatography, and quantitation by gas-liquid chromatography. Significant increases in progesterone production over incubated control values were found for PGE_1 ($p < .05$), PGE_2 ($p < .02$), and $PGF_2\alpha$ ($p < .01$). (Author)   003102

0144
SINGH, E.J.; CELIC, L.; SWARTWOUT, J.R.
Chromatographic separation of ovarian dermoid cyst lipids and prostaglandins.
Journal of Chromatography 63: 321-327. 1971.

Using Florisil column chromatography the ovarian dermoid cyst lipids were fractionated. Temperature-programmed gas chromatography showed the presence of carbon-14 to carbon-40 saturated hydrocarbons. The cholesteryl ester fraction had a high content of nyristic acid, 17.5%. There were no marked differences in the fatty acid composition of glycerides between mono-, di-, and triglycerides except for some differences in the content of saturated fatty acids. A glass fiber paper chromatographic method was developed for the separation and identification of prostaglandins. The following components were identified: PGA, PGB, PGE_1, PGE_2, $PGF_1\alpha$ and $PGF_2\alpha$. (Authors)   003361

0145
SMITH, W.L.; LANDS, W.E.M.
Stimulation and blockade of prostaglandin biosynthesis.
Journal of Biological Chemistry 246: 6700-6704. 1971.

An enzymic system of sheep vesicular gland which forms prostaglandins showed a time- dependent, concentration-dependent activation by phenol before full dioxygenase activity could be manifested. The activation process could be reversibly inhibited by O-phenanthroline. Aspirin and indomethacin did not instantly inhibit the dioxygenase, but acted in a time-dependent, concentration-dependent manner to block full activity of the synthetic system in an irreversible manner. The enzyme preparation was protected from the inhibitory action of these drugs by the presence of O-phenanthroline. (Authors)   003196

0146

SPELLACY, W.N.; BUHI, W.C.; HOLSINGER, K.K.

The effect of prostaglandin $F_2\alpha$ and E_2on blood glucose and plasma insulin levels during pregnancy. American Journal of Obstetrics and Gynecology 111: 239-243. 1971.

Blood glucose and plasma insulin were serially measured in 42 women who were receiving intravenous infusions of either prostaglandin E_2, prostaglandin $F_2\alpha$, or oxytocin. There were no significant changes in either the glucose or insulin levels as a result of the infusions. These results suggest that at the doses used the prostaglandins do not play a role in maternal carbohydrate metabolism and that they may be administered to pregnant women without producing significant carbohydrate metabolic alterations. (Authors)   003106

0147

SPEROFF, L.

Discussion of session on prostaglandins in female reproductive physiology.

In: Ramwell, P. and Shaw, J., eds. Prostaglandins. Annals of the New York Academy of Sciences 180: 513-517. 1971.

The highlights of this part of the conference are reviewed. The paradox is noted between the in vivo and in vitro effects of prostaglandins on ovarian tissue. The mechanism of the prostaglandin action in the induction of labor remains to be unraveled. An aspect of the clinical studies in pregnancy which is commented upon is the difficulty in achieving objectivity in evaluating results. A note of caution is stated in emphasizing that the prostaglandin is an experimental drug which still requires a great deal of investigation. Three areas which require concentrated effort are: development of a sensitive and specific assay, determination of optimum dosage and best system of delivery, and establishment of the safety of prostaglandins. It is, however, concluded that at this time no other agent holds as much promise as prostaglandins for safe fertility control. (ART)   002719

0148

SPEROFF, L.

The effect of prostaglandin $F_2\alpha$ infusion on free steriod levels in early pregnancy.

In: Sobrero, A.J. and Harvey, R.M. eds. Advances in Planned Parenthood. Vol. 7. (Proceedings of the ninth annual meeting of the American Association of Planned Parenthood Physicians, Kansas City, Missouri. April 5-6, 1971.) p. 151-154.

During the stepwise intravenous infusion of $PGF_2\alpha$, with final rate of 200 µg per minute, in 7 women 7 weeks pregnant, plasma progesterone levels did not change; estradiol and, especially, estriol levels declined sharply. An infusion of 50 µg per minute for 12 hours in a patients 11 weeks pregnant produced a gradual decline in progesterone and sharp falls in estrogen levels. Results were similar in a woman 20 weeks pregnant given 50 µg per minute for 6 hours on 2 consecutive days. The author concludes that $PGF_2\alpha$ does not have a luteolytic effect in early human pregnancy. (HC)   004655

0149

SPEROFF, L.

The prostaglandins -- enthusiasm and caution in equal doses.

Hospital Practice 8(2): 9. 1971.

Problems in evaluating the success of prostaglandins as abortifacients and oxytocics are outlined in an editorial stressing the need for research effort and financial support in the field of prostaglandin research. (RMS)   003294

0150
STURDE, H.C.; BOHM, K.
Das verhalten der sperma-prostaglandine unter gonadotropin-therapie. [Behavior of sperm prosta-
glandins under therapy with gonadotropin.]
Arzneimittel-Forschung 21: 986-989. 1971.

The present paper reports on the regular rise of the prostaglandin level in human seminal plasma after application of choriongonadotropine. The selected 24 subjects received 1000 I.U. choriongonadotro-pine (Primogonyl) 3 times a week over 6 weeks. Spermiograms including determination of sperm prostaglandin were taken before, during, and at the end of therapy, and once more after 4 weeks. 21 out of 24 treated men had inferior seminal quality (azoospermia, cryptospermia, oligospermia, hypospermia), which probably was the reason for the sterile marriage in 18 cases. While sperm volume, number of spermatozoa, and fructose concentration changed little if at all in 21 of the 24 patients, treatment with gonadotropine led to a steep continuous increase of sperm prostaglandins at first double, and later 2.5-fold of the initial value. In a previous study (Sturde, 1971) it was demonstrated that the production of prostaglandin in the glandulae vesiculares is stimulated by androgen. If under therapy with gonadotropine 21 of our patients showed such an increase of prostaglandins, regularly functioning Leydig cells can be assumed, which moreover have functioning capacity to spare. (Authors' modified)   003253

0151
STURDE, H.C.
Das verhalten der sperma-prostaglandine unter androgen-therapie. [Behavior of sperm prostaglandin
under therapy with androgen.]
Arzneimittel-Forschung 21: 1302-1307. 1971.

Sperm counts, fructose levels, and prostaglandin levels in the seminal fluid of men receiving androgen therapy for infertility were measured. Sperm counts and prostaglandin levels rose during the 12 weeks the men received 3 10-mg or 3 25-mg doses of androgen (mesterolone) intraorally. (The prostaglandin measured and the units used are not specified). It is suggested that the change in prostaglandin levels in patients on androgen therapy might prove useful in determining the success of the therapy, as it may reflect the effect of the androgen on Leydig-cells of the testes. (JRH)   003254

0152
STURDE, H.C.
Der einfluss des antiandrogens cyproteron auf die ejaculatbefunde junger manner einschliesslich der
sperma-prostaglandine. [The effect of cyproteron therapy on the semen and the sperm
prostaglandins production in young men.]
Archiv fur Dermatoligische Forschung 241: 86-95. 1971.

The semen of 22 young men after treatment with Cyproteron as an antiandrogen was analyzed. The daily oral dosage was 200-400 mg for 4-16 weeks. There were no differences in sperm volume, number of sperms per milliliter, and fructose concentration when compared to the normal range. However, after 6 weeks of Cyproteron therapy a steep increase of more than 3 times the baseline values (3.75 to 13.81) of the androgen-dependent sperm prostaglandins occurred. It is suggested that the inhibitory activity of androgens on the anterior lobe of the pituitary gland is abolished. In consequence more gonadotropines are produced, which stimulate the production of androgens by the Leydig cells. (Author's modified)   003372

0153

STURDE, H.C.; GLOWANIA, H.J.; BOHM, K.
  Vergeichende Ejaculatersuchungen bei Mannern aus sterilen and fertilen Ehen. [Comparative Ejaculate Investigations on Men from Sterile and Fertile Marriages]
  Arch fuer dermatologische Forschung 241: 426-437. 1971.

Sperm count and other ejaculate properties were compared from 100 men in fertile marriages to 100 men in sterile marriages. Prostaglandin content is mentioned as a potential factor which may influence fertility. (LJG)   002812

0154

SWEDIN, G.
  Biphasic mechanical response of the isolated vas deferens to nerve stimulation.
  Acta Physiologica Scandinavica 81: 574-576. 1971.

Isolated guinea pig vas deferens stimulated via the hypogastric nerve, were studied. It was found that 30-second stimulations of the hypogastric nerve produced an initial twitch (first phase) contraction followed by a slower (second phase) contraction that was often greater in amplitude than the twitch. PGE_1(1-5 ng/ml) decreased or abolished the twitch but had little or no effect on the second phase contraction. (JRH)   002873

0155

SWEDIN, G.
  Endogenous inhibition of the mechanical response of the isolated rat and guinea-pig vas deferens to pre- and postganglionic nerve stimulation.
  Acta Physiologica Scandinavica 83: 473-485. 1971.

Repeated periods with prolonged nerve stimulation (30 seconds) of the isolated vas deferens tended to depress especially the initial, rapid phase of subsequent contractions. The following findings led to the conclusion that the inhibition was induced by PG-like material released from the organ on nerve stimulation: 1) The inhibition was immediately abolished on washing of the organ; 2) it appeared faster in an organ bath of small volume than in one of a greater; 3) it was possible to transfer the substance from one bath to influence an organ in another; 4) the inhibitory agent, like exogenous PG, influenced both phases of contraction, the second, however, to a lesser extent; 5) the ganglionic transmission in the last part of the hypogastric nerve was found to be extremely sensitive to the inhibitory substance as to exogenous PG (.1-.4 ng/ml); 6) in vitro inhibition of PG synthesis led to a partial or total abolishment of the endogenous inhibition; 7) release upon nerve stimulation of PG-like material, mainly PGE_2, was established. It is suggested that this endogenous inhibiting process might play a modulating role in the nerve-induced mechanical activity of this organ also under in vivo conditions. (Author's modified)   003231

0156

SWEDIN, G.

Endogenous, possibly prostaglandin-mediated inhibition of the neuromuscular transmission in the vas deferens.

Journal of Pharmacy and Pharmacology 23: 994-995. 1971.

In this letter to the editor the author reported that prolonged nerve stimulation of isolated rat or guinea pig vas deferens causes the release of an endogenous inhibitory substance (most likely a prostaglandin), which mainly inhibits the twitch phase of the vas deferens contractions. It is suggested that this system may function in vivo to modulate nerve stimulation of this organ. (JRH)   002906

0157

THORBURN, G.D.; NICOL, D.H.

Regression of the ovine corpus luteum after infusion of prostaglandin $F_2\alpha$ into the ovarian artery and uterine vein.

Journal of Endocrinology 51: 785-786. 1971.

Ewes were prepared so that prostaglandin could be infused into either the uterine vein or the ovarian artery on the same side (other ovary removed). On Day 6, 7, 8, or 9 of the following estrous cycle the animals were infused with $PGF_2\alpha$ in either the artery or vein. Estrous behavior was judged by running the ewe with a vasectomized ram wearing a marking crayon. Venous blood was collected from the jugular vein and analyzed for progesterone. $PGF_2\alpha$ infused into either the ovarian artery (40 μg/h or 10 μg/h for 3 hours) or uterine vein (40 μg/h for 3 hours) caused a prolonged decrease in progesterone. Estrous behavior was suppressed but returned at 60 hours. Progesterone returned to normal after 3-4 days. PGE_2(40 μg/h for 6 hours) infused into the ovarian artery produced no change. Infusion of $PGF_2\alpha$ into the jugular vein did not produce a long-lasting effect. It was concluded that $PGF_2\alpha$ in physiological doses can cause regression of the corpus luteum in cyclic ewes. (JRH:) 003310

0158

TOPPOZADA, M.; BYGDEMAN, M.; WIQVIST, N.

Induction of abortion by intra-amniotic administration of prostaglandin $F_2\alpha$.

Contraception 4: 293-303. 1971.

Legal abortion was induced in 37 pregnant women in the second trimester by intraamniotic administration of prostaglandin $F_2\alpha$. The effect of different dose schedules on uterine contractility and clinical outcome was studied. Doses in the order of 5-25 mg were employed, demonstrating a marked and prolonged stimulatory effect on the uterus. The contractions became rather effective and fairly regular within 6 hours and were sustained for about 20 hours. The induction trial was successful in 89% of the cases with a mean induction-abortion interval of 28 hours. The possible mechanism of action and the results obtained by different doses are discussed. More experience is still required before a suitable dose schedule can be finally settled. This procedure seems to offer certain advantages over the currently used methods for termination of pregnancy in the second trimester. (Authors)   003094

0159

TOTHILL, A.; BAMFORD, D.; DRAPER, J.

Inhibition of prostaglandin release and the control of threatened abortion.

Lancet 2: 381. 1971.

The works of Vane, Smith, and Willis; Ferreira et al.; Beazley; and Sih et al. are briefly mentioned in this letter to the editor. The authors have shown in previous studies that in late pregnancy in rats PGE_2 can be detected and quantitated from uterine strips treated in vitro with catecholamines. It is now reported that the appearance of PGE_2 can be prevented under these circumstances by using a number of different psychotropic agents, of which desipramine appears to be the most potent. Others include tranylcypromine, phenelzine, and chlorpromazine. The standard treatment of threatened abortion relies upon the use of sedative drugs--for example, promazine--and it has always been assumed that their beneficial effect stems from their general sedative action. The work reported here raises the possibility that promazine and allied drugs may act by preventing the liberation of prostaglandin and thereby blocking its action on the uterus, and that their action may furthermore be potentiated by combining them with the anti-inflammatory drugs (aspirin, indomethacin, aspirin-like drugs). (ART) 002782

0160

TOTHILL, A.; RATHBONE, L.; WILLMAN, E.
Relation between prostaglandin E_2 and adrenaline reversal in the rat uterus.
Nature 233: 56-57. 1971.

The inhibitory response of the rat uterus to adrenaline in the presence of isoprenaline inhibition can be reversed in induced estrus but not in the progestational state, and it has been postulated that a motor substance accumulates during the isoprenaline inhibition and is released on addition of adrenaline. The same sequence of events has also been observed when adrenaline was used as the initial inhibitory agent. In an attempt to detect and identify the postulated motor substance, uterine horns or strips were set up in organ baths, and after addition of adrenaline it was found that a motor substance was released into the bath fluid from the strips of pregnant uterus. Dried extract from the bath fluid was tested and examined by paper chromatography with the AI and AII solvent systems of Green and Samuelsson. The biological activity behaved in a manner identical to that of a reference sample of authentic prostaglandin E_2 and the AII system showed it to be E_2. Other tests confirmed the identification. It is concluded that adrenaline reversal, wherever it occurs, is brought about by liberation of prostaglandin. (ART)   002742

0161

TYACK, A.J.; BAILLE, P.; MEEHAN, P.F.
In-vivo response of the human uterus to orciprenaline in early labour.
British Medical Journal 2: 741-743. 1971.

The effect of orciprenaline, a beta-adrenergic stimulant, on uterine activity in 10 women in early induced labor at term with intact membranes was studied. 9 had uterine contractions in response to intravenous oxytocin and 1 to PGE_2. No difference in orciprenaline effect was observed in the patient contracting in response to PGE_2 from those given oxytocin. The effect of the drug was clearly dose dependent. The effective dose varied between 10 and 20 μg/min. Tachyphylaxis was not observed. The only significant effects were tachycardia and increased pulse pressure in the mother and a smaller increase in heart rate in the fetus. (Authors' modified)   003347

0162

VENTURA, W.P.; FREUND, M.
Prostate and semen spasmogens: A new class of uterine stimulants.
Federation Proceedings 30: 476. 1971.

The authors report the isolation of a new class of compounds from the seminal fluid and prostate gland fluid of humans, rats, and guinea pigs which has uterine stimulatory properties. These compounds are not prostaglandin. (JRH)   003322

0163
VERMA, O.P.; HAWES, R.O.
The effect of semen on motility of the chicken oviduct.
Poultry Science 50: 199-203. 1971.

The in vitro effect of fresh, undiluted cock semen on the circular muscles of 4 regions of the hen reproductive tract was observed. Results were inconclusive, but the authors feel that semen does not show an oxytocinlike effect. Several papers on the presence and function of prostaglandins in seminal plasma are mentioned in discussion. (RMS)   002726

0164
WALLACH, D.P.; DANIELS, E.G.
Properties of a novel preparation of prostaglandin synthetase from sheep seminal vesicles.
Biochimica et Biophysica Acta 231: 445-457. 1971.

A system of improving the yield of prostaglandin E_2 biosynthesis by sheep seminal vesicular tissue is described. Microsomal synthetase is precipitated by acidification to pH 4.8-5.0 or by adding acetone to a concentration of 30% and centrifuging. An acetone-pentane powder of this microsomal precipitate retains full synthetase activity. It is suggested that synthetase is inhibited by one or more reaction products as the conversion of arachidonic acid to prostaglandin E_2 approaches asymptote after 10 minutes at 30 degrees at which point the enzyme may be removed and used to catalyze a second conversion. Cycling 10 times with fresh enzyme added to compensate for losses results in 3-fold increases in yields. (RMS)   002674

0165
WILLMAN, E.A.
The extraction of prostaglandin E_1 from human plasma.
Life Sciences 10: 1181-1191. 1971.

This paper describes a method of extraction of prostaglandins from plasma by a modification of that of Hickler. The phospholipid free plasma lipids were partitioned between hydrochloric acid, pH 3, and ethylacetate prior to silica gel chromatography. Consistent results were obtained in the range 2-400 ng/ml PGE_1 per ml plasma using few manipulations. (Author's modified)   003133

0166
WILSON, L., Jr.; CENEDELLA, R.J.; BUTCHER, R.L.; INSKEEP, E.K.
Endometrial prostaglandins during ovine estrual cycle.
Biology of Reproduction 5: 83. 1971.

Levels of prostaglandins (PG) $F_2\alpha$ and E_2 were measured in endometrium from each uterine horn of 35 mature ewes with corpora lutea on only one ovary. After the ewes were killed the endometria were dissected free, weighed, and stored at -20 degrees C. Prostaglandins were extracted, separated by thin layer chromatography, and measured by the gerbil-colon bioassay. The percentage of recovery was determined by using $PGF_1\alpha$-5,6-tritum, and the measured values were corrected to 100%. Ewes in 4 groups were killed on Days 3, 5, 11, or 14 of the estrual cycle, and the mean concentration (ng/g) for both uterine horns of $PGF_2\alpha$ was 33, 62, 75, and 201, and for PGE_2, 6, 3, 9, and 18, respectively. The levels of $PGF_2\alpha$ at Day 14 were different from all other days ($p<.05$). Ewes in a fifth group received a plastic intrauterine device (IUD) in each uterine horn at laparotomy on Day 2 of the estrual cycle and were killed on Day 5. They had a mean concentration (ng/g) for $PGF_2\alpha$ of 178 and for PGE_2 of 17. The concentration of $PGF_2\alpha$ was different for the Day 5 control (178 vs. 62; $p<.05$). Differences in prostaglandin $F_2\alpha$ levels were detected between uterine horns adjacent and opposite to a corpus luteum only in the Day 5 IUD-treated group (203 vs. 153; $p<.05$). (Authors' modified)   003053

0167
WILSON, L., Jr.; CENEDELLA, R.J.; BUTHCER, R.L.; INSKEEP, E.K.
Progesterone treatment on ovine endometrial prostaglandins.
Journal of Animal Science 33: 273. 1971.

2 experiments were designed to study the effects of progesterone treatment on levels of prostaglandins $F_1\alpha$, $F_2\alpha$, and E_2 in uterine endometrium of mature ewes. After the ewes were killed the endometria were dissected free, weighed, and stored at 20 degrees C. Prostaglandins were extracted, separated by thin layer chromatography, and measured by the gerbil-colon bioassay. The percent recovery was determined by using PGF1 a-5, 6-tritium, and the measured values were corrected to 100%. In the first experiment 16 ewes received 40 mg of progesterone (P) and 16 ewes received 1 ml corn oil (C) sc on Days 0 and 1 of the estrous cycle (estrus = Day 0). Half of each group were killed on Day 5 and the remainder on Day 9. The contents (ng) and concentrations (ng/g) of $PGF_1\alpha$ and $PGF_2\alpha$ were: Day 5C-$F_1\alpha$ (233, 17), $F_2\alpha$ (156, 11); Day 5P-$F_1\alpha$ (660, 89), $F_2\alpha$ (378, 32); Day 9C-$F_1\alpha$ (386, 32), $F_2\alpha$ (200, 17); and Day 9P-$F_1\alpha$ (373, 29), $F_2\alpha$ (324, 24). In the second experiment 2 groups of ewes were bilaterally ovariectomized on Day 4 of the estrous cycle and received either C (8 ewes) or 10 mg P/day on Day 4 and 8 (7 ewes). All ewes were killed on Day 9. The contents (ng) and concentrations (ng/g) of $PGF_1\alpha$ and $PGF_2\alpha$ were: C-$F_1\alpha$ (684, 72), $F_2\alpha$ (575, 64); P-$F_1\alpha$ (410, 37), $F_2\alpha$ (169, 14). Both content and concentration of $PGF_2\alpha$ were reduced ($p<.05$) by progesterone treatment. Over both experiments, PGE_2 was measurable in 12 of 46 samples and detectable in an additional 20 samples. (Authors)   003259

0168
WIQVIST, N.; BYGDEMAN, M.
Administration of prostaglandin for termination of pregnancy.
In: Abstracts, Seventh World Congress on Fertility and Sterility, Amsterdam, Excerpta Medica, 1971.
   (International Congress Series 234b) p. 117.

50-75 µg $PGF_2\alpha$ per minute intravenously for 7 to 8 hours aborted more than 90% of patients less than 8 weeks pregnant. Increasing the concentration to 100 µg per minute and the duration to 24 hours provided an 83% success rate in second trimester patients. Side effects of vomiting and diarrhea were most frequent and distressing at 100 µg per minute. 50 to 100 mg $PGF_2\alpha$ or 5 to 10 mg PGE_2 intravaginally produced marked uterine contractions. Intermittent intrauterine injections of 200 to 750 µg $PGF_2\alpha$ or 25 to 75 µg PGE_2 resulted in sustained intense uterine contractions, 19 of 26 patients aborted, and a low incidence of side effects was noted. (HC)   003495

0169
WIQVIST, N.; BYGDEMAN, M.; TOPPOZADA, M.
Induction of abortion by the intravenous administration of prostaglandin $F_2\alpha$.
Acta Obstetricia et Gynecologica Scandinavica 50: 381-389. 1971.

166 women in the first and second trimester of pregnancy were given $PGF_2\alpha$ intravenously at rates of 50, 75 or 100 μg per minute for 8 to 26 hours. The shorter was the duration of pregnancy, the higher the abortion success rate. At either low or high dosage nearly all abortions occurred between the tenth and twentieth hour of infusion. Abortion was successful in less than 30% of the women given the infusion at 50 or 75 μg per minute but 75% of the women in the 100 μg per minute group aborted. Parity was not a factor in determining success of abortion. Nearly all patients had side effects, especially vomiting and diarrhea. The authors suggest that the greater effectiveness of PG in early pregnancy is probably due to a greater intrauterine pressure. They conclude that for an acceptable success rate, high doses of $PGF_2\alpha$ must be used for protracted periods of time. However, the high incidence of side effects associated with this procedure represent an obstacle toward the practical and routine use of PG intravenously. Therefore, vacuum curettage is recommended for termination of early pregnancies and routes other than intravenous for administration of PG are preferable in second trimester abortion. (HC)   003406

0170

WIQVIST, N.; BYGDEMAN, M.; KIRTON, K.T.
  Non-steroidal antifertility agents in the female.
  In: Diczfalusy, E., and Borell, U., eds. Nobel Symposium 15, Control of human fertility. Stockholm, Almqvist and Wiksell, 1971. p. 137-155.

This paper delivered at a symposium deals mainly with the effects of prostaglandins on the human uterus during different phases of the menstrual cycle and pregnancy. The authors review their own and others' work which has shown that prostaglandins given intravenously stimulate uterine activity in all phases of pregnancy and during the menstrual cycle. It was also found that the threshold dose of PGE_1 is about the same in the pregnant and nonpregnant uterus, but the threshold dose for $PGF_2\alpha$ is much higher in the pregnant uterus than in the nonpregnant one. Unlike oxytocin, prostaglandins are highly stimulatory to the early pregnant uterus and can serve as efficient abortifacients at this stage. While $PGF_2\alpha$ has been shown to reduce peripheral progestin levels in animals, preliminary results with humans show no such effect in either pregnant women or nonpregnant women immediately after ovulation. (JRH)   003222

0171

ABDUL-KARIM, R.W.; BEYDOUN, S.N.
  Amniotic fluid: the value of prenatal analysis. First of two parts.
  Postgraduate Medicine 52: 147-149, 151, 153. August 1972.

This part of the review covers gross appearance, osmolality, electrolytes, pH, urea, uric acid, creatinine, proteins, amino acids, lipids, and prostaglandins which have all been measured in amniotic fluid during fetal development. Clinical applications of these data include lecithin/sphingomyelin ratio for predicting new born respiratory distress syndrome; creatinine for estimating fetal maturity; pigment or meconium staining indicating erythroblastosis, growth retardation, dysmaturity; and high osmolality correlating with poor perinatal outcome. Research is directed towards diagnosing amino acid disorders and predicting Apgar score by acid-base balance. PGFs have been shown to increase in labor, but no clinical applications of this fact have been made. (LJG)   004921

0172

AIKEN, J.W.

Aspirin and indomethacin prolong parturition in rats: evidence that prostaglandins contribute to expulsion of fetus.

Nature 240: 21-25. November 3, 1972.

Rats 18 to 21 days pregnant were given indomethacin .1 mg/kg or 1.0 mg/kg or aspirin 10, 30, or 100 mg/kg orally twice a day. 3 of 10 animals receiving the low dose and 8 of the 10 given the high dose of indomethacin had excessive vaginal bleeding, greater than 4 hour prolongation of parturition, and a high incidence of fetal mortality. 8 of 10 rats given 100 mg/kg and 1 of 10 given 30 mg/kg aspirin died. 3 animals receiving 30 mg/kg and 1 at 10 mg/kg had adverse effects similar to those described for indomethacin, especially a high number of stillbirths. Uteri taken from rats sacrificed during parturition and placed in tissue baths produced about 20 times more PGF-like substances than uteri from rats 18 to 19 days pregnant. The level of PGE-like substances did not increase. Uteri from pregnant animals treated with 1.0 mg/kg indomethacin released less PGF-like material than did uteri from untreated animals. Adrenalin added to the tissue bath relaxes the uterus; spontaneous activity returned in the untreated sample within 3 minutes but uteri from indomethacin-treated animals remained relaxed for more than 30 minutes. Aspirin 100 to 300 μg/ml or indomethacin .01 to 1.0 μg/ml added to the tissue bath reduced or blocked spontaneous contractions and release of both PGE and PGF-like materials. From these results the author concludes that factors other than prostaglandin may regulate time of onset of parturition but, once started, F prostaglandins are required for expulsion of the fetus. Further, use of antiinflammatory agents late in pregnancy should be contraindicated. (HC)   004513

0173

ALDERMAN, B.

Abortion with prostaglandins.

Lancet 2: 279. 1972.

The author reports achievement of 100% success in aborting 23 patients with either PGE_2 or $PGF_2\alpha$ given by the intrauterine route. In commenting on the problem of expulsion of the catheter, the author suggests the problem can be solved by inflating the balloon on the catheter to its full 30 ml capacity. Besides preventing premature expulsion of the catheter, the larger balloon may also help prevent leakage of prostaglandin through the cervix. (JRH)   004125

0174

ALTURA, B.M.; MALAVIYA, D. REICH, C.F.; ORKIN, L.R.

Effects of vasoactive agents on isolated human umbilical arteries and veins.

American Journal of Physiology 222: 345-355. 1972.

A variety of vasoactive substances known to be found in circulating blood was tested on longitudinal and helical strips of human umbilical arteries and veins in vitro. Of the substances tested, serotonin produced the greatest contraction. $PGF_2\alpha$ and PGA_1 (100 μM to 10 mM) produced from 45% to 90% of the contraction caused by serotonin, depending on which preparation was used (longitudinal or helical, artery or vein). Pyrilamine, phenoxybenzamine, atropine, and UML-491 failed to inhibit the action of PGA_1 (10 μg/ml) on these vessels. (JRH)   003785

0175
AMBACHE, N.; BUNK, L.P.; VERNEY, J.; ZAR, M.A.
Inhibitory nature of the adrenergic innervation in the guinea-pig vas deferens.
British Journal of Pharmacology 44: 359P-360P. 1972.

Twitches were elicited by 1 to 8 pulses (10 Hz, .2-1 ms duration) and recorded isometrically. All other details of experimental procedures are given by Ambache and Zar (1970, 1971), who presented evidence against motor transmission in the vas being adrenergic, and suggested that the adrenergic innervation might subserve an inhibitory function. This is corroborated by obtaining twitch inhibition with the indirectly acting sympathomimetics, tyramine and cocaine. Pretreatment with reserpine phosphate (.5 mg/kg sc 2 days, and 1.5 mg/kg ip 1 day before use) abolished inhibition by tyramine and cocaine without altering noradrenaline or PGE_2 inhibitions. This confirms that the mechanism of inhibition by tyramine or cocaine involves endogenous noradrenaline. If motor transmission were adrenergic, tyramine should have produced contractions in the normal vas deferens by releasing transmitter, but this was never seen in guinea pig preparations. The degree of inhibition produced by noradrenaline, tyramine, or cocaine remained unaltered after exposure for 3 hours to indomethacin or sodium meclofenamate (1-2 μg/ml), both of which abolish prostaglandin synthesis (Gryglewski & Vane, 1971). It is therefore unlikely that the noradrenaline inhibition is mediated by release of endogenous prostaglandin as proposed by Swedin (1971). Moreover, in guinea pig (and rabbit) vas deferens, phentolamine (2 μg/ml) antagonized noradrenaline inhibitions without affecting PGE_2 inhibitions. In addition, the motor transmission in other species (rat and Meriones vas deferens) was virtually unaffected by PGE_2 (2-500 ng/ml) but could be inhibited by tyramine or noradrenaline after the motor response to noradrenaline was abolished by phenoxybenzamine (1 μg/ml). In conclusion, the failure of tyramine to induce contractions, the ability of tyramine and cocaine to inhibit the motor transmission, and the loss of this ability after reserpine, all substantiate the inhibitory adrenergic function postulated previously. The existence of a prostaglandin link in noradrenaline inhibition seems improbable. (Authors' modified)   003751

0176
ANDERSON, G.G.; HOBBINS, J.C.; SPEROFF, L.
Intravenous prostaglandins E_2 and $F_2\alpha$ for the induction of term labor.
American Journal of Obstetrics and Gynecology 112: 382-386. 1972.

Prostaglandins E_2 and $F_2\alpha$ and synthetic oxytocin were studied for efficacy and side effects in a double-blind protocol for the induction of labor at term in 100 women. That an inducibility scoring index should be part of each clinical oxytocin trial is evidenced by the fact that erroneous conclusions would have been drawn if it had not been used. All patients who were classified as having 'easy inductions' were delivered independent of which drug they received; however, the 3 categories of graded 'difficult inductions' reflected success rates from 93.3%-40% depending upon difficulty of induction. Synthetic oxytocin and prostaglandin $F_2\alpha$ were found to be equally efficacious in the difficult groups, although more cases will be required to allow significant statistical analysis. Except for an increased incidence of innocuous uterine hypertonus, no significant side effects of prostaglandin infusions were noted. (Authors' modified)   003711

0177
ANDERSON, G.G.; HOBBINS, J.C.; SPEROFF, L.; CALDWELL, B.V.
Intravenous prostaglandins E_2 and $F_2\alpha$ and Syntocinon for the induction of term labor.
In: Southern, E.M., ed. The prostaglandins: clinical applications in human reproduction. (Brook Lodge Symposium, Augusta, Michigan.) Mount Kisco, New York, Futura Publishing Co., 1972. p. 85-94.

Labor was induced in 148 women at term by the incremental intravenous infusion of $PGF_2\alpha$ 2.5 μg to 40 μg per min, PGE_2.3 to 4.8 μg per min or oxytocin. The success rate was approximately 80% in each group. Those women classified as failures had Bishop scores less than 6. The IDT was 8.5 to 9 hours for women with Bishop scores of 0 to 6 and approximately 5 hours for women with higher Bishop scores whether given $PGF_2\alpha$ or oxytocin. There were no depressed infants delivered in this study. Uterine hypertonus was noted in 9 women given PG, 4 other women developed postpartum hemorrhage. There was no correlation between preinduction maternal plasma progesterone levels and Bishop scores. Plasma estriol levels dropped during delivery in 3 of 5 women given $PGF_2\alpha$. There was no significant difference in the mean maternal serum PGF levels between patients given $PGF_2\alpha$ and those given oxytocin. The authors conclude that oxytocin and $PGF_2\alpha$ given intravenously at term are indistinguishable in terms of efficacy or induction time. (HC)   004851

0178

ANDERSON, G.G.; HOBBINS, J.C.; RAJKOVIC, V.; SPEROFF, L.; CALDWELL, B.V.
Midtrimester abortion using intraamniotic prostaglandin $F_2\alpha$.
Prostaglandins 1: 147-155. 1972.

Prostaglandin $F_2\alpha$ ($PGF_2\alpha$) was instilled intraamniotically in varying amounts and at different intervals in 40 patients in the second trimester of pregnancy. Using an F prostaglandin radioimmunoassay as a clinical tool, a regimen was developed in which 40 mg of $PGF_2\alpha$ were instilled in a single injection. Of 35 patients so treated, 26 aborted completely and 9 partially. The results compare favorably with those reported with hypertonic saline. The mean treatment-abortion interval among the 26 patients who received the high dose was 22.3 hours. The side effect rate was low and acceptable, and no effect was noted on blood clotting mechanisms. (Authors)   003832

0179

ANDERSON, G.G.; HOBBINS, J.C.; RAJKOVIC, V.; SPEROFF, L.; CALDWELL, B.V.
Midtrimester abortion using intra-amniotic prostaglandin $F_2\alpha$ with intravenous Syntocinon.
In: Southern, E.M., ed. The prostaglandins: clinical applications in human reproduction. [Brook Lodge Symposium, Augusta, Michigan.] Mount Kisco, New York, Futura Publishing Co., 1972. p. 417-422.

40 women, 16 to 20 weeks pregnant, were given $PGF_2\alpha$ intraamniotically, 40 mg initially and another 10 to 20 mg dose 24 hours later if needed. Simultaneously, oxytocin was given intravenously by infusion at a rate of 66 mU per minute from time zero til case completion. All patients aborted; 37 were complete. Similarly, 36 of 40 women given only $PGF_2\alpha$ had complete abortions. The IAT averaged 23 1/2 hours for oxytocin group, compared to 23 hours for the women given only $PGF_2\alpha$. All patients experienced cramping, most had nausea and vomiting, diarrhea occurred in more women given oxytocin and $PGF_2\alpha$ than in those given $PGF_2\alpha$ alone. Uterine contractions were similar between groups as to onset, intensity, and form. Blood loss was greater in the women given both oxytocin and $PGF_2\alpha$. (HC)   004880

0180

ANDERSON, G.G.
Panel discussion Session IV labor induction.
In: Southern, E.M., ed. The prostaglandins: clinical applications in human reproduction. (Brook Lodge Symposium, Augusta, Michigan.) Mount Kisco, New York, Futura Publishing Co., 1972. pp. 223-239.

4 panelists review the Filshie and Craft studies on labor induction. In 100 cases induction was 98% successful, following oral prostaglandin E_2. In the second study, 29 of 32 primigravidae and 17 of 18 multigravidae delivered following oral PGE_2. Side effects were minimal in both studies. The disscussants spoke of varying dosage, PG metabolism, the relationship of uterine blood flow to parturition and induction of troublesome deliveries i.e., hydatidiform mole. (RAS)   004864

0181
ANDERSON, G.G.; HOBBINS, J.C.; SPEROFF, L.; CALDWELL, B.V.
The induction of therapeutic abortion using intravenous prostaglandin $F_2\alpha$.
Contraception 5: 303-311. 1972.

Intravenous prostaglandin $F_2\alpha$ was given to 42 patients to induce abortion. The overall success rate was about 50% using 3 separate protocols in which the dosage varied between 25 and 200 $\mu g/min$ and in which the duration of infusion varied from 6 to 24 hours. A high and unacceptable rate of side effects was associated with all dosage levels, but especially with the highest rate of infusion. There was no association of eventual abortion with duration of pregnancy, but there was a marked association with parity; multiparas aborted 2.5 times more frequently than primigravidas. It is concluded that intravenous prostaglandins given in accordance with the described protocols is not a clinical approach of terminating human pregnancy which offers any advantage over existing methods. (Authors)   003907

0182
ANONYMOUS
Clinical use of natural prostaglandins for fertility control and termination of pregnancy.
In: Bergstrom, S., Green, K., and Samuelsson, B., eds. [Proceedings of the] third conference on prostaglandins in fertility control, Stockholm, January 17-20, 1972. Stockholm, WHO Research and Training Centre on Human Reproduction, Karolinska Institutet [1972]. (Prostaglandins in Fertility Control 2) p. 11-16.

Various methods of inducing abortion with PGs are compared with each other and with other methods of inducing abortion. Intravenous administration is generally unacceptable because of the high incidence of side effects. Intrauterine or intraamniotic administration is superior to intraamniotic saline for inducing abortion in the second trimester because the success rate is similar, the incidence of side effects is lower and the induction interval is shorter. Intravaginal application of PGs needs more investigation and might prove to be a valuable method of postconceptional fertility control. A detailed comparison of preliminary results of an intrauterine and intraamniotic (2-dose levels) protocol is presented. The success rates (expulsion of the fetus) are 84% and 61%-96% respectively. (JRH)   004452

0183
ANONYMOUS
Discussion. Session II basic science.
In: Southern, E.M., ed. The prostaglandins: clinical applications in human reproduction.(Brook Lodge Symposium, Augusta, Michigan.) Mount Kisco, New York, Futura Publishing Co., 1972. p. 67-71.

Most of this discussion concerned attempts to demonstrate luteolysis in nonpregnant women by injecting PG and measuring hormone levels. 4 participants, R. Jewelewicz, S.M.M. Karim, C. Falman and G. Seegar-Jones, infused $PGF_2\alpha$ into small groups of women in luteal phase who experienced cramps and bleeding but normal menstruation at the expected time. Although conclusions regarding steroid and gonadotropin hormone were tentative because of inadequate sampling to reveal their pulsatile behavior, it was clear that $PGF_2\alpha$ was not luteolytic. (LJG)   004849

0184
ANONYMOUS

Discussion. Session 6. Therapeutic abortion.

In: Southern, E.M., ed. The prostaglandins: clinical applications in human reproduction. (Brook Lodge Symposium, Augusta, Michigan.) Mount Kisco, New York, Futura Publishing Co., 1972. p. 391-395.

The discussion following the sixth session on therapeutic abortion at the Brook Lodge Symposium on PGs had a central theme of fever and infection, but also a short presentation of a trial of iv $PGF_2\alpha$ or PGE_2 combined with oxytocin, and another presentation of the progesterone levels measured during iv PGE_2. When Beazley infused PGE_2 1 μg/minute, none of 7 women aborted; $PGF_2\alpha$ 10 μg/minute aborted 1 of 10; $PGF_2\alpha$ 10 μg/minute with oxytocin 64 μg/minute aborted 6 of 10; and PGE_2 1 μg/minute with oxytocin 64 μg/minute aborted 8 of 10. These were all complete abortions without any side effects. The participants felt that the fever, chills, and cold sensations during abortion induced by PGs were probably from hypothalamic stimulation, since the incidence of fever or endometritis possibly caused by infection was only about 1 or 2 cases in each series. There is a risk of infection when the membranes rupture without successful abortion: 2 discussants suggested using intravaginal PG or 15-methyl analog 48 hours after rupture of membranes. Filshie showed a graph of progesterone levels measured in 8 patients infused intravenously with constant 5 μg/minute PGE_2; progesterone did not differ before, during, or after PG infusion. (LJG)   004877

0185
ANONYMOUS

Discussion. Session 7. Therapeutic abortion.

In: Southern, E.M., ed. The prostaglandins: clinical applications in human reproduction. (Brook Lodge Symposium, Augusta, Michigan.) Mount Kisco, New York, Futura Publishing Co., 1972. p. 471-489.

In this lengthy discussion by 19 participants, several clinical reports are presented followed by researchers' opinions and comments. N. Lauerson found that complication rate was correlated with length of abortion interval and that oxytocin side effects could be decreased by alternatively infusing Ringer's solution and dextrose with oxytocin. I. Craft achieved higher success with PGE_2 at increasing doses, and chance of delivering a live fetus decreased with increasing dose. J. Elias tried 25 and 50 mg $PGF_2\alpha$ vaginal suppositories with 23 patients, which resulted in 19 abortions and 4 failures, with unacceptable side effects in most of the patients. R. Jewelewicz measured declines in HCG, human placental somatomamotropin, progesterone, and estrogens during $PGF_2\alpha$ infusions. These changes could be due to maternal infusion of fetal blood. F. Fuchs discussed the complication of cervical laceration. N. Wiqvist used a single extraamniotic injection of 15-methyl-$PGF_2\alpha$ for cervical dilatation on the day before vacuum aspiration. There was discussion on the incidence of lactation, and of chronoperiodicity affecting PG abortions. Birnhill reminded his colleagues that PGs must be considered in comparison with many other substances besides hypertonic saline that can be used to induce midtrimester abortion. (LJG)   004886

0186
ANONYMOUS

International Conference on prostaglandins.
Prostaglandins 2(5): November 1972.

Papers presented at the Vienna International Conference on Prostaglandins covered a wide range of subjects. Mid-trimester abortions have had successful results with both the intraamniotic and extraamniotic routes. Termination of early pregnancy by intrauterine, intravaginal or intravenous routes have been disappointing. A detailed explanation of PGA isomerase activity and the synthesis of PGC was presented. Other topics were renal blood flow and micropuncture, adrenal stimulation of aldosterone secretion by PGA_1, inhibition of gastric secretion and uterine contractility. Prostaglandins E_1 and E_2 increase sperm transport but additional research is needed into male reproduction. (RAS) 004576

0187
ANONYMOUS

Prostaglandin effect studied in cases of missed abortion.
Hospital Tribune World Service. September 18, 1972. p. 17.

$PGF_2\alpha$ administered intravenously with a simultaneous paracervical block was an effective procedure in 8 cases of missed abortion and 2 cases of intrauterine death. The paracervical block allowed for adequate dilation of the cervix and eliminated subsequent pain. (RAS)   004217

0188
ANONYMOUS

Prostaglandins and reproduction.
British Medical Journal 3: 337-338. August 5, 1972.

Prostaglandins and labor, prostaglandins and termination of pregnancy, and the use of prostaglandins for contraception in the future are discussed in this report of a symposium. (RAS)   004225

0189
ANONYMOUS

Prostaglandins go commercial.
Science News 102: 282. 1972.

The first commercially available prostaglandin has been approved in Britain for use to induce labor and terminate pregnancy. The products named Prostin E_2 and Prostin $F_2\alpha$ are manufactured by the Upjohn Company. (RAS)   004484

0190
ANONYMOUS

Prostaglandins speed induction time in therapeutic abortions.
Journal of the American Medical Association 220: 1556. 1972.

20 women, 13 to 20 weeks pregnant, were given 3 5 mg injections of $PGF_2\alpha$ intraamniotically at 10 minute intervals. 4 aborted within 24 hours, 10 more aborted after receiving another 15 mg 24 hours later. 21 women received 25 mg in 5 injections 10 minutes apart. 11 aborted within 24 hours and 8

after receiving another 25 mg 24 hours later. 5 women in the low dose group had fever. Premedication with prochlorperazine more than halved the incidence of nausea and vomiting. Multiparas aborted more easily and rapidly than primiparas. 80% of the women given hypertonic saline aborted within 48 hours and 97% within 72 hours. $PGF_2\alpha$ had the advantages of shorter IAT and none of the serious side effects associated with saline. (HC)   004714

0191
ANONYMOUS
Prostaglandins succeed in labor induction tests.
Medical Tribune and Medical News. September 18, 1972. p. 1, 20.

The success of labor induction in various clinical trials with prostaglandins is reported. Topics presented are 1) the effectiveness of PGE_2 administered orally, 2) a preliminary report of $PGF_2\alpha$ with high-risk patients, and 3) various studies of $PGF_2\alpha$ used in conjunction with oxytocin. (RAS)   004218

0192
ANONYMOUS
Prostaglandins.
British Medical Journal 4: 355-357. November 1972.

Prostaglandins are derivatives of 20-carbon, essential fatty acids. They are ubiquitous in mammalian tissue and have a wide range of pharmacological actions. Clinically, prostaglandins have been used to induce labor at term or to empty the uterus in cases of uterine death, severe rhesus isoimmunization, and hydatidiform mole. At higher dosages, prostaglandins have been used to induce abortions during the early and mid-trimester of pregnancy. Unpleasant, dose related, side effects, have been noted with the clinical use of these compounds. (JSL)   005015

0193
ANONYMOUS
Prostaglandins; new birth control hope or headache?
Family Planning Digest 1(2): 11-13. March 1972.

The article reviews present world-wide research into the clinical uses of prostaglandins, particularly as they relate to fertility control. Studies of administration of prostaglandins through various routes to induce labor, terminate pregnancy, or induce menstruation are mentioned. (RAS)   004112

0194
ANONYMOUS
Prostaglandins, therapeutic abortion and induced labor.
Research in Reproduction 4: 3. 1972.

The author cites the work of Anderson et al. (Contraception 5: 503, 1972) in which women 4 to 20 weeks pregnant were given 25 to 200 μg per minute $PGF_2\alpha$ by intravenous infusion for 6 to 24 hours. Abortion rate, complete and incomplete, was only 50%; side effects were many. However, the work of Vakhariya et al. (American Journal of Obstetrics and Gynecology 113: 212, 1972) was cited to illustrate the value of $PGF_2\alpha$ in labor induction. Prostaglandins were as effective as oxytocin. (HC) 004678

0195
ANONYMOUS
Recommended protocols for further studies.
In: Bergstrom, S., Green, K., and Samuelsson, B., eds. [Proceedings of the] third conference on prostaglandins in fertility control, Stockholm, January 18-20, 1972. Stockholm, WHO Research and Training Centre on Human Reproduction, Karolinska Institutet, [1972]. (Prostaglandins in Fertility Control 2) p. 17-25.

Recommended protocols for extraamniotic and intraamniotic administration of $PGF_2\alpha$ for the induction of abortion in the second trimester are given. Complete and detailed instructions are given in a step-by-step procedure for the administration of $PGF_2\alpha$, preparation and choice of patients, recording of side effects, successes, and contraindications. (HC)   004822

0196
ANONYMOUS
Research: experience with prostaglandins.
Family Planning Digest 1(5): 10. 1972.

In a study of human fallopian tubal fluid, it was reported that $PGF_2\alpha$ moved from the surface of the tubal mucosa to the lamina propria after ovulation. This suggests endocrine control and feedback mechanism. By its vasoconstrictive effect on blood vessels of the ovary $PGF_2\alpha$ could cause luteolysis. This has been demonstrated in animals but not in man. In another study 17 of 20 women 14 to 21 weeks pregnant aborted completely after given $PGF_2\alpha$ by the intraamniotic route. The initial dose was 25 mg with additional doses given every 10 minutes as needed up to a total of 70 mg. The IAT averaged 22 hours. Vomiting was the only side effect of consequence. Intravenous oxytocin produced abortions in 11 of 13 patients in whom PG induced contractions appeared inadequate. (HC)   004536

0197
ANONYMOUS
The clinical use of prostaglandins.
International Planned Parenthood Federation Medical Bulletin 6(4): 4. 1972.

This is a report of a meeting held at Upjohn Company, Kalamazoo, Michigan, June 12-14, 1972. Professor Karim of Uganda detailed his successful use of PGE_2 and $PGF_2\alpha$ intravenously in 1500 women for induction of labor. Severe gastrointestinal side effects were noted. Ian Craft of London and William Barr of Glasgow had some success with PGE_2 orally but Barr reported massive side effects. The intrauterine and intraamniotic routes were reported to have produced fewer side effects than intravenous administration while maintaining effectiveness for second trimester abortion. Vacuum aspiration is still the method of choice up to 12 weeks pregnancy. Intravaginal administration has the disadvantage of systemic absorption and concomitant serious side effects. 15-methyl derivatives were reported to have greater activity and fewer side effects than the parent compounds. Anne Wentz of Baltimore found no evidence of luteolysis using $PGF_2\alpha$ in 22 women suggesting that the use of PG as a contraceptive is still not practical. Members of the conference stressed caution and the need for more extensive clinical trials before the routine use of PG. (HC)   004306

0198
ANONYMOUS
The present state of prostaglandin research.
Nature 240: 1-2. November 1, 1972.

This report summarizes highlights of the international conference on PGs held in Vienna, September 24-28, 1972. Half of the sessions were devoted to reproduction. Trials of $PGF_2\alpha$ for early abortion achieved 80% success in 2 centers, but 20%-30% in all others, with no evidence of luteolysis. For asthma, PGE dilates bronchi as does isoprenaline, but evokes severe coughing; $PGF_2\alpha$ is 10,000 times more potent a vasoconstrictor in asthmatics than in normal subjects. 2 analogs of PGE_2 showed promise for treating gastric ulcers. Radioimmunoassay or gas-liquid chromatography and mass spectrometry are used to quantitate PGs. In plasma, an antioxidant or lung sampling may be necessary to circumvent biosynthesis of PGs by platelets. It was estimated that urinary excretion of PGE_2 is 2.5-3.3 μg/day in women but 6.5-46.7 in men; plasma levels of $PGF_2\alpha$ are no more than 2 pg/ml; and total PG production is up to 200 μg/day. New evidence showed that PG synthetase is consumed in its reaction. A new group called PGC with a 11,12 double bond was discovered. Studies of PG's mediation in inflammation (e.g., uveitis and burns) suggest that their antagonists may serve the pharmaceutical industry better than PGs themselves. (LJG) 004507

0199
ANONYMOUS
Use of prostaglandins is approved in the U.K.
Chemical and Engineering News 50: 21. October 23, 1972.

Use of prostaglandins in selected hospitals and clinics, primarily to induce labor and abortion, has been approved by the Committee on Drug Safety of the United Kingdom. (HC) 004489

0200
ARMSTRONG, D.T.; GRINWICH, D.L.
Blockade of spontaneous and LH-induced ovulation in rats by indomethacin, an inhibitor of prostaglandin biosynthesis.
Prostaglandins 1: 21-28. 1972.

Indomethacin, an inhibitor of prostaglandin synthesis, blocks ovulation in immature rats pretreated with pregnant mare serum gonadotropin (PMS) when given either at 0800, 1200, or 1600 hours on the second day after PMS treatment (the equivalent of proestrus in normally cycling adult rats). The drug also blocked ovulation in response to exogenous luteinizing hormone, whether the latter was administered 30 minutes before or 30 minutes after the inhibitor. Luteinization of follicles and signs of preovulatory progesterone secretion (loss of uterine lumen fluid) were not prevented when the inhibitor was given at 1600 hours or when exogenous LH was administered in addition to the inhibitor, indicating that the luteinizing and steriodogenesis actions of LH upon the ovary were not completely blocked. When the drug was administered before the critical period for LH secretion, follicular luteinization and signs of progesterone secretion were also prevented, suggesting an additional action of indomethacin at the level of the hypothalamic-pituitary axis, inhibiting LH secretion. (Authors' modified) 003744

0201
ARMSTRONG, D.T.; GRINWICH, D.L.; KENNEDY, T.G.
Role of prostaglandins in the steroidogenic and ovulatory responses of rabbit ovaries to luteinizing hormone.
In: Abstracts, 5th Annual Meeting of the Society for the Study of Reproduction, East Lansing, Michigan, June 26-29, 1972. p. 27-28.

The possible involvement of prostaglandins (PG) as mediators of ovarian responses to luteinizing hormone (LH) has been investigated in in vitro and in vivo studies with rabbits. PGE_2 and $PGF_2\alpha$ failed to increase in vitro synthesis of progesterone plus 20alpha-OH-pregn-4-en-3-one (20alpha-OH-P) in ovarian interstitial tissue slices when added to incubation media at concentrations of 10 $\mu g/ml$; replicate slices from the same ovaries responded to LH at concentrations of .05, .15, and .45 $\mu g/ml$, with 10-, 25-, and 43-fold increases, respectively. In in vivo studies, LH (50 μg NIH-LH-B7) was injected iv 1/2 hr. after iv injection of either indomethacin, an inhibitor of prostaglandin synthesis (10-40 mg/kg in phosphate buffer) or phosphate buffer vehicle. Mean level of 20alpha-OH-P observed in peripheral plasma obtained by heart puncture 1 hour after LH were increased by 112 ± 55 and 26 ± 12 ng/ml (18.1-fold and 9.4 fold) above levels observed before LH, in 10 indomethacin-treated and 5 phosphate buffer treated rabbits, respectively. None of 11 indomethacin treated and 9 of 10 vehicle treated control rabbits ovulated in response to this dosage of LH, as determined by flushing of oviducts or gross examination of ovaries 1 or 3 days after LH treatment. Indomethacin did not prevent luteinization of follicles in LH treated rabbits, as determined by gross and histologic examination of ovaries 8 days after treatment. These findings argue against a role of prostaglandins as mediators of the steroidogenic and luteinizing actions of LH in the rabbit ovary, but suggest an involvement in the process of ovulation. (Authors)   004049

0202

ARMSTRONG, D.T.; GRINWICH, D.L.; KENNEDY, T.G.

Role of prostaglandins in the steroidogenic and ovulatory responses of rabbit ovaries to luteinizing hormone.

Biology of Reproduction 7: 107. 1972.

PGE_2 and $PGF_2\alpha$ at 10 $\mu g/ml$ failed to increase progesterone synthesis when incubated with ovary slices; LH induced marked increases. The mean level of plasma 20alpha-hydroxy-pregnenone 1 hour after LH 50 μg intravenously was increased ninefold. In those animals given indomethacin 10-40 mg/kg intravenously 1/2 hour before LH, the level increased eighteen fold. 9 of the 10 animals given LH but none of the 11 indomethacin-pretreated animals ovulated. Indomethacin did not prevent the luteinization of follicles. These results do not indicate a role of prostaglandins as mediators of LH activity in the ovary but do imply an involvement in ovulation. (HC)   004266

0203

AULETTA, F.J.; CALDWELL, B.V.; VAN WAGENEN, G.; MORRIS, J.M.

Effects of postovulatory estrogen on progesterone and prostaglandin F levels in the monkey.

Contraception 6(5): 411-420. 1972.

Estrogens were given orally or intramuscularly in single or repeated doses during the postovulatory (luteal) period to previously mated monkeys. Plasma PGF levels measured by radioimmunoassay rose in more than half of the animals at the same time that progesterone levels rapidly declined. In most instances, values returned to normal whether or not estrogen was continued. The present studies do not establish any definite relationship between progesterone decline and prostaglandin F release. (Authors' modified)   004596

0204

BALLARD, C.A.; BALLARD, F.E.

Four years' experience with mid-trimester abortion by amnioinfusion.

American Journal of Obstetrics and Gynecology 114(5): 575-581. November 1, 1972.

3579 midtrimester patients received intraamniotic 20% saline infusion for abortion from 1967 to 1971. They were 10-48 years old, 30% under 20 years, and 41% were primigravidas. Protocols varied until 1969 with the first 360 patients; thereafter patients received a standardized treatment with a maximum of 50 ml amniotic fluid withdrawn and 200 ml 20% NaCl injected. Intravenous oxytocin was given if needed only after 48 hours. Now it is routinely given immediately after saline. Complications fell from 3.4% in the first 1814 patients to 2% after curettage became routine. The 3242 patients who were pregnant 16 weeks or more and subjected to the strict protocol had a success rate of 99.8% in a mean interval of 35.2 hours. Intraamniotic PGs were mentioned in the discussion. (LJG)   004831

0205
BALLARD, C.A.; QUILLIGAN, E.J.; JAKOBOVITS, A.
Pathological changes of prostaglandin.
Advances in the Biosciences (Suppl.) 9: 91. 1972.

The authors investigated macroscopic and microscopic fetal skin and placental changes in 25 patients with midtrimester abortion induced by intraamniotic prostaglandin $F_2\alpha$. The dose of prostaglandin was 25, 40, and 50 mg, repectively. The fetus and placenta were expelled within 1 to 2 days. Gross fetal changes consisted of bluish-black patches occasionally blending into massive subcutaneous hemorrhages under the skin in the typical cases (20 of 25), the most frequent location being the scalp, the buttock, frequently the trunk and/or extremities. Microscopically there were massive hemorrhages in the subcutaneous connective tissue and even in the superficial muscle layers. In the placenta there were 15 to 25 case areas of submembraneous hemorrhage. These frequently were grossly visible (several centimeters in length and width, and 3 to 5 mm thick) but never extending throughout the whole length of the cross section. Microscopically the membranes and the stroma edematous were thickened. The stroma cells showed vacuolization and pyknosis. The hemorrhagic areas underneath the membranous areas proved histologically to be the thrombin consisting of red blood cells and fibrin. Leukocytosis was remarkable in only 2 cases. The chorionic villi showed varying degrees of damage, occasionally only visibly outlined ('ghost villi'). Otherwise there were frequent vacuolization of the stroma cells in the villi. The histomorphological visible damages of the placentas explain the death of fetus. (Authors' modified)   004448

0206
BALLARD, C.A.; QUILLIGAN, E.J.
Therapeutic abortion in midtrimester using intra-amniotic instillation of prostaglandin $F_2\alpha$.
In: Southern, E.M., ed. The prostaglandins: clinical applications in human reproduction. (Brook Lodge Symposium, Augusta, Michigan). Mount Kisco, New York, Futura Publishing Co., 1972. p. 333-336.

20 women pregnant 14-20 weeks were given 25 mg $PGF_2\alpha$ by intraamniotic infusion and an additional 15 mg dose 6 hours later if necessary. 17 women aborted, 14 completely; most required both doses; the IAT averaged 15 1/2 hours. Only 3 patients had several episodes of severe vomiting and diarrhea. The authors noted that even though the initial dose is potentially dangerous, higher doses might be needed in women pregnant longer than 18 weeks. They suggested that the dose be determined by the amniotic fluid volume in each patient. (HC)   004872

0207
BARDEN, T.P.
Induction of preterm labor with prostaglandin $F_2\alpha$ in patients with premature rupture of membranes.
In: Southern, E.M., ed. The prostaglandins: clinical applications in human reproduction. (Brook Lodge Symposium, Augusta, Michigan). Mount Kisco, New York, Futura Publishing Co., 1972. p. 193-205.

13 women at term having premature rupture of membranes without spontaneous labor were given $PGF_2\alpha$ by intravenous infusion 2.5 μg/min for 60 min, doubled at 60-min intervals as needed. Total dose for delivery averaged 1.25 mg; IDT mean was 6 hours. There were 4 instances of hypertonus, 2 being associated with fetal heart rate deceleration. Apgar scores for the neonates averaged 9; all survived. 10 women having premature rupture of membranes prior to term were also given $PGF_2\alpha$ by intravenous infusion. Bishop score for these women averaged 5.3 compared with 7.5 for the other group. Total dose of $PGF_2\alpha$ was 1.25 mg. Mean IDT was 5 hours. 6 neonates had respiratory distress; 2 died. The author concluded that uterine activity in preterm $PGF_2\alpha$-induced cases was similar to term induction with amniotomy alone. (HC)   004861

0208
BARR, W.
Induction of labor by prostaglandins E_2.
In: Southern, E.M., ed. The prostaglandins: clinical applications in human reproduction. (Brook Lodge Symposium, Augusta, Michigan). Mount Kisco, New York, Futura Publishing Co., 1972. p. 219-222.

$PGF_2\alpha$ was given orally in a capsule to 50 women, 5 mg initially, 10 mg 30 minutes later, and additional 15 mg doses every 2 hours until labor was established. PGE_2 was given orally to another group of 50 women as a solution, .5 mg initially followed 30 minutes later by 1.0 mg doses until labor. 3 to 8 doses of $PGF_2\alpha$ were required, while with PGE_2 4 to 6 doses were sufficient. 33 of the women given $PGF_2\alpha$ delivered in less than 18 hours while with PGE_2, the number of successes was 32. With $PGF_2\alpha$, 84% of the patients had gastrointestinal side effects, mainly diarrhea. With PGE_2, complaints were few and minor. (HC)   004863

0209
BARR, W.; NAISMITH, W.C.M.K.
Oral prostaglandins in the induction of labour.
British Medical Journal 2: 188-191. 1972.

Prostaglandins E_2 and $F_2\alpha$ were administered by mouth to induce labor in 24 patients at or past term. The drugs were administered at 2-hourly intervals in doses ranging from .5 to 1.5 mg for prostaglandin E_2 and from 5 to 15 mg for prostaglandin $F_2\alpha$. Of the 10 cases in which prostaglandin E_2 was used, labor was successfully induced in 8 and there were no side effects. With prostaglandin $F_2\alpha$ labor was induced in 12 of 14 patients, 9 of whom had gastrointestinal disturbance, mostly of mild degree. With both drugs the infant was apparently unaffected and Apgar scores were satisfactory. Uterine hypertonus was not observed and the postpartum blood loss was within normal limits. (Authors)   003841

0210

BARTKE, A.; MERRILL, A.P.; BAKER, C.F.
Effects of prostaglandin $F_2\alpha$ on pseudopregnancy and pregnancy in mice.
Fertility and Sterility 23: 543-547. 1972.

The effects of $PGF_2\alpha$ on pseudopregnancy and pregnancy were studied in random bred laboratory mice. Subcutaneous administration of a water soluble derivative of $PGF_2\alpha$ ($1\text{-}PGF_2\alpha$ base 75% by weight) in doses ranging from 25-400 μg/day on Days 4-7 after mating significantly shortened pseudopregnancy. A single dose of 200 μg shortened pseudopregnancy when given on Days 4, 5, or 6, but had no effect when given on Days 2 or 3. In pregnant mice daily injections of 25 μg of this salt of $PGF_2\alpha$ on Days 4-7 did not interfere with gestation. However, the administration of 200 μg on Days 4 and 5, or 300 μg on Day 4, terminated pregnancy in 5 of 9, and 7 of 10 mice, respectively. In females in which pregnancy continued in spite of prostaglandin administration, the mortality of embryos was increased. Pregnancies which commenced, immediately after the treatment with prostaglandin were not affected. (Authors' modified)   004196

0211

BASHORE, R.A.
Studies concerning a radioimmunoassay for oxytocin.
American Journal of Obstetrics and Gynecology 113: 488-496. 1972.

There is no mention of prostaglandins in this paper presented at a symposium. However, in the discussion session after the paper was presented, Dr. R. C. Goodlin mentions that attempts have been made to develop blood tests for prostaglandins and other biologically active substances which may be involved in human labor. (JRH)   003997

0212

BASU, H.K; THELWALL-JONES, H.
Oral prostaglandin E_2 for induction of labor.
British Medical Journal 2: 527. 1972.

The authors object to the use of the term 'titration' with the use of oral prostaglandins for the induction of labor as it has been used by others. Their practice is to give .5 mg of PGE_2 and if vomiting or diarrhea does not appear the dose is increased to 1.5 or 2.0 mg given every 2 hours. They also report that dosages of more than 1.5 mg are often associated not only with vomiting and diarrhea but also with inefficient uterine activity. This result has also been reported by others and may be due to lack of absorption of PG after vomiting. The authors stress the need to closely monitor the fetus during induction of labor with prostaglandins. (JRH)   003958

0213

BATTA, S.K.; MUKERJEE, B.; SANTHAKUMARI, G.
Pharmacological studies on the uterine flushing in women with and without intrauterine devices.
Archives Internationales de Pharmacodynamie et de Therapie 196: 174-176. 1972.

The intrauterine fluid of women aged 25-40 who had IUDs for 1-7 years with no menstrual or gynecologic pathology, was compared to controls (IUD acceptors) by bioassay. Distilled water was injected into the uterus with a tuberculin syringe and polyethylene cannula, and aspirated after 15 seconds. The uterine flushings from controls and from IUD patients during the first week after menstruation depressed rat blood pressure; flushings from IUD patients in the second week after ovulation evoked a biphasic vasodepressor-vasopressor response; and flushings from IUD patients in the third week increased rat blood pressure. Flushings from controls sensitized the isolated rat uterus to oxytocin, and those from IUD patients desensitized uteri. Flushings from IUD patients sensitized guinea pig ileum to histamine 75-130%, and control flushings sensitized ileums 50-62.5% ($p < .05$). The authors conclude that the pharmacologic activity of the uterine fluid in these bioassays may be due to released prostaglandins, which are known to be present in menstrual fluid and to evoke similar biologic responses in other tissues. (LJG)   004380

0214
BEATTY, C.H.; BOCEK, R.M.
Prostaglandins: their effects on the carbohydrate metabolism of myometrium from rhesus monkeys.
Endocrinology 90: 1295-1300. 1972.

Slices of myometrium from pregnant rhesus monkeys (Macaca mulatta) were incubated in medium containing glucose-U-carbon-14 with and without PGE_1, PGE_2, and $PGF_1\alpha$. The addition of PGE_1 to the medium increased glucose uptake and the productions of lactate, lactate-carbon-14, and carbon-14-carbon dioxide with no change in oxygen consumption. About 68% of the total lactate produced in the control series arose from glucose-U-carbon-14. Although a significant decrease in glycogen levels at the end of 2 hours incubation was not demonstrated with PGE_1, some glycogenolysis must have occurred since lactate and lactate-carbon-14 production increased to the same extent, with no change in the percentage of lactate derived from glucose-U-carbon-14 and glycogen. The metabolic effects of PGE_2 and $PGF_1\alpha$ on myometrium were similar to those of PGE_1. Epinephrine did not affect the carbohydrate metabolism or oxygen consumption of the myometrium. In the series treated with both PGE_1 and epinephrine, glycogen levels at the end of incubation and incorporation of glucose-carbon-14 into glycogen were decreased compared with those of the control series. PGE_1 also increased glucose uptake and the production of lactate and lactate-carbon-14 of the myometrium from ovariectomized rhesus monkeys. There was no evidence of glycogen utilization by myometrium from ovariectomized monkeys during incubation. Hypoxia, like prostaglandins, increased the glucose uptake and conversion of glucose and glycogen to lactate with no change in the percentage of lactate arising from either source. (Authors)   003946

0215
BEAZLEY, J.M.; BRUMMER, H.C.; KURJAK, A.
Distribution of 9-H3-Prostaglandin $F_2\alpha$ in pregnant and non-pregnant subjects.
Journal of Obstetrics and Gynecology of the British Commonwealth 79: 800-803. September 1972.

Tritium-labelled $PGF_2\alpha$ was injected into 3 pregnant (16-20 weeks) women who were undergoing surgical abortion and 2 nonpregnant women who were being surgically sterilized to determine if pregnancy affects PG metabolism. Samples of blood and urine were collected at intervals after injection. Samples of skin, subcutaneous fat, rectus muscle, myometrium, fallopian tube, amniotic fluid, placenta, umbilicus, cord blood, fetal skin, fetal liver, and fetal lung were taken. The amount of label found in maternal subcutaneous fat and rectus muscle was highly variable, but there was no concentration of label in any other maternal tissue. In the fetal tissues, the liver shows consistently higher levels of radiolabel uptake than other tissues. There was no difference in the distribution or excretion of the radiolabelled $PGF_2\alpha$ between pregnant and nonpregnant women. (RAS)   004346

0216
BEAZLEY, J.M.
Induction of labor with prostaglandins: summary of present status.
In: Southern, E.M., ed. The prostaglandins: clinical applications in human reproduction. (Brook Lodge Symposium, Augusta, Michigan.) Mount Kisco, New York, Futura Publishing Co., 1972. p. 529-534.

The present status of the use of PGs for the induction of labor was summarized. Injected intravenously, PGE_2 seems more efficient in smaller doses than does $PGF_2\alpha$, especially in primigravida. It is reported by some that both PGs seem more effective than Syntocinon, a synthetic oxytocin, whereas others report that PGs and Syntocinon are equally effective. Many investigators suggest that $PGF_2\alpha$ is less flexible in its dose range than is Syntocinon. The efficacy of PGs has been found to improve after ruptured membranes. Transient hypertonus has often been reported with intravenous infusion. Oral administration appears successful for the induction of labor in both primigravida and multiparous women, although diarrhea and vomiting were noted in 10% of the patients in 1 study. Little is known at this point about the mode of action of PGs or about the effect of infused PGs upon fetal physiology. Further clinical investigations are anticipated. (MLH)   004890

0217
BEGUIN, F.; BYGDEMAN, M.; GREEN, K.; SAMUELSSON, B.; TOPPOZADA, M.; WIQVIST, N.
Analysis of prostaglandin $F_2\alpha$ and metabolites in blood during constant intravenous infusion of prostaglandin $F_2\alpha$ in the human female.
Acta Physiologica Scandinavica 86: 430-432. 1972.

An investigation was undertaken to explore the degree of individual variation and the induced biological effects of intravenously injected $PGF_2\alpha$. Kinetic studies have shown that $PGF_2\alpha$ is rapidly converted following injection to 15-keto-dihydro-$PGF_2\alpha$ and dihydro-$PGF_2\alpha$. In this investigation, 10 midpregnant women received infusions of $PGF_2\alpha$ 75 μg/minute for 10 hours to induce abortion. Peripheral venous blood (20 ml) was drawn before and 3 hours after the start of the infusion and left to clot overnight at 4 degrees C. The serum was isolated and 5-10 μg each of D4-$PGF_2\alpha$, D4-15-keto-dihydro-$PGF_2\alpha$, and D4-dihydro-$PGF_2\alpha$ were added; each carrier also contained a tritium-labeled tracer. The levels of the 3 compounds were measured by mass spectroscopy. The range of $PGF_2\alpha$ concentration was found to be 1.7-10.4 ng/ml serum; of dihydro-$PGF_2\alpha$, .9-4.5 ng/ml; and of 15-keto-dihydro-$PGF_2\alpha$, 55-143 ng/ml. The data demonstrated that there was considerable variation in the rate of the initial metabolic degradation and in the inactivation of exogenous $PGF_2\alpha$. It was also demonstrated that a high level of $PGF_2\alpha$ was usually associated with a high incidence of side effects, whereas the correlation to the mean uterine activity was less evident. The high levels of 15-keto-dihydro-$PGF_2\alpha$, 10-75 times those of $PGF_2\alpha$ suggested that it could be advantageous as an indicator for the release of $PGF_2\alpha$ into the circulation. The levels of this metabolite demonstrated also the rapid transformation of $PGF_2\alpha$. Biological effects of the metabolites in humans were unknown, but it was thought that they might be contributors to the uterine activity and side effects observed during administration of $PGF_2\alpha$. (MLH)   004767

0218
BEGUIN, F.; BYGDEMAN, M.; TOPPOZADA, M.; WIQVIST, N.
The response of the midpregnant human uterus to vaginal administration of prostaglandin suppositories.
Prostaglandins 1: 397-405. 1972.

The effect of a new form of vaginal prostaglandin suppositories on the contractility of the midpregnant uterus was studied in 20 volunteers admitted to the hospital for induction of legal abortion. 10 patients were given 50 mg $PGF_2\alpha$ and 10 other patients 20 mg PGE_2. The uterine contractions were recorded continuously for a period of 10 hours following the insertion of a single vaginal suppository. 50 mg $PGF_2\alpha$ induced a maximum uterine activity during the third hour followed by a rapid decline to reach a level close to zero by the tenth hour. 20 mg PGE_2 stimulated the uterus to achieve a maximum activity at the fifth hour. This high level of activity was sustained throughout the rest of the observation period. The incidence of vomiting and diarrhea showed a reasonable correlation to the level of uterine activity induced within the 2 groups. The potency of vaginal adminstration of 20 mg PGE_2 was also reflected by the fact that 7 out of the 10 cases in this group aborted within 10 hours following a single suppository as compared to only 1 out of 10 in the 50 mg $PGF_2\alpha$ group. These results indicate that vaginal administration of 20 mg PGE_2 is far more potent than 50 mg $PGF_2\alpha$. The possible role of the new prostaglandin suppositories for future clinical trials is discussed. (Authors) 003949

0219

BEHRMAN, H.R.; ORCZYK, G.P.; GREEP, R.O.
Effect of synthetic gonadotrophin-releasing hormone (Gn-RH) on ovulation blockade by aspirin and indomethacin.
Prostaglandins 1: 245-258. 1972.

To determine the site of the ovulation blocking action of indomethacin and aspirin, reversal of each drug's inhibitory effect by synthetic gonadotrophin-releasing hormone (Gn- RH) and luteinizing hormone (LH) was tested. Aspirin blockade of ovulation was consistently reversed by LH administration (10 μg/iv) 2 or 3 hours after drug injection and by Gn-RH (approximately .6 and .3 nm, sc) given twice either 3 and 4 or 2 and 3 hours after drug injection. In contrast, ovulation blockade by indomethacin was not reversed by Gn-RH, nor was it reversed by exogenous LH given at the time of the expected LH surge if the indomethacin had been administered 2 hours earlier. Treatment with either drug resulted in a reduction in pituitary and hypothalamic PGF levels. The dosage of and injection schedule for Gn-RH was also effective in reversing phenobarbital-blocked ovulation. All animals, including those that did not ovulate after indomethacin and Gn-RH or LH treatment, exhibited loss of uterine fluid on the day after proestrus. Aspirin blocked ovulation at the hypothalamic level, whereas indomethacin may act at the ovarian level by interfering with the rupture of the follicle and release of the ovum. (Authors)   003965

0220

BEHRMAN, H.R.; MOUDGAL, N.R.; GREEP, R.O.
Studies with antisera to luteinizing hormone in vivo and in vitro on luteal steroidogenesis and enzyme regulation of cholesteryl ester turnover in rats.
Journal of Endocrinology 52: 419-426. 1972.

It is briefly mentioned in the discussion section of this paper that others have shown that administration of prostaglandins causes an increase in 20alpha-dihydorprogesterone when luteal function is waning. (JRH)   003899

0221

BERGSTROM, S.

Clinical studies with prostaglandins.

Abstracts of papers presented at the National Meeting of the APhA Academy of Pharmaceutical Sciences 2(2): 12. 1972.

Prostaglandins have been used extensively since 1967 as oxytocic agents for the induction of labor at term and for the termination of pregnancy. Unlike oxytocin, which is most active during the latter periods, prostaglandins are fully active throughout pregnancy. Currently, studies on the use of the various prostaglandins, and dosages and routes of administration as abortifacients during various stages of pregnancy are in progress. (JSL)   005017

0222

BERGSTROM, S.; DICZFALUSY, E.; BORELL, U.; KARIM, S.; SAMUELSSON, B.; UVNAS, B.; WIQVIST, N.; BYGDEMAN, M.

Prostaglandins in fertility control. [Meeting summary]

Science 175: 1280-1287. 1972.

The authors review the material presented at a workshop conference on prostaglandins in fertility control which was sponsored by World Health Organization Research and Training Centre on Human Reproduction at Karolinska Institutet, Stockholm. It was agreed that a more unified approach to clinical prostaglandin research was needed. (JRH)   003738

0223

BHAGAT, B.; DHALLA, N.S.; GINN, D.; La MONTAGNE, A.E., Jr.; MONTIER, A.D.

Modification by prostaglandin $E_2(PGE_2)$ of the response of guinea-pig isolated vasa deferentia and atria to adrenergic stimuli.

British Journal of Pharmacology 44: 689-698. 1972.

PGE_2 exerted positive cardiostimulant effects on isolated guinea pig atria. The response was not altered by treatment of the animal with reserpine or by addition of propranolol to the organ bath. These results suggest that the cardiostimulatory actions of PGE_2 are not mediated through the release of catecholamines or stimulation of adrenoceptors. On the electrically driven atria, PGE_2 consistently exerted a cardiostimulant action, which was not appreciably altered by changes in calcium ion in the bathing medium. PGE_2 showed no effect on the transport of calcium by the fragments of heart sarcoplasmic reticulum. PGE_2 reduced the responses to both noradrenaline and tyramine in the isolated atria. The shifted dose-response curve was not parallel to the original. PGE_2 increased the contractor response of the isolated vas deferens to nerve stimulation or to direct electrical stimulation. PGE_2 antagonized the increase caused by by noradrenaline in contractor response of isolated vas deferens to direct electrical stimulation, whereas it affected the potentiation by noradrenaline differently when the vas deferens was contracting in response to nerve stimulation. In low concentration it inhibited and in large concentrations it slightly enhanced the potentiation by catecholamine. It is concluded that PGE_2 has actions on multiple sites. It has postjunctional as well as prejunctional effects on adrenergic neurones. (Authors' modified)   003974

0224

BHALLA, R.C.; SANBORN, B.M.; KORENMAN, S.G.

Hormonal interactions in the uterus: inhibition of isoproterenol-induced accumulation of adenosine 3':5'-cyclic monophosphate by oxytocin and prostaglandins.

Proceedings of the National Academy of Sciences (U.S.A.) 69(12): 3761-3764. December 1972.

The effect of isoproterenol, oxytocin, propranolol, PGE_1, PGE_2, and $PGF_2\alpha$ on tissue cAMP levels in ovariectomized rat uteri, myometrium, and endometrium were explored. The uterine horns were removed from rats which had been ovariectomized 5-8 days earlier. Uteri were stretched on glass rods to facilitate immersion in Eagle's minimal essential medium, or test substances. cAMP was determined by protein binding assay. Isoproterenol (beta adrenergic agent) increased cAMP in whole uteri, with a half maximal response of about 100 pmol per mg protein at about 5 μM isoproterenol for 5 minutes incubation. (PGE_1 and PGE_2 increased cAMP to 55.4 and 25.3 pmol per mg protein in cow endometrium). PGE_2 increased cAMP almost as much as isoproterenol at 50 μM PG, but $PGF_2\alpha$ was ineffective. When combined with isosproterenol, both PGs prevented the increase in cAMP with increasing concentration from .005 to .5 μM PG, but the inhibition decreased as concentration was raised to 50 μM PG. Qualitatively similar results were obtained with rat myometrium. Oxytocin alone or oxytocin with PGE_2 did not have any effect on cAMP content. When uteri were incubated with various combinations of theophylline, isoproterenol, propranolol, and PGE_2, cAMP was only inhibited about 25%, compared to 80% inhibition with all these except PGE_2. These data were taken as evidence of a single site for the beta-adrenergic receptor in uterine smooth muscle. (LJG)   004836

0225

BHALLA, R.C.; SANBORN, B.M.; KORENMAN, S.G.

Hormonal interrelationships in the uterus: oxytocin inhibition of catecholamine response.

In: Abstracts, Fourth International Congress of Endocrinology, Washington, D, C,. June 18-24, 1972. Amsterdam, Excerpta Medica, 1972. (International Congress Series No. 256) Abstract 550. p. 219.

Uteri from ovariectomized rats were incubated with isoproterenol, theophylline, $PGF_2\alpha$, PGE_2, estrogen and progesterone, to study effect on cAMP generation by protein binding assay. Isoproterenol 10^{-7} to 10^{-5}M produced a dose-dependent increase in cAMP, enhanced by theophylline 5 mM. PGE_2 5 x 10^{-7}M significantly stimulated cAMP with or without isoproterenol. $PGF_2\alpha$ at 1.4 x 10^{-5} M, estrogen or porgesterone or both, in vivo or in vitro did not affect cAMP or isoproterenol response. Oxytocin 1-50 mU per ml inhibited the cAMP release stimulated by isoproterenol. (LJG)   005013

0226

BLATCHLEY, F.R.; DONOVAN, B.T.

The effect of prostaglandin $F_2\alpha$ and prostaglandin E_2 upon luteal function and ovulation in the guinea-pig.

Journal of Endocrinology 53: 493-501. 1972.

The effect of treatment with $PGF_2\alpha$ on luteal function was examined in hysterectomized guinea pigs. Regression of corpora lutea was found to occur when .25 mg (or more) was injected intraperitoneally daily for 3 days, and ovulation usually ensued within 5 days after treatment. The administration of 1 mg $PGF_2\alpha$ daily for 7 days caused marked luteolysis, but ovulation did not occur. Ovulation was blocked in 3 of 4 intact female guinea pigs given 1 mg $PGF_2\alpha$ per day for 7 days from Day 15, but took place normally in 5 females injected with .25 mg per day. Treatment of hysterectomized guinea pigs with .62 mg adrenaline hydrochloride per day, 1.97 mg atropine sulphate per day or .52 mg histamine dihydrochloride per day did not cause luteal regression, while the intraperitoneal injection of .25 mg or 1 mg PGE_2 daily for 3 days was likewise ineffective. (Authors' modified)   004123

0227

BLATCHLEY, F.R.; DONOVAN, B.T.; HORTON, E.W.; POYSER, N.L.
The release of prostaglandins and progestin into the utero-ovarian venous blood of guinea-pigs during the oestrous cycle and following oestrogen treatment.
Journal of Physiology (London) 223: 69-88. 1972.

Blood was collected from the uteroovarian vein of guinea pigs on Days 3, 9, 10, 11, 12, 13, 14, and 15 of the estrous cycle, and from animals injected with estrogen over Days 4-6 and used on Day 7. The content of prostaglandin $F_2\alpha$ and E_2 was determined by bioassay and gas chromatography-mass spectrometry, and of progestin by competitive protein-binding assay. The concentration of prostaglandin $F_2\alpha$ on Day 3 was 14.1 ng/ml; it fell below detectable limits on Day 7, but after Day 11 the concentration rose to between 15.7-17.2 ng/ml. A value of 60.7 ng/ml was recorded on Day 15. The release of prostaglandin $F_2\alpha$ was increased in animals pretreated with estrogen but was below threshold levels in similarly treated hysterectomized females. Prostaglandin E_2 was not detectable in blood collected on Days 3, 9, 10, 11, and 12. From a threshold value of 1.1 ng/ml on Day 13, the concentration rose to 5.4 ng/ml on Day 14 and to 54.9 ng/ml on Day 15. The progestin concentration rose from 66 ng/ml on Day 3 to values of 80-180 ng/ml between Days 9-14. It fell to 45 ng/ml between Days 14 and 15. This work established that prostaglandin $F_2\alpha$ is released into the uteroovarian venous blood, and that its concentration fluctuates in the course of the estrous cycle and can be related to changes in luteal function. (Authors' modified)   004012

0228

BRAAKSMA, J.T.; BRENNER, W.E.; FISHBURNE, J.I., Jr.; STAUROVKSY, L.
Early myometrial effects of intrauterine - extraamniotic administration of prostaglandin $F_2\alpha$.
Advances in the Biosciences (Suppl.) 9: 84. 1972.

A method of administering prostaglandin $F_2\alpha$ ($PGF_2\alpha$) solution at various intrauterine-extraamniotic sites to induce abortion and to study the effect of $PGF_2\alpha$ on the myometrium is described. Four observations were noted during the study of women between 8 and 20 weeks gestation. Injection of $PGF_2\alpha$ in the fundus resulted in a greater myometrial response when compared to the response elicited by the same dose injected at lower uterine sites. The sensitivity of the myometrium to $PGF_2\alpha$ appeared to be greater in early pregnancies than in older pregnancies. The myometrium responded to $PGF_2\alpha$ by increasing the basal tonus more in early pregnancy than it did in later pregnancies. The abortifacient effect of $PGF_2\alpha$ appeared to be greater when administered high in the fundus rather than in the lower uterine segment. (Authors' modified)   004337

0229

BRENNER, W.E.; HENDRICKS, C.H.; FISHBURNE, J.I., Jr.; BRAAKSMA, J.T.; STAUROVSKY, L.G.
Dose-abortifacient response for intraamniotically administered prostaglandin $F_2\alpha$.
Pharmaceutical Science 2(2): 13. 1972.

A dose response study of intraamniotic $PGF_2\alpha$ was conducted in 132 women requiring therapeutic abortion. 15-75 mg $PGF_2\alpha$ was administered in 3 single-dose and 3 multiple-dose schedules; abortion within 24 hours was the end point. The dose response was determined by parity, gestational age, and whether PG was given in single or multiple doses. (LJG)   005016

0230

BRENNER, W.E.; HENDRICKS, C.H.; BRAAKSMA, J.T.; FISHBURNE, J.I.; STAUROVSKY, L.G.

Intra-amniotic administration of $PGF_2\alpha$ for induction of therapeutic abortion.

In: Southern, E.M. ed., The prostaglandins: clinical application in human reproduction. (Brook Lodge Symposium, Augusta, Michigan.) Mount Kisco, New York, Futura Publishing Co., 1972. p. 457-470.

61 women 14 to 22 weeks pregnant received $PGF_2\alpha$ intraamniotically by 1 of 4 dosage schedules; A) 3 5-mg doses at 10-15 minute intervals, 15 mg 24 hours later if necessary; B) initial 15 mg as above, additional 15 mg doses at 6, 24 and 30 hours if needed; C) 5 5-mg doses at 5 minute intervals, 25 mg 24 hours later if necessary; D) initial 25 mg as above, 25 mg at 6, 24 and 30 hours if needed. 14 of 20 women in Group A aborted, 9 completely; IAT averaged 28 hours. All 10 patients in Group B aborted, 6 completely, mean IAT was 22.8 hours. 19 of 21 women in Group C aborted, 13 completely; averaged IAT was 22.9 hours. All 10 in group D aborted, 7 completely; mean IAT was 17.7 hours. In Groups A and C multipara aborted faster. In Groups A and B women less than 16.5 weeks pregnant aborted faster than those greater than 16.5 weeks. Nearly all patients had pain; 20 to 30% of the women in each group had emesis. The authors conclude that the dosage schedule used for Group D most closely approximates the 'ideal.' (HC)   004885

0231

BRENNER, W.E.; HENDRICKS, C.H.; BRAAKSMA, J.T.; FISHBURNE, J.I.; KRONCKE, F.G.; STAU-ROVSKY, L.

Intraamniotic administration of prostaglandin $F_2\alpha$ to induce therapeutic abortion.

American Journal of Obstetrics and Gynecology 114: 781-787. November 15, 1972.

$PGF_2\alpha$ was administered intraamniotically to 4 women between 13 and 20 weeks gestation in order to compare the efficacy and tolerance of 2 dosage schedules. 5 mg $PGF_2\alpha$ was injected either transabdominally or transvaginally at 10-minute intervals, to a total dose of 15 mg in 20 patients and to a dose of 25 mg in 21 patients. In both cases, the injections were repeated after 24 hours if abortion had not yet occurred. The 25 mg dose schedule seemed most satisfactory with 50% of the patients aborting within 24 hours and 90% within 48 hours. In the 15 mg group only 20% of the patients had aborted within 24 hours and 70% had aborted within 48 hours. No differences were noted in the incidences or severity of pain, vomiting, or blood loss between the 2 groups; however, 5 cases of fever and 1 of bronchospasm were reported in the 15 mg group. Premedication with prochlorperazine in 11 patients of the 25 mg group seemed to lower the incidence of nausea and vomiting, but it was thought that this treatment might have decreased the effectiveness of $PGF_2\alpha$. Multiparous women were found to abort more frequently and in a shorter period of time than did nulliparous women under the same treatment plan. Intraamniotic administration of $PGF_2\alpha$ for the induction of abortion seems superior to hypertonic saline due to a shorter induction-to-abortion time. However a more effective rate of intraamniotic administration still seems possible; further study is needed. (MLH)   004152

0232

BRENNER, W.E.; BRAAKSMA, J.T.; FISHBURNE, J.I., Jr.; KRONCKE, F.G., Jr.,; HENDRICKS, C.H.; STAUROVSKY, L.G.

Intraamniotic prostaglandin ($F_2\alpha$) for inducing abortion.

Obstetrics and Gynecology 39: 628-629. 1972.

The efficacy and complications of the administration of intraamniotic prostaglandin $F_2\alpha$ ($PGF_2\alpha$) were studied in 27 healthy women between 8 to 20 weeks' gestation. 7 patients were monitored for continuous intrauterine pressure by the open end catheter technique. 6 of the 7 patients aborted within 18 hours without serious complications after a single dose of $PGF_2\alpha$ ranging from 2.5 to 50 mg (mean: 15 mg). 20 patients between 14 and 20 weeks' gestation received 15 mg of $PGF_2\alpha$ intraamniotically over a 20-minute period. If abortion did not occur within 24 hours, the same dosage was repeated. 16 patients aborted in a mean of 27.2 hours (range: 7-48 hours). 11 patients vomited, all requested analgesics, 10 had evidence of postpartum endometritis, and 1 experienced clinically evident bronchospasm. These results indicate that intraamniotic $PGF_2\alpha$ is effective with controllable complications. Improved efficacy is anticipated with higher doses. (Authors' modified)   003897

0233

BRENNER, W.E.

Intravenous prostaglandin $F_2\alpha$ for therapeutic abortion: the efficacy and tolerance of three dosage schedules.

American Journal of Obstetrics and Gynecology 113: 1037-1045. 1972.

Because dispute continues over the most effective intravenous prostaglandin $F_2\alpha$ dosage schedule associated with the fewest complications, the efficacy and complications of 3 dosage schedules were compared in 15 patients in the first and second trimester of pregnancy over an 18-hour observation period. In the first dosage schedule the dosage was increased from 25 to 200 μg/minute at regular intervals. In the second the dosage was increased based on uterine activity. In the third 25 μg/minute was given for the first 1/2 hour; then the dosage was increased to 50 μg/minute for the rest of the trial. All patients aborted with the low dosage schedule with fewer complications. No difference in uterine contractility was observed among subjects receiving the 3 dosage schedules or between those patients aborting and failing to abort. Although a relatively low dosage intravenous infusion is effective with few complications, its clinical usefulness is limited; however, it will remain an important research method. (Author's modified)   004139

0234

BRENNER, W.E.; HENDRICKS, C.H.; FISHBURNE, J.J.; BRAAKSMA, J.T.

Intravenous prostaglandin $F_2\alpha$ for therapeutic abortion: cardiovascular and respiratory effects.

In: Bergstrom, S., Green, K., and Samuelsson, B., eds. [Proceedings of the] third conference on prostaglandins in fertility control, Stockholm, January 17-20, 1972. Stockholm, WHO Research and Training Centre on Human Reproduction, Karolinska Institutet, [1972]. (Prostaglandins in Fertility Control 2) p. 156-163.

Cardiovascular and respiratory parameters were monitored during infusion of prostaglandin $PGF_2\alpha$ iv into 7 healthy women 12-16 weeks pregnant for therapeutic abortion. 45 minutes after premedication with 75 mg meperidine and 10 mg chlorpromazine im, $PGF_2\alpha$ was infused iv at 25 μg/minute for 30 minutes, 50 μg/minute for 1 hour, and 200 μg/minute for 1 hour. No significant or consistent cardiovascular changes occurred in direct arterial pressure or cardiac output. Vital capacity decreased significantly in 4 out of 7, somewhat during 50 μg/minute, but dramatically during 200 μg/minute infusion. First-second forced expiratory volume and maximum expiratory flow rate decreased, particularly during the 200 μg/minute dose, but returned to normal within 45 minutes after perfusion was discontinued. No clinical signs of bronchoconstriction were noted. The results suggest that $PGF_2\alpha$ might better suit patients with heart disease than saline abortion. In those with asthma, $PGF_2\alpha$ is not advisable, since other cases of respiratory distress are known to have occurred. (LJG)   004464

0235

BRENNER, W.E.; HENDRICKS, C.H.; FISHBURNE, J.J.; BRAAKSMA, J.T.

Prostaglandin $F_2\alpha$ administered vaginally for inducing therapeutic abortion.

In: Bergstrom, S., Green, K. and Samuelsson, B., eds. [Proceedings of the] third conference on prostaglandins in fertility control, Stockholm, January 17-20, 1972. Stockholm, WHO Research and Training Centre on Human Reproduction, Karolinska Institutet, [1972]. (Prostaglandins in Fertility Control 2) p. 170-174.

A trial of intravaginal prostaglandin $F_2\alpha$ (PGF$_2\alpha$) in 16 women 9-17 weeks pregnant is presented. 9 received 50 mg PGF$_2\alpha$ in 200 mg/ml solution injected in the posterior fornix at hourly intervals and 7 received 50 mg PGF$_2\alpha$ every 2 hours, for 24 hours or until abortion. 6 out of 9 in the 1-hourly group and 3 out of 7 in the 2-hourly group aborted; 4 of the total were incomplete abortions, including all 3 in the latter group. Twice as many women treated every hour required meperidine for pain as those treated every 2 hours. The 2-hourly schedule was inefficient in inducing abortion. (LJG)
004473

0236

BRENNER, W.E.; HENDRICKS, C.H.; BRAAKSMA, J.T.; FISHBURNE, J.I.

The abortifacient efficacy and tolerance of prostaglandin $F_2\alpha$ administered by the intraamniotic and intrauterine-extraamniotic routes.

In: Bergstrom, S., Green, K., and Samuelsson, B., eds. [Proceedings of the] third conference on prostaglandins in fertility control, Stockholm, January 17-20, 1972. Stockholm, WHO Research and Training Centre on Human Reproduction, Karolinska Institutet, [1972] (Prostaglandins in Fertility Control 2) p. 139-155.

20 women 14 to 20 weeks pregnant received 3 doses of 5 mg of PGF$_2\alpha$ intraamniotically at 10-minute intervals with most having an additional 15 mg dose 24 hours later. 14 women aborted, 9 completely, the IAT averaging 28 hours. Pain was significant in 9 women, 4 had emesis, 5 had blood loss greater than 400 ml, 5 had temperatures higher than 100. 10 other women 14 to 18 weeks pregnant received 25 mg PGF$_2\alpha$ intraamniotically; 1 required an additional dose 24 hours later. All women aborted, 6 completely, the IAT averaging 20 hours. Pain was evident in 4 cases, 3 women had emesis, 3 had blood loss greater than 400 ml. 9 women 9 to 17 weeks pregnant received 250 μg PGF$_2\alpha$ by the intrauterine route, 750 μg 2 hrs later, repeated every 2 hours as needed. Only 2 women aborted, 6 had pain, 2 had emesis, 4 had elevated body temperatures. The authors conclude that 25 mg intraamniotically provided the best results, 250 μg doses intrauterine being the least satisfactory protocol. (HC)   004453

0237

BRENNER, W.E.; HENDRICKS, C.H.; BRAAKSMA, J.T.; FISHBURNE, J.I., Jr.; STAUROVOSKY, L.G.

Vaginal adminstration of prostaglandin $F_2\alpha$ for inducing therapeutic abortion.

Prostaglandins 1: 455-467. 1972.

Because of the need for a safe, rapid, effective self-administered abortifacient, the vaginal administration of PGF$_2\alpha$ was investigated. 36 physically healthy women between 8 and 17 weeks gestation received 50 mg PGF$_2\alpha$ on either an every hour or every 2 hour dosage schedule for a total period of 24 hours. The effectiveness and tolerance of PGF$_2\alpha$ as the Tham salt was evaluated for 2 concentrations of solutions, tablets, and suppositories. The incidences of vomiting, diarrhea, fever, and pain were high, independent of the form or concentration administered whenever a dosage schedule resulted in adequate rates of abortion within the 24-hour period. (Authors)   004081

0238
BRUCE, J.E.F.
Abortion with extra-amniotic prostaglandins.
Lancet 2: 380. August 19, 1972.

The author suggested that when investigating the abortive effects of extraamniotic prostaglandins, mechanical stimulation should not be overlooked. He reported 100% success in inducing abortion in 28 patients by inserting a Foley catheter, overinflating a 30ml balloon with 80 ml of water, and giving high doses of intravenous oxytocin without prostaglandins. (RAS)   004180

0239
BRUMMER, H.C.
Detection and measurement of prostaglandins in biological fluids.
Acta Pharmaceutica Jugoslavica 22: 177-187. 1972.

Various methods of extraction, separation and determination of prostaglandins is presented. They all involve the precipitation of protein, followed by extraction into organic solvent at acid pH. Further separation of the different series and individual compounds by chromatographic and chemical techniques are mentioned. Bioassay on smooth muscle preparations, spectrophotometric measurement of ketonic prostaglandins, and the determination of $PGF_2\alpha$ by radioimmunoassay and the applications of these methods are discussed. The use of gas chromatography-mass spectrometry enzymatic and fluorometric techniques are mentioned. Comparison between techniques are made, radioimmunossay and gas chromatography appear to be the most sensitive and specific. (Author's modified)   005037

0240
BRUMMER, H.C.; COLLINS, W.P.
Factors affecting myometrial activity during pregnancy.
Journal of Obstetrics and Gynaecology of the British Commonwealth 79: 985-989. November 1972.

Myometrial strips were taken from uteri of women aborting by hysterotomy or delivering at term by caesarean. Sensitivity of this tissue to oxytocin and prostaglandins in vitro and progesterone content were determined. Plasma samples were taken at the same time as the fetus and placenta were removed and assayed for progesterone content. Myometrial concentration averaged 13 ng per 100 mg tissue for all patients while plasma progesterone levels averaged 4 μg per 100 ml for women 10 to 22 weeks pregnant and 17 μg per 100 ml for those women 30 to 41 weeks pregnant. The response of the myometrium to PGE_1 and PGE_2 was similar quantitatively and qualitatively. There was considerable variation in sensitivity between tissues but no correlation between the dose to elicit contraction and either myometrial or plasma progesterone level. This raises the question of the role of progesterone in pregnancy. Myometrial strips from multigravid patients at term were more sensitive than those from primigravida; sensitivity of term multigravida strips was greater than from multigravida 10-20 weeks pregnant. These results suggest that an earlier pregnancy leaves the uterus in a more excitable state. (HC)   004783

0241
BRUMMER, H.C.
Further studies on the interaction between prostaglandins and Syntocinon on the isolated pregnant human myometrium.
Journal of Obstetrics and Gynaecology of the British Commonwealth 79: 526-530. 1972.

Interaction between prostaglandins and Syntocinon is shown to be of 2 kinds. There may be potentiation, where an increased response to Syntocinon follows combination with a low dose of prostaglandin, and enhancement, which is of much longer duration. Potentiation can be produced by both E and F prostaglandins, whereas enhancement is confined to E prostaglandins. Varying the doses of prostaglandin E is shown to have little effect on the degree of enhancement produced. The physiological implications and clinical applications of these interactions are discussed. (Author) 004073

0242
BRUMMER, H.C.; GILLESPIE, A.
Seminal prostaglandins and fertility.
Clinical Endocrinology 1: 363-368. 1972.

Mean levels of prostaglandins E, A, and 19-hydroxy-A from seminal fluid, estimated spectrophotometrically, were comparable among 4 test groups. The samples came from 1) a group with unexplained infertility who were assessed normal, 2) men who were potentially infertile, 3) normal men whose wives had fertility problems, and 4) normal men who acted as controls. A smaller scatter of values for PGE was found in the group where infertility was unexplained. Rapid sample deterioration clouded the results of the experiment. (RAS)   004827

0243
BRUMMER, H.C.
Serum $PGF_2\alpha$ levels during late pregnancy, labour and the puerperium.
Prostaglandins 2(3): 185-194. 1972.

$PGF_2\alpha$ levels in the peripheral sera of women were radioimmunoassayed. Mean values of .59 ng/ml for late pregnancy, 7 ng/ml for early first stage, 1.9 ng/ml for late first stage, .6 ng/nl for second stages, .7 ng/ml for nonpregnant, and less than .4 ng/ml for postpartum was determined. There were no differences in $PGF_2\alpha$ values if labor was spontaneous or induced. (RAS)   004279

0244
BRUNTON, W.J.
Adenyl cyclase activity in rabbit oviduct: stimulation by isoproterenol and prostaglandins.
In: Abstracts, 5th Annual Meeting of the Society for the Study of Reproduction, East Lansing, Michigan, June 26-29, 1972. p. 26-27.

Earlier work suggested a role for beta-adrenergic agents and prostaglandins in oviductal secretion. Since there is evidence that cyclic AMP mediates the effects of these agents in other secretory tissues, this possibility was examined by determining the presence of adenyl cyclase in rabbit oviduct and the effect of these agents on its activity. Adenyl cyclase activity was detected in the epithelium of the rabbit oviduct. Activity required Mg^{++} and was stimulated by sodium fluoride. Prostaglandins E_1 and E_2 (10 µg/ml) or isoproterenol (10 µM) stimulated adenyl cyclase activity by 3-to 4-fold. These agents did not act by inhibiting oviductal phosphodiesterase, but this enzyme was inhibited by theophylline (.01 M). Prostaglandin stimulation was inhibited Ca^{++}. These results are consistent with the hypothesis that the effects of beta-adrenergic agents and certain prostaglandins on oviductal secretion are mediated by an increased synthesis of cyclic AMP. (Author)   004052

0245

BRUNTON, W.J.

Adenyl cyclase activity in rabbit oviduct: stimulation by isoproterenol and prostaglandins.

Biology of Reproduction 7: 106-107. 1972.

Adenyl cyclase activity in rabbit oviduct was stimulated 3- to 4-fold by 10^{-5}gm per ml PGE_1 or PGE_2. These agents did not inhibit phosphodiesterase. Prostaglandin stimulation was inhibited by calcium ions. The hypothesis is that the effect of prostaglandin on oviduct secretion is mediated by increased synthesis of cyclic AMP. (HC)   004269

0246

BYGDEMAN, M.; BEGUIN, F.; TOPPOZADA, M.; WIQVIST, N.

Further experience with intrauterine prostaglandin administration.

In: Southern, E.M., ed. The prostaglandins: clinical application in human reproduction. (Brook Lodge Symposium, Augusta, Michigan.) Mount Kisco, New York, Futura Publishing Co., 1972. p. 323-332.

$PGF_2\alpha$ and 15-methyl-$PGF_2\alpha$ were each tested by transabdominal intraamniotic and transcervical extraovular routes for first- and second-trimester abortion in a total of 151 women. 70 women, 20 first trimester and 50 second trimester, received 250-750 μg $PGF_2\alpha$ extraamniotically by Foley catheter. All the first trimester and 45 of 50 of the second trimester (93% overall) aborted within 48 hours. Those pregnant 7-8 weeks aborted within mean 6.2 hours; 9-12 weeks pregnant took 15.6 hours; the midpregnant took 24.2 hours. 16 women pregnant 13-16 weeks were given mean 540 μg 15-me-$PGF_2\alpha$ (range 200-1400 μg) and aborted in mean 14.8 hours (range 3-27.7), 50% completely. A single 350 μg dose of 15-me-$PGF_2\alpha$ evoked a uterine response comparable with 5 injections of $PGF_2\alpha$ of 500 μg. 65 women in 14-24 week gestation received 5-25 mg $PGF_2\alpha$ intraamniotically at 3-6 hour intervals; 30 of them were given 25 mg repeated every 24 hours. 29 of 30 aborted within 48 hours (97%), a mean interval of 26.8 hours. 18 women were injected only once with 5 mg 15-me-$PGF_2\alpha$, obtaining 94% success within 48 hours, or a mean interval of 21.6 hours. 6 of the 18 abortions were complete. Side effects for both trials of 15-me-$PGF_2\alpha$ were comparable with those from $PGF_2\alpha$. Blood $PGF_2\alpha$ levels were negligible in 1 woman who received intraamniotic $PGF_2\alpha$. 4 others were given 32 μCi of tritiated $PGF_2\alpha$ with their 25 mg intraamniotic injection. The half-life of $PGF_2\alpha$ in amniotic fluid was 6, 34, 35, and 38 hours, and urinary excretion was prolonged. These studies as well as simultaneous radiographic recordings of uterine activity supported the authors' impression that local administration of PG is safe and effective. (LJG)   004871

0247

BYGDEMAN, M.; BEQUIN, F.; TOPPOZADA, M.; WIQVIST, N.

Intra-amniotic administration of prostaglandin $F_2\alpha$ for the induction of second trimester abortion.

In: Bergstrom, S., Green, K. and Samuelsson, B., eds. [Proceedings of the] third conference on prostaglandins in fertility control, Stockholm, January 17-20, 1972. Stockholm, WHO Research and Training Centre on Human Reproduction, Karolinska Institutet, [1972]. (Prostaglandins in Fertility Control 2) p. 129-136.

To induce abortion with fewer side effects than obtained with systemic routes transabdominal intraamniotic infusion of prostaglandin $F_2\alpha$ ($PGF_2\alpha$) was investigated in 50 women pregnant 14-24 weeks. The amniotic cavity was punctured with an 18 guage thin-walled needle (subsequently removed) through which a polyethylene catheter was inserted for injecting $PGF_2\alpha$ and recording amniotic pressure. 6 dosage and time schedules were used, ranging from 5-10 mg at 3-6 hour intervals to 25 mg at 24 hour intervals, including the World Health Organization protocols of 5 mg in 3 doses every 10 minutes repeated after 24 hours, and 2 doses of 25 mg at 24 hour intervals. Abortion was successful in 46 (92%) of the patients. It was complete in all those given 25 mg or more. Pregnancies of 16 weeks or more needed a higher dose and had a longer interval to abortion. The most efficient dose was 25 mg at 24 hours interval, inducing abortion in a mean of 28.6 hours. Although 3 women who were given 40 mg within 12 hours aborted after 8, 15, and 20 hours, the risks with this schedule need more investigation. 26 women vomited, and 7 had diarrhea; pain during contraction could be controlled with analgesics. Recordings of intensity and frequency of uterine contractions and uterine tone are shown. 4 contractions in 10 minutes were usually observed within 6 hours, with maximal intensity within 10-20 hours. (LJG)   004472

0248
BYGDEMAN, M.; BEGUIN, F.; TOPPOZADA, M.; WIQVIST, N.; BERGSTROM, S.
Intrauterine administration of 15(S)-15-methyl-prostaglandin $F_2\alpha$ for induction of abortion.
Lancet 1: 1336-1337. 1972.

Clinical trials were performed on 31 midpregnant women to determine if the 15-methyl analogue of $PGF_2\alpha$ had any advantages over primary $PGF_2\alpha$ in the induction of abortion. Preliminary dose-response trials on 25 midpregnant women indicated that 15-methyl-$PGF_2\alpha$ was 5-10 times more potent than $PGF_2\alpha$ in stimulating uterine contractions and the response was somewhat prolonged. The prostaglandin was dissolved in dextran to produce a highly viscous solution preventing intravenous escape and vaginal leakage of the prostaglandin. The analogue was administered to 17 of the patients by the extraamniotic route and to 14 by the intraamniotic route. The extraamniotic route caused abortion in 12 patients after 1 injection, in 3 after 2 injections, and in the remaining 2 after 3-5 injections. The mean amount of 15-methyl-PGE2a required was 500 μg (about 1/10 the amount of $PGF_2\alpha$ needed) and the induction/abortion interval was 14.5 hours. In 5 of these patients the catheter was removed after giving them 1 comparatively large injection (200-750 μg). All of these patients aborted without receiving further prostaglandin. The advantages of this 1-shot procedure are obvious. In the intraamniotic trials, 1 5 mg injection was sufficient to induce abortion in all cases. The induction/abortion interval was 22 hours. The side effects in all of these trials were not severe and consisted of vomiting and occasionally diarrhea. The incidence of side effects was similar to those reported for $PGF_2\alpha$. (JRH)   004013

0249
BYGDEMAN, M.; BEGUIN, F.; TOPPOZADA, M.; WIDE, L.; WIQVIST, N.
Postconceptional fertility control by prostaglandin $F_2\alpha$.
In: Bergstrom, S., Green, K., and Samuelsson, B., eds. [Proceedings of the] third conference on prostaglandins in fertility control, Stockholm, January 17-20, 1972. Stockholm, WHO Research and Training Centre on Human Reproduction, Karolinska Institutet, [1972]. (Prostaglandins in Fertility Control 2) p. 175-181.

Prostaglandin ($PGF_2\alpha$) was given intravenously, intravaginally, or by intrauterine means to 34 women who had a positive pregnancy test by the Wide immunologic method 2 weeks after a missed period. 3 out of 7 (43%) women who were perfused iv with 68 μg $PGF_2\alpha$ for 5 hours and 10 out of 15 (67%) given 78 μg iv aborted, as judged by a negative pregnancy test 2 weeks later. 3 out of 8 given 125 mg $PGF_2\alpha$ in solution intravaginally aborted. 2 out of 4 (50%) patients given 500 μg $PGF_2\alpha$ by a single injection into the uterus aborted. Side effects of vomiting and diarrhea were associated with successful abortion, but menstruation-like bleeding occurred despite continued pregnancy in 11 out of 25 patients given $PGF_2\alpha$ iv or by the intravaginal routes. (LJG)   004474

0250

BYGDEMAN, M.

Session VIII: menstrual regulation and interception of pregnancy.

In: Southern, E.M., ed. The prostaglandins: clinical application in human reproduction. (Brook Lodge Symposium, Augusta, Michigan.) Mount Kisco, New York, Futura Publishing Co., 1972. p. 497-520.

This discussion contains 3 lengthy experimental reports, an exposition on the urgency of developing a method of inducing menstruation, and discussion by 10 others on related topics. W. Wiest reported on 5 groups of first and second trimester abortion patients whose decreasing progesterone and estrogen levels were shown to correlate with successful abortion. A. Wentz recorded progesterone, pregnanediol, estradiol, estrone, detailed side effects and temperature curves in 18 of 24 women infused with $PGF_2\alpha$ during menstrual cycles. 7 of the 18 cycles were shortened, and progesterone fell in 8 cycles. M. Embrey infused $PGF_2\alpha$ on days 23 or 24 into 7 normally cycling women, who had bleeding (no endometrial tissue) but menstruated at the expected time. C. Faimen infused $PGF_2\alpha$ into 7 women in luteal phase. In this group there was a slight decrease in progesterone in 4, a significant mean fall in estradiol, a fall in LH in 4, but no change in FSH. The only change in cycle length was a 5 day increase in the following cycle. Regarding menstrual induction, one participant suggested that pregnancy could be ruled out if a patient with missed menstruation has ferning cervical mucus. Others observed that it might be possible to induce menstruation or abortion by increasing endogenous intrauterine PG release, for example by causing uterine stretch as the Japanese do with the metreurynter. (LJG)   004888

0251

CALDWELL, B.V.; SPEROFF, L.; HOBBINS, J.C.; ANDERSON, G.G.

Aspirin as a contraceptive: the prostaglandins strike again.

In: Family Planning in the South. [Proceedings of a conference.] Atlanta, Georgia, Emory University Family Planning Program. 1972. p. 171-184.

The authors discuss the use of PGs as abortifacients using the intravenous, intravaginal, intrauterine and intraamniotic routes. Another topic discussed is the use of $PGF_2\alpha$ as a luteolytic and as an agent to induce menses. The third topic is the use of PG synthesis inhibitors, such as aspirin and indomethacin, as contraceptives. The authors conclude that PGs may be the method of choice for interrupting pregnancy in the second trimester and that prostaglandin inhibitors might become useful as contraceptives. (HC)   004688

0252

CALDWELL, B.V.; ANDERSON, G.G.; HOBBINS, J.C.; SPEROFF, L.

F prostaglandin levels in women receiving $PGF_2\alpha$ for therapeutic abortion.

In: Bergstrom, S., Green, K. and Samuelsson, B., eds. [Proceedings of the] third conference on prostaglandins in fertility control, Stockholm, January 17-20, 1972. Stockholm, WHO Research and Training Centre on Human Reproduction, Karolinska Institutet, [1972]. (Prostaglandins in Fertility Control 2) p. 182-188.

Various procedures for the administration of $PGF_2\alpha$ were studied to determine which provided the highest success rate of therapeutic abortion with the lowest side effect rate. $PGF_2\alpha$ levels in the blood were measured by radioimmunoassay. Intravenous infusion at a rate of 25 μg/minute in 4 patients led to a rise in the level of PGF within 10 minutes which reached 2-3 ng/ml by 30 minutes. In 1 patient the level reached 4 ng/ml when the infusion rate was doubled. Levels dropped off within 5-10 minutes after stopping the infusion, Half-life of $PGF_2\alpha$ could be calculated from this data to be 1-5 minutes.

This method was accompanied by high side effects. Intravaginal insertion of tablets (50 mg/tablet) every hour in 5 patients led to high PGF levels in the blood and unsatisfactory side effects in the 3 successful cases; levels fell and side effects disappeared after rupture of the membranes. Intravaginal insertion of 50 mg $PGF_2\alpha$ in solution in 1 patient produced blood levels of 5 ng/ml after 30 minutes and severe side effects. Intraamniotic administration of 40 mg $PGF_2\alpha$ in 30 patients led to a gradual rise in blood levels, suggesting a slow release of PGs from the amniotic cavity; there were no significant side effects with this method and the success rate was 100%. The intraamniotic route is therefore seen to be the most favorable. (MLH)   004457

0253

CALDWELL, B.V.; AULETTA, F.J.; GORDON, J.W.; SPEROFF, L.

Further studies on the role of prostaglandins in reproductive physiology.

In: Bergstrom, S., Green, K. and Samuelsson, B., eds. [Proceedings of the] third conference on prostaglandins in fertility control, Stockholm, January 17-20, 1972. Stockholm, WHO Research and Training Centre on Human Reproduction, Karolinska Institutet, [1972]. (Prostaglandins in Fertility Control 2) p. 217-233.

PG involvement in luteal function in sheep, rabbits, and monkeys is reviewed. The authors measured estradiol, progesterone, LH, and $PGF_2\alpha$ in sheep estrous cycle, then simulated the PGF peak on Day 14 in ovariectomized ewes by giving them progesterone from Days 1-11, and estradiol on Day 13. No PGF peak could be evoked without estrogen, or in sheep hysterectomized or immunized against estradiol. A model of the sheep cycle was formulated requiring estrogen for the $PGF_2\alpha$ peak and $PGF_2\alpha$ for the preovulatory LH surge. In rabbits, pseudopregnancy generally lasts 16 days, but is extended to 25-29 days by hysterectomy, immunization against $PGF_2\alpha$-BSA, or indomethacin, an inhibitor of PG synthesis. The authors studied the effect of pharmacologic doses of estrogen on luteal function in primates. 10-25 mg diethylstilbesterol for 1-6 days, or 25 mg Premarin for 1-3 days given in luteal phase caused PGF levels to rise and progesterone to decline. $PGF_2\alpha$ infused into the ovarian arteries stimulated progesterone production at low doses, but inhibited at high doses, and inhibition was reversed by chorionic gonadotropin, which simultaneously increased endogenous PGF 5-10 fold. Indomethacin studies suggest that PGs act on the pituitary and ovary. In vitro $PGF_2\alpha$ stimulates progesterone synthesis, possibly with LH via cAMP. In contrast, $PGF_2\alpha$ inhibits cholesterol synthetase and esterase in rat ovaries in vitro and may deplete precursors for steroidogenesis. (LJG)   004460

0254

CALDWELL, B.V.; TILLSON, S.A.; BROCK, W.A.; SPEROFF, L.

The effects of exogenous progesterone and estradiol on prostaglandin F levels in ovariectomized ewes.
Prostaglandins 1: 217-228. 1972.

F prostaglandin levels rose significantly on Day 14 of the sheep estrous cycle in intact ewes and in ovariectomized ewes treated with exogenous progesterone and estradiol as measured by radioimmunoassay. PGF did not show this significant elevation unless estradiol was injected following the sequence of progesterone injections. Low to undetectable PGF was found in hysterectomized animals and in animals immunized against estradiol. This evidence, combined with that reported by others, strongly favors the view that PGF is the 'luteolytic' factor of uterine origin in sheep that controls the normal duration of corpus luteum function. (Authors' modified)   003882

0255

CANTOR, B.; JEWELEWICZ, R.; WARREN, M.; DYRENFURTH, I.; PATNER, A.; VANDEWEILE, R.L.
Hormonal changes during induction of midtrimester abortion by prostaglandin $F_2\alpha$.
American Journal of Obstetrics and Gynecology 113: 607-615. 1972.

Prostaglandin $F_2\alpha$ was infused over 12 hours in 10 patients at 12-16 weeks gestation. There were 4 complete and 5 incomplete abortions. All patients developed significant side effects. Plasma levels of human chorionic gonadotropin, follicle-stimulating hormone, progesterone, free estrogens, and renin were determined serially. Results indicate functional damage of the placenta and fetus prior to clinical evidence of abortion. (Authors)   004020

0256

CARLSON, J.C.; RUGG, A.E.; GLEW, M.E.; BARCIKOWSKI, B.; McCRACKEN, J.A.
Luteolytic properties of prostaglandin $F_1\alpha$ in sheep.
In: Abstracts, 5th Annual Meeting of the Society for the Study of Reproduction, East Lansing, Michigan, June 26-29, 1972. p. 30-31.

The luteolytic effect of $PGF_1\alpha$ was studied in ewes with the left ovary autotransplanted to the neck with vascular anastomoses and compared with the luteolytic effects of $PGF_2\alpha$, PGE_1, and PGE_2. Each dose level of prostaglandin was infused in 7.38 ml of .9% saline/hr for 6 hours directly into the arterial supply of the transplanted ovary. Ovarian venous blood was sampled via the jugular vein during each infusion experiment. Progesterone was quantitated by GLC and LH by radioimmunoassay. In addition, the appearance and duration of estrus was established with a vasectomized ram wearing a crayon marking harness. $PGF_1\alpha$ in doses ranging from 2.5 μg/hr to 50 μg/hr caused complete luteal regression followed by estrous behavior. Progesterone secretion decreased to <5% of preinfusion levels within 24 hours and and LH peak appeared during estrus (from 2.6 ng/ml to 45 ng/ml, mean 4 ewes). For comparison $PGF_2\alpha$ was infused at doses ranging from 1.0 μg/hr for 6 hours into the transplants. At the lowest dose level (1.0 μg/hr) progesterone secretion fell to <50% of the control level within 6 hours. However, normal estrous behavior did not occur. Higher doses of $PGF_2\alpha$ caused complete luteal regression with progesterone secretion rate declining to <5% of preinfusion output followed by estrous behavior and an ovulatory LH peak. Control infusions of 7.38 ml of .9% saline/hr directly into the ovarian arterial supply were without effect. In addition, systemic infusions of $PGF_2\alpha$ (25 μg/hr) failed to cause any significant decrease in progesterone secretion rate from the ovary, indicating that the luteolytic effect of prostaglandins is mediated locally and not through the systemic circulation. In contrast to $PGF_1\alpha$ and $PGF_2\alpha$, the infusion of PGE_1 and PGE_2 directly into the ovary at levels up to 50 μg/hr for 6 hours did not induce estrus, suggesting that in the sheep the luteolytic properties of prostaglandins in vivo are restricted to the F series. (Authors)   004044

0257

CARLSON, J.C.; McCRACKEN, J.A.; FRIED, J.
Prostaglandin $F_1\alpha$ and its 7-oxa analogues-comparative luteolytic effects.
Advances in the Biosciences (Suppl.) 9: 101. 1972.

The luteolytic properties of $PGF_1\alpha$ and several of its 7-oxa synthetic analogues were studied in sheep with autotransplanted ovaries. PGs were infused in .9% saline directly into the arterial supply of the ovary. Ovarian blood flow, progesterone secretion rate, and LH levels were measured during the course of each experiment. Infusions of $PGF_1\alpha$ in doses ranging from 2.5 to 50 μg/hr for 6 hours induced complete luteal regression. Progesterone secretion dropped to <5% of control values within 24 hours. Behavioral estrus occurred in all animals by 48 hours postinfusion accompanied by an ovulatory LH peak in peripheral blood. Similar results were noted with $PGF_2\alpha$, but not PGE_1 or PGE_2. A control infusion of .9% saline failed to show any changes in the above parameters. In contrast to $PGF_1\alpha$, 6-hour infusions of (+)-7-oxa-$PGF_1\alpha$ (50 μg/hr) did not cause complete luteal regression, behavioral estrus, or an ovulatory LH peak. However, progesterone secretion fell to 26% of control values by the sixth hour then returned to control levels within 2 hours. Short term ovarian infusions of (-)-15-epi-7-oxa-$PGF_1\alpha$ (100 μg for 1 hour) showed a temporary increase in blood flow and progesterone output followed by a decrease to slightly less than preinfusion values. A similar but more sustained increase in blood flow occurred during infusion of similar amounts of (+)-15-epi-7-oxa-$PGF_1\alpha$. In addition (-)-7-oxa-$PGF_1\alpha$ (100 μg for 1 hour caused some decrease in progesterone secretion during infusion (71% of control) followed by a return to control levels. Blood flow however, remained essentially unchanged. These results indicate that the substitution of an oxygen atom in place of the carbon atom in position 7 in the prostaglandin molecule diminishes the potent luteolytic properties of $PGF_1\alpha$. (Authors' modified)   004313

0258
CARSTEN, M.E.
Prostaglandin's part in regulating uterine contraction by transport of calcium.
In: Southern, E.M., ed. The prostaglandins: clinical applications in human reproduction. (Brook Lodge Symposium, Augusta, Michigan.) Mount Kisco, New York, Futura Publishing Co., 1972. p. 59-66.

The electron microscopic appearance, ATP-dependent Ca^{++} binding and its inhibition by sodium azide and salyrgan, Ca^{++} release, and effects on both by PGE_2, $PGF_2\alpha$, and $PGF_1\beta$ were analyzed in sarcoplasmic reticulum (SR) purified by sucrose density gradient from pregnant cow and human myometrium. The Ca^{++} binding medium contained .5 mM ATP, .5 mM MgCl2, 2×10^{-5} M CaCl2, .01 M KCl, and .02 M imidazole buffer pH 7.0. After 1 minute incubation the SR took up 6.2 μmol Ca^{++} per g protein, uptake which was not inhibited by sodium azide (inhibitor or mitochondrial Ca^{++} binding) but was inhibited 90% by 3 mM salyrgan. PGE_2 and $PGF_2\alpha$ but not $PGF_1\beta$ at 100 μg per ml completely prevented Ca^{++} binding in bovine and human uterine SR. The effect of PGs on Ca^{++} release was tested by prolonging the uptake incubation for 2 minutes and then adding PG: 100 μg per ml $PGF_2\alpha$ caused release of all the bound Ca^{++} and PGE_2 75%. PGs may regulate uterine contraction by mobilizing bound calcium ions. (LJG)   004848

0259
CERINI, M.E.D.; CHAMLEY, W.A.; FINDLAY, J.K.; GODING, J.R.
Luteolysis in sheep with ovarian autotransplants following concurrent infusions of luteinizing hormone and prostaglandin $F_2\alpha$ into the ovarian artery.
Prostaglandins 2(6): 433-440. December 1972.

Ovariectomized ewes were infused with $PGF_2\alpha$ into the carotid artery at a rate of 5 μg per hour. Estradiol-17β was infused into the cannulated jugular vein at 10 μg per hour to stimulate LH release. $PGF_2\alpha$ did not effect LH secretion in series I (no estradiol), series II ($PGF_2\alpha$ infused at time of expected LH release), and series III ($PGF_2\alpha$ infused at time of estradiol administration). These results negate the notion that increased LH levels account for the luteolytic effect of $PGF_2\alpha$. (RAS)   004830

0260
CHALLIS, J.R.G.; HARRISON, F.A.; HEAP, R.B.; HORTON, E.W.; POYSER, N.L.
A possible role of oestrogens in the stimulation of prostaglandin $F_2\alpha$ output at time of parturition in a sheep.
Journal of Reproduction and Fertility 30: 485-488. 1972.

In a ewe sheep carrying twin fetuses $PGF_2\alpha$ concentration in uterine venous plasma, as detected by bioassay, remained less than 3.5 ng/ml; then rose sharply to approximately 25 ng/ml within 8 hours of lambing. This occurred 15 to 20 hours after estrogen levels began to rise and progesterone levels fell. No PGE_2 was detected. (HC)   004695

0261

CHAMLEY, W.A.; CHRISTIE, M.
Failure of prostaglandin $F_2\alpha$ to affect LH secretion in the ovariectomized ewe.
Prostaglandins 2(6): 465-470. December 1972.

4 sheep with ovarian autotransplants were given an intraarterial infusion of luteinizing hormone (LH) 10 μg per hour for 12 hours. Prostaglandin $F_2\alpha$ ($PGF_2\alpha$) was infused at 5 μg per hour for the final 6 hours. LH produced an increase in progesterone secretion. When $PGF_2\alpha$ was added, the rate fell. LH at very high concentrations failed to prevent the luteolytic effect of $PGF_2\alpha$. (HC)   004592

0262

CHAMLEY, W.A.; BROWN, J.M.; CAIN, M.D.; CERINI, J.C.; CERINI, M.E.D.; CUMMING, I.A.; GOD-ING, J.R.; KRAGT, C.
Luteolysis following intra-arterial infusion of prostaglandin $F_2\alpha$ directly into the ovine autotransplanted ovary.
Journal of Reproduction and Fertility 28: 153-155. 1972.

Whether $PGF_2\alpha$ is 'the' luteolysin depends upon the determination of the minimum effective dose, the isolation and quantification of $PGF_2\alpha$ in the uterine venous plasma, and the efficiency of its transfer from uterine vein to ovarian artery. Results are presented of experiments in which the minimum effective dose of $PGF_2\alpha$ for luteolysis has been investigated. In Experiment 1, 3 sheep were infused intraarterially at 40 μg/hr for 7 hours. Ovarian venous blood samples were taken every 2 hours for 110 hours, and these were assayed for plasma progesterone (P) and estradiol-17beta (E_2). Peripheral blood samples were taken for LH estimations. In all sheep, P secretion dropped by 50% of the control levels at 2.6 $\pm$ 1.4 hours ($\pm$ SD) after the start of infusion and fell to values less than 10% of the preinfusion levels. In 5 sheep, a major increase in E_2 secretion occurred approximately 35 hours from the start of infusion. In 4 of these animals, the pattern of E_2 secretion showed a biphasic peak with the second rise occurring immediately before the LH peak. Following the LH peak E_2 secretion returned to base line in both cases. All animals came into estrus 55 $\pm$ 8.4 hours from the start of infusion. In the 1 sheep which failed to show an LH peak, the initial rise in E_2 secretion occurred 25 hours from the start of infusion but no second peak occurred. In Experiment 2 (4 sheep), the rate of infusion of $PGF_2\alpha$ was reduced to 2 μg/hr and continued for 9.5 to 18 hr. A fall in P was observed in all animals. In 1 the initial level of plasma P was high (7 ng/ml) and the fall in P occurred as rapidly as in Experiment 1. In the other 3 sheep, the fall of P was more gradual; in 2 cases peripheral plasma P reached base line ($<$.5 ng/ml), and in the third animal P did not fall below 2 ng/ml. These experiments show that luteolysis can be produced by $PGF_2\alpha$ given at a rate as low as 2 μg/hr for 9.5 hours, and it seems likely that this infusion rate may approximate the minimum effective dose. These findings give further support to the role of $PGF_2\alpha$ as 'the' luteolytic hormone. (Authors' modified)   003716

0263

CHAMLEY, W.A.; BUCKMASTER, J.M.; CAIN, M.D.; CERINI, J.; CERINI, M.E.; CUMMING, I.A.; GODING, J.R.

The effect of prostaglandin $F_2\alpha$ on progesterone, oestradiol and luteinizing hormone secretion in sheep with ovarian transplants.

Journal of Endocrinology 55: 253-263. 1972.

3 ewes with ovarian autotransplants were given $PGF_2\alpha$ intravenously at a rate of 40 μg per hour for 4 hours; 3 others received $PGF_2\alpha$ at 10 μg per hour for 7 hours. 56 days later, 4 of these animals were given $PGF_2\alpha$ at a rate of 2 μg per hour for 9 1/2 to 18 hours. Ovarian vein blood samples were taken periodically for determination of progesterone, LH, and estradiol-17β levels. $PGF_2\alpha$ had no effect on ovarian blood flow. All animals given 10 or 40 μg per hour came into estrus. $PGF_2\alpha$ at either 10 or 40 μg per hour caused progesterone levels to fall to 50% of control levels in 2 1/2 hours and then further to 10% of control values. Estradiol-17β levels peaked 24 1/2 to 46 hours after infusion began and again at or just before estrus. LH surge was several hours after estrous. Infusing $PGF_2\alpha$ at 2 μg per hour resulted in a gradual fall in progesterone levels in all animals, a pattern resembling that seen in the normal estrous cycle. The author suggests 1) 2 μg per ml might be the minimal effective luteolytic dose; 2) the second estradiol peak may have been LH-induced; 3) $PGF_2\alpha$ did not exert its luteolytic effect by means of vasoconstriction. (HC)   004781

0264

CHANG, M.C.; HUNT, D.M.

Effect of prostaglandin $F_2\alpha$ on the early pregnancy of rabbits.

Nature 236: 120-121. 1972.

Adult rabbits were artificially inseminated and intravenously injected with HCG for induction of ovulation. They were injected subcutaneously on various days (3-10 and 21-23) after insemination with $PGF_2\alpha$. In 2 groups, $PGF_2\alpha$ was placed in a Silastic tube and implanted under the skin on Day 3 after insemination. The rabbits were killed in most cases from Day 9 to Day 14. The number of corpora lutea, uterine swellings (implantation sites), maternal placentae (deciduoma), fetal placenta with or without degenerated embryos, and living fetuses were counted. Results of this study indicate that subcutaneous injection of $PGF_2\alpha$ at a total dose of 5 mg/kg/day soon after ovulation for 1 day causes the disappearance of eggs from the tube and uterus, probably by stimulation of the contractility of these organs. Administration of $PGF_2\alpha$ at the same dose for 5 days when the eggs are in the uterus disturbs the development of the corpora lutea as well as that of the embryo and the normal process of implantation. Administration of 2 mg/kg for 3 days after implantation causes complete degeneration of embryos, which is probably due to some luteolytic effect. Abortion can also be induced by a dose of 2 to 5 mg/kg/day when subcutaneously injected 21 days after insemination. The effective dose, however, is close to a lethal dosage for the rabbit. (Authors' modified)   004136

0265

CHANNING, C.P.; CRISP, T.M.

Comparative aspects of luteinization of granulosa cell cultures at the biochemical and ultrastructual levels.

General and Comparative Endocrinology Suppl. 3: 617-625. 1972.

Sonic ultrastructural features of luteinization of rhesus monkey granulosa cell cultures are summarized. Results of experiments using porcine LH (.2 μg/ml) to stimulate and maintain granulosa cells in culture showed that LH elicited a stimulatory effect when added from Days 0-4, but had no effect if added from Days 4-8. Studies were cited in which PGE_1 and PGE_2, but not PGA or $PGF_2\alpha$, stimulated morphological luteinization and progestin secretion in rhesus monkey granulosa cell cultures. PGE_2 is also able to stimulate cAMP production in porcine granulosa cells and mouse ovaries. The author concluded that these observations, plus the ability of 7-oxa-13-prostynoic acid, a PG inhibitor, to inhibit the stimulatory effects of LH on cAMP production in mouse ovaries, give evidence that PG may be involved in the luteinization mechanism, perhaps somewhere between an LH receptor and adenyl cyclase. (MJN)   004942

0266
CHANNING, C.P.
    Effects of prostaglandins inhibitors, 7-oxa-13-prostynoic acid and eicosa-5, 8,11,14-tetraynoic acid, upon luteinization of rhesus monkey granulosa cells in culture.
    Prostaglandins 2(5): 351-367. November 1972.

To determine whether PGs mediate lutenization in monkey granulosa cells cultured with human chorionic gonadotropin (HCG) or luteinizing hormone (LH), 7-oxa-13-prostynoic acid, a PG antagonist, and eicosa-5,8,11,14-tetraynoic acid, an inhibitor or PG synthesis, were added into the culture medium. Neither agent inhibited progestin release by cultures from preovulatory untreated monkeys, although 50 μg per ml prostynoic stimulated progestin secretion slightly. 50 μg per ml prostynoic inhibited progestin secretion from cultures of monkeys treated with pregnant mare serum gonadotropin (PMSG) when the medium contained .1 μg per ml HCG or 1 μg per ml PGE_2. The threshold concentration of prostynoic was 10 μg per ml, against the stimulation of .01 μg per ml LH in the culture medium. Neither 1, 10, or 100 μg per ml eicosatetraynoic inhibited progestin release from PMSG treated monkey granulosa cells, but 100 μg per ml inhibited progestin release about 50% when .01 μg per ml LH was included in the medium. Morphology of cultured granulosa cells correlated with luteinization, but inhibition was accompanied by vacuolization and necrosis within 8-10 days, possibly due to inhibition of protein synthesis. Therefore these results must be interpreted with caution. (LJG)
    004578

0267
CHASALOW, F.; PHAARISS, B.
    Ovarian prostaglandin (PG) synthesis: Effect of aminoglutethimide (AG) and polyamines.
    Federation Proceedings 31: 545. 1972.

Ovarian homogenates from pregnancy (Day 8) Sprague-Dawley rats have been shown to incorporate tritiated arachidonic acid (H-AA) into PGs(1.2% of added H-AA incorporated into PG E_2 fraction). There is a 37% ($p < .05$) suppression of incorporation following injection of antibovine LH antiserum 1 hour prior to killing and the suppression is overcome by either simultaneous injection of 10 μg of LH or by inclusion of 10 μg/ml of LH in the homogenization media ($p < .001$). When AG (75mg/kg, ip) is injected 1 hour prior to killing, the incorporation of H-AA into PG is increased 70% ($p < .001$) and simultaneous anti-LH antiserum injection does not suppress incorporation. If AG (.3mM), or spermine (.5mM), or spermidine (.5mM) is included in the media, then the synthesis is also increased (18%, $p < .05$; 32%, $p < .005$; 80%, $p < .001$, respectively). Because of the endogenous PG precursors, the percent incorporation from exogenous H-AA suggests a physiologically significant amount of PG was formed (about 10 ng/ml). Since AG and spermine inhibit progesterone synthesis, this experiment suggests that progesterone may be involved in controlling PG synthetase activity in the ovary, providing support for a role for PGs in the regulation of ovarian steroidogenesis. (Authors)   003773

0268
CHASALOW, F.I.; PHARRISS, B.B.
Luteinizing hormone stimulation of ovarian prostaglandin biosynthesis.
Prostaglandins 1: 107-117. 1972.

Rat ovarian homogenates catalyze the synthesis of prostaglandin-like compounds from labeled arachidonic acid. When rats were injected with anti-LH antiserum this synthesis was suppresed. The suppression was overcome either by simultaneous LH injection or by addition of LH to the incubation media. This experiment is the first report of a tropic hormone sensitive prostaglandin synthesis system and provides evidence supporting a physiological role for prostaglandins in the ovary. (Authors)   003836

0269
CHATTERJEE, A.
The possible mode of action of prostaglandins. II. Failure of prostaglandin $F_2\alpha$ in the prevention of compensatory ovulation and ovarian hypertrophy following unilateral ovariectomy in rats.
Acta Endocrinologica 70: 786-790. 1972.

Unilateral ovariectomy in cyclic rats caused a 37.1% gain in weight of the remaining ovaries and doubling of the number of ovulated ova in the following cycle over that observed in their operated counterparts. $PGF_2\alpha$, a very consistent luteolytic agent assumed to act by restricting blood flow to the ovaries, failed to reverse the compensatory ovulation and hypertrophy of the surviving ovaries in the hemispayed rats even when a reasonably large dose of the drug was injected subcutaneously immediately following the operation on Day 1 and repeated again on Day 3 of the cycle. The surviving ovaries on the respective groups of the treated animals weighed (in mg) $41.33 \pm .5$ and 41.6 ± 1.1 as compared with their operated counterparts, i.e., $28.2 \pm .7$ and $28.75 + $ or $\cdot .5$. The possible mode of action of $PGF_2\alpha$ in the regulation of ovarian physiology is discussed. (Author's modified)   004146

0270
CHATTERJEE, A.
The possible mode of action of prostaglandins. I. Interruption of decidual reaction by $PGF_2\alpha$ and its prevention by using prolactin or progesterone.
Acta Endocrinologica 70: 781-785. 1972.

The administration of 500 μg $PGF_2\alpha$ subcutaneously on Day 5 of pseudopregnancy (day of uterine traumatization) caused luteolysis as well as lysis of the deciduoma between 72 to 96 hours. When the $PGF_2\alpha$ was postponed to Days 6, 7, or 8 of pseudopregnancy instead of Day 5, the regressive changes were found to be relatively less intense. Replacement of an exogenous prolactin (500 μg per rat) or hydroxyprogesterone caproate (1 mg per rat) for 4 consecutive days (Days 5-8) consistently prevented the luteolytic effects of $PGF_2\alpha$. The possible mode of action of $PGF_2\alpha$ is discussed. (Author's modified)   004147

0271
CHATTERJEE, A.
The possible mode of action of prostaglandins. IV. Reversal of the detrimental effects of prostaglandins E_2 in the termination of decidual growth in pseudopregnant rats by using prolactin or progesterone.
Prostaglandins 2(5): 417-425. November 1972.

Rats were given .5, 1.0 or 1.5 mg prostaglandins E_2(PGE_2) subcutaneously on Day 5 of pseudopregnancy. The animals were sacrificed on Day 9 and the ovaries, uteri and corpora lutea were weighed. Some animals also received prolactin or progesterone daily from Day 5 to Day 8. The low dose of PGE_2 had no effect. The middle dose induced bloody diestrus smears in 50% of the animals, much less decidual growth than in controls, luteolytic type corpora lutea and many follicles in the ovaries. With the high dose, all animals had bloody proestrus vaginal smears, decidual growth was terminated, ovarian weight was greater than controls and there was marked luteolysis, and well-developed follicles. All of the animals given the high dose exhibited estrous, mated and successfully ovulated. Prolactin or progesterone completely blocked the the detrimental effect of PGE_2 on decidual growth in the pseudopregnant rat. (HC)   004593

0272
CHESTER, R.; DUKES, M.; SLATER, S.R.; WALPOLE, A.L.
   Delay of parturition in the rat by anti-inflammatory agents which inhibit the biosynthesis of prostaglandins.
   Nature 240: 37-38. November 3, 1972.

Indomethacin 2 mg/kg, aspirin 200 mg/kg, and fenclozic acid 25 or 50 mg/kg given orally to rats on Days 20 and 21 of pregnancy delayed parturition by 16-35 hours. Cortisone 50 mg/kg subcutaneously did not delay parturition. Fenclozic acid and indomethacin, but not aspirin, caused an increase in stillbirths. $PGF_2\alpha$ 50 μg per rat given intraperitoneally 3 times on Day 20 prevented the delay in parturition and the increase in stillbirths produced by indomethacin. The authors suggested that a function of prostaglandins in late pregnancy in rats is to induce prepartum luteolysis; the delay in parturition caused by the antiinflammatory agents might be due to inhibition of the biosynthesis of the luteolytic agent resulting in delay of fall of progesterone blood levels. (HC)   004512

0273
CHRISTIE, G.A.
   Prostaglandins -- the point of no return.
   Medical Gynaecology and Sociology 6(3): 6-9. 1972.

The author traces the history of prostaglandin research, discusses the various chemical conformations, and mentions possible mechanisms of action. Isolation of PGs from menstrual fluid, which in vitro had stimulated uterine contraction, led to expanded research of PGs and human reproduction. Karim was the first person to use PGs via intravenous route to initiate abortion for women between weeks 9 and 22 of pregnancy. Since this initial effort various dosages and routes of adminstration have been tested and compared. Recent efforts have been toward a self-administered intravaginal suppository, to be used immediately following a missed menstrual period. Further research is needed prior to governmental approval for over-the-counter use of PGs. (RAS)   004817

0274
CONRADT, A.; UNBEHAUN, V.
   Behaviour of important metabolic parameters in maternal blood during prostaglandin $F_2\alpha$ infusion for therapeutic abortion.
   Advances in the Biosciences (Suppl.) 9: 97. 1972.

14 patients--6 intact pregnancies in Weeks 12-20 of gestation (Group 1), 3 hydatidiform moles, 3 missed abortions, 2 intrauterine fetal deaths (Weeks 32-34)--were continuously intravenously infused with prostaglandin $F_2\alpha$ (50-100 μg/min) until expulsion occurred or to a maximal dose of 80 mg. In all

cases the induction resulted in uterine contractions and, with the exception of 1 patient, in cervical dilation and complete or incomplete expulsion of the conceptus. The $PGF_2\alpha$-induced uterine contractions were readily blocked by intravenous application of a beta-adrenergic substance (Th 1165a). Before and during the infusion the following parameters were studied: glucose, lactate, pyruvate, free and esterified gycerol, triglycerides, 3-hydroxybutyrate, acetoacetate, free fatty acids, total bilirubin, aspartate aminotransferase, total and heat stable alkaline phosphatase, leucine arylamidase, potassium, inorganic phosphate. Some of the most striking data, obtained from the 6 patients of Group 1 were: Continous increase of maternal blood concentration of: glucose (78.3-96.2 mg%), pyruvate (.5-.81 mg%), lactate (8.82-6.88-9.24 mg%), total bilirubin (.45-.70 mg%), inorganic phosphate (3.4-4.5 mg%). Presently the mechanisms leading to these metabolic responses are not wholly understood. (Authors' modified)   004333

0275
CORLETT, R.C.; SRIBYATTA, B.; MISHELL, D.R.; BALLARD, C.; NAKAMURA, R.M., THORNEY-CROFT, I.H.
Termination of early gestation with vaginal prostaglandin $F_2\alpha$ tablets.
Prostaglandins 2(6):453-464. December 1972.

9 women less than 4 weeks pregnant were given 1 to 2 50-mg $PGF_2\alpha$ tablets intravaginally at intervals of 2 to 4 hours for 24 hours. Only 3 patients aborted, even though 7 patients developed vaginal bleeding. HCG, progesterone, and hydroxyprogesterone levels fell in only those patients that aborted. There was no consistent pattern for estradiol levels. Side effects, especially nausea, vomiting, diarrhea, and cramps, occurred in all but 2 patients. The results indicate that although absorption was adequate, individual variation in response precludes use of a fixed dose. The authors indicate that the mechanism of action for $PGF_2\alpha$ is not luteolytic but rather stimulation of uterine muscle contractions. (HC)   004586

0276
COUDERT, S.P.; PHILLIPS, G.D.; PALMER, M.; FAIMAN, C.
Prostaglandin F concentration in the peripheral blood of the ewe during the estrous cycle.
Prostaglandins 2(6): 501-509. December 1972.

PGF levels rose steadily in blood samples taken daily from the utero-ovarian vein of 8 ewes during an induced estrous cycle. In 6 of the animals the major peak of PGF was on Day 13, in 1 it was on Day 10, and 1 had a complex series of day-to-day peaks. Estrus was detected on Days 17 to 19. Other data to be published by these authors show that maximum serum LH levels were on Day 1 and progesterone levels remained stable. These data suggest that PGF may not have a direct luteolytic effect. (HC)   004589

0277
CRAFT, I.
Abortion: Use of prostaglandins and epidural analgesia.
Lancet 2: 41. 1972.

Using increasing doses of prostaglandin E_2 intraamniotically increased success rate of abortions while reducing the overall induction/abortion interval. Abortion occurred within 24 hours in 30 out of 34 patients given an injection of 20 mg PGE_2, and in the remainder after repetition at this time; the mean induction/abortion interval was approximately 14 hours. While these results are not ideal, they suggest that an acceptable and effective method may be developed using the more potent analogues given as a single administration by either route. One of the limitations of using PGE_2 in the doses used by others is the discomfort experienced by the patient as a result of induced uterine activity. To alleviate this, an epidural catheter is inserted immediately before the induction of abortion, with the same beneficial results as obtained in labor. (Author's modified)   004072

0278
CRAFT, I.
  Amniotomy and oral prostaglandin E_2 titration for induction of labour.
  British Medical Journal 2: 191-194. 1972.

The efficacy of oral prostaglandin E_2 used on a titration basis in association with amniotomy for the induction of labor was investigated in a series of 50 patients. Induction was successful in 29 out of 32 primigravid and 17 out of 18 multigravid patients. The mean induction-delivery intervals in successful cases were 10 1/2 and 6 hours, respectively. There were no significant effects on the fetuses. (Author)   003842

0279
CRAFT, I.
  Oral prostaglandins and amniotomy.
  British Medical Journal 2: 653. 1972.

The author discusses a paper by Barr and Naismith in which oral PGE_2 and $PGF_2\alpha$ were used in combination with amniotomy to induce labor in term pregnancies. The author suggests that his method, which involves amniotomy before administering the oral prostaglandin, is more effective than Barr and Naismith's procedure of performing the amniotomy after uterine contractions had started. The author reports a mean induction time for 16 multigravid patients of 7 hours 26 minutes with his protocol compared to 11 hours 9 minutes for 22 out of 24 patients with the Barr and Naismith procedure. The author reports also that amniotomy alone induced labor in 23 patients with a mean induction time of 12 hours 31 minutes. (JRH)   003985

0280
CSAPO, A.I.; et al.
  Discussion: Session III: Labor induction.
  In: Southern, E.M., ed., The prostaglandins: clinical applications in human reproduction. (Brook Lodge Symposium, Augusta, Michigan.) Mount Kisco, New York, Futura Publishing co., 1972. p. 159-168.

The first part of this discussion is a presentation by J. Sauvage of 1000 cases of monitored oxytocin-induced labor, who were given a test dose of oxytocin to assess inducibility; the remainder of the discussion by 8 others concerns largely doses and abnormal contraction patterns to PGs. Sauvage's 1000 patients were 33-42 weeks pregnant, 80% elective induction. Each was scored by the Friedman method and given 25-50 mU oxytocin iv. Contractions were recorded with an extraamniotic .8 ml microballoon before and for 30 minutes after oxytocin. Eventual successful induction was predicted by the degree of synchrony of the spontaneous contractions prior to oxytocin administration. Compared to control patients delivered by spontaneous labor, the induced group had fewer cesarean sections, fewer low forceps deliveries, less postpartum hemorrhage. In the discussion on PGs, it was noted that Karim successfully used lower doses of PG than those used in other trials. He continued to infuse a low dose, e.g. a maximum of 5 or 10 μg $PGF_2\alpha$ per minute, and he waited for labor to develop, rather than infuse a higher dose to achieve response. Another discussant volunteered that besides hypertonus, PGs often induce contractions of a higher frequency and lower amplitude than in normal labor. (LJG)   004858

0281

CSAPO, A.I.; KIVIKOSKI, A.; PULKKINEN, M.O.; WIEST, W.G.

First trimester abortions induced by the extraovular infusion of prostaglandin $F_2\alpha$.

Prostaglandins 1: 295-303. 1972.

A group of 12 patients, 10.8 ± .4 weeks pregnant, received by extraovular infusion 2-4 mg/ hour $PGF_2\alpha$. Complete abortion was provoked in 7 and incomplete abortion in 2 patients in 9.4 ± 1.5 hours by a total dose of 29.1 ± 3.9 mg $PGF_2\alpha$. 3 patients failed to abort. A rapid rise in RP was evident shortly after the start of $PGF_2\alpha$ infusion. This was followed by a gradual increase in IUP, which reached 49.9 ± 3.9 mm Hg within 3 hours. Plasma progesterone levels decreased from 31.4 ± 3.0 ng/ml to 22.3 ± 2.1 ng/ml (p<.05) and estradiol-17β levels decreased from 1.9 ± .2 ng/ml to 1.1 ± .2 ng/ml (p<.001) within this period in the 9 patients who successfully aborted. At the time of abortion plasma progesterone levels had decreased in these patients to 14.6 ± 2.4 ng/ml (p<.001) and estradiol-17β levels to .4 ± .1 ng/ml (p<.001). In those 3 patients who failed to abort, steroid levels remained unchanged. Because of early membrane rupture only 4 patients showed meaningful IUP values; these averaged 126.8 ± 4.5 mm Hg during advanced stages of clinical progress. The side effects were mild except in 2 patients, their vital signs and laboratory tests were normal. Additional, more penetrating studies will be necessary to elucidate fully the mechanism of action of $PGF_2\alpha$, and thus improve the therapeutic value of this compound as an abortifacient. (Authors)   003964

0282

CSAPO, A.I.; RUTTNER, B.; WIEST, W.G.

First trimester abortions induced by a single extraovular injection of prostaglandin $F_2\alpha$.

Prostaglandins 1: 365-371. 1972.

The abortifacient effect of a single extraovular dose of 10 mg $PGF_2\alpha$ has been successfully demonstrated without additional therapy in first trimester pregnant patients. The immediate response to $PGF_2\alpha$ was manifest as uterine pain, which tapered off after about 2 hours. Slight uterine bleeding and decreased levels of plasma estradiol-17β and progesterone measured at 2 hours suggested an initial insult of $PGF_2\alpha$ on the fetoplacental unit. Subsequent reappearance of uterine activity and progressive cervical dilatation signaled effective therapy. 6 patients out of 10 aborted completely in an average of 8.0 hours. The other 4 aborted their fetuses in 15.8 hours on the average but retained placentae. Plasma steroid levels in the former group were decreased still further at the time of diagnostic curettage (22 hours); however, steroid levels remained unchanged in the latter group during the interval between 2 and 22 hours. These results suggest the possibility that, following the successful provocation of fetoplacental insult by $PGF_2\alpha$, supplemental treatment with oxytocics might reduce instillation abortion time and increase the incidence of complete abortions. (Authors) 003983

0283

CSAPO, A.I.; KIVIKOSKI, A.; WIEST, W.G.

Massive initial prostaglandin impact in postconceptional therapy.

Prostaglandins. 2(2): 125-134. August 1972.

The abortifacient effect of an initial $PGF_2\alpha$ impact has been examined in 10 obstetrically normal first trimester pregnant patients. Sedated patients were given extraamniotically an average initial dose of

8.1 $\pm$ .8 mg PGF$_2\alpha$ during a 10-minute instillation. Side effects occurred occasionally but were minimal. Uterine contracture developed rapidly, reaching an average pressure of 83.2 $\pm$ 11.3 mm Hg in about 20 minutes, and then slightly declined with time. Superimposed on the contracture response were gradually increasing cyclic changes in intrauterine pressure that reached a magnitude of 129.8 $\pm$ 12.2 mm Hg by 10 hours after initial treatment. Initial therapy was augmented in some cases by an average of 4 mg PGF$_2\alpha$; only 4 patients required oxytocin supportive therapy. The patients aborted in an average 10.9 $\pm$ 2.0 hours. 7 aborted completely, 2 left behind small placental residues, and 1 retained the placenta for 11.5 hours. An abortion score (AbS) of 92 was obtained in the study, which is the highest in 6 consecutive studies using various methods of PGF$_2\alpha$ administration. Plasma estradiol-17β and progesterone levels decreased continuously during the instillation abortion time in the complete aborters, while the incomplete aborters showed lesser changes. It was concluded that massive intrauterine PGF$_2\alpha$ injection is a more efficacious and acceptable form of postconceptional therapy than protracted treatment. Such therapy appears to convert the refractory uterus into a spontaneously active and pharmacologically reactive organ by inducing vasoconstriction, myometrial stretch, and feto-placental insufficiency. (Authors' modified)   004221

0284

CSAPO, A.I.; KIVIKOSKI, A.; WIEST, W.G.
    Midtrimester abortions induced by intraamniotic prostaglandin F$_2\alpha$ treatment.
    Prostaglandins 1: 305-318. 1972.

A group of 10 patients, 16.2 $\pm$ .5 weeks pregnant, received intraamniotically 10 mg followed at 3 hours intervals by 5 mg PGF$_2\alpha$. The total dose of 31.5 $\pm$ 3.2mg PGF$_2\alpha$ successfully induced abortion in 15.1 $\pm$ 1.8 hours. Seven patients aborted completely and 3 incompletely. The rapid rise in RP was followed by a gradual increase in IUP and a continuing decrease in estradiol-17β and progesterone after a delay of about 6 hours. The systemic side effects were minimal, and the vital signs and laboratory tests revealed no significant changes. The case reports of 4 additional patients are presented, and the mechanism of the abortifacient action of PGF$_2\alpha$ therapy may favorably compete with methods currently used for midtrimester legal abortions. (Authors)   003963

0285

CSAPO, A.I.
    Moderator's summary.
    Journal of Reproductive Medicine 9(6): 327-330. December 1972.

The moderator summarized the reports given at the Brook Lodge Symposium, June 1972, on induction of labor with PGs. Because of the great variety of trials and results and 'individuality of reporting,' generalization was difficult. Most contributors considered PGE$_2$and PGF$_2\alpha$ potent labor inducers; several found them comparable with oxytocin in a few double-blind trials controlled for Bishop inducibility score. The majority occasionally observed increased resting pressure, lack of complete relaxation, hypertonicity, fetal bradycardia, prolonged contractions, prolonged second stage, need for cesarean section, and depressed newborns. The moderator commented on the desirability of monitored induction compared with spontaneous control of labor. He interpreted the events leading to labor as a phenomenon of slowly changing thresholds for the intrinsic myometrial stimulant, PG, competing with progesterone, its inhibitor. (LJG)   004573

0286
CSAPO, A.I.
Moderator's summary: labor induction.
In: Southern, E.M., ed. The prostaglandins: clinical applications in human reproduction. (Brook Lodge Symposium, Augusta, Michigan.) Mount Kisco, New York, Futura Publishing Co., 1972. p. 169-175.

The moderator attempted to summarize the reports on induction of labor with PGs, but because of the great variety of trials and results, and 'individuality of reporting,' generalization was difficult. Most contributors considered PGE_2 and $PGF_2\alpha$ potent labor inducers; several found them comparable to oxytocin, as evidenced by double-blind trials controlled for Bishop inducability score. The majority occasionally observed increased resting pressure, lack of complete relaxation, hypertonicity, fetal bradycardia, prolonged contractions, prolonged second stage, need for cesarean section, and depressed newborns. The moderator commented on the desirability of monitored induction compared to spontaneous control of labor. He interpreted the events leading to labor as a phenomenon of slowly-changing thresholds for the intrinsic myometrial stimulant, PG, competing with progesterone, its inhibitor. (LJG)   004859

0287
CSAPO, A.I.; WIEST, W.G.
On the mechanism of the abortifacient action of prostaglandin $F_2\alpha$.
Prostaglandins 1: 158-165. 1972.

Experiments were performed to determine the mechanism of the abortifacient activity of $PGF_2\alpha$ in rabbits. Either $PGF_2\alpha$, oxytocin, estradiol-17β or estradiol plus $PGF_2\alpha$ were given to pregnant rabbits that had been surgically altered as follows: unilateral placental displacement, bilateral placenta displacement, bilateral ovariectomy, or unilateral placental displacement plus bilateral ovariectomy. Circulating progesterone and estrogen levels were measured daily. It was found that procedures which resulted in decreased progesterone levels increased the sensitivity of the uterus to oxytocin and led to abortion (i.e., bilateral ovariectomy, bilateral placental displacement). Massive doses of $PGF_2\alpha$ given intravenously were ineffective in causing uterine contractions or increasing the sensitivity to oxytocin. However, intrauterine administration led to immediate contraction of the uterus and a delayed increase in sensitivity to oxytocin. It was also found that intrauterine $PGF_2\alpha$ caused a sharp decrease in progesterone levels but had little effect on estrogen levels. It was concluded that while large doses of $PGF_2\alpha$ (1000 μg) given intravenously are ineffective, smaller doses (500 μg) given by the intrauterine route are effective, perhaps by causing contractions that reduce mucosal blood flow, thus interfering with the communication between the placenta and the ovary that protects the pregnancy. (JRH)   003831

0288
CSAPO, A.I.
On the mechanism of the abortifacient action of prostaglandin $F_2\alpha$.
In: Southern, E.M., ed. The prostaglandins: clinical applications in human reproduction. (Brook Lodge Symposium, Augusta, Michigan.) Mount Kisco, New York, Futura Publishing Co., 1972. p. 337-365.

This report includes clinical trials of the author's PG impact theory, supported by further experiments on pregnant rabbits. The author used an arbitrary abortion score based on cervical dilatation and completeness of abortion. A score of 100 represented a complete abortion. 10 first trimester patients given 12 mg per hour iv $PGF_2\alpha$ scored 64; 10 given 500 μg per hour extraovular $PGF_2\alpha$ scored 44; 12

given 2-4 mg per hour extraovular $PGF_2\alpha$ scored 68. Cyclic uterine contractions eventually developed in all. Luteectomy in 30 women before 7 weeks of pregnancy caused progesterone withdrawal which was critical for evolution of cyclic uterine activity or response to oxytocin. Comparable studies in rabbits correlated $PGF_2\alpha$ threshold with progesterone level and showed that estrogen replacement could not prevent abortion. The author then proposed the hypothesis of 'initial impact' of a high initial $PGF_2\alpha$ dose intended to upset hormone regulatory balance. 10 first trimester patients were given 10 mg $PGF_2\alpha$ extraovularly and achieved an abortion score of 80. A group of 12 were given 10 mg $PGF_2\alpha$ intraamniotically, and 5 mg $PGF_2\alpha$ repeated every 3 hours as 'supportive therapy,' and scored 90. 10 patients received a single extraovular injection of mean 8.1 mg $PGF_2\alpha$, supported by oxytocin in 4 women; this resulted in a score of 92. The significance of the PG impact was thought to be the uterine contracture or possibly indirectly via ischemia, evoking endocrine imbalance. $PGF_2\alpha$ was proposed as the intrinsic myometrial stimulant, released by stretching of smooth muscle fibers. (LJG)   004873

0289
CSEPLY, J; CSAPO, A.I.
The effect of prostaglandin $F_2\alpha$ on the small arteries of the omentum uteri in the rat.
Prostaglandins 1: 235-238. 1972.

This paper presents preliminary results of studies on the effects of $PGF_2\alpha$ given topically to small arteries in the omentum uteri of the rat. The omentum uteri was placed under a microscope in a special chamber so that a small artery could be observed. $PGF_2\alpha$ and other vascular substances in various concentrations were applied locally, and their effects on the diameter of the artery recorded. The preparation was then washed with isotonic saline and the artery observed at 1 minute intervals until the preparation returned to normal. Norepinephrine (.1 μg) induced a vasoconstriction which disappeared completely 1 minute after washing. Oxytocin (50 mU) had a similar effect. $PGF_2\alpha$ produced vasoconstriction at doses as low as 10 μg or less. A dose of $PGF_2\alpha$ (100 μg), which had been previously shown to produce abortion when given intrauterinely, caused not only vasoconstriction, but complete stasis, which persisted up to 20 minutes after washing with saline. It is concluded that part of the abortifacient action of $PGF_2\alpha$ may be due to its persistent vasoconstrictor properties. (JRH) 003880

0290
DANDEKAR, P.; VAIDYA, R.; MORRIS, J.M.
The effect of coitus on transport of sperm in the rabbit.
Fertility and Sterility 23(10): 759-762. October 1972.

The major factor in sperm transport in the rabbit is sperm motility. However, prostaglandins in rabbit semen may stimulate uterine contractions. The coital act also plays a role. (HC)   004525

0291
DAVID, H.P.
Abortion in psychological perspective.
American Journal of Orthopsychiatry 42(1): 61-68. January 1972.

This article evaluates the literature and 2 studies currently in progress on psychological implications of abortion or denied abortion, discusses the relationship of abortion and contraceptive behavior, and predicts the impact of PGs if they become available as inducers of menstruation. Previous literature on psychological aspects of abortion is of poor scientific quality, rarely mentions the common emotional relief, and emphasizes postabortion psychoses, of which virtually none are documented.

Only 1 matched study has been done on denied abortion, but a large scale longitudinal study is being conducted in Czechoslovakia with meticulously matched groups and subgroups. Repeated abortions seem to be related to lack of contraceptive practice and are as high as 4.6 abortions per woman in an Armenian survey. A Swiss project is ongoing which will attempt to relate such psychologic factors as sadomasochiasm to repeated abortion seeking. Since some individuals and all societies now seem to need liberal abortion to prevent unwanted births, the most effective alternative is to perfect a self-administerd drug such as PGs to deal with unwanted pregnancy without the need for abortion. (LJG) 004533

0292

DAVIS, B.K.; CHANG, M.C.
Control of hamster fertility with prostaglandin $F_2\alpha$ implants.
Acta Endocrinologica 70: 97-103. 1972.

$PGF_2\alpha$ freely permeates polyacrylamide (PAA) during in vitro incubation in saline. In contrast, Silastic tubing was found to be comparatively impervious and subsequently was discounted as a possible implant medium for $PGF_2\alpha$ in these experiments. Subcutaneous and vaginal PAA implantation of 20, 75, and 100 μg $PGF_2\alpha$, 2 days after mating, significantly reduced (mainly at the higher levels) the number of female hamsters found pregnant upon autopsy 12 days p.c. Subcutaneous injection and vaginal douche were less effective means of administration. Local application of the drug to the vagina was associated with lower pregnancy rates than was parenteral administration. When PAA implants with .5 and 5 mg $PGF_2\alpha$ were given (sc injection) on the second day after mating and 5 mg $PGF_2\alpha$ was implanted an equal interval prior to mating, it became evident that the former approach was at least 10 times more effective in lowering the pregnancy rate in hamsters. This dependence on the time of $PGF_2\alpha$ implantation is attributed to rapid release of the drug from these implants and to the existence of a sensitive phase in the initial stages of pregnancy. (Authors' modified)   003935

0293

DEIS, R.P.; VERMOUTH, N.T.
Prolactin release induced by prostaglandin $F_2\alpha$ in pregnant rats.
In: Abstracts, Fourth International Congress of Endocrinology, Amsterdam, Excerpta Medica, 1972. (International Congress Series 256) Abstract 481. p. 191.

$PGF_2\alpha$ has been observed to induce lactogenesis as well as abortion in rats. Mean serum prolactin levels 8, 10, 12, 20, 22 and 24 hours following i.p. injection of $PGF_2\alpha$ on Day 18 of pregnancy were found to be significantly higher than control values. Prolactin release was found to be phasic, with peaks at 12 and 20 hours. In all $PGF_2\alpha$ treated rats, lactogenesis occurred 26 hours after the last prostaglandin injection and abortion was noted around Day 20 of pregnancy. Progesterone pretreatment was found to prevent abortion and had some ability in inhibiting the increase in the serum prolactin elicited by $PGF_2\alpha$. It is concluded that $PGF_2\alpha$ induces lactogenesis through stimulating prolactin release by decreasing progesterone levels. This would indicate a luteolytic ability for $PGF_2\alpha$. (JSL)   005061

0294

DEL CAMPO, C.H.; GINTHER, O.J.
Vascular anatomy of the uterus and ovaries and the unilateral luteolytic effect of the uterus: guinea pigs, rats, hamsters, and rabbits.
American Journal of Veterinary Research 33(12): 2561-2578. 1972.

Vascular anatomy of the uterus and ovary was compared in 3 species (guinea pig, rat, and hamster) which have a direct unilateral luteolytic pathway, to that of the rabbit, in which luteolysis is not unilateral. 8 guinea pigs, 6 rats, 6 hamsters, and 8 rabbits were injected with red and blue latex, fixed, dehydrated, cleared, and the reproductive vascular system studied with drawings, photographs and measurements. The chief differences were: 1) rabbits had a less prominent branch of the uterine artery extending along the uterine horn to the ovary; 2) uterine venous blood drained in a common trunk (uteroovarian vein) and cranially (based on increasing diameters) in the 3 species with unilateral luteolysis, but in rabbits, uterine venous blood drained caudally and the uterine tip was further from the ovary; 3) the ovarian blood supply (uterine artery and ovarian artery) in the 3 unilateral species were coiled around the uterine and uteroovarian veins, but the rabbit ovarian artery was not as close to that part of the ovarian vein comparable to the uteroovarian vein in the other species. (LJG) 004786

0295
DENKO, C.W.; MOSKOWITZ, R.W.; HEINRICH, G.
Interrelated pharmacologic effects of prostaglandins and bradykinin.
Pharmacology 8: 353-360. 1972.

The pharmacologic interaction of prostaglandin and bradykinin was tested by studying in vitro the contractions of isolated rat uterine muscle. 100 ng $PGF_2\alpha$, 100 ng PGE_2, 500 ng PGB_2, or 10 ng PGE_2 were each combined with 2, 1, 1, and 1 ng of bradykinin, respectively; the onset of contraction was more rapid than when each drug was tested separately. A second experiment in which the subliminal dose of each compound, when added sequentially, again evoked a synergistic response. The authors speculated on how this interaction may be involved in the inflammation process. (RAS)   004481

0296
DJERASSI, C.
Fertility control through abortion.
Bulletin of the Atomic Scientists 28(1): 9-14, 41-45. January 1972.

In placing abortion in the context of world fertility control, the author considers contraceptive behavior of cultural groups, legal abortion, incidence of abortion, likelihood of new abortifacients being developed, and types of conceivable abortifacients, i.e. those interfering with embryonic implantation or development, and the oxytocic abortifacients, specifically PGs. Contraception as a form of preventive medicine tends to be practiced not at all by very low socioeconomic-cultural groups, in the form of abortion by higher groups, and as true contraception by those at the highest cultural level. Legal abortion was permitted in 1930 in Scandinavia, after World War II in Japan, and in 1950 in eastern Europe. It was calculated that 40 million or 8% of the world's fertile women are aborted per year. Thus the need for safer self-administered abortion is apparent, but regulatory agencies, politics, priorities and interested economic groups in wealthy nations will delay the process of developing a new method for 10-16 years, enough time for 1 billion people to be born. An abortifacient interfering with embryonic development is unlikely to be introduced, mainly because it would have to be tested for teratologic effects, a time consuming process, for which the proper animal model is essentially unavailable. PGs show promise by intravenous and intrauterine routes for midtrimester abortion (vacuum aspiration will probably always be better for early abortion). Their disadvantages include high cost, side effects such as hypotension, need for extensive testing by international protocols for midtrimester abortion, lack of a route for self administration in postcoital use and high incidence of side effects for use as a menses inducer. (LJG)   004530

0297

DOUGLAS, R.H.; GINTHER, O.J.

Effect of prostaglandin $F_2\alpha$ on length of diestrus in mares.

Prostaglandins 4(2): 265-268. October 1972.

$PGF_2\alpha$ (1.25, 2.50, 5.00, and 10.00 mg) was injected subcutaneously into 12 pony mares (in groups of 3) on Day 6 of diestrus to determine if it had any luteolytic effects. The $PGF_2\alpha$ caused a significant shortening of diestrus at all 4 dose levels, but mean length of diestrus did not differ among the 4 amounts. Another experiment was performed with lower doses (.1-1.25 mg) in an attempt to determine the minimal effective dose. These results were inconclusive. The $PGF_2\alpha$ treatment had no effect on the length of the next diestrus. (RAS)   004401

0298

DREHER, E.; LIPPERT, T.H.

Induction of abortion by intravenous infusions of prostaglandin $F_2\alpha$.

Archiv fur Gynaekologie 213: 48-53. 1972.

The induction of abortion in 12 patients during the twelfth to eighteenth week of pregnancy was attempted by the intravenous administration of $PGF_2\alpha$ as a continuous infusion for 10 hour periods on 2 consecutive days. The concentration of $PGF_2\alpha$ was increased from 25 μg/minutes to 100 μg/minutes by the third or fourth hour. Uterine activity, respiratory rate, pulse, temperature and blood pressure were measured hourly. Unpleasant side effects were recorded at 2 hour intervals. Abortion was induced in 2 cases, the first after 18 hours, 50 minutes of total infusion time (94.5 mg total dose) and the second after 20 hours (93 mg total dose). The most common unpleasant side effects noted were diarrhea and stomach pain. (JSL)   004844

0299

EAGLING, E.M.; LOVELL, H.G.; PICKLES, V.R.

Interaction of prostaglandin E_1 and calcium in the guinea-pig myometrium.

British Journal of Pharmacology 44: 510-516. 1972

PGE_1 increased the responses of guinea pig myometrium in a low calcium medium to added Ca^{++}, acetylcholine, vasopressin, Ba^{++}, and Sr^{++}. The concentration of PGE_1 used (50 pg/ml) was clearly below the threshold for direct spasmogenesis. In the presence of PGE_1 the doses necessary for half-maximal contractions were decreased by factors of 2.6 for Ca^{++}, 2.4 for acetylcholine, and 3.7 for vasopressin. The responses to Ba^{++} or Sr^{++}, though studied less extensively, were found to be affected in much the same manner. The K^+ depolarized myometrium in a low Ca^{++} medium contracts in response to added Ca^{++}. These responses also were increased by low concentrations of PGE_1, but the effective concentrations of PGE_1, was indistinguishable from that for direct spasmogenesis. Possible mechanisms for the interactions of PGE_1 and Ca^{++} in the myometrium are discussed. It is tentatively suggested that these findings may be relevant to the physiological control of human myometrium. (Authors' modified)   003855

0300

ECKSTEIN, P.

Recent research on the mode of action of intra-uterine devices in primates.

[Symposium on the use of non-human primates for research on problems of human reproduction, Sukhumi, U.S.S.R., December 13-17, 1971.] Acta Endocrinologica Suppl. 166: 364-380. 1972.

New data since 1970 on the mechanism of action of IUDs in women, primates, and other laboratory species are collected and treated under such topics as effect of IUD surface area, heavy metals, endometrial repsonse, deciduation, possible luteolysis, embryotoxicity, and composition of luminal fluid. The most significant findings included the following: IUD effectiveness is correlated with surface area. Heavy metals can increase effectiveness without increasing side effects. Polymorphs and macrophages taken from IUDs actively phagocytize inert particles. By double transfer of mouse embryos, sojourn in an IUD-containing uterine horn for 2-4 hours reduced survival from 40%-50% to 10%-15%. Uterine flushings from baboons were not toxic to mouse embryos in culture. Protein content, reducing sugar, and proteinase activity were higher in IUD uteri. PGs were shown by bioassay to be increased in uterine fluid from IUD patients, but no increase was detected in baboons with IUDs. (LJG)   004838

0301
EL-MAHGOUB, S.
Prostaglandins and reproduction.
Ain Shams Medical Journal 23(1): 89-98. January 1972.

This review covers various topics of PGs in reproductive physiology briefly, emphasizing human data and commenting on possible clinical applications. Topics treated are history, nomenclature, chief tissue sites of synthesis and degradation, male fertility, in vitro and in vivo contractility of nonpregnant and pregnant uterus and oviduct, luteolysis, progesterone synthesis, induction of labor, abortion and menstruation. It was considered unlikely the PGs would be used for labor (because of side effects) or abortion (on moral grounds) in Egypt, but if a monthly contraceptive for self administration was developed, it might be of profound significance. (LJG)   004842

0302
ELIAS, J.A.
Experience with prostaglandins E_2 and $F_2\alpha$ for induction of labor.
In: Southern, E.M., ed. The prostaglandins: applications in human reproduction. [Brook Lodge Symposium, Augusta, Michigan.] Mount Kisco, New York, Futura Publishing Co., 1972. p. 121-128.

Using a double-blind protocol, 30 patients, 38 to 42 weeks pregnant, all with intact membranes received either PGE_2 or $PGF_2\alpha$ by intravenous infusion. Rates were .3 μg per minute for PGE_2 or 2.5 μg per minute $PGF_2\alpha$, doubled after 30 minutes and again at 1, 2 and at 6 hours, if necessary. Labor was successfully induced in all patients. The IDT intervals averaged 11.8 hours for PGE_2 and 10.5 for $PGF_2\alpha$ with some correlation existing between IDT and inducibility. The total dose for PGE_2 averaged 1.8 mg and 5.6 mg for $PGF_2\alpha$. The only side effect noted was phlebitis. Fetuses were normal with regard to Apgar scores, pH and cardiology. Thus, there were no significant differences between PGE_2 and $PGF_2\alpha$ or between primiparas and multiparas. In another study labor was induced successfully in 15 or 17 patients at term, all with intact membranes, using PGE_2 orally, .5 mg initially, then 1.0 mg every 2 hours, increased to 1.5 mg every 2 hours if necessary. The average total dose was 3.5 mg, the IDT averaged 8.7 hours. Vomiting occurred in only 2 patients; all fetuses were normal. In 3 inductions for intrauterine deaths at 28 to 37 weeks, the mean total oral dose of PGE_2 was 16 mg and the IDT 29 hours. (HC)   004855

0303

ELIASSON, R.

Prostaglandins and reproduction -- a general survey.

[Abstracts of] First International Planned Parenthood Publication, South-East Asia and Oceania Regional Medical and Scientific Congress, Sydney, Australia, August 14-18, 1972. p. 14.

The author mentions the chemical composition, areas of function, and potential use for fertility control of prostaglandins. (RAS)   004178

0304

ELLINGER, J.V.; KIRTON, K.T.

Ovum transport in rabbits injected with prostaglandin E_1 or $F_2\alpha$.

In: Abstracts, 5th Annual Meeting of the Society for the Study of Reproduction, East Lansing, Michigan, June 26-29, 1972. p. 25.

Ovum transport in the rabbit is generally considered to be regulated by ovarian steroids. However, the prostaglandins have been shown to alter oviduct motility in several species. The present study was undertaken to investigate possible effects of some prostaglandins on ovum transport in the rabbit. Ovulation was induced in Dutch-belted rabbits by intravenous injection of 100 IU of human chorionic gonadotrophin (HCG); $PGF_2\alpha$ or PGE_1(5 mg subcutaneously) was administered 13 hours postovulation. Animals were sacrificed 6 hours after treatment, the oviducts removed, divided into 3 equal segments, and flushed with saline. All ova from control animals (27/27) were found in the middle third of the oviduct, while in PGE_1 and $PGF_2\alpha$ treated animals, 10 of 29 and 0 of 27 ova, respectively, were found in the middle segment. The majority of ova in prostaglandin-treated animals (19/29 for PGE_1; 27/27 for $PGF_2\alpha$) were found either in the distal third of the oviduct, the uterus, or were not recovered. Unrecovered ova were assumed to be in the uterus or to have been transported through the reproductive tract; prostaglandin treatment hastened the arrival of ova at the uterus by approximately 40 hours. PGE_1 or $PGF_2\alpha$ (5 mg subcutaneously) administered 4 hours postovulation had no effect on the location of ova in the oviduct when examined 5 hours after treatment; however this treatment resulted in an early loss of cumulus cells from the ova. It is concluded that both PGE_1 and $PGF_2\alpha$ can hasten the passage of ova through the rabbit oviduct. The premature removal of cumulus cells from the ovum may be the initial step in accelerated egg transport. (Authors)   004050

0305

ELLINGER, J.V.; KIRTON, K.T.

Ovum transport in rabbits injected with prostaglandin E_1 or $F_2\alpha$.

Biology of Reproduction 7: 106. 1972.

Rabbits pretreated with HCG were given single subcutaneous injections of 5 mg PGE_1 or $PGF_2\alpha$ 13 hours postovulation. In control animals all ova were in the middle third of the oviduct while in prostaglandin-treated animals most ova were in the distal third of the oviduct, the uterus, or were not recovered. Administered 5 hours postovulation, prostaglandins had no effect on the location of the ova. It is concluded that both PGE_1 and $PGF_2\alpha$ can hasten the passage of the ova through the rabbit oviduct. (HC)   004268

0306

ELLIS, L.C.; HARGROVE, J.L.; JOHNSON, J.M.; SEELEY, R.R.

Prostaglandins and the dual endocrine role of the testis.

Research in Reproduction 4: 2. 1972.

Gonadotropins increase the synthesis of both androgens and PGs in rabbit testes, but PG synthesis only is diminished by adrenalectomy. Exogenous PGs and endogenous PG-like compounds can induce contractions in quiescent testes, but this response cannot be duplicated by acetylcholine, serotonin, or epinephrine. (Seasonal changes in the spontaneity of testicular contractions in vitro imply higher neural control of PG synthesis and testicular responses). Synthesis of PGs and androgens are divided into 4 functional units by exogenous serotonin and melatonin. Serotonin inhibits androgen synthesis, increases PG synthesis, and increases tonus and rate testicular contractions. Melatonin inhibits all of these effects. Lipid peroxidation, where unsaturated fatty acids (dihomo-gamma-linolenic, arachiodonic, and 5,8,11,14,17-eicosapentaenoic acids) are converted into either malonaldehyde or PGs, are important biochemical functions in the testes. (MJN)   004485

0307
ELLIS, L.C.
Rat testicular prostaglandin synthesis and its relationship to androgen synthesis.
Federation Proceedings 31: 295. 1972.

Lipid peroxidation, prostaglandin synthetase activity, and androgen synthesis were studied in rat testicular preparations from rats of various ages and after hypophysectomy, adrenalectomy, and injections of gonadotropins. Lipid peroxidation and androgen synthesis had the same distribution with respect to the seminiferous tubules and parenchyma of the testis, and both increased concomitantly with age and development of the gonad. Both functions were markedly reduced by hypophysectomy as was prostaglandin synthetase activity. LH increased both androgen synthesis and prostaglandin synthetase activity. Adrenalectomy decreased prostaglandin synthetase activity without appreciable effects on androgen synthesis. Biological compounds with antioxidant and radioprotective properties affected all 3 phenomena. Melatonin inhibited androgen synthesis, but increased lipid peroxidation. Serotonin inhibited androgen synthesis and lipid peroxidation, but increased prostaglandin synthetase activity. Ascorbic acid, hydrogen peroxide, and vitamin E all affected the above 3 phenomena similar to serotonin and melatonin. These data plus the recent observations in our laboratory that prostaglandins modulate rabbit testicular contractions suggest that the testis is a dual endocrine organ. (Author)   003769

0308
ELLSWORTH, L.R.; ARMSTRONG, D.T.
Ability of prostaglandin E_2 to induce luteinization in transplanted ovarian follicles in the rat.
Proceedings of the Canadian Federation of Biological Societies 15: Abstract 403. 1972.

Ovarian follicles removed from diestrus rats were incubated with 25 μg per ml PGE_2 or 10 μg per ml LH for 2 to 2 1/2 hours. The follicles were then homotransplanted under the kidney capsule in hypophysectomized, ovariectomized recipients. Some animals received only PGE_2-exposed follicles, others had PGE_2 and LH-exposed follicles. 5 days later follicles had transformed into corpora lutea. These results suggest that PGE_2 may be involved in the luteinizing action of LH. (HC)   004718

0309
EMBREY, M.P.; HILLIER, K.
Abortion with extra-amniotic prostaglandins.
Lancet 2: 654-655. September 23, 1972.

Other letters on the Foley catheter and the effect of the mechanical stimulation on the uterus are discussed with the suggestion that inflating the balloon too much may delay delivery and inflating too little may cause premature extrusion of the catheter. The authors found that inflation of the balloon to 40 ml in 2 patients at 18 weeks' gestation did not cause significant uterine activity during an 18-hour period, especially when compared with the activity created by administration of extraamniotic prostaglandins. (JRH)   004249

0310

EMBREY, M.P.

Extra-amniotic prostaglandins.

In: Bergstrom, S., Green, K., and Samuelsson, B., [Proceedings of the] third conference on prostaglandins and fertility control, Stockholm, January 17-20, 1972. Stockholm, WHO Research and Training Centre on Human Reproduction, Karolinska Institutet, [1972]. (Prostaglandins in Fertility Control 2) p.109-117.

Prostaglandin (PG) E_2 and $PGF_2\alpha$ were given by 14-16 French gauge Foley catheter extraamniotically to 94 patients for therapeutic abortion. The catheter was filled with saline containing 1 mg/ml ampicillin. PGE_2 was given at a 50 μg initial dose, then 200 μg at 1 or 2 hour intervals; $PGF_2\alpha$ was given at 250 μg, then 750 μg at 1 or 2 hour intervals. 66 pregnancies (70%) were aborted within 24 hours, 82 (87%) in 36 hours, and 88 (94%) in 48 hours. Abortion interval was shorter for multigravidae (mean 19.1 hr) and more successful (90%), than for primigravidae (24.6 hours and 86%). 61 patients received a mean 11800 μg of $PGF_2\alpha$; 33 received a mean 2300 μg of PGE_2, and experienced a shorter mean abortion interval of 19.5 hours, compared with 24.1 hours for $PGF_2\alpha$. No differences were noted with 1 or 2 dose intervals, incidence of vomiting or of complete abortions, between the 2 PGs. 25 patients vomited once or twice; 2 had mild diarrhea; 40 had tachycardia of 10/min or more; 22 had increased blood pressure 20 mm Hg or less; 38 aborted completely, judged clinically or by uterine evacuation. (LJG)   004470

0311

EMBREY, M.P.; HILLIER, K.; MAHENDRAN, P.

Induction of abortion by extra-amniotic administration of prostaglandins E_2 and $F_2\alpha$.

British Medical Journal 3: 146-149. 1972.

The use of prostaglandins E_2 and $F_2\alpha$ administered by extraamniotic instillation for the induction of abortion, was studied in 94 patients in the first and second trimesters of pregnancy. Abortion was successfully induced in 87% of patients within 36 hours and in 94% within 48 hours. The mean abortion time was 22.4 hours. In 60% of patients abortion was complete. Though the differences were not statistically significant, multigravid patients aborted more quickly on the average than primigravidas, while the mean abortion time in PGE_2-treated patients was less than in those receiving $PGF_2\alpha$. No serious complications occurred. Some side effects were observed. Occasional vomiting was the most common symptom, but the incidence of side effects was lower than with alternative routes of administration. A leucocytosis was often noted, but there were no significant instances of infection. The method has proved a safe and effective means of terminating pregnancies in the second trimester. (Authors' modified)   004120

0312

EMBREY, M.P.

Intra-amniotic prostaglandins.

In: Bergstrom, S., Green, K. and Samuelsson, B., eds. [Proceedings of the] third conference on prostaglandins in fertility control, Stockholm, January 17-20, 1972. Stockholm, WHO Research and Training Centre on Human Reproduction, Karolinska Institutet, [1972]. (Prostaglandins in Fertility Control 2) p. 137-155.

A trial of transabdominal intraamniotic prostaglandin (PGF$_2\alpha$) to induce abortion in 12 women is reported. PGF$_2\alpha$ was administered by a fine polythene catheter through a Touhay needle under local anesthesia. 12 women received 3 5 mg injections at 10 minute intervals, repeated in 24 hours if necessary. Abortion time was 10-24 hours in 3, 24-36 hours in 4, 36-48 hours in 1, and over 48 hours in 4; 8 of these had complete abortions. 4 other patients received 25 mg PGF$_2\alpha$, repeated in 24 hours if necessary. 1 of these aborted in 24 hours, 3 within 48, 1 failed to abort, and 2 of the 4 were imcomplete. (LJG)   004471

0313

EMBREY, M.P.
   Moderator's summary.
   In: Southern, E.M., ed. The prostaglandins: clinical applications in human reproduction. (Brook Lodge Symposium, Augusta, Michigan.) Mount Kisco, New York, Futura Publishing Co., 1972. p. 491-494.

The moderator summarized and commented on 7 presentations of results from trials of PGs as abortifacients. He felt that intravenous and vaginal routes produced unacceptable side effects, but neither intrauterine nor intraamniotic routes permit injection of PG. Of these 2 local routes, he preferred the intrauterine for its simplicity, relative innocuousness, and applicability to women pregnant 11-14 weeks. In general, he approved of those reports which included their 'mishaps,' comparisons of PGF$_2\alpha$ with PGE$_2$, comparisons of PG inductions with traditional methods, and diurnal variability in success rates or hormone levels. (LJG)   004887

0314

EMBREY, M.P.; HENDRICKS, C.H.; BRENNER, B.; QUILLIGAN, E.J.; BALLARD, C.; BYGDEMAN, M.; WIQVIST, N.
   Results of a WHO preliminary study on the termination of pregnancy using intrauterine administration of prostaglandins.
   In: Bergstrom, S., Green, K., Samuelsson, B., eds. [Proceedings of the] third conference on prostaglandins in fertility control, Stockholm, January 17-20, 1972. Stockholm, WHO Research and Training Centre on Human Reproduction, Karolinska Institutet, [1972] (Prostaglandins in Fertility Control 2) p. 107-108.

The protocol for intrauterine PGF$_2\alpha$, designed by participants at a WHO meeting in Stockholm, August 25-29, 1971, to be applied to 100 patients in each center by February 1972, is described. PGF$_2\alpha$ 250 μg/ml is to be injected through a 14-French 30 ml Foley catheter just above the internal os, 250 μg at 0 hours, 750 μg at 2 hours, and 750 μg every 2 hours until abortion. If adverse reaction occurs, the next dose would be 500 μg 4 hours later, repeated every 2 hours. If the patient does not tolerate 400 μg, the same procedure would be followed with 25 μg doses. In case of lack of progress, doses are to be increased by 250 μg up to 750 μg. If abortion does not occur within 30 hours, pregnancy is to be terminated by other appropriate means. (LJG)   004451

0315

EMBREY, M.P.; HILLIER, K.

Therapeutic abortion by extra-amniotic administration of prostaglandins.

In: Southern, E.M., ed. The prostaglandins: clinical applications in human reproduction. [Brook Lodge Symposium, Augusta, Michigan.] Mount Kisco, New York, Futura Publishing Co., 1972. p. 381-390.

70 women were given PGE_2 by the extraamniotic route, 200 μg initially, dose repeated at 2-hour intervals until abortion occurred or up to 36 hours. Another 93 women were given $PGF_2\alpha$, 750 μg initially, repeated at 2-hour intervals as needed. 15% of the women were in the first trimester; the others were in the second trimester of pregnancy. 85% of the women given $PGF_2\alpha$ aborted; the mean IAT was 25 hours. With PGE_2, the success rate was 93% and the IAT averaged 19 1/2 hours. 63% of all the abortions were complete. 25% of the patients experienced vomiting. Blood loss exceeded 300 ml in only 13 women. Total WBC and neutrophil counts increased and lymphocyte count decreased during treatment. In other studies reported by the authors, intravenous infusion by oxytocin 80 mU/minute 6 hours after the PGE_2 injection resulted in a decrease of from 5 to 7 1/2 hours in the average IAT. (HC)   004876

0316

ESKIN, B.A.; SEPIC, R.; AZARBAL, S.; SLATE, W.G.

In vitro fertility responses of cervical mucus treated with prostaglandin ($F_2\alpha$).

Obstetrics and Gynecology 39: 628. 1972.

Little research has been done on the use of prostaglandins to improve fertility. However, there are many reports on the effectiveness of prostaglandins (PG) in terminating pregnancy (abortion and induction of labor). Previous studies revealed a subnormal quantity of prostaglandins in the semen of infertile males. Changes in cervical mucus resulting from treatment with prostaglandin have not been considered. The purpose of this study was to determine whether fertility could be improved by applying $PGF_2\alpha$ to cervical mucus as well as by comparing the characteristics of mucus treated with prostaglandin with control patterns. Using dose-response curves, 250 pg of $PGF_2\alpha$ proved to be most effective on the mucus. The technique employed in this preliminary research involves: 1) collection of cervical mucus from presumably ovulating women on Cycle Days 11-13, 2) use of semen samples from normal males, 3) determination of the changes in the characteristics of cervical mucus when $PGF_2\alpha$ is added, 4) measurement of the penetration and motility of spermatozoa in cervical mucus of untreated controls with physiologic saline added or with $PGF_2\alpha$ in physiologic saline, and 5) preparation of sperm-mucus slides and tube test using methods similar to those of Kunitake and Davazan. The results of this study, as compared to controls, were: 1) increased spermatozoa penetration (2-3X), 2) increased Spinnbarkeit of cervical mucus (2X), 3) increased spermatozoa drive (2X), and 4) increased spermatozoa motility (1.5X). From these preliminary observations, it seems that $PGF_2\alpha$ has some definitive, although selective, effects on the physical traits of cervical mucus. In addition there is evidence of increased penetration and motility of spermatozoa after cervical mucus is treated with $PGF_2\alpha$. Although these preliminary findings seem promising, before clinical investigations or appraisal can be instituted further in vitro research with prostaglandins is necessary. (Authors) 003894

0317

FERRIS, T.F.; STEIN, J.H.; KAUFFMAN, J.

Uterine blood flow and uterine renin secretion.

Journal of Clinical Investigation 51: 2827-2833. November 1972.

Uterine blood flow and renin secretion were examined in nephrectomized pregnant rabbits by studying hemodynamics and the effects of hemorrhagic hypotension, angiotensin, propranolol, isoproterenol and norepinephrine. Blood flow was measured with [85]Sr or [141]Ce labeled microspheres 19 μm in diameter in 24-28 day pregnant Australian white rabbits 24 hours after nephrectomy, under Nembutal, anesthesia. In 40 rabbits mean blood flow was 32.4 ml per minute or 4.7% of cardiac output (measured by dye dilution). Plasma renin activity (PRA) in uterine vein was 994 ng per 100 ml per hour, and 823 in carotid vein. Bilateral ligation of the uterine artery lowered uterine blood flow to 1.95% of cardiac output, increased uterine vein PRA from 1434 to 4430 (p<.001), and carotid artery PRA from 1009 to 2300 (p<.01). Hypotension caused by withdrawing 50 ml blood increased uterine vein PRA from 913 to 3638 (p<.001), and carotid artery PRA from 774 to 1730 (p<.01), while percentage uterine blood flow remained constant. Angiotensin, 10 ng per kg per minute, iv, increased uterine blood flow from 4.1% to 8.4% (p<.005), and this increase was blocked by prior iv propranolol. Isoproterenol, .5 μg per minute, increased uterine flood flow from 3.5 to 6.4% (p<.02). Norepinephrine had no effect. These experiments verified the uterus as an extrarenal site of renin secretion, and showed that renin may help regulate uterine blood flow. The possible release of renin and PGs as vasodilators by ischemic uterus was mentioned. (LJG)   004839

0318
FISHBURNE, J.I., Jr.; BRENNER, W.E.; BRAAKSMA, J.T.; HENDRICKS, C.H.
Bronchospasm complicating intravenous prostaglandin F$_2\alpha$ for therapeutic abortion.
Obstetrics and Gynecology 39: 892-896. 1972.

Investigations using PGF$_2\alpha$ for therapeutic abortion are currently in progress in several centers. This case report describes an asthmatic attack during intravenous administration of PGF$_2\alpha$ to a patient with a history of bronchial asthma. Moderately severe bronchoconstriction was noted clinically during infusion of 200 μg/minute. The symptoms cleared rapidly soon after the drug was discontinued. The literature on respiratory effects of prostaglandins is reviewed. (Authors)   004100

0319
FISHBURNE, J.I., Jr.; BRENNER, W.E.; BRAAKSMA, J.T.; STAUROVSKY, L.G.; MUELLER, R.A.; HOFFER, J.L.; HENDRICKS, C.H.
Cardiovascular and respiratory responses to intravenous infusion of prostaglandin F$_2\alpha$ in the pregnant woman.
American Journal of Obstetrics and Gynecology 114 (6): 765-772. November 1972.

11 women premedicated with meperidine and prochlorperazine, 12 to 18 weeks pregnant, were given PGF$_2\alpha$ at 25 μg per ml by constant intravenous infusion at 25 μg per minute for 30 minutes, then 50 μg per minute for 1 hour followed by another premedication, then 200 μg per minute for 1 hour. 50 μg per minute was restarted until 21.5 hours or until abortion occurred. Cardiovascular and respiratory systems were monitored. Only 5 patients aborted within the 24-hour infusion period, 2 of which were complete. There were no significant cardiovascular changes. Respiratory parameters, vital capacity, forced expiratory volume, maximal mid expiratory flow rate, and maximal expiratory flow rate decreased by 15.5% to 26.9% during maximal infusion. Results suggest that PGF$_2\alpha$ increases airway resistance probably by producing bronchoconstriction. (HC)   004490

0320

FREE, M.J.; JAFFE, R.A.

Factors affecting blood flow and pressures in testes of conscious rats.

In: Abstracts, 5th Annual Meeting of the Society for the Study of Reproduction, East Lansing, Michigan, June 26-29, 1972. p. 47-48.

A miniature friction flowmeter and techniques for the cannulation of small blood vessels have been used to determine the effects of fluid dynamic changes and biologically active substances on blood flow and pressures in testes of conscious rats. Norepinephrine and epinephrine (.03-.15 μg) given locally decreased testis blood flow, while acetylcholine (1.0-10.0 μg) increased it slightly. Single local (.01-1.0 μg) of prostaglandins E_1, E_2, and $F_2\alpha$ markedly increased testis vein pressure and reduced blood flow and lateral pressure in the testis artery. $PGF_2\alpha$ was the most potent venoconstrictor, while $PGE_1$was the most potent inhibitor of blood flow. Catecholamines and prostaglandins injected into the spermatic vein resulted in a paradoxical fall in lateral pressure in the testis artery at dose levels that had no effect on central blood pressure. Serotonin was even more potent in this respect, reducing testis artery pressure by as much as 75%. (Authors' modified)   004043

0321

FRIED, J.; MEHRA, M.M.; GAEDE, B. J.

Novel selective inhibitors of human placental PG-15-dehydrogenase.

Advances in the Biosciences (Suppl.) 9: 18. 1972.

Rac-prostanoic acid (I) and racemic 7-thia-13-prostynoic acid (II) were synthesized and found to be the most potent inhibitors of human placental prostaglandin 15-dehydrogenase known. The enzyme preparation used was that described in Proc. Nat. Acad. Sci., USA, 69: 533, as well as a more highly purified enzyme. Altogether 7 inhibitors were tested and their I[50] determined, using 19 μM PGE_1as a substrate. The most active inhibitors, I and II, had I[50]=15 and 13 μM, respectively. II showed competitive inhibition, Ki=2.8 μM (Km=7.7 μM), whereas I and the others showed mixed inhibition. II was tested by Kuehl and Humes for its binding to their prostaglandin 'receptor' preparation (Proc. Nat. Acad. Sci. USA, 69: 480)and was found to possess an affinity to the receptor equal to that of 7-oxa-13-prostynoic acid. The latter when tested as a PGE_1dehydrogenase inhibitor showed approximately one tenth the activity of II. II therefore exhibits selectivity as a 15-dehydrogenase inhibitor. (Authors' modified)   004385

0322

FRISCH, A.W.

International Conference on Prostaglandins.

London, United States Office of Naval Research, December 28, 1972. (ONR London Conference Report C-20-72) 15 P.

A review of the International Conference on Prostaglandins held in Vienna in 1972 is presented. Short summaries of the papers are given. Prostaglandins hold tremendous potential clinical value in neuromuscular responses, blood clotting, burns, shocks, and pregnancy. The mechanisms are not fully understood, but they act as mediators of many physiological and pathological responses. (RAS) 004744

0323
FUCHS, A.-R.; MOK, E.; SUNDARAM, K.
  Prostaglandin effects in pregnant rats.
  In: Abstracts, 5th Annual Meeting of the Society for the Study of Reproduction, East Lansing,
    Michigan, June 26-29, 1972. p. 29.

Prostaglandins increase ovarian steroid synthesis in vitro but have luteolytic effects in vivo, at least in
some species. This study was, therefore, undertaken to obtain more detailed information on the in
vivo effects. Over 200 rats were used at various stages of gestation. The compounds were
administered intravenously as an infusion of 6 to 7 hours' duration. Blood samples were collected
serially through an indwelling catheter in the right heart. At autopsy 1-4 days later the uterus was
inspected and the ovaries examined for 20-alpha-hydroxysteroid dehydrogenase (20-alpha-ase)
activity. PGE_1(.5 mg/rat) was without effect regardless of the day of administration (4-22). By
contrast, $PGF_2\alpha$ effectively terminated pregnancy, but only during a limited period of gestation. On
Days 4 to 8, $PGF_2\alpha$ had no effect in doses up to 2 mg/rat, and no 20-alpha-ase activity appeared. On
Days 9 to 12, 250 μg $PGF_2\alpha$ per rat was 100% effective. In these rats plasma progesterone levels
declined rapidly during the infusion and 20-alpha-ase activity was evident within 24 hours. On Day 14,
$PGF_2\alpha$ resulted in a partial or complete resorption in 40% of the rats, whereas on Days 16 and 18
$PGF_2\alpha$ was without effect on gestation, but 20-alpha-ase activity appeared in all rats. Exogenous
progesterone or LH prevented the effects of $PGF_2\alpha$ on Day 10. Prolactin, alone or with estradiol-17β,
was ineffective. The results support the contention that $PGF_2\alpha$ interferes with the action of LH on the
ovary. (Authors)   004045

0324
FUCHS, A.-R.; MOK, E.; SUNDARAM, K.
  Prostaglandin effects in pregnant rats.
  Biology of Reproduction 7: 108. 1972.

$PGF_2\alpha$ given intravenously to rats as an infusion over 6 to 7 hours had no effect in doses up to 2 mg
per rat on Days 4-8 of gestation. On Days 9-12, 250 μg terminated pregnancy in all animals, and
progesterone levels declined during the infusion. On Day 14 partial or complete resorption occurred in
40% of the rats, whereas on Days 16 and 18 the drug had no effect on gestation. PGE_1.5 mg per rat
had no effect regardless of day of administration. LH prevented the effects of $PGF_2\alpha$. (HC)   004264

0325
FUCHS, A.-R.; MOK, E.; SUNDARAM, K.
  Prostaglandin effects on luteal function in pregnant rats.
  In: Abstracts, Fourth International Congress of Endocrinology, Washington, D.C., June 18-24, 1972.
    Amsterdam, Excerpta Medica, 1972. (International Congress Series No. 256) Abstract 474. p.
    188.

Given as an intravenous infusion over a 7-hour period, PGE .5 mg had no effect on progesterone levels
in the blood or on 20-alpha-hydroxy steroid dehydrogenase activity of the ovaries of pregnant rats.
$PGF_2\alpha$, .25 mg or more per rat, given on Day 10 terminated pregnancy in all rats, and strong enzyme
activity was noted. When $PGF_2\alpha$ was given on Day 14, resorptions occurred in nearly 1/2 of the
animals treated, but enzyme activity was still obtained. Progesterone levels declined during the
infusion to nonpregnant levels in the animals aborting. Injections of progesterone or LH blocked the
effects of $PGF_2\alpha$; prolactin alone or with estradiol was ineffective. The authors contended that $PGF_2\alpha$
interferes with the action of LH on the corpus luteum. (HC)   005064

0326
FUCHS, A.-R.
Uterine activity during and after mating in the rabbit.
Fertility and Sterility 23(12): 915-923. December 1972.

For eliciting uterine contractions in estrous or estrogen-treated ovariectomized rabbits, the threshold dose of $PGF_2\alpha$ was 10 to 25 μg per animal intravenously. After mating, prostaglandin sensitivity remained until Day 6 or 7. $PGF_2\alpha$ also elicited contractions in phenoxybenzamine-pretreated animals. The response of the uterus to $PGF_2\alpha$ differed from that seen after mating or after oxytocin injection. (HC)   004646

0327
FUCHS, A-R.
Prostaglandin effects on rat pregnancy. I. Failure of induction of labor.
Fertility and Sterility 23: 410-416. 1972.

Infusions of prostaglandins E_1, E_2, and $F_2\alpha$, as well as of other uterine- activating substances, were given intravenously to pregnant rats at or near term in an attempt to induce labor. Uterine activity was monitored by means of previously inserted intrauterine balloons. All infusions were given from 12 noon to 4 p.m., starting on Day 21 of gestation and repeated daily until parturition occurred. Induction of labor was considered successful if delivery started during the infusion and the expulsion of the litter was completed during or within 30 minutes of the infusion. Parturition was not induced in any of the 14 prostaglandin-treated rats. The dosages used varied from 200 to 950 μg/day/rat; the infusion rates varied from 1 to 5 μg/min. Induction of labor was successful in 8 out of 10 rats receiving oxytocin infusions (1-2 mU/min) and in all 7 rats receiving oxytocin together with arginine-vasopressin (both 1-2 mU/min). Infusions of vasopressin alone or of norepinephrine (.1-.5 μg/min) did not induce labor. Neonatal mortality was much higher in PGE_1- and norepinephrine-treated rats than in other groups. Uterine activity elicited by the 3 prostaglandins was more variable and not as strong as that induced by oxytocin. Besides, the prostaglandin-induced activity was usually stronger on Day 21 than on Day 22, which is in contrast to the oxytocin-induced uterine activity. (Author) 003987

0328
GILLESPIE, A.
Factors affecting the dose of prostaglandin E_2 and Syntocinon required to induce labour.
Journal of Obstetrics and Gynaecology of the British Commonwealth 79: 135-138. 1972.

The dose of prostaglandin E_2 or Syntocinon required to induce labor in 300 women showed no correlation with maternal weight, parity, or gestation. (Author)   003814

0329
GILLESPIE, A.; BRUMMER, H.C.; CHARD, T.
Oxytocin release by infused prostaglandin.
British Medical Journal 1: 543-544. 1972.

Plasma oxytocin levels were measured serially in 22 women receiving prostaglandin E_2 or $F_2\alpha$ intravenously for the induction of labor. Oxytocin was detected in the plasma of 19 of the 22 women; positive levels were found in 60 (43%) of 139 plasma samples, an incidence similar to that in the late first stage of spontaneous labor. Oxytocin was found in the maternal plasma, even when the fetus was dead, and in the plasma of 2 men receiving prostaglandin infusions. This indicates that prostaglandins stimulate the pituitary directly and suggests that this mechanism may play a part in the oxytocic action of infused prostaglandins. (Authors)   003787

0330
GILLESPIE, A.
Prostaglandin-oxytocin enhancement and potentiation and their clinical applications.
British Medical Journal 1: 150-152. 1972.

The pharmacological phenomena of enhancement and potentiation of uterine response occur when combinations of prostaglandins and oxytocin are given serially and simultaneously to a patient. Employing these phenomena allows small doses of the drugs to achieve the same effects as a large dose given alone. In a pilot study of the use of the combination of prostaglandin and oxytocin for the induction of midtrimester abortion, 7 of 9 women were aborted within 48 hours. Side effects attributable to prostaglandin were eliminated or reduced in severity. (Author's modified)   003786

0331
GILLETT, P.G.
Prostaglandins and therapeutic abortion.
Journal of Reproductive Medicine 8: 329-334. 1972.

The history, chemistry, and possible biological significance of the prostaglandins are briefly reviewed. Results of a recent clinical trial in which $PGF_2\alpha$ was administered by continuous intravenous infusion to induce therapeutic abortion are summarized, and the significance of the plasma $PGF_2\alpha$ and serum progesterone values are discussed. In addition, the side effects and possible methods by which they might be circumvented are outlined. (Author's modified)   004192

0332
GILLETT, P.G.; KINCH, R.A.H.; WOLFE, L.S.; PACE-ASCIAK, C.
Therapeutic abortion in the second trimester by intra-amniotic prostaglandin $F_2\alpha$.
In: Southern, E.M., ed. The prostaglandins: clinical applications in human reproduction. [Brook Lodge Symposium, Augusta, Michigan.] Mount Kisco, New York Futura Publishing Co., 1972. p. 373-380.

20 women, 15 to 18 weeks pregnant, were given $PGF_2\alpha$ intraamniotically, 5 mg during the initial 5-minute period followed by 20 mg during the next 5 minutes. Patients were medicated for analgesia and anti-nausea. All women aborted, only 6 were complete, the mean IAT was 29 1/2 hours. Nausea and vomiting occurred in 13 patients. Plasma levels of $PGF_2\alpha$, monitored in 3 patients, were higher 4-8 hours after $PGF_2\alpha$ injection. Levels of $PGF_2\alpha$ in the amniotic fluid were 4-fold higher 30 minutes after injection and the mean half-life was 12 1/2 hours. The authors suggest that higher doses or the addition of an oxytocin infusion may improve the results since the method described is not better than the hypertonic saline plus oxytocin protocol. (HC)   004875

0333
GILLETT, P.G.; KINCH, R.A.H.; WOLFE, L.S.; PACE-ASCIAK, C.
Therapeutic abortion with the use of prostaglandin $F_2\alpha$: a study of efficacy, tolerance, and plasma levels with intravenous administration.
American Journal of Obstetrics and Gynecology 112: 330-338. 1972.

Recent reports indicate that certain prostaglandins are effective abortifacients through their oxytocic activity. In this study, 10 women, 10 to 15 weeks pregnant, received $PGF_2\alpha$ by intravenous infusion for termination of pregnancy. Complete abortion was achieved in 9 cases with a mean induction-abortion interval of 24 hours 41 minutes, and a mean drug dose of 93.5 mg. An infusion rate of 50 μg/min was sufficient to induce uterine activity adequate for abortion. The incidence of side effects at this dosage level was high. Electrocardiogram changes and pyrexia were more pronounced at higher infusion rates. Plasma progesterone levels fell only after abortion. Mean plasma $PGF_2\alpha$ levels before infusion were .40 ng/ml and increased significantly with increases in infusion rate to a mean level of 3.00 ng/ml at 22 μg/min. Mean plasma $PGF_2\alpha$ levels double the preinfusion levels appeared sufficient to initiate adequate uterine activity. We conclude that although intravenous $PGF_2\alpha$ is an effective abortifacient, this method of delivery has limited clinical usefulness. (Authors' modified) 003712

0334

GINTHER, O.J.; MECKLEY, P.E.
Effect of intrauterine infusion on length of diestrus in cows and mares.
Veterinary Medicine/Small Animal Clinician 67: 751-754. 1972.

The authors note that studies by other investigators suggest that the substance produced by the uterus which causes regression of the corpus luteum is prostaglandin. They also cite studies in which exogenous prostaglandin interfered with luteal function or caused abortion in a number of animal species. (HC)   004674

0335

GODING, J.R.; BECK, C.; BROWN, J.M.; CERINI, J.C.; CERINI, M.E.D.; CHAMLEY, W.A.; CUM-
MING, I.A.; FELL, L.R.; FINDLAY, J.K.; HEARNSHAW, H.; JONAS, J.; SALAMONSEN, L.
Gonadotrophins in the oestrous cycle.
    In: [Abstracts of] First International Planned Parenthood Federation South-East Asia and Oceania
        Regional Medical and Scientific Congress, Sydney, Family Planning Association of Australia,
        August 1972. p. 9-10.

The authors report that in pregnant sheep there is evidence of a fetal gonadotropin which inhibits the action of $PGF_2\alpha$ and resembles human LH rather than ovine LH. (JRH)   004181

0336

GODING, J.R.; CAIN, M.D.; CERINI, J.; CERINI, M.; CHAMLEY, W.A.; CUMMING, I.A.
Prostaglandin $F_2\alpha$ 'the' luteolytic hormone in the ewe.
Journal of Reproduction and Fertility. 28: 146-147. 1972.

Infusions were made of $PGF_2\alpha$ (Alza) in saline into the uterine vein on the same side as the ovary bearing a CL in 6 sheep on Day 7 to Day 8 of the estrous cycle. The contralateral ovary was excised. The criteria for luteolysis were: 1) progesterone secretion falling to base line levels, and 2) ovarian morphology (5 days after infusion) showing luteal regression and fresh ovulations. 1 ewe was infused with 1000 μg/hr for 3 hours, another 200 μg/hr for 3 hours, and another 20 μg/hr for 9 hours, while the remaining 3 ewes received 20 μg/hr for 7 hours. All sheep showed luteolysis except 1 ewe infused with 20 μg/hr for 7 hours. Control experiments with saline infusion into the uterine vein of 3 sheep for 7 hours and peripheral infusion (jugular vein) of 200 μg $PGF_2\alpha$/hr for 3 hours into 1 sheep, showed no evidence of luteolysis. These experiments support the proposition that $PGF_2\alpha$ can exert its luteolytic action by a countercurrent mechanism and suggest that the minimum effective dose is in the region of

20 μg/hr for 7 hours. In a further series of experiments $PGF_2\alpha$ in saline was infused (4.2 ml/hr) into the uterine horn on the side of the CL in sheep on Days 7 to Day 8 of the cycle. Peripheral progesterone levels were measured, ewes were observed for estrous behavior and ovaries were examined 5 days after infusion. 5 control ewes recieved saline infusions and luteolysis did not occur. Luteolysis, overt estrus, and ovulation followed infusion of 200 μg/hr for 3 hours (1 ewe). 2 ewes received infusions of 50 μg/hr for 9 hours. Neither ewe mated, but 1 exhibited luteolysis; at ovarian examination this ewe had 1 corpus albicans and 2 large follicles. 1 ewe which received 20 μg/hr for 9 hours showed no evidence of luteolysis or mating. It was concluded that $PGF_2\alpha$ may have a practical application to synchronize estrus and/or ovulation in the ruminant by administration into the uterus or upper vagina. (Authors' modified)   003717

0337
GODING, J.R.; CUMMING, I.A.; CHAMLEY, W.A.; BROWN, J.M.; CAIN, M.D.; CERINI, J.C.; CERINI, M.E.D.; FINDLAY, J.K.; O'SHEA, J.D.; PEMBERTON, D.H.
Prostaglandin $F_2\alpha$, 'the' luteolysis in the mammal?
Gynecologic Investigation 2(1-6): 73-97. 1972.

The evidence for $PGF_2\alpha$ being the luteolysin in sheep, evaluations in 10 other species, plus applications in artificial insemination are reviewed. Sheep experiments have shown that $PGF_2\alpha$ is luteolytic when infused iv at 2 μg per hour. $PGF_2\alpha$ was identified at 25 ng per ml on Day 15 in uterine vein by gas liquid chromatography and mass spectrometry. The route taken by PG from uterus to ovary could be from uteroovarian vein to ovarian artery because they are intimately interlaced at the ovarian hilus. Tritiated $PGF_2\alpha$ can cross into the ovarian artery with a recovery of about 1% and a lag of 30 minutes; this is down a concentration gradient since no $PGF_2\alpha$ is detectable in peripheral circulation. Luteolysis can be achieved by infusing 20 μg per hour into the uterine vein (with an estimated final concentration of 20 ng per ml), or 2 μg per hour into the ovarian artery. In normal cycling sheep whose uteroovarian blood was analyzed by radioimmunoassay every 3 hours, the $PGF_2\alpha$ peaks were correlated with troughs of progesterone. Luteolytic mechanisms in cow, sow, guinea pig, and possibly hamster and rabbit resemble those in sheep. Except for possible control by prolactin in rats, the mechanism is unknown in goat, horse, monkey and human. This knowledge has been applied, with a view toward synchronization of estrus in cows and ewes, in experiments using intrauterine instillation of $PGF_2\alpha$. (LJG)   004841

0338
GORDON, E.S.
The prostaglandins. Physiologic actions and clinical potential.
Postgraduate Medicine, October 1972. p. 75-79.

This discussion of PGs praises their physiologic effects and possible uses in reproduction and hypertension. PGE_2 and $PGF_2\alpha$ were shown to be responsible for the uterine contractions of menstruation in 1963. PGE_2 is the more potent oxytocic, inducing labor or abortion when infused at .005-.01 μg/kg/minute. Contractions begin in 30 minutes, resemble normal labor (which is accompanied by PG surges in maternal blood during contractions), and result in delivery after 5-6 hours without fetal problems, complications, or side effects. Intravenous PG 25-100 μg/minute will abort 94%-100% of patients, with a few mild gastrointestinal side effects. These can be eliminated by administering PG by transcervical catheter hourly into the extraovular space, or by intravaginal tampon. Men with functional infertility have semen PGE levels of less than 15 μg/ml, compared with 55.2, the mean in normal men. PGAs and some PGEs also appear to be ideal for treating essential hypertension. They decrease systolic and diastolic pressure, increase cardiac output, decrease renal resistance, and are the most potent natriuretic agents known. (LJG)   004553

0339
GRANSTROM, E.
Metabolism of prostaglandin $F_2\alpha$ in female subjects.
In: Bergstrom, S., Green, K., and Samuelsson, B., eds. [Proceedings of the] third conference on prostaglandins in fertility control, Stockholm, January 17-20, 1972. Stockholm, WHO Research and Training Centre on Human Reproduction, Karolinska Institutet, [1972] (Prostaglandins in Fertility Control 2) p. 107-108.

3 experiments were conducted on metabolites of $PGF_2\alpha$ in female urine and plasma. First [9beta-3H]-$PGF_2\alpha$ was injected iv and analyzed in urine collected for 5-6 hours. Lipophilic (Amberlite XAD-2), silicic acid, and reversed phase partition chromatography yielded 4 peaks. After making appropriate derivatives of 3 of these peaks for gas-liquid chromatography and mass spectrometry, the author characterized a total of 14 peaks by further analysis. These included: (III) 7alpha,9alpha-dihydroxy-ketodinorprosta-3-en-18-dioic acid, and its corresponding menocarboxylic acid; (I) a series of C16 compounds with 2 or 3 of 5-, 7- and/or 9alpha-hydroxy groups, 11- or 13-keto groups, or their delta or gamma lactones; (IV) a 5alpha,7alpha-dihydroxy-11-keto)tetranor, omega-dinor)prosta-1, 14-dioic acid and its lactones. The second experiment was iv injection of [9beta-3H]-$PGF_2\alpha$; the plasma withdrawn 10 minutes later contained $PGF_2\alpha$, 9alpha,11alpha,15-tri-hydroxyprost-5-enoic (dihydro-$PGF_2\alpha$), and 70% of the material appeared as 9alpha,11alpha-dihydroxy-15-keto-prost-5-enoic (15-dehydro-$PGF_2\alpha$). In the third study, injections of both [14C]- and [3H]-$PGF_2\alpha$ or 15-dehydro-$PGF_2\alpha$ showed the time course of the above process. $PGF_2\alpha$ was 97% metabolized within 1.5 minutes, and the half-life of the dehydro-$PGF_2\alpha$ was 7 minutes. (LJG)   004462

0340
GRANSTROM, E.
On the metabolism of prostaglandin $F_2\alpha$ in female subjects: structures of two metabolites in blood.
European Journal of Biochemistry 27: 462-469. 1972.

Several tritium-labeled products appeared in blood after intravenous injection of [9beta-tritium]-prostaglandin $F_2\alpha$ into a female subject. The major metabolite was identified as 9alpha,11alpha,dehydroxy-15-oxoprost-5-enoic acid and one of the minor products was 9alpha,11alpha,15-trihydroxyprost-5-enoic acid. Data on the disappearance from blood of [9beta-tritium]-prostaglandin $F_2\alpha$ and of [9beta-tritium]9alpha,11alpha-dihydroxy-15-oxoprost-5-enoic acid after intravenous administration are given. (Author's modified)   004028

0341
GRANSTROM, E.
On the metabolism of prostaglandin $F_2\alpha$ in female subjects: structures of two C14 metabolites.
European Journal of Biochemistry 25(3): 581-589. 1972.

[9beta-3H]-$PGF_2\alpha$ and unlabeled $PGF_2\alpha$ were injected iv into women and metabolites were isolated from pooled urine collected for up to 8 hours. 35 μg of the labeled $PGF_2\alpha$ and 5-50 mg of the unlabeled PG were administered. After lipophilic chromatography (Amberlite XAD-2) and silicic acid chromatography, 95% of the radioactivity was recovered. Reversed phase chromatography yielded 4 peaks, 2 of which were already identified: the other 2 peaks were analyzed in this study. Each was subjected to mass spectrometry after appropriate purification by thin layer and silicic acid chromatography and conversion to acetates or trimethylsilyl ethers. One peak was shown to be 5alpha,7alpha-dihydroxy-11-keto-(tetranor,omega-dinor)-prosta-1,14-dioic acid and the other 5alpha,7alpha,11-trihydroxy-(tetranor,omega-dinor)-prosta-1,14-dioic acid. Both occurred as delta lactones and the trihydroxy metabolite was also identified as its gamma lactone. It was suggested that

112

these structures were formed by 2 beta oxidations at the omega end of $PGF_2\alpha$, a pathway considered unique in humans. (LJG)   005006

0342
GREEN, K.; BEGUIN, F.; BYGDEMAN, M.; TOPPOZADA, M.; WIQVIST, N.
   Analysis of prostaglandin $F_2\alpha$ and metabolites following intravenous, intra-amniotic and vaginal administration of prostaglandin $F_2\alpha$.
   In: Bergstrom, S., Green, K., and Samuelsson, B., eds. [Proceedings of the] third conference on prostaglandins in fertility control, Stockholm, January 17-20, 1972. Stockholm, WHO Research and Training Centre on Human Reproduction, Karolinska Institutet, [1972]. (Prostaglandins in Fertility Control 2) p. 189-200.

Blood levels of prostaglandin ($PGF_2\alpha$), 9alpha,11alpha-dihydroxy-15-keto-prost-5-enoic acid, and 9alpha,11alpha,15-trihydroxy-prost-5-enoic acid were determined during intravenous, vaginal, or intraamniotic administration of $PGF_2\alpha$ in 8 midpregnant women. Uterine activity, tone, and side effects were recorded. Intravenous perfusion of 75 µg/minute $PGF_2\alpha$ for 10 hours in 5 patients raised $PGF_2\alpha$ in the blood from 1.2-5.7 ng/ml, in a plateau-shaped curve. Trihydroxyprostenoic acid rose to 1.2 ng/ml, and dihydroxyketoprostenoic rose to 28-33 ng/ml in a patient infused with $PGF_2\alpha$ for 5.5 hours. Dihydroxyketoprostenoic acid rose to 42-56 ng/ml in another patient infused for 10 hours. In a patient given 25 mg intraamniotic $PGF_2\alpha$, no $PGF_2\alpha$ was detectable in the blood, but in a patient given 50 mg in a vaginal suppository, a peak in $PGF_2\alpha$ was detected 2 hours later. Uterine activity and tone as well as frequency of vomiting and diarrhea tended to follow blood levels of $PGF_2\alpha$. (LJG)   004465

0343
GREEN, K.; SAMUELSSON, B. CARLSON, J.C.; McCRACKEN, J.A.
   Prostaglandin $F_2\alpha$ identified as a luteolytic hormone in sheep.
   In: Abstracts, Fourth International Congress of Endocrinology. Washington, D. C., June 18-24, 1972. Amsterdam, Excerpta Medica, 1972. (International Congress Series No. 256) Abstract 475. p. 189.

Ovarian and uterine transplantation, cross circulation and uterine vein ligation studies indicated that $PGF_2\alpha$ is a luteolytic factor in uterine vein blood in sheep. Intraarterial (but not systemic) infusions of $PGF_2\alpha$ into the transplanted ovary mimicked this luteolytic factor. Uterine venous plasma from individual sheep was examined for $PGF_2\alpha$ during luteal regression and on other cycle days. [D8]-$PGF_2\alpha$ and [3H]-$PGF_2\alpha$ were added to each sample as carrier and tracer respectively, and measured by gas liquid chromatography and mass spectometry. Luteal function was assessed by determining progesterone in ovarian venous plasma by gas liquid chromatography. Coincident with the fall in progesterone during luteolysis there was a 10-20 fold rise (2.0 ng per ml to 42.0 ng per ml) in $PGF_2\alpha$ in uterine vein plasma. Venous blood from the uterus was sampled from sheep bearing utero-ovarian transplants. On cycle days 15 and 16 the concentrations of $PGF_2\alpha$ (ng per ml) were as follows: 0 hour, <2 ng; 8 hours, 15 ng; 16 hours, 13 ng; 0 hours, 21 ng; 8 hours, 3 ng; 16 hours, <2 ng; 0 hours, <2 ng. By infusing 25 µg $PGF_2\alpha$ per hour (the calculated release rate during luteolysis) into the uterine vein in situ, premature luteal regression was induced consistently in the adjacent ovary, but not when infused systemically. This not only verified the counter current exchange of $PGF_2\alpha$ between the uterine vein and the ovarian artery, but also confirmed the role of $PGF_2\alpha$ as a luteolytic hormone in the sheep. (Authors' modified)   005009

0344

GRINWICH, D.L.; KENNEDY, T.G.; ARMSTRONG, D.T.

Dissociation of ovulatory and steroidogenic actions of luteinizing hormone in rabbits with indomethacin, an inhibitor of prostaglandin biosynthesis.

Prostaglandins 1: 89-96. 1972.

The possible involvement of prostaglandins as mediators of ovarian responses to luteinizing hormone (LH) has been investigated in in vivo studies with the aid of indomethacin (1-[p-chloro-benzoyl]-5-methoxy-2-methylindole-3-acetic acid), an inhibitor of prostaglandin biosynthesis. LH (50 μg NIH-LH-B7) was injected iv 1/2 hour after iv injection of either indomethacin (20 - 40 mg/kg) or phosphate buffer vehicle. Mean levels of progesterone and 20alpha-hydorxypregn-4-en-3-one (20alpha-OH-P) observed in peripheral plasma obtained by cardiac puncture 1 hour after LH were increased by 5.35 ± 1.48 and 26.0 ± 11.6 ng/ml, respectively, in phosphate buffer-treated controls, and by 5.60 ± 4.47 and 100.6 ± 37.1 ng/ml, respectively, in indomethacin-treated rabbits. The apparently greater increase in 20alpha-OH-P in indomethacin-treated rabbits could be attributed almost entirely to the larger amounts of interstitial tissue in the latter animals, as indicated by essential disappearance of this difference when covariance analyses were performed to correct for this difference in ovarian interstitial tissue weight. None of 12 indomethacin-treated and 10 of 11 vehicle-treated control rabbits ovulated in response to this dosage of LH, as determined by flushing of oviducts to recover ova, and by gross examination of ovaries 1 or 3 days after LH treatment. Indomethacin did not prevent luteinization of follicles in LH-treated rabbits, as determined by gross and histologic examination of ovaries 8 days after treatment. These findings argue against a role of prostaglandins as mediators of the acute steroidogenic and luteinizing actions of LH in the rabbit ovary, but suggest an involvement in the process of ovulation. (Authors)   003838

0345

GRINWICH, D.L.; KENNEDY, T.G.; ARMSTRONG, D.T.

Inhibition of the ovulatory action of luteinizing hormone by indomethacin, an inhibitor of prostaglandin synthesis.

Proceedings of the Canadian Federation of Biological Societies 15: Abstract 404. 1972.

Indomethacin 10-40 mg/kg was given intravenously to rabbits 1/2 hour before intravenous injection of LH 50 μg. LH-induced ovulation was blocked in all animals. Ova were in the follicles, and there was evidence of granulosa cell luteinization 3 days later. By Day 8 corpora lutea were present. Indomethacin did not block the increase in serum progesterone levels induced by LH. These results suggest an involvement of prostaglandin in the ovulatory but not in the luteinizing or steroidogenic effect of LH. (HC)   004719

0346

GUSTAVII, B.; BRUNK, U.

A histological study of the effect on the placenta of intra-amniotically and extra-amniotically injected hypertonic saline in therapeutic abortion.

Acta Obstetrica et Gynecologica Scandinavica 51: 121-125. 1972.

The authors found extensive decidual cell damage when hypertonic saline was used to induce abortion. Since others have reported the presence of prostaglandins in decidual cells and because of the known role of prostaglandins in labor, spontaneous abortion, and induction of therapeutic abortion, the authors suggest that the release of endogenous prostaglandins may be involved in the mechanism by which hypertonic saline induces abortion. (JRH)   004092

0347
GUSTAVII, B.
Labour: a delayed menstruation?
Lancet 2: 1149-1150. 1972.

A model is proposed for the mechanism initiating contractions in the pregnant uterus, according to which labor may be regarded as delayed menstruation. An increase of labilizing factors (estrogen, progesterone) and/or decrease in stabilizing factors, perhaps a stabilizing protein, that influence decidual plasma and/or lysosomal membranes lead to increased lysosome permeability. The resultant release of lysosomal hydrolytic enzymes to the cytoplasm initiates intracellular processes resulting in prostaglandin synthesis and release. The prostaglandin induces uterine contraction and decidual cell stress leading to further permeability of decidual lysosomal membranes and the cycle is complete. (HC)   004487

0348
GUSTAVII, B.; GREEN, K.
Release of prostaglandin $F_2\alpha$ following injection of hypertonic saline for therapeutic abortion: a preliminary study.
American Journal of Obstetrics and Gynecology 114(8): 1099-1100. December 15, 1972.

Extraamniotic injection of hypertonic saline in 3 women, 17 to 22 weeks pregnant, caused all 3 to abort in 32 to 54 hours. The concentration of $PGF_2\alpha$ in amniotic fluid increased 24 hours after injection, before uterine contractions were effective, and continued to rise. The authors note that the question of whether the release of PG is cause or result of increased uterine activity needs further research. (HC)   004648

0349
GUTIERREZ-CERNOSEK, R.M.; MORRILL, L.M.; LEVINE, L.
Prostaglandin $F_2\alpha$ levels in peripheral sera of man.
Prostaglandins 1: 71-80. 1972.

Antibodies to $PGF_2\alpha$ and $PGF_1\alpha$, which discriminate between these 2 prostaglandins and the metabolic products 15-keto-$PGF_2\alpha$ and 13,14-dihydro-15-keto-$PGF_2\alpha$ were used to determine $PGF_2\alpha$ levels in peripheral sera of man. The procedure for extraction of serum and the efficiency and reproducibility of this extraction are presented. 76 samples of peripheral sera were analyzed by radioimmunoassay. These values are presented. In general, sera from males were found to contain larger amounts of $PGF_2\alpha$ than sera from females. (Authors)   003742

0350
GUTIERREZ-CERNOSEK, R.M.; ZUCKERMAN, J.; LEVINE, L.
Prostaglandin $F_2\alpha$ levels in sera during human pregnancy.
Prostaglandins 1: 331-337. 1972.

Prostaglandin $F_2\alpha$ levels in serum of pregnant women throughout gestation were determined by radioimmunoassay. Significant differences in $PGF_2\alpha$ concentration were found in the serum of women at various stages of gestation. The amounts of $PGF_2\alpha$ in the serum increased gradually as pregnancy progressed and reached a peak during the second trimester. However, during the latter part of the third trimester, $PGF_2\alpha$ levels approximated those found in the serum of nonpregnant women. (Authors)   003960

0351
GUTKNECHT, G.D.; JOHNSTON, J.O.
   Direct luteolytic effect of locally administered prostaglandin $F_2\alpha$ in the pregnant hamster.
   In: Abstracts, 5th Annual Meeting of the Society for the Study of Reproduction, East Lansing, Michigan, June 26-29, 1972. p. 30.

   The luteolytic dose of prostaglandin $F_2\alpha$ ($PGF_2\alpha$) to terminate pregnancy was determined in 146 hamsters treated either subcutaneously or locally into the uterus or ovarian bursa on Day 4 after coitus. $PGF_2\alpha$ when placed bilaterally into the ovarian bursa was 25 times more active in terminating pregnancy than when instilled bilaterally into the proximal end of the uterine horn cavity, and 100 times more active in terminating pregnancy than when injected subcutaneously. Statistically, the minimal effective dose of $PGF_2\alpha$ to terminate pregnancy in 50% of the treated hamsters was .4 μg after placement into the ovarian bursa, 10 μg after uterine instillation, and 39 μg after subcutaneous injection. In another group of Day 4 pregnant hamsters, 10 μg of $PGF_2\alpha$ was placed unilaterally into either the right or left ovarian bursa while vehicle was placed into the respective contralateral bursa. At 24 hours post-treatment, the ovaries were removed and the total ovarian progesterone concentration was determined by the protein binding assay. The ovaries treated with $PGF_2\alpha$ contained significantly less ($p < .02$) progesterone than their contralateral controls (29.2 $\pm$ 3.4 vs. 50.7 $\pm$ 5.6 ng/ovary). These data add evidence to the concept that $PGF_2\alpha$ has a direct effect on the ovary and/or corpora lutea in the pregnant hamster. (Authors)   004047

0352
GUTKNECHT, G.D.; DUNCAN, G.W.; WYNGARDEN, L.J.
   Inhibition of prostaglandin $F_2\alpha$ or LH induced luteolysis in the pseudopregnant rabbit by 17β-estradiol.
   Proceedings for the Society of Experimental Biology and Medicine 139: 406-410. 1972.

   Administration of prostaglandin $F_2\alpha$ ($PGF_2\alpha$) or luteinizing hormone (LH) induces luteolysis in the rabbit. The present study was designed to compare the effect of estrogen treatment on the luteolytic action of $PGF_2\alpha$ and LH in pseudopregnant Dutch Belted rabbits. $PGF_2\alpha$ (500 μg administered subcutaneously twice a day on Days 9 through 12 of pseudopregnancy) reduced ($p < .001$) corpora lutea weights by Day 14 and lowered peripheral plasma progestin levels on Days 11 ($p < .001$) and 14 ($p < .05$). LH (50 μg intravenously on Day 9) comparably reduced corpora lutea weights, although plasma progestin levels increased due to LH-induced ovulations. In animals receiving 17β-estradiol (20 μg/day subcutaneously on Days 9 through 12) plus LH, corpora lutea weights did not differ significantly from controls. Corpora lutea in rabbits receiving 17β-estradiol plus $PGF_2\alpha$ were heavier ($p < .05$) than in animals receiving $PGF_2\alpha$ alone but less ($p < .05$) than controls. This luteotropic effect of estradiol was also reflected by the presence of significant ($p < .05$) plasma progestin concentrations on Days 11 and 14. A 5-fold increase in the dose of $PGF_2\alpha$ did not completely overcome the luteotropic effect of 17β-estradiol. (Authors)   003810

0353
GUTKNECHT, G.D.; SOUTHERN, E.M.
   Present status of prostaglandins in pregnancy termination.
   Journal of Reproductive Medicine 8: 209-210. 1972.

   The authors review the history of $PGF_2\alpha$ and PGE_2 in terminating pregnancy in humans and describe some of the current areas of prostaglandin research in population control. (JRH)   004011

0354

HAOUR, F.; COHEN, M.; BERTRAND, J.

Stimulative effects of prostaglandins, theophylline and glucose on the in vitro release of human chorionic somatomammotrophin (HCS).

In: Abstracts, Fourth International Congress of Endocrinology, Washington, D.C., June 18-24, 1972. Amsterdam, Excerpta Medica, 1972. (International Congress Series No. 256) Abstract 267. p. 107

Release of human chorionic somatomammotropin (HCS) by placenta was increased by PGE_1, PGE_2, theophylline, and glucose in vitro. Placenta fragments taken 1 hour after delivery of term pregnancies were preincubated in Krebs-Ringer-2% glucose for 45 minutes, then incubated with test compounds for 2 hours while samples were radioimmunoassayed for HCS every 30 minutes. PGE_1 and PGA_2, 10^{-5} to $10^{-4}M$, doubled HCS in the medium after 1 hour. $PGF_2\alpha$ was ineffective. After 2 hours control incubations produced 100-150 ng HCS per mg tissue in medium and 288 ng per mg tissue. Theophylline, $2.2 \times 10^{-2}M$, stimulated HCS release two-to threefold in 30 minutes. A glucose concentration of .4% gradually increased HCS concentration to significant levels after 1 hours, but in 4% glucose, HCS did not rise. Thus local factors may regulate HCS production in human placenta. (LJG)   005007

0355

HARGROVE, J.L.; SEELEY, R.R.; JOHNSON, J.M.; ELLIS, L.C.

The obligatory role of prostaglandins in the regulation of spontaneous smooth muscle contraction of the testes and uterus in vitro.

Advances in the Biosciences (Suppl.) 9: 135. 1972.

This investigation was undertaken to ascertain what relationship exists between endogenous prostaglandins and spontaneous contractions of rabbit testes and rat uteri in vitro. Contractions were recorded with a sensitive myotransducer and a polygraph with the muscle suspended in oxygenated Tyrodes solution (35 degrees C) in a water-jacketed muscle warmer. Spontaneous contractions were observed in rabbit testes and in both castrated and castrated, estrogen-treated rat uteri. Replacing the bathing media with prewarmed, preoxygenated Tyrodes solution resulted in disappearance of contractions in all preparations after several changes. Bathing inactive preparations in the original bathing media reestablished normal rhythmic contractions. Extraction of this media at pH 3 with ethyl acetate yielded a residue that induced rhythmical contractions in inactive prepartions. Subjecting the residue from the uteri to thin-layer chromatography against authentic prostaglandin standards resulted in smooth muscle-stimulating substances that corresponded with PGE_2 and $PGF_2\alpha$. Acetylcholine and serotonin would not but epinephrine would stimulate the inactive testis. Extracting the media from epinephrine-stimulated preparations did not give an active residue. We concluded that prostaglandins were obligatory for spontaneous contractions of these preparations in vitro, and perhaps in vivo. (Authors' modified)   004284

0356

HARPER, M.J.K.; SKARNES, R.C.

Inhibition of abortion and fetal death produced by endotoxin or prostaglandin $F_2\alpha$ [in mice].

Prostaglandins 2(4): 295-309. October 1972.

A series of experiments were performed on mice pregnant 16 days to study the mechanism by which $PGF_2\alpha$ and bacterial endotoxin induce abortion, fetal death, and maternal diarrhea. Large doses of a serotonin inhibitor (cyproheptadine) reduced $PGF_2\alpha$-induced fetal death, but small doses seemed ineffective. Large doses of norepinephrine, histamine, l-dopa, bradykinin, and oxytocin had little or no effect on abortion, diarrhea, or fetal death. The results suggest that endogenous prostaglandins are the agents responsible. (RMS)   004404

0357
HARRISON, F.A.; HEAP, R.B.; HORTON, E.W.; POYSER, N.L.
Identification of prostaglandin $F_2\alpha$ in uterine fluid from nonpregnant sheep with an autotransplated ovary.
Journal of Endocrinology 53: 215-224. 1972.

In 4 sheep with an autotransplanted ovary, cyclic estrous behavior was abolished due to the persistence of a corpus luteum in the transplant. Fluid recovered from the uterus was viscous, odorless, and sterile, and was present in amounts of 10 to 1900 ml. Uterine fluid contained prostaglandin $F_2\alpha$, the total amount recovered ranging from 150 ng to 7.6 mg (15-4000 ng/ml). The identity of prostaglandin $F_2\alpha$ was confirmed by gas chromatography-mass spectrometry. No prostaglandins of the E series were found. Biopsy specimens of the uterine wall showed that the uterine and glandular epithelium comprised tall, columnar cells typical of the luteal phase of the normal cycle. The uterine glands were considerably distended and stromal edema was appreciable. These findings are discussed in the context of the hypothesis that the uterine luteolytic factor (luteolysin) is prostaglandin $F_2\alpha$. (Authors)   003994

0358
HASPELS, A.A.; LUIGIES, J.H.H.
Induction of abortion by intravenous and intrauterine administration of prostaglandin $F_2\alpha$.
In: Southern, E.M., ed. The prostaglandins: clinical applications in human reproduction. (Brook Lodge Symposium, Augusta, Michigan.) Mount Kisco, New York, Futura Publishing Co., 1972. p. 433-441.

10 women in the first trimester of pregnancy received $PGF_2\alpha$ by intravenous infusion at rates of 25-200 µg/minute. Total dose for each patient was 75 mg. 7 women aborted, 4 completely. Nearly all had nausea and vomiting; some had diarrhea. 4 other women received repeated 375 to 750 µg doses by the intrauterine route; total dose averaged 7.5 mg. 3 women aborted, 2 completely; none had any side effects. 3 women pregnant 14-20 weeks given 25 or 50 mg $PGF_2\alpha$ intraamniotically in 5 mg increments, aborted, 2 completely; none had serious side effects. (HC)   004882

0359
HEDQVIST, P.; EULER, U.S. von
Prostaglandin controls neuromuscular transmission in guinea-pig vas deferens.
Nature New Biology 236: 113-115. 1972.

Guinea pig vasa deferentia were isolated and mounted in a small organ bath with platinum electrodes in the wall. The vas deferens was stimulated with biphasic pulses at a frequency of 10 pulses/sec, a duration of 1 millisecond and at supramaximal voltage. Recordings of mechanical responses of the vas deferens as well as rat stomach strips used for bioassay for PG activity were made isotonically. Thin layer chromatography was also performed. Bioassay was carried out on rat stomach strips and on electrically stimulated guinea vasa deferentia. Transmural stimulation, 25 pulses of 10 pulses/sec, 1-millisecond duration and at supramaximal voltage, resumed each minute caused reproduceable contractions of the vas deferens preparation. Addition to the incubation medium of PGE_1 or PGE_2 (2 ng/ml) markedly inhibited the effector response to nerve stimulation. When the vas deferens was subjected to 20 minutes of intermittent nerve stimulation, the bath fluid contained small amounts of prostaglandin-like material. The appearance of such material was markedly increased when the preparation was continously stimulated at 10 pulses/sec for 20 minutes. Meanwhile the effector response fell off quite rapidly. Bath fluid collected from a vas deferens so stimulated inhibited the neuromuscular transmission of another preparation subjected to intermittent nerve stimulation. No PG activity was obtained in bath fluid collected in absence of nerve stimulation. Nor was any activity recovered in the seminal fluid collected from vas deferens, indicating that nerve activity is a prerequisite for PG release in measurable quanitities. When 5,8,11,14-eicosatetraynoic acid (ETA) (2 μg/ml), an inhibitor of PG synthesis, was added, the effector response to intermittent nerve stimulation progressively increased to reach after approximately 10 minutes a plateau which was maintained largely unchanged after the preparation was washed. On the other hand, the effector response to a standard dose of norepinephrine (.6 μg/ml) was not potentiated by ETA, indicating the drug must have facilitated the neuromuscular transmission by a prejunctional action, probably by inhibition of PG synthesis. Our results provide strong support for the view that endogenous PGs of the E type modulate the effector response to nerve activity in sympathetically innervated tissues. (Authors' modified)   003812

0360
HEDQVIST, P.
Prostaglandin induced inhibition of neurotransmission in the isolated guinea pig seminal vesicle.
Acta Physiologica Scandinavica 84: 506-511. 1972.

The contractile response of the guinea pig seminal vesicle to postganglionic nerve stimulation is inhibited by PGE_1 and PGE_2 in low doses (1-20 ng/ml) and potentiated by high doses (100-200 ng/ml). PGE_1 and PGE_2 consistently increase the contractile response to exogenous noradrenaline (.6 μg/ml). The potentiation by high doses of PGE_1 of the effector response to nerve stimulation is antagonized by SC 19220, an inhibitor of prostaglandin action. The effect of PGE_1 and PGE_2 on the neuroeffector system of the seminal vesicle may be explained by a dual action -- inhibition of noradrenaline release from the nerve terminals and potentiation of the effector response to the noradrenaline released. An endogenous prostaglandin-mediated regulation of the smooth muscle response to nerve activity is suggested. (Author's modified)   004077

0361
HEDQVIST, P.; EULER, U.S. von
Prostaglandin-induced neurotransmission failure in the field-stimulated, isolated vas deferens.
Neuropharmacology 11: 177-187. 1972.

The contractile response of the guinea pig vas deferens to postganglionic nerve stimulation is inhibited by PGE_1 in low doses and potentiated by high doses. PGE_1 consistently increases the contractile response to exogenous noradrenaline, the potentiation progressively increasing with the dose of PGE_1. Similar effects were obtained in the rabbit, cat, and rat vas deferens preparation. PGE_2 was equiactive with PGE_1, while $PGF_2\alpha$ was 100-1000 times less active. The effect of PGE_1 and PGE_2 on the neuroeffector system of the vas deferens preparation may be explained by a dual action -- inhibition of the release of NA from the nerve terminals and potentiation of the effector response to the NA released. The inhibitory action of PGE_1 and PGE_2, which was most marked at low impulse frequency and short pulse duration, was not blocked by polyphloretin phosphate or SC-19220, nor was it affected by addition of acetylcholine, noradrenaline, phenoxybenzamine, propranolol or tranylcypromine. The potentiation by high doses of PGE_1 of the effector response to nerve stimulation, as well as the potentiated response to exogenous noradrenaline (both effects claimed to be post-junctional), were greatly or completely antagonized by SC-19220. (Authors)   003886

0362

HENDRICKS, C.H.; BRENNER, W.E.; EKBLADH, L.; BROTANEK, V.; FISHBURNE, J.I.
Efficacy and tolerance of intravenous $PGF_2\alpha$ and PGE_2.
American Journal of Obstetrics and Gynecology 111: 564-579. 1972.

PGE_2 and $PGF_2\alpha$ were investigated by a random double-blind iv technique to 1) compare their abortifacient and oxytocic activity, 2) determine the complications and their frequency in therapeutic dosages, and 3) derive an infusion dosage with maximum effectiveness and minimum complications. 10 gravidas, 7-20 weeks pregnant, desiring therapeutic abortion, were infused with PGE_2 and $PGF_2\alpha$ at 2.5-20 and 25-200 μg/minute, respectively, in progressively increasing doses for 12 hours. Both were effective abortifacients; nausea, emesis, and fever (temperature $>$ 100) complicated $PGF_2\alpha$ administration and increased in frequency and severity with progressive infusion rates. 60% of patients infused with 5 μg/minute of PGE_2 developed transient phlebitis, which increased in intensity with higher rates of infusion. The percentage of patients developing a minimum amount of uterine activity observed with abortion increased as rates of infusion were increased. No one infusion rate of prostaglandin would produce a maximum rate of abortion with minimum complications. To attain this goal, it is recommended that the individual patient receive a progressively increasing rate of infusion until adequate uterine activity is achieved. (Authors' modified)   004185

0363

HENDRICKS, C.H.
Prostaglandins and therapeutic abortion: summary of present status.
In: Southern, E.M., ed. The prostaglandins: clinical applications in human reproduction. (Brook Lodge Symposium, Augusta, Michigan.) Mount Kisco, New York, Futura Publishing Co., 1972. p. 523-527.

The status of PG as an abortifacient, the concern of almost half of the papers given at the Brook Lodge Symposium, is summarized. Most studies have concerned midtrimester abortion, where the doses at various routes of administration have been explored. Probably the best combination is the intraamniotic route because it permits a low incidence of side effects, low dose, and a high efficacy of over 90% within 32 hours. In first trimester, PGs have not surpassed vacuum aspiration, which takes about 2-3 minutes. PGs would be useful if an oral, intravaginal, or single-dose parenteral method could be developed. It is urgent that a method be devised for using PG for inducing menstruation within 1-2 weeks of amenorrhea, but there are many practical and pharmacologic obstacles to overcome, such as lack of a method of diagnosing early pregnancy, failure of many women to date their cycles, and severity of PG side effects by the intravaginal route. (LJG)   004889

0364

HENZL, M.R.
Luteolysis in humans and subhuman primates.
Research in Prostaglandins 1(5): 1-3. May 1972.

The corpus luteum functions in humans for 12-14 days in menstrual cycles and 8-10 weeks in pregnancy. In pregnant monkeys the corpus luteum secretes progesterone in 3 surges: Days 9-11, Days 29-33, and after 50 days of pregnancy. Evidence from monkeys, baboons, and hypophysectomized humans suggests that luteinizing hormone is required to maintain functioning corpora lutea. Estrogens and/or synthetic progestagens can inhibit luteal function, possibly by blocking 3-beta-ol-steroid dehydrogenase, or perhaps by affecting feedback with luteinizing hormone. A uterine luteolytic factor, possibly endometrial prostaglandin (PG), has been postulated. It is possible to abort early pregnancy with $PGF_2\alpha$ in rhesus monkeys and in humans; luteolysis has been offered as the mechanism, but progesterone measurements contradict this theory. Luteolysis does not occur when $PGF_2\alpha$ is given to rhesus monkeys in the menstrual luteal phase or when PGE_2or $PGF_2\alpha$ is given to humans in the luteal phase. 2 special cases demonstrate that PGs can suppress ovarian progesterone syntheses: ovarian organ culture and the local injection of PG into the ovarian artery of sheep or rhesus monkeys. (LJG)   003932

0365
HENZL, M.R.; NORIEGA, L.; AZNAR, R.; ORTEGA, E.; SEGRE, E.
The uterine effects of vaginally administered prostaglandin E_2.
Prostaglandins 1: 205-215. 1972.

A group of 15 women with normal menstrual cycles was treated with prostaglandin E_2(PGE_2) during the luteal phase of the cycle (Day 20-22). A solution of PGE_2in normal saline or polyethylene glycol was applied vaginally in amounts ranging from 25 μg to 2 mg, either as a single dose or as a series of repeated doses, and uterine activity was monitored. A single dose of 1 mg distinctly increased uterine activity; 8 to 10 mg given in divided doses within 45 minutes changed the pattern of luteal phase uterine activity into intensive and regular contractions. The uterine effects of PGE_2were manifest in 20 to 40 minutes after administration; the effect lasted for 2 to 4 hours. The systemic reactions and side effects which occur during intravenous infusion of PGE_2were absent. (Authors)   003883

0366
HERTELENDY, F.; FRAWLEY, T.F.
Induction of premature oviposition in the coturnix by prostaglandins, unsaturated fatty acids and phospholipase A.
Advances in the Biosciences (Suppl.) 9: 136. 1972.

Intravenous oxytocin causes premature egg laying in the domestic hen. In view of the oxytocic effect of prostaglandins (PG), we examined the avian uterus as a simple model to study the mode of action of PG. We chose as end point the expulsion of an egg within 30 minutes after the intrauterine injection of a test compound. Four experimental approaches were taken on: 1) the response to PG; 2) the effect of PG precursors; 3) the availability of endogenous PG precursors using phospholipase A; and 4) the presence of intrinsic phospholipase activity using exogenous phospholipids. PG dissolved in ethanol-saline was given to coturnix quail having a palpable egg in the uterus. Of 152 attempts, 122 inductions were successful. The minimum effective doses (50% or greater success) were: E_1-.005 μg, E_2-.1 μg, $F_2\alpha$-1 μg, A_1-5 μg, and $F_1\beta$ and F2b-10 μg. The mean induction time for all PG tested was 5.1 minutes except for $F_2\alpha$ (21.1 minutes). The PG effect was inhibited by epinephrine and isoproterenol (1-5 μg) but not by theophylline or dibutyryl cyclic AMP (1-100 μg). Linoleic and linolenic acids (5 μl) induced oviposition in 7/11 and 12/13 cases respectively in 23.3 minutes. Arachidonic acid was 100 times more potent (.05 μl). The effect of these PG precursors was inhibited by intrauterine administration of aspirin (5 mg) or indomethacin (.1 mg). Phospholipase A (1 μg protein) induced egg laying in 8/8 attempts. Simultaneous injection of indomethacin (.1 mg) blocked this effect. Lecithin or isolecithin (.1-1 mg) caused expulsion of the egg and indomethacin (.1 mg) inhibited the response. The avian uterus is highly sensitive to exogenous PG and can synthesize PG locally from exogenous and endogenous precursors, indicating a probable physiologic role for PG in the initiation of uterine activity leading to egg laying. (Authors' modified)   004251

0367

HERTELENDY, F.

Prostaglandin-induced premature oviposition in the coturnix quail.

Prostaglandins 2(4): 269-279. October 1972.

PGE_1(1.0-.001 μg), PGE_2(10.0-.01 μg), $PGF_2\alpha$ (15.0-.1 μg), PGA_1(10.0-1.0 μg), $PGF_1\beta$ (10.0-1.0 μg), $PGF_2\beta$ (10.0-5.0 μg), arachidonic acid, phospholipase A, and indomethacin were given by intrauterine injection to Japanese quail (Coturnix corturnix japonica) to determine their effects on oviposition. All of the PGs tested caused premature oviposition with the order of effectiveness being $PGE_1 > PGE_2 > PGF_2\alpha > PGA_1 > PGF_1\beta = PGF_2\beta$. As little as 5 ng of PGE_1 induced premature oviposition. The higher doses of PGE_1, PGE_2, and $PGF_2\alpha$ caused sedation and diarrhea. Both arachidonic acid (10-100 μg) and phospholipase A (1 μg protein) caused pre- premature oviposition. Indomethacin (.1 mg) completely blocked the effects of arachidonic acid and phospholipase A but had no effect on PGE_1. The sensitive avian uterus may serve as a model in future studies. (JRH)   004402

0368

HILLIARD, J.

Letters to the editors.

Prostaglandins 1: 180-181. 1972.

The author discusses 4 papers on indomethacin which appeared in a previous issue of Prostaglandins (February 72). He is especially interested in the effects of this blocker of prostaglandin biosynthesis on ovulation, steroidogenesis, pseudopregnancy, and pregnancy in rabbits. The author poses several interesting questions about the possible role of prostaglandins in the final stage of ovulation and the continuation of pregnancy through stimulation of a pituitary gonadotropin like LH, which in turn maintains estrogen production. Several experiments are suggested which might clarify these problems. (JRH)   003878

0369

HILLIER, K.; EMBREY, M.P.

High-dose intravenous administration of prostaglandin E_2 and $F_2\alpha$ for the termination of midtrimester pregnancies.

Journal of Obstetrics and Gynaecology of the British Commonwealth 79: 14-22. 1972.

Intravenous prostaglandins E_2 and $F_2\alpha$ were administered for the induction of abortion in 20 patients. A steeply increasing high-dose regime was employed. Side effects were closely monitored and blood and urine analyses carried out before, during, and after infusion. Compared with a low-dose regime previously reported, the efficacy of the method was disappointing while the incidence of toxic effects was high. Leucocytosis and ketonuria were common findings. In 3 patients curtailment of infusion of $PGF_2\alpha$ was necessitated by persistent vomiting and diarrhea while in 2 patients receiving PGE_2a 'phlebitic' type reaction prevented further infusion. In these 5 patients prostaglandin therapy was continued by the intrauterine route without disturbing side effects. (Authors)   003715

0370

HILLIER, K.

Oxytocin release by infused prostaglandin.

British Medical Journal 2: 46. 1972.

The author disputes the claim of others that the release of endogenous oxytocin during prostaglandin infusion is caused by direct action of the prostaglandin in the pituitary gland. (JRH)   003860

0371

HILLIER, K.; DUTTON, A.; CORKER, C.S.; SINGER, A.; EMBREY, M.P.

Plasma steroid and luteinizing hormone levels during prostaglandin $F_2\alpha$ adminstration in luteal phase of menstrual cycle.

British Medical Journal 4: 333-336. 1972.

An intravenous infusion of prostaglandin $F_2\alpha$ (12.5-250 μg per min) was administered to 4 volunteers in the mid-late luteal phase and 3 in the early luteal phase of the menstrual cycle. Frequent measurements of plasma progesterone, estrogens, and LH showed that administration of high doses depressed plasma progesterone levels in the late luteal phase and caused concomitant side effects. Levels of progesterone in the early luteal phase were unaffected. In both phases estrogen and LH levels were little altered. In 2 subjects hourly progesterone levels measured throughout the day at a similar time in a subsequent control menstrual cycle, showed an appreciable variation in 1 but steady levels in the second. This variation may contribute to the magnitude of the fall in progesterone noted during the infusion of prostaglandins. (Authors)   004828

0372

HINGORANI, V.; GANESH, K.

Induction of abortion by extra-amniotic administration of prostaglandin $F_2\alpha$; a preliminary report.

Contraception 6(5): 353-359. November 1972.

$PGF_2\alpha$ was given to 20 patients, 8-22 weeks pregnant, via the extraamniotic route at an initial dose of 500 μg, supplemented at 2-hour intervals with 750-μg doses until abortion occurred or for 30 hours. Complete evacuation occurred in 3 first and and 3 second trimester patients. Another 10, 5 first and 5 second trimester, were easily terminated vaginally. 2 had vomiting and 1 had pyrexia. The average time for induction of abortion was 19 hours, and the average dose was 6.3 mg. (HC)   004597

0373

HINMAN, J.W.

Devoloping applications of prostaglandins in obstetrics and gynecology.

American Journal of Obstetrics and Gynecology 113: 130-138. 1972.

The stated purpose of this review is to present a clear up-to-date summary of the clinical results with prostaglandins in obstetrics and gynecology. It contains sections on induction of labor, induction of abortion, and prostaglandins in contraception. Suggestions are made about areas that need more research. The bibliography includes 69 entries. (JRH)   003951

0374

HORTON, E.W.

Biological significance of the prostaglandins.

In: Horton, E.W., ed. Monographs on endocrinology, Vol. 7. New York, Springer. 1972. p. 179-190.

Seminal PGE was found in lower concentrations in men in infertile marriages than in men of proven fertility although there was no difference in the concentrations of other PGs between the 2 groups. The mechanism by which PGE compounds aid conception is unknown. PGE_1 was found to stimulate or enhance smooth muscle contractions of ejaculation. PGs were found also to act upon the female reproductive tract by affecting sperm transport or by affecting contractions of the fallopian tube to delay transport of the ovum into the uterine cavity until fertilization had taken place. PGE_2 and $PGF_2\alpha$ have been identified in menstrual fluid, human endometrium, amniotic fluid, and the decidua, and have been implicated in menstruation, in some cases of spontaneous abortion, and in the uterine contractions of parturition. It was found that PGs are released into the circulation from various organs on chemical or nerve stimulation, but many PGE and PGF compounds are removed in the lungs. PGE_1 was found to antagonize pressor and vasoconstrictor substances and to block the lipolytic action of adrenaline and other lipolytic hormones. It was concluded that PGE_1 affects the enzyme adenyl cyclase, sometimes inhibiting the stimulant hormone but sometimes mimicking its action. Although much study has been done, little is known about the mechanism which determines which PG is released at a particular site in response to a particular stimulus. PGE_1 was found to effect a change in membrane permeability to ions which secondarily activate adenyl cyclase. (MLH)   004909

0375
HORTON, E.W.
Female reproductive tract smooth muscle.
In: Horton, E.W., ed. Monographs on endocrinology, Vol. 7. New York, Springer, 1972. p. 87-103.

The effects of PGs on the female reproductive tract smooth muscle were discussed. PGs extracted from human seminal plasma and administered intravaginally to nonpregnant women seemed to cause an increase in uterine motility at the time of ovulation but seemed to have no effect at other phases in the menstrual cycle. The pregnant human myometrium seemed to be especially sensitive to PGs. It was shown that PGs could induce both abortion and labor, with PGE_2 reported to be more active than $PGF_2\alpha$. High success rates were recorded when PGs were injected intravenously to induce abortion in the first and second trimesters; increased dosages were related to the appearance of side effects including dysmenorrhea, nausea, and diarrhea. Intrauterine administration of PGs was also found to be effective at doses 1/10 as high as those required for intravenous infusion, and with no side effects. Intravaginal administration seemed similarly effective and was thought to carry less danger of infection. Whether intravaginal PGE_2 is absorbed into the circulation or whether it acts locally is unknown. The clinical value and safety of PGs for the induction of parturition is as yet undecided. The use of PGs for contraceptive purposes is being studied. Contractions of human myometrial strips in vitro were decreased in amplitude and frequency by both whole semen and by PGE and PGA compounds, though the PGA compounds were 10-30 times less potent than the PGE. $PGF_1\alpha$ and $PGF_2\alpha$ stimulated contractions. Hormonal status seemed to affect the sensitivity of the tissue. PGE_1 and PGE_2 caused contractions of the uterine end of the human fallopian tube in vitro and relaxed the distal end. All parts of the tube were relaxed by PGE_3 and contracted by $PGF_1\alpha$ and $PGF_2\alpha$. Rat, guinea pig, and rabbit uteri were found to contract in vitro with PGE_1; guinea pig uterus in vivo contracted, but both rabbit uterus and oviduct in vivo were inhibited. Rat uterus was found to be more sensitive to $PGF_1\alpha$ and $PGF_2\alpha$ than to PGE_1 and PGE_2, but the sensitivity could be increased by prior ovariectomy of the animal. (MLH)   004900

0376
HUNT, W.L.; NICHOLSON, N.
Studies on semen from rabbits injected with H-3-thymidine and treated with prostaglandins E_2 and $F_2\alpha$.
Fertility and Sterility 23(10): 763-768. October 1972.

Prolonged daily administration of 1 mg of PGE_2 or $PGF_2\alpha$ to male rabbits did not affect their fertility, the quantity of semen, the percentage of motile spermatozoa, the percentage of live spermatozoa, or the measurable sperm output for a 61-day test cycle. However, the pattern of radioactive sperm cells in the ejaculates indicated that PGE_2 and $PGF_2\alpha$ caused a decrease of 2 days in the interval from the injection of tritium-labeled thymidine to the arrival of labeled spermatozoa. The total amount of radioactivity per ejaculate was calculated by multiplying the disintegrations per minute per quantity of ejaculate containing 30×10^6 spermatozoa by the total quantity of the ejaculate. (Authors' modified) 004524

0377
ITO, H.; KATAYAMA, T.; TAKAGISHI, H.; MOMOSE, G.
Contraction of seminal vesicle and prostaglandin E_1.
Prostaglandins 1: 327-330. 1972.

It was studied how PGE_1 would affect the responses of isolated human seminal vesicles to adrenalin. PGE_1 in the final concentration of 1.3 µg/ml suppressed the contraction of human seminal vesicle that would have occurred in reaction to adrenalin added 1 minute later. When the concentration of PGE_1 was increased to 6.7 µg/ml, the inhibitory action was further enhanced. The meanings of this phenomenon were discussed. (Authors)   003961

0378
JACOBSON, H.I.; BULLOCK, D.W.; KEYS, P.L.
Effect of prostaglandin $F_2\alpha$ on estrogen receptor in corpus luteum and uterus.
In: Abstracts, Fourth International Congress of Endocrinology, Washington, D. C., June 18-24, 1972. Amsterdam, Excerpta Medica, 1972. (International Congress Series No. 256) Abstract 472. p. 188.

To clarify the sequence of events during luteolysis induced by $PGF_2\alpha$, Dutch-belted rabbits were mated to sterile males on Day 9 were treated with $PGF_2\alpha$, 400 µg per kg in 15% gelatin or with vehicle. Uteri and corpora lutea (CL) were weighed 24 hours later and luteal and uterine estrogen receptor concentrations assayed. Blood for progesterone assay was collected before injection and at autopsy. Mean CL weight was 9.12 mg in $PGF_2\alpha$ treated and 11.28 in controls (p<.01). CL and uterine estrogen receptor concentrations were 3.64 and 27.76 fmol per mg in treated rabbits and 14.26 and 17.10 fmol per mg in controls. Plasma progesterone decreased 26% and 7% in treated and control groups. Thus in 24 hours $PGF_2\alpha$ significantly reduced CL weight and estrogen receptor concentrations, possibly reduced progesterone output and raised uterine estrogen receptor concentration compared to controls. (Authors' modified)   005008

0379
JACOBSON, H.I.; KEYES, P.L.; BULLOCK, D.W.
Regulation of uterine estrogen receptor by luteal progesterone in the pseudopregnant rabbit.
In: Abstracts, 5th Annual Meeting of the Society for the Study of Reproduction, East Lansing, Michigan, June 26-29, 1972. p. 28-29.

Induction of pseudopregnancy in rabbits results in a decline in the uterine concentration of available cytoplasmic estrogen receptor (ER), which rises again as the animal returns to the estrous state (Lee et al., 1971. Science, 173, 1032-1033). We have recently found that early termination of pseudopregnancy by injection of a luteolytic dose of prostaglandin $F_2\alpha$ (PG) on Day 9 causes reversion of uterine ER to estrous levels within 24 hours. On Day 10, uterine ER concentration in PG-treated animals was 27.8 ± 5.5 fmole/mg (mean $\pm$ SD) (n=5) compared to 17.7 ± 3.4 fmole/mg in pseudopregnant controls (n=5). To determine whether this effect was due to loss of luteal progesterone, Dutch Belted rabbits were mated to vasectomized males (Day 1) and ovulation was confirmed by laparotomy on Day 2. On Day 9 the animals were randomly assigned to 1 of 3 treatments: sham operation (n = 2), surgical ablation of corpora lutea (lutectomy, n = 4), or lutectomy plus progesterone (4 mg sc in oil, n = 2). At autopsy 24 hours later (Day 10) uterine ER concentration was 13.6 ± 1.7 fmole/mg in the sham-operated group, 29.8 ± 3.2 fmole/mg after lutectomy, and 14.4 ± 2.7 fmole/mg in lutectomized rabbits treated with progesterone. Elevation of uterine ER is thus caused either by PG-induced luteolysis or by lutectomy, and can be prevented in lutectomized rabbits by administration of progesterone. We conclude that the depression of uterine ER concentration during pseudopregnancy in the rabbit is caused by progesterone of luteal origin. (Authors)   004046

0380

JAFFE, B.M.; PARKER, C.W.

Extraction of PGE from human serum for radioimmunoassay.

In: Bergstrom, S., Green, K., and Samuelsson, B. [Proceedings of the] third conference on prostaglandins and fertility control, Stockholm, January 17-20, 1972. Stockholm, WHO Research and Training Centre on Human Reproduction, Karolinska Institutet, [1972]. (Prostaglandins in Fertility Control 2) p. 69-82.

8 extraction procedures for preparing serum for radioimmunoassay of prostaglandin (PG) E_1 were compared: ethanol, chloroform, ethyl acetate, and ethyl acetate with silicic acid chromatography, each of these with and without prior petroleum ether extraction. Sera from 5 persons were allowed to clot, centrifuged, and frozen. Half of each was extracted with neutral petroleum:ether (1:1). 1 ml portions contained .15 to 40 ng/ml added PGE_1, and 3 or 8 ng/ml tritiated PGE_1. The immunoassay used a rabbit anti-PGE_1-keyhole limpet hemocyanin conjugate. Ethanol extracts of lyophilized serum, suspended in phosphate buffered saline with 1% gelatin (GPBS), were dialyzed 24 hours against GPBS. 19.1% of added PGE_1 was recovered without, and 28.3% with, petroleum ether extraction (before dialysis), but the dialysis did not yield the expected recovery. For chloroform extraction, serum was titrated to pH 2.5 with 2 N HCL, and 3 volumes of chloroform added. The dried organic phase was dissolved in GPBS. Recovery was only 8.1 and 15.4% without and with petroleum ether extraction; thus PGE_1 was probably converted to PGA. After adding 3 volumes of ethyl acetate: isopropanol:.1 N HCL (3:3:1), then 3 volumes of water and 2 of ethyl acetate, the organic phase was dried and dissolved in GPBS. Recovery was linear, 39.4% without, and 51.7% with prior petroleum ether, but was not comparable with recovery with buffer alone. When serum was extracted as above with ethyl acetate, and chromatographed on silicic acid columns by the method of Behrman, 40.2% of PGE_1 was recovered without, and 65.7% with, prior petroleum ether extraction, and recovery was commensurate with serum content plus added PGE_1. The 5 sera contained 292 to 680 pg/ml PGE_1 by this immunoassay. (LJG)   004468

0381

JARABAK, J.

Human placental 15-hydroxyprostaglandin dehydrogenase.

Proceedings of the National Academy of Sciences 69: 533-534. 1972.

Normal term human placentas are a rich source of 15-hydroxyprostaglandin dehydrogenase. The enzyme is extremely labile, and partial purification could be achieved only after stabilization with glycerol. The instability of the enzyme and its Km for NAD are indications that it is different from the 15-hydroxyprostaglandin dehydrogenase isolated from swine lung. Human placental tissue should provide a very useful source from which large amounts of highly purified 15-hydroxyprostaglandin dehydrogenase may be obtained. (Author's modified)   003944

0382
JEWELEWICZ, R.; CANTOR, B.; DYRENFURTH, I.; WARREN, M.P.; VANDE WIELE, R.L.
  Intravenous infusion of prostaglandin $F_2\alpha$ in the mid-luteal phase of the normal human menstrual
    cycle.
  Prostaglandins 1: 443-451. 1972.

  The luteolytic effect of $PGF_2\alpha$ was studied in 3 normal women. Plasma levels of estrogens,
  progesterone, and gonadotropins were measured daily. Following a control cycle, an infusion of 100
  $\mu g/min$ of $PGF_2\alpha$ over an 8 hour period was given. 2 subjects received the $PGF_2\alpha$ on the ninth and 1
  on the fifth postovulatory day. The steroid patterns during the control and treatment cycles were not
  significantly different and the length of the luteal phases was identical. To accentuate luteal activity, 2
  subjects received, in subsequent cycles, 10,000 IU HCG 48 hours prior to the $PGF_2\alpha$ infusion. In these
  cycles, despite the $PGF_2\alpha$ the steroids continued to rise and then gradually declined in normal fashion.
  The luteal phase was not curtailed. All subjects experienced nausea, vomiting, painful uterine
  contractions, staining, and temperature elevations above 100 degrees F during the infusion. It is
  concluded that in the human $PGF_2\alpha$ at the given dose was not luteolytic. (Authors)   004080

0383
JONES, G.S.; WENTZ, A.C.
  The effect of prostaglandin $F_2\alpha$ infusion on corpus luteum function.
  American Journal of Obstetrics and Gynecology 114: 393-404. October 1, 1972.

  The effect upon menstrual function of an 8-hour infusion of 25 mg of $PGF_2\alpha$ was tested in 21 normal
  volunteer women between the ages of 21 and 40 years. The infusions were given between the
  fourteenth and the twenty-seventh days of the luteal phase of the cycle, Day 14 being arbitrarily
  assigned as the day of ovulation. 13 of 21 women showed a shortening of the luteal phase by at least 1
  day. 1 of 2 women who received $PGF_2\alpha$ prior to ovulation failed to ovulate, and 2 additional patients
  had anovulatory cycles in the immediate postinfusion cycle. The steroid parameters, as measured by
  serum progesterone, urinary pregnanediol, and in selected cases, estradiol, failed to show a consistent
  change. Measurements of serum progesterone and estradiol during the infusion suggested an initial
  depression and subsequent elevation of steroid values. The conclusion was that an 8-hour infusion of
  $PGF_2\alpha$ in the woman fails to disrupt corpus luteum function. (Authors' modified)   004226

0384
JONES, R.L.
  Functions of prostaglandins.
  Pathobiology Annual 2: 359-380. 1972

  PG physiology was reviewed up to 1971, including sperm transport, luteolysis, menstruation,
  parturition, natriuresis, gut fluid transfer, hypertension, essential fatty acid deficiency, lipolysis, and
  inflammatory response. Many of the normal physiologic actions of PGs were discussed in terms of
  substrate release by phospholipase, substrate availability, or essential fatty acid deficiency. Some
  other roles for PG considered likely were: relaxation of myometrium after coitus due to interaction of
  semen PGE and the oxytocin surge of orgasm; uterine contraction in labor, abortion, and
  menstruation determined by mixtures of PGs in amniotic fluid, possibly secreted by decidual cells;
  PGA is possibly the natriutetic hormone acting by redistributing renal blood flow or by inhibiting
  sodium transport; negative feedback control of lipolysis by PGE_1. PGs may function in pathologic
  conditions such as dysmenorrhea, essential fatty acid deficiency, essential hypertension, diarrhea of
  cholera and carcinoid tumors, polymorphonuclear leukocyte migration, and inflammation. (LJG)
  005028

0385

JONSSON, H.T., Jr.; SHELTON, V.L.; BAGGETT, B.
Stimulation of adenyl cyclase by prostaglandins in rabbit corpus luteum.
In: Abstracts, 5th Annual Meeting of the Society for the Study of Reproduction, East Lansing, Michigan, June 26-29, 1972. p. 27.

Studies reported by Marsh (1970) and by Kuehl et al. (1970) have recently indicated that prostaglandins E_1 and E_2 may play an important, but as yet unexplained, role in the action of LH on adenyl cyclase (AC) in bovine and mouse ovaries, respectively. Present studies provide additional evidence for such stimulatory activity in the rabbit corpus luteum. Pseudopregnancy was induced in mature female New Zealand white rabbits with ovine LH. Homogenates of 4 and 10 day corpora lutea (CL) were incubated with ATP-alpha^{32}p. Incubations were with about .14 mg protein in a total volume of .32 ml. Levels of labeled 3',5'-AMP isolated chromatographically provided a measure of luteal AC activity in the presence of stimulatory levels of PG and/or LH. The results were as follows: 1) A rough dose-response relationship existed between PG level and AC activity. 2) This response was maximal at 2.0 g/incubation, but was present at levels as low as .03 g PGE_1. 3) Marked stimulation was obtained with PGE_1 and PGE_2, but was greater for PGE_1 in both 4 and 10 day CL. 4) Combinations of maximally stimulating amounts of both LH and PG produced greater stimulation than did either one added alone. 5) LH singly or in combination with PG gave greater stimulation in the 4 day than in the 10 day CL. 6) Fluoride produced a marked (30-fold) stimulation of AC, but this response was unaltered by added PG. 7) Slight stimulation was consistently observed with $PGF_1\alpha$ and $PGF_2\alpha$. These results support the hypothesis that PG is involved in the regulation of the function of the corpus luteum by an action on adenyl cyclase. The fact that LH causes a further stimulation of this enzyme beyond that obtained with PG alone suggests that PG is not a mediator of the stimulatory effect of LH. (Authors)   004051

0386

JONSSON, H.T., Jr.; SHELTON, V.L.; BAGGETT, B.
Stimulation of adenyl cyclase by prostaglandins in rabbit corpus luteum.
Biology of Reproduction 7: 107. 1972.

Homogenates of corpora lutea from LH-treated pseudopregnant rabbits were incubated with ATP in the presence of prostaglandins and/or LH. The levels of AMP isolated from the medium provided a measure of adenyl cyclase (AC) activity. PGE_1 stimulated AC activity in a dose-related manner from .03 to 2.0 μg. PGE_1 was more potent than PGE_2. $PGF_1\alpha$ and $PGF_2\alpha$ had a slight stimulant effect. Maximal doses of LH and PG together produced greater stimulation than either one alone. It was suggested that prostaglandins may play a role in regulating corpus luteum function by an action on AC and that they do not mediate the stimulant effect of LH. (HC)   004267

0387

JUBIZ, W.; FRAILEY, J.; CHILD, C.
Physiologic significance of prostaglandins of the E group (PGE).
In: Abstracts, Fourth International Congress of Endocrinology, Washington, D. C., June 18-24, 1972. Amsterdam, Excerpta Medica, 1972. (International Congress Series No. 256) Abstract 482. p. 191.

PGEs were determined in human plasma by double-antibody radioimmunoassay, using rabbit antibodies against pig gamma globulin-PGE_1-conjugates having cross reaction with PGA, B and F groups of less than 5%. Normal plasma levels were 150-1400 pg per ml, with a circadian peak between midnight and 0400. PGEs were highest in the middle of the menstrual cycle. In pregnancy or on oral contraceptives PGE levels were low when measured directly in plasma, but above normal when plasma was extracted, suggesting presence of estrogen induced binding proteins. (LJG)   005010

0388

JUBIZ, W.J.; FRAILEY, J.; BARTHOLOMEW, K.
   Prostaglandins of the F group (PGF). Assessment by radio-immunoassay of their role in human physiology.
   Clinical Research 20: 178. 1972.

We have developed a radioimmunoassay to investigate the role of PGF in human physiology. Antibodies were generated in rabbits by repeated injections of $PGF_2\alpha$ coupled to porcine gamma globulin. When tracer amounts of tritiated $PGF_2\alpha$ are incubated with $PGF_2\alpha$ standards and anti-$PGF_2\alpha$ antiserums (final dilution 1: 1500), a sensitivity of 25 pg is obtained. The assay is able to detect circulating PGF levels in 50 μl of unextracted plasma. Antibodies are very specific for PGF. As compared with $PGF_2\alpha$, only $PGF_1\alpha$ cross-reacts signficantly (30%). Prostaglandins of all the other groups (A,B,E) and the precursors and metabolites of PGF give negligible cross-reactivity ($<.1\%$). The method is simple, reliable, and reproducible. PGF levels in normal subjects at 8 a.m. range from .25- 1.30 ng/ml. Values are slightly lower in females than in males. Striking circadian variation with highest levels between midnight and 4 a.m. has been observed in normal subjects. During the menstrual cycle in normal females PGF levels are highest at midcycle. Low levels are obtained in women on oral contraceptives and during normal pregnancy. We feel that PGF plays a role in human physiology. Their role in human diseases is yet to be determined. (Authors' modified)   003819

0389

JUNGMANNOVA, C.; HAVRANEK, F.; HODR, J.
   The effect of prostaglandin $F_2\alpha$ on the placental vessels in vitro.
   Journal of Reproductive Medicine 9(2): 79-80. August 1972.

The dose response of the resistance of placental arteries to prostaglandin $F_2\alpha$ ($PGF_2\alpha$) was recorded electronically. 12 marginal chorial arteries from term deliveries were perfused with Krebs-Ringer solution, and given .001 μg of norepinephrine in a .1 ml volume, then .1 ml $PGF_2\alpha$ at doses of .0001 to 100 μg, followed by the same dose of norepinephrine. Vasoconstriction was recorded as a mean of 113 sq. mm for .0001 μg $PGF_2\alpha$. Significant increases in vasoconstriction occurred at doses of .01 to 10 μg and ranged from 234.1 to 351 sq. mm (p=.05;p=.01 at .1 μg). The response was variable at 100 μg. The constriction recorded during infusion of norepinephrine averaged 307.91 sq. mm at the beginning and 312.4 sq. mm at the end of the PG experiments. (LJG)   004213

0390

KAMMERMAN, S.; CANFIELD, R.E.; KOLENA, J.; CHANNING, C.P.
   The binding of iodinated HCG to porcine granulosa cells.
   Endocrinology 91(1): 65-74. 1972.

Intact, viable porcine granulosa cells were incubated with radioiodinated HCG. At concentrations greater than 10 μg per ml, PGE_2inhibited by approximately 20% the uptake of HCG by the cells. The limited ability of $PGE_2$to compete for uptake of HCG demonstrates specificity of receptor sites on granulosa cells for the gonadotrophic hormone. (HC)   004519

0391
KARIM, S.M.M.

Abortifacient action of prostaglandins.

In: [Abstracts of] First International Planned Parenthood Federation South-East Asia and Oceania Regional Medical and Scientific Congress, Sydney, Family Planning Association of Australia, August 1972. p. 28.

The author reviews the use of prostaglandins for the induction of abortion in first and second trimester pregnancies. It is suggested that intrauterine or intraamniotic routes offer the highest success rates with the fewest side effects. The use of intravaginal prostaglandin is discussed as a method of inducing menstruation as a once a month contraceptive. It is reported that preliminary trials with some synthetic analogues show promise of overcoming the major drawbacks of natural prostaglandins given intravaginally, which are the necessity of 1-6 intravaginal doses over 3 days and unpleasant side effects. (JRH)   004176

0392
KARIM, S.M.M.; SHARMA, S.D.

Oral administration of prostaglandin E_2 for the induction and acceleration of labor.

In: Southern, E.M., ed. The prostaglandins: clinical applications in human reproduction. (Brook Lodge Symposium, Augusta, Michigan.) Mount Kisco, New York, Futura Publishing Co., 1972. p. 207-217.

1000 women at or near term were given .5-2.0 mg PGE_2 orally every 2 hours for the induction or acceleration of labor. Of the women given PGE_2 for induction of labor, 90% delivered vaginally within 48 hours. Labor was accelerated in 95% of the women treated, nearly all delivering within 24 hours. Side effects noted were nausea and vomiting, diarrhea and hemorrhage. The authors concluded that these results confirm previous reports of the efficacy and safety of oral PGE_2 as well as the convenience of the method to the patient and the medical and nursing staffs. (HC)   004862

0393
KARIM, S.M.M.

Physiological role of prostaglandins in the control of parturition and menstruation.
Journal of Reproduction and Fertility (Suppl.) 16: 105-119. 1972.

The following evidence supports the suggestion that prostaglandins have a role in spontaneous abortion and labor. Prostaglandins E_2 and $F_2\alpha$ are present in amniotic fluid and in maternal circulation in appreciable quantities only during spontaneous abortion and labor. Both these prostaglandins have a stimulant action on the pregnant myometrium in vitro and in vivo. Prostaglandins E_2 and $F_2\alpha$ administered by different routes can be used to terminate pregnancy at all stages of gestation. Intravenous infusion of 10% ethanol inhibits spontaneous and prostaglandin-induced uterine activity. Alcohol has no effect on oxytocin-induced uterine activity. There is also evidence implicating prostaglandins in the process of menstruation. Prostaglandins are present in menstrual fluid and in disintegrating endometrium. Prostaglandin-like activity is present in peripheral blood at the time of menstruation. Prostaglandins E_2 and $F_2\alpha$ can be used to induce menstruation. (Author's modified)
003931

0394

KARIM, S.M.M.

Prostaglandins in fertility control.

In: Potts, M., and Wood, C., eds. New concepts in contraception. Oxford, Medical and Technical Publishing Co., Ltd., 1972. p. 125-142.

Pharmacologic induction of abortion and of menstruation are reviewed, considering historical attempts briefly and PGEs, PGFs, and analogs in detail. Nonmedical abortifacients usually contain no active ingredients except purgatives or emetics. Trials of other drugs as abortifacients, such as aminopterin, have produced unacceptable toxic and teratologic effects. The first trials with $PGF_2\alpha$ and PGE_2iv were conducted in 1970. Later investigation determined intravenous PG is effective but entails infusion for 12-24 hours with constant surveillance, and high incidence of nausea, vomiting, diarrhea, and sometimes fever and vasovagal response. As an alternative, intrauterine, intraamniotic, or intravaginal routes offer fewer side effects, but still require uterine monitoring and repeated doses. The intraamniotic route can only be used in second trimester. The PG analog 15(S)15-methyl-PGE_2 may be more convenient for local administration because it is 400 times more potent and lasts 4 times as long as PGE_2. As with abortion, no nonmedical or medical drug is known to be effective in inducing menstruation. The potential use of PG as a menstrual inducer was discussed. Preliminary trials indicated that natural PGs in up to 5 doses can induce menstruation but with severe uterine pain, gastrointestinal side effects, and heavy bleeding, especially if the period is delayed over 7 days or if tissues are retained. 15(S)-methyl-PGE_2successfully induced menstruation in 60% of 60 women with 1 intravaginal 25 mg dose and caused fewer side effects. (LJG)   005062

0395

KARIM, S.M.M.; MACINTOSH, D.

Recent advances in prostaglandin research.

In: Horrobin, D., and Gunn, A., eds. The international handbook of medical science, 2nd ed. Oxford, Medical and Technical Publishing Co. Ltd., 1972. p. 17-45.

Potential uses of prostaglandins in all clinical fields are mentioned. The interpretation of varying results, the ubiquity of prostaglandins in mammalian tissue, and the the development of new analogs initiate a volume of speculation on their activity. Nonetheless, knowledge of some physiological effects of prostaglandins has found valuable clinical application in obstetrics. In several centers prostaglandins $E_2$and $F_2\alpha$ are routinely employed for the termination of second trimester pregnancy and for the induction of labor. Potential applications of prostaglandins in other clinical areas have been indicated but at the present time their greatest promise is in the field of fertility control. Population increase without a concomitant increase in food production is a problem facing many countries. The failures with and hazards attending the use of contraceptive procedures currently available are well known; prostaglandins may offer a safe and highly effective means of contraception through their ability to induce menstrual bleeding, administration being required only once a month. (MJN)   005082

0396

KARIM, S.M.M.; SHARMA, S.D.; FILSHIE, G.M.

Termination of pregnancy with 15 methyl analogues of prostaglandins $E_2$and $F_2\alpha$.

In: Southern, E.M. ed. The prostaglandins: clinical applications in human reproduction. (Brook Lodge Symposium, Augusta, Michigan.) Mount Kisco, New York, Futura Publishing Co., 1972. p. 307-321.

79 women pregnant 7 to 36 weeks were given 15(S)15-methyl $PGF_2\alpha$ (free acid) or 15(S)15-methyl PGE_2methyl-ester intramuscularly, intravaginally, intraamniotically or orally. Doses of 250 to 500 μg

of the $PGF_2\alpha$ analog or 25 to 50 μg of the PGE_2derivative were required to produce adequate uterine contractions. The analogs are, therefore, 40 to 400 times more potent than the parent compounds. 3 women 13 to 16 weeks pregnant and 5 women 7 to 10 weeks pregnant given 25 μg doses of 15 methyl PGE_2intramuscularly at 8 hour intervals aborted with 14 to 18 hours; all were complete except 2 in the latter group. 6 women in the second trimester given 50 μg doses intravaginally every 8 hrs aborted within 20 hrs, all except 1 completely. Of 42 women 13 to 22 weeks pregnant given the drug intraamniotically every 10 hrs, 38 aborted after 1 or 2 50 μg doses while the other 4 women required 3 or 4 doses; 34 abortions were complete. Given intramuscularly or intravaginally most women experienced cold and shivering; nausea and vomiting were noted when the drug was given intraamniotically. 2 women 7 to 9 weeks pregnant and 3 women 13 to 15 weeks pregnant were given 1 to 3 intramuscular injections of .25 or .5 mg 15-methyl $PGF_2\alpha$ at 8 hour intervals. All aborted within 28 hours, 3 completely. Abortion was complete in 5 of 6 women 13 to 18 weeks pregnant who were given 1 to 3 .5 mg doses $PGF_2\alpha$ analogue intravaginally every 8 hours. Another group of 6 women in the second trimester were given 1 to 3 .5 mg doses $PGF_2\alpha$ analogue by the intraamniotic route every 10 hours. All aborted within 32 hrs, 5 completely. Vomiting and/or diarrhea occurred in all but 1 woman given the drug intramuscularly or intravaginally; there were virtually no side effects when the drug was given intraamniotically. 500 μg 15-methyl $PGF_2\alpha$ or 50 μg 15-methyl PGE_2orally caused nausea and diarrhea in all 8 non-pregnant women tested. 6 women at term were given 1 to 4 doses of 20 or 40 μg 15-methyl PGE_2orally at 8 hr intervals. All expelled the fetus and placenta within 25 hrs; 2 had hypertonus. The authors conclude that 15-methyl $PGF_2\alpha$ had advantages over the parent compound only when given intraamniotically for the termination of second trimester pregnancy. 15-methyl PGE_2produced fewer side effects than the parent compounds, is safer if inadvertently absorbed systemically, and might be useful for first trimester abortion given as a single intramuscular injection. (HC)   004870

0397
KARIM, S.M.M.; SHARMA, S.D.
Termination of second trimester pregnancy with 15 methyl analogues of prostaglandins $E_2$and $F_2\alpha$. Journal of Obstetrics and Gynaecology of the British Commonwealth 79: 737-743. August 1972.

Women 13 to 22 weeks pregnant were given 15 methyl analogues of PGE_2or $PGF_2\alpha$ by the intraamniotic, intravaginal or intramuscular route. With 15 methyl $PGF_2\alpha$ the uterine stimulant effect lasted for 8 hours when given by intramuscular injection or vaginal instillation and 12 hours when given by intraamniotic injection. Doses used were 250 to 500 μg. Diarrhea and vomiting occurred in most patients given the drug by the intramuscular or intravaginal routes but not in those women given the drug by the intraamniotic route. 15 methyl PGE_2methyl ether was 10 times more potent than $PGF_2\alpha$ methyl ether, duration of action was the same, cold and shivering or pyrexia occurred in half of the patients. Abortion was complete in 25 or the 30 patients given either $PGF_2\alpha$ or PGE_2. Severity of pain associated with uterine contractions and cervical dilatation during abortion appeared to be less with the methyl derivatives than with the parent compounds. The methyl analogues had greater potency and longer duration of action than the parent compounds. Greater side effects with 15 methyl $PGF_2\alpha$ would limit its usefulness but 15 methyl PGE_2may have some advantages over its parent compound. (HC)   004740

0398

KARIM, S.M.M.; SHARMA, S.D.; FILSHIE, G.M.

Termination of second trimester pregnancy with intra-amniotic administration of prostaglandins E_2 and $F_2\alpha$.

In: Southern, E.M., ed. The prostaglandins: clinical applications in human reproduction. (Brook Lodge Symposium, Augusta, Michigan.) Mount Kisco, New York, Futura Publishing Co., 1972. p. 403-416.

PGE_2 and $PGF_2\alpha$ were administered intraamniotically to 110 women between 13 and 26 weeks gestation in order to evaluate 5 different dose schedules for the termination of second trimester pregnancies. Results over a period of 48 hours following injection were: 5 mg PGE_2 given to 10 women (Group 1) as a single intrammniotic injection resulted in 9 successes; 25 mg $PGF_2\alpha$ given to 10 women (Group II) as a single injection resulted in 8 successes; 5 mg PGE_2 every 10 hours given to 40 women (Group III) for a maximum of 4 doses resulted in successful abortion in 38 cases and complete in 31, the mean treatment-abortion interval being 14 hours. 30 women (Group IV) were treated with 25 mg $PGF_2\alpha$ every 10 hours, maximum of 4 doses. In 28 abortion was successful and in 21 it was complete; the mean treatment-abortion interval was 17 hours. A mixture of 2.5 mg PGE_2 and 12.5 mg $PGF_2\alpha$ was given every 10 hours to 10 women (Group V); abortion occurred in all patients and was complete in 8, the mean interval being 14 hours. 10 women with intrauterine fetal death before Week 28 of gestation (Group VI) were administered a dose of 5 mg PGE_2 every 10 hours. All cases were successful within 24 hours and 8 of the 10 were complete; the mean treatment-abortion interval was 10 hours. Side effects reported in this study were nausea in 12 patients, vomiting in 32, headache in 1, and pyrexia in 5, the latter related to a longer treatment-abortion interval. None of the side effects interfered with the treatment, and none seemed significantly different in incidence in the various treatment groups. Problems associated with this treatment were accurate localization of the amniotic cavity and loss of uterine activity following early rupture of the membranes. It was concluded that repeated doses of the PGs every 10 hours in contrast to single doses led to increased abortifacient success (95% success in this study) and a decreased treatment-abortion interval; PGE_2 appeared slightly more effective than $PGF_2\alpha$. (MLH)   004879

0399

KARIM, S.M.M.; FILSHIE, G.M.

The use of prostaglandin-E_2 for therapeutic abortion.

Journal of Obstetrics and Gynaecology of the British Commonwealth 79: 1-13. 1972.

Prostaglandin E_2 was infused in 139 women recommended for therapeutic termination of pregnancy. Abortion occurred in 130 women and was complete in 105. Uterine activity was measured in 67 patients before, during, and after prostaglandin infusion. In 6 selected cases, hematological, biochemical, hormonal, and renal function tests were in normal ranges. Nausea or vomiting occurred in 50 patients, diarrhea in 9. Erythema appeared in 33 women, developing into phlebitis in patients infused longer than 24 hours. Mild pyrexia was seen in most patients, 4 experienced headache, 1 experienced blurred vision, 1 showed tachycardia. Symptoms were mild and controllable by normal means. Blood transfusion was required in 1 out of 4 patients in which hemorrhage occurred. 3 genital tract and 1 urinary tract infections were experienced. (RMS)   003714

0400
KARIM, S.M.M.
The use of prostaglandins in abortion.
In: Lewit, S., ed. Abortion techniques and services. [Proceedings of a conference held at New York, N.Y., June 3-5, 1971] Amsterdam, Excerpta Medica, (International Congress Series No. 255) 1972. p. 68-77.

Trials of PGE_1, PGE_2, and $PGF_2\alpha$ for abortion and induction of late menstruation in Uganda from 1968 to 1971 are summarized. The abortion studies were discussed in relation to variables influencing side effects and efficacy. PGE_2 is 10 times more active than $PGF_2\alpha$. The route of administration determines side effects to some extent; more phlebitis results from iv route and fewer systemic effects from intrauterine route. Low doses, i.e. 5 μg per minute PGE_2 iv, in the author's opinion, are more effective than high doses, up to 20 μg per minute, used in drug tolerance studies, which may result in hypertonus or tachyphylaxis without cervical dilatation. Efficacy is directly related to duration of administration. Induction-to-abortion interval lengthens with length of gestation. Small trials were described of intravaginal, intraamniotic, and intrauterine routes of PG administration for abortion. PGE_2 5-7.5 mg, $PGF_2\alpha$ 50 mg, and a PG analogue .5-1 mg repeated 1-3 times were also used intravaginally to induce menstruation from 3 days before to 21 days after expected menstruation in 69 women. Bleeding occurred in all of 69 cases but was prolonged, indicating retained tissue, in several of 15 women who received PGs 7-21 days after expected menstruation. (LJG)   004807

0401
KELLER, P.J.; RUPPEN, M.; GERBER, C.; SCHMID, J.
Placental function in prostaglandin induced labour.
Journal of Obstetrics and Gynaecology of the British Commonwealth 79: 804-806. September 1972.

The use of $PGF_2\alpha$ to induce labor was studied in order to determine possible effects on the function of the human placenta. 3 primigravid and 5 multiparous women with prolongation of pregnancy of 9-17 days received $PGF_2\alpha$ intravenously at a rate of 2.5-10 μg/minute to a total dose of .9-9.8 mg. There was no uterine hypertonus; all cases were successfully terminated with a mean duration of labor of 8 hours 25 minutes. 5 ml blood samples were drawn every hour before and during the PG infusion until parturition. 4 pregnant women at term served as controls, and blood sampling was done at the same intervals. Placental function was assessed by estimation of the serum activity of human chorionic somatomammotrophin (HCS), human chorionic gonadotrophin (HCG), progesterone, and of heatstable alkaline phosphatase (HSAP). For PGF induced labor there was no significant change in any of the 4 parameters from the control group; the values were stable during labor and dropped after delivery. Therefore use of $PGF_2\alpha$ appeared unobjectionable to placental function. (MLH)   004345

0402
KIRSHEN, E.J.; NAFTOLIN, F.; RYAN, K.J.
Intravenous prostaglandin $F_2\alpha$ for therapeutic abortion.
American Journal of Obstetrics and Gynecology 113: 340-344. 1972.

The use of prostaglandin $F_2\alpha$ in increasing (25-250 μg per minute) and constant (50-100 μg per minute) doses via intravenous infusion in a group of 15 women undergoing therapeutic abortion is described. The constant lower-dose regimen was superior to the graded dose program from the standpoint of both obtaining abortion and minimizing side effects. Although no serious side effects were encountered, troublesome nausea, vomiting, and diarrhea were frequent. The occurrence of prolonged infusion times (which could not be forecast from the gestational age or parity of the patients), an overall failure rate of 42%, and the accompanying side effects obviate this form of treatment as an alternative to suction evacuation of the uterus for routine therapeutic abortion. (Authors)   004032

0403
KIRTON, K.T.; FORBES, A.D.
Activity of 15(S)15-methyl prostaglandin E_2 and $F_2\alpha$ as stimulants of uterine contractility.
Prostaglandins 1: 319-325. 1972.

Experiments were performed to compare the biological potency of 15(S)15-methyl-$PGF_2\alpha$ and 15(S)15-methyl-PGE_2 with natural $PGF_2\alpha$ and PGE_2 in pregnant rhesus monkeys. It was found that the 15-methyl analogues have greater potency and a longer duration of response than the parent compounds as judged by the rate and amplitude of contractions in the monkey uterus. Only 3.5 mg (Day 27) or 5.0 mg (Day 130) of 15(S)15-methyl-$PGF_2\alpha$ or .4 mg (Day 130) of 15(S)15-methyl PGE_2 were required to terminate pregnancy. Only .08 μg/min of 15(S)15-methyl-PGE_2 for 45 minutes caused maximal contractions. 10 times as much PGE_2 was required to produce the same amount of contraction. The increased potency of the 15-methyl analogues may be due to the inability of prostaglandin dehydrogenase to degrade these compounds, or it could be due to increased affinity of the uterine receptors for these analogues. (JRH)   003962

0404
KIRTON, K.T.; CORNETTE, J.C.; BARR, K.L.
Characterization of antibody to prostaglandin $F_2\alpha$.
In: Bergstrom, S., Green, K., and Samuelsson, B. [Proceedings of the] third conference on prostaglandins and fertility control, Stockholm, January 17-20, 1972. Stockholm, WHO Research and Training Centre on Human Reproduction, Karolinska Institutet, [1972]. (Prostaglandins in Fertility Control 2) p. 60-68.

Antibody was produced against prostaglandin (PG) $F_2\alpha$-bovine serum albumin conjugate, which was synthesized by Lieberman's method, in goats and rabbits. Specificity of the antibody, determined by inhibition assays with labeled $PGF_2\alpha$ and unlabeled related PGs, showed 17% cross reaction with $PGF_1\alpha$, but less than 1% cross reaction with 15 keto, and 13,14-dehydro-$PGF_2\alpha$, and other endogenous PGs. Affinity, tested by double antibody technique, yielded an association constant of 2.75 x 10^{-9}M; and a rapid equilibrium was obtained. The radioimmunoassay of serum showed a coefficient of variation of 10%-12% with serum, and serial dilutions of serum or PG standards produced concentration curves with comparable slopes. In human serum, $PGF_2\alpha$ ranged .5-2 ng/ml and injected $PGF_2\alpha$ had a half-life of less than 3 minutes. (LJG)   004467

0405
KIRTON, K.T.; WYNGARDEN, L.J.
In vivo alteration of uterine cAMP by prostaglandin.
Advances in the Biosciences (Suppl.) 9: 108. 1972.

The mechanism of prostaglandin action is associated with cyclic AMP formation in many tissues. They (PG) have pharmacologic effects on uterine smooth muscle of most species, however the mechanism of this action remains to be determined. Cyclic AMP concentrations were measured in hamster uterine tissue at 30 minutes after prostaglandin injections. Tissue was quickly dissected and frozen on dry ice. cAMP was quantitated by the method of Gilman et al. The injection of 500 μg of PGE_1 or PGE_2 in estrus significantly increased levels (p<.05) from control values of 445 to 639 or 642 p moles/100 mg protein, respectively. Injections of $PGF_2\alpha$, or of PGE_1 or E_2 in pseudopregnant animals failed to alter cAMP concentrations significantly. These results will be compared to prostaglandin-induced alteration of adenyl cyclase measured in vitro following in vivo treatment, and prostaglandin-stimulated uterine contractility. The in vivo results reported here are qualitatively similar, but quantitatively different from in vitro effects of prostaglandins on uterine cAMP reported earlier by Harbin and Clauger. (Authors' modified)   004319

0406
KIRTON, K.T.
Prostaglandins and reproduction in sub-human primates.
In: Karim, M.M., ed. The prostaglandins. Progress in research. New York, Wiley-Interscience, 1972. p. 47-70.

Specific topics reviewed are the effects of prostaglandins in nonpregnant animals in vitro, effects in nonpregnant animals in vivo for both subprimate and primate, the effects of prostaglandins in early and late stages of pregnancy, and the use of prostaglandins as abortifacients. (RAS)   004110

0407
KIRTON, K.T.; GUTKNECHT, G.D.; BERGSTROM, K.K.; WYNGARDEN, L.J.; FORBES, A.D.
Prostaglandins and reproduction.
In: Southern, E.M., ed. The prostaglandins: clinical applications in human reproduction. (Brook Lodge Symposium, Augusta, Michigan.) Mount Kisco, New York, Futura Publishing Co. 1972. p. 37-46.

The authors summarize their work on corpus luteum function in rhesus monkeys and PG binding in golden hamsters, as well as a wide selection of other evidence, to show that PGs may act directly on smooth muscle cells of the uterus. $PGF_2\alpha$ has been shown to be luteolytic in sheep, cows, many laboratory rodents, but not in primates or women. The best approximation of luteolysis in primates was the shortening of the menstrual cycle by 2-9 days in rhesus monkeys given 2 sc injections of $PGF_2\alpha$ 15 mg on Days 22 and 23, while serum progesterone fell about 2 ng per ml lower than in controls and ovarian blood flow decreased significantly for 1.5-2 hours. In other experiments on golden hamsters, uterine strips were incubated with labeled PGE_1 or $PGF_2\alpha$ with various concentrations of unlabeled PGs. A linear relationship between bound and unbound PGs resulted, suggesting a single group of PG receptors with a concentration of 5 x 10^{-10}M. Some of the other data considered indicative of a direct action of PGs on uterine smooth muscle are failure of $PGF_2\alpha$ to stimulate adenyl cyclase, failure of adrenergic blocking agents to inhibit PG induced contraction, and direct effect of PGs on release of calcium ions by uterine sarcoplasmic reticulum. (LJG)   004846

0408
KIRTON, K.T.
The role of prostaglandins in reproduction in sub-human primates.
In: Bergstrom, S., Green, K., and Samuelsson, B., eds. [Proceedings of the] third conference on prostaglandins in fertility control, Stockholm, January 17-20, 1972. Stockholm, WHO Research and Training Centre on Human Reproduction, Karolinska Institutet, [1972]. (Prostaglandins in Fertility Control 2) p. 208-216.

This review centers on the role of PGs in the reproductive physiology of female monkeys. PGs and LH stimulate adenyl cyclase in corpus luteum, resulting in progesterone synthesis. Results on whether PGs mediate LH action in vitro are ambiguous, and PGs definitely inhibit steroidogenesis in vivo in laboratory rodents and sheep. In monkeys and women, $PGF_2\alpha$ caused a temporary decrease in plasma progestin level and vaginal bleeding, but no change in menstrual interval. In rhesus monkeys pregnant less than 25 days, $PGF_2\alpha$ rapidly decreased progestin level, resulting in abortion, but after 30-40 days of pregnancy, abortion is complete before progestin decreases to negligible levels. The mechanisms are not known in detail; probably PGs compromise corpus luteum progestin synthesis in early pregnancy, but PGs cause uterine contractions in later pregnancy, which eventually disrupt placental progestin synthesis. (LJG)   004463

0409
KIRTON, K.T.
The role of prostaglandins in reproductive physiology.
Abstracts of paper presented at the 13th National Meeting of the APhA Academy of Pharmaceutical Sciences 2(2): 10. 1972.

PGs are known to affect activity of the myometrium, to modulate the cyclic AMP system in hormone action, and to occur in high concentrations in seminal fluid and other reproductive tissues. Specific roles of PGs include control of parturition in sheep and women and luteolysis in sheep and cows. Less direct evidence, usually based on exogenous administration of PGs, implicates PGs in oviduct motility and ovum transport. (LJG)   005089

0410
KLOCK, F.K.; JUNG, H.
Investigations to the prostaglandin release by human myometrium.
Advances in the Biosciences (Suppl.) 9: 127. 1972.

The presence of various prostaglandins in nearly all mammalian organs, including the human, has become present knowledge. Some tissues release prostaglandin E and F spontaneously or on stimulation. This has been demonstrated for tissues such as the brain, the diaphragm, the stomach, the spleen, and the lungs. The present investigation is concerned with a study on release of prostaglandins from human myometrium under the influence of passive stretching in an in vitro system. Prostaglandins were extracted from the Ringer solution (perfusate) and from homogenized muscle strips as well. They were assayed by the method of Horton and Main. PGE and PGF were separated by the thin-layer chromatography technique according to Green and Samuelsson. PGE was quantitatively determined by the bioassay using the rat stomach fundus and PGF using the rabbit jejunum. The spontaneously acting myometrium releases very small quantities of prostaglandin E and prostaglandin F in vitro. After stretching in the perfusion fluid a significant increase is observed in the concentration of PGE and decrease of PGF. The increase in concentration of PGE and the decrease in PGF suggests an enzymatic mechanism rather than an autooxidative conversion of unsaturated fatty acids to prostaglandins. (Authors' modified)   004293

0411
KOCHENOUR, N.; ENGEL, T.; HENRY, G.; DROEGEMUELLER, W.
Midtrimester abortion produced by intra-amniotic prostaglandin $F_2\alpha$ augmented with intravenous oxytocin.
American Journal of Obstetrics and Gynecology 114: 516-519. October 15, 1972.

10 healthy women in midtrimester were given single intraamniotic injections of $PGF_2\alpha$ in 10 mg or 15 mg dose. Oxytocin at a rate of 25 to 33 mU/minute was administered intravenously 10 minutes later. 8 women aborted; the IAT was 17 hours for the 15 mg group and 18 1/2 for the 10 mg group. Only 1 patient had a brief episode of nausea and vomiting. The method may be superior to hypertonic saline because of a shorter abortion time and milder side effects. (RAS)   004258

0412
KOERING, M.J.
    Light and electron microscopic examination of the ovaries of pregnant rabbits treated with
        prostaglandin $F_2\alpha$ ($PGF_2\alpha$).
    Anatomical Record 172: 348-349. 1972.

    Ovaries from control and $PGF_2\alpha$-treated pregnant rabbits were examined by light and electron
    microscopy. 18 rabbits were randomly assigned in pairs to 1 of 3 groups (days 8, 14, and 21 of
    pregnancy). 1 animal of each pair was treated with .25 mg/kg of $PGF_2\alpha$ b.i.d. for 3 days and the other
    animal received the appropriate volume of the vehicle. On the fourth day, the ovaries were removed
    and each was divided so that 1 part was prepared for light microscopy and the other portion fixed for
    electron microscopy. The ovarian observations were correlated with the state of the uterus, which was
    examined and prepared for light microscopy and also with peripheral progestins level before and after
    $PGF_2\alpha$ administration. Results indicate that the dosage of $PGF_2\alpha$ administered caused corpus luteum
    degeneration as the luteal cells decreased in size, had a crenated nucleus, and became surrounded by
    connective tissue fibers. The luteal cell cytoplasm showed a reduction of organelles, and the lipid
    droplets were of various sizes compared with the controls. The interstitial cells of the treated ovary
    appeared similar to the control ovary. Patches of degenerating cells were identified, but such
    variability is normal in the rabbit. The blood supply was not as abundant or distinct as was
    characteristic of the corpus luteum. The majority of the cells had many lipid droplets, mitochondria
    with a dense matrix, and an abundance of scattered tubular smooth endoplasmic reticulum. The
    effect of $PGF_2\alpha$ on the uterus varied depending on the stage of pregnancy. The Day 21 animals
    aborted, the Day 14 animals resorbed, and variable results were observed in the Day 8 animals. Some
    of the Day 8 blastocysts showed signs of resorption, while others appeared normal. (Author's
    modified)   003957

0413
KOLENA, J.; CHANNING, C.P.
    Stimulatory effects of LH, FSH and prostaglandins upon cyclic 3',5'-AMP levels in porcine granulosa
        cells.
    Endocrinology 90: 1543-1550. 1972.

    Incubation of porcine granulosa cells with 2 μg/ml FSH or LH resulted in a 3- to 5- fold augmentation
    of cyclic 3',5'-AMP content. When gonadotropins were added in the presence of 3 mM aminophylline,
    a 10-fold stimulation of cyclic 3',5'-AMP was found. The stimulatory effect of LH was greater than that
    of FSH. The effect of maximal doses of LH and FSH were not additive, whereas the effects of
    submaximal doses were additive. LH stimulated cyclic 3',5'-AMP within 5 minutes and the levels
    remained elevated for 3 hours. No increase in progestin secretion was found until after 3 hours. After
    this time, the content of cyclic in cells dropped and secretion of progestin increased further. The
    addition of prostaglandin E_1 and E_2 stimulated the formation of cyclic 3',5'-AMP in granulosa cells. The
    stimulatory effect of prostaglandin together with gonadotropins was more than additive and was not
    inhibited in the presence of 7-oxa-13-prostynoic acid or inhibitors of protein synthesis. These results
    strongly suggest that gonadotropic induction of luteinization is mediated by cyclic 3',5'-AMP and that
    prostaglandins may not be mediators of gonadotropic action upon cyclic 3',5'-AMP formation in
    porcine granulosa cells. (Authors)   004131

0414
KORDA, A.; SHEARMAN, R.P.; SMITH, I.D.
    Termination of pregnancy by intra-uterine prostaglandin $F_2\alpha$.
    Australian and New Zealand Journal of Obstetrics and Gynaecology 12: 166-169. 1972.

14 women 12 to 19 weeks pregnant were given $PGF_2\alpha$ either 1) by transcervical extraovular administration of 375 to 750 μg at hourly or second hourly intervals to total doses of 9.25 to 37.5 mg or 2) by the intraamniotic route to total doses of 30 to 45 mg (30 mg plus 15 mg after 24 hours). Of the 7 transcervical patients, 4 abortions were complete; 3 patients developed pyrexia related to genital tract infection and 5 patients had nausea. Using the intraamniotic route, all patients aborted completely; 5 patients had nausea, but there were no infections. (HC)   004797

0415

KORENMAN, S.G.; SANBORN, B.M.; BHALLA, R.C.
Adenyl cyclase and the cyclic AMP responsive systems in the uterus.
Advances in Experimental Medicine and Biology 36: 241-262. 1972.

This paper contains 2 different sets of experiments: cAMP content of ovariectomized rat uteri incubated in vitro with hormones and prostaglandins, and detailed kinetic studies of a cAMP binding and protein kinase from bovine endometrium. Estradiol did not affect cAMP content (protein binding assay) 2-30 minutes after ip injection into ovariectomized rats or during 30 minutes of incubating their uteri in vitro, but 5 μM isoproterenol increased cAMP from 15 pmol per mg to 250 pmol. Oxytocin 1 mU per ml to 50 mU per ml inhibited the cAMP increase due to isoproterenol by 50%. PGE_2 stimulated cAMP but $PGF_2\alpha$ was inactive. Both inhibited cAMP increase evoked by $10^{-5}M$ isoproterenol at low doses, but PGE_2 did not inhibit at 5 x $10^{-5}M$. The cAMP receptor with protein kinase activity was purified from fresh bovine endometrium cytosol fraction. The cAMP-binding and kinase activities were separated and recombined. The receptor and combined forms had single order Scatchard plots with equilibrium association constants of 3 and .6-1 x 10^{-8} per M. The receptor-cAMP complex had simple first order dissociation kinetics and linear second order association. But the kinetics of interaction of receptor-kinase and cAMP concentration made the second order association kinetics of the receptor and cAMP much more complex. (LJG)   004914

0416

KORENMAN, S.G.; BHALLA, R.C.; SANBORN, B.M.
Independent stimulatory and inhibitory prostaglandin effects in the uterus.
Advances in the Biosciences (Suppl.) 9: 128. 1972.

The beta-adrenergic effector isoproterenol produces a dose-related stimulation of castrated rat uterine adenyl cyclase associated with inhibition of uterine contraction. This effect is maximally inhibited by both PGE_2 and $PGF_2\alpha$ at a .5 μM concentration as determined by suppression of hormone-stimulated levels of cAMP. PGE_2 stimulates uterine cAMP with a minimal effective dose of .5 μM and a maximum at .1mM. $PGF_2\alpha$ was ineffective at concentrations as high as .1mM. In contrast to the beta-adrenergic effect, PGE_2-induced cAMP accumulation was inhibited neither by propranolol nor by oxytocin. It was concluded that there are probably 2 species of adenyl cyclase receptors in the uterus affected by PGE_2--an inhibitory one associated with beta-adrenergic activation and stimulation of uterine contraction and a stimulatory one located elsewhere and related to an unknown function. (Authors' modified)   004292

0417

KRESNADI, S.; HARPER, M.J.K.; LLOYD, C.W.
Comparative effectiveness of estrogens as antifertility agents when administered to rats by different routes.
Endocrinology 90: 834-838. 1972.

In the introduction of this paper it is mentioned that the intravaginal application of antifertility substances, such as prostaglandins, has been investigated in the hope of achieving a local effect on reproductive tissues and avoiding systemic side effects. In the discussion section the results of these experiments with estrogen given intravaginally to rats are compared with the results of others on prostaglandins given by the same route. Previous experiments have indicated that to achieve antifertility effects in humans, rabbits, and sheep by this route enough prostaglandin must be given to produce systemic effects. However, results of these experiments with estrogen suggest that the intravaginal route may produce local effects if the antifertility substance is given at a precise time in the reproductive cycle and in a vehicle which produces a steady sustained uptake of the substance. (JRH)   003890

0418
KUEHL, F.A.; HUMES, J.L.; HAM, E.A.; CIRILLO, V.J.
Cyclic AMP and prostaglandins in hormone action.
Intra-Science Chemistry Reports 6(1): 85-95. 1972.

The authors examined the mode of action of PGs in 2 hormone systems: cAMP production by mouse ovary and PG binding by rat epididymal lipocytes or their homogenates. cAMP release was studied kinetically: dose response curves generated Vmax of 2×10^{-6}M for PGE_1 and 2×10^{-4}M for $PGF_2\alpha$. At maximal doses combined PGs were not additive. cAMP and progesterone synthesis both responded to PGs in the order $PGE_2 > PGE_1 > PGA_1 > PGF_2\alpha$. From this it was proposed that PGs stimulate progesterone synthesis via cAMP. The PG derivative 7-oxa-13-prostynoic acid inhibited cAMP release stimulated by PGE_2 and PGE_1 by mouse ovary competitively with a Ki of 3×10^{-5}M, and it blocked cAMP release stimulated by LH with a Ki of 5×10^{-5}M. This prompted studies of PG receptors which were more easily accomplished with rat lipocyte suspensions or homogenates than with mouse ovaries. [3H]-PGE_1 was bound and released in 90% yield. Unlabeled PGE_1, 7-oxa-13-prostynoic, 7-oxa-15-hydroxy-13-prostynoic acid and arachidonic acid displaced some of the bound [3H]-PGE_1, but palmitic, oleic, linoleic acids, testosterone, progesterone, polyphloretin phosphate, and SC 19220 did not. These competition studies suggested that there is a PG receptor. (LJG)   005053

0419
KUEHL, F.A., Jr.; HUMES, J.L.; CIRILLO, V.J.; HAM, E.A.
Cyclic AMP and prostaglandins in hormone action.
Advances in Cyclic Nucleotide Research 1: 493-502. 1972.

The relationship of PGs to intracellular concentrations of cyclic AMP and thus to hormone action was investigated. PGE_1 and PGE_2 were found in in vitro studies to mimic the action of luteinizing hormone (LH) in increasing cyclic AMP levels in the mouse ovary in a dose-related manner. Both PGs stimulated this tissue to a similar maximum value, double that attained by LH, suggesting that both PGs were activating the same receptor site-adenyl cyclase complex. PGE_2 seemed the more potent in this assay, and $PGF_2\alpha$ appeared to be ineffective. In various tissues from hypophysectomized rats and normal rats and mice, PGs consistently increased cyclic AMP formation, with the PGE type being the most active in all instances. It was suggested that both PGs and LH act via a common receptor site, the premise of which requires that the effects of maximum stimulatory amounts of each are not additive. Studies have been undertaken to investigate this point, but without homogeneous cell populations any conclusions are at best tentative. A report suggested the existence of a group of compounds structurally related to the PGs which selectively antagonize the smooth muscle effects of PGs, and 1 such compound, 7-oxa-13-prostynoic acid [I(d)] was investigated. It was found to inhibit competitively the stimulatory effect of PGE-induced cyclic AMP formation in the mouse ovary and was thought to act at a site common to both PGE_1 and PGE_2. I(d) was found also to block the effect of LH in increasing

AMP formation. Since AMP levels are directly related to progesterone production, it appears that PGs as well as LH may play an essential intermediate role in the action of many hormones. (MLH) 004793

0420

KUFHNE, D.

Neuere ergebnisse bei der entwicklung and anwendung der hormonalen kontrazeption. [New results from the development and application of hormonal contraception.]

Deutsche Gesundheitswesen 27: 488-494. 1972.

The author discussed the precautions that should be followed when prescribing steroidal contraceptives. In the final section, the author mentions prostaglandins as being a possible birth control substance that could avoid many of the problems of steroids. However, investigation into the mechanism of prostaglandin action on uterine contractility is needed. (JRH)   004230

0421

LeMAIRE, W.J.; SPELLACY, W.N.; SHEVACH, A.B.; GALL, S.A.

Changes in plasma estriol and progesterone during labor induced with prostaglandin $F_2\alpha$ or oxytocin. Prostaglandins 2(2): 93-101. 1972.

In a double blind study, 12 women received oxytocin for the induction of labor at term, and 20 subjects received intravenous $PGF_2\alpha$. During the infusions, plasma progesterone and estriol levels were measured and compared with preinfusion levels of these steroids. From the analysis of the data, it was concluded that neither the infusion of oxytocin nor $PGF_2\alpha$ per se alters the plasma levels of either progesterone or estriol in term pregnant subjects. (Authors' modified)   004223

0422

LeMAIRE, W.J.; SHAPIRO, A.

Effect of prostaglandin $F_2\alpha$ infusion on the corpus luteum of the human cycle.

Advances in the Biosciences (Suppl.) 9: 115. 1972.

The study was carried out to evaluate the effect of prostaglandin $F_2\alpha$ ($PGF_2\alpha$) on the human corpus luteum of the menstrual cycle. Six volunteers were selected and an infusion of $PGF_2\alpha$ at a rate of 25-50 μg per minute was carried out on different days (5 to 9 days) after ovulation. The infusion was continued for 8 hours. Throughout the entire cycle, plasma samples were obtained for steroids and gonadotropins as well as during a control cycle. During the infusions of $PGF_2\alpha$ 2 hourly blood samples were obtained. In none of the subjects was there a change of the plasma progesterone levels during the infusion period in spite of uterine cramps and vaginal spotting in some subjects. The progesterone levels during the cycle in which $PGF_2\alpha$ was infused was not different from the progesterone levels during a control cycle. The length of the cycle was not altered. Plasma gonadotropins were also measured but at the time of submission of this abstract, the results have not been analyzed. It is concluded from the progesterone data that $PGF_2\alpha$ at the rate of infusion used and administered on Days 5 to 9 postovulatory does not affect the human corpus luteum of the menstrual cycle. (Authors' modified)   004288

0423

LeMAIRE, W.J.; SHAPIRO, A.G.

Prostaglandin $F_2\alpha$: its effect on the corpus luteum of the menstrual cycle.
Prostaglandins 1: 259-267. 1972.

Prostaglandin $F_2\alpha$ ($PGF_2\alpha$) was infused into a peripheral vein of 6 female volunteers during the luteal phase of their cycle. At an infusion rate of 25 to 46 μg/min, no effect was observed on peripheral plasma progesterone levels and luteal phase length. Vaginal spotting was a common occurrence. It is concluded from this study that at the days and rates of infusion used, $PGF_2\alpha$ is not luteolytic in the nonpregnant woman. (Authors)   003966

0424

LABHSETWAR, A.P.

Luteolytic and ovulation-inducing properties of prostaglandin $F_2\alpha$ in pregnant mice.
Journal of Reproduction and Fertility 28: 451-452. 1972.

$PGF_2\alpha$ has been shown to have luteolytic effects in many laboratory species. It also induces ovulation in rats. Experiments were performed to determine if it also has these properties in mice. Inbred ICI mice were injected subcutaneously once a day with $PGF_2\alpha$ in buffered saline on Days 4 through 6 of pregnancy (vaginal plug taken as Day 1). All animals were killed on Day 8 and examined for the number of implantation sites and tubal ova, and the ovaries were examined histologically for corpora lutea. It was found that 100 μg $PGF_2\alpha$ daily (about 3 mg/kg) terminated pregnancy in all animals and a high proportion showed evidence of a second ovulation. The results indicate that $PGF_2\alpha$ has a luteolytic effect in mice and can induce ovulation. Experiments with different doses of prostaglandin indicated that, with the exception of hamsters, mice are the most sensitive species to the luteolytic effects of $PGF_2\alpha$ on a dose per unit of weight basis. (JRH)   003846

0425

LABHSETWAR, A.P.

Antifertility properties of a new prostaglandin--ICI 74205.
Prostaglandins 2(5): 375-392. November 1972.

.5 mg/kg/day of $PGF_2\alpha$ and .10 mg/kg/day of ICI-74205 derivative administered subcutaneously was the minimum effective dose for terminating pregnancy in hamsters. 20 mg/kg/day $PGF_2\alpha$ and 1 mg/kg/day ICI-74205 derivative administered orally was needed for 100% abortion rate. Therefore, ICI-74205 was 5 and 20 times more potent as an antifertility agent than $PGF_2\alpha$ by subcutaneous and oral routes, respectively. Chemically ICI-74205 differs from $PGF_2\alpha$ in having 2 extra carbon atoms in the lower side chain. (RAS)   004580

0426

LABHSETWAR, A.P.

Effects of prostaglandin $F_2\alpha$ on some reproductive processes of hamsters and rats.
Journal of Endocrinology 53: 201-213. 1972.

In an attempt to characterize the endocrine profile of prostaglandin $F_2\alpha$ ($PGF_2\alpha$) in relation to the female reproductive system, the compound (racemic form) was administered to hamsters and rats in various reproductive states. The prostaglandin terminated pregnancy when given once a day either subcutaneously (50 μg/hamster) or orally (1.5-2 mg/hamster) from Days 4 to 6 of pregnancy inclusive, or as a single subcutaneous injection (50 μg/animal) on Day 4. In the rat, higher (500 μg/injection) and more frequent (twice daily) sc injections were required to get even fetal resorption. Concomitant administration of progesterone (4 mg/animal) in either species protected pregnancy. Prostaglandin $F_2\alpha$ terminated pregnancy without interfering with the Pontamine blue reaction, suggesting that its antifertility effects were not mediated by inhibition of implantation. In both hamsters and rats the prostaglandin markedly reduced the size of deciduomata, which could be restored to normal by administration of progesterone. Prostaglandin $F_2\alpha$ delayed passage of zygotes through the fallopian tubes in a proportion of rats but failed to accelerate egg transport in rats and hamsters. Furthermore, it caused a marked histological degeneration of the corpora lutea and induced formation of a fresh set of corpora lutea in pseudopregnant, pregnant, and pseudopregnant-hysterectomized hamsters. These deleterious effects of prostaglandin were accompanied, in hamsters, by the appearance of freshly ovulated tubal ova. Most of the endocrine effects of $PGF_2\alpha$ observed in this study can be accounted for by its luteolytic property. (Author)   003993

0427
LABHSETWAR, A.P.
   Effects of prostaglandins on luteolysis and some other reproductive processes of small laboratory
      animals.
   Research in Prostaglandins 2(1): 1-3. July 1972.

The role of prostaglandins in the reproductive process is discussed in relation to their possible use in fertility control. Luteolysis by prostaglandins involves both a local effect of vasoconstriction or vasodilation and a central effect altering LH and perhaps other luteotrophic hormone secretions. Antifertility effects of $PGF_2\alpha$ were reported in rats, hamster, mice, and rabbits. PGE_2 and E_1 also caused antifertility in hamsters. Both PGs, $F_2\alpha$ and E_2, were associated with a marked decline in circulating progesterone level. Direct effects of PGE_2 on the uterus (stimulating smooth muscle contraction) may be another antifertility aid. (IC)   004101

0428
LABHSETWAR, A.P.
   Evidence for luteolytic effects of prostaglandin E_2 in hamsters and rats.
   In: Abstracts, Fourth International Congress on Endocrinology. Amsterdam, Excerpta Medica, 1972.
      (International Congress Series 256) p. 188.

PGE_2, 150 μg/day/animal, given on Days 4 through 6 of pregnancy terminated pregnancy, markedly decreased serum progesterone and caused degeneration of corpora lutea in hamsters. The same results were obtained in rats using 1 mg PGE_2/animal twice a day on Days 4 through 7 of pregnancy. Administration of progesterone maintained pregnancy but did not prevent luteolysis. PGE_2 also suppressed decidual growth and decreased progesterone levels in pseudopregnant hamsters. The author suggests that PGE_2 is luteolytic. (HC)   005096

0429
LABHSETWAR, A.P.
   New anti-fertility agent -- an orally active prostaglandin-ICI 74,205.
   Nature 238: 400-401. 1972.

A 22 carbon derivative of $PGF_2\alpha$, ICI 74,205 (2 carbons added to the end of the methyl side chain) was compared to $PGF_2\alpha$ for its ability to abort hamsters. When given by the oral route the derivative was 20 times more potent than natural $PGF_2\alpha$ given on Days 4-6 of pregnancy. No side effects were noted with ICI 74,205, nor was it found to have estrogenic, antiestrogenic, androgenic, or anabolic activity.

This derivative was also more effective than $PGF_2\alpha$ in producing abortions when given subcutaneously. Its abortifacient action could be blocked with progesterone. The derivative only had 35% and 1% the smooth muscle stimulating activity of $PGF_2\alpha$ and PGE_1, respectively, when tested on uterine strips in vitro. These experiments show that lengthening the methyl side chain results in a preferential enchancement of oral activity of the compound. (JRH)   004148

0430

LABHSETWAR, A.P.

Prostaglandin E_2: analysis of effects on pregnancy and corpus luteum in hamsters and rats.
Acta Endocrinologica (Suppl. 170) 71: 3-32. 1972.

Effects of PGE_2 on some reproductive processes of hamsters and rats were studied. Anti-fertility doses of PGE_2 caused morphological degeneration of corpus luteum. induced fresh ovulations during the course of the treatment, decreased peripheral progesterone levels and inhibited development of deciduomata in response to trauma. Both pregnancy and decidual growth could be maintained by simultaneous administration of exogenous progesterone. PGE_2 also decreased plasma progesterone concentration in pseudopregnant-hysterectomized hamsters. PGE_2 produced no demonstrable effects on egg transport in hamsters but induced delayed implantation in rats. The anti-fertility doses of PGE_2 but not of $PGF_2\alpha$ inhibited implantation of a proportion of blastocysts and reduced the distance between the adjacent implantation sites. When PGE_2 and $PGF_2\alpha$ were given together to pregnant rats in doses which were virtually ineffective individually, the combined treatment caused loss of pregnancy. Administration of PGE_2 to near term hamsters produced premature littering but oxytocin proved inactive. Anti-fertility doses of PGE_2 produced diarrhea in rats but not in hamsters. (Author's modified)   004745

0431

LABHSETWAR, A.P.

Prostaglandin E_2: Evidence for luteolytic effects.
Prostaglandins 2(1): 23-31. July 1972.

PGE_2 has been examined in laboratory animals for possible antifertility and luteolytic properties. Daily subcutaneous injections from Days 4-6 into hamsters (100-200 μg once a day) and from Days 4-7 (500-1000 μg twice daily) resulted in termination of pregnancy. The treatment caused degeneration of corpora lutea and a sharp decrease in the serum concentration of progesterone. Exogenous administration of progesterone protected the pregnancy, but did not prevent degeneration of corpora lutea. It is concluded that PGE_2 has antifertility and luteolytic properties although it is less potent in this respect than $PGF_2\alpha$. (Author's modified)   004169

0432

LAMPRECHT, S.A.; ZOR, U.; TSAFRIRI, A.; LINDNER, H.R.

Functional relationships between prostaglandin E_2 and luteinizing hormone in the rat ovary.
Israel Journal of Medical Sciences 8: 170. 1972.

Prostaglandin E_2 (PGE_2) is known to mimic the in vitro action of luteinizing hormone (LH) on ovarian tissues with respect to stimulation of the production of cyclic AMP (cAMP) protein kinase activity; induction of ovum maturation, and luteinization and progesterone synthesis. Hence it has been proposed that prostaglandins may be obligatory mediators of the action of LH on the ovary. However, neither the addition of 7-oxa-13-prostynoic acid, an antagonist of the biological action of prostaglandins, nor indomethacin, an inhibitor of prostaglandin biosynthesis, to the incubation medium was able to block the stimulatory action of LH on cAMP production by the intact ovary. Addition of crude Vibrio cholerae enterotoxin to the incubation medium casued a slow increase in the rate of cAMP formation by the ovary, similar to the known effect of the toxin on gut and thyroid. Whole ovaries or isolated follicles incubated with LH for 18 hours and then thoroughly washed became refractory to further stimulation by LH, as judged by cAMP formation, but remained fully responsive to PGE_2. It is widely held that many of the biological actions of cAMP are mediated by activation of protein kinase(s) by being bound to an inhibitory subunit of the enzyme. Incubation of intact ovaries with either LH or PGE_2 resulted in reduced binding of exogenous cAMP by a protein fraction of the 2,000 X g supernatant from the incubated ovaries. This reflected an increased generation of endogenous cAMP (about tenfold) and consequent saturation of the cAMP-binding regulatory subunit of protein kinase, resulting in a four- to fivefold increase in the activity of this enzyme. (Authors) 003853

0433
LANDS, W.E.M.; LeTELLIER, P.; ROME, L.; VANDERHOEK, J.
Alternate modes of inhibiting the prostaglandin synthetic capacity of sheep vesicular gland preparations.
Federation Proceedings 31: 476. 1972.

2 early steps in converting cellular fatty acids to prostaglandins are 1) release from ester precursors and 2) attack by oxygen to initiate the oxidative cyclization. Some acids released can directly inhibit as well as serve in an enzyme-catalyzed destruction of the oxygenase as described for the soybean enzyme (Proc. N.Y. Acad. Sci. 180,107(1971)). The rate of destruction differs for each acid and depends on the presence of oxygen. Rate constants K2 ranged from .1/min for 18:2 to greater than 1 for 20:3 and 20:4 (with 18:3 and 20:2 intermediate). Acids containing (n-3) double bonds were poor substrates for this oxygenase and caused no enzyme destruction, but did serve as inhibitors with KI values similar to the KM values of the substrates (about .1 mM). Thus, the composition of tissue esters may regulate the capacity for prostaglandin formation by providing both reversible and nonreversible inhibitors as well as substrates. A variety of synthetic compounds inhibit prostaglandin formation in vitro, and may act by blocking the dioxygenase reaction (J. Biol. Chem. 246,6700(1971)). These exogenous inhibitors (drugs) can also be classified as reversible (e.g., oxyphenbutazone) and nonreversible (e.g.,indomethacin) in a manner similar to the endogenous inhibitors (fatty acids). (Authors)   003756

0434
LAUDERDALE, J.W.
Effects of $PGF_2\alpha$ on pregnancy and estrous cycle of cattle.
Journal of Animal Science 35: 246. 1972.

$PGF_2\alpha$ was given either intravenously or subcutaneously as either a single or multiple injection to cattle 40 to 120 days pregnant. Total doses of 45 and 150 mg produced abortions in 20 of 20 cows 2-7 days after treatment. 3 of 6 animals given 15 and 30 mg aborted within 14 days. 10 of 14 aborted cows became pregnant again 13-33 days after aborting. 30 mg given subcutaneously 2-4 days postestrus produced estrus in 15-19 days; those animals given the drug 6-9 or 13-16 days postestrus had a 2-4 day interval to estrus. (HC)   004491

0435
LEHMANN, F.; PETERS, F.; BRECKWOLDT, M.; BETTENDORF, G.
Plasma progesterone levels during infusion of prostaglandin $F_2\alpha$ in the human.
Prostaglandins 1: 269-277. 1972.

Progesterone levels were recorded daily in 1 nonpregnant human volunteer and in several patients hospitalized for therapeutic abortion (10-18 week gestation). In the nonpregnant volunteer $PGF_2\alpha$ was infused intravenously on Day 21 of 2 menstrual cycles. The dosage levels were 1 mg in 4 hours and 25 mg in 5 hours. Control progesterone levels in this female were also recorded in 2 cycles, 1 before and 1 after the experimental cycles. The progesterone levels in patients being aborted with $PGF_2\alpha$ infusions were recorded as was the levels in 1 patient aborted by dilation and curettage. Progesterone levels dropped sharply 72 hours after the 25 mg dose of $PGF_2\alpha$ in the nonpregnant woman but the 1 mg dose cycle was not different from the untreated control cycles. There was a pronounced decline in progesterone levels in patients being aborted with $PGF_2\alpha$ and this decline generally continued after infusion was stopped. In 1 patient there was an initial decline, but progesterone levels stabilized and did not decline in spite of of continued infusion. This patient failed to abort and had to be aborted by dilation and curettage. The rate of progesterone decline in patients aborted with $PGF_2\alpha$ and the patient aborted by dilation and curettage were almost identical. It was concluded that large doses of $PGF_2\alpha$ may be luteolytic in the nonpregnant human, but the abortifacient action of this prostaglandin can be attributed to its effect on the uterus and consequent disruption of the placenta. Its effect on progesterone synthesis cannot be determined from this data. (JRH)   003967

0436
LEHMANN, F.; PETERS, F.; BRECKWOLDT, M.; BETTENDORF, G.
Plasmaprogestins during infusion of prostaglandin $F_2\alpha$.
Acta Endocrinologica (Suppl.) 159: 61. 1972.

This paper reports the effect of $PGF_2\alpha$ (infusion into the cubital vein) on the plasma concentration of progesterone during therapeutic abortion with this substance in 12 patients (Weeks 10-26 of gestation). Blood samples were taken before, during, and after the infusion. Infusion rate ranged between 80-100 μg/min until abortion was complete, either with or without curettage. Average time for complete abortion was 10 hours. Side effects (vomiting, diarrhea) could be overcome with atropin, paspertin iv and 1% morphine solution (20 drops every 2-3 hours orally). Plasma progesterone levels were estimated by a CPB method using 3% dog plasma. Results did not show any uniform reaction on the $PGF_2\alpha$ infusion. In some cases a dramatic decrease of the plasma progesterone level could be noticed, in others progesterone values did not drop until complete abortion had taken place. These findings did not correlate with the weeks of gestation. These results support the assumption that $PGF_2\alpha$ during pregnancy (Weeks 10-26) and in the described concentration acts directly on the smooth muscle of the uterus, and does not act by luteolysis of the corpus luteum as discussed by Behrmann and Kirton, which only could have been expected some 80 hours later according to McCracken et al. (Authors' modified)   003736

0437
LEHMANN, F.F.; PETERS, F.; BRECKWOLDT, M.; BETTENDORF, G.
Luteolysis after infusion of prostaglandin $F_2\alpha$ in man.
In: Abstracts, Fourth International Congress of Endocrinology, Washington, D.C., June 18-24, 1972.
Amsterdam, Excerpta Medica, 1972. (International Congress No. 256) Abstract 477. p. 189.

1 or 25 mg PGF$_2\alpha$ was infused intravenously in women on different days of the menstrual cycle. The low dose caused a marked decrease in plasma estrogen and progesterone levels, whereas the high dose blocked steroid synthesis and induced menstruation. (HC)   005002

0438
LESLIE, D.C.; LAUFE, L.E.
The evaluation of intra-amniotic prostaglandin F$_2\alpha$ in the management of early midtrimester pregnancy termination.
In: Southern, E.M., ed. The prostaglandins: clinical applications in human reproduction. [Brook Lodge Symposium, Augusta, Michigan.] Mount Kisco, New York Futura Publishing Co., 1972. p. 451-455.

5 women, 14 to 17 weeks pregnant, received PGF$_2\alpha$ intraamniotically, 10 mg initially, additional doses given as needed. Another group of 5 women received PGF$_2\alpha$, 20 mg initially, then 2 injections of 10 mg at 10 to 15 minute intervals. 2 patients from each group aborted within 12 to 19 hours. In all patients, rhythmic contractions occurred every 1 1/2 to 2 minutes and subsided in tone and frequency after 6 to 8 hours. Plasma progesterone levels dropped markedly after abortion. Nausea and vomiting occurred in all patients, diarrhea and flushing in 5 and 4 women respectively. The authors suggest that additional PGF$_2\alpha$ 6 to 8 hr after the initial dose should increase the 24-hr success rate. (HC)   004884

0439
LESLIE, D.C.; LAUFE, L.E.
The evaluation of intra-amniotic prostaglandin F$_2\alpha$ in the management of early midtrimester pregnancy termination.
Journal of Reproductive Medicine 9(6): 453-455. December 1972.

PGF$_2\alpha$ administered intra-amniotically to induce abortion has been studied in 10 women who were in the first half of the second trimester of pregnancy. A group of 5 received 10 mg of PGF$_2\alpha$, administered through a catheter which had been inserted into the amniotic cavity. Additional PGF$_2\alpha$ was administerd at the discretion of the attendant. The other group of 5 received an initial dose of 20 mg followed by 2 more 10 mg injections of PGF$_2\alpha$ at 10 to 15 minute intevals. 4 of the 10, 2 from each group, aborted within 12 to 19 hours and another 4 showed signs of impending termination at the end of the 24 hour test period. All patients experienced systemic side effects, presumably from the absorption of PGF$_2\alpha$ from the amniotic sac. (JSL)   005000

0440
LEVINE, L.; CERNOSEK, R.M.G.; POLET, H.; GERSHMAN, H.
Prostaglandins: serologic specificities and estimation in biological fluids.
In: Bergstrom, S., Green, K., and Samuelsson, B., eds. [Proceedings of the] third conference on prostaglandins in fertility control, Stockholm, January 17-20, 1972. Stockholm, WHO Research and Training Centre on Human Reproduction, Karolinska Institutet, [1972]. (Prostaglandins in Fertility Control 2) p. 107-108.

Methods of producing and measuring prostaglandin (PG) antibodies, studies of their specificity, and results of their use in bioassay of PGs are summarized. For specificity studies, PG antigens were synthesized by linking PGs to polylysine by an amide bond to the PG carboxyl, and complexing the polylysyl PG to succinyl hemocyanin. Antibodies were produced in monkey, rabbits, and guinea pigs. 1000-fold increase in binding specificity (by slope of binding curves) resulted after multiple injections. Rabbit anti-PGE$_1$ was an IgG, 7s antibody, as judged by gel filtration. Although hyperimmune sera from rabbit, guinea pig, and monkey could distinguish the class of PG (degree of unsaturation) antisera of all types (PGA, E, F, etc.) were active toward PGB. Binding of PGB_1 and anti-PGB_1 was measured by equilibruim dialysis: association constants and standard free energies are tabulated. Less than 1% cross reaction between anti-$PGF_2\alpha$, and free energies of -1.8 to -4.8 kcal were observed with 10 other PGs. For bioassay, a 100-fold increase in sensitivity was achieved by using polyvalent antisera made with I-125-tyrosine polypeptide conjugated with the PG hapten. The monovalent antiserum was adequate, however, for a new bioassay in which the immune complex was collected on nitrocellulose filters and counted by scintillation. All solutions had to be controlled for salt, pH, and protein concentration. The nitrocellulose filter bioassay was used to measure the conversion of PGA_1 to PGB_1 by serum from rats and other species, and to measure the level of $PGF_2\alpha$, as well as recovery of added $PGF_2\alpha$, in human serum. Serum $PGF_2\alpha$, levels were .25-.49 ng/ml in most females and .5 or more in most males, after precipitation with methylal-ethanol and dialysis against tris-gelatin solution. (LJG) 004466

0441
LIEHR, R.A.; MARION, G.B.; OLSON, H.H.
Effects of prostaglandin on cattle estrus cycles.
Journal of Animal Science 35: 247. 1972.

2 groups of 3 heifers each were injected with 500 μg of $PGF_2\alpha$ either contralaterally or ipsilaterally into the uterine horn on Day 5 of estrous cycle. Cycle length remained 20 days for the 2 test groups and control. 5 heifers received 6 ng contralaterally and 5 heifers were injected ipsilaterally on Day 9 in a second experiment. Prostaglandin levels dropped to non-detectable levels for ipsilateral treatment but remained high for contralateral treatment. Cycle length was 11.4 days ipsilaterally and 15.2 contralaterally. (RAS)   004212

0442
LIGGINS, G.C.; GRIEVES, S.A.; KENDALL, J.Z.; KNOX, B.S.
The physiological roles of progesterone, oestradiol-17β and prostaglandin $F_2\alpha$ in the control of ovine parturition.
Journal of Reproduction and Fertility (Suppl.) 16: 85-103. 1972.

Experiments were performed to determine the role of progesterone, estrogen, and prostaglandins on parturition in sheep and any interaction between them. In 1 set of experiments, blood samples were taken from the uterine vein and the maternal jugular vein before and after induction of labor by fetal infusion of ACTH or dexamethasone. In another set of experiments ewes were killed in various stages of late pregnancy and samples of the myometrium, maternal and fetal cotyledons were assayed for prostaglandin content. During a further series of experiments $PGF_2\alpha$ (.5-16 μg/min) was infused into the systemic aorta, the amounts of $PGF_2\alpha$ in vena cava and jugular vein plasma determined and the amniotic pressure continuously recorded. PGE, PGE_2, and $PGF_1\alpha$ were not detected in any plasma or tissue. $PGF_2\alpha$ was only detected in jugular plasma when it was being infused into the aorta. Small amounts of $PGF_2\alpha$ were found in the myometrium and in maternal and fetal cotyledons. The amount of $PGF_2\alpha$ in the myometrium and maternal cotyledons increased sharply when ewes entered labor. $PGF_2\alpha$ did not alter progesterone levels, but it did cause an increase in estrogen levels. It was also found that a rise in fetal corticosteroid levels caused an increase in synthesis of $PGF_2\alpha$ in the maternal cotyledons within 24 hours. In sheep no evidence has been found that $PGF_2\alpha$ can stimulate uterine activity in pregnancy. It seems unlikely from these experiments that either progesterone or estrogen levels are the prime determinant of the onset of parturition. It is unclear whether $PGF_2\alpha$ can fill the requirements. (JRH)   003930

0443
LOUIS, T.M.; HAFS, H.D.; MORROW, D.A.
    Estrus and ovulation after uterine PGF$_2\alpha$ in cows.
    Journal of Animal Science 35: 247-248. 1972.

    PGF$_2\alpha$ 5 mg was infused into the uterine horn of cows on Day 11 of estrus and again on Day 15.
    Following each infusion, luteal diameter fell in 48 hours from 2.7 cm to .4 cm after Day 11 and to .9
    cm after Day 15; estrus began at 68 to 73 hours and ovulation at 94 to 99 hours. LH levels in jugular
    blood serum averaged .9 ng/ml at start of infusion and peaked at 7.3 ng/ml 69 hours later. Results
    indicated that PGF$_2\alpha$ is luteolytic in cows and may be an effective method of inducing ovulation. (HC)
        004566

0444
LOUIS, T.M.; HAFS, H.D.; MORROW, D.A.
    Estrus and ovulation after PGF$_2\alpha$ in cows.
    Journal of Animal Science 35: 1121. 1972.

    The uterine horn contralateral to the corpus luteum was infused once with 5 mg prostaglandin F$_2\alpha$
    (PGF$_2\alpha$ THAM salt) in .5 ml saline in each of 5 cows on Day 11 of the estrus cycle. Cows were observed
    twice daily for signs of estrus and ovarian activity was monitored by rectal palpation. Luteal diameter
    averaged 2.3 + or - .2 cm at PGF$_2\alpha$ infusion; it fell to 1.6 ± .2 cm at 24 hours and to .9 ± .2 cm at 48
    hours. Estrus began at 75 ± 9 hours and ovulation occurred at 99 ± 12 hours. Blood serum was
    taken at 12-hour intervals from regular cannulae. LH (RIA) averaged .9 ± .2 ng per ml at PGF$_2\alpha$ and
    peaked (average 10.6 ± 3.6 ng per ml) at 78 ± 6 hours. None of these responses differed
    significantly from responses after infusion of 5 mg of PGF$_2\alpha$ into the ipsilateral horn on Days 7, 11 or
    15. In a second experiment, 6 heifers were infused once into the vagina with 30 mg of PGF$_2\alpha$ in 1.5 ml
    saline on Day 11 of the cycle. Luteal diameter averaged 2.5 ± .1 cm at infusion and 2.2 ± .1 cm at 24
    hours; it fell to 1.5 ± .2 cm at 48 hours and .6 ± .4 cm at 96 hours. Estrus began at 117 + or - 18
    hours and ovulation occurred at 138 ± 20 hours. LH averaged .7 ± .1 ng per ml at infusion and
    peaked (average 5.8 ± 2.0 ng per ml) at 128 ± 19 hours. Thus, the luteolytic response to vaginal
    PGF$_2\alpha$ may be delayed 1 or 2 days relative to that from uterine PGF$_2\alpha$. Control estrus cycles before
    and after uterine PGF$_2\alpha$ averaged 20.2 ± 1.2 and 20.8 ± .7 days, respectively. Control cycles after
    vaginal PGF$_2\alpha$ averaged 21.2 ± .3 days. (Authors)   004845

0445
LUKASZEWSKA, J.; WILSON, L., Jr.; HANSEL, W.
    Luteotropic and luteolytic effects of prostaglandins in the hamster.
    Proceedings of the Society for Experimental Biology and Medicine 140: 1302-1307. 1972.

    PGF$_2\alpha$ was administered to pseudopregnant hamsters, and its effects were compared with the effects
    of lipid extracts of hamster uteri. Relatively high doses of PGF$_2\alpha$ (10 µg) injected intraperitoneally
    caused pseudopregnant hamsters to return to estrus after 2 to 4 days. Single injections of 50 and 100
    ng of PGF$_2\alpha$ into the ovarian bursae depressed corpus luteum weights. However, 2 ng of PGF$_2\alpha$
    injected into the ovarian bursa caused a large increase in luteal tissue progesterone. The luteolytic
    activity of the uterus is found in the lipid rather than in the saline extract of the uterus. The PGF
    content of the pseudopregnant hamster uterus increases sharply just prior to corpus luteum
    regression, which coupled with its luteolytic properties suggests a possible physiological role for PGFs
    in luteolysis. However the PGs might not exert a direct effect, since injection into 1 ovarian bursa
    causes luteal regression in both ovaries. (JRH)   004308

0446
McCRACKEN, J.A.; GLEW, M.E.; CARLSON, J.C.
Prostaglandin transport by a counter current mechanism.
Federation Proceedings 31: 546. 1972.

Transplantation of the ovary or uterus, cross-circulation experiments, and uterine blood vessel ligation studies in sheep suggested that a luteolytic factor passed from the uterine vein directly into the adjacent ovarian artery. Tritium labelled $PGF_2\alpha$ infused into the uterine vein resulted in higher amounts of radioactivity in the adjacent ovarian artery than in the iliac artery (Rec. Prog. Horm. Res., 27: 537, 1971). Long-term infusion experiments have shown that the rate of transfer of unchanged tritiated $PGF_2\alpha$ is relatively slow and that a dynamic equilibrium is not reached until about 2 hours from the start of the infusion. After stopping the infusion, $PGF_2\alpha$ persists for some time in the ovarian artery, since in this mechanism the lungs, the main site of $PGF_2\alpha$ catabolism, are bypassed. The amount of infused tritiated $PGF_2\alpha$ transferred in this way was in the range of 5% to 10%. These experiments were repeated with unlabelled $PGF_2\alpha$ in amounts calculated to simulate the measured physiological concentrations of $PGF_2\alpha$ in the uterine vein during luteolysis. Premature luteal regression was induced consistently, thus confirming not only the countercurrent exchange mechanism, but also the identity of $PGF_2\alpha$, as a luteolytic hormone from the uterus of the sheep. (Authors)   003763

0447
McCRACKEN, J.A.; CARLSON, J.C.; GLEW, M.E.; GODING, J.R.; BAIRD, D.T.; GREEN, K.; SAMU-ELSSON, B.
Prostaglandin $F_2\alpha$ identified as a luteolytic hormone in sheep.
Nature New Biology 238: 129-134. August 2, 1972.

Experiments leading to a conclusion that prostaglandin (PG) $F_2\alpha$ controls regression of the corpus luteum in sheep are summarized. Sheep, as well as guinea pig, rat, hamster, and cow, require a local uterine factor for luteolysis. As shown by transplanting the ovary to the neck, and crossing the blood flow of a donor on Cycle Day 15 with a recipient, there was a 50% decrease in progesterone secretion in the recipient, a good indicator of luteolysis. $PGF_2\alpha$ caused luteolysis when infused into the artery of an autotransplanted ovary for 3-6 hours at 25 μg/hour, but not if PG were administered systemically. The route taken by the luteolytic factor from the uterus to the ovary was evidently from the utero-ovarian vein to the ovarian artery by a countercurrent transfer, since no luteal regression occurs if these vessels are physically separated. A test of the $PGF_2\alpha$ and the countercurrent theories was performed by infusing tritium-labeled $PGF_2\alpha$ into the uterine vein and collecting ovarian arterial blood. Tritium was associated with the $PGF_2\alpha$ spot from thin layer chromatography, and peaked 30 minutes after the infusion was stopped. Further measurements of the $PGF_2\alpha$ showed that it varies from a resting level of 2 ng/ml to 20 ng/ml in the uterine vein plasma during luteolysis. 25 μg/hour for 3 hours into the uterine vein are required to induce luteolysis 5 days early, or 1-2 μg/hour into the artery of a transplanted ovary. Thus about 5% of uterine $PGF_2\alpha$ may reach the ovary by countercurrent flow. (LJG)   004096

0448
McCRACKEN, J.A.; CARLSON, J.C.; GREEN, K.; SAMUELSSON, B.
Prostaglandin $F_2\alpha$ identified as a luteolytic hormone in sheep using GLC/mass spectrometry.
Advances in the Biosciences (Suppl.) 9: 100. 1972.

Cross circulation experiments between sheep bearing uteroovarian transplants and sheep with ovarian transplants indicated that a luteolytic factor was present in uterine vein blood at the time of luteal regression. Intraarterial (but not systemic) infusions of $PGF_2\alpha$ into the transplanted ovary exactly mimicked the uterine vein luteolytic factor. Uterine venous plasma from both normal sheep and sheep with uteroovarian transplants was examined for $PGF_2\alpha$ during luteal regression and at other times of the cycle. (5,6,8,9,11,12,14,15) D8-$PGF_2\alpha$ and (17,18) tritiated $PGF_2\alpha$ were added to each sample as carrier and tracer, respectively, prior to measurement and identification by GLC/mass spectrometry. The functional state of the corpus luteum was determined by the concentration of progesterone in ovarian venous plasma using GLC. Coincident with the fall in progesterone concentration during luteal regression, plasma concentration of $PGF_2\alpha$ increased by at least 10 fold (from less than 2.0 ng/ml to greater than 20.0 ng/ml). In 2 sheep bearing uteroovarian transplants, the release of $PGF_2\alpha$ from the uterus during luteal regression was determined in serial samples from the same individual animal. Results were determined by recording blood flow and measuring the A-V difference in $PGF_2\alpha$ concentration across these preparations. The release of $PGF_2\alpha$ from the uterus rose from less than .5mcg/hr to greater than 12.5mcg/hr during luteolysis. Premature luteal regression was induced by infusing the calculated release rate of $PGF_2\alpha$ during luteal regression into the uterine vein in situ (but not when infused systemically). This not only confirmed the countercurrent transfer of $PGF_2\alpha$ from the uterine vein to the adjacent ovarian artery but also the role of $PGF_2\alpha$ as a luteolytic hormone. (Authors' modified)   004330

0449

McCRACKEN, J.A.

Prostaglandins and luteal regression - a review.

Research in Prostaglandins 1(4): 1-4. 1972.

The role of prostaglandins in luteal regression, especially in sheep, is reviewed. (JRH)   004186

0450

McCRACKEN, J.A.

The role of prostaglandins in corpus luteum regression.

In: [Abstracts of] First International Planned Parenthood Federation South-East Asia and Oceania Regional Medical and Scientific Congress, Sydney, Family Planning Association of Australia, August 1972. p.16.

The author reviews the evidence that $PGF_2\alpha$ is the uterine luteolytic factor in sheep and perhaps other animals. (JRH)   004174

0451

McCRACKEN, J.A.

The role of prostaglandins in luteal regression.

In: Bergstrom, S., Green, K. and Samuelsson, B., eds. [Proceedings of the] third conference on prostaglandins in fertility control, Stockholm, January 17-20, 1972. Stockholm, WHO Research and Training Centre on Human Reproduction, Karolinska Institutet, [1972]. (Prostaglandins in Fertility Control 2) p. 234-258.

Experiments leading to a conclusion that prostaglandin ($PGF_2\alpha$) controls regression of the corpus luteum in sheep are summarized. Sheep, as well as guinea pig, rat, hamster, and cow, require a local uterine factor for luteolysis. As shown by transplanting the ovary to the neck and crossing the blood flow of a donor on Cycle Day 15 with a recipient, there was a 50% decrease in progesterone secretion in the recipient, a good indicator of luteolysis. $PGF_2\alpha$ caused luteolysis when infused into the artery of an autotransplanted ovary for 3-6 hours at 25 μg/hour, but not if PG were administered systemically.

The route taken by the luteolytic factor from the uterus to the ovary was evidently from the utero-ovarian vein to the ovarian artery by a countercurrent transfer, since no luteal regression occurred if these vessels were physically separated. A test of the $PGF_2\alpha$ and the countercurrent theories was performed by infusing tritium-labeled $PGF_2\alpha$ into the uterine vein and collecting ovarian arterial blood. Tritium was associated with the $PGF_2\alpha$ spot from thin layer chromotography, and peaked 30 min after the infusion was stopped. Further measurements of $PGF_2\alpha$ showed that it varies from a resting level of 2 ng/ml to 42 ng/ml in the uterine vein plasma during luteolysis. 25 μg/hour for 3 hours into the uterine vein are required to induce luteolysis 5 days early, or 1·2 μg/hour into the artery of a transplanted ovary. Thus about 5% of uterine $PGF_2\alpha$ may reach the ovary by countercurrent flow. (LJG) 004475

0452

MALNASI, Z.; LEHRNER, J.; SCHWARZ, J.

Immune reaction manifesting with anterior uveitis and urogenital disease: prostaglandins as haptens.
Acta Chirurgica Academicae Scientiarum Hungaricae 13(4): 375-383. 1972

Exacerbation of iridocyclitis in a male patient by Raveron, an aqueous nonprotein prostate extract prescribed for prostatits, and in a female during menstruation prompted experiments designed to see if PGs were involved; indirect fluorescent antibody techniques were used. Antigens appeared brilliant green when bound antibodies from patients' sera were layered with FITC-labeled sheep anti-human IgG-IgM antiserum and viewed in a UV fluorescent microscope. Serum from both patients contained antibodies againinst human and rabbit iris, against epithelial cells of prostatic acini if tissue was mounted in water at pH 7.0, against prostatic secretion at pH 4.5, and against the contours of smooth muscle cells in pupil and prostate. The pH dependence and intracellular location of fluorescence were considered evidence that PGs were specific haptens in this disorder. (LJG) 005749

0453

MALOFIEWJEW, M.; LAUDANSKI, T.; STRACZKOWSKI, W.

Wplyw prostaglandyny A_2(PGA)2 na aktywnosc skurczowa myometrium szczura. [The effect of prostaglandin A_2(PGA_2) upon the contractility of rat myometrium.]
Ginekologia Polska 43(11): 1289-1294. 1972.

Isometric and isotonic contractions, spontaneous or under the influence of PGA_2, of isolated Wistar rat uterus were recorded on a kymograph; uteri were in Tyrode's solution with isotonic magnesium at 36 degrees C. .001 and .005 μg PGA_2 increased basal tone, frequency, and height of transient isotonic contractions. The height of contractions in response to .1 IU oxytocin appeared just after adding PGA to the bath. PGA_2 had no effect on isotonic contractions after bradykinin. Isometric contractions against 5 g appeared to decrease in number after PGA_2. (LJG) 004910

0454

MANDL, J.P.

The effect of prostaglandin E_1 on rabbit sperm transport in vivo.
Journal of Reproduction and Fertility 31: 263-269. 1972.

A total of 75 μg PGE_1 was added to the inseminate placed intravaginally in rabbits. The numbers of spermatozoa found in the uteri and oviducts of treated animals sacrificed 1/2 or 2 hours later were greater than in controls. However, this difference was significant only for those killed 2 hours later (p = .014). (HC) 004611

0455
MANN, T.
Advances in male reproductive physiology.
Fertility and Sterility 23(10): 699-707. October 1972.

The recent advances in our knowledge of spermatozoa, male testicular and hypothalamic hormones, accessory organ secretions, pharmacology, and artificial insemination are reviewed. Specific topics included are insect feromones, testicular sperm collection, migration of the kinoplasmic droplet, staining of the Y-bearing sperm with quinacrine and of the acrosome with eucrisin, role of 5alpha-dihydrotestosterone in action of testosterone, daily LH and testosterone rhythms, sex-specific structure of the hypothalamus, quantities of PGEs in human semen, and analysis of semen by split-ejaculate method. (LJG)   004518

0456
MARAZ, A.; MECS, E.
The influence of prostaglandin E_1 on the blast transformation of lymphocytes from pregnant and normal humans and from humans with tumors.
Advances in the Biosciences (Suppl.) 9: 72. 1972.

The optimal conditions for 3H thymidine incorporation into PHA (phytohaemagglutinin) stimulated lymphocytes were determined in normal and pregnant humans and in humans with tumors. The effect of PGE_1, added in different concentrations, was examined on the blast transformation of such cells. PGE_1, given in a certain concentration, markedly increased the incorporation of thymidine into lymphocytes from normal (non-pregnant) humans and humans without tumors; by contrast there was only a faint stimulation by PGE_1 in the case of lymphocytes from pregnant humans and humans with tumors. (Authors' modified)   004431

0457
MARLEY, P.B.
An attempt to inhibit the uterine luteolysin in the guinea-pig.
Journal of Physiology 222: 169P-170P. January 1972.

In order to determine whether $PGF_2\alpha$ is the luteolytic hormone which terminates the life span of the corpus luteum in the guinea pig, indomethacin, which blocks the synthesis of $PGF_2\alpha$ in vitro, was administered in toxic doses (40 mg/kg/day). Treatment was started from 1-10 days after vaginal opening had occurred. The length of the estrous cycles in 2 animals had no effect and shortened the cycle in 2 others. The mean lengths of the estrous cycles before, during, and after treatment were reported as 17.6, 15.5, and 17.3 days, respectively. Animals started on Day 11, showed a significant lengthening of about 4 days whether administered twice daily by stomach tube (17.9, 22.0, and 20.0 days) or added to the drinking water (17.7, 21.3, and 19.7 days). There was no significant effect with treatment started on Day 15, nor with a single daily dose of 15 mg/kg/day started on Day 11, nor with indomethacin in the drinking water starting on Day 7. The author felt that the lengthening when it occurs is short in relation to the cycle itself and thus it is unlikely that $PGF_2\alpha$ is the uterine luteolysin in the guinea pig. (MJN)   004567

0458
MARLEY, P.B.
Effects of prostaglandin $F_2\alpha$, E_2 and E_1 on fertility in mice.
Nature New Biology 235: 213. 1972.

The effect of $PGF_2\alpha$, PGE_2, and PGE_1 on pregnancy in mice was measured. The prostaglandins were injected subcutaneously with 1 or 2 mg/10 ml/kg body weight with some mice receiving up to 3 injections in a 24-hour period. Only $PGF_2\alpha$ and PGE_2 exhibited antifertility action with $PGF_2\alpha$ being more effective. The litter sizes of the mice were not changed, only the number of pregnancies, suggesting an all-or-none effect. A series of experiments with ovarietomized and hypophysectomized mice injected with progesterone, LH, or pituitary extracts were performed to determine the mechanism of $PGF_2\alpha$ actions. The experiments suggest that $PGF_2\alpha$ functions by inhibiting the action of gonadotropic hormones in the ovary. (JRH)   003745

0459

MARSH, J.M.; LE MAIRE, W.J.

Effect of gonadotropins and prostaglandins on cyclic AMP accumulation and steroidogenesis in the human corpus luteum in vitro.

In: Abstracts, Fourth International Congress of Endocrinology, Washington, D. C. June 18-24, 1972. Amsterdam, Excerpta Medica, 1972. (International Congress Series No. 256) Abstract 476. p. 189.

PGE_2 or HCG stimulated the synthesis of progesterone and cAMP accumulation in incubated slices of human corpora lutea. Human LH also stimulated cAMP accumulation but $PGF_2\alpha$ had no effect. This response was due to a stimulation of adenyl cyclase. Corpora from pregnant women responded much less than tissue from the menstrual cycle. The authors conclude that the stimulation by PG on steriodogenesis in the human corpus luteum is mediated by cAMP. (HC)   005060

0460

MATHER, E.C.; DALE, H.E.

Seminal plasma effects on bovine endometrial respiration.

Journal of Reproduction and Fertility 28: 197-206. 1972.

The authors report the isolation of a water soluble substance from bull seminal plasma which reduces the oxygen uptake of bovine endometrium in vitro. This substance was not a prostaglandin. (JRH) 003811

0461

MIDWINTER, A.; BOWEN, M.; SHEPHERD, A.

Continuous intrauterine infusion of prostaglandin E_2 for termination of pregnancy.

Journal of Obstetrics and Gynaecology of the British Commonwealth 79: 807-809. September 1972.

25 women pregnant 10-17 weeks were given an initial dose of 100 μg of PGE_2 followed by an infusion (50 μg/ml, 1.5 ml/hour) into the extraovular space. The mean IAT was 15 1/2 hours; 8 of the 24 abortions were incomplete. Most patients experienced a dull ache, some had painful contractions, and 9 patients vomited. There was no correlation between total dose (mean of 1380 μg) or IAT and age, parity, or length of gestation. (HC)   004892

0462

MILLER, A.W.F.; CALDER, A.A.; MACNAUGHTON, M.C.

Termination of pregnancy by continuous intrauterine infusion of prostaglandins.

Lancet 2: 5-7. 1972.

Prostaglandins (E$_2$or F$_2\alpha$) were introduced into the extraamniotic space by continuous slow infusion with a Palmer pump to produce abortion in 60 patients in the midtrimester of pregnancy. In 50 cases (83%) abortion occurred with prostaglandins only. In 10 cases an intravenous infusion of oxytocin (Syntocinon) was used in addition to prostaglandins, but in no case was abdominal hysterotomy required to empty the uterus. Of 52 patients receiving prostaglandins E$_2$, 47 (90%) aborted with prostaglandins only in a mean infusion time of 15.75 hours (range: 6-30 hours). There were no significant complications attributable to prostaglandins. (Authors)   004071

0463
MOGHISSI, K.S.; RANGARAJAN, N.S.; LACROIX, G.E.
Induction of labor with prostaglandin F$_2\alpha$ and oxytocin: a matched study.
In: Southern, E.M., ed. The prostaglandins: clinical applications in human reproduction. [Brook Lodge Symposium, Augusta, Michigan.] Mount Kisco, New York, Futura Publising Co., 1972. p. 115-119.

PGF$_2\alpha$, 2.5 μg per minute, or oxytocin, 1.25 mU per minute, was given intravenously to 2 groups of women at term, group 'A' having cervical scores of less than 7, and group 'B' with scores of 8 or more. Rates were doubled at 15 to 30 minute intervals as needed. In group 'A' 9 of 10 women given oxytocin delivered, the IDT being 493 minutes; 5 of 8 women in this group given PGF$_2\alpha$ delivered, the IDT being 307 minutes, and 2 had prolonged contractions. In group 'B' all 10 women given oxytocin delivered, IDT was 292 minutes; 11 of 12 women given PGF$_2\alpha$ delivered, the IDT was 346 minutes, and 6 had prolonged contractions. The mean total doses used for oxytocin were 8.2 mU per minute in group 'A' and 5.6 mU per minute in group 'B'. For PGF$_2\alpha$ the values were 13 μg per minute in group 'A' and 15 μg per minute in group 'B'. With PGF$_2\alpha$ there were 2 cases of fetal bradycardia. The authors conclude that PGF$_2\alpha$ was not superior to oxytocin for labor induction. (HC)   004854

0464
MOGHISSI, K.S.
Prostaglandins in reproduction.
Obstetrics and Gynecology Annual. 297-337. 1972.

Basic prostaglandin structure, synthesis and assay techniques are discussed. Prostaglandin effects on smooth muscle, gastrointestinal tract, cardiovascular system, nervous system are reviewed. An extensive discussion of prostaglandins in reproduction covers male genitourinary tract, isolated myometrium and cervix, in vivo studies on contractility, and induction of labor and termination of pregnancy. Prostaglandin mode of action and potential uses are hypothesized. (RAS)   004734

0465
NAFTOLIN, F.; KIRSHEN, E.J.; RYAN, K.J.
Therapeutic abortion utilizing local application of PGF$_2\alpha$.
In: Southern, E.M., ed. The prostaglandins: clinical applications in human reproduction. (Brook Lodge Symposium, Augusta, Michigan). Mount Kisco, New York, Futura Publishing Co., 1972. p. 423-431.

Of 8 women pregnant 10-12 weeks given 8-16 50 mg PGF$_2\alpha$ tablets intravaginally at 1- to 2-hour intervals, only 1 aborted but 6 had nausea, vomiting, and diarrhea. 3 other women given 50 mg PGF$_2\alpha$ in liquid form intravaginally every 1 to 2 hours for 24 hours failed to abort, while all experienced side effects. 6 women were given 200 μg PGF$_2\alpha$ by the intrauterine route every 1 to 4 hours; 3 aborted, 2 completely. All patients had nausea and vomiting. 20 women pregnant 14-19 weeks were given PGF$_2\alpha$

intraamniotically 15-30 mg initially and supplementary doses if needed. 18 aborted, 15 completely; 10 had nausea and vomiting. (HC) 004881

0466

NAISMITH, W.C.M.K.; BARR, W.; MacVICAR, J.

Induction of labour by simultaneous intravenous administration of prostaglandin E_2 and oxytocin. British Medical Journal 4: 461-462. November 25, 1972.

10 patients were given a simultaneous intravenous infusion of oxytocin at a rate of .66 mU/minute, doubled every 15 minutes until adequate uterine response was produced, and PGE_2 at a rate of .5 μg/minute; another 10 patients received oxytocin and saline. The addition of PGE_2 significantly reduced the mean maximum oxytocin rate of infusion from 71 to 17 mU/minute ($p < .025$). The duration of labor was shortened (nonsignificantly) from 591 to 453 minutes. There were no significant side effects. (HC) 004612

0467

NAKAJIMA, A.; NISHIMURA, T.

The effect of prostaglandin $F_2\alpha$ and E_1 on the electrical activity of the rabbit fallopian tube. Acta Obstetrica et Gynaecologica Japonica 19(1): 40-46. 1972.

The effect of $PGF_2\alpha$ and PGE_1 on the electrical activity of the rabbit fallopian tube was investigated with the use of an intracellular recording technique. Female rabbits were pretreated with 50 μg of estradiol daily for 4 days, then 5 mg of progesterone for 5 days. The fallopian tube was removed and mounted in saline without stretch. Crystalline PGs were dissolved in 50% ethanol and added to the bath to maintain necessary concentrations. In the ampulla a plateau-type action potential was recorded, while in the isthmus bursts of spike discharge were obtained. The membrane potential was the same at both sites. $PGF_2\alpha$ exhibited an excitatory effect, enhancing spike activity and depolarizing the membrane potential in both the ampulla and the isthmic portion of the oviduct. PGE_1, on the other hand, inhibited spontaneous contraction of the oviduct and hyperpolarized the membrane potential. Adrenergic blocking agents such as Hydergin, Regitine, and Inderal seemed to have no effect on the response of the fallopian tube to PGs. It was concluded that PGs have an inherent ability to stimulate the cell membrane and do not act through alpha and beta receptors in the membrane. Further study is needed. (MLH) 004085

0468

NAKANO, J.

Prostaglandins: their pathophysiological roles and therapeutic applications. Transactions of the Association of Life Insurance Medical Directors of America 56: 184-206. 1972.

The biochemistry and pharmacology of PGs were discussed. The PGs are 20-carbon, hydroxylated fatty acids and are analogues of a hypothetical compound, prostanoic acid. They are present in almost all tissues and are found in many species of animals. Newly synthesized PGs are rapidly metabolized after exerting their biological actions, rather than stored in the body. They are found to be rapidly inactivated by lung, liver, and kidney tissue, and slightly inactivated in plasma, heart, and brain, due to the presence of a PG dehydrogenase which is differentially active. PG precursors such as dihomo-gamma-linoleic acid and arachidonic acid are stored as a moiety of the phospholipids in the cell membrane; the PGs are released through unknown mechanisms by seemingly non-specific stimuli and factors. Recent studies indicate a close relationship between PGs and cyclic AMP formation or action; in some tissues PGs have been found to inhibit the AMP system thereby reducing the action of various hormones, and others PGs have been found to be stimulatory, causing multiple pharmacological actions. PGs are thus thought to play a variety of physiological roles, both in inducing pathological conditions and regulating normal functions. Efforts are being made to discover and develop PG synthesis inhibitors or PG receptor antagonists to block the pathogenic effects. Because of their hormonal relationships, possible therapeutic uses of PGs have been considered in the induction of abortion and labor, in contraception, in the prevention of platelet aggregation in blood storage, and in the treatment of bronchial asthma. Early clinical trials have revealed side effects and further study is necessary. (MLH)   004801

0469
NALBANDOV, A.V.
Letters to the Editors.
Prostaglandins 1: 181. 1972.

I was interested in all the indomethacin papers, but I was especially happy that 3 of the papers correctly, I am sure, conclude that the prostaglandin synthesis inhibitor works at the ovarian and not at the hypothalamic level. The old work of Ferrando and Nalbandov (Endocrinology 85:38, 1969) has been extended by Kao and Nalbandov (Endocrinology, in press) to show that alpha site blocking drugs and drugs which block catecholamine synthesis or those which deplete it also block ovulation in rabbits and chickens. All of this work was done by the direct injection of the drugs into the lumena of follicles (rabbits) or into the follicle wall (chickens). Here there is no doubt that the effect is local, since neighboring follicles are not affected. This technique has been developed by Jones and Nalbandov (Biology of Reproduction, in press) and would probably be very suitable for the local injection of prostaglandins. Then there could be no doubt that the stuff acts locally and not somewhere else, such as hypothalamic level. All 4 papers on indomethacin suffer from the defect that you can not disentangle local from peripheral effects. I recommend intrafollicular injection very highly. In animals with small follicles (guinea pigs or rats) drugs can be injected into the ovarian stroma and can be shown to diffuse throughout the ovary affecting one but not the untreated ovary. Try it, you'll like it. (Author's modified)   003879

0470
NATHAN, A.H.
Prostaglandins: a status report.
Chemtech, September 1972. p. 540-546.

The history and present status of PG research was discussed. The structure of the first 2 PGs, PGE_1 and $PGF_1\alpha$, were mapped by gas chromatography followed by mass spectroscopic analysis of the fractions. Other PG structures have since been elucidated as have their differences in activity and their metabolic pathways. It was found that PGs could be formed biologically from the naturally occurring 20-carbon unsaturated fatty acids, gamma homolinolenic and arachidonic acids and this reaction served as the basis for an important preparative measure. It was found later that PGs are produced and stored in Plexaura homomalla, a sea whip which belongs to a group of corals. The PGs stored in this animal account for up to 1.5% of its weight; this became another major source. Analogues of PGs have been developed with enhanced activities, affecting blood pressure and muscle contraction among other physical properties. PGs have been found capable of inducing abortion and labor by stimulating contractions of the uterus. Other possible reproductive applications are in fertility control and in the induction of menstruation following missed periods. PGs may be effective in treating asthma and ulcers; they have been found also to have a direct relationship with cyclic AMP. It has been suggested that $PGF_2\alpha$ in combination with certain hormones may cause reversion of cancer cells to normal cells and that PGs may have a relationship with the substances involved in the transmission of nerve impluses. (MLH)   004755

0471

NATHANIELSZ, P.W.; ABEL, M.; SMITH, G.W.

Initiation of parturition in the rabbit by intra-aortic infusion of prostaglandin $F_2\alpha$.

Journal of Endocrinology 55: 617-618. 1972.

$PGF_2\alpha$ was infused into the aorta of rabbits at rates of 1.125 ng to 75 μg per hour from Day 21 until delivery. At all dosages delivery time was reduced. The authors hypothesize that $PGF_2\alpha$ first causes luteolysis, then is oxytocic by directly stimulating the myometrium. (HC)   004835

0472

NISWENDER, G.D.; MENON, K.M.J.; JAFFE, R.B.

Regulation of the corpus luteum during the menstrual cycle and early pregnancy.

Fertility and Sterility 23: 432-442. 1972.

In this review of factors regulating the corpus luteum in humans the authors briefly state some of the evidence that $PGF_2\alpha$ may be the luteolytic factor. (JRH)   003988

0473

NOVY, M.J.

Distribution of ovarian blood flow in rabbits as measured by radioactive microspheres.

In: Abstracts, 5th Annual Meeting of the Society for the Study of Reproduction, East Lansing, Michigan, June 26-29, 1972. p. 24-25.

Quantitative measurements of the distribution of blood flow to the corpus luteum and the interstitial tissue of the ovary have not previously been reported. We measured regional blood flow to the ovaries, uterus, and other organs in unanesthetized estrous and pseudopregnant rabbits (Days 7-9) by nuclide-labeled microspheres and systemic blood flow by dilution of indocyanine green dye. The mean blood flow to the ovary in estrous rabbits was 2.0 ± 1.4 (SD) ml/min/gm (n = 11). Ovarian blood flow increased to 10.4 ± 8.2 (SD) ml/min/gm (n=6, p<.001) during the luteal phase. A positive regression of corpus luteum weight on ovarian blood flow was noted. The partition of radioactive microspheres indicated that greater than 75% of total ovarian blood flow is distributed to the corpora lutea in pseudopregnant rabbits. The effect of intravenous injections of $PGF_2\alpha$ (150 μg/kg) on ovarian and regional blood flows was studied. In estrous rabbits, an initial increase in the fraction of systemic blood flow to the ovaries was followed by a decline toward control levels 30 minutes after injection. Changes in total ovarian blood flow and corpus luteum blood flow were unpredictable in pseudopregnant rabbits. Ovarian interstitial blood flow, however, increased following the injection of $PGF_2\alpha$ in pseudopregnant rabbits. A significant increase in blood flow to the uterus, large intestine, and kidneys was noted in estrous and pseudopregnant rabbits after injection of $PGF_2\alpha$ (p<.01). It is concluded that blood flow to the corpus luteum and the interstitial tissue of the ovary is regulated independently during the estrous cycle. (Author)   004055

0474

NUTLER, D.O.

Prostaglandins.

Journal of the Medical Association of Georgia 61: 112-113. March 1972.

The author mentioned PG nomenclature and potential areas of clinical use. The physiologic action of PGs may be in their ability to act as local metabolic and vascular regulator substances. Recent investigations into calcium kinetics and cAMP suggests PGs may act to modulate or control hormonal messages across the cell membrane. Clinically PG offers an attractive method for pregnancy termination in second trimester and may be developed for use in antihypertensive therapy. (RAS) 004615

0475
O'GRADY, J.P.; KOHORN, E.I.; GLASS, R.H.; CALDWELL, B.V.; BROCK, W.A.; SPEROFF, L.
Inhibition of progesterone synthesis in vitro by prostaglandin $F_2\alpha$.
Journal of Reproduction and Fertility 30: 153-156. 1972.

A technique is described for measuring the effect of $PGF_2\alpha$ on the synthesis of progestin by the rabbit corpus luteum in vitro. The ovaries were removed on the 10th day after mating. Each corpus luteum was dissected out, cut in half, and placed in a modified organ bath apparatus. The solution to be tested was added to the incubation medium. After a 6-hour incubation period, the amount of progestin in the tissue and incubation medium was measured by the radioimmunoassay method. It was found that 10 μg/ml of $PGF_2\alpha$ added to the incubation medium caused about a 50% reduction in progestin production during the incubation period. (JRH)   004127

0476
O'GRADY, J.P.; CALDWELL, B.V.; AULETTA, F.J.; SPEROFF, L.
The effects of an inhibitor of prostaglandin synthesis (indomethacin) on ovulation, pregnancy, and pseudopregnancy in the rabbit.
Prostaglandins 1: 97-106. 1972.

Ovulation, induced by mating or with exogenous gonadotropins, was blocked by indomethacin given during a critical time period. A single concomitant injection of 20 mg/kg was sufficient to block LH-induced ovulation. Additional treatment (8 mg/kg every 12 hours for 48 hours) was necessary if HCG were administered instead of LH, probably reflecting the longer half-life of HCG. Coitus-induced ovulation was not affected if indomethacin (20 mg/kg) was given at the time of mating, but a single sc injection of 8 mg/kg given 8 hours after mating inhibited ovulation. The block was presumably at the ovarian level, since in all cases where indomethacin was effective, the ovaries were characterized by large hemorrhagic follicles.in which the ova were retained. Considerable luteinization of the follicles was observed in association with normal progesterone levels, suggesting that indomethacin affected only the physical process of ovulation. Indomethacin treatment was also found to lengthen pseudopregnancy to the same extent as produced by hysterectomy, and when administered from the day of mating indomethacin was associated with fetal resorption in pregnant animals. (Authors' modified)   003837

0477
O'SULLIVAN, D.A.
Prostaglandin publication activity soars.
Chemical and Engineering News 50: 23. October 23, 1972.

The recent surge in prostaglandin research has prompted the need for this data to be communicated quickly and distributed worldwide. The Worcester Foundation for Experimental Biology in collaboration with George Washington University recently published a volume of 99 abstracts concerning prostaglandins in fertility control. Shortly, the University plans to publish a bibliography and index of the world literature. The journal 'Prostaglandins,' originated by members of Yale's Department of Obstetrics and Gynecology, has begun publication. The journal will publish recent research articles and allow for opinions, criticisms, and announcements. In Paris 'Opportunities in Prostaglandins,' printed by Dynachim, covers the chemical synthesis and properties of prostaglandins. (RAS) 004486

0478

OKAMURA, H.; YANG, S.L.; WRIGHT, K.H.; WALLACH, E.E.
    The effect of prostaglandin $F_2\alpha$ on the corpus luteum of the pregnant rat. An ultrastructural study.
    Fertility and Sterility 23: 475-483. 1972.

Pregnant rats were treated with $PGF_2\alpha$ and their corpora lutea were investigated morphologically. Pregnancy was terminated between Days 7 and 11 in all 8 animals receiving 1 mg intramuscularly for 2-4 consecutive days. Lutein cells from rats treated with $PGF_2\alpha$ showed massive accumulation of lipids. Smooth endoplasmic reticulum decreased and free ribosomes and rough endoplasmic reticulum increased. Mitochondrial changes were not significant. The luteolytic effect of $PGF_2\alpha$ was confirmed ultrastructurally. Its mode of action in luteolysis is discussed and compared to morphologic changes seen in natural corpus luteum regression. (Authors' modified)   004115

0479

ORCZYK, G.P.; BEHRMAN, H.R.
    Ovulation blockage by aspirin or indomethacin-in vivo evidence for a role of prostaglandin in gonadotrophin secretion.
    Prostaglandins 1: 3-20. 1972.

Inhibitors of prostaglandin (PG) synthesis, aspirin and indomethacin, were used to assess the role of PG in gonadotrophin secretion. Both drugs reduced plasma PGF content, and indomethacin reduced pituitary and hypothalamic concentration of PGF measured by a radioimmunoassay procedure reported herein. Chronic and acute administration of either indomethacin or aspirin blocked ovulation but indomethacin was effective at 1/30 the dose of aspirin. Injection of either LH or a mixture of PGE_2 and $PGF_2\alpha$ at the time of the expected ovulatory surge of LH was effective in reversing the blockage of ovulation that occurred after a single injection of indomethacin administered about 3 hours before the expected ovulatory surge of LH. However, LH did not reverse the blockage of ovulation produced when indomethacin was administered chronically beginning 30 hours before the expected ovulatory LH surge. These data support the hypothesis that prostaglandins play a functional role in regulating the release of LH necessary for ovulation in the rat. (Authors)   003743

0480

OSHIMA, K.; SASADA, M.; MATSUMOTO, K.; ISHII, K.; MATSUBAYASHI, K.
    Absorption of orally-administered prostaglandin E_2 and uterine contractility and body temperature in monkeys.
    Hormones 3: 278. 1972.

PGE_2 was administered orally to 3 types of monkeys at a dose of 1 mg/kg body weight. Increases in uterine contraction, body temperature, and concentration of PGE_2 in blood were noted. Blood concentration was highest 60-90 minutes after treatment, followed by a gradual decline. (RAS) 004210

160

0481
OSTER, G.
   Conception and contraception.
   Natural History 81: 46-53, 76-77. 1972.

   The reproductive physiology of women and current and experimental contraceptive methods are
   included in this feature article directed to the popular audience. Prostaglandins are mentioned as an
   early abortifacient, with hopeful anticipation that an analogue will be developed to be taken orally or
   intravaginally for missed menstruation. (LJG)   004832

0482
PACE-ASCIAK, C.; WOLFE, L.S.; GILLETT, P.G.; KINCH, R.A.
   Disappearance of prostaglandin $F_2\alpha$ from human amniotic fluid after intraamniotic injection.
   Prostaglandins 1: 469-477. 1972.

   Levels of prostaglandin $F_2\alpha$ in the amniotic fluid from 5 women at various times after single
   intraamniotic administration of 25 mg $PGF_2\alpha$ for induction of abortion in the second trimester are
   reported. A slow disappearance of $PGF_2\alpha$ was observed, with amniotic fluid levels of 13%-49% of the
   original dose still present 12-18 hours after administration of the drug. A metabolite identified as
   15-keto-13,14-dihydro-$PGF_2\alpha$ by sodium borodeuteride reduction and mass spectrometry appeared at
   8-12 hours. No obvious correlation could be found between $PGF_2\alpha$ levels in the amniotic fluid and the
   induction-abortion interval, gestational age, or fetal weight. The metabolite is likely of fetal or placental
   origin. (Authors)   004082

0483
PALOMAKI, J.F.; LITTLE, A.B.
   Surgical management of abortion.
   New England Journal of Medicine 287: 752-754. October 12, 1972.

   Current concepts of abortion are discussed including patient preparation and the techniques of
   dilatation and curettage, hypertonic solutions, hysterotomy and hysterectomy. The side effects
   associated with the use of $PGF_2\alpha$ by the oral, intravaginal or intravenous route is considered by the
   author to have delayed clinical application. Intrauterine instillation, however, may be successful and
   have fewer side effects. (HC)   004554

0484
PARK, M.K.; DYER, D.C.
   Effect of prostaglandin antagonists on isolated human umbilical arteries.
   Federation Proceedings 31: 546. 1972.

   3 compounds known to block the effects of prostaglandins were studied on isolated human umbilical
   arteries. The arteries were suspended in a series of muscle baths containing modified Krebs solution,
   maintained at 37 degrees C. and aerated with 95% oxygen - 5% carbon dioxide. All muscles were
   placed under 1 g tension and responses were recorded isotonically. 1-acetyl-2-(8-chloro-10, 11-
   dihydrodibenz)(b,f)(1,4) oxazepine-10-carbonyl) hydrazine (SC-19220) at a concentration of .1 mM
   shifted the dose response curve (DRC) to PGE_2 to the right 3-fold and the DRC to 5-hydroxytryptamine
   (5-HT) 1.8-fold to the right at the ED50 to 5-HT. Polyphloretin phosphate (PPP) at a concentration of
   100 μg/ml shifted the DRC to PGE_2 to the right 9-fold and the DRC to 5-HT to the right 3.5-fold. 7-oxa-
   13-prostynoic acid (EC-I-148) exhibited a strong intrinsic activity at a concentration of 100 ng/ml and

higher. A concentration of EC-I-148 (10 ng/ml) which did not produce contractions also did not block responses produced to PGE_2 or 5-HT. Propylene glycol, the solvent for SC-19220, was observed to antagonize contractions to both PGE_2 and 5-HT thereby indicating that solvent controls are necessary when evaluating compounds such as SC-19220. We conclude that both PPP and SC-19220 are not specific antagonists of PGE_2. (Authors)   003760

0485

PARK, M.K.; RISHOR, C.; DYER, D.C.
  Vasoactive actions of prostaglandins and serotonin on isolated human umbilical arteries and veins.
  Canadian Journal of Physiology and Pharmacology 50: 393-399. 1972

Cumulative responses to prostaglandins $E_1(PGE_1)$, $E_2(PGE_2)$, $F_1\alpha$ ($PGF_1\alpha$) and $F_2\alpha$ ($PGF_2\alpha$) were obtained on isolated human umbilical arteries and veins. All 4 prostaglandins produced contractions. $PGF_2\alpha$ was the most active prostaglandin on umbilical arteries, while PGE_2 and $PGF_2\alpha$ were equiactive and more potent than PGE_1 or $PGF_1\alpha$ on umbilical veins. 5-hydroxytryptamine (5-HT) was at least 100 times more potent than the prostaglandins. SC-19220, a prostaglandin antagonist, in a high concentration was found to moderately antagonize the vasoactive effect of PGE(2) and to slightly antagonize contractions to 5-HT. Also propylene glycol, the solvent for SC-19220, was observed to antagonize contractions to both PGE_2 and 5-HT, thereby indicating that solvent controls are necessary when evaluating compounds such as SC-19220. (Authors' modified)   003996

0486

PEDERSEN, P.H.; LARSEN, J.F.; SORENSEN, G.
  Induction of labour with prostaglandin $F_2\alpha$ in missed abortion, fetus mortuus, and anencephalia.
  Prostaglandins 2(2): 135-141. 1972.

7 cases of missed abortions, 6 cases of intrauterine death, and 4 with anencephalic fetuses all had labor sucessfully induced with intravenous $PGF_2\alpha$, mean dosage 42 mg, and 36 mg, and 53 mg respectively. Side effects of nausea and diarrhea were mild. The induction-termination ranged from 3 to 12 1/2 hours. (RAS)   004220

0487

PHARRISS, B.B.; TILLSON, S.A.; ERICKSON, R.R.
  Prostaglandins in luteal function.
  Recent Progress in Hormone Research 28: 51-89. 1972.

A review of recent investigations supports $PGF_2\alpha$ as a uterine luteolytic factor. Different mechanisms for the action were hypothesized. Prostaglandins seem interrelated to the action of LH stimulation of luteal steroidogenesis. Additional experiments on the role of prostaglandin in the ovary are presented. (RAS)   004237

0488

PIKE, J.E.
  Recent advances in prostaglandin research.
  Abstracts of paper presented at the 13th National Meeting of the APhA Academy of Pharmaceutical
    Sciences 2(2): 7. 1972.

Clinical researchers in prostaglandins are now investigating renal-cardiovascular related problems, induction of labor, therapeutic abortion, fertility control, bronchodilation and gastric antisecretory activity. The 6 primary prostaglandins are the PGE and PGF structures of the 1, 2, and 3 series. (RAS)
005097

0489
PION, R.J.; WABREK, A.J.; WILSON, W.B.
Innovative methods in the prevention of the need for abortion.
In: Lewit, S., Ed. Abortion techniques and services. Amsterdam, Excerpta Medica, 1972. p.14-16.

All of the current abortion techniques are used at a limited gestational age, but notably after a positive diagnosis of pregnancy. It is feasible to test clinically several modes of terminating a possible pregnancy within 2 weeks of missed menstruation, such modes as electrical, ultrasonic, mechanical, thermal, or chemical stimulation of the endometrium. The only chemical method commonly mentioned is PG, but it is not known whether PG can induce menstruation, particularly by self-administration. If a method of inducing menstruation is found, it should be well publicized, perhaps as a means of preventing the need for abortion. (LJG)   005019

0490
PION, R.J.; HALE, R.W.; REICH, L.
Vaginal administration of prostaglandins and early abortion.
In: Southern, E.M., ed. The prostaglandins: clinical applications in human reproduction. (Brook Lodge Symposium, Augusta, Michigan.) Mount Kisco, New York, Futura Publishing Co., 1972. p. 367-372.

10 women 7 to 8 weeks pregnant received hourly 50 mg doses of $PGF_2\alpha$ intravaginally. Another 8 women 10 to 17 days pregnant were given 25 mg doses at 4 hour intervals. In the first group the total dose given was 600 mg for most women; all had vaginal bleeding and a dilated cervix but only 4 aborted. In the second group the total dose varied from 100 to 400 mg; all women had a dilated cervix, only 2 aborted. Side effects, such as vomiting and diarrhea, appeared fleetingly. The authors conclude that PG is inferior to suction curettage for first trimester abortion. (HC)   004874

0491
PORTER, J.F.; SHENNAN, A.T.; SMITH, S.
Plasma kinin and kininogen levels in women during pregnancy and in labour.
Journal of Reproduction and Fertility 30: 247-354. 1972.

The introduction to this article mentions that the increase in the oxytocic activity of maternal blood during labor cannot be attributed to oxytocin, but is likely due, in part, to prostaglandins. (RAS)
004117

0492
PORTOGHESE, P.S.
The once-a-month birth control pill of the future.
Minnesota Pharmacist, March 1972. p.8-9.

The author traces the history of prostaglandin research from the 1930's to the present interest in a self-administered menstrual inducer. The author mentions the early work of Karim for induction of labor and abortions. Additional research into the various routes of administration, and dosages are required to fully evaluate potential clinical use of prostaglandins. (RAS)   004559

0493
POYSER, N.L.
Production of prostaglandins by the guinea-pig uterus.
Journal of Endocrinology 54: 147-159. 1972.

Guinea pig uteri were obtained on selected days of the estrous cycle and were either homogenized in ethanol, homogenized in Tyrode's solution and then incubated, or homogenized in Krebs' solution containing indomethacin and then incubated. All homogenates and incubates were subjected to column chromatography, and the fractions obtained were bioassayed on the rat fundus. PGs were identified using gas chromatography and mass spectrometry. PG level in the guinea pig uterus was significant only on Day 14 of the estrous cycle. Incubation increased PG content on all days of the cycle, highest levels being attained on Days 14 and 15. $PGF_2\alpha$ content was 5-10 times higher than PGE_2. Incubation in the presence of indomethacin reduced PG levels by more than 50%. The authors concluded that their data support the view that the uterine luteolytic hormone in the guinea pig is $PGF_2\alpha$. (HC)   004834

0494
PUCK, T.T.; WALDREN, C.A.; HSIE, A.W.
Membrane dynamics and the action of debutyryl adenosine 3':5'-cyclic monophosphate and testosterone on mammalian cells.
Proceedings of the National Academy of Sciences (U.S.A.) 69(7): 1943-1947. July 1972.

The process termed 'reverse transformation' by which epitheloid cells in culture take on a fibroblast form when cultured with dibutyryl cAMP and testosterone or PGs was observed by time-lapse cinematography. Epithelial cells (S3 clones of HeLa cells or CHO-K1 Chinese hamster ovary cells) appeared under time-lapse as compact cells actively extruding about 10-40 knobs, each extension and retraction taking about 15 seconds. The first effect of .2mM dibutyryl cAMP and 15 μM testosterone was total cessation of the knob extension within 15 minutes, with only a slow ruffling of the membrane. By 5-8 hours, cells had elongated into fibroblast form, extruding knobs for 1 hour after mitosis only. Permanent fibroblasts (V79 Chinese hamster cells) were transformed from 8% to 85% epithelioid cells by adding 2-13 μM cytochoasin B or .27 μM cholcemid; this transformation could be prevented or reversed by 1 mM dibutyryl cAMP plus 15 μM testosterone. It was proposed that these processes are general and reversible although the fibroblast and knobbed epithelial are the most organized form of the microtubular microfibrillar system. (LJG)   004963

0495
PULKKINEN, M.O.
Induction of labour. Recent developments.
Annales Chirurgiae et Gynaecologiae Fenniae 61: 47-51. 1972.

Recent developments in the field of regulation of uterine contractility and induction of labor are reviewed. Estrogens, hypertonic solutions, oxytocin, and amniotomy are discussed. The author states that PG in first trimester has a 95% success rate but side effects occur. Use of PG in second trimester is not considered promising. The iv route has side effects; extraovular administration has not been perfected. For abortions during second trimester, 10 mg initially followed by 1 or 2 doses of 5 mg $PGF_2\alpha$ by the intraamniotic route is best. For induction of labor between 34 and 38 weeks, $PGF_2\alpha$ at an

164

infusion rate of 5 μg/minute may be more effective than oxytocin. Another method for inducing uterine contractions is volume increase (stretch), which, among other effects, produces an increase in PG synthesis. (HC)   004541

0496
RAO, V.S.N.; SHARMA, P.L.
Potentiating effect of d-INPEA on prostaglandin (PGF$_2\alpha$ and PGE$_2$)-evoked contractions of the isolated rat uterus.
European Journal of Pharmacology 20: 363-365. 1972.

PGF$_2\alpha$ and PGE$_2$-induced contractions of isolated rat uterus in an organ bath were potentiated 4 and 10 times respectively with the addition of d-INPEA (1 x 10^{-5}g/ml). The results were consistent with all experiments; however, an increase in PGE$_2$ concentration was needed to reproduce initial contractions. The authors suggested that if in in vivo studies the potentiating action of d-INPEA continues, clinicians may be able to use a smaller dose of prostaglandin for induction of labor or abortion, and thereby reduce side effects. (RAS)   004696

0497
RATNAM, S.S.; CHENG, M.C.E.; NG, A.
Place of prostaglandins in termination of pregnancy.
In [Abstracts of] First International Planned Parenthood Federation South-East Asia and Oceania Regional Medical and Scientific Congress, Sydney, Family Planning Association of Australia, August 1972. p. 29.

The authors review the history of the use of prostaglandins for the termination of pregnancy. Their view of prostaglandins for this purpose is generally unfavorable, however, PGE$_2$ is more effective than PGF$_2\alpha$. (JRH)   004177

0498
RATNER, A.; PEAKE, G.T.
The role of a prostaglandin (PG) receptor site in pituitary growth hormone (GH) and prolactin (PL) release.
In: Abstracts, Fourth International Congress of Endocrinology, Washington, D.C., June 18-24, 1972. Amsterdam, Excerpta Medica, 1972. (International Congress Series No. 256) Abstract 476.

PGE$_1$, PGE$_2$, and PGA at .1 mM, theophylline, or dibutyryl cyclic AMP had no effect on PL release from incubated pituitary explants. The PGs did increase GH release by 2-3 times and cyclic AMP by 3-15 times. At 1 μM PGE$_1$ increased GH release but decreased PL release. 7-oxa-13-prostynoic acid inhibited GH release stimulated by PGE$_1$. Added alone, prostynoic acid inhibited GH and PL release, but failed to inhibit the two- to threefold increase in GH release found following 6.7 mM theophylline or 20 mM dibutyryl cyclic AMP. (RAS)   004713

0499
RAZ, A.
Binding of prostaglandin E$_2$(PGE$_2$) to human plasma albumin and its effect on the physiological activity of PGE$_2$ in vitro and in vivo.
Israel Journal of Chemistry 9: 38BC. 1972.

PGE$_2$added to human plasma in vitro became bound to albumin. Albumin-bound PGE$_2$did not cause contraction of the gerbil colon in vitro but did cause a lowering of blood pressure when given intravenously or intraarterially to rats. (HC)   005059

0500
RING, A.
Clinical experience with prostaglandin F$_2\alpha$ for induction of labor.
In: Southern, E.M., ed. The prostaglandins: clinical applications in human reproduction. [Brook Lodge Symposium, Augusta, Michigan.] Mount Kisco, New York, Futura Publishing Co., 1972. p. 129-133.

Labor was successfully induced in 38 of 42 women given PGF$_2\alpha$ by intravenous infusion. The rate was 2.5 μg per minute increased to 5 μg per minute after 30 minutes and then to 10 μg per minute if necessary. The average IDT was 5 1/2 hours and the total dose averaged 2.5 mg. Side effects included nausea, vomiting, diarrhea, and phlebitis. 10 to 15 hours after PGF$_2\alpha$ administration was terminated, there were significant increases in leucocyte count, SGOT, SGPT and blood urea. The fetuses delivered were normal in terms of Apgar scores, pH values, and cardiology. Labor was also induced in another 13 patients with intrauterine fetal deaths between weeks 28 and 34. The dosage of PGF$_2\alpha$ was 25 mg and the IAT 7 hours. Side effects were more frequent and more intense. The marked difference in PGF$_2\alpha$ dosage in the 2 studies might be due to either the presence of dead rather than live fetuses or gestational age. (HC)   004856

0501
RISLEY, P.L.; STAHL, P.
Effects of prostaglandins on hamster seminal vesicle responses to catecholamines and acetylcholine.
Biology of Reproduction 6: 224-233. 1972.

Effects of prostaglandins E$_1$, E$_2$, A$_1$, F$_1\alpha$ on contractile responses of hamster seminal vesicle smooth muscles to stimulation by epinephrine, norepinephrine, and acetylcholine are compared. Hamster seminal vesicles provide suitable preparations for comparisons of physiological actions of these pharmacological agonists. Pairs of seminal vesicles from a single animal respond synchronously in the muscle bath preparation, and were used in these tests. Potentiation of the responses to these stimulants occurred in the presence of all the prostaglandins tested. The relative effectiveness of the potentiation was estimated on a μg/ml dose basis to approximate a 1:2:5:10 ratio for E$_1$, E$_2$, A$_1$, and F$_1\alpha$ respectively, for epinephrine and norepinephrine. PGE$_2$was about as effective as PGE$_1$in potentiating responses to acetylcholine stimulation. These results approximate and confirm those reported by Naimzada (1969) for potentiating actions of these PGs on the responses of guinea pig seminal vesicles to epinephrine and norepinephrine or to nerve stimulation. Further studies are in progress relating to the actions of prostaglandins in the regulation of neuromuscular function in male reproductive organs. (Authors' modified)   003905

0502
ROBERTS, C.D.
Aspirin and human seminal prostaglandins.
Lancet 1: 1070. 1972.

The author uses different statistical methods to analyze the data of Collier and Flower on the effects of aspirin on PGE and PGF content of human seminal fluid. The authors of the original paper stated that the seminal fluid volume was normal in both experimental and control groups. However, as a result of his reanalysis of the data, the author feels that the best explanation of the results he obtained would be that there is a significant difference in volume. He also found a significant difference in the amount of prostaglandins present in the 2 time periods measured both in experimental and control groups. He suggests further clinical tests in which the total amount of prostaglandin is measured rather than the concentration and that the action of the time period on treatment be considered in the statistical model. (JRH)   003915

0503

ROBERTS, G.; MOTTRAM, R.F.; PARRY, H.; BLOOM, A.
Cyanosis due to intravenous prostaglandin $F_2\alpha$.
Lancet 2: 425-426. 1972.

The authors report the development of cyanosis in the forearm of a patient being infused intravenously with $PGF_2\alpha$ (5 μg/min) for therapeutic abortion. The infusion cannula was moved to the other arm and the infusion resumed at 40 μg/min. The cyanosis appeared in this arm in 1 minute. The symptoms disappeared 15 minutes after termination of the infusion. Further investigations with this patient 2 months later showed a similar cyanosis upon infusion of $PGF_2\alpha$ (5-40 μg/min). Platelet aggregation studies with thrombin, collagen, and ADP revealed no differences in platelet behavior before or after infusion. This patient had a history of rubella arthropathy 3 years prior to these experiments. The authors suggest that $PGF_2\alpha$ should be contraindicated in patients with a history of peripheral vascular disease. (JRH)   004138

0504

ROBERTS, G.; GOMERSALL, R.; ADAMS, M.; TURNBULL, A.C.
Therapeutic abortion by intra-amniotic injection of prostaglandins.
British Medical Journal 4: 12-14. 1972.

27 patients, premedicated with 10 mg of morphine and 5 mg perphenazine, 14 to 22 weeks pregnant, were given a single intraamniotic injection of 25 mg $PGF_2\alpha$ (5 mg/ml) or 3 1.0-mg doses of PGE_2at 10-minute intervals. Only 6 patients given $PGF_2\alpha$ alone aborted, only 2 were complete; the IAT averaged 19 1/2 hours. Another 6 patients were given supplemental intravenous oxytocin infusions after 24 to 48 hours with an eventual average IAT of 59 hours; 3 of these were complete. 2 patients failed to abort. With PGE_2alone 11 patients aborted, 8 completely, the IAT averaging 18 1/2 hours. The other 2 patients were given supplemental oxytocin and aborted in 48 hours; 1 was complete. Severe abdominal pain was noted by all patients, 4 patients vomited, most had mild pyrexia. Uterine activity was also monitored. Initially, contractions were very frequent and of small amplitude, but as abortion proceeded, frequency decreased and intensity increased to between 88 and 110 mm Hg. (HC) 004550

0505

ROTHWELL, R.O.
Response of rabbit myometrium to prostaglandin $F_2\alpha$, monitoring spontaneous and electrical field stimulation activity.
Texas Reports on Biology and Medicine 30: 232-233. 1972.

The estrogen-dominated rabbit uterus in vitro was more sensitive to stimulation by $PGF_2\alpha$ than the progesterone-dominated uterus. The stimulatory effect of $PGF_2\alpha$ was antagonized by electrical field stimulation. The author suggests that the effect of $PGF_2\alpha$ may be dependent upon the presence of some endogenous substance, such as noradrenaline, which is depleted by electrical field stimulation. (HC)   004515

0506
ROTHWELL, R.O.
Response of rabbit myometrium in vitro to prostaglandin $F_2\alpha$, monitoring spontaneous and electrical field stimulation activity.
Texas Medicine 68: 68-74. September 1972.

The effects of $PGF_2\alpha$ on the response of rabbit myometrium under differing hormonal conditions were studied. Part of the uterine horn was removed from virgin female New Zealand white rabbits. The tissues were impaled and suspended in saline. One group was allowed to contract spontaneously and the second was stimulated with an electric field. In each experimental situation PG and oxytocin were added separately to investigate responses. It was found that estrogen-dominated tissues had more inherent spontaneous activity initially and were more responsive to electrical stimulation than were progesterone-dominated tissues. Mechanisms were not understood. With time the estrogen group lost activity when electrically stimulated to a greater extent than did the progesterone group. When $PGF_2\alpha$ was added to the saline solution the estrogen group was more responsive and the same was true when oxytocin was added, with and without electrical stimulation. It was suggested that progesterone may depress the muscle's ability to respond. Finally, electrical field stimulation abolished the responsiveness of all tissues to $PGF_2\alpha$ and decreased the effects of oxytocin. The effect of PG may be mediated through the release and utilization of norepinephrine. (MLH)   004960

0507
ROWSON, L.E.A.; TERVIT, R.; BRAND, A.
The use of prostaglandins for synchronization of oestrus in cattle.
Journal of Reproduction and Fertility 29: 145. 1972.

Prostaglandin $F_2\alpha$ injected nonsurgically into the ipsilateral uterine horn of cattle induced regression of the corpus luteum in almost every case when administered at a dose level of .5 mg/day on 2 consecutive days of the cycle, between Days 5 and 16. Synchronization of estrus was very exact, most animals showing heat on the morning of the 3rd day after treatment. The same treatment from Days 1 to 4 of the cycle was ineffective. A single dose of .5 mg gave intermediate results. The fertility of eggs recovered from cows given PMSG 1 day before the prostaglandin treatment was normal, and following transfer of these eggs pregnancy rates were normal, even when they were transferred to recipients which had themselves been synchronized by prostaglandin administration. The intrauterine administration of prostaglandin $F_2\alpha$ would appear to be an extremely efficient method of synchronizing estrus in the cow, and has the advantage that normal fertility occurs at the immediate post-treatment estrus in contrast to the progestagen method. (Authors' modified)   003919

0508
SAKSENA, S.K.; HARPER, M.J.K.
Level of F prostaglandin (PGF) in peripheral plasma and uterine tissue of cyclic rats.
Prostaglandins 2(6): 511-517. December 1972.

Levels of PGF in peripheral plasma and uterine tissue of rats throughout the estrous cycle were determined. From a mean of 2.6 ng per ml on Day 1 (estrous), plasma levels increased significantly to 4 to 5 ng per ml on successive days. Uterine concentration averaged 149 to 185 ng per gm tissue for

Days 1 through 3 of the cycle but dropped to 85 ng per ml on Day 4. These data suggest that ovarian steroid levels play an important role in regulating prostaglandin synthesis and release. (HC)   004590

0509
SAKSENA, S.K.; HARPER, M.J.K.
   Levels of F prostaglandin (PGF) in uterine tissue during the estrous cycle of hamster: effects of
      estradiol and progesterone.
   Prostaglandins 2(5): 405-411. November 1972.

Hamsters were sacrificed on different days of the estrous cycle and the prostaglandin F ($PGF_2\alpha$) content of their uteri measured. Some animals were treated with estradiol- 17β or progesterone for 2 days prior to sacrifice. Mean uterine concentrations of PGF were 6.8, 16.3, 9.6, and 7.6 to 19.6 ng per gm tissue on Days 1 through 4, respectively. Estradiol or progesterone markedly increased $PGF_2\alpha$ concentration. The levels of $PGF_2\alpha$ in uterine tissue do not correlate with levels in uterine vein blood previously reported by these authors. They suggest that the blood levels more accurately monitor physiological changes and that differences in uterine content of $PGF_2\alpha$ may be simply a reflection of uterine weight changes. (HC)   004594

0510
SAMUELSSON, B.
   Endogenous synthesis of prostaglandins.
   In: Bergstrom, S., Green, K., and Samuelsson, B., eds. [Proceedings of the] third conference on
      prostaglandins in fertility control, Stockholm, January 17-20, 1972. Stockholm, WHO Research
      and Training Centre on Human Reproduction, Karolinska Institutet, [1972]. (Prostaglandins in
      Fertility Control 2.) p. 1-17.

Major urinary metabolites of PGE_1 and PGE_2 were traced with labeled PGE and intermediaries in man, rat, and guinea pig, in order to estimate total PGE synthesis. The main metabolic pathway in man was C15 hydroxylation, reduction of the C13 double bond, usually 2 steps of beta oxidation, or sometimes omega oxidation, resulting in 7alpha-hydroxy-5,11-diketotetranorprosta-1,16-dioic acid. In man 50% of radioactivity from iv 3H-PGF2 appears in urine in 2 hours. Men and women excreted .4-3.8 µg per kg metabolite per 24 hours. Rats produce a tetranor-PGE_1 and tetranor-PGB_1 by doing beta oxidation first and omitting the 15-dehydrogenases step, in addition to the hydroxy-diketo-tetranorprostadioic acid found in human urine. Total tetranor-PGE_1 ranged from 2.04-3.43 µg per kg per 24 hours in rats, and increased 2-2.5 fold after acute cold stress. Guinea pigs excreted mainly 5beta,7alpha-dihydroxy-11-ketotetranor-prostanoic acid, by following the same series of steps as do humans but finally reducing the C6 keto group. 5-10 µg per kg was produced in 24 hours, which could be reduced 98% by feeding 50 mg indomethacin daily for 3 days. (LJG)   004458

0511
SATOH, K.; RYAN, K.J.
   Prostaglandins and their effects on human placental adenyl cyclase.
   Journal of Clinical Investigation 51: 456-458. 1972.

Prostaglandins increased adenyl cyclase activity in human term placental homogenates in a dose-dependent manner during 10-minute incubation periods. The potency of prostaglandins examined was demonstrated to be, in ascending order, prostaglandin $F_1\alpha < A_2, F_2\alpha, B_2 < A_1 < E_2 < E_1$. Although no specific trophic or regulating factors for placental function have been described as yet, it is possible that prostaglandins which are synthetized in decidual tissue could play such a physiological role. (Authors)   003719

0512

SCHENKEL-HULLIGER, L.; DESAULLES, P.A.

The steroidogenic response of hamster ovaries to prostaglandins and luteinizing hormone, comparison of in vivo and in vitro administration.

Advances in the Biosciences (Suppl.) 9: 107. 1972.

The in vitro steroidogenic effect of prostaglandins on the ovary has been described for different species of rodents. The discrepancy between this 'luteotropic' effect in vitro from the known 'luteolytic' effect of the prostaglandins in vivo has so far not been elucidated. Ovaries from cyclic hamsters were incubated in vitro; progestin and estrogen concentrations were determined in the incubation fluid. Steroid secretion of ovaries from animals pretreated with prostaglandin E_2 or with $F_2\alpha$ or with LH was compared with that of ovaries exposed to the hormones in vitro. Prostaglandins added to incubated ovaries have a stimulating effect: yet the steroid secretion shows a different time response pattern compared with that found after exposure to LH. If administered in vivo 1 to 2 hours before the animal is sacrificed, prostaglandins induce a small and transitory increase in steroid secretion of incubated ovaries. Thus the primary action of prostaglandins on the hamster ovary seems to consist in a stimulation of steroidogenesis whether applied in vivo or in vitro. (Authors' modified) 004317

0513

SCHER, J.; DAVEY, D.A.; BAILLIE, P.; FRIEND, J.; FRIEND, D.M.

Comparison of prostaglandin $F_2\alpha$ and oxytocin in the induction of labour.

South African Medical Journal 46: 2009. December 1972.

43 women 36 to 43 weeks pregnant received intravenously either oxytocin or $PGF_2\alpha$, the latter initially at 5 μg per minute with increments at 30 minute intervals to 40 μg per minute. Labor was induced more often in the women given PG, especially those with a ripe cervix. The times for initial and adequate contractions were comparable, as was the pattern of labor. The authors conclude that $PGF_2\alpha$ intravenously was more effective, than oxytocin for the induction of labor, although the differences were not statistically significant. The authors suggest that hypertonus can be avoided by using 30 minute intervals between dose increments. (HC)   005065

0514

SEELEY, R.R.; HARGROVE, J.L.; JOHNSON, J.M.; ELLIS, L.C.

Modulation of rabbit testicular contractions by prostaglandins, steroids and some pharmacological compounds.

Prostaglandins 2: 33-40. 1972.

Rabbit testes were suspended in Tyrode solution. Spontaneous contractions developed within 1 hour, or they could be induced with PGE_1 (70.0 nM). Progesterone, pregnenolone, and testosterone inhibited both spontaneous and PGE_1 induced contractions. When those steroids were added to the bath before PGE_1 they blocked its stimulation. The authors conclude that prostaglandins may function in modulating the capsular motility in vivo. (JRH)   004170

170

0515
SELLNER, R.G.; WICKERSHAM, E.W.
    Effects of prostaglandins on steroidogenesis by rabbit ovarian tissue in vitro.
    In: Abstracts, 5th Annual Meeting of the Society for the Study of Reproduction, East Lansing,
        Michigan, June 26-29, 1972. p. 28.

Following reports that prostaglandins of the A, E, and F series stimulate steroidogenesis by bovine
luteal tissue in vitro, an investigation was undertaken to determine the effects of prostaglandins E_1, E_2,
and $F_2\alpha$ on progesterone production by rabbit ovarian tissue during in vitro incubation. Sliced
ovarian tissue from 9 sexually mature New Zealand white rabbits was incubated with each of the 3
prostaglandins at a concentration of 1.0 μg/ml KRB. Each of the 3 prostaglandins was used in each of
the 9 incubations. After 2 hour incubations, tissue samples were analyzed for progesterone content by
a method involving homogenization with NaOH, extraction with ethyl acetate, purification by thin-layer
chromatography, and quantitation by gas-liquid chromatography. Significant increases in progesterone production over incubated control values were found for $PGE_1(p<.05)$, $PGE_2(p<.02)$ and $PGF_2\alpha$
$(p<.01)$. (Authors)   004048

0516
SELLNER, R.G.; WICKERSHAM, E.W.
    Effects of prostaglandins on steroidogenesis by rabbit ovarian tissue in vitro.
    Biology of Reproduction 7: 107-108. 1972.

Ovarian tissue from rabbits was incubated with 1.0 μg/ml PGE_1, PGE_2, and $PGF_2\alpha$ for 2 hours.
Significant increases in progesterone production were found for each prostaglandin. (HC)   004265

0517
SEPPALA, M.; KAJANOJA, P.; WIDHOLM, O.; VARA, P.
    Prostaglandin-oxytocin abortion: a clinical trial on intra-amniotic prostaglandin $F_2\alpha$ in combination
        with intravenous oxytocin.
    Prostaglandins 2(4): 311-319. October 1972.

Women pregnant 14-20 weeks were given either 25 mg $PGF_2\alpha$ intraamniotically and a second injection
24 hours later if needed, or the above and an intravenous infusion of oxytocin starting 3 hours after
the first PG injection. 18 of 21 women given only $PGF_2\alpha$ aborted, 9 completely. The mean IAT was
26.6 hours; the average dose was 36.6 mg. 35 of 36 women given $PGF_2\alpha$ and oxytocin aborted, 19
completely. The IAT was reduced to 17.3 hours, and the dose of $PGF_2\alpha$ was decreased to 26.4 mg.
Primigravid women aborted more slowly than multigravid, especially those given only $PGF_2\alpha$.
Incidence of nausea and vomiting was similar, whereas women given only PG2a had an elevation of
leucocytes. Since most women given $PGF_2\alpha$ and oxytocin aborted within 24 hours, the authors
suggested that this method improves the clinical management of abortions. (HC)   004405

0518
SEPPALA, M.; VARA, P.
    Prostaglandin-oxytocin enhancement.
    British Medical Journal 1: 747. 1972.

We report here preliminary results on the abortifacient effect of a combined treatment with $PGF_2\alpha$ and
oxytocin. The combined treatment, administered extra-amniotically by means of a Foley catheter
introduced between the fetal membranes and the uterine wall through the cervical canal, was given to

32 women admitted to hospital for termination of midtrimester pregnancies. A dose of .5 mg $PGF_2\alpha$ (Astra) was given hourly to 20 women. Higher doses (.75-2.0 mg) were used if the uterine contractions were weak or lacking. 9 women received a mixture containing .5 mg of $PGF_2\alpha$ and 1 I.U. of oxytocin, and 3 women received 3 I.U. of oxytocin alone. The results are shown in the Table. 2 out of the 3 failures in the prostaglandin group could be successfully treated with an intravenous oxytocin infusion (5 I.U./hour, which caused uterine contractions when prostaglandin had no more effect. Hysterotomy was needed in one case, where a bicornuate uterus was found at the operation. Extra-amniotic oxytocin alone caused uterine contractions in all 3 cases included in the trial, though abortion was achieved in only one. The 2 failures in the oxytocin group were treated with extra-amniotic prostaglandin. In the total series of 32 cases, hysterotomy was not required in 31 (97%) if both prostaglandin and oxytocin were used. $PGF_2\alpha$ and oxytocin have a contractile effect on the uterus in vivo during the mid-trimester. Whether the effects of these 2 drugs are additive or potentiative is being studied. (Authors' modified)   003809

0519
SETTY, B.S.; KAR, A.B.
Prostaglandins and 'functional' sterility in male rats.
Current Science 41: 64-65. 1972.

The possibility that PGE_2 or $PGF_2\alpha$ may induce 'functional' sterility in the male was studied. Castrated rats were given testosterone propionate (1 mg, intramuscularly) to maintain potency and libido. In addition, each rat in a group of 7 received daily, subcutaneous injections of 100 μg of PGE_2. Another group received 100 μg of $PGF_2\alpha$, and a third received the vehicle alone. Fertility, as measured by successful mating, was tested from the third day of drug treatment. Rats were sacrificed on the eighth day and spermatozoa from the epididymus and vas deferens, examined microscopically. No significant differences were found in fertility or spermatozoal number, morphology and motility between prostaglandin treated and control groups. (JSL)   005021

0520
SHAIKH, A.A.
Regulation of menstrual cycle and termination of pregnancy in the monkey by estradiol and $PGF_2\alpha$.
Prostaglandins 2(3): 227-233. September 1972.

4 monkeys (Macaca fascicularis) were given daily injections of 12 μg estradiol for 3 days starting on Day 18 or 19 of the menstrual cycle, and from Day 3 of the estradiol treatment they were given daily injections of 15 mg $PGF_2\alpha$. On the third, fourth, or fifth day of the treatment with $PGF_2\alpha$ all of the monkeys showed menstrual bleeding. Thus, in every case there was shortening of the luteal phase of the cycle. Injections of $PGF_2\alpha$ alone for 5 days did not have this effect. In the pregnant monkey, treatment with 15 mg $PGF_2\alpha$ for 5 days did not terminate pregnancy, whereas only 3 injections of the same dose of $PGF_2\alpha$ caused abortion in another monkey pretreated with 12 μg estradiol for 3 days. It was postulated that 1) administration of estrogen may cause a block of gonadotrophin release by a negative feedback mechanism, 2) estrogen may also cause the release of endogenous prostaglandins, 3) the corpus luteum because of lack of gonadotrophins may become more vulnerable to exogenously administered and/or endogenously secreted prostaglandins, and 4) estrogen may act directly on the ovary and make the luteal cells more sensitive to prostaglandins (for example, by causing changes in vascular permeability). (Authors' modified)   004273

0521
SHANKLIN, D.R.
Doctors, drugs, and the FDA.
Journal of Reproductive Medicine 9(5): 203-205. November 1972.

The increasing complexities of medical practice and the need for increased communication between patients, doctors, regulatory agencies, and journals which hopefully result in more responsible behavior on the part of all was discussed. With regard to diuretics, this journal convened 2 symposiums, culminating in the announcement that it considers diuretics and and appetite depressants contraindicated in pregnancy. The Brooks Lodge Symposium on PGs will be published in this journal, with the aim of informing doctors about their use when the FDA approves PGs for marketing. (LJG)   005022

0522
SHEARMAN, R.; SMITH, I.; KORDA, A.
Second trimester termination by intra-uterine prostaglandin $F_2\alpha$: clinical and hormonal results with observations on induced lactation and chronoperiodicity.
In: Southern, E.M., ed. The prostaglandins: clinical applications in human reproduction. [Brook Lodge Symposium, Augusta, Michigan.] Mount Kisco, New York Futura Publishing Co., 1972. p. 443-450.

24 women in second trimester pregnancy received $PGF_2\alpha$ intramniotically, 30 mg initially, 15 mg 24 hours later, and another 15 mg dose 18 hours later if necessary. Another group of 7 women were given $PGF_2\alpha$ by transcervical extraovular route, 375 to 750 μg doses at hourly or 2-hourly intervals. All patients were premedicated to minimize pain and gastrointestinal side effects. All patients aborted, 19 of 24 in the first group and 4 of 7 in the second group being complete. The IAT averaged 27 1/2 hours for those in the first group and 35 1/2 hours for the others. Nausea and vomiting occurred frequently; skin flushing, dyspnea, and headache were also noted. 3 of the transcervical patients developed genital tract infections. Nearly all women were lactating when examined 72 to 96 hours after abortion. Plasma estrogens fell during $PGF_2\alpha$ treatment, progesterone levels concomitantly rose. (HC) 004883

0523
SHERMAN, A.I.; VAKHARIYA, V.R.
An evaluation of prostaglandin $F_2\alpha$ for the induction of labor at term.
In: Southern, E.M., ed. The prostaglandins: clinical applications in human reproduction. (Brook Lodge Symposium, Augusta, Michagan.) Mount Kisco, New York, Futura Publishing Co. 1972. p. 95-106.

A double-blind comparison of iv Pitocin and $PGF_2\alpha$ was conducted on 100 consecutive patients selected for induction of term labor and assigned randomly into equal groups by easy and difficult inductions as indicated by Bishop score. The Bishop score was modified by doubling the cervical dilation factor. Drugs were administered by increasing iv infusion rate every 30 minutes from .8002 mU Pitocin per minute or 2 μg $PGF_2\alpha$ per minute to a maximum of 16.01 mU Pitocin per minute or 40 μg $PGF_2\alpha$ per minute. All easy inductions (Bishop scores above 6) were successful. There were 2 failures with Pitocin, and 3 with PG, all with Bishop scores below 3 and all of whom delivered after Pitocin infusion the next day or spontaneously 10 or 12 days later. 83.3 and 83.7% of $PGF_2\alpha$ and Pitocin patients developed first contractions within 30 minutes. Complications included 1 cord compression requiring cesarean section in a PG patients, and 2 breech presentations in the Pitocin group. There were no differences in mean contraction amplitude and frequency, but 9 incidents of uterine hypertonus occurred, 8 of them in the PG group. Apgar scores in patients with hypertonus were all 8-10 at 1 minute and 9-10 at 5 minutes. 3 infants with Apgar scores of 5 or below at 1 minute had unrelated problems. In conclusion, it was felt that the modified Bishop score was a good indicator, that 10 μg PGF$_2\alpha$ is comparable to 32 or 40 mU Pitocin rather than the 4 mU used in this series, and in general that PGF$_2\alpha$ is safe and effective for inducing term labor. (LJG)   004852

0524
SJOSTRAND, N.O.
  A note of the dual effect of prostaglandin E$_1$ on the responses of the guinea-pig vas deferens to nerve
    stimulation.
  Experientia 28: 431-432. 1972.

Experiments were performed to determine the effect of PGE$_1$ on membrane potentials in the guinea pig vas deferens. PGE$_1$(10-200 ng/ml) caused a depolarization of the smooth muscle cells. It also decreased the magnitude of evoked excitatory junction potentials (EJP). The preparations varied greatly in their sensitivity to PGE$_1$ in both respects. There was no correlation between the amount of depolarization and the amount of inhibition of EJP. In sensitive preparations a higher frequency of stimulations was required to achieve an action potential. However, in preparations that showed a high degree of depolarization the action potential, once achieved, was often larger and sometimes shorter in duration and the contraction of the organ became larger. Atropine had no effect on the response to PGE$_1$. It was concluded that the inhibitory effect on the motor response of the vas deferens can be explained electrophysiologically on the basis of the inhibition of EJP. The potentiation found by some workers could be due to the lowering of membrane potentials causing a propagation of action potentials in the tissue. It is possible that the effects of PGE$_1$ on EJPs and membrane potentials are due to their opening ion channels in the membranes. (JRH)   003917

0525
SKARNES, R.C.; HARPER, M.J.K.
  Relationship between endotoxin-induced abortion and the synthesis of prostaglandin F.
  Prostaglandins 1: 191-203. 1972.

The results of this study provide direct evidence for the synthesis of F prostaglandins in the uterine endometrium in response to parenteral administration of bacterial endotoxin. This finding, together with indirect evidence on similarities in response of pregnant mice to either endotoxin or PGF$_2\alpha$, and the effects on this response of agents which block prostaglandin action or synthesis lead us to conclude that the abortifacient action of endotoxin is mediated by PGF$_2\alpha$. The data suggest that diarrhea produced by endotoxin is also a manifestation of prostaglandin synthesis, probably by cells in the intestinal mucosa. The finding of large amounts of PGF or its metabolites in the urine of male as well as pregnant female mice implies that endotoxemia evokes a generalized synthesis of F prostaglandins. Intrauterine fetal death appears not to be due to a direct effect of prostaglandin on the uterus, but to the effect of other substances (possibly serotonin) released by endotoxin or exogenous prostaglandin. (Authors)   003884

0526
SMITH, A.P.
  Side-effects of prostaglandins.
  Lancet 2: 655. September 23, 1972.

The author points out that PGF$_2\alpha$ causes bronchoconstriction or bronchospasm when given as an aerosol or intravenously. He urges caution in the use of PGF$_2\alpha$ in patients with a history of bronchial asthma. (JRH)   004248

0527
SMITH, I.D.; SHEARMAN, R.P.; KORDA, A.R.
Lactation following therapeutic abortion with prostaglandin $F_2\alpha$.
Nature 240: 411-412. December 15, 1972.

80 women, 12 to 24 weeks pregnant, were given $PGF_2\alpha$ intraamniotically, 30 mg initially with supplemental 15 mg doses if necessary. 75% of these women lactated following abortion. Of 24 patients aborted by hysterotomy and 12 by suction curette, only 1 lactated. The onset of lactation was earlier in the more advanced pregnancies. Milk samples from these patients appeared to be similar in content to that obtained from women following full-term pregnancy. (HC)   004599

0528
SORGEN, C.D.; GLASS, R.H.
Lack of effect of prostaglandin $F_2\alpha$ on the fertilizing ability of rabbit sperm.
Prostaglandins 1: 229-233. 1972.

Caput and cauda sperm incubated with prostaglandin $F_2\alpha$ were inseminated into the oviduct of recipient which were given 75 IU of HCG (APL, Ayerst) at the same time. None of the 18 eggs exposed to caput sperm and all of 13 eggs exposed to cauda sperm were fertilized. Capacitated ejaculate sperm incubated with prostaglandin $F_2\alpha$ were inseminated into the oviduct of recipient given 75 IU HCG 12-13 hours earlier. Of 74 eggs recovered, 53 were fertilized. The results are consistent with those found utilizing sperm without prostaglandin treatment. (Authors)   003881

0529
SOUTHERN, E.M.; PATEL, N.C.
Prostaglandins and their clinical applications in human reproduction.
In: Southern, E.M., ed. The prostaglandins: clinical applications in human reproduction. (Brook Lodge Symposium, Augusta, Michigan.) Mount Kisco, New York, Futura Publishing Co., 1972. p. 535-545.

The use of PGs in therapeutic abortion and in the induction of labor was outlined, with emphasis on $PGF_2\alpha$ since it was available in advance of PGE_2. With intravenous administration, it was found that increasing the infusion rate of $PGF_2\alpha$ beyond 50 μg/minute is without benefit in regard to stimulating the uterus for response to abortion. The incidence of side effects by this route suggested that effective alternative routes had to be found; lower dose levels and routes of less systemic absorption were thus necessary. Local administration directly into the uterine cavity seemed favorable. Intraamniotic administration has since been found to have a high success rate. Vaginal application by aqueous solution or in tablets is still in the exploratory stage. As studies have progressed, gestational age has been found to be a critical factor in success rates. No maternal mortality has ever been attributed to the use of PGs. Side effects have consisted chiefly of excessive contractile patterns, such as hypertonus, which are thought to be dose-related. Knowledge of the clinical approach of choice is expected to advance rapidly. (MLH)   004891

0530
SPELLACY, W.N.; GALL, S.A.
Prostaglandin $F_2\alpha$ and oxytocin for term labor induction.
In: Southern, E.M., ed. The prostaglandins: clinical applications in human reproduction. (Brook Lodge Symposium, Augusta, Michigan). Mount Kisco, New York, Futura Publishing Co., 1972. p. 107-113.

In a double-blind study 115 women 36 to 44 weeks pregnant received $PGF_2\alpha$ 2.5 μg per minute while 107 women were given oxytocin (.5 mU per minute) intravenously. The rate of infusion was doubled at 30, 60, 120 and 360 minutes if necessary. With $PGF_2\alpha$ 67% of the women with a Bishop score of 0 to 6 delivered while 87% of those with scores of 7 to 13 delivered. In the oxytocin group, 53% of the women with low scores delivered as did 90% of those with high scores. Length of labor averaged 386 minutes for $PGF_2\alpha$-treated women and 425 minutes for those given oxytocin. 9 cesarean sections were required in the $PGF_2\alpha$ group because of fetal distress. Side effects in the $PGF_2\alpha$ group included uterine hypertonus, fetal bradycardia, nausea, vomiting, and hot flashes while fetal bradycardia was the only side effect occurring in more than 10% of the women given oxytocin. All fetuses were normal with regard to weight and Apgar scores. The authors conclude that PG offers no increased efficacy for labor induction. (HC)   004853

0531

SPEROFF, L.; CALDWELL, B.V.; BROCK, W.A.; ANDERSON, G.G.; HOBBINS, J.C.
Hormone levels during prostaglandin $F_2\alpha$ infusions for therapeutic abortion.
Journal of Clinical Endocrinology and Metabolism 34: 531-536. 1972.

Plasma levels of unconjugated estrone, 17β-estradiol, progesterone, 17-hydroxyprogesterone, human chorionic gonadotropin (HCG), and human chorionic somatomammotropin (HCS) were measured in patients 7-20 weeks pregnant receiving prostaglandin $F_2\alpha$ infusions for therapeutic abortions. The results indicate that prostaglandin $F_2\alpha$ does not exert a luteolytic effect in terminating pregnancy of 7 weeks or more duration. Significant progesterone changes were not seen prior to abortion, and no effect was seen on 17-hydroxyprogesterone levels. There was a gradual decline in estradiol levels during prostaglandin $F_2\alpha$ infusion, while significant falls were seen in estriol levels preceding any changes in estradiol or progesterone. There were no significant changes in HCG levels, while HCS levels declined during the infusion. The data do not rule out the possibility of a luteolytic effect for prostaglandin $F_2\alpha$ in the first few weeks of pregnancy. (Authors)   003888

0532

SPEROFF, L.
Research on abortifacients.
In: Lewit, S., ed. Abortion techniques and services. Amsterdam, Excerpta Medica, 1972. p. 78-83.

$PGF_2\alpha$, used to induce abortions in women pregnant 7-20 weeks, had no ability to lower plasma levels of unconjugated progesterone and 17-hydroxyprogesterone. The mechanism of $PGF_2\alpha$ action in the termination of pregnancy is therefore thought to involve its oxytocic effects rather than a luteolytic activity. $PGF_2\alpha$ infusion did have a marked ability to lower the plasma estriol, and to a lesser extent, the estradiol concentrations. The mechanism of this action is not known. (JSL)   005020

0533

SPILMAN, C.H.; HARPER, M.J.K.
Effect of prostaglandin on oviduct motility in conscious rabbits.
Biology of Reproduction 7: 106. Abstract 19. 1972.

Intravenous injections of 25 or 50 μg per animal of PGE_1 or PGE_2 produced a decrease in tone and frequency of spontaneous activity of the oviduct. $PGF_1\alpha$ and $PGF_2\alpha$ at 50, 100 or 200 μg per animal intravenously increased tone, amplitude and frequency of contractions. PGE_2 .5 or 1.5 mg subcutaneously suppressed, while $PGF_1\alpha$ 5 mg or $PGF_2\alpha$ 7.5 mg increased oviduct activity. E and F prostaglandins were mutually antagonistic suggesting their involvement in the normal regulation of oviduct function. (HC)   004270

0534
SPILMAN, C.H.; HARPER, M.J.K.
Effect of prostaglandins on oviduct motility in conscious rabbits.
In: Abstracts, 5th Annual Meeting of the Society for the Study of Reproduction, East Lansing, Michigan, June 26-29, 1972. p. 25-26.

A silicone balloon attached to polyethylene tubing was placed in each isthmus of the oviducts of 14 New Zealand white rabbits. Oviduct motility was recorded in these unanesthetized animals with a Grass polygraph. Spontaneous oviduct motility was suppressed following an intravenous (iv) injection of prostaglandin (PG)E_1 or E_2 at doses of 25 or 50 μg/animal. Usually, a large decrease in basic tone and a decrease in the frequency of contraction followed the administration of PGE_1 or E_2. These effects lasted an average of 7 minutes. Occasionally a high dose of PGE_1 or PGE_2 (50 or 100 μg) was followed first by a slight contraction and then a suppression of spontaneous activity. Often E prostaglandins completely abolished the spontaneous activity of the oviduct for a short period of time. $PGF_1\alpha$ and $F_2\alpha$ had no effect in animals with no spontaneous activity, indicating that their action may be related to other tubal regulatory mechanisms. In animals with spontaneous activity, $PGF_1\alpha$ and $F_2\alpha$ at doses of 50, 100, or 200 μg/animal iv evoked a sustained increase in tone lasting an average of 5.5 minutes. In some cases this sustained contraction was followed by a slight suppression of spontaneous activity, while in others it was followed by an increase in amplitude and frequency of contraction. PGE_1 and E_2 abolished contractions induced by $PGF_1\alpha$ or $F_2\alpha$. Conversely, contractions were induced by $PGF_1\alpha$ or $F_2\alpha$ during a period of suppressed activity caused by PGE_1 or E_2. Spontaneously occurring sustained contractions were abolished by the injection of either PGE_1 or E_2. Subcutaneous injection of PGE_2 (.5 or 1.5 mg) suppressed oviduct activity for 20 to 40 minutes. $PGF_1\alpha$ (5 mg) and $PGF_2\alpha$ (7.5 mg) so increased the frequency and amplitude of contraction for 1 to 1.5 hours. Since E and F prostaglandins have opposite effects, they may be involved in the normal regulation of oviduct function, the F prostaglandins causing tubal occlusion and the E prostaglandins abolishing it. (Authors)   004054

0535
SPILMAN, C.H.; DUBY, R.T.
Prostaglandin mediated luteolytic effect of an intrauterine device in sheep.
Prostaglandins 3(2): 159-168. September 1972.

An intrauterine device was inserted into 1 uterine horn of 6 sheep on Days 2-4 of the estrous cycle. $PGF_2\alpha$ rose from 34.4 to 216.6 ng/g in the endometrium in closest proximity to the IUD. Radioimmunoassay of uterine vein plasma PGF recorded an increase of 1.5 vs 15.6 ng/ml after IUD placement. These effects of an IUD were abolished by indomethacin treatment. Corpus luteum (CL) development and/or maintenace were inhibited by an IUD. Total progesterone content and CL weight were greater in indomethacin-treated animals than in controls. These results support the role of PGF as the luteolytic substance causing regression of the CL during the estrous cycle and suggest that the IUD produces its antifertility effects by stimulating the production of $PGF_2\alpha$ from the endometrium. (LJG)   004281

0536
STAHL, P.
> Comparative responses of hamster vas deferens and seminal vesicle to the effect of prostaglandins E_1
> and E_2 on stimulation by epinephrine.
> Prostaglandins 2(6): 491-500. December 1972.

Contractions of hamster vas deferens and seminal vesicle preparations in vitro were elicited by epinephrine. Added to the bath at concentrations of .1 or 1.0 ng per ml, PGE_1 inhibited most seminal vesicle responses. The drug had either no effect or augmented contractions at 10 ng per ml and it augmented nearly all epinephrine-induced contractions at concentrations of 100 or 1000 ng per ml. PGE_2 augmented contractions of most seminal vesicle preparations at concentrations of .1 or 1.0 ng per ml, had no effect on nearly all preparations at 10 or 100 ng per ml, but then augmented all epinephrine-induced contractions at concentrations of 500 or 1000 ng per ml. The vas deferens response to epinephrine was not altered by PGE_1 at concentrations of .1 to 100 ng per ml, responses were augmented 20% at 500 ng per ml, 50% at 1000 or 2000 ng per ml, and 80% to 90% at approximately 4000 ng per ml. PGE_2 augmented contractions of vas deferens 45% at concentrations as low as .001 ng per ml. No inhibition of vas deferens contractions was found at any dose of either PGE_1 or PGE_2. The PGE_2 effect was maintained despite several washings of the tissue. This was not true for PGE_1. The author suggests a role of PG in the sequential activity of the smooth muscles of the male reproductive system during ejaculation. (HC)   004588

0537
STEVENSON, P.M.; THOMAS, P.
> The effect of prostaglandin $F_2\alpha$ on the utilization of (6, 14-C)glucose by the superovulated rat ovary.
> Journal of Endocrinology 53: [30-31] 1972.

Glucose utilization by whole homogenates of the superovulated rat ovary was studied using 14-C tracer techniques. $PGF_2\alpha$ (.3 $\mu g/ml$) reduced total glucose utilization to about 1%. The effect of $PGF_2\alpha$ on glucose metabolism was concentration dependent, .1 $\mu g/ml$ causing greater inhibition than either 1.0 or .01 $\mu g/ml$. The distribution of label from the [6, 14-C] glucose, used in the presence and absence of $PGF_2\alpha$ did not change. (Authors' modified)   004379

0538
STRICKLER, R.C.
> Abortion with prostaglandins.
> Lancet 2: 539. September 9, 1972.

The author suggests that the mechanical stimulation of the Foley catheter may have contributed to the successful use of extraamniotic prostaglandins to induce abortion. He reports successful induction of abortion with a Vorhees bag and no prostaglandin. (JRH)   004128

0539
SWARTWOUT, J.R.; SINGH, E.J.; BOSS, S.
> Prostaglandins in human female reproductive tract.
> Obstetrics and Gynecology 39: 629. 1972.

Intensified interest in the physiologic and pharmocologic roles of prostaglandins creates the need for a reliable method for their analysis. Since these potent hormones occur only in trace amounts in cervical mucus, menstrual blood, myometrium, endometrium, ovarian tissue, and ovarian dermoid cyst, a glass fiber, paper chromatographic method must be considered as a method for quantification. It is reported that prostaglandins in human seminal fluid are important in normal fertility. Therefore, several methods are investigated for the separation, identification, and quantification of primary and dehydrated prostaglandins in the female reproductive tract. (Authors' modified)   003895

0540

SYMONDS, E.M.; FAHMY, D.; MORGAN, C.; ROBERTS, G.; GOMERSALL, C.R.; TURNBULL, A.C.
    Maternal plasma oestrogen and progesterone levels during therapeutic abortion induced by intra-
        amniotic injection of prostaglandin $F_2\alpha$.
Journal of Obstetrics and Gynaecology of the British Commonwealth 79: 976-980. November 1972.

10 patients 14 to 22 weeks pregnant were given 25 mg of $PGF_2\alpha$ as a single intraamniotic injection. Blood samples were taken periodically from 40 hours prior to injection until 1 hour after removal of the uterine contents. Samples were extracted and fractionated using column chromatography, and levels of estradiol and total unconjugated estrogens and progesterone were determined by radioimmunoassay. Plasma progesterone levels paralleled those of the estrogens. Before, as well as after, administration of $PGF_2\alpha$ steroid levels showed considerable fluctuation between samples. Levels fell either after the placenta was delivered or within 6 hours of injection. These results suggest that the action of PG is one of direct myometrial stimulation rather than through direct effect on estrogen and progesterone biosynthesis. (HC)   004792

0541

TAKEGUCHI, C.; SIH, C.J.
    A rapid spectrophotometric assay for prostaglandin synthetase: application to the study of non-
        steroidal antiinflammatory agents.
    Prostaglandins 2(3): 169-184. September 1972.

A rapid spectrophotometric assay for the quantitation of PG synthetase activity has been devised. The principle of this method entails the measurement of the arachidonic acid-dependent formation of adrenochrome from L-epinephrine during PG biosynthesis by bovine seminal vesicle microsomes. The validity of this assay was established by examining the substrate specificity of a variety of unsaturated fatty acids, their methyl esters, and the effect of inhibitors on PG synthetase. The results were in good agreement with those obtained by other assay methods. It was found that naphthalenediols, in particular 2,7-naphthalenediol, were potent inhibitors of PG synthetase at concentrations (I.D.50, 2 x 10^{-6}M) comparable with indomethacin. (Authors' modified)   004280

0542

TAN, W.C.; PRIVETT, O.S.
    Analysis of prostaglandins in rat vesicular glands.
    Lipids 7(9): 622-624. 1972.

Rat vesicular glands were frozen, ground to a fine powder, extracted with cold acetone, and PGE content assayed by UV absorption. Valid results can be obtained for the concentration of endogenous prostaglandins using this technique where homogenization as normally applied fails. Analysis of the PGE fraction via thin-layer chromatography revealed that the major component was PGE_2. Powder incubated with buffer and arachidonic acid gave a low yield of PGE_2. (HC)   004516

0543

TAYLOR, G.S.; EINHORN, V.F.

The effect of prostaglandins on junction potentials in the mouse vas deferens.
European Journal of Pharmacology 20: 40-45. 1972.

The authors examined the effect of prostaglandins on the amplitude of the excitatory junction potentials produced in the isolated mouse vas deferens in response to transmural stimulation of sympathetic nerve endings. PGE_1 or PGE_2 infused at a concentration of 10^{-9}gm per ml produced a gradual reduction in amplitude of the junction potentials. At 10^{-10}gm per ml PGE_1 or PGE_2 did not alter the resting membrane potential but action potentials and resultant contractions were no longer recorded. All parameters were reduced at 10^{-5}gm per ml. All values returned to normal 10-15 minutes after washout of PGE. $PGF_2\alpha$ reduced all parameters at 10^{-11}gm per ml. In the presence of higher calcium concentrations, PGE_2 had no effect; lowering the concentration increased the tissue sensitivity. The results suggest that PG acts by reducing the output of sympathetic transmitter. (HC)
004711

0544

TAYLOR, P.L.; THOMAS, P.; STEVENSON, P.M.

The function of malate in ovarian steroidogenesis.
Journal of Endocrinology 53: [30]. 1972.

The complete oxidation of fatty acid depends on a supply of malate. The equilibrium of ovarian mitochondrial malic dehydrogenase is predominantly in the direction of malate, but this equilibrium is changed by acetyl-CoA. We have found that malic enzyme is inhibited by both cyclic-AMP and $PGF_2\alpha$ but the activity of malic dehydrogenase is unaffected by these agents and by steroid hormones. It is postulated that malate is important in ovarian metabolism because it allows the oxidation of fatty acids which supply the electrons for pregnenolone formation from cholesterol. (Authors' modified)
004236

0545

TCHILINGUIRIAN, N.G.O.

Comparison of prostaglandin $F_2\alpha$ and oxytocin in the induction of labor in high risk pregnant women.
In: Southern, E.M., ed. The prostaglandins: clinical applications in human reproduction. (Brook Lodge Symposium, Augusta, Michigan.) Mount Kisco, New York, Futura Publishing Co., 1972. p. 179-192.

$PGF_2\alpha$ was given to induce labor in 10 high-risk patients, and 14 others received $PGF_2\alpha$ or oxytocin in a double-blind trial. Indications in the double-blind group were diabetes, essential hypertension, postmaturity, and toxemia. The patients were aged 15-42, weight 90-230 pounds, parity 0-4, Bishop score 0-13, including ruptured membranes, and without previous cesarean section or hysterotomy. Drugs were infused intravenously and stepwise ranging from 2.5-40 μg/minute $PGF_2\alpha$ and 1-16 mU/minute oxytocin. Results are presented (with primigravidas and multigravidas combined) as ranges in easy-, moderate-, and difficult-to-induce classes, as medians, and as graphs called mean eyeballed laborgrams. There were 7 out of 7 successful deliveries in the PG group and 1 out of 7 successes in the oxytocin group. The PG group had 1 episode of vomiting, 1 of uterine hypertonus, 1 of uterine incoordination, and 1 severe variable deceleration of fetal heart rate. The oxytocin group had 1 case of vomiting and 1 of severe late deceleration. Maternal and fetal vital signs and Apgar scores were all normal. (LJG)   004860

0546

THIERY, M.; VROMAN, S.; KETS, H. VAN; DEROM, R.
Fetal effects of prostaglandins.
European Journal of Obstetrics and Gynecology 4: 125-129. 1972.

Labor was induced in 25 women with $PGF_2\alpha$ given by intravenous infusion. The rate schedule was 2.5 μg/minute for the first 30 minutes, then 5.0 μg/minute for 30 minutes, 10 μg/minute for 60 minutes, 20 μg/minute for 4 hours, and if necessary, 40 μg/minute. Amniotomy was performed 90 minutes after PG infusion was started. Blood samples were taken from the mother and the baby at birth for determination of acid-base status. Labor was satisfactorily induced at the 10 μg/minute level in 2 women, at the 20 μg/minute level in 20 women, and at the 40 μg/minute level in 2 women. No serious side effects were observed, and all fetuses were delivered within 10 hours. Transient uterine hypertonus was observed in 2 patients, vomiting occurred in 2 women, and 5 fetuses had bradycardia. There was a significant difference ($p < .05$) between the mean pH and base excess of the PG babies and their normal controls. (HC)   004893

0547

THIERY, M.; VROMAN, S.; VANDERHEYDEN, K.; HEMPTINNE, D., DE; DEROM, R.; KETS, H. VAN; MARTENS, G.
The fetal effect of prostaglandin $F_2\alpha$ applied in the elective induction of labor at term.
In: Southern, E.M., ed. The prostaglandins: clinical applications in human reproduction. (Brook Lodge Symposium, Augusta, Michigan). Mount Kisco, New York, Futura Publishing Co., 1972. p. 135-158.

In a double-blind study, 50 multiparae at term were given incremental intravenous infusions of $PGF_2\alpha$ 2.5-40 μg/min or equivalent oxytocin. Low amniotomy was performed 90 minutes after the start of the infusion. The IDT was 7 hours or less in all but 1 woman given $PGF_2\alpha$; the mean dose of $PGF_2\alpha$ was 5.5 mg. There were 14 clinical abnormalities such as vomiting, uterine hypertonicity, or fetal bradycardia, but these were not considered $PGF_2\alpha$- or oxytocin-specific. There were 5 $PGF_2\alpha$ and 1 oxytocin neonate with abnormal Apgar scores, 3 of whom had had bradycardia during labor. The pH of the umbilical artery blood fell outside the limits of normality for 3 PG and 1 oxytocin neonates, all of whom had had bradycardia. 25 nulliparae at term were given $PGF_2\alpha$ intravenously 2 μg/minute doubled at 30-minute intervals as needed. Amniotomy was performed 60 minutes before the infusion began. The mean dose used was 4 mg, the IDT averaged 6 1/2 hours. Uterine hypertonus and fetal bradycardia were noted in 2 patients. 4 infants had a low Apgar score at birth; none had abnormal blood pH values. The authors concluded that, provided uterine hyperactivity is avoided, the fetal condition is not endangered by the intravenous infusion of $PGF_2\alpha$ to induce labor at term. (HC) 004857

0548

THIERY, M.; VROMAN, S.; DEROM, R.; KETS, H., VAN
The fetal effects of prostaglandin $F_2\alpha$.
In: Saling, S., and Dudenhausen, J.W., eds. Perinatale Medizin -- Vol. 3. Stuttgart, Georg Thieme, 1972. p. 295-301.

22 healthy term pregnant women, aged 16-40, weighing 45-115 kg, parity 2-4, were infused with $PGF_2\alpha$ iv and compared with normal spontaneous deliveries of 174 multiparae with respect to recorded uterine activity, fetal heart rate, Apgar score, pH of cord blood, mean base excess, and mean fetal-maternal difference in excess lactate. The fixed $PGF_2\alpha$ schedule was 2.5 μg/minute for 30 minutes, 5 μg for 30 minutes, 10 μg for 1 hour, 20 μg for 4 hours, and 40 μg for 4 hours; membranes

were ruptured and uterine motility and fetal heart rate were recorded 90 minutes after starting infusion. There were 2 incidents of hypertonus in first stage related to PG, 3 incidents of persistent bradycardia in second stage questionably related to PG, and 1 fetal bradycardia and 1 severe variable deceleration considered unrelated to PG. The study group had 5 Apgar scores below 8 at 1 minute and 5 below 8 at 4 minutes, but all infants developed normally for the first week. The mean pH, mean base excess, and fetal-maternal difference in excess lactate were not considered significantly different from controls. (LJG)   004986

0549

THORBURN, G.D.; NICOL, D.H.; BASSETT, J.M.; SHUTT, D.A.; COX, R.I.

Parturition in the goat and sheep: changes in corticosteroids, progesterone, oestrogens and prostaglandin F.

Journal of Reproduction and Fertility (Suppl.) 16: 61-84. 1972.

The corpus luteum is the main source of progesterone during late pregnancy in goats, while the placenta is the main source in sheep. Therefore, experiments were performed to determine if an increase in fetal corticosteroid levels would induce parturition in goats as it does in sheep. It was found that infusion of synthetic ACTH into the fetal goat led to an increase in fetal corticosteroid levels, which was followed by a decrease in maternal progesterone levels and premature birth. It has been suggested that the increase in fetal corticosteroids might cause an increase in synthesis and release of PGF by the placenta, which would in turn cause the decrease in progesterone production due to the known luteolytic properties of PGF. However in goats this does not seem to be true, since the decrease in progesterone precedes the surge in PGF levels in the uterine vein plasma. As in sheep the levels of PGF increased sharply during the 24-hour period before birth, but unlike the goat the sheep exhibits considerable PGF in the uterine vein plasma 2 to 3 weeks before birth. While these results (from only 1 animal) would seem to indicate that PGF is not responsible for luteal regression in goats, the surge in PGF immediately before birth indicates that PGF may play an important role in the onset of labor by stimulating myometrial activity in sheep and goats. (JRH)   003929

0550

THORBURN, G.D.; COX, R.I.; CURRIE, W.B.; RESTALL, B.J.; SCHNEIDER, W.

Prostaglandin F and progesterone concentrations in the utero-ovarian venous plasma of the ewe during the oestrous cycle and early pregnancy.

In: [Abstracts of] First International Planned Parenthood Federation South-East Asia and Oceania Regional and Scientific Congress, Sydney, Family Planning Association of Congress: August 1972. p. 17.

The aim of this study was to obtain a detailed description of the temporal changes in the concentration of these hormones around the time of luteal regression. 4 ewes were prepared with indwelling catheters in both uteroovarian veins 10 days after estrus, and blood samples were collected at 3-hour intervals for 6 to 23 days thereafter. PGF was measured by radioimmunoassay. A series of peaks in PGF concentration was detected between Days 13-17 of the estrous cycle. These peaks (5-22 ng/ml) were of short duration (<6 hours) and increased in frequency as estrous approached. Values were usually below .2 ng/ml on other days. Transient but marked decreases in progesterone concentration followed each PGF peak. The recovery in progesterone concentration became progressively less until low levels were reached on Day 15. In ewes in which CL regression did not occur (e.g., early pregnancy), there were varying levels of PGF present on Days 13-14, but in contrast to the cycling ewes, the series of peaks of PGF on Days 15-16 were entirely absent. These results indicate that the PGF concentration peaks on Days 13-14 initiate CL regression and that the peaks on Days 15-16 are necessary to complete luteolysis. These results provide further support for the hypothesis that $PGF_{2}\alpha$ is the ovine luteolytic factor. (Authors' modified)   004175

0551
THORBURN, G.D.; COX, R.I.; CURRIE, W.B.; RESTALL, B.J.; SCHNEIDER, W.
Prostaglandin-F concentration in utero-ovarian venous plasma of ewe during oestrus cycle.
Journal of Endocrinology 53: 325-326. 1972.

This paper describes in detail the changes in PGF concentration in utero-ovarian venous plasma of conscious ewes during the estrous cycle. Polyvinyl catheters (1.5 mm OD) were inserted into the right and left utero-ovarian veins via uterine vein branches in 2 Merino ewes 10 and 11 days after estrus. Uteroovarian venous samples were collected every 2-3 hours from the day after surgery. Concentrations of PGF were measured by a radioimmunoassay. With frequent sampling we detected a complex series of peaks in the concentration of PGF between Days 13 and 17 of the estrous cycle. These peaks were of short duration and increased in frequency as estrus approached. Very low concentrations were measured at other times except on Days 2 and 3 in samples from 1 ewe. In this same ewe, which had a corpus luteum (CL) in each ovary, there was striking similarity in the concentrations of PGF on each side. The early PGF peaks (Day 13) coincide with the early histological signs of luteal regression. Marked cytological changes and the major decrease in progesterone occur at the time of the later, more frequent PGF peaks (Day 15-16). Thorburn and Nicol have shown that short (3-hour) infusions of $PGF_{2}\alpha$ into the uterine vein result in only a temporary decrease in progesterone levels, so these later peaks (Day 16) may be required to complete luteolysis. Preliminary results from pregnant ewes show no similar series of peaks during this time (Days 15-16). The demonstration of appreciable concentrations of PGF in the utero-ovarian vein during the time of luteal regression further supports the hypothesis that $PGF_{2}\alpha$ is the ovine luteolytic factor. (Authors' modified)   003995

0552
THORBURN, G.D.; HALES, J.R.S.
Selective reduction in blood flow to the ovine corpus luteum after infusion of $PGF_{2}\alpha$ into a uterine vein.
Proceedings of the Australian Physiological and Pharmacological Society 3: 145. 1972.

Infusion of $PGF_{2}\alpha$ 40 μg/hour for 6 hours into the uterine vein of ewes reduced blood flow to the ovary. The authors suggested that $PGF_{2}\alpha$ induced corpus luteum regression by a selective reduction in capillary blood flow to the corpus luteum. (HC)   005075

0553
TOM, W.K.C.; SRIBYATTA, B.; THORNEYCROFT, I.H.; MISHELL, D.R., Jr.
Fertility regulation with cyclic luteal phase vaginal administration of prostaglandin $F_{2}\alpha$.
Contraception 6(6): 479-488. December 1972.

Of the 10 nonpregnant women used in this study, 4 were given doses of 3000 IU and 6000 IU HCG on consecutive days one week postovulation. During their next menstrual cycle, 2 of these women received 100 mg $PGF_{2}\alpha$ intravaginally in 2 divided doses 4 hours apart 1 week postovulation, while the other 2 were given $PGF_{2}\alpha$ according to the same schedule on 2 consecutive days. All 4 women were then given HCG as described above for 2 days. Another group of 2 women were given only $PGF_{2}\alpha$ while a group of 4 were given $PGF_{2}\alpha$ and HCG without HCG pre-treatment during their previous cycle. HCG increased the mean length of time from LH peak to menses from 15 to 17 1/2 days and the mean summation levels of progesterone from 114 to 250 ng. $PGF_{2}\alpha$ given to these women during their next cycle did not further increase these parameters. In the 2 women given only PGF$_2\alpha$, there was no change in the length of the luteal phase and no decrease in progesterone levels. The 4 women given HCG and PGF$_2\alpha$, but not HCG during their previous cycle, had the mean length of the luteal phase prolonged from 14 to 19 days and the mean progesterone level increased from 99 to 271 ng. All 10 women experienced adverse side effects such as diarrhea, vomiting. Since PGF$_2\alpha$ failed to have a luteolytic effect in this study, vaginal administration during the luteal phase of the menstrual cycle will not be an effective method of fertility control. (HC)   004647

0554

TOM, W.K.C.; THORNEYCROFT, I.H.; NAKAMURA, R.M.; MISHELL, D.R. Jr.
Intra-vaginal prostaglandin F$_2\alpha$ as a luteolytic agent.
Advances in the Biosciences (Suppl.) 9: 116. 1972.

10 women were treated with 100 mg of intravaginal prostaglandin F$_2\alpha$ (PGF$_2\alpha$) as the THAM salt in the form of lactose tablet for 1 or 2 days in the midluteal phase of the menstrual cycle. 8 of the women were also treated with injections of human chorionic gonadotrophin (HCG) in order to mimic HCG production from an implanting embryo. Serum luteinizing hormone (LH) and progesterone levels were measured by radioimmunoassay and were correlated with the onset of the menstrual period. In all 10 women, PGF$_2\alpha$ in the dosages given did not shorten the time interval between the LH peak and the onset of menses when compared to their untreated control cycles or to previous cycles artificially lengthened by HCG. In addition, there was no significant difference in serum progesterone between PGF$_2\alpha$ treatment cycles and control cycles with or without HCG. During treatment, 9 of the 10 women had objectionable side effects including diarrhea, abdominal cramping, pain, pelvic pressure, nausea, and vomiting. These data suggest that intravaginal PGF$_2\alpha$ in the dosage and formulation used was systemically absorbed but had no luteolytic effects. (Authors' modified)   004304

0555

TOPPOZADA, M.; BYGDEMAN, M.; WIQVIST, N.
Prostaglandin administration for induction of mid-trimester abortion in complicated pregnancies.
Lancet 2: 1420-1421. December 30, 1972.

26 midtrimester patients, 14 of whom had systemic disorders and 12 of whom had pelvic abnormalities, were treated with prostaglandins. 18 were given PGF$_2\alpha$, 5 were given 15-methyl-PGF$_2\alpha$, and 3 were given PGE$_2$. The route was intrauterine for 11, intraamniotic for 9, intravenous and vaginal for 3 each. Abortion was successful in all cases. IAT and incidence of side effects were reported to be consistent with earlier results in uncomplicated pregnancies. (HC)   004614

0556

TOPPOZADA, M.; BEGUIN, F.; BYGDEMAN, M.; WIQVIST, N.
Response of the midpregnant human uterus to systemic administration of 15(S)-15-methyl-
     prostaglandin F$_2\alpha$.
Prostaglandins 2(4): 239-249. 1972.

15 or 20 μg of 15-methyl PGF$_2\alpha$ or 200 μg of PGF$_2\alpha$ as a single intravenous injection was the threshold dose for uterine stimulation in women 13 to 24 weeks pregnant. A single dose of 50 μg of the methyl analogue provided a slower initiation of tonus elevation but a more sustained effect than the parent compound at 200 μg intravenously. At 1.0 mg intramuscularly the methyl analogue produced marked sustained contractions; 1.5 mg was associated with frequent side effects. A constant intravenous infusion of the methyl analogue at a rate of 5 μg/minute provided the same degree of uterine activity

associated with 75 μg/minute of PGF$_2\alpha$. 8 of 10 women given the infusion of the methyl analogue aborted; 3 of these were complete. The IAT averaged 9 hours; vomiting and diarrhea occurred in 7 patients. With the parent compound, only 2 of 16 patients aborted; the incidence of side effects was the same. The enhanced potency of the 15-methyl analogue might be due to its slower metabolism. (HC)   004600

0557
TRETHEWIE, E.R.
   Aspirin and seminal prostaglandins.
   Lancet 2: 282. 1972.

Dr. Charles DeWitt Roberts (May 13, p. 1070) confirms statistically that aspirin reduces the concentration of prostaglandins E and F in seminal fluid. This mechanism is in accord with the effects of acetylsalicylic acid. This acid inhibits the release of histamine and S.R.S. (once thought to be prostaglandin) in the antigen/antibody reaction of anaphylaxis (now presumably IgE). I believe that aspirin inhibits the release substances by interfering with antibody groupings. Further, I have shown that aspirin also reduces the output of histamine in venom injury and that blood -- probably its protein constituents -- plays an essential part in the reaction. Spector and Willoughby also implicated protein. It is possible that aspirin interferes with a membrane lipoprotein precursor releasing prostaglandin E and F. (Author)   004126

0558
TSAFRIRI, A.; LINDNER, H.R.; ZOR, U.; LAMPRECHT, S.A.
   In-vitro induction of meiotic division in follicle-enclosed rat oocytes by LH, cyclic AMP and
      prostaglandin E$_2$.
   Journal of Reproduction and Fertility 31: 39-50. 1972.

Enlarged follicles removed from rat ovaries before 1400 on the day of proestrus progressed from the dictyate stage of the first meiotic division to mature oocytes, if insulin and PGE$_2$, luteinizing hormone (LH), human chorionic gonadotropin (HCG), or follicle stimulating hormone (FSH) were added to the medium, or if dibutyryl cAMP were injected into the antrum. Follicles were cultured on steel grids in Eagle's medium with 20% fetal calf serum; meiosis was detected by interference contrast and phase contrast microscopy. Criteria for meiosis were: 1) disappearance of germinal vesicle and nucleolus; 2) perivitelline space; 3) extrusion of first polar body; 4) dyads. LH, .5-1 μg per ml; brought on maturation in 82% of oocytes, but most (61%) remained tetraploid. 5-10 μg LH, with 5 μg per ml insulin, evoked maturation in 72-79% of oocytes. HCG, 5-10 I.U. per ml, FSH, 10-20 μg per ml, and PGE$_2$, 1.4 X $10^{-6}$to 2.8 x 10^{-5}M, also induced maturation. Prolactin, progesterone, 20alpha-dihydroprogesterone, estradiol-17β, and PGF$_2\alpha$ were ineffective. Estradiol, progesterone and cyanoketone (cyanotrimethyl androstenone), an inhibitor of steroidogenesis, did not inhibit maturation induced by LH. LH and PGE$_2$increased cAMP in isolated follicles, but only dibutyryl cAMP, 6 μg per follicle injected into the antrum, caused maturation. The authors propose that cAMP and possibly PGs are involved in maturation of oocytes. (LJG)   004598

0559
TSAFRIRI, A.; LINDNER, H.R.; ZOR, U.; LAMPRECHT, S.A.
   Induction of meiotic division in follicle-enclosed ova in culture.
   Israel Journal of Medical Sciences 8: 170. 1972.

Enlarged graafian follicles taken from rats on the day of proestrus were placed in tissue culture. The ability of various hormones and other compounds to induce resumption of meiosis in the enclosed oocytes was measured. LH, chorionic gonadotropin, and FSH induced resumption of meiosis, while prolactin, progesterone, and estrogen were ineffective. LH was found to cause an increase in cAMP production in isolated follicles, PGE_1 stimulated cAMP formation and also evidenced resumption of meiosis. $PGF_2\alpha$ was only partly effective and linolenic acid had no effect. Indomethacin did not prevent induction of meiosis by LH. Exogenous application of cAMP or dibutyryl cAMP had no effect, but microinjection of dibutyryl cAMP into the follicle caused resumption of meiosis. (JRH)   003902

0560

TSAFRIRI, A.; LINDNER, H.R.; ZOR, U.; LAMPRECHT, S.A.

Physiological role of prostaglandins in the induction of ovulation.

Prostaglandins 2(1): 1-10. July 1972.

The administration of prostaglandin E_2 to adult rats at .7, 1.0 or 1.5 mg/rat on the afternoon of proestrus in which the preovulatory surge of LH was prevented by Nembutal induced ovum maturation in 58%, 70%, and 90% and ovulation in 42%, 60%, and 81% of the animals, respectively; the incidence of persistent uterine distension was reduced by the prostaglandin treatment, suggesting that ovarian progesterone secretion was stimulated. Injection of indomethacin, an inhibitor of prostaglandin synthesis, on its own at 14.30 on the day of proestrus (5-10 mg/rat) prevented follicular rupture in 78-89% of the animals, but maturation of the oocytes retained in the follicles was unimpaired. A dose of 1 mg indomethacin was partially effective in inhibiting ovulation. Concomitant treatment with indomethacin and Nembutal prevented both follicular rupture and ovum maturation. Administration of LH at a dose level adequate to induce ovulation in Nembutal-blocked rats (2.5 μg/rat), failed to overcome the indomethacin-induced block of ovulation, but prostaglandin E_2 brought about follicular rupture in most of the indomethacin-treated animals. Both indomethacin and PGE_2 animals showed signs of respiratory distress. It is concluded that 1) indomethacin, under the conditions studied, does not block LH release, but exerts its antiovulatory action directly on the follicle: it prevents follicular rupture, but not ovum maturation; 2) prostaglandins have an essential role in the mechanism by which LH brings about follicular rupture; 3) though prostaglandins E_2 is able to induce ovum maturation, prostaglandins are not indispensible for this action of LH. (Authors' modified)   004167

0561

TURMAN, E.J.

Improve reproductive performance of beef cattle in the southern plains.

In: U.S. Department of Agriculture Research Work Unit. Project Abstract. Accession No. 22740. 1972.

Objectives: Improve reproductive performance of beef cows through the use of biologically active chemicals. Approach: Observe the superovulatory response of beef cows and heifers to repeated subcutaneous injections of PMS, alone or in combination with HCG, and administered at different levels and at different times during the estrual cycle. Determine the relationship of response to PMS and age, weight, postpartum interval, and lactational status of the cows. Observe the relationship between circulating levels of biological components and response to PMS. Evaluate, if possible, the synthetic releasing hormones, prostaglandins, and other potentially effective chemicals as agents for improving reproductive performance. (Author)   004819

0562

TYLER, E.T.; LEVIN, M.L.; ELLIOTT, J.
  Current status of contraceptives in population control.
  International Surgery 57(5): 406-409. May 1972.

The authors state that future developments in the field of contraception and birth control include the use of prostaglandins as early abortifacients. They note that prostaglandins given intravenously will abort pregnancy in the first trimester, but they also note that there is no substantive information regarding contraceptive activity or prevention of implantation by prostaglandins. (HC)   004511

0563

VAKHARIYA, V.; SHERMAN, A.I.
  Prostaglandins in human reproduction.
  Michigan Medicine, September 1972. p. 777-784.

The role of PGs in human reproduction was reviewed. PGs are 20-carbon fatty acids containing a cyclopentane ring based on a prostanoic acid skeleton, and are found in many human tissues and fluids, primarily in seminal plasma. High concentrations of $PGF_2\alpha$ and lower concentrations of PGE_1, PGE_2, and $PGF_1\alpha$ have been discovered in amniotic fluid and decidua of pregnant women during labor, and detectable amounts of $PGF_2\alpha$ have been found in peripheral venous blood during labor. The actual role of the PGs is unknown, though it was suggested that they may sensitize the uterine muscle to circulating oxytocin. The use of PGs to induce labor has been investigated. Intravenous administration was seen as the best route; success rates varied, and uterine hypertonus has been a reported side effect in some cases. PGs have been used also to induce abortion. When they were injected intravenously, side effects including nausea and vomiting and (less frequently) elevated temperature and phlebitis at the infusion site were reported. Intraamniotic administration seemed to avoid these side effects and is therefore favored. The mechanism of action is unresolved. Other possible roles of PGs in the physiology of reproduction were discussed. (MLH)   004739

0564

VAKHARIYA, V.R.; SHERMAN, A.I.
  Prostaglandin $F_2\alpha$ for induction of labor.
  American Journal of Obstetrics and Gynecology 113: 212-222. 1972.

In 100 uncomplicated pregnancies in multiparas between 36 and 43 weeks of gestation, labor was induced electively with oxytocin and prostaglandin $F_2\alpha$ in a double blind study. All patients were examined prior to induction, and a pelvic score for 'inducibility' was determined. They were divided into 2 categories--'easy' and 'difficult'--by an arbitrary division from the inducibility score. Blood chemistry and urinalysis were studied on each patient prior to and on the day following the induction for possible alterations. Mothers' vital signs, frequency and intensity of uterine contractions, and fetal heart tones were monitored throughout the labor, and the majority of the patients were subjected to electronic fetal monitoring. This study compares the results of the use of oxytocin and prostaglandin $F_2\alpha$, both for their effectiveness and safety in each category. Prostaglandin $F_2\alpha$ compares very favorably with oxytocin as a drug for the induction of labor in multiparas at term. In the dose necessary to induce labor, prostaglandin $F_2\alpha$ was found to be safe for both mother and infant. (Authors)   003914

0565
VANE, J.R.; WILLIAMS, K.I.
Prostaglandin production contributes to the contractions of the rat isolated uterus.
British Journal of Pharmacology 45: 146P. 1972.

We have now studied the effect of indomethacin upon spontaneous contractions of the rat isolated uterus and those elicited by agonists. Uterine horns from virgin rats of the Wistar strain were superfused at 10 ml/minute with Krebs solution at 37 degrees C or bathed with De Jalon's solution at 35 degrees C in a 15 ml bath. Submaximal contractions of the uteri were induced by oxytocin or $PGF_2\alpha$; indomethacin (.25-1 μg/ml) was then added to the bathing fluid. Whereas the effects of $PGF_2\alpha$ were relatively unchanged (dose ratio 1.6 $\pm$ .5 [mean $\pm$ S.E.M., 12 experiments]) the activity of oxytocin was reduced (dose ratio 5 $\pm$ 1, 18 experiments). Prostaglandin output into the bathing fluid (15 ml organ bath Krebs solution, 37 degrees C) was also measured, using uteri from rats which were 17-22 days pregnant. Bath fluid was withdrawn at 15-minute intervals. PG-like activity was assayed in terms of $PGF_2\alpha$. PG-like activity (2.1-6.5 ng/g/ml of fluid over 15 minutes) was present in the bath, and the output was maintained over a 3-hour period. Indomethacin (1-4 μg/ml) reduced the output to undetectable amounts within 45 minutes. At the same time spontaneous activity of the uteri was abolished. These results suggest that intramural generation of PG in the rat isolated uterus contributes to the maintenance of spontaneous activity and to the contractions induced by oxytocin. (Authors' modified)   004001

0566
VARAVUDHI, P.; CHOBDIENG, P.
Biological evidence for the direct stimulating effect of $PGF_2\alpha$ on the release of pituitary luteolytic agents of pseudopregnant rats.
Prostaglandins 2(3): 199-205. September 1972.

Pseudopregnancy was induced in rats by electrical stimulation of the uterine cervix until vaginal cornification ceased, designated as Day LO. The first appearance of leucocytes in the vaginal smear was designated L1, the second L2, etc. On Days LO, L4, L7, and L9, $PGF_2\alpha$ was administered via sterotoxic implantation into either the anterior pituitary (AP) or the median eminence (ME). Termination of corpus luteum function was determined by the reappearance of nucleated or cornified epithelial cells in the smear. ME implant did not show significant shortening of pseudopregnancy, whereas the AP did. Durations of the leucocytic smear were 11.8 and 10.8 days compared with 13.5 without $PGF_2\alpha$. In hysterectomized pseudopregnant animals, again ME implant showed no significant shortening, whereas the AP implants did (17.7 and 12.9 compared with 18.7 without $PGF_2\alpha$). It was concluded that $PGF_2\alpha$ may be one of the uterine factors capable of inducing luteolysis in rats directly and stimulating adenohypophyseal tissue to release luteolytic agents of LH in nature. (MJN)   004277

0567
VERMOUTH, N.T.; DEIS, R.P.
Prolactin release induced by prostaglandin $F_2\alpha$ in pregnant rats.
Nature New Biology 238: 248-250. 1972.

This study was undertaken to see if $PGF_2\alpha$ can induce the release of prolactin when injected into pregnant rats. A solution of $PGF_2\alpha$ was injected intraperitoneally in 2 doses of .3 mg each on Day 18 of pregnancy at 0800 and 1200 in some rats and 1600 in other groups. The mean serum prolactin concentration at 8, 10, and 12 hours after the last dose of $PGF_2\alpha$ is significantly greater than the control values (p<.001) with a peak at 12 hours. After 16 hours, however, prolactin values are similar to the control rats. A second rise in prolactin level is observed 20, 22, and 24 hours after $PGF_2\alpha$ with a maximum at 20 hours. To see if the release of prolactin is due to a decrease of plasma progesterone concentration, we tried to prevent the peak of serum prolactin seen 12 hours after the administration of $PGF_2\alpha$ by injecting progesterone 14 and 4 hours before this peak. The dose of progesterone (5 mg) administered prevented the release of prolactin only partially but significantly. In another group injected with a single dose of progesterone (10 mg) 4 hours before the second peak of prolactin level, a partial but significant block was observed. $PGF_2\alpha$ induced lactogenesis 26 hours after the last dose of the drug. Abortion was also observed around Day 20 of pregnancy. These results suggest that $PGF_2\alpha$ administration is followed by increased circulating levels of prolactin. This may account for the lactogenic action of $PGF_2\alpha$. Furthermore, the possibility that $PGF_2\alpha$ releases other related hormones must be considered. A very important finding is that a single dose of progesterone administered to pregnant rats 18 hours after $PGF_2\alpha$ prevented the abortive action of $PGF_2\alpha$. (Authors' modified) 004151

0568

VIRUTAMASEN, P.; WRIGHT, K.H.; WALLACH, E.E.
Effects of prostaglandins E_2 and $F_2\alpha$ on ovarian contractility in the rabbit.
Fertility and Sterility 23: 675-682. 1972.

Intraarterial injection of PGE_2 (5 µg/kg) into the dorsal aorta of pentobarbital-anesthetized rabbits caused an inhibition of spontaneous ovarian contractions which lasted up to 4 minutes. A similar amount of $PGF_2\alpha$ stimulated tone, frequency, and/or amplitude. Isoproterenol given by the same route after the $PGF_2\alpha$ greatly reduced the PG-stimulated increase in contractility. Only 1 of 12 ovaries developed spontaneous contractions in vitro but addition of $PGF_2\alpha$ (2-5 µg) to the organ bath initiated contractions after a delay of 3-30 minutes. PGE_2 inhibited contractions initiated by either $PGF_2\alpha$ or norepinephrine. Phenoxybenzamine stopped contractions started by $PGF_2\alpha$ when added to the organ bath. When the experiments were performed on rabbits pretreated 9 1/2-14 hours previously with human chorionic gonadotropin to induce ovulation, the responses were similar to the nonovulatory experiments except the response to $PGF_2\alpha$ was more variable. When given intravenously the effects of the PGs (5-10 µg/kg) were much less prominent on ovarian contractions. The authors conclude that these results support the possible role of prostaglandins in ovulation at a local level. (JRH)   004200

0569

VROMAN, S.; THIERY, M.; YO LE SIAN, A.; DEPIERE, M.; VANDERHEYDEN, C.; DEROM, R.; KETS, H. VAN; BROUCKAERT, J.
A double blind comparative study of prostaglandin $F_2\alpha$ and oxytocin for the elective induction of labor.
European Journal of Obstetrics and Gynecology 4: 115-123. 1972.

Term labor was induced in 50 consecutive multiparas given either $PGF_2\alpha$ or oxytocin iv in a random double-blind protocol with dose equivalency of 2.5 µg PG to 1 mU oxytocin. Patients were selected with parity of 1-6, uneventful prenatal course, vertex presentation, and intact membranes. They were infused with 5% glucose for 1 hour. Dose rates were doubled at .5, 1, 2, and 6 hours or until labor was satisfactory, usually stopping at 16 mU oxytocin or 40 µg $PGF_2\alpha$. Membranes were ruptured 90 minutes after starting induction. There were no differences between the 2 drugs in duration of induction, although amniotomy seemed to reduce induction interval. There were 8 abnormal events during monitoring of the PG group of which 2 fetal bradycardia and 2 uterine hypertonus may have been drug related. There were 6 in the oxytocin group, all nonspecific. 6 Apgar scores of 1 minute and 1 score of 5 minutes were 7 or below; the mean pH of cord blood was significantly lower in the PG group; the mean base deficit was significantly greater in both groups than in spontaneous deliveries (both $p < .05$) but all were within normal limits. The only maternal side effects were vomiting in 2 patients in each group. There was no change in intraocular pressure. $PGF_2\alpha$ and oxytocin were considered equally effective, but PG induction should be monitored carefully. (LJG)   004985

0570

WALTMAN, R.; TRICOMI, V.; PALAV, A.B.
Mid-trimester hypertonic saline-induced abortion: effect of indomethacin on induction/abortion time.
American Journal of Obstetrics and Gynecology 114(6): 829-831. November 15, 1972.

15 midtrimester patients were given hypertonic saline and indomethacin. The mean IAT was 70 1/2 hours, whereas for those patients previously given only saline the IAT averaged 36 hours. Since both saline and prostaglandins increase uterine motility and since indomethacin has been shown to inhibit prostaglandin biosynthesis, the results of this study may be evidence for saline exerting its abortifacient effect through a prostaglandin-mediated mechanism. The use of indomethacin for preventing premature labor is suggested. (HC)   004746

0571

WENTZ, A.C.; JONES, G.S.; GRAEBER, J.
Effect of infused prostaglandin $F_2\alpha$ on hormonal levels during early pregnancy.
American Journal of Obstetrics and Gynecology 114(7): 908-913. December 1972.

8 women, 11 1/2 to 18 1/2 weeks pregnant, were given $PGF_2\alpha$ by intravenous infusion at a rate of 25 μg per minute for 30 minutes, followed by 50 μg per minute for 23 1/2 hours and then 100 μg per minute for additional 12 hours if necessary. 2 patients aborted placental tissue but had no fetuses; 5 patients aborted, the IAT averaging 16 1/2 hours; the infusion was terminated in 1 patient because of fever and vomiting. The patients with missed abortions had no estriol in the serum, their estradiol and progesterone levels were initially low and decreased 50% during $PGF_2\alpha$ infusion. In those patients that had successful abortions, serum estriol, estradiol, and progesterone levels fell 50% prior to abortion. In the patient that failed to abort, estriol levels dropped during the first 12 hr but then rose again during the next 12 hr; estradiol values remained unchanged; progesterone levels dropped the first 8 hr rose sharply and then plateaued during the remaining 16 hr. $PGF_2\alpha$ appears to have an effect on placental steroidogenesis. Absolute levels are not crucial for labor induction. Decreased steroid production may be a result of placental anoxia caused by vascular impairment resulting from uterine contractions. (HC)   004645

0572

WENTZ, A.C.; CUSHNER, I.M.; AUSTIN, K.; SHAMS, M.
Intra-amniotic administration of prostaglandin $F_2\alpha$ for abortion.
American Journal of Obstetrics and Gynecology. 113: 793-803. July 15, 1972.

Prostaglandin $F_2\alpha$ was administered intraamniotically to 20 women for midtrimester abortion. 17 patients aborted completely, 1 incompletely, and 2 patients failed to abort after 32 hours. The average medication-abortion time was 22 and 8 minutes. The only significant side effect was vomiting. Intravenous oxytocin was evaluated as an ancillary means of inducing more rapid abortion. Intraamniotic prostaglandin $F_2\alpha$ was shown to be an effective oxytocic agent, but side effects may limit its clinical usage. (Authors' modified)   004145

0573
WENTZ, A.C.; KING, T.M.
Intramyometrial prostaglandin $F_2\alpha$.
American Journal of Obstetrics and Gynecology 114: 112-114. 1972.

A rhesus monkey at 139 days gestation was anesthetized with Fluothane and the abdominal cavity opened. Either 20% saline or $PGF_2\alpha$ (1-5 mg/ml) was injected in small amounts (.2-2 ml) into several sites on the anterior myometrium. The saline caused immediate blanching followed by hyperemia of the injection site. The $PGF_2\alpha$ caused no noticeable change, but uterine contractions started while the abdomen was being closed. 7 days later laparotomy was performed. The saline injection sites showed full thickness myometrial necrosis upon histologic examination. The $PGF_2\alpha$ sites showed no noticeable tissue damage. It is concluded that extravasation of even small quantities of hypertonic saline along the injecting needle can cause severe permanent damage to the myometrium, while $PGF_2\alpha$ does not have this disadvantage. (JRH)   004214

0574
WENTZ, A.C.; KING, T.M.
Myometrial necrosis after therapeutic abortion.
Obstetrics and Gynecology. 40(3): 315-320. September 1972.

Extensive myometrial necrosis occurred after the injection of hypertonic saline to terminate pregnancy when prostaglandin $F_2\alpha$ failed to induce abortion. Myometrial necrosis was produced in pregnant rhesus monkeys by the injection of hypertonic saline. Termination of pregnancy by this technique is contraindicated in a contracting or hypertonic uterus. (Authors' modified)   004250

0575
WHO PROSTAGLANDIN TASK FORCE
Phase IIb clinical trials comparing intra-amniotic prostaglandin $F_2\alpha$ and 15-Methyl $PGF_2\alpha$.
In: Bergstrom, S., ed. Report from meetings of the prostaglandin task force steering committee, Chapel Hill, June 8-10, 1972, Stockholm, October 2-3, 1972, Geneva, February 26-28, 1972. Stockholm, WHO Research and Training Centre on Human Reproduction, Karolinska Institutet, 1973. (Prostaglandins in Fertility Control 3) p. 65-69.

The WHO Prostaglandin Task Force Committee rejected the intravenous use of $PGF_2\alpha$ as a method of terminating pregnancies, whereas the intrauterine method was considered promising. The committee designed this phase of the clinical research protocol to compare the safety and effectiveness of intrauterine (intraamniotic) instillation of 40 mg of $PGF_2\alpha$ to 2.5 mg of 15-me-$PGF_2\alpha$ for termination of second trimester pregnancies. The protocol includes the selection of patients, contraindications to the use of prostaglandins, side effects, complications, and instructions for using the protocol form for clinical trial of prostaglandin. (AS)   006528

0576
WHO SCIENTIFIC GROUP
Advances in fertility control.
WHO Chronicle 20: 12-19. 1972.

This is a shortened version of the 48-page report by the same title, the work of the World Health Organization Scientific Committee on Methods of Fertility Regulation. Most of the text concerns mode of action, effectiveness, side effects, and contraindications of oral and injectable contraceptives and

IUDs. Side effects of oral contraceptives, classified into common, subjective, and rare side effects, are chiefly ascribed to elevated corticosteroids; little detail is supplied about thromboembolism. For IUDs a distinction is made between leucocytic infiltration (the probable mode of action) and pelvic inflammatory disease (a complication). New methods of fertility control and the use of PGEs and PGFs to abort pregnancies by iv, intrauterine, or intravaginal routes are discussed. Best results with PGs are achieved in the first 2 months of pregnancy; incomplete abortion is more common later. (LJG) 004508

0577

WILKS, J.W.; HUNTER, K.K.; NORLAND, J.F.
Prostaglandin $F_2\alpha$: synthesis by the rabbit ovary, uterus and uterine cervix.
In: Abstracts, 5th Annual Meeting of the Society for the Study of Reproduction, East Lansing, Michigan, June 26-29, 1972. p. 24.

Implication of the prostaglandins in the physiology of the ovary and uterus prompted this investigation on the capability of the reproductive tissues to synthesize prostaglandin $F_2\alpha$ ($PGF_2\alpha$). Tissue slices of ovary, uterus, and uterine cervix from estrous and pseudopregnant (Day 9) rabbits were superfused with Krebs-Ringer bicarbonate buffer in vitro at 37 degrees C. Samples of buffer were collected at 10 minute intervals from superfusion flasks and assayed for $PGF_2\alpha$ by radioimmunoassay. Tissue $PGF_2\alpha$ concentrations prior to incubation (but after slicing) were 699, 80, and 206 ng/gm tissue for ovary, uterus, and cervix, respectively. The mean rates for release of $PGF_2\alpha$ into the buffer over a 4 hour incubation period were 52, 71, and 30 ng/gm tissue/hr for ovarian, uterine, and cervical tissue, respectively. $PGF_2\alpha$ concentrations increased in all tissues during incubation, indicating that prostaglandin concentrations in the buffer reflected synthesis and not depletion of tissue $PGF_2\alpha$. Uterine tissue from estrous rabbits synthesized greater ($p<.005$) quantities of $PGF_2\alpha$ than did tissue from pseudopregnant animals; in contrast, pseudopregnancy enhanced ($p<.025$) $PGF_2\alpha$ synthesis by the cervix. Synthesis of $PGF_2\alpha$ was similar for ovarian tissue from both estrous and pseudopregnant rabbits; interstitial tissue released more $PGF_2\alpha$ into the buffer than did corpora lutea (53 vs. 36 ng/gm tissue/hr, $p<.01$). The rate of $PGF_2\alpha$ release was relatively constant throughout a 4 hour incubation except for uterine tissue from estrous animals, which showed an increasing release rate. The data indicate that the ovary and uterus can synthesize $PGF_2\alpha$ de novo and that $PGF_2\alpha$ synthesis can be altered by pseudopregnancy or estrus. (Authors)   004053

0578

WILKS, J.W.; HUNTER, K.K.; NORLAND, J.F.
Prostaglandin $F_2\alpha$: synthesis by the rabbit ovary, uterus, and uterine cervix.
Biology of Reproduction 7: 105. 1972.

Reproductive tissue slices from estrous or pseudopregnant rabbits were incubated and samples of buffer solution assayed periodically for $PGF_2\alpha$. Concentrations increased with time, suggesting synthesis rather than depletion of tissue stores. Uterine tissue from estrous animals synthesized more $PGF_2\alpha$ than tissue from pseudopregnant animals; the opposite was true for uterine cervical tissue. Synthesis by ovarian tissue was similar for both groups. (HC)   004271

0579

WILKS, J.W.; FORBES, K.K.; NORLAND, J.F.
Synthesis of prostaglandin $F_2\alpha$ by the ovary and uterus.
In: Southern, E.M., ed. The prostaglandins: clinical applications in human reproduction. (Brook Lodge Symposium, Augusta, Michigan.) Mount Kisco, New York, Futura Publishing Co., 1972. p. 47-58.

PGF$_2\alpha$ release into incubation medium was observed with pseudopregnant and normal rabbit and rhesus monkey ovary and uterus. PGF$_2\alpha$ was radioimmunoassayed in samples of superfusion medium taken every 10 minutes, and in tissues before and after 4 hours' incubation in Krebs-Ringer bicarbonate with 2 mg per ml glucose. For the first hour PGF$_2\alpha$ fell in the medium but from then on it was biosynthesized and released, since tissue levels did not decrease. Dutch-belted and New Zealand rabbit uterine slices released 110 ng PGF$_2\alpha$ per gm tissue per hour with an increasing rate; slices from pseudopregnant rabbits on Day 9 released 49 ng per gm per hour at a constant rate. Neither LH, FSH, or prolactin affected PGF$_2\alpha$ released by rabbit ovaries, nor did LH affect corpora lutea, nor progesterone or estradiol affect rabbit uterus. Monkey ovary, myometrium, and endometrium released 66, 85, and 293 ng per gm per hour over 8 hours. LH but not FSH or prolactin stimulated PGF$_2\alpha$ release by ovary. Estradiol enhanced PGF$_2\alpha$ release by myometrium; progesterone did not; neither altered release by endometrium. Thus biosynthesis of PGF$_2\alpha$ with some information on its hormonal control was demonstrated in vitro. (LJG)   004847

0580

WILLIAMS, C., 3rd
    Prostaglandins - miracle or medicine show?
    Science Digest 72: 10-15. September 1972.

Although recent publication on potential clinical prostaglandin use has been encouraging, the author called attention to the less publicized side effects (bradycardia, bronchoconstriction, pyrexia, inflammation, etc.) which might limit use. The author suggested that philosophical questions concerning abortion vs. fertility control would be raised by the development of a self-administered monthly prostaglandin menstrual inducer. Additional research into prostaglandin mode of action and clinical test trials are needed prior to FDA approval. (RAS)   004682

0581

WILLIS, A.L.; JOHNSON, M.; RABINOWITZ, I.; WOLF, P.L.
    Prostaglandin F2 may induce sickle-cell crisis.
    New England Journal of Medicine 286: 783-784. 1972.

Recently, PGE$_2$ and PGF$_2\alpha$ have been used in human patients to induce labor or abortion. PGE has also been administered to human beings in aerosol form as a bronchodilator. We believe that use of PGs in human beings may be premature, since the role of PGs in pathologic processes has not been sufficiently examined. That systemic administration of PGs is fraught with risk is now becoming more apparent. PGEs are potent promoters of inflammation and fever. PGE$_2$ in minute concentrations renders normal human erythrocytes more rigid and enhances platelet aggregation, and PGF$_2\alpha$ may induce bronchospasm. We have now shown in vitro that PGE$_2$ in concentrations as low as 10 ng/ml can induce sickling of erythrocytes from patients with sickle cell anemia and that this effect can be prevented or reversed by means of urea, a known inhibitor of sickling. Comparable results were not observed with erythrocytes from normal persons. Erythrocytes from fresh heparinized blood were suspended in isotonic saline, and the sickling process followed by light microscopy with the use of wet-mount and Wright-Giemsa stained preparations. Shape changes of the erythrocyte and effects on hemoglobin were also observed by means of light scattering and absorption spectrophotometry. Our findings so far do not exclude the possibility of qualitative differences between the effects of different prostaglandins, such as seen in rat skin and platelets. There is still the possibility, therefore, that administration of specific prostaglandin analogues or prostaglandin antagonists will be of clinical use in future treatment of sickle cell crisis. At present, however, we do urge that administration of PGE$_2$ be absolutely contraindicated in patients with sickle cell anemia or trait and that in general, prostaglandin administration to human subjects be viewed with more caution. (Authors' modified)   003700

0582

WILSON, L., Jr.; BUTCHER, R.L.; CENEDELLA, R.J.; INSKEEP, E.K.
Effects of progesterone on endometrial prostaglandins in sheep.
Prostaglandins 1: 183-190. 1972.

The effects of progesterone on endometrial levels of prostaglandins were investigated in intact and ovariectomized ewes. In the first experiment 17 ewes were ovariectomized bilaterally on Day 4 of the estrous cycle. 8 ewes served as controls and 9 ewes received 10 mg progesterone on Days 4 through 8. All ewes were killed on Day 9. The concentration (ng/g) and content (ng/uterus) of $PGF_2\alpha$ were lower (p<.05) in the progesterone treated ewes (18 and 204, respectively) compared to the controls (64 and 575, respectively). In the second experiment, 20 intact ewes received 40 mg progesterone in corn oil and 20 intact ewes received corn oil only on Days 0 and 1 of the estrous cycle. 10 ewes in each group were killed on Day 5 and 10 on Day 9. A significant increase (p<.05) in $PGF_2\alpha$ content, but not concentration, was detected due to progesterone treatment. $PGF_2\alpha$ did not vary with day or the interaction of day with progesterone treatment. No differences in $PGF_1\alpha$ content or concentration were detected in either experiment. The results suggest that progesterone by itself is not responsible for the previously reported increase in levels of prostaglandin in the endometrium of ewes at Day 14 of the estrous cycle. The luteolytic effect of progesterone given early in the estrous cycle could result in part from increased levels of $PGF_2\alpha$. (Authors' modified)   003885

0583

WILSON, L., Jr.; CENEDELLA, R.J.; BUTCHER, R.L.; INSKEEP, E.K.
Levels of prostaglandins in the uterine endometrium during the ovine estrous cycle.
Journal of Animal Science 34(1): 93-99. 1972.

$PGF_2\alpha$, $PGF_1\alpha$, and PGE_2 were measured in the endometrium from each uterine horn of 35 mature ewes with corpora lutea on only one ovary. The assay procedure combined a thin layer chromatography technique for seperation and gerbil colon bioassay to determine percent recovery of the labelled prostaglandin. Ewes in 4 groups were killed on Day 3, 5, 11, or 14 of the estrous cycle, and the mean concentration (ng/g) for both uterine horns of $PGF_2\alpha$ was 32, 60, 76, and 202, and the mean content (ng) was 261, 344, 311, and 1244, respectively. The levels of $PGF_2\alpha$ at Day 14 were different from all the other days (p<.05). Ewes in a fifth group received a plastic intrauterine device (IUD) in each uterine horn at laparotomy on Day 2 of the estrous cycle and were killed on Day 5. They had a mean concentration (ng/g) and content (ng) for $PGF_2\alpha$ of 179 and 1050, respectively. The concentration and content of $PGF_2\alpha$ were different from the Day 5 controls. No differences in endometrial weights were detected among the treatments. Differences in $PGF_2\alpha$ levels were not detected between uterine horns adjacent or opposite to a corpus luteum. $PGF_1\alpha$ and PGE were detectable but often could not be quantified. PGE_1 was not detected. The authors suggest that $PGF_2\alpha$ may be an active luteolytic substance. (Authors' modified)   004153

0584

WILSON, L., Jr.; BUTCHER, R.L.; INSKEEP, E.K.
Prostaglandin $F_2\alpha$ in ovine uterus during early pregnancy.
In: Abstracts, 5th Annual Meeting of the Society for the Study of Reproduction, East Lansing, Michigan, June 26-29, 1972. p.23.

2 studies examined levels of $PGF_2\alpha$ in endometrium of pregnant and nonpregnant ewes. In Experiment I, 7 pregnant and 7 nonpregnant ewes were laparotomized on Day 13 postestrus, then heparinized, and 40 ml each of jugular and uterine venous blood were collected from each ewe. After sacrifice uterine horns were removed and flushed with saline; the eluted material was observed for the

presence of embryos. Endometrium was dissected from the remainder of the uterus and both were weighed and frozen. Prostaglandins were extracted by solvent partitioning and thin-layer chromatography and quantified by gerbil colon bioassay. Endometrial $PGF_2\alpha$ content (ng) and concentration (ng/g) were greater (p$<$.05) in pregnant (2375 and 152, respectively) than in nonpregnant ewes (1478 and 89). This effect was reflected in an increase (p$<$.05) in $PGF_2\alpha$ (ng/ml) in uterine venous plasma of pregnant (4.7) compared to nonpregnant ewes (1.6). No differences were found in levels of $PGF_2\alpha$ in the remainder of the uterus, and it was not detectable in jugular plasma. In Experiment II, samples were taken as above from pregnant and nonpregnant ewes at Days 11, 12, 13, 14, 15, and 16 and from pregnant ewes at Day 18 postestrus. Endometrial levels of $PGF_2\alpha$ (ng) averaged 144, 448, 969, 1711, 3507, 2130, and 3508, respectively, for the 7 pregnant groups and 337, 751, 1059, 1144, 1526, and 1318, respectively, for the 6 nonpregnant groups, with 2 to 4 ewes per group. (Authors) 004042

0585

WILSON, L., Jr.; BUTCHER, R.L.; INSKEEP, E.K.
Prostaglandin $F_2\alpha$ in ovine uterus during early pregnancy.
Biology of Reproduction 7: 105. 1972.

On day 13 postestrus, jugular and uterine venous blood and endometrial samples were taken from 7 pregnant and 7 nonpregnant ewes. Prostaglandins were extracted by thin layer chromatography and quantified by gerbil colon bioassay. Endometrial $PGF_2\alpha$ content (ng) and concentration (ng/gm) were greater (p$<$.05) in pregnant (2375 and 152, respectively) than in nonpregnant ewes (1478 and 89). There was also an increase (p$<$.05) in $PGF_2\alpha$ (ng/ml) in uterine venous plasma of pregnant (4.7) compared with nonpregnant ewes (1.6). No differences were found in levels of $PGF_2\alpha$ in jugular plasma. In another experiment using 2 to 4 ewes per group, $PGF_2\alpha$ endometrial levels were higher in the nonpregnant ewes on days 11 and 12, similar on day 13, then higher in pregnant animals on days 14, 15, and 16 postestrus. (HC)   004272

0586

WILSON, L., Jr.; BUTCHER, R.L.; INSKEEP, E.K.
Prostaglandin $F_2\alpha$ in the uterus of ewes during early pregnancy.
Prostaglandins 1: 479-481. 1972.

$PGF_2\alpha$ has been postulated to be the uterine luteolytic factor. If this is true, then one might expect the levels of $PGF_2\alpha$ to be lower during early pregnancy. However, in the present study levels of $PGF_2\alpha$ were found to be significantly higher in both content (ng) and concentration (ng/gm) in the endometrium of Day 13 pregnant ewes (2375, and 152, respectively) compared to Day 13 nonpregnant animals (1478, and 89, respectively). These results imply that the luteotropic effects of the embryo must occur by some mechanism other than inhibition of production or release of the uterine luteolytic factor. (Authors)   004083

0587

WIQVIST, N.; BYGDEMAN, M.; TOPPOZADA, M.
Current developments in the use of prostaglandins for induction of abortion.
Research in Prostaglandins 2(3): 1-4. November 1972.

The experience gained from trials of PGs as abortifacients from December 1971 is updated. Combining oxytocin with intravenous PGE_2 was attempted to reduce side effects. A successful trial of intravaginal $PGF_2\alpha$ could not be repeated in 2 centers. Most of the work has been on extraamniotic and intraamniotic administration. Extraamniotic doses are usually 250-750 μg PGF$_2$α or 50-200 μg PGE$_2$every 2 hours, totaling about 10 injections, with over 80%-90% success rates. Continuous infusion offered no advantage; a single 10 mg dose was not well tolerated. A large variety of doses and single or multiple injections by intraamniotic route have been reported without definite consensus on the optimal combination. The 15-methyl-PGF$_2$α analog appeared advantageous; it offered more gradual onset of contractions, more potent and prolonged oxytocic effects, and uterine pain was the predominant side effect. PGF$_2$α was also used for cervical dilation in late first-trimester vacuum aspiration. (LJG)   004812

0588

WIQVIST, N.; BEGUIN, F.; BYGDEMAN, M.; TOPPOZADA, M.
Extra-amniotic administration of prostaglandin for induction of abortion.
In: Bergstrom, S., Green, K. and Samuelsson, B., eds. [Proceedings of the] third conference on
   prostaglandins in fertility control, Stockholm, January 17-20, 1972. Stockholm, WHO Research
   and Training Centre on Human Reproduction, Karolinska Institutet, [1972]. (Prostaglandins in
   Fertility Control 2) p. 118-128.

The main part of this paper contains results of descriptive studies on extraamniotic abortion with recordings of uterine motility and radiographic tracings of PGF$_2$α in the uterus; the appendix contains pooled results from 3 centers (Karolinska Institute, Oxford University, and University of North Carolina) of extraamniotic PGF$_2$α abortions. The local action of PGF$_2$α was demonstrated by injecting 15 IU and 10 IU doses of oxytocin into the extraovular space of a midpregnant woman, without effect. 500 μg PGF$_2$α given about 1 hour later initiated regular contractions. 4 ml of radioopaque dye was applied extraovularly in volunteers; it remained in a pool around the Foley catheter. 500 μg PGF$_2$α mixed with dye, however, evoked contractions in the lower segment around the balloon, which forced the dye to spread, and strong coordinated contractions ensued. Low and high fundal applications of PGF$_2$α were also compared. The high injections, done through a No. 5 French gauge infant-feeding tube, caused the medium to spread like a cap over the fundus. Lower doses were effective, but the high injection is more difficult technically. The uterine response resembled that seen in iv injections, being short in duration, and was probably due to intravenous leakage. Comparing 10 women given 250 μg PGF$_2$α into the fundus with 10 given 500 μg into the low uterine segment, there was no obvious difference in uterine tone, frequency, or intensity of contractions. Low instillation of PGF$_2$α, mean 5.5 mg dose, to 8 women pregnant 7-8 weeks, 12 women pregnant 9-12 weeks, and 47 women in second trimester, aborted all successfully except 5 failures in the midtrimester group. Earlier pregnancies were aborted more rapidly than the mean 23.4 hours in the midpregnant group. (LJG)   004461

0589

WIQVIST, N.; BEGUIN, F.; BYGDEMAN, M.; FERNSTROM, I.; TOPPOZADA, M.
Induction of abortion by extra-amniotic prostaglandin administration.
Prostaglandins 1: 37-53. 1972

Abortion was induced by intermittent extraamniotic administration of prostaglandin F$_2$α in 70 women. It was found that the best results were obtained by separate doses of 250-750 μg PGF$_2$α given in volumes of 1-6 ml every 1-4 hours via a self retaining Foley catheter. A radiological study revealed that a small volume of PGF$_2$α solution injected into the lower uterine segment induces local contractions which force the pool of solution upwards to spread between the uterine wall and the fetal membranes. This mechanism is believed to facilitate the development of forceful, reasonably coordinated contractions that result in expulsion of the conceptus. In the second trimester, abortion occurred within a mean period of 24 hours in 88% of the cases. The incidence of generalized side effects in terms of vomiting and diarrhea was very low compared to that during continuous intravenous infusion of the compound. In 10 late first or early second trimester pregnancies PGF$_2\alpha$ or PGE$_2$was injected extraamniotically during a limited period to dilate the cervix as a pre-operative measure. The stimulating action of prostaglandins on uterine contractility induced a cervical dilatation that made subsequent instrumental evacuation a simple procedure. This form of pre-operative treatment may be utilized to avoid complications in 'border line cases' where the size of the uterus renders primary evacuation hazardous. (Authors' modified)   003740

0590
WIQVIST, N.; BYGDEMAN, M.
Prostaglandins.
Lebanese Medical Journal/Journal Medical Libanais 25(1-2): 127-142. 1972.

The activity of PGs on uterus in vitro and in vivo and for inducing labor and abortion is summarized, mainly from Swedish work. In vitro PGFs stimulate activity in nonpregnant myometrium, and both PGEs and PGFs stimulate isolated pregnant myometrium. However, in vivo PGEs and PGFs stimulate contractions in pregnant and nonpregnant uterus. PGF$_2\alpha$ is luteolytic in sheep, laboratory rodents, and rhesus monkeys, but not in women. PGE$_2$has induced labor in 50% and 95% of 2 series of patients. 100 μg PGE$_2$evoked stronger contractions lasting 3 times as long as those produced by 1 IU oxytocin in a single iv injection. Successful abortion was more likely early in gestation, required at least 100 μg PGF$_2\alpha$ per minute iv, but produced gastrointestinal side effects about every 2 hours. A much lower PG dose and lower incidence of side effects resulted from extraovular administration. Some findings provide a theoretical background for the potential use of PGs for once-a-month contraceptives or for missed menstruation, if side effects can be controlled. (LJG)   005079

0591
WIQVIST, N.; BEGUIN, F.; BYGDEMAN, M.; TOPPOZADA, M.
Recent aspects on systemic administration of prostaglandin.
    In: Southern, E.M., ed. The prostaglandins: clinical applications in human reproduction. (Brook Lodge Symposium, Augusta, Michigan.) Mount Kisco, New York, Futura Publishing Co., 1972. p.295-306.

Blood levels of PGF$_2\alpha$ in 5 midpregnant women given an intravenous infusion of 75 μg per minute for 10 hours ranged between 1.7 and 10.4 ng per ml. Incidence of vomiting and diarrhea and degree of uterine activity correlated with blood levels of PGF$_2\alpha$. Uterine activity in 20 midpregnant women given 20 mg PGE$_2$as a vaginal suppository was much greater than in 10 women given a 50 mg PGF$_2\alpha$ vaginal suppository. 9 of the PGE$_2$group and 2 given PGF$_2\alpha$ aborted. Incidence and duration of vomiting and diarrhea were greater with PGE$_2$. Given as a single intravenous injection, 10 μg 15-methyl PGF$_2\alpha$ elicited a detectable level of uterine contractility in 5 of 8 midpregnant women, approximately 1/20 the dose needed with the parent compound. With 15-methyl PGF$_2\alpha$ the rise in tonus was slower and duration was prolonged. Intravenous infusion of 15-methyl PGF$_2\alpha$ 5 μg per minute for 10 hours in 7 midpregnant women resulted in abortion in 6, the IAT averaging 8 1/2 hrs. The incidence of vomiting and diarrhea was similar to that noted in 10 women given the parent compound at 75 μg per minute. (HC)   004869

0592
WIQVIST, N.; BEGUIN, F.; BYGDEMAN, M.; TOPPOZADA, M.
Vaginal administration of prostaglandin.
    In: Bergstrom, S., Green, K., and Samuelsson, B. [Proceedings of the] third conference on prostaglandins in fertility control, Stockholm, January 17-20, 1972. Stockholm, WHO Research

and Training Centre on Human Reproduction, Karolinska Institutet, [1972]. (Prostaglandins in Fertility Control 2) p. 164-165.

Suppositories containing 50 mg PGF$_2\alpha$ or 20 mg PGE$_2$in a lipid base were administered intravaginally to 2 groups of 8 midpregnant women. The average magnitude of contraction for PGE$_2$was 50-60 mm of Hg compared to 25 mm for PGF$_2\alpha$. PGE$_2$also had a greater frequency and duration of contraction. 5 of 8 women aborted with PGE$_2$, but only 1 of 8 in the PGF$_2\alpha$ cases. Nausea and diarrhea were common side effects in both groups. (RAS)   004459

0593

WITTING, W.C.; WORK, B.A., Jr.; LAROS, R.K., Jr.

Uterine activity response to constant infusion of prostaglandin F$_2\alpha$ in term human pregnancy: a preliminary report.

In: Southern E.M., ed. The prostaglandins: clinical applications in human reproduction. (Brook Lodge Symposium, Augusta, Michigan.) Mount Kisco, New York, Futura Publishing Co., 1972. p. 77-84.

30 term pregnant women with intact, and 30 with artificially or spontaneously ruptured membranes were infused with a constant dose of PGF$_2\alpha$, (5 women in each group, .5, 1, 2, 4, 8, and 16 μg per minute). Parity ranged from 0 to 5; 30 were 0 or 1. Uterine activity, recorded transabdominally or transcervically was higher in those with ruptured membranes, and consistently increased by PG doses above 4 μg per minute, without relation with Bishop score. The report covers the preliminary results from the first 43 patients. 2 infants had Apgar scores of 7 or less at 1 minute and 1 had a score of 6 at 5 minutes. 6 of 43 women had hypertonus or hypersystole during 8 or 16 μg per minute PGF$_2\alpha$. Some impressions gained in this trial were that 4 μg per minute was the optimum PGF$_2\alpha$ dose; amniotomy markedly increased success; PGF$_2\alpha$ may be more effective than oxytocin in patients with an unripe cervix; some women had effective labor with 'inadequate' contraction pattern. (LJG)   004850

0594

WOLFE, L.; MAMER, O.; ROSTWOROWSKI, K.

Prostaglandin levels in human body fluids.

Clinical Research 20: 925. 1972.

PGs were isolated from plasma, CSF, and amniotic fluid, purified on columns and TLC, and measured using gas chromatography-mass spectrometry. Levels of PGF$_2\alpha$ in normal female plasma were at the barely detectable level of .3 ng/ml. Levels in male plasma averaged 1.3 ng/ml. Amniotic fluid from women pregnant 15-18 weeks contained approximately 3.6 ng/ml. Normal CSF levels were 1.0 ng/ml but were 5-6 ng/ml in patients with focal cerebral seizures. (HC)   004966

0595

WOLFE, L.S.; PACE-ASCIAK, C.

Measurement of prostaglandin F$_2\alpha$ concentration in plasma during clinical evaluation in therapeutic abortion.

In: Bergstrom, S., Green, K. and Samuelsson, B., eds. [Proceedings of the] third conference on prostaglandins in fertility control, Stockholm, January 17-20, 1972. Stockholm, WHO Research and Training Centre on Human Reproduction, Karolinska Institutet, [1972]. (Prostaglandins in Fertility Control 2) p. 201-207.

A method for measuring the concentration in the plasma of PGF$_2\alpha$, used for inducing late first and early second-trimester abortion, was described. 20 ml of blood was withdrawn before, during, and

198

after induction with $PGF_2\alpha$. The plasma was twice extracted; the PGs (with 3,4 tetradeutero $PGF_2\alpha$ as carrier) were eluted from a small silicic HA column, were further purified by thin-layer chromatography, and were converted to methyl esters and then to trimethylsilyl ethers. They were introduced into the LKB-9000 mass spectrometer, which detected $PGF_2\alpha$ peaks. Quantities of $PGF_2\alpha$ in the plasma could be calculated from a standard line when values were higher than .3 ng/ml blood. When the $PGF_2\alpha$ was injected intravenously, plasma levels could be directly related to the presence of side effects; when injected intraamniotically, plasma levels remained relatively constant, indicating that diffusion from the amniotic sac is slow. Problems with this method of $PGF_2\alpha$ measurement arose with the use of a deuterated carrier molecule and with the lengthy purification process. Radioimmunoassay may give slightly higher plasma levels. (MLH)   004454

0596
ZOR, U.; LAMPRECHT, S.A.; KANEKO, T.; SCHNEIDER, H.P.G.; McCANN, S.M.; FIELD, J.B.; TSAFRIRI, A.; LINDNER, H.R.
  Functional relations between cyclic AMP, prostaglandins, and luteinizing hormone in rat pituitary and ovary.
  Advances in Cyclic Nucleotide Research 1: 503-520. 1972.

This review covers the work of the authors and others on the role of PGs and cAMP in LH release by LH-RF, and on LH action and protein kinase activity in ovaries in neonatal and adult rats in vitro. The authors used hypothalamic extract as a source of LH-RF and saw elevated adenyl cyclase, cAMP and LH. PGE_1 and theophylline increased cAMP but not LH. In other pituitary systems discrepancies between cAMP and LH release have been observed in Ca^{++} free medium and in high K^+ medium, and in adrenalectomized rats. Neonatal rat pituitaries can respond to adult hypothalamic extracts, but not vice versa. In cow, rabbit, and rat systems in vitro, cAMP can evoke luteinization or LH can increase cAMP production, and in rat, cow and mouse, PGs stimulate cAMP release. The authors could not elicit ovum maturation by cAMP alone in vitro, although PGs added to the cultured follicles stimulated cAMP content. Newborn rat ovaries could not respond to LH with cAMP synthesis, but PGE_2 increased cAMP 6.5 fold. Both LH and PGE_2 stimulate ovarian protein kinase after the first week of life in rats. Discrepancies in these studies may be explained by use of crude hypothalmic extract and whole pituitaries rather than pure components or other experimental conditions. Until these discrepancies are resolved, it is premature to assert that cAMP is involved in LRH action, the authors conclude. (LJG)   004802

0597
ZOR, U.; LAMPRECHT, S.A.; TSAFRIRI, A.; LINDNER, H.R.
  Functional relationships between luteinizing hormone, prostaglandin E_2 and cyclic adenosine 3',5'-monophosphate (cAMP) in the rat ovary during postnatal development.
  In: Abstracts, Fourth International Congress of Endocrinology. Amsterdam, Excerpta Medica. 1972. (International Congress Series Number 256) Abstract 478. p. 190.

Ovaries from rats up to 10 days of age were unresponsive to LH but responded to PGE_2 with increased cAMP formation. cAMP stimulated only slightly protein kinase activity of ovarian homogenates using ovaries from animals less than 7 days old but had a marked effect on samples taken from animals older than 12 days. Either LH or PGE_2 markedly increased cAMP content and protein kinase activity of ovaries from 28-day old rats. The authors conclude that ovarian enzyme activity is acquired during the second week of postnatal development. (HC)   005085

0598

ABDEL-AZIZ, A.; BAKRY, N.

The reactivity of the rat uterus to sympathomimetics during the oestrous cycle and pregnancy and after parturition.

Journal of Reproduction and Fertility 35: 217-223. 1973.

Spontaneous contractions of isolated rat uteri were inhibited by isoproterenol, epinephrine, and norepinephrine. Reactivity did not change during the estrous cycle. During pregnancy and 1 to 3 days after parturition, reactivity was markedly decreased. 4 to 7 days after parturition, reactivity approached that obtained with nonpregnant uteri. The authors suggested that these changes in reactivity may be due to the appearance of excitatory adrenergic receptors during periods of high estrogen levels. Other factors discussed included binding, inactivating enzymes, progesterone levels, and presence of prostaglandins. (HC)   006919

0599

ABEL, M.; TAUROG, J.; NATHANIELSZ, P.W.

A comparison of the luteolytic effect of $PGF_2\alpha$ and cortisol in the pregnant rabbit.

Prostaglandins 4(3): 431-440. September 1973.

Serial plasma progesterone measurements were made by immunoassay on pregnant rabbits given intraarterial cortisol or $PGF_2\alpha$ starting on Day 21 of gestation. Rabbits infused with saline had a brief drop in progesterone, but recovered and delivered at term. Parturition before term exhibited threshold and latency for both agents. A rate of 1.125 ng per hour was required for $PGF_2\alpha$, and 80 μg per hour for cortisol, to induce labor in about 4 days. At higher doses, progesterone fell more sharply until no further decrease in latency to parturition occurred at 2.25 μg per hour for $PGF_2\alpha$ and .3 mg per hour for cortisol. Infusion of cortisol 1200 μg per hour for brief time periods, such as 3 or 6 hours, produced a more gradual fall in progesterone than infusion for 12 or 24 hours at this dose. The authors suggest that $PGF_2\alpha$ may be an intermediate in the process of inducing labor by cortisol. (LJG)   005446

0600

ABEL, M.; SMITH, G.W.; NATHANIELSZ, P.W.

Prostaglandin-induced parturition in the rabbit; quantitative aspects of the route and duration of administration.

Journal of Endocrinology 58(1): xiv. 1973.

Earlier investigation showed that $PGF_2\alpha$ induces parturition when infused intraaortically in pregnant rabbits at a continuous dosage of 2.25 μg per hour with delivery occurring in 35.8 ± 4.4 hours. When same amounts of $PGF_2\alpha$ were infused into the inferior vena cava delivery was more variable. In 6 rabbits infused intravenously with $PGF_2\alpha$, 2 animals delivered at 240 and 280 hours after infusion. The others delivered at 39, 69.5, 69.5 and 77.5. The latency to delivery in the intravenously infused was significantly longer ($p < .05$) in each case than the intraarterial ones. When intraaortically $PGF_2\alpha$ infusion was a short period (2-8 hours) and stopped, the mean latency to delivery was 93.1 ± 16.7 hours (n = 18). Results on 3 animals submitted to different infusion regimes suggest that delivery is accompanied by a fall in plasma progesterone concentrations from 18.9 to 2.6 ng per ml. (Authors' modified)   005519

0601

AHREN, K.; PERKLEV, T.

Effects of PGE_1 and 7-oxa-13-prostynoic acid on the isolated prepubertal rat ovary.

Advances in the Biosciences 9: 717-721. 1973.

Addition of 7-oxa-13-prostynoic acid (7-oxa-13-PA) to isolated whole ovaries from prepubertal rats stimulated ovarian glycolysis in a way similar to that previously found for luteinizing hormone (LH) and prostaglandins (PGs). PGE_1 but not 7-oxa-13-PA increased the rate of amino acid uptake and amino acid incorporation into protein of the isolated ovaries. A high concentration of 7-oxa-13-PA (100 ug/ml) blocked the effect of PGE_1, but this concentration of 7-oxa-13-PA inhibited in itself protein synthesis in the ovary. Lower concentrations of 7-oxa-13-PA did not inhibit ovarian protein synthesis and did not block the effect of PGE_1. (Authors)   006490

0602

AKKAPEDDI, M.K.; HALPERN, D.; DAVIS, R.H.; BALIN, H.

Cervical hydrogel dilator: a new delivery system for prostaglandins.

Abstract Symposium on Controlled Release of Biologically Active Agents. Birmingham, Alabama, Southern Research Institute. April 19-20, 1973. p.165-176

The administration of PGs by means of a cervical hydrogel dilator was investigated in rabbits. With this method the PG is released from the matrix of a swellable hydrophilic polymer inserted at the cervix uteri where it affects both cervical dilation and controlled release of the drug at the uterus. Polymers used in this study were a,omega-polyethylene glycol diacrylates (PEGDA) and dimethacrylates (PEGDMA) and a-acrylamidoglucose; after dialysis in water they were dried in air and vacuum, allowed to absorb PG solutions, and were dried by N2 to predetermined cylindrically shaped rods. PGE_2 and $PGF_2\alpha$ used for this purpose had been converted to tromethamine salts. The rods were inserted into 12-day pregnant New Zealand White rabbits, 1 rod into 1 of the cervices of the 2 uterine horns. Each animal was observed over a period of 24-72 hours for abortion effect. The polymers were thought to be biocompatible with rabbit cervical tissue because there was no evidence of tissue damage or local reaction. Excellent abortion was noted with the equivalent of 1.5 mg $PGF_2\alpha$ incorporated into the polymer, especially when the latter had good swellability. It was concluded that PGs are released at the uterus in sufficient levels, while cervical dilation due to polymer swelling has a synergistic effect inasmuch as the effective dose levels are 1/20 that of other routes of administration. The results with PGE_2 were similar, though less effective. It was found also that the rate of release of the drug may be constant or may increase with time until either the drug is exhausted or the polymer has attained equilibrium swelling, whichever is earlier. This route of administration of PGs was considered favorable. (MLH)   006926

0603

ALAM, N.A.; CLARY, P.; RUSSELL, P.T.

Depressed placental prostaglandin E_1 metabolism in toxemia of pregnancy.

Prostaglandins 4(3): 363-370. September 1973.

Metabolism of PGE_1 by placenta from normal and toxemic pregnancies was estimated by thin layer chromatography. Fresh placentas were homogenized in .1 M phosphate buffer, centrifuged at 15,000 g X 15 minutes, and the supernatant was incubated with tritiated PGE_1. The reaction mixture was extracted and chromatographed on thin layer with standard PGA_1 and PGE_1. 3.3 gm of normal placenta metabolized all of the substrate, 5 ng of PGE_1, in 5 minutes, but eclamptic placenta used only 12% of the substrate. 18 normal placentas produced mean 6.2 ng per gm metabolites in 5 minutes (range 3.6-9.7): 10 preeclamptic placentas (blood pressure 140/90 or more) produced mean

2.4 (range 1.4-3.7): 6 postpartum eclamptic placentas produced mean .5 (range .2-1.3); 2 postpartum eclamptic produced mean 1.8 ng metabolites (range .8-2.9). The products of the reaction were presumably PGA_1 and 15-ketoPGE1, which were not separated by this system. The authors suggest mechanisms by which the placenta may regulate blood pressure during pregnancy. (LJG) 005439

0604

ALDERMAN, B.; THELWALL-JONES, H.

Application of the potentiating effect of prostaglandin E_2 and oxytocin to induced second trimester abortion.

Journal of Obstetrics and Gynaecology of the British Commonwealth 80: 1021-1024. November 1973.

Extraamniotic PGE_2 was combined with intravenous oxytocin to induce abortion in 28 consecutive patients 11-22 weeks pregnant, 12 of whom were primiparae and 16 multiparae. The first 7 patients received only PGE_2 by #14 Foley catheter, 2 ml of a 100 μg per ml solution, then 300 μg hourly until abortion. They aborted in mean 21.3 hours (range 13-33), after a mean dose of 3.6 mg (range 2.6-4.5). The rest were given the same dose of PGE_2 plus iv oxytocin simultaneously at a constant rate of 100 mU. per minute. They aborted in mean 10.1 hours (range 5.5-18.3), after receiving mean PGE_2 dose of 2.14 mg (range 1.2-3.4) and oxytocin 62.4 U. (range 33-109.5). All uteri were explored under general anesthesia within 24 hours, and incidence of complete abortion was not reported. There were no significant differences in abortion interval by gestational age or parity. The first 16 were anesthetized, and all who required it for pelvic pain, received pethidine, which may have contributed to the 57% incidence of vomiting. (LJG)   005827

0605

ALLEN, W.R.; ROSSDALE, P.D.

A preliminary study upon the use of prostaglandins for inducing oestrus in non-cycling thoroughbred mares.

Equine Veterinary Journal 5(4): 137-140. October 1973.

38 nonpregnant, noncycling mares were given 1 or 2 doses of 100-300 μg of ICI-79939, a synthetic analog of $PGF_2\alpha$, by intrauterine infusion. 34 of the animals returned to estrus and ovulated. All were covered, 12 became pregnant, and 9 produced a normal, live foal. The authors concluded that mares not showing regular cycles during the breeding season are in prolonged diestrus rather than anestrus. (HC)   006969

0606

ALLEN, W.R.; ROWSON, L.E.A.

Control of the mare's oestrus cycle by prostaglandins.

Journal of Reproduction and Fertility 33: 539-543. 1973.

It was suggested that uterine luteolysin reaches the ovaries through general circulation in the mare, rather than by direct pathway from the uterine horn to the ovary as it does in the cow, sheep, and pig. This study describes a new synthetic analogue of $PGF_2\alpha$ (ICI-79939) which exhibits a high luteolytic potency in the mare. 14 Welsh pony mares were given ICI-79939 at various stages of diestrus for a total of 40 estrus cycles. The results indicate that ICI-79939 is highly luteolytic when administered between the fourth and thirteenth days of diestrus. This compound has 200 times the luteolytic potency of natural $PGF_2\alpha$, and small amounts are sufficient to induce luteolysis (less than 100 μg). This finding may be of significance to horse breeders. (IC)   006434

0607

ALSAT, E.; CEDARD, L.

The stimulatory action of the prostaglandins on the production of oestrogens by the human placenta perfused in vitro.

Prostaglandins 3(2): 145-153. February 1973.

Human placentas were perfused for 2 to 2 1/2 hours in a medium containing testosterone. Estrone and estradiol content was measured fluorometrically at 30-minute intervals. $PGE_2$1.5 or 3×10^{-6}M added to the bath for 1 hour doubled the estrone content. $PGF_2\alpha$ at 3×10^{-6}M markedly increased the estradiol content. $PGF_2\alpha$ at 8×10^{-6}M and PGE, at 2×10^{-6}M markedly elevated both estrone and estradiol levels. (HC)   006569

0608

AMOROSO, E.C.; HARRISON, F.A.; HEAP, R.B.; POYSER, N.L.

The production of prostaglandin $F_2\alpha$ by the uterus of the sheep.

Journal of Endocrinology 57(1): p59. 1973.

The accumulation of large volumes (to 1900 ml) of uterine fluid containing high concentrations of $PGF_2\alpha$ (up to 4 μg/ml) in sheep having a transplanted ovary and maintenance of plasma progesterone levels similar to those of the luteal phase of the normal sheep estrous cycle led to 2 experiments to test the hypothesis that progesterone is involved in the stimulation of fluid accumulation and $PGF_2\alpha$ production in the sheep uterus. 1) A ewe having 1 ovary removed was given progesterone injections daily for 115 days during which no behavioral estrus was observed and after which 188 ml of uterine fluid containing 15.6 μg $PGF_2\alpha$ were aspirated. 2) To utilize endogenous secretion of progesterone during pregnancy, 5 sheep had an endometrial septum constructed in 1 uterine horn before mating, and 4 had normal pregnancies. At laparotomies performed 109-149 days after mating, 127-180 ml fluid containing 28-1500 ng $PGF_2\alpha$/ml were collected. Fluid collection after lambing also contained high concentrations of $PGF_2\alpha$. Conclusions are that progesterone plays an important role in accumulation of fluid rich in $PGF_2\alpha$ in the uterine lumen of sheep. (IC)   006454

0609

AMY, J.J.; KARIM, S.M.M.; SIVASAMBOO, R.

Intra-amniotic administration of prostaglandin 15(S)15-methyl-E_2methyl ester for termination of pregnancy.

Journal of Obstetrics and Gynaecology of the British Commonwealth 80: 1017-1020. November 1973.

Investigation on 2 different dosages of intraamniotic administration of prostaglandin 15(S)15-methyl-PGE_2methyl ester was reported. 18 of 20 patients at 14-22 weeks gestation were successfully aborted after 1 intraamniotic injection of 100 μg of this compound. The remaining 2 patients required a second injection 24 hours after the first. The mean induction-abortion interval was 16.5 hours. Side effects were limited to a rise in temperature (1.0 degrees C to 1.5 degrees C) in 3 patients and vomiting in 3 patients. Of 10 patients given 200 mg. of this compound, 9 aborted and 1 patient required 2 injections. A temperature elevation of 1.6 degrees C in 1 patient and vomiting in 2 patients were noted. The mean induction-abortion interval was 17 hours. This difference from the induction abortion-interval in the other group was nonsignificant. (AS)   005821

0610
ANDERSON, G.G.; SPEROFF, L.
Clinical use of prostaglandins in reproduction.
In: Ramwell, P.W., ed. The prostaglandins, Vol. 1. New York, Plenum Press, 1973. p. 365-389.

The use of PGE_2 and $PGF_2\alpha$ for term labor and abortion are considered with emphasis on the first published trial, the authors' experience, and hormone measurements. Term labor was first induced by Karim with .05 μg per kg per minute $PGF_2\alpha$ in 29 of 35 women. Threshold doses are estimated to be 2 μg per minute PGE_2 and 40 μg per minute $PGF_2\alpha$. Several double blind trials have not revealed differences between oxytocin and PG, including one study carefully controlled for inducibility in multiparae, using 2.5-40 μg $PGF_2\alpha$, .3-4.8 μg PGE_2, and 1-16 mU oxytocin per minute iv. Hormone assays have shown a decrease in estriol in PG induction but an increase with oxytocin. The first trials of PG for abortion were by Karim and by Roth-Brandel published in 1970, by iv and sc routes. Karim has done a wide variety of trials with the highest success rates, perhaps related to his smaller patients and longer infusions. In the author's trial with iv $PGF_2\alpha$ in 43 patients, half had complete or incomplete abortions, with side effects correlated with dose rate. Of the iv, sc, vaginal, oral and local routes, the author prefers intraamniotic, first used in 1968, because it requries 1/10 the dose and has limited side effects. Hormone assays have not shown a decline in hydroxyprogesterone, an indicator of human corpus luteum steroidogenesis, but there are decreases in estriol, estradiol, human chorionic somatomammotropin during abortion. (LJG)   006911

0611
ANDERSON, G.G.
Induction of term labor with intravenous $PGF_2\alpha$; a review.
Prostaglandins 4(5): 765-774. November 1973.

Since the first trials of $PGF_2\alpha$ and PGE_2 for induction of term labor in 1971, clinicians have questioned whether PGs are more effective or safer than oxytocin. 6 recent studies have employed the Bishop score of inducibility, double-blind protocols, and standardized dose regimens, and have resulted in comparable efficacy for PG and oxytocin. A few reports suggest that $PGF_2\alpha$ may be more effective than oxytocin at any stage of gestation and in missed abortion, intrauterine death, molar pregnancy, anencephaly, and possibly very low Bishop scores. $PGF_2\alpha$ has been used in small numbers of high risk patients and patients with ruptured membranes and unfavorable cervices. In a double-blind study 23 newborns were evaluated extensively until 72 hours old without finding any differences between $PGF_2\alpha$ and oxytocin induction. The only significant drawbacks with $PGF_2\alpha$ are uterine hypertonus, which can be reversed by stopping the PG infusion, and a narrower range of safety between adequate and over response. Unlike oxytocin, PGs have no antidiuretic action. (LJG)   005647

0612
ANDERSON, G.G.; HOBBINS, J.C.; SPEROFF, L.
Intravenous prostaglandins E_2 and $F_2\alpha$ for the induction of term labour.
In: Bossart, H., Cruz, J.M., and Huber, A., eds. Perinatal medicine. (Third European Congress on Perinatal Medicine, Lausanne, 1972.) Bern, Hans Huber, 1973. p. 368.

A total of 100 women had term labor induced by iv PGE_2 .3-4.8 μg/minute with 83% success, $PGF_2\alpha$ 2.5-40 μg/minute with 76% success, or oxytocin 1-16 mU/minute with 85% success. PGs and oxytocin were equally efficacious, even with comparable Bishop inducibility scores; for example, difficult inductions with a score of 2 took a mean 10 hours, but easy inductions with a score of 12 took 4 hours to deliver. PGs and oxytocin were also equally safe with respect to maternal vital signs, fetal heart rate, and neonatal condition, although the PG group had more episodes of innocuous uterine hypertonus or tachysystole. The infusion-to-delivery interval averaged 1 hour shorter in the PG inductions. PGF levels in maternal circulation increased 4-fold during labor over preinduction levels, and cord blood levels were significantly higher than maternal levels. PGF levels were higher in the PGF group, as compared with the oxytocin group, but the curves of PGF and oxytocin groups were alike. (LJG)   005095

0613

ANDERSON, G.G.; HOBBINS, J.C.; RAJKOVIC, V.; GOLDSTEIN, L.; SPEROFF, L.; CALDWELL, B.V.
Midtrimester therapeutic abortion using intraamniotic $PGF_2\alpha$.
Advances in Biosciences 9: 539-543. 1973.

82 patients received an initial intraamniotic injection of 40 mg $PGF_2\alpha$ except for a few who received the 40 mg over the first hour. A second group of 40 patients received 40 mg initially with a concomitant infusion of 66 mU/minute of Syntocinon, and 13 received an initial 40 mg followed by 20 mg in 6 hours. 28 patients received a second dosage at 24 hours if they did not appear to be progressing toward abortion. The results obtained show that there was no significant difference in the treatment-abortion interval or in the success rate between the group that received the initial high dose and the group that also received Syntocinon. However, the treatment-abortion interval in the group that received an additional 20 mg in 6 hours appeared to be significantly shorter. (AS)   006791

0614

ANDERSON, G.G.; HOBBINS, J.C.; SPEROFF, L.; CALDWELL, B.V.
The use of prostaglandins for the induction of labor.
Journal of Reproductive Medicine 10(3): 121-124. March 1973.

The authors review the use of prostaglandins for labor induction. Early studies differed in protocol, conduct of study, analysis of results, and use of different prostaglandins with and without other agents. Preassessment of the patients' inducibility and double-blind well-controlled studies can be expected to lead to better data. Intravenous $PGF_2\alpha$ is considered as effective as oxytocin in inducing labor and will initiate contractions at any time during pregnancy. Intravenous PGE_2 is 10 times more potent than $PGF_2\alpha$ but has no advantage in terms of time of labor or success rate. Oral PGE_2 is equal to oxytocin for inducing labor. 15-methyl PGE has both the activity and the side effects of the parent compound, can be used in lower doses, and has a longer duration of action. Early alarm about side effects was based on much larger doses that were being used for abortion. Future developments in this area may come from studies using low oral doses of PGE to advance the patient's inducibility and studies in which prostaglandins are used together with oxytocin, as well as the development of better prostaglandin derivatives. (HC)   006780

0615

ANONYMOUS
A system for controlled delivery of prostaglandins for inducing abortions.
Chemical and Engineering News 51(18): 16. April 30, 1973.

A syposium on controlled release of biologically active agents sponsored by Southern Research Institute was told of preliminary results of in vivo tests on rabbits, using partially deacetylated cellulose acetate capsules of 2 mm diameter containing $PGF_2\alpha$. The capsules release the drug in a controlled manner and cause smooth muscle movement associated with abortion in the rabbit oviducts. (Author's modified)   006527

0616

ANONYMOUS

Discussion.

In: Jacomb, R.G. ed. The use of prostaglandins E_2 and $F_2\alpha$ in obstetrics and gynaecology. (Proceedings of the Upjohn Prostaglandins Symposium, London, September 21, 1972.) Miami, Florida, Symposia Specialists, 1973. p. 99-106.

A discussion among 6 prostaglandin researchers touched upon various subjects: 1) the problem of platelet adhesiveness stimulated by prostaglandins during abortion, 2) the relationship of mechanical stimulation of a catheter to the stimulating effect of the prostaglandin, and 3) the merits of intraamniotic vs. extraamniotic administration of prostaglandins. The chairman, A.C. Turnbull, suggested that prostaglandins are the method of choice for midtrimester terminations. (RAS) 006958

0617

ANONYMOUS

Intrauterine administration of prostaglandin $F_2\alpha$ for induction of abortion: intraamniotic method.

In: Bergstrom, S., ed. Report from 1972 meetings of the prostaglandin task force steering committee, Chapel Hill, June 8-10, 1972, Stockholm, Ocotober 2-3, 1972, Geneva, February 26-28, 1972. Stockholm, WHO Research and Training Centre on Human Reproduction, Karolinska Institutet, 1973. (Prostaglandins in Fertility Control 3) p. 27-28.

An intraamniotic abortion method for pregnancies over 14 weeks using injections of 25 mg $PGF_2\alpha$ into the amniotic sac through a catheter is described. The catheter remains in situ after the preliminary injection for another injection of 25 mg after 6 hours unless abortion has already occurred or complications arise. The physician should manage incomplete abortions at his discretion but not before 48 hours following initiation of $PGF_2\alpha$ method. (IC)   006446

0618

ANONYMOUS

Intrauterine administration of prostaglandin $F_2\alpha$ for induction of abortion: extraamniotic method.

In: Bergstrom, S., ed. Report from 1972 meetings of the prostaglandin task force steering committee, Chapel Hill, June 8-10, 1972, Stockholm, October 2-3, 1972, Geneva, February 26-28, 1972. Stockholm, WHO Research and Training Centre on Human Reproduction, Karolinska Institutet, 1973. (Prostaglandins in Fertility Control 3) p.29-30.

To induce abortion prostaglandin $F_2\alpha$ solution (250 μg/ml) is injected into the lower segment of the uterus between the uterine wall and the fetal membranes through a Foley catheter. The catheter is introduced through the cervix and the balloon prefilled to maintain the catheter tip within the lower uterine segment. The catheter remains in situ to allow periodic injections of $PGF_2\alpha$. Initial dosage is 250 μg + 500 μg at 15 min then 750 μg at 2 hour intervals. Abortion should take place within 36 hours. (IC)   006447

0619

ANONYMOUS

Intrauterine administration of prostaglandin $F_2\alpha$ or 15-methyl $PGF_2\alpha$ for induction of abortion.

In: Bergstrom, S. ed. Report from meetings of the prostaglandin task force steering committee, Chapel Hill, June 8-10, 1972, Stockholm, October 2-3, 1972, Geneva, February 26-28, 1972. Stockholm,

WHO Research and Training Centre on Human Reproduction, Karolinska Institutet, 1973. (Prostaglandins in Fertility Control 3) p. 79.

A suggested protocol for the use of $PGF_2\alpha$ as an abortifacient is presented. 40 mg of $PGF_2\alpha$ or 2.5 mg of the 15-methyl analog is injected intraamniotically. Analgesics, anti-emetics and antibiotics are administered as needed. Results are considered successful if abortion occurs within 48 hours. (HC) 006789

0620
ANONYMOUS
Prostaglandin on the US market.
Nature 246(5434): 444. December 21/28, 1973.

This report announces the introduction to the United States market of Prostin $F_2\alpha$ for the induction of abortion during the second trimester of pregnancy. Only university medical centers will be sold the drug, and it will be administered intraamniotically. (HC)   006933

0621
ANONYMOUS
Prostaglandin used as postcoital abortifacient.
Ob. Gyn. News 8(3): 39. Feb. 1, 1973.

10 healthy pregnant women were given $PGF_2\alpha$ to determine its effectiveness as an abortifacient under special conditions. 8 aborted when their levels of human chorionic gonadotrophin (HCG) showed a significant drop according to Lt. Cdr. Donald R. Tredway, MC, USN, at the meeting of the Armed Forces District meeting of the American College of Obstetricians and Gynecologists. No patient was more than 14 days past her expected menses, and all of them received the agent intravaginally by intermittent instillation during a 48-hour period. The total dosage ranged from 200 to 600 mg. Serum HCG and progesterone levels were measured by radioimmunoassay just before the treatment was started at frequent intervals for 48 hours, and at 1 week and 2 weeks after treatment. All patients experienced significant side effects, mainly diarrhea and vomiting. (Authors' modified)   006284

0622
ANONYMOUS
Prostaglandins and abortion.
People 1(1): 28-29. October 1973.

The author traces briefly the historical development of the use of PGs for abortion. In initial studies the drugs were given intravenously, but unpleasant side effects occurred. The intrauterine and intraamniotic routes have been used successfully. 1 dose of $PGF_2\alpha$ given intraamniotically has produced abortion within 12 hours with few side effects in second-trimester pregnancies. (HC) 006720

0623
ANONYMOUS.
Prostaglandins and the uterus.
Lancet 2(7833): 829-830. October 13, 1973.

Current prospects for use of PGs in a once-a-month contraceptive, as a midtrimester abortafacient, and for induction of labor are surveyed. PGs will probably not become the hoped-for monthly contraceptive because they are not luteolytic in humans and do not evoke true menstruation. Only a few clinicians favor PGs for first trimester abortion or for cervical dilatation adjunctive to surgical evacuation because of the unacceptable side effects. For midtrimester abortion, PGs are safer than hysterotomy or hypertonic saline, and concomitant oxytocin can raise success rates and lower required dose of PG. For midtrimester abortion, major research emphasis is on doses, routes, combined therapy, and actual frequency of side effects and complications. PGs are not used so commonly for induced labor as oxytocin although PGs are less damaging to fetus, placenta, and mother's water balance. PGs may be the choice in cases of fetal death, preeclampsia, missed abortion, hydatidiform mole, or cardiac disease. The merits of iv, oral, vaginal, and intrauterine routes are being debated and will probably depend on the individual circumstances. (LJG)   005467

0624

ARCHER, D.F.; PETRILLI, E.S.
Alterations in biosynthesis of progesterone in human corpora lutea of pregnancy by prostaglandins.
Advances in the Biosciences 9: 669-672. 1973.

Human corpora lutea obtained at midpregnancy were incubated with prostaglandin $F_2\alpha$ and prostaglandin E_1 in vitro in concentrations of 50 $\mu g/ml$. Tissue progesterone as well as the incorporation of carbon 14 labelled sodium acetate into progesterone were analyzed. Procedural losses were corrected by the addition of progesterone-7-alpha-tritium as internal standard. The results indicate that prostaglandin $F_2\alpha$ in the concentrations used in this study significantly decreased the incorporation of carbon 14 labelled sodium acetate to progesterone while having little significant effect on the endogenous levels of progesterone. The preliminary data available for prostaglandin E_1 does not show a statistically significant difference between progesterone levels and the incorporation of carbon 14 labelled sodium acetate into progesterone. (Authors)   006484

0625

ARCHER, D.F.
Effects of prostaglandin $F_2\alpha$ on cholesterol biosynthesis by human corpora lutea of pregnancy.
Biology of Reproduction 9: 95. 1973.

Corpora lutea from women 14 to 18 weeks pregnant were incubated with carbon-14-sodium acetate and various concentrations of $PGF_2\alpha$. There was no significant effect of $PGF_2\alpha$ on cholesterol content of carbon-14-sodium acetate or biosynthesis suggesting no significant alteration of endogenous cholesterol or of biosynthesis of cholesterol. (HC)   005523

0626

AREF, I.; HAFEZ, E.S.E.; KAMAR, G.A.R.
Postcoital prostaglandins, in vivo oviductal motility, and egg transport in rabbits.
Fertility and Sterility 24(9): 671-676. September 1973.

$PGF_2\alpha$ and PGE_2 were given sc to rabbits 12 or 24 hours after coitus, and their effect on oviduct motility and egg transport were examined. Oviduct motility was recorded by indwelling microballoon, placed 3 days before mating, and eggs were flushed from 4 separate oviduct segments and the uterus, and counted. 12 mg $PGF_2\alpha$, given 24 hours after coitus, stimulated amplitude and duration of contractions for 12 hours after a 15-20 minute delay; 6 mg $PGF_2\alpha$ evoked an intermediate response; 3 mg stimulated motility for 3-4 hours after a 2-2.5 hour delay. 12 mg PGE_2 decreased motility for 3-4 hours after a 10-15 minute delay, which was followed 5 hours later by a 2-3 hour compensatory stimulation. 3 mg PGE_2 inhibited motility for 1-2 hours, followed by .5-1 hour stimulation. In untreated rabbits 89% of eggs were found in the third oviduct segment (third from fimbrial end). 12 mg $PGF_2\alpha$ given 24 hours after mating sent 63% of eggs to the fourth segment and 36% into the uterus 24 hours later; 4 uterine eggs were unfertilized and 6 were 16-32 cell morulae. 3 or 12 mg PGE_2 did not stimulate egg transport into the uterus. Knowledge of the developmental fate of eggs whose transport was stimulated by PG would be useful in formulating a PG postcoital contraceptive. (LJG)   005500

0627

ARMSTRONG, D.T.; MOON, Y.S.; GRINWICH, D.L.
Possible role of prostaglandins in ovulation.
Advances in the Biosciences 9: 709-715. 1973.

Indomethacin, an inhibitor of prostaglandin biosynthesis, was effective in blocking luteinizing hormone (LH)-induced ovulation in rabbits when administered intravenously 30 min before, at the same time as, or as late as 5 hr after the gonadotropin. Luteinization was not prevented in those follicles in which ovulation was blocked, and the acute steroidogenic response of the interstitial tissue was not interfered with. Intrafollicular injections of indomethacin 5 hr after LH similarly blocked ovulation, but not luteinization. Control follicles injected with the phosphate buffer vehicle underwent normal ovulation and luteinization. These findings provide indirect evidence in support of a role for prostaglandins, acting at the follicular level in the process of ovulation. (Authors' modified)   006489

0628

ASH, R.W.; HEAP, R.B.
The induction and synchronization of parturition in sows treated with I.C.I. 79,939, an analogue of prostaglandin $F_2\alpha$.
Journal of Agricultural Science 81: 365-368. 1973.

Compound ICI (Imperial Chemical Industries) 79,939 was given in divided doses of 50-250 μg, totaling mean 530 μg im, to 2 Landrace sows, 1 Large White sow, and 7 Large White gilts on days 109-111 of pregnancy to induce delivery. 79 normal sows and 68 gilts from the same herd delivered in a mean 114 $\pm$ .2 days, and only 1 of the 147 delivered on Day 111. The treated pigs delivered in about 26 hours, a mean gestation length of 111 $\pm$ .2 days. Piglets were usually born within 3-5 hours, placentae expelled within 4 hours of the first piglet. No adverse clinical signs were observed. The piglets weights at birth and at 3 weeks were normal. 22 piglets died, 15 of which were from gilts which failed to lactate; 77 survived to 3 weeks; and 20 were used experimentally so that details of their survival were unknown. (LJG)   006420

0629

AULETTA, F.J.; SPEROFF, L.; CALDWELL, B.V.
Prostaglandin $F_2\alpha$ induced steroidogenesis and luteolysis in the primate corpus luteum.
Journal of Clinical Endocrinology and Metabolisn 36(2): 405-407. 1973.

Monkeys pretreated with FSH, LH, and HCG were given $PGF_2\alpha$ at progressively increasing concentrations by intravenous or intraarterial infusion. In 2 animals, $PGF_2\alpha$ up to 1 μg/minute increased progesterone concentration in the ovarian vein. At 50 μg/minute, progesterone concentration and output decreased markedly. In 1 animal given $PGF_2\alpha$ intraarterially, low doses had both stimulatory and inhibitory effects. 1 animal given 10 ng/minute intraarterially on Day 22 had an increase in progesterone concentration. An increased infusion of 50 μg/minute markedly decreased progesterone concentration and output in this animal. $PGF_2\alpha$ had no effect on ovarian venous blood flow and caused no histological change in ovarian structure. (HC)   006781

0630
BABEJ, M.; SANDOW, J.; KIRCHER, E.
Interaction of prostaglandins and LH-RH in vitro.
Acta Endocrinologica (Suppl.) 173: 81. 1973.

It was shown in previous work that pituitary cAMP was stimulated after in vitro incubation with prostaglandins and that the pituitary adenyl cyclase system was also stimulated by natural and synthetic releasing hormones (GH-RH, LH-RH) with the simultaneous release of pituitary hormones. A test model used was the effect of synthetic LH/FSH-RH on the gonadotropin release from isolated rat anterior pituitaries to see if prostaglandins also stimulated release of pituitary hormones. Pituitary halves were incubated in KRBG medium, hormone content of the medium was determined, and PGE_2, PGA_2, and $PGF_2\alpha$ were added. The modulating influence on the effect of a standard dose of LH-RH was investigated. Results were that the effects of PGE_2 and $PGF_2\alpha$ on FSH release by LH-RH in vitro are reciprocal and vary with increasing doses of PG. The LH release is also reciprocally influenced by PGA_2 and $PGF_2\alpha$. (IC)   006456

0631
BADRAOUI, M.H.H.; BONNAR, J.; HILLIER, K.; EMBREY, M.P.
Blood coagulation changes during mid-trimester abortion induced by prostaglandin $F_2\alpha$.
British Medical Journal 4(5889): 375-378. November 17, 1973.

Serial studies on the coagulation system were made during second-trimester abortion induced by extraamniotic, intraamniotic, vaginal, and intravenous prostaglandin $F_2\alpha$. No significant changes were found in the prothrombin time, partial thromboplastin time, or level of fibrin-fibrinogen degradation products. An increase in the activity of Factor X occurred with all routes of administration; the activity of Factor VIII increased during extraamniotic, vaginal, and intravenous administration; Factor V activity increased during extraamniotic and vaginal administration; and the activity of Factor VII-X complex increased slightly at the time of abortion with all methods. The findings suggest that though prostaglandin induction of second-trimester abortion produces changes in the coagulation system, the effects are much less than those which accompany induction of abortion by hypertonic saline or abdominal delivery in late pregnancy. (Authors)   005826

0632
BADRAOUI, M.H.H.; BONNAR, J.; HILLIER, K.; EMBREY, M.P.
Coagulation changes during termination of pregnancy by prostaglandins and by vacuum aspiration.
British Medical Journal 1: 19-21. January 6, 1973.

750 μg doses of $PGF_2\alpha$ were instilled at 2-hour intervals into the extraamniotic space of 12 women pregnant 14-20 weeks. Prothrombin time decreased and the levels of coagulation factors rose during uterine contractions and at the time of abortion. No such changes occurred in patients pregnant 8-12 weeks having abortions by vacuum aspiration. The authors concluded that these changes are probably related to the physiological changes which normally occur in the hemostatic system in the second trimester. (HC)   006882

0633

BAIRD, D.T.; COLLETT, R.A.; FRASER, I.S.; KELLY, R.W.; LAND, R.B.; WHEELER, A.G.
Progesterone secretion from the ovary in the ewe following infusion of uterine venous plasma.
Journal of Reproduction and Fertility 35: 13-22. 1973.

The luteolytic activity of uterine venous plasma collected from donor ewes at different stages of the estrous cycle was tested by observing the change in the rate of progesterone secretion from the CL of recipient ewes with the ovary autotransplanted to the neck. Infusion of uterine venous plasma decreased progesterone secretion 20% in the first hour, but about 15% in the next 4-5 hours in ewes given plasma from donors in cycle Days 14-46. Mean progesterone secretion rate during infusion decreased 9.6% with Day 10 donor plasma and 33% with Day 15 plasma. The mean $PGF_2\alpha$ concentration (1.38 ng/ml) as measured by combined gas chromatography/mass spectrometry in uterine venous plasma collected around the time of luteal regression was greater than that in plasma collected on Days 9, 10, or 11 (.47 ng/ml). Cannulation of the uterine vein on the same side as a corpus luteum prevented 3 donors from going into estrous within 18 days. These findings confirm the increase in luteolytic activity of uterine venous plasma collected at the time of expected luteal regression. (Authors' modified)   005510

0634

BALLARD, C.A.; QUILLIGAN, E.J.
Intraamniotic prostaglandin $F_2\alpha$ for midtrimester abortion.
Advances in the Biosciences 9: 551-554. 1973.

Prostaglandin $F_2\alpha$ was administered by the intraamniotic route to 30 patients for induction of legal abortion. Ten patients received 15 mg followed 24 hr later with an additional 15 mg if abortion had not occurred. Twenty patients received 25 mg followed with a repeat 25 mg in 24 hr. Forty-eight hr without abortion was considered a failure. The success rate was dose dependent, only 40% aborting with the lower dose as opposed to 80% with the higher dose. There was no difference in gastrointestinal side effects in the two groups, and the majority of patients had only mild symptoms. (Authors)   006472

0635

BARCIKOWSKI, B.; SAKSENA, S.K.; BARTKE, A.
Androgenic regulation of plasma prostaglandin F levels in the rat.
Journal of Reproduction and Fertility 35(3): 549-551. 1973.

Levels of PGF in plasma of rats were decreased 1 to 2 weeks after castration. Daily treatment with testosterone 100 or 250 μg subcutaneously for 7 days prevented this decrease. The authors conclude that androgen-dependent tissues may be an important source of PG. (HC)   005760

0636

BARCIKOWSKI, B.; CARLSON, J.C.; WILSON, L.; McCRACKEN, J.A.
The effect of endogenous and exogenous estradiol-17β on the release of prostaglandin $F_2\alpha$ from the ovine uterus.
Biology of Reproduction 9: 70. 1973.

Samples of blood from the utero-ovarian vein were collected periodically throughout the estrous cycle of 6 ewes. Estrogen peaked in plasma on Day 3-4 and on Day 14 coincidental with a 3-8.5 ng per ml $PGF_2\alpha$ peak. On Day 15 larger $PGF_2\alpha$ peaks (9-23 ng per ml) along with a fall in progesterone occurred. Infusion of .5-1 ng per minute estrogen for 6 hours into the arterial supply of the autotransplanted uterus during the late luteal phase resulted in a marked release of $PGF_2\alpha$ from the uterus. Given systemically, estrogen did not have this effect. The authors suggest that both estrogen and progesterone play a role in the synthesis and/or release of $PGF_2\alpha$ and that this effect is mediated directly on the uterus and not through some indirect mechanism such as the pituitary. (HC)   005522

0637

BARR, W.

Oral prostaglandin E_2 in the induction of labour.

In: Jacomb, R.G., ed. The use of prostaglandins E_2 and $F_2\alpha$ in obstetrics and gynecology. (Proceedings of the Upjohn Prostaglandin Symposium, London, September 21, 1972.) Miami, Florida, Symposia Specialists, 1973. p. 21-24.

28 women pregnant 39-41 weeks, almost all primigravidae, after amniotomy were given PGE_2 orally 1 mg initially followed by 1.5 or 2 mg doses at 2-hour intervals. 11 of the 23 primigravidae delivered within 12 hours; the average dose was 5.2 mg. The remaining patients delivered with a supplementary infusion of oxytocin. Side effects were common; 18 women vomited. The authors concluded that this method is less effective than intravenous oxytocin for inducing labor in primigravid patients. (HC) 006959

0638

BARTKE, A.; MUSTO, N.; CALDWELL, B.V.; BEHRMAN, H.R.

Effects of a cholesterol esterase inhibitor and of prostaglandin $F_2\alpha$ on testis cholesterol and on plasma testosterone in mice.

Prostaglandins 3(1): 97-104. January 1973.

Mice were injected subcutaneously with 50, 100 or 200 μg $PGF_2\alpha$ 2 times a day for 3 1/2 days. The level of esterified cholesterol in the testes increased significantly; there was no change in level of free cholesterol or in the weight of the testes or seminal vesicles. Testosterone levels in plasma were significantly reduced, probably due to reduced steroidogenesis in the testis. (HC)   006787

0639

BATTA, S.K.; LABHSETWAR, A.P.

Do prostaglandins stimulate LH release and thereby cause luteolysis?; reply.

Prostaglandins 4(5): 626-628. November 1973

The author (S. K. Batta) repeats results on intracarotid PGs because in his opinion A.P. Labhsetwar (Prostaglandins 3: 729, 1973) had cited them incorrectly. In spayed female rats pretreated with estradiol benzoate, PGE_1 increased LH at 2-30 minutes and FSH at 6-8 minutes; $PGF_2\alpha$ increased LH at 6-8 minutes, decreased LH to negligible values at 30 minutes, and increaaed FSH at 2-60 minutes. In male rats PGE_1 increased LH at 5-10 minutes. A.P. LABHSETWAR replied below that in his estradiol benzoate treated female rats, LH rose within minutes, and rose significantly but not as high after injecting the vehicle, phosphate buffer pH 7.4. FSH, however, did not rise in plasma after $PGF_2\alpha$ was injected into the carotid or third ventricle. He felt that he had not misquoted BATTA'S work, and agreed that PGs stimulate gonadotropin secretion, whatever the physiologic significance of these observations. (LJG)   005631

212

0640

BATTA, S.K.; PIVA, F.; MARTINI, L.
Effect of prostaglandins E_1, E_2, and $F_2\alpha$ on gonadotrophin-induced ovulation in immature female rats.
Advances in the Biosciences 9: 723-730. 1973.

PMS induced ovulation in 28 of 36 rats and the number of ova shed per animal was 25.6. PGE_1 given at 10 or 50 μg per rat increased the number of ova shed to 38.3 and 49.5 per animal, respectively, while the percent of induced ovulations remained significantly unchanged. $PGF_2\alpha$ at 10 μg or 50 μg per rat lowered the percent of animals ovulating but did not effect the number of ova shed per ovulating rat. Nembutal totally blocked PMS-induced ovulation. In a second series of experiments, PMS and HCG in combination produced ovulation in 93.9% of the rats with 33.6 ova shed per ovulating animal. Nembutal given 52 hours after PMS did not reduce significantly the number of animals who ovulated, but only 19 ova were found for each rat. 10 μg PGE_1 and 10 μg $PGF_2\alpha$ had 32% and 80% animals ovulating and 8.8 and 41.2 ovas shed per animal. (RAS)   006641

0641

BEATTY, C.H.; BOCEK, R.M.; YOUNG, M.K.
Effect of oxytocin and epinephrine on the adenylate cyclase activity of myometrium from pregnant rhesus monkeys.
Hormone and Metabolic Research 5(3): 213-215. 1973.

The effects of oxytocin and epinephrine with or without caffeine on adenyl cyclase activity of myometrial slices from term pregnant rhesus monkeys (average gestation 165 days, range 145-160) were examined. Adenyl cyclase activity was determined from the formation of carbon-14 labeled cAMP from labeled ATP formed by incubation with adenine-8-carbon-14. 90-150 mg myometrial slices were incubated for 60-90 minutes with labeled adenine, followed by incubation in fresh Krebs buffer with test compounds for 10 minutes. 2 mM caffeine stimulated adenyl cyclase from 876 to 1240 dpm per 100 mg, a 42% increase (p<.005), and 1×10^{-7}M epinephrine increased activity further to 1763 dpm, another 42% increase.(p<.005). In a second series, 1 mM caffeine heightened adenyl cyclase from 765 to 884 dpm per 100 mg, a 16% increase (p<.02), and 2.5 or 5×10^{-8}M oxytocin further enhanced activity to 1220 and 1216, a 38% increase (p<.01). Without caffeine, epinephrine and oxytocin increased adenyl cyclase 15% and 14%, respectively (p<.05). The results with oxytocin were contrary to expectations, but they may be explained by recalling that PGE, like oxytocin, stimulates cAMP and mobility in rat and in pregnant human uterus. (LJG)   006592

0642

BEATTY, C.H.; YOUNG, M.K.; BOCEK, R.M.
Effect of prostaglandins on myometrial adenyl cyclase activity during the menstrual cycle.
Biology of Reproduction 9: 67-68. 1973.

Adenyl cyclase activity of myometrial slices from ovariectomized monkeys during an artificial menstrual cycle was studied using carbon-14 labeled material. On Day 8, control level of cAMP was 114 dpm x 10^{-2} per gm wet weight of tissue; PGE_2 or $PGF_2\alpha$ at 2.8 μM increased the level 185% and 54%, respectively. Results were similar on Day 18, early luteal phase. On Day 27, control cAMP levels were 155 dpm x 10^{-2} per gm tissue, and PGE_2 and $PGF_2\alpha$ increases were 300% and 58%, respectively. (HC)   005521

0643
BEAZLEY, J.M.
Prostaglandins and their use for induction of labour -- the intravenous route.
In: Jacomb, R.G., ed. The use of prostaglandins E_2 and $F_2\alpha$ in obstetrics and gynecology.
(Proceedings of the Upjohn Prostaglandins Symposium, London, September 21, 1972.) Miami,
Florida, Symposia Specialists, 1973. p. 11-19.

The author minimized hypertonicity, i.e., increased myometrial tone, as a consequence of using PGs
intravenously. He noted that myometrial sensitivity to PG varies with parity and gestational age,
primigravida being less sensitive than multipara, and PG doses for inducing labor being 1/10 those
needed for promoting midtrimester abortions. PGE_2.5-2 μg/minute or $PGF_2\alpha$ 5-10 μg/minute will
usually induce labor within 12 hours. For best results the drug should be titrated using a constant
infusion pump, thereby reducing the likelihood of side effects. Either PGE_2or $PGF_2\alpha$ is at least as
effective as oxytocin for inducing labor. The author concluded that PGs will not readily replace
oxytocin for routine intravenous use in the induction of labor. (HC)   006954

0644
BEAZLEY, J.M.
Treatment with prostaglandins.
Clinics in Endocrinology and Metabolism 2(3): 411-422. November 1973.

The uses of $PGE_2$and $PGF_2\alpha$ in obstetrics and gynecology are discussed, and the relationship between
parity or gestational age and myometrial sensitivity is described. The danger of a sharp initial rise in
myometrial tone, i.e., hypertonus, following iv infusion of PG is minimized by the author. Results of
studies using PGs by intravenous, intrauterine, and intraamniotic routes for the termination of early
pregnancies are cited. The author noted that side effects are a limitation with the intravenous route
and that the intraamniotic has advantages over intrauterine administration. Also, $PGE_2$is more
effective and has fewer side effects than $PGF_2\alpha$. For induction of labor, intravenous PGE is again more
effective than PGF, and each is as effective as oxytocin. The oral route may have considerable
practical value. 15-methyl analogs are more potent than the parent compounds but may also be more
toxic. PGs have also been used to potentiate oxytocin for pregnancy termination. (HC)   006941

0645
BEDWANI, J.R.
Effects of hormones, prostaglandins and cyclic-AMP on the synthesis of progesterone by the human
    placenta in vitro.
Journal of Reproduction and Fertility 34: 141-145. 1973.

Human placentae were incubated with $PGF_2\alpha$, PGE_1, human chorionic gonadotropin (HCG), sheep
prolactin, vasopressin, adrenalin, cyclic AMP, or dibutyryl cyclic AMP, and tritiated cholesterol to
measure progesterone synthesis. Fresh, term placentae were chilled in saline and cut in 200-μm
slices. The incubation medium contained human serum albumin, 250 μg per ml; glucose-6-phosphate,
4mM; NADP, .65 mM or the optimal concentration of .25 mM; tritiated 7-alpha cholesterol, .09-45 μC;
per ml; all in 20 ml Krebs phosphate buffer, pH 7.4. Progesterone synthesis, determined by
extraction, column chromatography, and thin-layer chromatography, was linear for 2 hours after a 15-
minute lag. None of the added substances affected production of progesterone except 5 mM cyclic
AMP plus 10 mM theophylline, which significantly inhibited progesterone synthesis ($p < .05$). (LJG)
006449

0646
BELL, W.R.; WENTZ, A.C.
Abortion and coagulation by prostaglandin. Intra-amniotic dinoprost tromethamine effect on the coagulation and fibrinolytic systems.
Journal of the American Medical Association 225(9): 1082-1084. August 27, 1973.

Alterations observed in the blood coagulation mechanism in women undergoing abortion by the intraamniotic administration of hypertonic sodium chloride led to the study of the coagulation and fibrinolytic systems in 20 patients undergoing the termination of second-trimester pregnancies (gestation periods ranging from 13.5 to 21 weeks) by the intraamniotic administration of $PGF_2\alpha$. An initial administration of 30 mg of $PGF_2\alpha$, made by way of a catheter inserted through the abdominal wall, was followed by 25 mg at 6 or 8 hours, and again at 24 and 32 hours if necessary. 18 of the 20 patients aborted, with an average time of 16 hours. Prothrombin, thrombin, and euglobin times, as well as plasminogen levels, were normal and did not change during $PGF_2\alpha$ treatment. Changes in the platelet count, fibrinogen concentration, and plasma levels of fibrinogen-fibrin degradation products were noted during and following the abortion procedure but did not exceed the range of normal values. A rise in the white blood cell count to approximately twice that of normal was observed. Vomiting in 12 patients was the only significant side effect observed. (JSL)   006918

0647
BERGSTROM, S.
Introduction.
Advances in the Biosciences 9: 1-5. 1973.

This was the opening address at the International Prostaglandin Conference held in Vienna in 1972. The speaker traced the testing of prostaglandin, spoke of the need to compare various assay methods, and suggested the importance that prostaglandin antagonists and inhibitors may play in living organisms. The largest number of papers presented dealt with human reproduction. (RAS)   006790

0648
BERGSTROM, S.; WILSON, R.
Phase III clinical trials of intra-amniotic prostaglandin $F_2\alpha$ versus hypertonic saline (trial no. 101) and of extra-amniotic prostaglandin $F_2\alpha$ (trial no. 102).
In: Bergstrom, S., ed. Report from meetings of the prostaglandin task force steering committee, Chapel Hill, June 8-10, 1972, Stockholm, October 2-3, 1972, Geneva, February 26-28, 1972. Stockholm, WHO Research and Training Centre on Human Reproduction, Karolinska Institutet, 1973. (Prostaglandins in Fertility Control 3) p. 12-26.

After introducing the overall structure of the WHO Prostaglandin Task Force, the details of patient selection, randomization, contraindications, reporting of side effects and complications, follow up, returning data forms, and processing data are summarized for the Phase III Clinical Trial of $PGF_2\alpha$ for midtrimester abortion. About 20 centers will compare hypertonic saline and intraamniotic $PGF_2\alpha$ in women 15-20 weeks pregnant (Trial No. 101), or extraamniotic $PGF_2\alpha$ in women 13-20 weeks pregnant (Trial No. 102). Sample reporting forms for trial results and a clinical log are included. (LJG)   006734

0649
BERGSTROM, S.
Summary of round-table discussion on fertility regulation.
Advances in the Biosciences 9: 843-854. 1973.

Intraamniotic administration of PGs has been used to terminate pregnancies after the fourteenth week of gestation when the amniotic sac can be easily punctured. A dose of 25 mg of $PGF_2\alpha$ repeated after 6 hours if necessary, or a single dose of 40-50 mg, is required, the increased dose being associated with a high occurrence of side effects. The induction-abortion period is significantly shorter with intraamniotically injected PGs than with hypertonic saline; side effects such as vomiting and diarrhea occur with greater frequency but dangerous side effects are fewer. Extraamniotic administration which can be used in the thirteenth-fourteenth week of pregnancy induces a local stimulatory effect on the myometrium. Repeated injections of $PGF_2\alpha$ (250-750 μg) or PGE_2(50-200 μg) at 2-hour intervals are required. This method is 90% successful in the second trimester with a mean induction-abortion interval of 20-24 hours. Vaginal administration seems favorable because of its simplicity and possible utilization on a self-administration basis. PGs are introduced in the form of a concentrated solution or as tablets, in which case the dose must be repeated every 2-4 hours to maintain a high level of uterine activity; suppositories which are effective for 10 hours have been tried clinically. Individual responses to vaginal administration of PGs are unpredictable and side effects are as high as with intravenous administration. The possible use of PGs in interfering with fertilization and implantation is being considered. (MLH)   006639

0650

BLACKBURN, M.G.; MANCUSI-UNGARD, H.R., Jr.; ORZALESI, M.M.; HOBBINS, J.C.; ANDERSON, G.G.
Effects on the neonate of the induction of labor with $PGF_2\alpha$ and oxytocin.
American Journal of Obstetrics and Gynecology 116(6): 847-853. July 1973.

A double-blind study was designed at the Yale-New Haven Hospital to evaluate the clinical and biochemical alterations in the neonate after using $PGF_2\alpha$ and oxytocin to induce labor. 23 infants were used, 11 born after induction of labor with $PGF_2\alpha$, 12 with oxytocin. At half hour intervals from birth, body temperature, respiration, systolic blood pressure, heart rate, and arterial samples for pH, $PCO2$, $PO2$, base excess, and lactate content measurements were made. Glucose, serum sodium, CBC, hemoglobin, and plasma $PGF_2\alpha$ levels were also recorded. Indications for induction, duration of labor, mode of delivery, and any pertinent observations were recorded. Evaluation of the results indicated no significant differences between the 2 groups. (IC)   006432

0651

BOLOGNESE, R.J.; CORSON, S.L.
Abortion of early pregnancy by the intravaginal administration of prostaglandin $F_2\alpha$.
American Journal of Obstetrics and Gynecology 117: 246-250. 1973.

A study was made using 12 women to determine if intravaginal $PGF_2\alpha$ could be used to terminate early pregnancy. 4 patients who received 50 mg per ml solution (total dose 135 to 600 mg/12 hours) failed to abort and required dilatation and suction evacuation; 2 patients experienced vaginal bleeding. 6 women receiving 200 mg per ml (total dose 300 to 700 mg/15.5 hours) solution successfully aborted. 2 attempts were unsuccessful. All patients experienced vaginal bleeding. Vomiting and diarrhea were common. Progesterone and HCG levels were inconsistent. It is suggested that if the side effects could be reduced, intravaginal $PGF_2\alpha$ may be a reliable self-administered abortifacient. (IC)   005504

0652

BOLOGNESE, R.J.; CORSON, S.L.; MEROLA, J.C.L.

Prostaglandin PGE_2as an intravaginal abortifacient during late first, and early second trimester pregnancy.

Obstetrics and Gynecology 41(4): 640. April 1973.

PGE_2suppositories were inserted intravaginally to abort patients 13-16 weeks pregnant. 24-40 mg PGE_2every 2 hours (total dose range: 80-240 mg) aborted 12 out of 13 patients in a mean 12 hours' (range: 8-26 hours) time to abortion. Most required suction dilatation and curettage for retained placenta, including 1 undiagnosed, partly retained molar pregnancy. Side effects included gastrointestinal complaints, headaches, and fevers up to 103 degrees F. for the duration of PG treatment. (LJG)   006442

0653

BOLOGNESE, R.J.; CORSON, S.L.

The effect of vaginally administered prostaglandin $F_2\alpha$ on corpus luteum function.

American Journal of Obstetrics and Gynecology 117(2): 240-245. September 15, 1973.

7 nonpregnant women took $PGF_2\alpha$ intravaginally on Postovulatory Day 10-11 and gave daily blood samples for progesterone and LH assays from Menstrual Cycle Day 10 until day of menses during 1 test cycle and 2 control cycles. They received from 125 to 400 mg $PGF_2\alpha$ Tham salt, at 50 or 200 mg/ml, in divided doses over a 4-hour to 2-day period, starting on Day 10 after ovulation. 24 and 48 hours later each was given 3000 and 6000 IU of human chorionic gonadotropin (HCG). A transient decrease in serum progesterone, about 2-4 ng/ml, occurred in 5 women, and progesterone did not rebound after HCG. Length of luteal phase ranged from 7 to 17 days (mean 12.1) in control cycles and from 17 to 23 days in the PG cycle in 5 women, but lasted 5 and 8 days in 2 others. Side effects included 2 cases of vomiting and diarrhea, 2 of vaginal bleeding, and 1 of vaginal burning. Individual chemical profile was not affected. Thus, $PGF_2\alpha$ did not impair corpus luteum function. (LJG) 005503

0654

BOLT, D.J.

Reduced luteolytic effect of $PGF_2\alpha$ by hysterectomy or HCG in ewes.

Journal of Animal Science 37(1): 302. 1973.

$PGF_2\alpha$, 10 or 40 mg intramuscularly on Day 10, markedly reduced the mean weight and progesterone content of the corpus luteum of ewes. The effect of PG was much less in animals hysterectomized 1-3 hours before the injection. HCG also prevented the lytic effect of PG. (HC)   006934

0655

BOWEN-SIMPKINS, P.

The induction of second trimester abortions using an intra-amniotic injection of urea and prostaglandin E_2.

Journal of Obstetrics and Gynaecology of the British Commonwealth 80: 824-826. September 1973.

31 patients with gestation periods varying from 15 to 22 weeks were injected with 80 gm of urea followed by 5 mg of PGE_2. Abortion was successful in 30 of the cases, the mean injection-to-abortion interval being 10 hours and 30 minutes. Abortion was complete in 25 cases. Painful uterine contractions occurred within 2 hours of the injection; membrane rupture followed 2-4 hours later. Side

effects including diarrhea, nausea, and vomiting were reported by less than half of the patients. There was no association drawn between the injection-to-abortion interval and either maternal parity or gestational age. The advantages of this method are that it avoids the use of intravenous infusions, the injection-to-abortion interval is short, and the combined urea and PG solution costs 1/3 as much as PG used alone. (MLH)   005820

0656

BRADLEY-WATSON, P.J.; BEARD, R.J.; CRAFT, I.L.
Injuries of the cervix after induced midtrimester abortion.
Journal of Obstetrics and Gynaecology of the British Commonwealth 80: 284-285. March 1973.

A large cervical tear extending through the cervix and about 3 cm up the posterior uterine wall was noted in a patient undergoing termination of a 20-week gestation. 75 mg of $PGF_2\alpha$ had been administered intraamniotically, followed after 24 hours by another 75 mg. Incomplete abortion occurred 17 hours after the second injection; the placenta was removed by cord traction 4 hours later. Analgesia had been necessary 10 hours after the first dosage. (RMS)   005526

0657

BRANDA, L.A.; VAILLANCOURT, P.; KOMINKOVA, E.
Effect of oxytocin on the perfused human placenta in vitro.
American Journal of Obstetrics and Gynecology 117(8): 1116-1125. 1973.

The effect of oxytocin on umbilical circulation through the placenta was studied using an in vitro perfusion procedure. Human placentas were obtained from normal term pregnancies. Oxytocin administered into one umbilical artery as a single dose in a range from 6 to 3600 mU had no ability to significantly alter the pressure or rate of flow through the other artery or the umbilical vein. When oxytocin was administered as a continuous infusion for 5-10 minutes at concentrations ranging from .5 to 200 mU/ml of perfusate, a decrease in the arterial pressure and increase in the venous flow rate was noted. This effect was most pronounced at the low oxytocin concentrations. Up to 20 mU/ml of oxytocin perfused into the maternal side of the placenta had no ability to alter umbilical arterial pressure or venous flow. This may indicate that oxytocin in the maternal circulation does not enter the fetal circulatory system. The rapid degradation of oxytocin by placental oxytocinase may explain this observation. (JSL)   006920

0658

BRENNER, W.; BYGDEMAN, M.
Results with intra-amniotic administration.
In: Bergstrom, S., ed. Report from meetings of the prostaglandin task force steering committee, Chapel Hill, June 8-10, 1972, Stockholm, October 2-3, 1972, Geneva, February 26-28, 1972. Stockholm, WHO Research and Training Centre on Human Reproduction, Karolinska Institutet, 1973. (Prostaglandins in Fertility Control 3) p. 33-36.

5 collaborative trials of different doses of intraamniotic $PGF_2\alpha$ or its 15-methyl analogue for midtrimester abortion are tabulated. First McGill and Southern California compared 25 mg $PGF_2\alpha$, repeated 6 hours later if needed, to 200 ml 20% NaCl in a total of 249 women. The most notable difference was the lower failure rate and 19.5 hour mean abortion time with PG compared to 27.8 hours with saline. 3 other centers had comparable data with 25 mg intraamniotic $PGF_2\alpha$. Southern California and Karolinska tried a single 40 mg dose of $PGF_2\alpha$ and North Carolina 50 mg $PGF_2\alpha$. The mean results of 40 mg compared to 50 mg were 64% and 75% abortion in 24 hours, 92% and 96%

abortion within 48 hours, 21.9 and 19.5 hours mean abortion interval. Single intraamniotic injections of 15-methyl-PGF$_2\alpha$ of 1, 2.5, or 5 mg aborted 46%, 96%, and 95% of patients in 48 hours, with abortion times of 20.1, 19.9, and 18.4 hours in 13, 22, and 39 women given the respective doses. Side effects of vomiting and diarrhea occurred in up to 59% at the 2.5 mg dose which was considered an acceptable frequency. (LJG)   006726

0659
BRENNER, W.E.; FISHBURNE, J.I.; McMILLAN, C.W.; JOHNSON, A.M.; HENDRICKS, C.H.
Coagulation changes during abortion induced by PGF$_2\alpha$.
American Journal of Obstetrics and Gynecology 117(8): 1080-1087. December 15, 1973.

Coagulation changes related to the defibrinogenation syndrome usually seen in hypertonic-saline-induced abortions were monitored in 6 women undergoing midtrimester abortion by transabdominal intraamniotic PGF$_2\alpha$. The 6 women, aged 18-22, 13-19 menstrual weeks' gestation, were given 50 mg PGF$_2\alpha$-Tham salt, repeated 24 hours later if necessary. They aborted within 8.5-38.6 hours, 3 completely and 3 incompeltely. Blood samples were taken from 4 days before PG injection until after abortion to assess fibrinogen decay by the selenium-75-selenomethionine method; other factors were tested every 4 hours from 4 hours befors PG injection until abortion. Mean levels of Factor V, VIII, and fibrinogen increased when expressed as percent of abortion time. No fibrin degradation products were seen. No significant changes were detected in hematocrit, platelets, prothrombin, thrombin clotting, partial thromboplastin times, Factor V, Factor VIII, and fibrinogen in any subject 4 hours postabortion or in any patient who did not abort. The authors contrasted these results with those in 5 women having midtrimester abortion by 200 ml 20% hypertonic saline. These women had significant increases in fibrin degradation products, decreases in platelets, fibrinogen, fibrinogen survival, and Factors V and VIII. The authors commented that a clinically significant defibrinogenation is rare in saline abortions, but PGF$_2\alpha$ may be a safer method for cases of intrauterine fetal death if defibrinogenation syndrome is already present. (LJG)   005807

0660
BRENNER, W.E.; DINGFELDER, J.R.; HENDRICKS, C.H.; STAUROVSKY, L.
Induction of therapeutic abortion with a single dose of intraamniotically administered prostaglandin F$_2\alpha$.
Prostaglandins 4(4): 485-498. 1973.

To determine the abortifacient effectiveness and complications of a single intraamniotic injection of 50 mg PGF$_2\alpha$ over 10 minutes, 40 women 13-26 weeks pregnant requesting termination of pregnancy were studied. The subjects received the PGF$_2\alpha$ with no additional oxytocics or surgical intervention until they aborted or at the end of the 48-hour trial. 25 patients (Group I) received 10 mg prochlorperazine intramuscularly to alleviate nausea or vomiting when they requested. 15 subjects (Group II) were given 10 mg prochlorperazine 30 minutes before PGF$_2\alpha$ and at 6-hour intervals until abortion. 77% of the patients aborted within 24 hours and 95% within the 48-hour period. Mean induction-to-abortion time was 19.1 hours. There were 68% complete abortions, 28% imcomplete, and 5% failures within the 48 hours. There were no serious complications. Group II patients had significantly less episodes of vomiting than Group I. No significant differences in the mean abortion times, cumulative abortion rates, or intraamniotic pressures were noted between the groups. Conclusions were that a single 50-mg dose of PGF$_2\alpha$ results in practicable abortion rates and that vomiting can be controlled with prochlorperazine without significantly altering the abortifacient or oxytocic effect of PGF$_2\alpha$. (Authors' modified)   005528

0661
BRENNER, W.E.; HENDRICKS, C.H.; BRAAKSMA, J.T.; FISHBURNE, J.I., Jr.; STAUROUSKY, L.G.; HARRELL, L.C.
   Induction of therapeutic abortion with intraamniotically administered prostaglandin $F_2\alpha$: a comparison of three repeated-injection dose schedules.
   Obstetrics and Gynecology 41(4): 633. April 1973.

To determine which intraamniotic $PGF_2\alpha$ dose schedule was most effective with minimum complications, 71 midtrimester patients had therapeutic abortion induced by 1 of 3 dose schedules. 25 mg, repeated at 6, 24, and 30 hours to those who had not yet aborted, produced the lowest complication rate and the fewest incomplete abortions. With 50 mg repeated in 24 hours if necessary, or 25 mg repeated at 6, 12, 24, and 30 hours if necessary, rates of incomplete abortions and complications were higher. Rates for expulsion of the placenta and hemorrhage after abortion were calculated, and their use in determining a logical time for surgical intervention was suggested. (LJG)
006443

0662
BRENNER, W.E.; HENDRICKS, C.H.; FISHBURNE, J.I.; BRAAKSMA, J.T.; STAUROVSKY, L.G.; HARRELL, L.C.
   Induction of therapeutic abortion with intra-amniotically administered prostaglandin $F_2\alpha$: a comparison of three repeated-injection dose schedules.
   American Journal of Obstetrics and Gynecology 116(7): 923-930. 1973.

$PGF_2\alpha$ was administered intraamniotically to 71 patients between 10 and 23 weeks gestation in order to compare the effectiveness of 3 dose schedules. 22 women in Group I received 25 mg initially and again at 6, 24, and 30 hours if abortion had not occurred. 24 women in Group II received 50 mg in the initial injection which was repeated after 24 hours. 25 women in Group III received 25 mg initially and again at 6, 12, 24, 30, and 36 hours. There were no statistically significant differences in abortion rates among the 3 groups. There were no differences in effectiveness in nulliparous patients, but Group I and Group III schedules were found to be superior in multiparous patients. Incomplete abortion appeared to be less frequent in Group I than in Groups II or III. Gestational age seemed to have no effect on abortion rates or on completeness of abortion. It was concluded that the Group I dose schedule was most satisfactory, in terms of greater effectiveness and lower incidence of incomplete abortion. In addition, leukocyte changes were smaller and minor complications were less frequent with this dose schedule. Side effects reported in this study included vomiting and pain and less frequently diarrhea, excessive blood loss, fever, endometritis, and cervical trauma. Further study is anticipated. (MLH)   006743

0663
BRENNER, W.E.; HENDRICKS, C.H.; FISHBURNE, J.I.; STAUROVSKY, L.; BRAAKSMA, J.; TAFT, R.
   Intraamniotic prostaglandin $F_2\alpha$ dose-twenty-four-hour abortifacient response.
   Journal of Pharmaceutical Sciences 62(8): 1278-1282. 1973.

132 women received intraamniotic $PGF_2\alpha$ following 6 dose schedules, 3 single dose and 3 multidose. Single injections of 15 mg, 25 mg, and 50 mg aborted 20%, 40%, and 45% women successfully. Multiple dose procedures of 15 mg followed by 15 mg 6 hr. later, 25 mg with additional 25 mg 6 hr. later and 25 with identical dose at 6 and 12 hr. later if needed, aborted 70%, 68% and 60%, respectively, of injected women. 73% women of 16 weeks gestation or longer aborted with a single dose of 50 mg $PGF_2\alpha$. The low dose multiple had 100% success with gestation less than 16 weeks. Vomiting was most common in patients with a high multiple dose schedule and pain was most

frequent in high single dose. Side effects did not correlate with dose, parity or gestational age. (RAS)
005584

0664

BRENNER, W.E.; HENDRICKS, C.H.; DINGFELDER, J.; STAUROVSKY, L.
Laminaria augmentation of intra-amniotic prostaglandin $F_2\alpha$ for the induction of mid-trimester abortion.
Prostaglandins 3(6): 879-894. June 1973.

To determine if augmentation of intraamniotic $PGF_2\alpha$ with laminaria would improve midtrimester abortion rates, 43 women of ages 16-34 and 10-23 weeks gestation were divided into 2 groups, depending upon whether laminaria were used (21 patients, Group I) or not used (22, Group II). Therapeutic abortion was attempted using 25 to 100 mg of $PGF_2\alpha$ at 5 mg/ml in 25-mg doses at 6, 24, and 30 hours. Patients with laminaria had a mean abortion time of 14.6 hours; 95% aborted within 24 hours of initial $PGF_2\alpha$ injection, and all aborted within 24.5 hours of the initial 25 mg $PGF_2\alpha$. Patients without laminaria had a mean abortion time of 18.9 hours; only 68% aborted within 24 hours and 1 failed to abort within 48 hours. There were no significant differences in frequency or severity of complications or in uterine contractility between the 2 groups. 3 Group I patients required curettage, 1 patient did not abort; 5 Group II patients required curettage. Augmenting the $PGF_2\alpha$ intraamniotic abortion method with laminaria appears practicable. (IC)   006467

0665

BRENNER, W.E.; DINGFELDER, J.R.; STAUROVSKY, L.G.; HENDRICKS, C.H.
Vaginally administered $PGF_2\alpha$ for cervical dilatation in nulliparas prior to suction curettage.
Prostaglandins 4(6): 819-836. December 1973.

Cervical dilatation was compared in 20 untreated nulliparas to that in 40 nulliparas given 50 mg $PGF_2\alpha$ by vaginal suppository 3 hours before mechanical dilation in a preliminary study to determine whether PG would decrease longterm side effects, such as cervical incompetence. Patients, 6 to 12 weeks gestation, aged 14-25, received 50 mg meperidine and 5 mg diazepam iv, and paracervical block of 20 ml 1% lidocaine. After 10 minutes the cervix was stabilized with a single-toothed tenaculum and progressively dilated from #19 to #35 (mm circumference) Pratt dilator. The circumference which met initial resistance was accepted as clinically significant dilatation. All PG patients were dilated to at least #25 (8 mm diameter) and 30% were dilated to #35 (11.1 mm diameter), compared to 30% and 0% of the untreated group. The proportion of PG treated women dilated sufficiently for suction curettage without needing further dilation at 6 weeks gestation was 100%, at 8 weeks 100%, at 9 weeks 90%, 10 weeks 53%, 12 weeks 30%. Among controls 100% were dilated sufficiently at 6 weeks, 30% at 8 weeks, 10% at 9 weeks, 5% at 10 weeks and 0% at 12 weeks. Aside from diarrhea and vomiting in 43% and uterine pain in 5% of treated patients, 1 (3%) PG patient and 3 (15%) controls suffered minor cervical trauma and 1 PG patient's uterus was perforated. (LJG)   005675

0666

BRODY, M.J.
A review: modulation of autonomic transmission by prostaglandins.
Population Report: Prostaglandins G(3): 25-29. December 1973.

Inhibition of effector responses to catecholamines by PGE's have been observed in the heart, spleen, vas deferens and blood vessels, although the effects of PGE's are not uniformly depressent. PG's are found not only to alter the ability of autonomic effectors to respond to transmitter but also significantly to influence the responses of these effectors to activation of their nerve supply. The mechanism of action in depressing responses to nerve stimulation appears to involve inhibition of release of adrenergic transmitter. There is additional direct evidence that PG's depress the nerve terminal liberation of norepinephrine. The specificity of PG action has been found to vary with species and with differences in concentration and quality of the PG's. The PG's have the potential to facilitate as well as depress autonomic transmission. There is more than ample indirect evidence that PG's are formed or released directly in the anatomic region of the autonomic neuroeffector junction. Stimulation of sympathetic nerves or infusion of norepinephrine has been demonstrated to cause release of PGE's, with smaller amounts of PGA and PGF, from the spleen, kidney, vas deferens and seminal vesicles, heart, and adipose tissue. Inhibition of PG synthesis followed by activation of adrenergic innervation is shown to yield greater effector organ response and also to increase responsiveness to norepinephrine. PGF's can increase vascular responses to norepinephrine in concentrations which have no effect on the basal contractile activity of vascular smooth muscle. Neurogenically evoked autonomic responses have also been demonstrated to be facilitated by PGE's and more significantly by PGF's. Data indicates that the modulating role of PG's may well involve the balance between the depressant effects of PGE's and the facilitating effects of PGF's. It seems plausible to postulate that the modulation of autonomic transmission by PG's with opposite effects could involve affinity of the PG's for a similar receptor site on or in the nerve terminal. (MLH)   006978

0667
BROWN, A.A.; HAMLETT, J.D.; HIBBARD, B.M.; HOWE, P.D.
　　Induction of labour by amniotomy and intravenous infusions of oxytocic drugs: a comparison between prostaglandins and oxytocin.
　　Journal of Obstetrics and Gynaecology of the British Commonwealth 80: 111-115. February 1973.

To determine if PGs have sufficient advantages over oxytocin in inducing labor to warrant changing present practices, 170 patients pregnant 37 or more weeks received PGE_2, $PGF_2\alpha$, or oxytocin intravenously immediately after amniotomy. Pairs of patients receiving oxytocin and PG were matched for age, parity, and gestation period. The outcome of induction was considered unsuccessful if labor failed to progress because of poor uterine action or if infusion was discontinued because of side effects. All inductions with oxytocin were successful. PGE_2 was unsuccessful in 6% (3 patients) and $PGF_2\alpha$ was unsuccessful in 12.5% (4 patients). Delivery occurred within 12 hours in 83% PGE_2 cases, 69% $PGF_2\alpha$ patients, and 88% and 89% of the respective oxytocin controls. Most PG patients had effective contractions with .5 μg per minute PGE_2 and 5 μg per minute $PGF_2\alpha$. 6 of the 53 PGE_2 patients required .75 μg per minute and 4 of the 32 $PGF_2\alpha$ patients required 7.5 μg per minute. No case of operative delivery was attributable to PGs or oxytocin. The condition of infants was satisfactory. Side effects included inflammatory reaction occurring at the infusion site in 10 PG patients. Only 1 $PGF_2\alpha$ patient suffered excessive vomiting. No patient had significant intestinal colic or diarrhea. Uterine hypertonus was not observed. It was concluded that low amnitomy prior to PG infusion makes lower dosages effective, thus reducing incidence of uterine hypertonus and other side effects. At present, the authors saw no practical reason for substituting PGs for oxytocin in induction of labor as PGs sometimes have unpleasant side effects. (IC)   005525

0668
BRUMMER, H.C.; CRAFT, I.L.
　　Prostaglandin $F_2\alpha$ and labour.
　　Acta Obstetricia et Gynecologica Scandinavica 52(3): 273-275. 1973.

$PGF_2\alpha$ was measured by radioimmunoassay in 10 ml blood samples taken at random from 58 women during labor, 23 women during the first postpartum week, and 7 women throughout labor. The random samples produced the following curve: .7 ng/ml $PGF_2\alpha$ at 0-4 cm cervical dilation, 1.8 at 5-9 cm, .5 at second stage, .35 at delivery, and .25 postpartum. All values from the 7 women followed throughout labor were within 1 standard deviation of the random samples. Peak $PGF_2\alpha$ levels were associated with high uterine activity and cervical dilatation, regardless of whether labor was spontaneous or induced by artificial rupture of membranes. (LJG)   005665

0669
BRUMMER, H.C.
Serum $PGF_2\alpha$ levels during human pregnancy.
Prostaglandins 3(1): 3-5. January 1973.

In 128 samples from 28 women serum $PGF_2\alpha$ levels averaged .62, .39, and .45 ng per ml during trimesters I, II, and III, respectively. $PGF_2\alpha$ levels were at their lowest, rather than highest, levels during weeks 17 to 24, contradicting a previous report (Prostaglandins 1: 331, 3318 1972). (HC) 005763

0670
BRUMMER, H.C.
Vasectomy and seminal prostaglandins.
Fertility and Sterility 24(2): 131-133. February 1973.

Mean seminal fluid levels of PGE, PGA, and 19-hydroxy PGA were significantly higher in 11 men tested 12 and 16 weeks after vasectomy. The authors conclude that these compounds do not arise from the testes but they offer no explanation for the increase in PG levels. (HC)   005755

0671
BRUNCK, U.; GUSTAVII, B.
Lability of human decidual cells: in vitro effect of autolysis and osmotic stress.
American Journal of Obstetrics and Gynecology 115(6): 811-816. March 15, 1973.

Decidual and trophoblastic cells from evacuation abortions in 64 women 9-12 weeks pregnant were fixed in hypoosmolar or normoosmolar solution and allowed to autolyze to compare stability of the cells and their lysosomes. Decidual cells had a dense granular cytoplasm with distinct acid phosphatase granules, but decidual cells fixed in hypotonic solution appeared vacuolated and ballooned, either devoid or diffusely stained for acid phosphatase. Trophoblastic cells appeared similar in the 2 fixatives except for few tiny vacuoles in the hypotonic fixative. After 1 hour of autolysis, decidual cells were distinctly vacuolated and swollen, stained diffusively for acid phosphatase, and had vacuolated, fragmented, pyknotic nuclei. Trophoblast cells were still rather well preserved after 8 hours of autolysis, showing occasional brush borders and acid phosphatase granules. Experimental data on the presence of prostaglandins is mentioned only in discussion; this study indicates that decidual cells have fragile lysosomal membranes and supports the authors' view that decidual cells may induce abortion or labor by releasing prostaglandins. (LJG)   006441

0672

BUCKLE, J.W.; NATHANIELSZ, P.W.

The effect of low doses of prostaglandin $F_2\alpha$ infused into the aorta of unrestrained pregnant rats: observations on induction of parturition and effect on plasma progesterone concentration.

Prostaglandins 4(3): 443-457. September 1973.

Uterine pressure, progesterone level, and labor were observed in pregnant rats infused with $PGF_2\alpha$. 12 controls and 12 study rats were operated upon on Days 17 or 18 of pregnancy to insert a cannula from the left carotid to the descending aorta just above the ovarian arteries and implant a pressure sensitive radio-transducer in the uterus. The controls received saline with heparin. The study rats received 4.67 μg $PGF_2\alpha$ per hour for 11.5 hours starting late in Day 18, and also several 10 minute infusions of saline and .1 μg PGF per minute for testing uterine sensitivity to PG. Blood was sampled twice daily for immunoassay of progesterone. Rats given PG delivered dead fetuses in 40-41 hours, 60 hours before normal liveborn controls ($p < .001$). Progesterone fell with a half life of 42.5 hours in controls, 20 hours in rats infused 1 day after operation (Day 18), and 16.5 hours in rats infused immediately (Day 17). Uterine activity in all animals showed regular contractions 16-24 hours after surgery, then was variable, followed by strong contractions during delivery, a quiet phase of about 16 hours, and resumption of rhythmic activity with strong contractions during suckling. (LJG) 005448

0673

BULLARD, P.D.; HERRICK, C.N.; HINDLE, W.H.; HALE, R.W.; PION, R.J.

Histopathologic changes associated with prostaglandin induced abortion.

Contraception 7(2): 133-144. February 1973.

10 patients pregnant less than 10 weeks received $PGF_2\alpha$ by intravenous infusion. 3 women were given the drug at a rate of 25 μg/minute which was increased over the next 12 hours to 200 μg/minute, and 7 patients received 50 μg/minute for 12 hours. The 3 women given the higher rate all had painful uterine contractions, vaginal bleeding, nausea, vomiting, diarrhea, headache, and flushing; none aborted. The other 7 patients had only vaginal bleeding and painful uterine contractions; 4 aborted completely and 1 was incomplete. All renal and liver function studies performed showed no change. 8 patients had leucocytosis, 5 had increased cortisol levels, and 4 had significant increases in progesterone levels. In all cases, regardless of mode of termination, microscopic examinations of products of conception were similar. Each specimen showed marked hemorrhage, decidual and placental necrosis, and intervillous fibrin deposition. Fetal tissue showed little significant change. Since the specimen from suction curettage had no hemorrhage or necrosis, the authors suggested that these changes may be secondary to induced uterine contractility and subsequent anoxia. The authors concluded that $PGF_2\alpha$ intravenously does not compete with early suction abortion. (HC) 005756

0674

BYGDEMAN, M.; BEGUIN, F.; TOPPOZADA, M.; WIQVIST, N.

Intrauterine administration of prostaglandin $F_2\alpha$ for induction of abortion.

Advances in the Biosciences 9: 525-531. 1973.

Extraamniotic administration of $PGF_2\alpha$ is an efficient method for induction of abortion particularly useful in the 13th-14th week of gestation. The need of an indwelling catheter and repeated injections comprise the major drawbacks of the method. Three different dose schedules were used for the intraamniotic administration of $PGF_2\alpha$. With 15 mg of $PGF_2\alpha$ repeated after 24 hr, an abortion could be induced in approximately 50% of the cases compared to a success rate of 97% if 25 mg was used. If the interval between the injections was decreased to 6 hr, the success rate remained high, and, in addition, the induction abortion interval was significantly reduced. This advantage has, however, to be weighed against the increased incidence of side effects. Injection of ^{3}H-PGF$_2\alpha$ together with 25 mg PGF$_2\alpha$ revealed a slow disappearance rate of radioactivity from the amniotic fluid and a low concentration of PGF$_2\alpha$ in the plasma. Moreover, there was a slow metabolism of the compounds in the amniotic fluid. (Authors)   006470

0675
BYGDEMAN, M.; WIQVIST, N.
The relation between prostaglandin and oxytocin action in the pregannt myometrium
In: Josimovich, J.B., ed. Uterine contraction--side effects of steroidal contraceptives. New York, Wiley-Interscience, 1973. (Volume 1 of Problems of Human Reproduction) p. 171-178.

Uterine sensitivity to oxytocin or prostaglandin varies considerably. Nonpregnant and early pregnant uterus will not respond to oxytocin but small doses of PG will cause contractions. As pregnancy progresses, the uterus becomes more sensitive to either PG or oxytocin stimulation. PGE$_2$infused at 2-8 μg/min and oxytocin at 50 μg/min had a different contractile response. Oxytocin induced coordinated contractions whereas prostaglandin induced irregular contractions. At term PGE$_2$or PGF$_2\alpha$ induced coordinated contractions, however, oxytocin contractions appear to have more intensity. Oxytocin inability to stimulate contraction at midpregnancy indicates a separate mechanism of action for the 2 compounds. Prostaglandin and oxytocin given in combination result in an additive effect on uterine contractility. (RAS)   006316

0676
CALDER, A.; EMBREY, M.P.
Prostaglandins and the unfavourable cervix.
Lancet 2(7841): 1322-1323. December 8, 1973.

The authors reported that the continuous infusion of PGE$_2$via the intrauterine route was safe and effective for inducing labor in high-risk and difficult cases for 40 women at or near term. The authors were also successful in inducing labor in 7 women with very low inducibility prospects, the mean induction/delivery interval averaging 13 hours. There was no fetal distress or uterine hypertonus. (HC)   006914

0677
CALDWELL, B.V.; AULETTA, F.J.; SPEROFF, L.
Prostaglandins in the control of ovulation, corpus luteum function, and parturition.
Journal of Reproductive Medicine 10(3): 133-138. March 1973.

This review summarizes the critical roles of PGs in ovulation, of PGF$_2\alpha$ in luteolysis, and theories of PGs' role in parturition. In rabbits (induced ovulators), ovulation is prevented if indomethacin is injected 8 hours after mating, but not earlier. Indomethacin may act in rats (spontaneous ovulators) directly on the ovary or on secretion of gonadotropins and hypothalamic releasing factor. When ovulation is blocked, luteinization proceeds as usual but the ovum is retained. Much evidence from various nonprimate species exists to support the view that PGF$_2\alpha$ produced by the uterus caused luteolysis. In cycling sheep there is an LH surge followed 12-20 hours later by ovulation, a rise in progesterone production which plateaus from Days 8-14, and then an increase in estrogen and PGF$_2\alpha$ the latter causing a fall in progesterone in 24 hours. In rabbits and hamsters, indomethacin or immunization against PGF$_2\alpha$ prolongs lifespan of corpora lutea. PGF$_2\alpha$ has not caused luteolysis in primates, but estrogen does so and induces PGF$_2\alpha$ secretion. No proof is available that either endogenous oxytocin or PGs initiate labor, although they are probably involved in the chain of events leading to parturition. (LJG)   005662

0678
CARLSON, J.C.; BARCIKOWSKI, B.; McCRACKEN, J.A.
Prostaglandin $F_2\alpha$ and the release of LH in sheep.
Journal of Reproduction and Fertility 34: 357-361. 1973.

The effect of $PGF_2\alpha$ on circulating luteinizing hormone (LH) and estradiol in cyclic and anestrous ewes was observed by perfusing $PGF_2\alpha$ at .75 to 100 μg per hour in the carotid, and sampling blood from the jugular vein. 3 cyclic ewes between Cycle Days 5-10 had control LH levels of .5-2.1 ng per ml during saline infusion, and peak LH levels of 1.8, 6.2, and 11.7 during $PGF_2\alpha$ infusion of 6 μg per hour or more. A second LH peak of 15 ng per ml was observed 10.5 hours after $PGF_2\alpha$. Estradiol peaks of 10 and 70 pg per ml occurred 1 and 8.5 hours after $PGF_2\alpha$ in 1 sheep, compared with control values of less than 8 pg per ml. $PGF_2\alpha$ in doses of 5 to 100 μg per hour had no effect on LH in 4 anestrous ewes (March to July) or on 2 ovariectomized ewes. The authors discussed evidence for their suggestion that hypothalamic PG synthesis may be involved in release of LH by the pituitary. (LJG) 006450

0679
CARSTEN, M.E.
Prostaglandins and cellular calcium transport in the pregnant human uterus.
American Journal of Obstetrics and Gynecology 117(6): 824-832. November 15, 1973.

The effects of PGE_2, $PGF_1\beta$, and inhibitors on ATP-dependent calcium binding and calcium release were quantitated by atomic absorption in sarcoplasmic reticulum (SR), and mitochondria from pregnant human uterus. The cellular fractions were isolated in the 15,000-40,000 x g pellet, and separted by sucrose density. Calcium binding was monitored for 1 minute in a medium containing .5 mM ATP, .5 mM MgCl2, 2×10^{-5}M CaCl2, .01 M KCl in .02 M imidazole buffer pH 7. Calcium release was assessed by adding PG for 2 minutes and estimating calcium after 1 minute. The SR vesicles were about 1000 Angstroms in diameter (range 500-2000), and behaved alike regardless of gestational age from 6 weeks to term. They contained 31.8 nmol intrinsic calcium per ml protein, bound 5.2 nmol calcium in presence of ATP, in comparison with 161.4 nmol per ml ATP dependent binding in the mitochondrial fraction. Sodium azide 5 or 10 mM, and dicumarol 60 μM, specific inhibitors of calcium binding in mitochondria, did not inhibit calcium binding by SR, but 3 mM Salyrgan, a sulfhydryl reagent, and 1 mM EGTA, a calcium chelator, both inhibited calcium binding. PGE_2 but not $PGF_1\beta$ inhibited ATP dependent calcium binding and enhanced calcium release in proportion to concentration from 25-200 μg per ml. The findings suggest that PGs exert their effect on uterine contractility through transport of calcium. (LJG)   005505

0680
CARSTEN, M.E.
Prostaglandins and cellular calcium transport in the non-pregnant uterus.
Federation Proceedings 32(3): 390Abs. 1973.

The binding of calcium by suspensions of sarcoplasm reticulum from uteri of non-pregnant cows was inhibited by PGE_2 or $PGF_2\alpha$ at concentrations of 1 to 1000 ng per ml; $PGF_1\alpha$ was ineffective. Prostaglandins added after maximum calcium binding had occurred caused a dose-dependent release. These data suggest that the uterus-stimulating effect of PG is based on a calcium displacement mechanism. (HC)   006786

0681

CARSTEN, M.E.

Sarcoplasmic reticulum from pregnant bovine uterus. Prostaglandins and calcium.

Gynecologic Investigation 4(2): 95-105. 1973.

The effect on calcium binding, calcium release, and reversible calcium binding of PGE_2, $PGF_2\alpha$, and $PGF_1\beta$ in purified sarcoplasmic reticulum from third-trimester pregnant cows was measured by an atomic absorption technique. 100-200 $\mu g/ml$ PGE_2 or $PGF_2\alpha$ inhibited calcium binding 100% in a 2-minute incubation with ATP and magnesium. $PGF_1\beta$ was without effect. After a 2-minute incubation, 50 $\mu g/ml$ PGE_2 led to release of 75%-90% of bound calcium, and 100 $\mu g/ml$ $PGF_2\alpha$ to 50%-70%. An experiment was conducted to compare the chelator EGTA (ethylene glycol-bis-[beta-amino-ethylether]-N,N-tetraacetic acid), which removed 42% of the intrinsic calcium, with PGs. EGTA 1 mM removed 10.3 nmol intrinsic calcium per mg protein, and 100 $\mu g/ml$ PGE_2 removed 5.8 nmol calcium. Both treated sarcoplasmic reticulum fractions bound 1.7 nmol calcium per mg protein during incubation with ATP. The EGTA chelation and rebinding process was reversible: 1.5 nmol calcium per mg protein was bound in the second incubation. The renewed binding was greater in the presence of ATP. The author discussed the possibility that PGs may interact with calcium to increase force of uterine contractions. (LJG)   005772

0682

CENEDELLA, R.J.; CROUTHAMEL, W.G.

Effect of aspirin upon male mouse fertility.

Prostaglandins 4(2): 285-290. August 1973.

16 male Swiss Webster mice were individually housed with 4 females for 12 days, then given 50 mg/kg aspirin orally twice daily for 7 days, and mated for 12 more days of continued aspirin treatment. Aspirin treatment raised plasma salicylate level to 13.9 $\mu g/ml$ 2 hours after the last dose and to 7.9 $\mu g/ml$ 18 hours after. Control matings yielded a wide range of fertility, about 3-15 placentae per female (mean 10.5). During aspirin treatment, fertility increased generally to mean 11.7 placentae per female and variation decreased (range about 10-14). By analysis of variance with multiple comparisons between drug effect and individual mice, fertility increased significantly in 4 initially subfertile males ($p < .05$) and decreased in 1 male from 15 to 10 placentae per female. (LJG)   005530

0683

CERINI, J.C.; CAIN, M.D.; CHAMLEY, W.A.; CUMMING, I.A.; FINDLAY, J.K.; GODING, J.R.

Luteolysis in the ewe: a study using a radioimmunoassay for prostaglandin F.

Journal of Reproduction and Fertility 32(2): 326-327. 1973.

A simplification of the Upjohn method using rabbit antiserum against $PGF_2\alpha$ coupled to bovine serum albumin was developed to assay PGF series in .1 ml ovine plasma without prior plasma extraction or purification. $PGF_2\alpha$ standards were nitrogen dried in tubes and .04 ml tris-HCl buffer added. Sample plasma or arterial plasma from a hysterectomized ewe were added with immune rabbit serum and $PGF_2\alpha$ in tris buffer, then shaken and incubated. Aliquots of .2 ml were applied to 35-mm columns (2.5 ml) Sephadex G25 Fine to isolate protein-bound tracer. Toluene was used as scintillant. Other PGs were assayed in quadruplicate and amounts required to displace 50% of tritiated $PGF_2\alpha$

compared. Relative to $PGF_2\alpha$ (100%) the following reactivities were obtained: $PGF_2\alpha$ THAM salt, 77%; $PGF_1\alpha$, 32%; PGE_1, .08%; PGA_2, .0002%; PGB_2, .01%. During $PGF_2\alpha$ infusion in 2 ewes with ovarian autotransplants, the concentration of hormone in the uteroovarian vein reached 4.5 nm per ml. Before and after infusion PGF was detectable (less than .2 ng per ml). Progesterone fell following infusion and animals returned to estrus. Concentrations of PGF in 1 ewe under slight anaesthesia were: carotid artery--undetectable; ovarian artery--.2 ng per ml; uteroovarian vein--3.9 ng per ml. The effect of trauma upon release of PGF was investigated by subjecting the uterus to 10-minute massage. No significant increase in uteroovarian vein PGF occurred, but concentration fell to 1 ng per ml when treatment ceased. (Authors' modified)   006444

0684

CERINI, M.E.D.; CHAMLEY, W.A.; FINDLAY, J.K.; GODING, J.R.
   Luteolysis in sheep with ovarian autotransplants following concurrent infusions of luteinizing hormone and prostaglandin $F_2\alpha$ into the ovarian artery.
   Prostaglandins 3(4): 399-404. 1973.

LH was administered at 10.0 μg per hour for 6 hours to determine if the luteolytic action of subsequent $PGF_2\alpha$ administration in 4 ovarian autotransplanted sheep could be overcome. $PGF_2\alpha$ was administered at 5 μg per hour during the subsequent 6 hours while maintaining the LH infusion. LH alone caused an increase of progesterone secretion, which fell from 800-1200 μg per hour to below 180 μg per hour after 6 hours of $PGF_2\alpha$ infusion and continued to fall thereafter. In 1 sheep given 5 μg of $PGF_2\alpha$ per hour and no LH, progesterone levels fell from 300 μg per hour to less than 20 μg per hour at the end of the infusion. All animals came into estrus within 2 days of treatment. Thus luteolysis in response to $PGF_2\alpha$ at levels just over twice the minimum dose known to cause luteolysis was not prevented by LH vastly in excess of levels reported during the luteal phase of the cycle, leading the authors to conclude that the survival of the corpus luteum in early pregnancy is not due to increased plasma LH. (RMS)   006439

0685

CHALLIS, J.R.G.; DAVIES, I.J.; RYAN, K.J.
   The relationship between progesterone and prostaglandin F concentrations in the plasma of pregnant rabbits,
   Prostaglandins 4(4): 509-516. October 1973.

Progesterone and total PGF were radioimmunoassayed in plasma from 7 rabbits to document their temporal relationships throughout pregnancy. The rabbits were bled at 1-2 day intervals throughout gestation. PGF was extracted, chromatographed on silicic acid, and separated by dextran-charcoal. Mean progesterone increased from 4.9 ng per ml on Day 3 to a broad maximum of 20.7 on Day 13, then declined to 12.8 on Day 29, and fell precipitously from Days 29-32 (delivery). Mean total PGF showed an apparent peak on Day 6, with variations in individuals, averaging below 300 pg per ml until Day 21. PGF then increased until Day 30 ($p < .001$, for Days 27-30 combined, vs Days 15-21 combined). PGF values were variable after delivery. There was an increase in PGF just before the rapid decline in progesterone on Days 29-32 in 5 of 7 rabbits. (LJG)   005531

0686

CHAMLEY, W.A.; CHRISTIE, M.
   Failure of prostaglandin $F_2\alpha$ to affect LH secretion in the ovariectomized ewe.
   Prostaglandins 3(4): 405-412. April 1973.

To determine whether $PGF_2\alpha$ affects luteolysis in sheep by altering luteinizing hormone (LH) secretion, the authors infused $PGF_2\alpha$ and monitored LH secretion in ovariectomized sheep. $PGF_2\alpha$ in saline with 200 U heparin per ml was infused at 5 μg per hour into an exterior carotid loop; and PGE_2 in saline with 5% ethanol at 10 μg per hour into the cannulated jugular vein; 10-ml blood samples were taken from the jugular vein; LH was determined by solid-phase radioimmunoassay. 1) In 4 sheep which received saline from Hours 3-6 and $PGF_2\alpha$ from Hours 6-11, LH fluctuated widely between 5 and 25 ng per ml, as usual in ovariectomized sheep. 2) In 3 sheep given $PGF_2\alpha$ during Hours 11-21, the time of the expected LH surge resulting from PGE_2 infused from Hours 1-5, LH rose to 60, 71, and 77 ng per ml. LH rose to 78 and 95 in 2 controls given PGE_2 only. 3) In 2 sheep given $PGF_2\alpha$ from Hours 2-12 before and during PGE_2 infusion (Hours 8-12), LH rose to 90 and 165 ng per ml. 4) In a sheep given $PGF_2\alpha$ from Hours 2-12 during and after PGE_2 infusion (Hours 2-6) LH rose to 120 ng per ml. Thus LH secretion in response to PGE_2 was not affected by $PGF_2\alpha$. (LJG)   006440

0687
CHAMLEY, W.A.; CERINI, J.C.; GODING, J.R.
Luteal function in sheep with ovarian autotransplants given concurrent infusions of prolactin and prostaglandin $F_2\alpha$ into the ovarian artery.
Prostaglandins 4(5): 711-716. November 1973.

4 sheep with ovarian autotransplants were infused with prolactin and $PGF_2\alpha$, and compared to 1 control given $PGF_2\alpha$ only. Prolactin, 10 or 100 μg per hour, was infused into the carotid (ovarian) artery for 6 hours after a 2.5 hour control infusion, then $PGF_2\alpha$ 5 μg per hour was added for 6 more hours. During prolactin, progesterone fluctuated from 103.5-424.0 μg per hour, within normal limits. Following $PGF_2\alpha$ progesterone was suppressed below 30 μg per hour in the control, fell slightly in 1 ewe given prolactin 10 μg per hour, and fell distinctly in the 2 ewes given 100 μg per hour. Plasma prolactin fluctuated widely in all sheep, but ovarian blood flow did not change. Thus prolactin alone failed to overcome the luteolytic effect of $PGF_2\alpha$. (LJG)   005642

0688
CHANG, M.C.; HUNT, D.M.; POLGE, C.
Effects of prostaglandins on sperm and egg transport in the rabbit.
Advances in the Biosciences 9: 805-810. 1973.

An increased rate of sperm transport, but not faster capacitation, was demonstrated by insemination of rabbits, with semen suspended in 0.01 % PGE_1 or PGE_2, but not in $PGF_2\alpha$. Such an effect was not shown by subcutaneous injection of 0.5 mg or 1 mg/kg of PGE_1, PGE_2, or $PGF_2\alpha$ before insemination. The transportation of eggs was found to be severely disturbed by subcutaneous injection 12 hr after ovulation of 1-5 mg/kg of PGE_1 or $PGF_2\alpha$, but not PGE_2. (Authors)   006501

0689
CHANNING, C.P.
The interrelationship of prostaglandins, cyclic 3',5'-AMP and ovarian function.
Research in Prostaglandins 2(5): 1-4. March 1973.

This review combines evidence from studies of adrenal and ovary to reach an understanding of prostaglandins (PG) and cyclic AMP (cAMP) in ovum maturation, ovulation, and progesterone synthesis by granulosa cells of the follicle. Luteinizing hormone (LH), dibutyryl cAMP and PGs can stimulate (possibly sequentially) meiosis in cultured rat follicles, probably by inducing granulosa cells to release cAMP into the fluid. PGs are essential for completion of ovulation, induced by LH via cAMP as second messenger. Luteinization (progesterone release) in cultured bovine or rabbit corpora lutea can proceed with LH alone, but if PGs are added, the amount of progesterone or cAMP released is greater than additive. PGs will inhibit progesterone release in cultured rabbit corpora lutea if incubated 6 hours or more. Cultured granulosa cells from monkey, pig, or cow will synthesize progesterone in response to added cAMP, and pig granulosa cells will generate cAMP when exposed to LH. The notion that cAMP is essential is strengthened by use of aminophylline, imidazole, and the recovery of more cAMP from larger follicles. The author feels that PGs may not be essential in the action of LH on adenyl cyclase in all these systems, except ovulation. PGs might act at the rate limiting step of cholesterol synthetase or esterase in progesterone synthesis. (LJG)   006438

0690
CHAPMAN, F.
   Prostaglandins.
   Off Our Backs 3(10):1. September 1973.

The author cited the various clinical uses of prostaglandins, with emphasis on their ability to induce labor or terminate pregnancy. The author suggested that physicians and women who may be prescribed prostaglandins fully review their side effects prior to administration. Additionally, the author hoped the FDA would carefully review and test prostaglandins prior to their commercial distribution. All such information should be available to the public. (RAS)   005689

0691
CHATTERJEE, A.
   Possible mode of action of prostaglandins: 5-differential effects of prostaglandin $F_2\alpha$ before and after
      the establishment of placental physiology in pregnant rats.
   Prostaglandins 3(2): 189-199. February 1973.

$PGF_2\alpha$ in a single injection of 2 mg per kg sc was given to pregnant rats on Day 10, 11, 12 or 13, and supplemented by 5 mg ovine prolactin sc or 5 mg hydroxyprogesterone caproate sc from Days 10-12, to study luteolysis and resorption of pregnancy. $PGF_2\alpha$ given on Day 10 caused complete fetal resorption and total placental resorption. On Day 11 partial placental resorption and total fetal resorption was evident. On Day 20, the controls and the rats treated on Day 1 had 95.8 and 91.8% surviving fetuses; those treated on Day 12 had 62.5%. Hydroxyprogesterone or prolactin was given to rats treated with $PGF_2\alpha$ on Day 11 and completely prevented luteolysis and fetal resorption, as shown by fetal, placental, ovarian and luteal weights, and number of surviving fetuses at examination on Day 20 and at term. (LJG)   006564

0692
CHATTERJEE, A.
   Some studies on the effects of prostaglandin $F_2\alpha$ and indomethacin in the physiology of
      pseudopregnancy and pregnancy in rats.
   Proceedings of the Indian National Science Academy 39(3,Pt.3): 408-419. June 1973.

The mechanism of regulation of luteal function by $PGF_2\alpha$ was studied using pregnant, pseudopregnant, or unilaterally ovariectomized rats. $PGF_2\alpha$ 2 mg/kg injected subcutaneously had no effect on the compensatory ovarian hypertrophy or on the increased number of ova noted in unilaterally ovariectomized rats. Given on Day 5 of pseudopregnancy, $PGF_2\alpha$ induced estrus, increased ovarian weight, and markedly reduced uterine weight. Injected on Day 6, proestrus was induced and uterine weight was moderately reduced. Prolactin or progesterone given on Days 5 through 8 blocked the

PGF$_2\alpha$ effects. Injected on Days 10 or 11 of pregnancy, PGF$_2\alpha$ caused resorption of fetuses and placentae and significant decreases in the weight of ovaries and corpora lutea. PGF$_2\alpha$ injected on Day 12 caused a decrease in placental weight and fetal survival. Either prolactin or progesterone injected on Day 10 through 12 again blocked the PGF$_2\alpha$ effects. Indomethacin injected on Days 8, 9, or 10 prolonged pseudopregnancy by 8-9 days. Injected on Day 2 of pregnancy, there were no implantation sites; on Days 5, 8, or 12, fetal survival rate was reduced. This effect of indomethacin was blocked by chorionic gonadotropin or prolactin. The authors concluded that PGF$_2\alpha$ acts through the hypothalamic-pituitary complex. (HC)   006947

0693

CHATTERJEE, A.

The possible mode of action of prostaglandins: VI--failure of prostaglandin F$_2\alpha$ in the interruption of pregnancy in rats having pituitary heterotransplant under the kidney capsule.
Prostaglandins 4(6): 915-922. December 1973.

Since prolactin is the only pituitary hormone capable of supporting pregnancy in hypophysectomized rats and the only pituitary hormone not released by hypothalamic stimulation, abortion by PGF$_2\alpha$ was attempted in heterotransplanted pregnant rats. Pituitaries were transplanted under the kidney capsules in cycling rats, which were then impregnated, and PGF$_2\alpha$ 2 mg/kg was injected on Pregnancy Day 10. Weights of corpora lutea checked by midventral incision on Day 20 were 5.0 mg in 6 sham-operated controls, 2.9 mg in 8 sham controls aborted with PGF$_2\alpha$, but 5.1 mg in 10 pituitary-grafted rats given PGF$_2\alpha$. On Day 20, none of 69 implanted fetuses had survived in 8 sham controls given PG, but 78 (97.5%) fetuses survived in 10 grafted rats, of which 66 (94.3%) were delivered alive, weighing mean 5.8 gm compared with 100% survival and 5.8 gm in 6 sham controls. Histologic studies showed luteolysis in sham-operated rats given PGF$_2\alpha$ and a number of functioning lactotrophs in the pituitary grafts. (LJG)   005683

0694

CHATTERJEE, A.

The possible mode of action of prostaglandins. III. Ovarian response to exogenous gonadotrophins in the presence of PGF$_2\alpha$ or PGE$_2$ in immature rats.
Endokrinologie 61(2): 307-311. 1973.

Ovulation was induced in immature rats by an intraperitoneal injection of PMS (Gestyl) followed in 48 hours by HCG. PGF$_2\alpha$ and PGE$_2$ injections, 250 μg subcutaneously daily for 3 days or 500 μg daily for 2 to 3 days, were given concurrently with the gonadotrophins. Autopsies conducted 24 hours after the HCG injection revealed no inhibitory effect of PG on ovarian or uterine weight. The authors suggested that the luteolytic effect of PG may be indirect rather than at the ovarian level. (HC)   006168

0695

CHAUDHURI, G.

Release of prostaglandins by the [rat uterus containing an] I.U.C.D.
Prostaglandins 3(6): 773-784. June 1973.

PGF$_2\alpha$, PGE$_1$, and PGE$_2$ were estimated in rat uterine horns containing a silk suture, in control horns, and in uteri of rats given indomethacin, an inhibitor of PG synthesis. A 2.5-cm silk thread was inserted through a midventral incision. Rats were mated 2 weeks later, and pregnancy was dated from the day sperms were detected in the vagina. Both uterine horns from 20 rats were flushed with 2 ml Krebs solution, or horns from 38 rats were perfused for 1 hour with 35 ml Krebs solution gassed with 95%

oxygen and 5% carbon dioxide. PG activity was estimated by extracting the acidified perfusate with ethyl acetate, dissolving the residue in Krebs solution, and testing it on rat stomach strip, rat colon, and chick rectum, all treated with mepyramine, hyoscine, phenoxybenzamine, propranolol, and methysergide. Those solutions which contained PG activity were again extracted and chromatographed on thin layer, and assayed again on rat stomach and colon. PGE and F activities on Days 3-6 of pregnancy were higher in IUD horns than in controls. PGE_2 activity ranged from 8-12 ng in control horns and 22-45 in IUD horns with the 2-minute flush, and from 9-32 ng in control and 23-110 ng in IUD horns with the 1-hour perfusion method. Indomethacin 2 mg/kg sc twice daily did not reverse the contraceptive action of the intrauterine threads, as judged by number of implantations. (LJG) 006463

0696
CHEAH, S.F.; KHAIRUDDIN, Y.
Prostaglandin $F_2\alpha$ for induction of labor.
Medical Journal of Malaysia 27(3): 211-216. March 1973.

$PGF_2\alpha$ was infused iv in 17 Malaysian women after amniotomy to induce labor at term. Dose rates ranged from 2-6 μg/minute in 2 μg/ml of 5% dextrose. Patients were followed by 10 minute tocography recordings, and a clinical exam was given every 30 minutes and a vaginal exam every 6 hours. 12 of 17 were delivered without assistance within 5-24 hours, 9 of these within 12 hours. 4 of the 5 failures were subsequently delivered within 3-8 hours after changing to iv Syntocinon. All infants were normal. Uterine activity began from 30 minutes to 5.5 hours, was normal in successful deliveries, but irregular with increased tone of 5-15 cm between contractions in unsuccessful cases; however, no hypertonicity or fetal distress occurred. Side effects reported were sweating, vomiting, phlebitis, postpartum hemorrhage, and hyperventilation. (LJG)   006396

0697
CLARK, K.E.; RYAN, M.J.; BRODY, M.J.
Effects of prostaglandins E_1 and $F_2\alpha$ on uterine hemodynamics and motility.
Advances in the Biosciences 9: 779-782. 1973.

The dog uterus perfused in situ at constant flow was used to study the effects of prostaglandins on uterine vascular resistance (UVR) and contractile activity (UCA) and on adrenergic control of these parameters. In nonpregnant dogs, neither PGE_1 nor $PGF_2\alpha$ (1 μg/min) altered UCA. $PGF_2\alpha$ did not affect UVR, whereas PGE_1 produced a marked decrease in UVR which persisted long after the infusion was terminated. $PGF_2\alpha$, but not PGE_1, significantly enhanced vasoconstrictor responses to uterine nerve stimulation and intraarterial norepinephrine. In pregnant dogs near term, intraarterial infusion of large doses of $PGF_2\alpha$ and PGE_1 failed to alter UCA, whereas these uteri were sensitive to oxytocin, raising the possibility that the abortifacient activity of PGs may not be caused by a direct effect on uterine smooth muscle. It may be concluded from these data that the facilitating effect of $PGF_2\alpha$ on adrenergic transmission and the long-lasting vasodilator effect of PGE_1 make these agents candidates for physiologic regulators of uterine blood flow. (Authors' modified)   006497

0698
CLARK, K.E.; RYAN, M.J.; VAN ORDEN, D.E.; FARLEY, D.; VAN ORDEN, L.S. BRODY, M.J.
Role of prostaglandins in estrogen-induced uterine hyperemia.
Pharmacologist 15: 209. 1973

Ovariectomized rats were pre-treated for 2 days with indomethacin. 2 hours before sacrifice, estradiol-17β was administered. Both uterine blood volume and PGF content were increased by the estrogen. Both effects were significantly inhibited by indomethacin. (HC)   005575

0699
CLITHEROE, H.J.
Cyclic variations in endometrial prostaglandins during the human menstrual cycle.
Acta Endocrinologica Suppl. 177: 317. 1973.

Endometrial samples were obtained from 110 patients and assayed for PG content. Early to late follicular phase endometrium had 10 to 50 ng $PGF_2\alpha$ per gm wet weight tissue. Postovulatory, early secretory endometrium had similar amounts, while in mid and late secretory endometrium the $PGF_2\alpha$ content was 80 to 250 ng per gm tissue. $PGE_2$was also present, but only $PGF_2\alpha$ increased postovulatory. (HC)   005576

0700
COCEANI, F.; OLLEY, P.M.
The response of the ductus arteriosus to prostaglandins.
Canadian Journal of Physiology and Pharmacology 51(3): 220-225. 1973.

Effects of various PGs (PGE_1, PGE_2, $PGF_1\alpha$) on lamb ductus arteriosus were investigated under aerobic and anaerobic conditions. $PGE_1$and E_2markedly relaxed the anoxic ductus over a dose range of 1 nM to 1 mM, whereas they had little or no effect on the tissue after exposure to oxygen. Papaverine relaxed the ductus before and after exposure to oxygen. $PGE_1$and $E_2$are less active on anoxic tissue depolarized by excess potassium. The greater effectiveness of PGs on the anoxic ductus suggests a role for these compounds in the regulation of vessel tone during fetal life. (Authors) 006459

0701
CONRAD, J.T.; ONWUDIWE, F.
Dose-response effects of the prostaglandins, $PGE_1$and $PGF_2\alpha$, upon the isometric tension of segments of the ampulla, isthmus and uterus of the estrous rabbit.
Prostaglandins 4(1): 47-55. July 1973.

Dose response of isotonic contractions of uterus, isthmus, and ampulla segments from estrous New Zealand rabbits to $PGE_1$and $PGF_2\alpha$ were recorded. All rabbits had had 1 litter, were fed low fat diet. Their tissues were allowed to develop regular contractions for 20 minutes (unstimulated control), then tested with 1 dose of PG. Results were expressed as a percent change from control. The drugs were applied in concentrations from $10^{-5}$to 10^{-8}g per cc. $PGF_2\alpha$ stimulated peak responses at 10^{-7}g per ml Ringers solution for the oviduct and $10^{-8}$for the uterus. There was a gradient with the uterus contracting 220%, the isthmus 130%, and the ampulla 60% more frequently than unstimulated controls. Peak amplitude of contraction was about 80% higher in all tissues. PGE_1depressed both frequency and amplitude of contraction, with increasing dose from $10^{-8}$to 10^{-5}g PG per ml, in isthmus and ampulla. The uterus however, tended to lose its sensitivity to inhibition by $PGE_1$with increasing dose. (LJG)   005779

0702

CORLETT, R.C.; SRIBYATTA, B.; THORNEYCROFT, I.H.; MISHELL, D.R., Jr.; BALLARD, C.; NAKA-MURA, R.M.

Abortifacient activity of vaginally administered prostaglandin $F_2\alpha$.
Advances in the Biosciences 9: 575-579. 1973.

9 women 33 to 40 days past expected menstruation were given a total dose of 200 to 1100 mg $PGF_2\alpha$ in the form of tablets containing 50 mg $PGF_2\alpha$, inserted vaginally at intervals over a 24-hour period. Although vaginal bleeding occurred in 3 1/2 to 16 hours in 7 patients, only 3 aborted. Progesterone, but not estrogen, levels fell following a decline in HCG levels in the patients that aborted. There were no changes in steroid levels in the women that did not abort. All but 2 patients had 1 or more side effects such as nausea, vomiting, diarrhea, chills, or cramps. The results indicate that $PGF_2\alpha$ acts by stimulating uterine muscle interfering with blood supply and causing a fall in HCG which in turn causes a fall in steroid hormone levels. (HC)   006475

0703

CORSON, S.L.; BOLOGNESE, R.J.; MEROLA, J.

Intraamniotic prostaglandin $F_2\alpha$ to induce midtrimester abortion.
American Journal of Obstetrics and Gynecology 117(1): 27-34. September 1973.

To investigate the success of $PGF_2\alpha$ in inducing abortion, 100 volunteers (ages 14-42, gravida 16-22 weeks) were given initial $PGF_2\alpha$ dosages of 20-40 mg with subsequent dosage following 5 different protocols. Oxytocic agents were used only in absence of uterine contractility, or if membranes ruptured just prior to administering $PGF_2\alpha$, to assess PG action alone. Laminaria was used in 3 cases having an unusually rigid cervix. An extensive review of previous studies involving $PGF_2\alpha$ as an abortifacient, which discuss dosages and side effects, is presented and comparisons made with the saline method. In this study 65 patients were aborted completely with intraarterial $PGF_2\alpha$, and 30 incompletely aborted. 40 patients experienced vomiting; 8 had diarrhea. No statistically significant linear regression existed between time to abort and patient age, parity, or gestational age. The authors agreed that an initial dose of 40 mg produces better results than smaller ones. $PGF_2\alpha$ has the disadvantage of causing more vomiting and diarrhea than hypertonic saline methods, but this may be reduced by 15-methyl analogues. (IC)   006461

0704

COUDERT, S.P.; FAIMAN, C.

Effect of prostaglandin $F_2\alpha$ on anterior pituitary function in man.
Prostaglandins 3(1): 89-95. January 1973.

5 men each received $PGF_2\alpha$ intravenously at .05 µg/kg/minute, .20 µg/kg/minute, 2.0 µg/kg/minute or vehicle alone for 30 minutes, with treatments at weekly intervals. Serum FSH and TSH levels were not altered, LH levels decreased with time in all groups. Growth hormone (GH) levels were not affected by the lower doses, but rose after the high dose was given. Cortisol levels initially decreased in all groups but rose sharply in the high dose group. At the high dose, all patients reported discomfort. The GH response may have been a nonspecific effect of stress; similarly the cortisol increase may be due to increased ACTH release, again a stress response. In conclusion, $PGF_2\alpha$ is not a useful agent to study pituitary reserve in man. (HC)   006783

234

0705
COX, R.I.; THORBURN, G.D.; CURRIE, W.B.; RESTALL, B.J.; SCHNEIDER, W.
Prostaglandin F group (PGF), progesterone and estrogen concentrations in the uteroovarian venous plasma of the conscious ewe during the oestrous cycle.
Advances in the Biosciences 9: 625-630. 1973.

Hormone measurements were carried out in detail with plasma from blood samples collected at 3 hourly intervals from sheep prepared with indwelling catheters in their uteroovarian veins. A peak of ovarian estradiol-17β was found on day 10-11 of the estrous cycle on the day before the first observed peak in PGF and is consistent with estrogen initiating uterine PGF secretion. The PGF pattern showed a series of peaks (5-22ng/ml) each of about 3-hr duration on days 13-17. Transient marked decreases in progesterone followed each PGF peak until luteolysis was complete. (Authors)   006479

0706
CRAFT, I.
Induction of abortion by combined intra-amniotic urea and prostaglandin E_2 or prostaglandin E_2 alone.
Lancet 1(7816): 1344-1346. June 1973.

15 women, 16 weeks to 20 weeks pregnant, given 1 to 2 doses of 10 mg PGE_2 intraamniotically aborted, 4 completely. IAT averaged approximately 27 hours; vomiting occurred in 3. Another 15 women given 140 ml of urea intraamniotically before the PG injection also aborted, 9 completely. IAT averaged 10 1/2 hours, vomiting occurred in 7. The authors conclude that the combined urea/PG method is advantageous in that it is effective within 24 hours of a single initiating procedure without the need of prolonged nursing supervision or supplementary intravenous oxytocin infusion. (HC)
006913

0707
CRAFT, I.
Intra-amniotic prostaglandin E_2 and $F_2\alpha$ for induction of abortion: a dose-response study.
Journal of Obstetrics and Gynaecology of the British Commonwealth 80(1): 46-47. January 1973.

Intra-amniotic prostaglandins were administered in patients 15 and 21 weeks pregnant. Initially a single intra-amniotic injection of 5 mg of PGE_2 or 25 mg of $PGF_2\alpha$ was given. During this study, the dose of prostaglandin was increased to a maximum of 20 mg of PGE_2 and 100 mg of $PGF_2\alpha$. The effectiveness of the lower doses used was disappointing. However, with the highest dose abortion resulted within 24 hours in 88% of those given 20 mg of PGE_2 and in 100% of those given 100 mg of $PGF_2\alpha$. A dose-response relationship was noted. The mean injection-abortion interval for patients having 20 mg of PGE_2 was 12 hours 57 minutes. For those receiving 100 mg $PGF_2\alpha$ it was 17 hours and 23 minutes. Mild vomiting and diarrhea occurred. (AS)   006792

0708
CRAFT, I.
Intra-amniotic urea and prostaglandin E_2 for abortion: a clinical study to determine the efficacy of using a variable prostaglandin dosage.
Prostaglandins 4(5): 755-763. November 1973.

Urea 80 g followed immediately by either 5 mg PGF2 in 10 women, or 10 mg $PGF_2\alpha$ in 20, was instilled intraamniotically to induce midtrimester abortion. The standard procedure included estimation of gestational age and placental location by ultrasound, local anesthesia with 1%

Lignocaine, amniocentesis of 30-150 ml fluid by cannula, oral antibiotics, liberal analgesics, and vaginal exam every 4 hours during the day. 80 ml sterile Hartman's solution was added to urea (80 g Ureaphil) in a sterile bottle, resulting in 140 ml solution. After instilling the urea, 5 or 10 mg PGE_2 in alcohol was injected, and the cannula removed. The 20 women given 10 mg aborted in a mean 10 1/2 hours (range 3 1/2-23 1/2), with complete abortion in 11 (55%), diarrhea in 4 (20%), and vomiting in 14 (70%). The 5 mg group aborted in mean 11 hours (range 5 1/2-18 1/2), with complete abortion in 6 (60%), diarrhea in 1 (10%), and vomiting in 4 (40%). The 10 mg group differed in having more primigravidae, and lower mean volume of amniotic fluid aspirated. Time to abortion was not significantly different, but 5 mg caused fewer side effects. This technique has the advantages of a single injection, less medical supervision, and expense, and less incidents of side effects. (LJG) 005646

0709
CRAFT, I.
Intraamniotic prostaglandin E_2 and urea for abortion.
Lancet 1(7806): 779. April 7, 1973.

5 patients pregnant 17-20 weeks, given 10 mg PGE_2 and 80 gm urea by the intraamniotic route aborted completely; the IAT averaging 8 hours. Another group of 5 patients given PGE_2 alone had imcomplete abortions, the IAT averaging 22 1/4 hours. (HC)   006795

0710
CRAFT, I.
Intraamniotic prostaglandins and urea for abortion.
Lancet 2: 207. July 28, 1973.

After successful therapeutic midtrimester abortion with 10 mg PGE_2 and 80 g urea intraamniotically, the author reports a trial of 5 mg PGE_2 plus urea, compared with the 10-mg dose. 8 patients, including 4 primigravidae, mean age 22.1 years, mean gestation 19 weeks, given 5 mg PGE_2, experienced 5 (62.5%) complete abortions in mean 11.5 hours (range: 7.3-18.5 hours). In comparison, 10 mg PGE_2 was given to 19 patients, including 15 primigravidae, of mean age 20.4, mean gestation 18 weeks; 10 of these patients (52.6%) aborted completely in a mean 10.5 hours (range: 3.75-23.5 hours). Those given 5 mg PGE_2 suffered 2 episodes of vomiting and 1 of diarrhea, while the 10-mg group had 14 incidents of vomiting and 4 of diarrhea. 20 mg of $PGF_2\alpha$ is also effective for inducing abortion. (LJG) 006445

0711
CRAFT, I.
Oral prostaglandin E_2 and amniotomy for induction of labor.
Advances in the Biosciences 9: 593-598. 1973.

The efficiacy of using oral prostaglandin E_2 in association with amniotomy for induction of labor was investigated in a series of 80 unselected patients. Increasing doses were given to stimulate an early onset of optimal uterine activity. Induction was successful in 43 out of 50 primigravid and 28 out of 30 multigravid patients. The mean induction-delivery intervals in successful cases were 10 1/4 and 6 1/4 hr, respectively. There were no significant effects on the fetuses. (Author)   006478

0712
CRAFT, I.
Prostaglandins and convulsions.
Lancet 2(7842): 1389. 1973.

The purpose of this letter is to refute a letter published previously (Lancet, Nov.3, p.1003) in which it had been asserted that $PGF_2\alpha$ injected intraamniotically for middle-trimester abortion was associated with atypical EEG patterns and epileptic convulsions. Some patients had a previous history of convulsions; the incidence of epileptic seizures in the patients studied was similar to that in the general population; and a dose-dependent relationship was not observed. (HC)   005758

0713
CRAFT, I.
The use of prostaglandin pessaries prior to vaginal termination.
Prostaglandins 3(3): 377-381. March 1973.

10 subjects pregnant 8 to 12 weeks were given 100-mg pessaries of $PGF_2\alpha$ inserted 2 times within 24 hours for vaginal termination of pregnancy. Cervical dilatation was increased 4 to 8 mm in 5 subjects, 4 of whom had vaginal bleeding. All patients had uterine cramps, 5 had diarrhea, 5 felt flushed. The procedure has limitations in that it was less successful in primiparous patients and in that side effects were common. The authors concluded that $PGF_2\alpha$ has no local specific cervical softening activity. (HC)   005568

0714
CRAFT, I.L.
Intra-amniotic prostaglandin E_2 and $F_2\alpha$ for induction of midtrimester abortion.
In: Jacomb, R.G., ed. The use of prostaglandins E_2 and $F_2\alpha$ in obstetrics and gynecology. (Proceedings of the Upjohn Prostaglandins Symposium, London, September 21, 1972.) Miami, Florida, Symposia Specialists, 1973. p. 81-91.

Women pregnant 15-21 weeks were given single doses of 5-20 mg PGE_2 or 25-100 mg $PGF_2\alpha$ intraamniotically. None of 5 women given 25 mg $PGF_2\alpha$ and only 2 of 10 women given 5 mg PGE_2 aborted. Amniocentesis did not improve the outcome. 8 of 15 women given 10 mg PGE_2 and 31 of 35 given 20 mg aborted within 24 hours. Vomiting and diarrhea were noted in the highdose group. 9 of 12 women given 50 mg $PGF_2\alpha$, 14 of 16 given 75 mg, and all 8 given 100 mg aborted within 24 hours. Vomiting occurred but the incidence was less than that noted with PGE_2; none of the women had diarrhea. The author noted that these doses differ from those reported by others and suggested that there is variation in sensitivity to prostaglandins in patients of different ethnic groups. He concluded that the method is quick and easy to perform with a high success rate, a single injection can be used, IAT is short, and hospital stay is brief. The disadvantages include high doses required, high degree of discomfort, diarrhea, and vomiting, possibly due to extramembranous leakage of prostaglandins and intrauterine sepsis and hemorrhage, both highly unlikely consequences. (HC)   006953

0715
CRAFT, I.L.; SCRIVENER, R.; DEWHURST, C.J.
Prostaglandin $F_2\alpha$ levels in the maternal and fetal circulations in late pregnancy.
Journal of Obstetrics and Gynaecology of the British Commonwealth 80(7): 616-618. July 1973.

Since infused prostaglandins were reported to stimulate oxytocin release and oxytocin release by the human fetus is maximal at delivery (higher concentrations found in fetal than maternal circulation), suggesting that it may be a stimulus to uterine activity, a study was undertaken to determine $PGF_2\alpha$ concentrations in maternal and fetal blood at normal delivery and elective cesarean. $PGF_2\alpha$ was significantly higher in the umbilical artery and vein at normal delivery and cesarean than in maternal blood. There was no significant arteriovenous difference in the umbilical cord as there is in oxytocin levels. Although results indicate that there is $PGF_2\alpha$ production in the fetoplacental unit, there is no indication that this source of $PGF_2\alpha$ contributes to initiation or maintenance of labor. (IC)   006452

0716

CRAFT, I.L.; FERGUSSON, I.L.C.; SMITH, B.; YOUSSEFNEJADIAN, E.
Sex steroid hormone levels in plasma following intra-amniotic injection of urea and prostaglandin E_2.
Journal of Obstetrics and Gynaecology of the British Commonwealth 80(12): 1095-1099. December 1973.

In 6 patients pregnant 15-22 weeks, given 80 gm urea and 10 mg PGE_2 intraamniotically, the IAT averaged 10 1/2 hours. Plasma levels of progesterone, 17-hydroxyprogesterone, estrone, estradiol, and estriol all fell, the sharpest decline being progesterone and estriol within 1 hour after injection. The authors concluded that urea increases the effectiveness of PGE_2 by causing fetal cellular dehydration and that PGE_2 indirectly reduces steroid levels, particularly progesterone, by impairing the fetoplacental unit, disrupting placental function, and causing fetal death. (HC)   005759

0717

CSAPO, A.I.; CSAPO, E.F.; FAY, E.; HENZL, M.R.; SALAU, G.
The delay of spontaneous labor by Naproxen in the rat model.
Prostaglandins 3(6): 827-837. June 1973.

Naproxen [d-2-(6-methoxy-2-naphthyl) proprionic acid] inhibits PG synthesis in vitro. Since PGs were recognized as intrinsic myometrial stimulants and implicated in the initiation of labor, it was of interest to examine the efficacy of Naproxen in delaying the onset of labor. Naproxen (20 mg/kg/day or more) was administered to 345 pregnant rats when delivery was in progress (Day 21 of pregnancy) and resulted in a high incidence of prolonged and interrupted labor, with consequent adverse effects. The onset of labor was effectively delayed without serious effects on pregnant animals or fetuses when Naproxen was given at a daily dose of 5-15 mg/kg beginning 3 days before term. Pregnancy was delayed in 98% of the animals and 88% of the fetuses remained undelivered until autopsy, 24 hours after control rats delivered spontaneously. Naproxen is concluded to effectively prolong pregnancy and delay labor in rats. (IC)   006464

0718

CSAPO, A.I.; PULKKINEN, M.O.; KAIHOLA, H.L.
The effect of luteectomy-induced progesterone-withdrawal on the oxytocin and prostaglandin response of the first trimester pregnant human uterus.
Prostaglandins 4(3): 421-429. 1973.

A single intravenous dose of 250 mU oxytocin or 100 μg $PGF_2\alpha$ resulted in slight, transient contractions in 9 normal first trimester pregnant patients. The normal early pregnant human uterus appeared refractory to stimulants. Luteectomy-induced rapid and continued P-(progesterone) withdrawals in 5 of the 9 patients making these uteri responsive organs. The conversion was noted by advanced cyclic IUP during the oxytocin response and PG response within 24-48 hours after luteectomy. Abortion occurred in 60.6 ± 7.6 hours. Conversion was delayed in 2 patients whose P-withdrawal was slow and who aborted after 140.5 ± 13.5 hours. In 2 patients the luteectomy-induced P-withdrawal was transient and they failed to abort. The present study demonstrated that during normal early pregnancy, when progesterone levels are high and myometrial stretch moderate, the uterus cannot respond to stimulants with advanced cyclic IUP. This substantiates the evidence that effective abortifacient therapy with PG depends on the degree of P-withdrawal it provokes, since it is the degree of P-withdrawal which controls the level of cyclic IUP induced by a given dose of PG. (Authors' modified)   005454

0719
CSAPO, A.I.; MOCSARY, P.; NAGY, T.; KAIHOLA, H.L.
The efficacy and acceptability of the 'prostaglandin impact' in inducing complete abortion during the second week after the missed menstrual period.
Prostaglandins 3(2): 125-139. February 1973.

22 women pregnant for 9-14 days were aborted by a single intrauterine injection of 5 mg $PGF_2\alpha$. 12 were sedated, given the $PGF_2\alpha$, and monitored for 7 hours by intrauterine pressure recording and blood samples for estradiol and progesterone; 10 received PG only. Sedation consisted of 100 mg pethidium HCl iv, 10 mg diazepam iv, 4 mg lidocain, and 6.5 mg thiethylperazin dimaleate suppository. The first uterine response was an increased sustained contracture (mean 14 mm Hg in mean 20 minutes). The sedated patients began bleeding in mean 144 minutes and the untreated group in 234 minutes; bleeding lasted 2-21 days (mean 8 days). Abortion probably took place 3 days later, but definitely not during observation. Outcome of treatment was complete abortion and negative pregnancy test at 14 days in 20 of the patients (2 required curretage). Side effects included nausea and vomiting in 3 sedated patients and in 10 nonsedated, 3 changes in blood pressure including 1 vascular collapse, 2 headaches, 3 uterine pain, 1 hyperventilation. Progesterone fell from 19.17 to 10.69 ng/ml, and estradiol from .47 to .27 ng/ml in 24 hours. According to the author's PG impact theory, exogenous PG could have caused a gradual progesterone withdrawal, upset endocrine balance, and lowered uterine threshold to oxytocin. (LJG)   005569

0720
CSAPO, A.I.
The prospects of PGs in postconceptional therapy.
Prostaglandins 3(3): 245-289. March 1973.

This review describes the function of the pregnant human uterus conceived as an 'intrinsically active, but endogenously suppressed organ'; about 'PG impact' and 'biphasic action' of $PGF_2\alpha$; discusses clinical trials and prospects for using PGs as abortifacients in comparison with other procedures. Supportive data include animal experiments and clinical trials, shown in 32 illustrations. Oxytocics, by lowering membrane potential of myometrial cells, can induce either contracture from ineffective overstimulation, or cyclic intrauterine pressure, leading to delivery when administered at 60 mm Hg at term, or higher pressures before term. PG release by stretched guinea pig and rabbit uteri provides evidence that PG may be the intrinsic myometrial stimulant. Animal experiments, luteectomy in women, and the progesterone withdrawal data show that progesterone controls the uterine threshold for PG. Clinical infusion of PGs abortion, up to 12,000 μg per hour, show the process to be time dependent, probably involving endocrine imbalance, vasoconstriction, local contraction, and stretching to enable the uterus to produce sufficient intrauterine pressure. Thus the 'PG impact' method, a single extraovular injection of 10 mg doses of $PGF_2\alpha$ with sedation and oxytocin support, evolved. The author believes that this procedure may ultimately produce fewer complications than dilatation and curettage or hypertonic saline, and might be amenable to self-administration after a missed period. (LJG)   005567

0721
CSAPO, A.I.
The regulatory interplay of progesterone and prostaglandin $F_2\alpha$ in the control of the pregnant uterus.
In: Josimovich, J.B., ed. Uterine contraction--side effects of steroidal contraceptives. New York, Wiley-Interscience, 1973. (Volume 1 of Problems of Human Reproduction) p. 223-255.

The author presents data, largely from recordings of rabbit uterine activity, showing how the interrelationships between progesterone, uterine volume, pressure, and $PGF_2\alpha$, uphold his model of an intrinsically active uterus inhibited by the volume/progesterone ratio. In rabbits intrauterine pressure reaches effective levels for labor only after volume increases, progesterone declines (about Day 26) and volume/progesterone ratio increases abruptly (Day 30-32). In ovariectomized rabbits, uterine activity increases precociously unless progesterone is supplied exogenously. Rats show this behaviorally, since in progesterone deficiency they destroy some fetuses and maintain pregnancy by retaining the placentae. Evolving uterine motility was also demonstrated in women in whom luteectomy before Day 49 resulted in eventual abortion. The role of uterine volume is shown by the phenomena of missed abortion, where labor fails to occur when uterine contents cease growing, and that of spontaneous labor which evolves 3 weeks after progesterone plateaus. $PGF_2\alpha$, especially if given topically, can cause midtrimester abortion if progesterone falls to a critical level, and intrauterine tension rises sufficiently to evoke effective wall tension. The $PGF_2\alpha$ dose required for abortion in rabbits was inversely proportional to progesterone level. Uterine sensitivity to electric stimulation was quantitively related to progesterone level, providing the mechanistic link between action of intrinsic $PGF_2\alpha$, progesterone level, and uterine motility. $PGF_2\alpha$ was presumably released by stretch in myometrial fibers. In clinical terms, these considerations imply that a minimal abortion time of 6-12 hours is needed to lower progesterore level and develop effective uterine contractions. (LJG)
006310

0722
CSAPO, A.I.; CSAPO, E.F.; FAY, E.; HENZL, M.R.; SALAU, G.
The role of estradiol 17β in the activation of the uterus during premature labor and the effect of Naproxen, an inhibitor of prostaglandin synthesis.
Prostaglandins 3(6): 839-846. June 1973.

52 pregnant rats were ovariectomized on Day 16 and treated daily with placebo; estradiol dipropionate, 1 μg im; Naproxen, d-2-(6-methoxy-2-naphthyl) proprionic acid (an inhibitor of PG synthesis), 25 mg/kg; or both estrogen and Naproxen. The rats were allotted to groups by litter size observed during ovariectomy, checked frequently for premature delivery and weight of fetuses and placentae, and killed on Day 21. The 15 control rats delivered 21% of their fetuses prematurely, and retained 44% of these placentae, which weighed a mean of .61 g, compared to normal .53 g for Day 21. The estrogen-treated rats delivered 94% of their fetuses prematurely, retained 11% of these placentae, weighing a mean .48 g. Naproxen-treated rats retained all their fetuses until Day 21. Rats given both drugs delivered 50% of their fetuses early. Both Naproxen groups had placentae of normal weight. The authors conclude that ovariectomy-induced estrogen and progesterone deficiency promote premature delivery. This is enhanced when normal estradiol level is restored due to resultant enhancement of PG synthesis as indicated by retention of fetuses in Naproxen-treated rats. (LJG)
006465

0723

CURRIE, W.B.; THORBURN, G.D.

Induction of premature parturition in goats by prostaglandin $F_2\alpha$ administered into the uterine vein.
Prostaglandins 4(2): 201-214. August 1973.

Premature parturition was induced in 7 attempts in 5 Saanen goats by infusing $PGF_2\alpha$ via indwelling catheter into the uterine vein on the same side as the corpus luteum; peripheral progesterone and uteroovarian vein PGF were measured from infusion until delivery. The polyethylene catheters were placed into a small uterine vein near the tip of the horn near the ovary with the corpus luteum and in the uteroovarian veins 1-2 cm proximal to the iliac vein junction. Goats received 20, 15, 10, and 5 mg progesterone im daily before PG infusion. $PGF_2\alpha$ was infused at 1.88 (for 3 hours) and .94 (for 6 hours) μg/minute without luteolysis or delivery. Subsequent infusions of 1.88, 3.75, 7.5, and 15 μg/minute for 6 hours (total doses of .55-5.4 mg) induced delivery within 30 to 34 hours. Only 1 fetus survived, born 11 days premature; the others were 19-25 days premature. PGF in uteroovarian vein plasma, estimated by microcolumn chromatography and radioimmunoassay, rose during infusion from less than .2 ng/ml to 38 ng/ml in the goat infused with 15 μg/minute, and to 2-4 ng/ml in 2 others. PGF fell slightly, then gradually rose to 55-62 ng/ml during second stage labor. Progesterone fell from about 4-7 during infusion to less than 1 ng/ml during delivery. (LJG)   005570

0724

CURRIE, W.B.; WONG, M.S.F.; COX, R.I.; THROBURN, G.D.

Spontaneous or dexamethasone-induced parturition in the sheep and goat: changes in plasma concentrations of maternal prostaglandin and fetal oestrogen sulphate.

Memoirs of the Society for Endocrinology 20: 95-118. 1973.

PGF, progesterone, glucocorticoids, estrone, estradiol-17alpha, estradiol-17β, and the estrogen sulfo-conjugates were measured by protein binding assays at frequent intervals until birth in pregnant and fetal sheep and goats and in pregnant sheep induced with dexamethasone. Blood was sampled from fetal carotid artery and jugular vein, maternal uteroovarian veins by surgically inserted catheters. In ewes progesterone kept a constant ratio of 5:1 (10 and less than 3 ng per ml) in uteroovarian vein and jugular vein. In fetal lambs, corticosteroids rose during the last 10 days with rapid increase up to 185 μg per ml during labor and 400 μg per ml at birth. Fetal lambs had estradiol-17alpha-sulfate, and estrone-sulfate of 2-4 ng per ml rising to 3-11 ng the day before and 40-50 ng at birth. In ewes PGF remained below 2.5 ng per ml, began to rise 25 hours before delivery, but rose markedly in second stage of labor (head at cervix). Estradiol-17β ran parallel to PGF, at about 20 pg per ml during pregnancy and up to 150 pg per ml at delivery. In goats, the fetal corticosteroids began to rise 2 weeks before birth, more sharply 4-5 days before, and reached 400 μg per ml during birth. Progesterone fell and PGF peaked 24 hours before delivery. In induced delivery (50 μg per hour dexamethasone for 50-62 hours) no fetal corticosteroid response took place, conjugated estrogens rose normally, but maternal estrogens and PGF rose less than in normal delivery. All neonates died within 36 hours of respiratory failure. It was suggested that the increase in fetal corticosteroids precedes other events and may cause the maternal fall in progesterone, that the increase in estrogens may be initiated by the fetus, that dexamethasone may act as substrate for these estrogens, that free estrogens may stimulate PGF release. The chief differences between sheep and goats are that progesterone originates in placenta in sheep but in corpus luteum in goats; and that goats have higher estrogen levels throughout pregnancy, but the final estrogen increment is higher in sheep, resulting in comparable estrogen to progesterone ratios. (LJG)   006291

0725
CUSHNER, I.M.
Prostaglandins: do they have any place in abortion?
In: Lewit, S., ed., Advances in Planned Parenthood, Vol. 8. [Proceedings of the tenth annual meeting of the American Association of Planned Parenthood Physicians, Detroit, Michigan, April 1972] Amsterdam, Excerpta Medica, 1973. p. 192-195.

The author reviews the 26 months since Karim and Filshie first reported successful use of intravenous PGF$_2\alpha$ as an abortifacient in relation to changes in abortion methods, the increased number of abortion patients, and potential developments of new abortifacients. Prostaglandins reduce hospital stay as compared to saline abortion; however, prostaglandin side effects require close medical supervision. Suction curretage is the method of choice for first trimester, whereas prostaglandins are the preferred method for second trimester. During the trouble period of 13-16 weeks, prostaglandins presently are the only method available to terminate pregnancy. Prostaglandin research will hopefully culminate with the development of a self-administered analogue for immediate menstrual induction. (RAS)   006968

0726
CUSHNER, I.M.
The current status of the use of prostaglandins in induced abortion.
American Journal of Public Health 63(3): 189-190. March 1973.

The results of current research in trials of PGs as abortifacients, and the safety, cost and availability of their application are summarized. PGF$_2\alpha$, PGE$_2$ and PGE$_1$ are abortifacient by all routes except orally. The mechanism is probably their oxytocic properties, since little evidence exists that PGs are luteolytic in women. PGs offer no advantage over suction abortion in the first 12 weeks, but may prove useful in Weeks 13-15 when no other method is advised. The advantages of PGs are shorter injection-to-abortion interval, less than 24 hours with PGE$_2$; less risk of uterine perforation since only about 25% of pateints require curettage for incomplete abortion, no chance of coagulopathy or intravascular saline injection. Costs would be reduced in proportion to hospital stay: 1 day for PGs, 2-3 days for saline, 1 week for hysteromtomy. But skilled personnel are still required in case of incomplete abortion, for care of PG side effects such as vomiting or fever, and for complications of hemorrhage and infections that may accompany any separation of trophoblast or placenta. (LJG)   005572

0727
DE PAIVA, C.E.N.; CSAPO, A.I.
The effect of prostaglandin on the electric activity of the pregnant uterus.
Prostaglandins 4(2): 177-188. August 1973.

Simultaneous unilateral recordings of intrauterine pressure and electrical activity by external macroelectrode were made daily from Days 17-22 in untreated rats, rats given PGF$_2\alpha$ and ovariectomized rats supplemented with estradiol dipropionate 1 μg per day im. Rats were surgically rendered unilaterally pregnant with 3-5 fetuses, and implanted with 2 loop electrodes, a recording microballoon of .8 ml volume, and a catheter for instilling extraovular PG. The controls showed small asynchronous spikes and contractions, gradually developing by Day 21 into synchronously propagated electric activity always accompanied by contractions. In ovariectomized rats the process was similar, but accelerated, until abortion on day 18. The initial response to 50-100 μg PGF$_2\alpha$ was a multitude of continuous isolated spikes and a high level resting pressure for several hours. On subsequent days when the rats were given 25, 10 and 5 μg PGF$_2\alpha$, an accelerated normal evolution took place, culminating in abortion on Day 19. The author comments throughout the article on the simultaneous behavior of PGs and progesterone from comparable rats in other experiments. (LJG)   005573

0728
DEERY, D.J.; HOWELL, S.L.
Rat anterior pituitary adenyl cyclase activity: GTP requirement of prostaglandin E_1 and E_2 and synthetic luteinising hormone-releasing hormone activation.
Biochimica et Biophysica Acta 329(1): 17-22. 1973.

Rat anterior pituitary adenyl cyclase was stimulated by PGE_1 and PGE_2, and the concentration response curves, effect of sodium ions, role of GTP, and stimulation by luteinizing hormone-releasing hormone (LH-RH) were explored. 10-20 μg protein of male rat anterior pituitary homogenate was incubated with .01 mM [32P]-ATP, 1 mM EDTA, 5 mM magnesium chloride, 3 mM theophylline, .1% bovine plasma albumin, 25 mM phosphocreatine, and 1 mg/ml creatine phosphokinase in a total volume of 100 μl of 25 mM Tris pH 7.6. The reaction was linear for 15 minutes. The optimum stimulation, 175%, was recorded at 1 mM PGE_1, and 170% at .01-.1 mM PGE_2. Addition of .1 mM GTP further enhanced adenyl cyclase activity in presence of .1 mM PGE_2 and at all PGE_1 concentrations from 1 μM to .01 M. Added LH-RF .2-200 ng/ml had no effect on adenyl cyclase activity, but in the presence of .1 M GTP, LH-RF at 20 ng/ml almost doubled the enzyme activity. .01 mM sodium fluoride increased adenyl cyclase 600%, but increasing concentrations of GTP .01 mM to .01 M reduced the activation to 400%. When ATP (substrate) concentration was 2 mM, neither GTP nor LH-RH activated rat pituitary adenyl cyclase. (LJG)   005588

0729
DEL CAMPO, C.H.; GINTHER, O.J.
Vascular anatomy of the uterus and ovaries and the unilateral luteolytic effect of the uterus: horses, sheep, and swine.
American Journal of Veterinary Research 34(3): 305-316. 1973.

The vascular systems of the reproductive tracts of 3 mares, 15 ewes, and 6 sows were injected with red and blue latex, removed, and studied from photographs and drawings. All 3 species have a large common vein draining the ovary, oviduct, and uterus, an anastomosis between veins of the left and right sides, and an ovarian artery. In sheep and swine, which have direct unilateral luteolysis from the uterus, 1 branch of the ovarian artery was convoluted around the ovarian branches of the ovarian vein. Swine had a more complicated system of convolutions of the ovarian artery over 2 or 3 channels of the ovarian vein. Horses have systemic rather than unilateral luteolysis and their ovarian artery only passed obliquely over a branch of the ovarian vein. These anatomic characteristics of sheep and swine are compatible with the postulated luteolytic role of $PGF_2\alpha$. (LJG)   006259

0730
DEMERS, L.M.; BEHRMAN, H.R.; GREEP, R.O.
Effects of prostaglandins and gonadotrophins on luteal prostaglandin and steroid biosynthesis.
Advances in the Biosciences 9: 701-707. 1973.

Steroid, protein, and prostaglandin biosynthesis by rat corpora lutea in long-term organ culture was examined. Luteinizing hormone increased steroid output in this system while $PGF_2\alpha$ produced a marked inhibition of progesterone, 20a-Ol, and protein synthesis. LH increased PGF synthesis by these cultured CL, while surprisingly PGF2 caused a marked increase in the production of PGF. These results, particularly the ability of corpora lutea to synthesize PGF, raises interesting possibilities for the regulation of luteal function through control of intrinsic PG synthesis. (Authors)   006488

0731

DEMERS, L.M.; GABBE, S.G.; VILLEE, C.A.; GREEP, R.O.
  Human chorionic gonadotropin-mediated glycogenolysis in human placental villi: a role of prostaglandins.
  Biochimica et Biophysica Acta 313: 202-210. 1973.

$PGE_2$and human chorionic gonadotropin (HCG) induced glycogenolysis with increased cAMP and activated glycogen phosphorylase in placental villi of 18-20 weeks gestation. Incubations consisted of 100 mg fresh placenta. In 2 ml 199 medium, at 37 degrees C. in 95% oxygen to 5% carbon dioxide for 30 minutes with 50 I.U. HCG per ml or 10 μg PGE_2per ml. Glucose-6-PO4 dependent and independent forms of glycogen synthetase were unaffected by HCG. HCG activated glycogen phosphorylase significantly ($p<.001$), and doubled the ratio of active to inactive phosphorylase (a/b), without changing the amount of inactive b form. HCG simultaneously decreased glycogen from 165.2 to 139.1 mg per 100 ml, and increased cAMP from .246 to .448 pM per mg. PGE_2caused a higher activation of glycogen phosphorylase a, a greater degree of glycogenolysis, and increased cAMP to .612 pM per mg ($p<.001$). The authors proposed a second messenger role of PGE_2in the action of HCG on human placenta. (LJG)   005667

0732

DHONT, M.; THIERY, M.; LEPOUTRE, L.; VERMEULEN, A.; VANDEKERCKHOVE, D.
  Maternal serum levels of human chorionic gonadotropin (HCG), human chorionic somatomammotropin (HCS), progesterone (P) and estradiol (E) before and during labor induced at term by orally administered prostaglandin E_2.
  International Research Communications System (73-6) 15-18-9. June 1973. 1 p.

The hormonal parameters of placental function were investigated by assaying levels of HCG, progesterone, estradiol, and human chorionic somatomammotropin (HCS) following oral administration of $PGE_2$to induce labor. PGE_2doses varied from .5 to 8.5 mg (mean 2.2), and the interval between first dose and delivery varied from 2 hours 18 minutes to 22 hours 15 minutes (mean 6 hours 43 minutes). Blood samples were taken before induction, 30 and 150 minutes following initial dose of PGE_2, and immediately following delivery. Except for progesterone, which was determined by a competitive-binding method, all samples were radioimmunoassayed. No change in the concentration of HCG, HCS, or estradiol was noted. The slight decrease in progesterone compares favorably with data from spontaneous labor. No negative effects on placental function could be surmised. (RAS) 006897

0733

DICKEY, J.F.; HENDRICKS, D.M.; HILL, J.R.
  Gonadal hormones in $PGF_2\alpha$/PMS treated heifers.
  Journal of Animal Science 37(1): 307. 1973.

78 heifers were allotted to the following treatments: 1) control; 2) PMS on Day 16 of cycle; 3-5) $F_2\alpha$ on Days 3-5, 9-10, or 16-17 of cycle, respectively; 6-8) $F_2\alpha$ plus PMS given 2 days, before $F_2\alpha$; and 9-11) $F_2\alpha$ plus PMS given on same day (Groups 1 and 2: 12 heifers each; Groups 3-11: 6 heifers each). 2 mg. of $F_2\alpha$ were placed in the body of the uterus and 1600 IU of PMS was injected sc. Jugular vein plasma was collected every day after treatment with $F_2\alpha$ until estrus and every other day after mating

until Day 16. Plasma was assayed for progesterone (P) and total estrogens (E) by RIA. Mean E levels on Day -1 or 0 (estrus) for Groups 1-11 were 14, 27, 14, 11, 11, 15, 26, 20, 19, 29, and 17 pg/ml, respectively. The estrogen pattern prior to and after mating in $F_2\alpha$-treated heifers (Groups 3,4 and 5) were similar to the control's. Estrogen levels were much higher before mating in the heifers treated with PMS or PMS plus $F_2\alpha$. Groups 2, 9, 10, and 11 had higher E levels on Days 6 to 8 after mating than did Group 1 (13, 27, 21, and 28 vs. 6 pg/ml). Other groups had levels mating similar to those of Group 1. P patterns before and after mating in Groups 3, 4, and 5 ($F_2\alpha$ alone) were similar to the control pattern. The levels (in ng/ml) on Day 2 before estrus and Day 14 after estrus were for Group 3, 2.7 and 5.6 for Group 4, 2.3 and 5.2; for Group 5, 1.6 and 6.3 and for Group 1, 2.6 and 5.8. The levels were for the other groups: 4.5 and 26.0 (Group 2), 6 and 9.8 (6), 3.9 and 12.5 (7), 7.5 and 14.1 (8), 1.4 and 6.2 (9), 2.7 and 14.6 (10), 3.2 and 12.1 (11). (Authors' modified)   005979

0734
DIEHL, J.R.; DAY, B.N.
Effect of prostaglandin $F_2\alpha$ on luteal function in swine.
Journal of Animal Science 37: 307-308. 1973.

Luteolytic effects of $PGF_2\alpha$ in normal cycling and pregnant gilts were investigated. Blood samples were collected twice daily from indwelling jugular catheters, and plasma progesterone levels were measured by competitive protein binding. 11 cycling gilts received 2 mg or 5 mg $PGF_2\alpha$ injected into the lumen of the uterine horns on Day 10 of the cycle or injected intramuscularly on Day 12 of the cycle; plasma progesterone levels were stable. 7 pregnant gilts received 5 mg $PGF_2\alpha$ injected intramuscularly at 25 to 30 days of gestation; progesterone levels dropped by 12 hours following treatment. Abortion occurred by 28 hours in 4 of the 7, and estrus was exhibited 3 days later. These results suggested that luteolytic action by $PGF_2\alpha$ in swine occurs only during early gestation. (MLH)   005818

0735
DIEHL, J.R.; GODKE, R.A.; KILLIAN, D.B.; DAY, B.N.
The induction of parturition in swine with prostaglandin $F_2\alpha$.
Biology of Reproduction 9: 104. Abstract 110. 1973.

2.1 mg of $PGF_2\alpha$ was infused intravenously during a 10-hour period in 5 gilts 5 to 7 days before expected date of parturition. The interval from onset of infusion to birth of the first pig averaged 29 hours in PGF-treated compared to 79 hours for control animals. Progesterone levels dropped to 1.3 ng per ml in PGF-treated animals compared to levels of 5.3 ng per ml in the controls 10 hours after infusion. Corticoid levels were elevated in 3 of 4 controls but showed little change in the PGF-treated animals. (HC)   005574

0736
DIMOND, P.A.
Prostaglandins - fact sheet.
Washington, D.C., National Institute of Child Health and Human Development, May 1973. 3 p.

This prostaglandin fact sheet prepared for the Center for Population of the National Institute of Child Health and Human Development summarizes the organizations interest in prostaglandins. The Center for Population does not support research of prostaglandins for abortifacients; however, efforts to synthesize an analogue that would reduce side effects are being attempted. Additional research into the effects of prostaglandins on ovarian function in nonpregnant women and prostaglandin interaction with LH, whereby PGs control pituitary secretion of gonadotropins, are now under investigation. Prostaglandins thus far have not shown a luteolytic effect in man. (RAS)   005502

0737

DJERASSI, C.

Directions and potentials of contraceptive research.

International Journal of Health Services 3(4): 583-590. 1973.

Future birth control methods (abortifacients, pills, IUDs, sterilization, and male contraceptives, were evaluated, concentrating on their applicability to fast growing countries and the practical impediments to their being developed. All new methods are handicapped by the low priorities and strict regulations of the wealthy countries that can afford costly research and development. A chemical abortifacient, especially a luteolytic drug, is the most needed breakthrough. The chief drawbacks are lack of appropriate animal models, 10 years or more of lead time needed, and once it appears, resistance from the surgical abortionists. PGs are the most likely chemical abortifacients to be marketed because their teratologic effects are least likely. Very little research has been done to adapt steroid contraceptives to developing countries. If a significantly better IUD were found, the past disillusioned IUD users and prejudiced affluent groups would have to be convinced. Developing a male pill is exceedingly unlikely. A chemical female tubal sterilant or a local agent to decapacitate sperm are conceivable. (LJG)   006928

0738

DONALD, I.

A review of procedures in induction of labour.

In: Jacomb, R.G., ed. The use of prostaglandins E_2 and $F_2\alpha$ in obstetrics and gynecology. (Proceedings of the Upjohn Prostaglandin Symposium, London, September 21, 1972.) Miami, Florida, Symposia Specialists, 1973. p. 5-9.

The author reviewed methods employed for labor induction from as far back 1595, noting in all instances a reluctance to hasten labor. Even after the need for surgical intervention was recognized, the possibility of complications from premature intervention such as sepsis, deleterious effect on the fetus, and psychological damage to the mother has made physicians cautious. Prior to inducing labor, the prime standard to be met is certainty of the child's maturity or imminent danger of fetal death in utero. To secure delivery within 12-18 hours after membrane rupture, the author suggested the use of an oxytocin drip. However, the ideal drug, hopefully prostaglandin, would initiate labor and remove the need for surgical intervention. (HC)   006960

0739

DOUGLAS, R.H.; GINTHER, O.J.

Effect of prostaglandin $F_2\alpha$ in ewes and pony mares.

Journal of Animal Science 37(1): 308. 1973.

The dose response of interval from treatment to estrus in ewes given $PGF_2\alpha$ im on Day 8 after estrus ranged from 9.5 days in controls to 3.8 days in ewes given 8 mg $PGF_2\alpha$. Luteal weights on Day 12 were 725 mg in controls, 352 mg in ewes given 2 mg im, 105 in ewes given 6 mg im, and 96 in ewes given 2 mg intrauterine. The dose response of interval from treatment to estrus in pony mares given $PGF_2\alpha$ sc on Day 6 of diestrus ranged from 14.7 days in controls to 2.3 days in mares given 10 mg $PGF_2\alpha$; 1.25 mg was the minimal effective dose. Mean interval from treatment to estrus in mares given 1.25 mg $PGF_2\alpha$ sc fell from 19 days when mares were treated on Day 1 to a minimum of 3 days when treated on Day 7 and increased to 4 days when treated on Day 13. Abortion occurred in 3 out of 7 mares pregnant 40-120 days when given 1.25 mg $PGF_2\alpha$, and in 4 out of 8 mares given 2.5 mg. (LJG) 005900

0740
DOUGLAS, R.H.; GINTHER, O.J.
Luteolysis following a single injection of prostaglandin $F_2\alpha$ in sheep.
Journal of Animal Science 37(4): 990-993. 1973.

Groups of 4 ewes were given $PGF_2\alpha$ intramuscularly (im) on Day 1 of estrus, the doses being 1, 2, 4, 6, or 8 mg. Length of estrous cycle and interval from treatment to return to estrus were significantly reduced only in the 6 and 8 mg groups. In another study, groups of 4 animals were given 2 or 6 mg $PGF_2\alpha$ im or 2 mg by the intrauterine route (iu) with vehicle being given concomitantly by the alternate route. All treatments significantly reduced mean corpora lutea weight, 2 mg iu being as effective as 6 mg im. 3 of the 4 animals in each of these groups and only 1 ewe in the 2 mg im group returned to estrus and had regressed corpora lutea. The authors note that the 2-3 day interval from $PGF_2\alpha$ treatment to estrus is comparable with results obtained using the drug by intravascular infusion. The luteolytic effect of $PGF_2\alpha$ was more pronounced when given locally (iu) than when given systemically (im). (HC) 006021

0741
DUNN, M.V.; HUMPHRIES, N.G.; JUDKINS, G.R.; KENDALL, J.Z.; KNIGHT, G.W.
The effect of prostaglandin $F_2\alpha$ antibody on gestation length in the rat.
Prostaglandins 3(4): 509-514. April 1973.

Rabbit antiserum, equally specific for $PGF_1\alpha$ and $PGF_2\alpha$ when injected on day 17 of pregnancy in rats, prolonged gestation and delivery. 3-3.5 ml freeze-dried antiserum of 1:3300 to 1:22,800 titer was reconstituted about double strength in water for injection into 6 rats. Duration of gestation in 6 treated rats was mean 23.8 days (range 23.3-25.2 days) compared to mean 22.6 (range 22.0-23.8) in 7 controls. (p<.01). Vaginal bleeding from 1.5 to 58 hours beforehand indicated that delivery was prolonged. Treated rats delivered mean 4.8 viable fetuses (range 2-6), but controls delivered 6.2 (range 4-9) (not signficant). These results support a role for $PGF_2\alpha$ in the events leading to parturition in the rat. (LJG) 005746

0742
DUNN, M.V.; HUMPHREYS, N.G.; JUDKINS, G.R.; KENDALL, J.Z.; KNIGHT, G.W.
The effect of prostaglandin $F_2\alpha$ antibody on gestation length in the rat.
New Zealand Medical Journal 78(501): 368. October 24, 1973.

Premature parturition can be induced in the rat by the administration of $PGF_2\alpha$. On the other hand, inhibitors of prostaglandin synthesis (indomethacin, aspirin, and fenclozic acid) prolong pregnancy and labor. In the present study antibodies specific for PGF1 and $PGF_2\alpha$ were obtained from immunized rabbits. The antibody titre ranged from 1:3300 to 1:22,000. 13 nulliparous Wistar rats were separated from the male 6-8 hours after observed mating (designated Day 0 of pregnancy). Antisera were injected intraperitoneally into 6 rats on Day 17.75 of pregnancy. 7 control rats were similarly injected with serum from a nonimmunized rabbit. Postpartum the antibody titre was low. Intraperitoneal administration of $PGF_2\alpha$ antibody lengthened gestation from 22.6 ± .6 days to 23.8 ± .9 days (p<.01). Prenatal vaginal bleeding was a common feature suggesting that labor was also prolonged. These observations lend support to a role for prostaglandin $F_2\alpha$ in the physiology of parturition in the rat. (Authors' modified)   005813

0743
EDWARDS, R.G.
Studies on human conception.
American Journal of Obstetrics and Gynecology 117(5): 587-601. November 1, 1973.

The author's extensive research on the physiology, endocrine relationships, and timing of follicular maturation and ovulation; harvesting, fertilizing, and culturing human eggs in vitro; and clinical applications are surveyed. Methods of controlling growth of preovulatory follicles by administering clomiphine and human chorionic gonadotropin (HCG) were sought. A correlation between follicles 1.5 cm or larger and urinary estrogens was detected. Mild hyperfolliculation could be induced for collection of ova by laparoscopy, or ovulation could be precisely timed at 36 hours after 5000 IU HCG given on Day 12 or 13 after a course of clomiphine from Days 3-8. Endocrine relationships are appearing in studies of follicular fluids. In rabbit follicles, HCG causes transient estrogen and cAMP synthesis followed just before ovulation by a burst of $PGF_2\alpha$, which may induce collagenase needed for follicular rupture. In human follicular fluid (100 follicles) estrogen and progesterone levels were correlated; $PGF_2\alpha$ has been measured in only 4 follicles, with variable results from 4.2-28 ng/ml. Studies on fertilization and cleavage of human ova in vitro have centered on finding proper media for capacitation and culture and have resulted in a few blastocysts grown for up to 147 hours. Clinical applications included oligomenorrhea (18 pregnancies in 132 treated cycles), but there have been no results so far in treating oligospermia or occluded oviducts. The following topics are being investigated: trisomy in zygotes (none encountered), male and female sperm antibodies (few true cases seen), and reimplantation of zygotes fertilized in vitro (no pregnancies to date). (LJG)   005668

0744
EINARSSON, S.; VIRING, S.
Effect of boar seminal plasma on the porcine uterus and the isthmus part of oviducts in vitro.
Acta Veterinaria Scandinavica 14(4): 639-641. 1973.

Boar seminal plasma, seminal vesicle secretion, filtrate (less than 1000 mol. wt.) and 1:5 concentrate (from ultrafiltration) were tested on longitudinal strips of uterus or oviduct isthmus from 9 gilts 24 hours after estrus. Spontaneous activity in aerated Tyrode solution at 37 degrees was about 1 contraction per minute. Seminal plasma, seminal plasma concentrate, and seminal vesicle secretion decreased amplitude and frequency when added to the organ bath. Concentrate and seminal vesicle secretion increased tonus and amplitude of oviduct strips. Filtrate had no effect. (LJG)   005721

0745
EINER-JENSEN, N.
Decreased endometrial blood flow and plasma progesterone level after instillation of 10 μg prostaglandin $F_2\alpha$ into the lumen of the uteri of rhesus monkeys.
Prostaglandins 4(4): 517-522. October 1973.

The effect of 10 μg $PGF_2\alpha$ injected into the uterine cavity of 3 rhesus monkeys, once weekly for 4 weeks, on endometrial blood flow and plasma progesterone were measured. Blood flow estimates were made by injecting .1 ml 133-Xenon in saline, 2mCi per ml, and monitoring from a Geiger-Muller tube taped to the abdomen. When 10 μg $PGF_2\alpha$ and .2 ml 133-Xenon saline was given 15 minutes after the control test, mean blood flow decreased 39% in 11 tests (p<.01), lasting at least 15 minutes. Plasma progesterone was equal to or less than 2.5 mg per ml in 4 control samples, but decreased 11, 14, 33 and 33% 20 hours after intrauterine $PGF_2\alpha$. 5 instances of low uterine blood flow, possibly conditioned by pain experienced during previous PGF injections, were observed before PG in Weeks 3 and 4. The low blood flow was presumably vasoconstriction mediated by autonomic nervous system. (LJG)   005532

0746

ELGER, W.; HASAN, S.H.; FRIEDREICH, E.

'Uterine' and 'luteal' effects of prostaglandins (PG) in rats and guinea pigs as potential abortifacient mechanisms.

Acta Endocrinologica (Kobenhavn) (Suppl.) 173: 46. 1973.

The abortifacient mechanisms of PGE_2, $PGF_2\alpha$, and PGF2delta were studied in rats and guinea pigs by determining serum progesterone with radioimmunoassay and observing uterine ovum transport. Rats injected with PGs 1.0 mg sc daily were aborted. Progesterone rose after PGF2delta, but fell significantly below controls after Day 3, even on lower doses without abortion. PGE_2 did not affect progesterone level. Simultaneous progesterone treatment did not reverse the abortifacient action of PGE_2, and only partly that of PGF2delta. Guinea pigs were given PG on 2 consecutive days in midpregnancy and were aborted by both PGs with a rapid decline in serum progesterone. The authors concluded that these PGs act mainly on uterine smooth muscle, since their effect on progesterone was minor. (LJG)   006462

0747

ELIAS, J.A.

Maternal and foetal safety during induction of labour with prostaglandins.

In: Jacomb, R.G., ed. The use of prostaglandins E_2 and $F_2\alpha$ in obstetrics and gynecology. (Proceedings of the Upjohn Prostaglandins Symposium, London, September 21, 1972.) Miami, Florida, Symposia Specialists, 1973. p. 35-39.

Labor was successfully induced in all 30 women given PGE_2 or $PGF_2\alpha$ intravenously and in 26 of 30 women given PGE_2 orally. There were no significant changes in maternal pulse or blood pressure; diarrhea and vomiting did not occur in any of the women treated intravenously and in only a small number given the drug orally. None of the patients had hyperpyrexia, and the minor fluctuations in uterine tone noted were similar to those seen in spontaneous labor. In all but 2 instances fetal acid-base status was in the normal range; 7 fetuses had heart rate abnormalities. Apgar scores were not affected by PG treatment; pediatric assessment at birth revealed no abnormalities. The author concluded that the use of PG for labor induction is as safe as any in existence. (HC)   006961

0748

ELIAS, J.A.

Prostaglandin F vaginal pessaries for midtrimester abortion.

Advances in the Biosciences 9: 581-584. 1973.

Termination of midrimester pregnancies by $PGF_2\alpha$ vaginal pessaries is described. Of the 25 patients studied, 21 (84%) were successfully aborted, with an average induction-abortion interval of 18.4 hr. Side effects (diarrhea and vomiting) were troublesome in 10 patients. The advantages of this method are discussed. (Author)   006476

0749

ELIASSON, R.; BYGDEMAN, M.; ENEROTH, P.
Effects of hormones on prostaglandins in human semen.
Acta Physiologica Scandinavica 5396: 19. 1973.

Prolonged treatment of infertile males with gonadotropins or androgens had no significant effect on PG content of seminal fluid analyzed chemically. (HC)   006163

0750

ELIASSON, R.; LINDHOLMER, C.; JOHNSEN, O.
Effects of human seminal plasma on some functional properties of the human spermatozoa.
Acta Physiologica Scandinavica Suppl. 396: 107. 1973.

Oxygen consumption of human sperm was significantly higher in salt solutions than in whole semen. Vesicular fluid had negative effects on sperm motility and survival. Prostatic fluid had a motility promoting effect and protected the sperm against the effects of vesicular fluid. A better understanding of the relationship between accessory genital glands and sperm may lead to the development of reversible male contraceptives that do not interfere with spermatogenesis. (HC)   006165

0751

EMBREY, M.P.; HILLIER, K.; CALDER, A.A.
Early abortion induced with prostaglandin $F_2\alpha$ and a prostaglandin analogue I.C.I. 74,205.
Lancet 2(7837): 1100. November 10, 1973.

The efficacy of intrauterine-administered $PGF_2\alpha$ and I.C.I. 74,205 (racemic 20-ethyl-$PGF_2\alpha$) in terminating early pregnancy was studied in 15 women with confirmed pregnancies. The duration of amenorrhea varied from 38 to 64 days. 7 patients were treated with $PGF_2\alpha$ and 8 with I.C.I. 74,205. .5-1.0 mg of prostaglandin, up to a maximum of 6 mg, was instilled into the uterus by way of a catheter at 1- to 2-hour intervals. The maximum duration of treatment was 12 hours, except for 1 patient who received 11 mg of I.C.I. 74,205 over a 22-hour period. Patients were sent home after 24 hours. 6 of the 7 pregnant patients receiving $PGF_2\alpha$ aborted in 21 days, and the seventh, within 28 days. Of the 8 receiving I.C.I. 74,205, 5 aborted by Day 21 and 1 more within 28 days. While both prostaglandins were found to be effective abortifacients, $PGF_2\alpha$ acted slightly faster and with fewer incomplete abortions than I.C.I. 74,205. Side effects observed included vomiting, hot flushes, and uterine cramps. (JSL)   006884

0752

EMBREY, M.P.; WIQVIST, N.
Extra-amniotic administration.
In: Bergstrom, S., ed. Report from meetings of the prostaglandin task force steering committee, Chapel Hill, June 8-10, 1972, Stockholm, October 2-3, 1972, Geneva, February 26-28, 1972. Stockholm, WHO Research and Training Centre on Human Reproduction, Karolinska Institutet, 1973. (Prostaglandins in Fertility Control 3) p. 9-11.

Ongoing collaborative trials of extraamniotic $PGF_2\alpha$ 250-750 µg every second hour, of extraamniotic PGE_2 and iv oxytocin, and of extraamniotic 15(S)15-methyl-$PGF_2\alpha$ 500-800 µg in a single injection, for midtrimester abortion are updated in this brief report. 3 centers (Oxford, Karolinska, New Delhi) had similar results, 60.7% success with extraamniotic $PGF_2\alpha$ in 24 hours and 83.4% in 36 hours, with few side effects, in a total of 206 patients. 4 others (North Carolina, Bombay, Chandigarh, Ljubljana) had lower success rates in a total of 41 patients. 145 patients received PGE_2 and oxytocin at Oxford; 80% of these had aborted in 18 hours compared to 36 hours with PGE_2 alone in another group of 70. 82.9% of 24 early midtrimester patients given 15-methyl-$PGF_2\alpha$ aborted within 36 hours, with mean induction interval of 13.7 hours. Side effects and complications were comparable to those seen with repeated intraamniotic $PGF_2\alpha$. (LJG)   006737

0753
EMBREY, M.P.; HILLIER, K.; MAHENDRAN, P.
Extraamniotic prostaglandin administration for the induction of abortion.
Advances in the Biosciences 9: 507-513. 1973.

Prostaglandins E_2 and $F_2\alpha$ were administered by extraamniotic instillation to 163 patients in the first and second trimesters of pregnancy. Abortion was successfully induced in 88% of patients within 36 hr and in 94% within 48 hr. The mean abortion time was 22.2 hr. 63% of the abortions were complete. On average, multigravidas aborted more quickly than primigravidas, and the mean abortion time in PGE_2-treated patients was less than in those receiving $PGF_2\alpha$. In a separate trial, the use of intravenous oxytocin in addition to extraamniotic prostaglandins resulted in a substantial decrease in the mean abortion time. (Authors)   006469

0754
ENGEL, T.; GREER, B.; KOCHENOUR, N.; DROEGEMUELLER, W.
Midtrimester abortion using prostaglandin $F_2\alpha$, oxytocin, and laminaria.
Fertility and Sterility 24(8): 565-568. August 1973.

The protocol used in this study consisted of a laminaria inserted into the cervix the evening before the operation, intravenous infusing of oxytocin for 1 hour the next morning, amniocentesis, then a single intramniotic injection of 15 mg $PGF_2\alpha$ and oxytocin infusion continued until abortion occurred. All 20 patients pregnant 16-20 weeks aborted. The median IAT for primagravidae was 14 1/2 hours and for multiparous patients 6 1/2 hours. The only toxicity was vomiting in 3 patients. The authors concluded that this protocol fulfills their 3 strict criteria for success: single $PGF_2\alpha$ injection, abortion within 24 hours, and minimal side effects. (HC)   006883

0755
ERICSSON, R.J.
Prostaglandins (E_1 and E_2) and reproduction in the male rat.
Advances in the Biosciences 9: 737-742. 1973.

Prostaglandins PGE_1 and PGE_2 inhibit spermatogenesis when given subcutaneously to mature rats at 2.0 mg/kg body weight b.i.d. for 15 days. Seminiferous tubules are not equally affected; some have polykaryocytes and others have primarily exfoliated immature germ cells, while still other tubules appear normal. The severity of spermatogenic inhibition is greater with PGE_1. Epididymides of rats from both groups contain sperm with decreased motility, immature germinal cells, and sperm separated into head and tail segments. PGE_1 causes a significant weight loss ($p<.01$) in the testes, epididymides, seminal vesicles, and ventral prostates; PGE_2 significantly reduces weights ($p<.05$) of the 2 accessory glands. Mean testosterone blood levels approach statistically significant lower concentrations in PGE_1 rats. The mechanism by which these pharmacologic amounts of PGE_1 and PGE_2 alter the male reproductive system is unknown. Body stress is evident, with some resultant deaths. Changes in the vascular system brought about by prostaglandins could theoretically result in the observed disturbance of the male reproductive tract--specifically, blood vessels of the pampiniform plexus. (Authors' modified)   006491

0756
ESKIN, B.A.; AZARBAL, S.
Effects of $PGF_2\alpha$ on periovular cervical mucus.
Advances in the Biosciences 9: 731-735. 1973.

A total of 25 healthy women in their reproductive years (age range: 19-30) and regularly menstruating, who permitted cervical mucus to be taken, were used in this study. Prostaglandin $F_2\alpha$ was added to the spermatozoa-cervical mucus system during the periovular period. A tube-slide technique was employed followed by determination of motility, penetration, and drive of the spermatozoa through the treated mucus. The results indicated that there is a significant increase ($p < .001$) in all these modalities when the pharmacologic and physiologic level (250 ng/cc) of $PGF_2\alpha$ is employed. The greatest effects were seen on cervical mucus derived from subjects within 24 hours prior to ovulation, as recorded by increased motility, drive, and penetration of spermatozoa. These findings are highly significant for improving fertility. (Authors' modified)   006796

0757
ESKIN, B.A.; AZARBAL, S.; SEPIC, R.; SLATE, W.G.
In vitro responses of the spermatozoa-cervical mucus system treated with prostaglandin ($F_2\alpha$).
Obstetrics and Gynecology 41(3): 436-439. 1973

Cervical mucus obtained from women shortly before or after midcycle was incubated for 1 hour with spermatozoa and 250 picograms $PGF_2\alpha$ or saline. Examination under a microscope showed that $PGF_2\alpha$ increased motility, drive and penetration of the spermatozoa through the mucus. The effectiveness of $PGF_2\alpha$ was greatest just before or at the time of ovulation. The authors suggest that fertility responses are improved in the presence of physiological doses of $PGF_2\alpha$. (HC)   005606

0758
EULER, U.S., von
Some aspects of the actions of prostaglandins.
Archives Internationales de Pharmacodynamie et de Therapie 202: 295-307. 1973.

The history, isolation, nomenclature, and structure of prostaglandins are presented. Early research (1960s) demonstrated prostaglandin ability to stimulate uterus contractility in vitro. Further in vivo studies led to prostaglandins' ability to work as an abortifacient. Prostaglandins' luteolytic action has been clearly demonstrated in sheep. Additional subjects of prostaglandin research are mentioned: action on the cardiovascular system, respiratory tract, gastrointestinal tract, blood platelets, adipose tissue, and nervous system. (RAS)   006681

0759
FAVIER, J.; RIETVELD, W.J.
The effect of prostaglandin E_1 on the pregnant human uterus.
American Journal of Obstetrics and Gynecology 115(1): 33-36. January 1, 1973.

The effect of PGE_1 on the responsiveness of the pregnant human uterus to nonspecific stimuli was investigated. 35 women of 24 to 43 weeks gestation were administered intravenously PGE_1 at a rate of

.25 μg/minute up to .50 μg/minute. Before administration of the PG, in cases where no uterine contractions were present and in cases of regular contractions, application of stimuli did not change the activity pattern; in cases of small irregular contractions the stimuli produced some coordinated contractions. In the 14 cases of no uterine activity, the administration of PG resulted in fluctuations in basal tone and contractions upon increased stimulation. In some cases the activity decreased in amplitude and in frequency after a few hours continual PG but oxytocin showed a favorable effect. In the 12 cases of irregular uterine contractions, PGE_1 resulted in an increase in amplitude of the contractions and rupturing the membranes. In the 9 cases of regular uterine contractions, PGE_1 had little or no effect, but the administration of oxytocin showed improvement. Side effects of the administration of PGE_1 were nausea and vomiting in 6 of the 35 patients and phlebitis at the locus of the infusion catheter in 4. It was concluded that PGE_1 may exert a sensitizing effect on the myometrium but that the sensitivity to oxytocin, an endogenous stimulus, may dominate the effect of nonspecific stimuli. (MLH)   006788

0760
FAWKE, L.
Thymoxamine on the guinea-pig vas deferens: an effect unrelated to alpha-receptor blockade.
Archives Internationales de Pharmacodynamie et de Therapie 204(2): 316-325. 1973.

Thymoxamine, in concentrations of 5-125 ng per ml, reduced the amplitude of contraction of the innervated or transmurally stimulated guinea pig vas deferens. Both vasa from 1 animal were used, 1 acting as control, and all drugs were added to the 100 ml organ bath 1 minute before contraction. Inhibition by thymoxamine was not due to alpha-adrenoceptor blockade, and was seen in the dose response curve before alpha-blockade became apparent. The inhibition was not influenced by propranolol, 500 ng per ml, acetylsalicylic acid, 500 μg per ml, or hyoscine, 100 ng per ml, indicating that neither beta receptors nor PGs were involved. (Author's modified)   005664

0761
FENNER-CRISP, P.A.
Effects of prostaglandins $E_2(PGE_2)$ and $F_2\alpha$ ($PGF_2\alpha$) on decidual response and corpora lutea in the rat.
Federation Proceedings 32: 804Abs. 1973.

Pseudopregnant rats given PGE_2 or $PGF_2\alpha$ 200 μg/day subcutaneously on Days 2 through 7 showed structural changes in corpora lutea indicative of luteolysis. These included increase in lipid content, decrease in cell size, decrease in size and number of organelles, and increase in intercellular space. The decidual response was markedly decreased by PG; this effect was partially blocked by concomitant treatment with progesterone. (HC)   006785

0762
FILSHIE, G.M.
A comparison of dose schedules for intra-amniotic prostaglandin 15 me E_2(S) methyl ester.
In: Jacomb, R.G., ed. The use of prostaglandins E_2 and $F_2\alpha$ in obstetrics and gynecology. (Proceedings of the Upjohn Prostaglandin Symposium, London, September 21, 1972.) Miami, Florida, Symposia Specialists, 1973. p. 93-97.

36 women pregnant 9-24 weeks received 50, 75, or 100 μg doses of 15-methyl PGE_2 methyl ester intraamniotically at 10-hour intervals. All patients aborted, the mean IAT being 17 1/2, 18 1/2, and 9 hours for the 50, 75, and 100 μg groups, respectively. The longer the gestation period, the more prolonged was the IAT. Nausea or vomiting, cold, and pyrexia were side effects noted. The author

concluded that the 100 μg dose shortens the IAT without increasing side effects and that the danger of cardiovascular disturbances with the methyl ester is less than with the parent compound. (HC)
006956

0763
FITZPATRICK, R.J.; SHARMA, S.C.
Prostaglandin F in the vena cava blood during the oestrous cycle of the ewe.
Journal of Physiology 234(2): 57P-59P. July 1973.

PGF was determined in posterior vena cava and jugular vein plasma collected twice daily for the last 10 days before estrus in 5 normally cycling ewes. An indwelling catheter was introduced via the saphenous vein to the junction of the posterior vena cava with the uterine veins for sampling. PGF activity remained below 100 pg/ml except for a peak of 1260-2705 pg/ml 72-84 hours before estrus and a transient increase 24-36 hours before estrus. The 2 peaks coincided with increases in PGF in jugular vein plasma from less than 25 to a maximum of 434 pg/ml. As the fall in progesterone has been noted 72 hours before estrus, these data support the luteolytic role of $PGF_2\alpha$ in sheep. (LJG)
005666

0764
FLOWER, R.J.; CHEUNG, H.S.; CUSHMAN, D.W.
Quantitative determination of prostaglandins and malondialdehyde formed by the arachidonate oxygenase (prostaglandin synthetase) system of bovine seminal vesicle.
Prostaglandins 4(3): 325-341. September 1973.

Quantitative radiochemical and spectrophotometric methods were used for the stimultaneous determination of PGE_2, $PGF_2\alpha$, PGD2, and malondialdehyde (MDA) by action of PG synthetase. Several differences were observed in optimal incubation conditions for formation of each product. Several nonsteroidal anti-inflammatory agents inhibited the formation of all 4 products to the same degree, but benzydamine and phenylbutazone differentially inhibited the formation of 1 or more products. The results indicate that more than 1 enzyme or site is involved in the last step of the synthetase reaction (breakdown of cyclic endoperoxide intermediate). The spectrophotometric determination of MDA formation provides a convenient assay of PG synthetase activity. (Authors' modified)   005436

0765
FORD, S.P.; PEXTON, J.E.; WILSON, L. Jr.; BUTCHER, R.L.; INSKEEP, E.K.
Effects of estradiol-17β on uterine PGF in ewes.
Journal of Animal Science 37(1): 311. Abstract 324. 1973.

The effects of .5 mg estradiol-17β injected im on cycle Days 4 and 5 or 9 and 10 in mature ewes were observed on PGF in uterine vein blood, on progesterone in jugular and uterine vein plasma, and on weights of corpora lutea, endometrium and uterus. Ewes were killed with pentobarbital 18 hours after the second injection (Day 6 or 11). PGF (by radioimmunoassay) averaged 8.4 ng per ml in uterine vein in 8 ewes treated on Days 10-11, but .9 ng per ml in all others (p<.01). The 8 sheep treated on Days 4-5 had corpora lutea weighing mean 430 mg, and jugular plasma progesterone of mean 1.5 ng per ml; compared to mean 703 mg and mean 3.0 ng per ml in 8 sheep treated on days 9-10 (p<.01). Endometrial weights were greater in estradiol treated ewes than in 16 controls: 27.99 g and 17.53 g, respectively (p<.01). (LJG)   006024

0766

FREDHOLM, B.; HEDQVIST, P.

Increased release of noradrenaline from stimulated guinea pig vas deferens after indomethacin treatment.

Acta Physiologica Scandinavica 87(4): 570-572. 1973.

In a study of mechanisms controlling sympathetic neurotransmission, indomethacin, a strong PG-synthesis inhibitor, was observed to increase noradrenaline from electrically stimulated guinea pig vas deferens. 15 male guinea pigs were killed and the vas deferens removed, incubated in radiolabeled Tyrodes solution, rinsed, and superfused. Cation exchange column chromatography showed that noradrenaline constituted more than 90% of the radioactivity retained in the organ and more than 70% in the superfusate. Administration of 10 ng/ml PGE_1 or E_2 markedly and reversibly depressed release of noradrenaline in response to transmural stimulation of the vas deferens, which had caused increased efflux of radioactivity. Infusion of indomethacin did not change resting efflux of radioactivity. When indomethacin infusion was stopped, increased noradrenaline flow in response to stimulation occurred. Infusion of PGEs reversed the potentiating effect of indomethacin on noradrenaline release and blocked the effect when given before indomethacin. It is suggested that indomethacin increases noradrenaline release from sympathetic nerves by inhibition of local PG formation. This study and previous ones support the view that locally formed PGs serve the function to modulate sympathetic neuroeffector transmission. (IC)   006468

0767

FREE, M.J.; TILLSON, S.A.

Secretion rate of testicular steroids in the conscious and halothane-anesthetized rat.

Endocrinology 93(4): 874-879. 1973.

Whole blood testosterone from rat femoral artery and testis vein and testis blood flow were measured as affected by PGE infusion and halothane anesthesia. Blood was collected by catheter, testosterone was determined by radioimmunoassay after extraction and column chromatography, and testis blood flow rate was measured by a miniature friction flow meter. In 14 conscious rats mean testis blood flow was .234 ml per g testis per minute, testosterone concentration was .23 ng per ml in femoral artery blood, and 18.2 in testis vein, giving a secretion rate of 4.2 ng per g testis per minute. In 8 halothane-anesthetized rats blood flow was .245, testosterone was .49 in femoral artery, and 28.5 in testis vein, and secretion rate was 6.7 (n.s.). Testosterone secretion rates fell with time after anesthesia. .3-10 μg PGE_2 infused into the testis capsule or testis vein over 30 minutes reduced testis blood flow in 6 rats from .241 to .110 ml per g per minute (p<.005) and hence reduced secretion rate, but did not significantly affect testosterone concentration. In all rats, testosterone secretion rate was proportional to blood flow. (LJG)   005607

0768

FREID, N.D.; TREDWAY, D.R.; MISHELL, D.R., Jr.

Termination of early pregnancy with prostaglandin E_2 vaginal suppositories.

Contraception 8(3): 255-263. September 1973.

10 women pregnant less than 6 weeks received 20 mg PGE_2 suppositories vaginally at 2-hour intervals, with the total dose being 80-120 mg. 7 women aborted, 5 completely, the IAT being 24 hours or less in all but 1 of the patients. HCG levels fell in those 5 patients who aborted completely but rose in the 3 failures and in 1 of the 2 incompletes. Progesterone levels also fell in the patients who aborted completely, but they remained relatively stable in the 3 failures and in 1 of the 2 women who aborted incompletely. All patients had side effects including nausea, diarrhea, abdominal cramps,

vaginal bleeding, and temperature elevation. The authors concluded that PGE_2 is no more effective than $PGF_2\alpha$ when given vaginally for termination of early pregnancy and that the incidence of side effects limits the usefulness of this method. (HC)   005604

0769
FUCHS, A.-R.
Parturition in rabbits and rats.
Memoirs of the Society for Endocrinology 20: 163-185. 1973.

The parameters involved in labor in rabbits and rats are reviewed, including progesterone, estrogen, uterine stretch, oxytocin, vasopressin, and PGs (which did not achieve great significance in the author's results). Rats and rabbits both depend on the corpus luteum for their progesterone supply, but differ in that rats have moderate assynchronous uterine activity and sensitivity to oxytocin throughout gestation, while the rabbit uterus is quiescent. Rabbits require estrogens for luteal function during gestation and need a systemic stimulant (probably not PG) to initiate labor. Rats release oxytocin to stimulate labor, and vasopressin as a consequence, proved by quantitating the peptides in the neurohypophysis. In rats, recordings of uterine activity from both ends of the uterus simultaneously show that contractions are random until Day 22, when spontaneous or oxytocin (but not $PGF_2\alpha$) induced synchronous contractions can be propagated. Uterine stretch may be a factor, since the uterus stops growing a few days before term, while fetus weight doubles. This stretch is much greater, but insufficient to coordinate synchronous contractions, however, in rats whose pregnancy is prolonged by progesterone treatment. Ovariectomy in rats at different stages of gestation can cause resorption of all or some fetuses, sporadic delivery and destruction of some with retained placentas, or postponement of delivery. Thus estrogens or other ovarian hormones such as relaxin are required to maintain pregnancy. 3 PGs were ineffective oxytocics in rats at term, but $PGF_2\alpha$ induced premature delivery when given on Days 18-20, after causing a 30% decline in progesterone level. (LJG)   006945

0770
FUCHS, A.-R.; FUCHS, F.
Possible mechanisms of the inhibition of labor by ethanol.
   In: Josimovich, J.B., ed. Uterine contraction -- side effects of steroidal contraceptives. New York, Wiley-Interscience, 1973. (Vol. 1 of Problems of Human Reproduction.) p. 287-300.

The authors discussed a hypothesized mechanism of action for the inhibition of labor by ethanol, namely, neurohypophysial inhibition. Ethanol delays parturition by inhibiting oxytocin release in animals and in the human. This effect has led to its successful use in the treatment of premature labor. Since the presence of oxytocin in maternal blood during labor has not been confirmed, other possible mechanisms of action for ethanol have been suggested. PGE_2 and $PGF_2\alpha$ are either the agents triggering labor or are released following myometrial contractions induced by neurohypophysial hormones. Ethanol inhibits uterine activity induced by PG but not that caused by oxytocin. Ethanol may also act by causing the release of inhibitory substances such as epinephrine, or by acting directly on the myometrium. The authors concluded that the main action of ethanol in labor is inhibition of oxytocin release. On the basis of this conclusion they proposed a sequence of events leading to the onset of labor. (HC)   006311

0771
FUCHS, A.-R.; MOK, E.
Prostaglandin effects on rat pregnancy: II. Interruption of pregnancy.
Fertility and Sterility 24(4): 275-283. April 1973.

The effects throughout rat gestation of intravenous and intrauterine infusions of PGE_1, PGE_2, $PGF_2\alpha$, oxytocin, vasopressin, prolactin, luteinizing hormone, estradiol, and progesterone were assessed. Drugs were administered by indwelling catheter at doses of 500 μg PGE_1, 250 or 1000 μg PGE_2; 50 to 2000 μg $PGF_2\alpha$, 200 or 1000 mU oxytocin, 200 or 1000 mU vasopressin, 50 mg LH, 4 mg prolactin, .2 or 2 μg estradiol, and 2 mg progesterone. Rats were killed 1-5 days later and checked for implantations. PGE_1 was ineffective, and PGE_2 only terminated pregnancy in 2 out of 4 rats which received 1 mg on Day 10. $PGF_2\alpha$ at 125 μg per rat iv caused resorption of embryos on Days 10-12, and premature delivery when given on Days 18 and 20. Larger intrauterine doses of $PGF_2\alpha$ caused resorption on Day 10. Doses of 2 mg $PGF_2\alpha$ over 6-7 hours stimulated uterine activity, but activity did not appear above normal during premature delivery. Oxytocin induced 33% resorption and oxytocin combined with vasopressin 50% resorption on Day 10. 2 mg progesterone im for 2 days and 50 μg LH prevented the abortive effect of $PGF_2\alpha$. PGEs and high $PGF_2\alpha$ doses caused central nervous depression and diarrhea. (LJG)   005605

0772
GALLANT, S.; BROWNIE, A.C.
The in vivo effect of indomethacin and prostaglandin E_2 on ACTH and DBCAMP-induced steroidogenesis in hypophysectomized rats.
Biochemical and Biophysical Research Communications 55(3): 831-836. 1973.

The question of PG participation in ACTH stimulation of corticosteroid secretion was examined by injecting hypophysectomized rats with indomethacin, ACTH, PGE_2, or dibutyryl cyclic AMP (dbcAMP), and assaying plasma for corticosterone. Female rats were hypophysectomized, given 2 mg indomethacin ip 4 and 20 hours later. 200 μg PGE_2 or vehicle were given 24 hours after operation then either saline, 8 U. ACTH or 7 mg dbcAMP were administered iv 15 minutes before decapitation. Corticosterone was assayed fluorometrically. Indomethacin or PGE_2 alone or combined did not affect corticosterone secretion. ACTH multiplied corticosterone level about 10 fold in 3 experiments and dbcAMP increased it about 5 fold. Indomethacin inhibited 50% of the ACTH stimulation ($p < .001$), but did not prevent the action of dbcAMP. PGE_2 given with indomethacin and ACTH partly restored the response to ACTH. Thus PGs may modulate action of ACTH in adrenal steroidogenesis. (LJG) 005951

0773
GANS, P.
Effects of prostaglandins and oxytocin on the uterine motility of the living rat under various hormonal conditions.
Advances in the Biosciences 9: 783-788. 1973.

The prostaglandins E_1, omega-homo-E_1, and $F_2\alpha$ caused uterine stimulation in the in vivo rat uterus. A distinct influence of estrone and progesterone on the sensitivity of the uterus to the prostaglandins was demonstrated. Estrone inhibited the prostaglandin effects, while this inhibition was completely or partly counteracted by progesterone. This was demonstrated in castrated rats after administration of estrone or estrone and progesterone. Castrated rats which were given no hormones were most sensitive to the prostaglandins. Under all experimental circumstances PGE_1 was more active than $PGF_2\alpha$. The activity of omega-homo-PGE_1, which was only tested in castrated animals without hormone treatment, equals that of PGE₁. Contrary to the results obtained with prostaglandins, the uterus-stimulating effect of oxytocin was enhanced by estrone, while progesterone did not block this effect. In untreated intact rats the sensitivity to PGE_1; varied with the stage of the estrus cycle, the sensitivity being less in estrus than in diestrus. (Author's modified)   006498

0774

GEVERS, R.H.; FAVIER, J.; RIETVELD, W.J.

Indications and clinical use of prostaglandin E₁ for induction of labour.

In: Bossart, H., Cruz, J.M., and Huber, A., eds. Perinatal medicine. (Third European Congress on Perinatal Medicine, Lausanne, 1972.) Bern, Hans Huber, 1973. p. 367.

Labor was induced with PGE₁ only if the uterus was judged insensitive to oxytocin. PGE₁ was administered in 11 cases of intrauterine death and in 22 cases of premature induction (32-37 weeks) for dysmaturity, toxemia, diabetes, or Rh incompatibility. Uterine activity was first recorded transcervically to test uterine responses. If the uterus was active or if it responded to an increase in pressure caused by infusing 5 ml of saline, it was considered sensitive to oxytocin, and oxytocin induction was begun. Without response, PGE₁ was infused at .25-.5 µg/minute. The PG was flushed every 30-60 minutes to see if normal labor had begun, and if so, oxytocin was substituted for PG. If basal tone increased after an increase in PGE₁ dose, Alupent was administered. PGE₁ induction was successful in 11 case of intrauterine death at 24-33.5 weeks and in 22 live births. Duration of labor was usually short, Apgar scores were usually good, and total PGE₁ dose was between .125-.25 mg. (LJG)   005093

0775

GIANNINA, T.; BUTLER, M.; SAWYER, W.; STEINETZ, B.

On the mechanism of prostaglandin F₂α-induced abortion in hamsters.

Biology of Reproduction 9: 104-105. 1973.

12.5 µg of PGF₂α subcutaneously aborted hamsters on Day 5 of pregnancy. Plasma progesterone levels dropped 88% 3 hours after PGF₂α administration. Progesterone injections blocked the abortifacient action of PGF₂α. Pregnancy was also maintained by injections of 17alpha-ethyl-19-nortestosterone but progesterone levels were still depressed. Pregnancy was maintained in 37% of a group of hypophysectomized hamsters treated with prolactin and FSH, although progesterone levels dropped 70% 3 hours post-PGF₂α injection. The authors conclude that a sustained drop of 85% to 90% in progesterone level is necessary for PGF₂α to induce abortion in all hamsters. (HC)   005608

0776

GILLESPIE, A.

Interrelationship between oxytocin (exogenous and endogenous) and prostaglandins.

Advances in the Biosciences 9: 761-766. 1973.

Endogenous oxytocin was present in the plasma of women receiving intravenous prostaglandin E₂ or F₂α in an incidence similar to that found in late first stage labor. The phenomena of 'enhancement' and 'potentiation' of the response of the human pregnant uterus to exogenous oxytocin after treatment with prostaglandin have been demonstrated in vivo. Employing these phenomena allows termination of pregnancy without the side effects usually associated with prostaglandin administration. (Author)   006494

258

0777
GILLESPIE, A.; BEAZLEY, J.M.; BALLARD, R.M.
The use of prostaglandin-oxytocin combinations for the termination of pregnancy.
International Research Communications System (73-8) 10-26-1. August 1973. 1 p.

Women in midtrimester pregnancy received intravenous infusion of PGE_2 1 μg/minute, or $PGF_2\alpha$ 10 μg/minute, alone or in combination with oxytocin. Only 1 of 17 women given a PG alone aborted, whereas 14 of 20 given oxytocin and a PG aborted. The IAT averaged 32 1/4 hours, and the mean doses were 2 mg and 32.5 mg for PGE_2 and $PGF_2\alpha$, respectively. Doubling the rates of infusion in 5 women decreased the average IAT to approximately 14 hours, and the doses of PGE_2 and $PGF_2\alpha$ were 1.5 and 18.2 mg, respectively. The authors suggested that alterations in the dosage regime should increase the abortion rate and decrease the IAT. (HC)   006929

0778
GILLETT, P.G.; KINCH, R.A.H.; WOLFE, L.S.; PACE-ASCIAK, C.
Induction of abortion by intraamniotic $PGF_2\alpha$: a comparison of dose schedules.
Advances in the Biosciences 9: 545-550. 1973.

20 women pregnant 14-20 weeks were given 5 mg $PGF_2\alpha$ intraamniotically in the first 5 minutes followed by 20 mg during the next 5 minutes. This procedure was repeated in 6 hours with 18 patients. Demerol was used for analgesia and Phenergan for nausea and vomiting. All patients aborted, 7 completely; the IAT averaged 25 hours. Most patients vomited and 7 had diarrhea. Results of this study indicate that the second dose increased both the efficacy and side effects of the induction dose. (HC)   006797

0779
GIMENO, M.F., de; GIMENO, A.L.; LIMA, F.; BORDA, E.
Effect of polyphloretin-phosphate, an inhibitor of prostaglandins, on the spontaneous and induced motility of isolated rat uterus.
Acta Physiologica Latino-Americana 23(2): 105-112. 1973.

The isometric developed tension (IDT) of rat uterine strips was recorded under several conditions. Spontaneous activity of strips taken from rats in natural estrus progressively decreased. Addition of polyphloretin phosphate (PPP), a specific PG antagonist, did not alter the decremental curve. Strips from ovariectomized rats did not lose their spontaneous motility; addition of PPP to the medium caused a significant decrement in activity. Addition of 2.5 μg/ml PGE or .18 μg/ml $PGF_2\alpha$ to the medium after spontaneous activity disappeared restored activity. Addition of PPP at the same time as PG blocked the PG-induced stimulation. These results were obtained with uteri from rats in natural estrus or from ovariectomized animals. Using thin-layer chromatography, only $PGF_2\alpha$ was isolated from ovariectomized rat uterine strips; only PGE was isolated from rats in natural estrus. The authors concluded that spontaneous motility of uterine strips from ovariectomized rats is associated with the presence of $PGF_2\alpha$. They further suggested that the presence of PGE in uteri may not be associated with the functional activity exhibited by the spontaneously active tissue obtained from rats in estrus. (HC)   005899

0780
GLEESON, A.R.; THORBURN, G.D.
Plasma progesterone and prostaglandin F concentrations in the cyclic sow.
Journal of Reproduction and Fertility 32(2): 343-344. 1973.

PGF, uterine progesterone, and venous progesterone were measured in 3 cycling sows every 3 hours for 7-9 Days with 1.5 mm polyvinyl catheters inserted into the jugular, uteroovarian, and uterine veins. Results from 1 sow are plotted, showing progesterone peaks of about 500-800 ng per ml in uteroovarian vein plasma and about 50 ng per ml in jugular vein plasma on Cycle Day 13, falling to less than 10 ng per ml in 2-3 days. PGF rose from undetectable levels on Day 11 to about 8 ng per ml on Day 13 or 14. (LJG)   006778

0781

GODING, J.R.; BUCKMASTER, J.M.; CERINI, J.C.; CERINI, M.E.D.; CHAMLEY, W.A.; CUMMING, I.A.; FELL, L.R.; FINDLAY, J.K.; JONAS, H.
Gonadotrophins in the ovine oestrous cycle.
Journal of Reproduction and Fertility, Suppl. 18: 31-37. 1973.

This review describes current understanding of the behavior of estradiol, progesterone $PGF_2\alpha$, LH, prolactin, and FSH in the sheep estrous cycle. There are independent stages of follicle development throughout the cycle. After preovulatory LH surge and appearance of vaginal mucus, estrus and sexual receptivity last up to 2 days. Ovulation occurs 20-30 hours after estrus. 13-16 days after ovulation, luteal regression is accompanied by fall in progesterone and, after a variable delay, surges in estradiol take place at 4-6 day intervals. The corpus luteum secretes progesterone from Day 4-6 after ovulation, with abrupt fluctuations, approaching a plateau. $PGF_2\alpha$ increases intermittently in uterine vein blood, the peaks accompanying throughs in progesterone secretion, and accelerating until luteolytic levels of $PGF_2\alpha$ reach the ovarian arterial circulation. Basal LH level (3-5 ng per ml) is under negative feedback control of 1-2 week response time, by estradiol, such that minimal LH levels are maintained for secondary sex characteristics without preventing short-term perturbations in estradiol secretion. The preovulatory LH surge of 200 ng per ml is due solely to the preovulatory estradiol wave. Prolactin has a preovulatory surge controlled by estradiol and is also involved in maintaining the corpus luteum. FSH surges with LH and exhibits its own erratic profile before ovulation, probably induced by estrogen acting at the hypothalamic level. FSH fluctuates throughout the luteal phase, independent of the corpus luteum. (LJG)   005657

0782

GRAN, L.
On the effect of a polypeptide isolated from 'Kalata-Kalata' (Oldenlandia affinis DC) on the oestrogen dominated uterus.
Acta Pharmacologica et Toxicologica 33(5-6): 400-408. 1973.

Agueous extract and 4000 mol wt peptides from Oldenlandia affinis DC, an African herb, were tested on isotonic contractions of rat uteri and isometric contractions of rabbit and human uterine strips, and in rats and rabbits by ip, iv, and oral routes. The extract produced contractions similar to serotonin, but the peptide evoked stepwise contractions, at 10 μg per ml, comparable to .08 mU oxytocin in rats. Progesterone (20 μg per ml) inhibited contractions, but acetyl salicylic acid, an inhibitor of PG synthesis, had no effect. At 10-20 μg per ml Kalata peptide stimulated rabbit longitudinal strips but inhibited circular strips. At 20 μg per ml the peptide increased tone and contractions of human uterine strips. Rats given 4-400 μg per kg peptide showed inconsistent uterine contractions, with a threshold of 20 μg per ml in ip fluid, but no dose response. Rabbits showed contractions with 100 μg per kg iv. Both species died of fibrillation at doses of .6-2 mg per kg. No uterine or circulatory effects appeared orally. It was recommended that this drug be discontinued in primitive medicine and that it not be tested clinically as an oxytocic, since lethal doses were only slightly higher than effective doses for uterine contractions. (LJG)   005726

0783

GRANSTROM, E.; GREEN, K.; BYGDEMAN, M.; TOPPOZADA, M.; WIQVIST, N.

Metabolic and quantitative studies in connection with intraamniotic administration of prostaglandin $F_2\alpha$ for induction of therapeutic abortion.

Life Sciences 12 (Pt 1): 210-229. 1973.

The fate of $PGF_2\alpha$ and its metabolites was followed in amniotic fluid, plasma, and urine in 8 women given 25 mg $PGF_2\alpha$ intraamniotically for abortion at 16-23 weeks pregnancy. The $PGF_2\alpha$ contained 32 μCi of tritium. $PGF_2\alpha$ and metabolites were analyzed by extraction, ion exchange, countercurrent and gas liquid chromatography, and mass spectrometry. $PGF_2\alpha$ disappeared slowly from amniotic fluid, such that the half life of 6 to 38 hours was inversely proportional to fluid volume (120-580 ml). The main metabolite was identified as 9alpha,11alpha-dihydroxy-15keto-prost-5-enoic acid. A second peak was heterogeneous and contained too small amounts to be characterized. 24 hours later 8-24% of the radioactivity was excreted in 3 women's urine, and 43% within 18 hours in 1 woman. Plasma $PGF_2\alpha$ level ranged from .1-.4 ng per ml, much too low to explain gastrointestinal side effects in 3 women. The maximum plasma levels of the metabolite ranged from 1.48-14.8 ng per ml. Placental and fetal 15-hydroxyprostanoate dehydrogenase may metabolize the bulk of intraamniotic $PGF_2\alpha$. (LJG) 006740

0784

GRANSTROM, E.

Metabolism and analysis of $PGF_2\alpha$ given by the intra-amniotic route.

In: Bergstrom, S., ed. Report from meetings of the prostaglandin task force steering committee, Chapel Hill, June 8-10, 1972, Stockholm, October 2-3, 1972, Geneva, February 26-28, 1972. Stockholm, WHO Research and Training Centre on Human Reproduction, Karolinska Institutet, 1973. (Prostaglandins in Fertility Control 3) p. 62-64.

The work of the author and of 3 other laboratories quantitating $PGF_2\alpha$ and its metabolites in various body fluids after intraamniotic injection are summarized and compared. The author found that preinjection levels could only be determined by deuterium-labeled $PGF_2\alpha$ carrier and mass spectrometry, but after 25 mg $PGF_2\alpha$ was injected, $PGF_2\alpha$ could be measured by simple gas liquid chromatography. In amniotic fluid 13-49% of the $PGF_2\alpha$ remained up to 18 hours later, while 13-21% was converted to 15-keto-13,14-dihydro-$PGF_2\alpha$. This slow disappearance was inversely proportional to amniotic fluid volume but was variable in individuals. In 4 women who received tritiated $PGF_2\alpha$, 12% was metabolized to 15-keto-13,14-dihydro-$PGF_2\alpha$ after 10 hours and 17% fell into 2 peaks (not analyzed). $PGF_2\alpha$ half-life was computed as 5, 13, 18, and 19 hours. Plasma $PGF_2\alpha$ levels were about .1 ng per ml before injection, and rose to up to .4 ng per ml. The $PGF_2\alpha$ metabolite was 40 pg per ml before injection and usually 3-5 ng per ml after injection. In 3 of 4 women, 8-24% of the $PGF_2\alpha$ was excreted in urine in 24 hours, mainly as 5alpha,7alpha-dihydroxy-11-ketotetranor-prosta-1,16-dioic acid. Another author reported .4-.7 ng per ml $PGF_2\alpha$ before injection without any increase. In contrast 2 others found a slow increase to 2 ng per ml (by radioimmunoassay), and spikes of 3-7 ng per ml. (LJG)　006724

0785

GREATHOUSE, T.R.

Induction of multiple ovulation and advancement of puberty in beef cattle.

Washington, D.C., U.S. Dept. of Agriculture, Research Work Unit/Project Abstract No. 0062592. August 3, 1973.

Procedures were investigated for inducing controlled multiple ovulations in beef cattle and for advancing puberty and maintaining pregnancy in yearling beef heifers. 27 Hereford or Hereford-crossbred cows received intrauterine infusions of $PGF_2\alpha$ to induce corpus luteum regression. 12 were implanted laterally behind the shoulder with 50 μg of gonadotropin relasing hormone (GnRH) and 15 with 12 μg of follicle stimulating hormone (FSH). 72 hours later the cows were artificially inseminated and injected with 200 μg of GnRH intramuscularly; they were inseminated again at 90 hours. 10 cows in each group ovulated following treatment, the rate in the GnRH group being 1.2 and in the FSH group 2.1 Yearling heifers in both groups failed to respond to treatment; multiparous cows had the best responses. It was observed that GnRH could induce multiple ovulations when its release from the capsule could be properly regulated. (MLH)   004820

0786

GREEN, K.; GRANSTROM, E.

Literature data on endogenous levels of prostaglandins in humans measured by bioassay, radioimmunoassay and gas chromatography - mass spectrometry.

In: Bergstrom, S., ed. Report from meetings of the prostaglandin task force steering committee, Chapel Hill, June 8-10, 1972, Stockholm, October 2-3, 1972, Geneva, February 26-28, 1972. Stockholm, WHO Research and Training Centre on Human Reproduction, Karolinska Institutet, 1973. (Prostaglandins in Fertility Control 3) p. 55-61.

Data on endogenous levels of primary prostaglandins in peripheral venous blood as measured by radioimmunoassays range from non-detectable 100-200 picograms per ml to several nanograns per ml. The high values assayed most likely come from a combination of factors: 1) unspecific quantitation techniques 2) assay sensitivity not high enough 3) formation of primary prostaglandins during collection and handling of blood. Measuring an initial metabolite (15-keto-13,14-dihydro-PGE_2 or $PGF_2\alpha$) would give a better indication of primary endogenous levels. The metabolite is found 10 to 75 times higher than that of the primary prostaglandin. With gas chromatographic-mass spectrometric techniques the level of 15-keto-13,14-dihydro-$PGF_2\alpha$ has been found to be about 25 picograms per ml of plasma. (Authors' modified.)   006727

0787

GRIEVES, S.A.; KENDALL, J.Z.; LIGGINS, G.C.

Measurement of prostaglandin $F_2\alpha$ ($PGF_2\alpha$) and its 15-keto metabolities in the human menstrual cycle.

New Zealand Medical Journal 78(501): 367-368. October 24, 1973.

The authors felt that because of the rapid metabolism of $PGF_2\alpha$ a radioimmunoassay for 15-keto-$PGF_2\alpha$ better estimates $PGF_2\alpha$ in peripheral blood. $PGF_2\alpha$, 15-keto-$PGF_2\alpha$, and progesterone were measured in normal and anovulatory menstrual cycles. The 15-keto-$PGF_2\alpha$ antibody from New Zealand white rabbits was used in a similar method to that used for $PGF_2\alpha$, with extraction and silica gel column chromatography. Plasma $PGF_2\alpha$ was .5-1 ng per ml with little cyclic variation. 15-keto-$PGF_2\alpha$ ranged from 0-3.5 ng per ml, highest in the late luteal phase. (LJG)   005851

0788

GUSTAVII, B.

Studies on the mode of action of intra-amniotically and extra-amniotically injected hypertonic saline in therapeutic abortion.

Acta Obstetricia et Gynecologica Scandinavica Suppl. 25: 5-22. 1973.

This review concentrates on 6 papers by the author, which provide evidence that the uterine decidua is the target tissue controlling the process of labor in the model of therapeutic abortion by saline injection in midpregnancy. The introduction also examines the following hypotheses: progesterone withdrawal, increased uterine volume, fetal death, and oxytocin release. 150-200 ml extraamniotic 20% saline (observed by x-ray) spread rapidly around the amniotic sac and separated the membranes from the decidual lining, then diffused into the amniotic fluid to raise its salinity to 4.4%. Since 5% saline by intraamniotic route failed to induce abortion, saline does not act on fetus or fetal membranes. Extraamniotic 20% saline produced necrosis, hemorrhage, and fibrin deposits in the villi beneath the membranes in a thin band, and intraamniotic saline produced a wider zone. The decidual cells after both treatments had pyknotic or disintegrating nuclei and vacuolated or dissolved cytoplasm, the damage increasing with time between saline and uterine evacuation. Most of the placenta and all of the myometrium were intact. The decidual tissues proved less stable when fixed in improper osmotic pressure or after autolysis, with respect to morphology and dispersion of acid phosphatase granules, than trophoblast tissues. Intraamniotic $PGF_2\alpha$ levels were measured in 3 women before and after extraamniotic hypertonic saline; $PGF_2\alpha$ increased 24 hours after saline and later increased further. Thus PG released from decidual cells may be a factor in saline abortion. (LJG)
005655

0789

GUTIERREZ-CERNOSEK, R.M.; LEVINE, L.
Serological aspects of prostaglandins.
Journal of Reproductive Medicine 10(3): 125-132. March 1973.

After a brief summary of PG structure, nomenclature, and metabolism, this review covers specificity of rabbit antibodies against $PGF_1\alpha$, $PGF_2\alpha$, and 15-keto-$PGF_2\alpha$ and their use in immunoassay of serum levels in men, and in women both pregnant and in labor. Rabbit antibodies against $PGF_2\alpha$ coupled with poly-L-lysine are sensitive to nanogram quantities of $PGF_2\alpha$ and are specific to the extent that 22 times as much $PGF_1\alpha$ is needed for 50% inhibition of binding. $PGF_2\alpha$ antibodies recognize 1) the cyclopentane ring, 2) 9-hydroxyl alpha or beta orientation, 3) 5-6 double bond, 4) 15-hydroxyl group, and 5) 13-14 double bond. $PGF_1\alpha$ antibodies are as sensitive, but less specific: $PGF_2\alpha$ cross reacts appreciably. Monkey antibodies against 15-keto-$PGF_2\alpha$ are specific for the 15, 13-14, and 11 sites, with very little cross-reaction from $PGF_2\alpha$ or 13,14-dihydro-15-keto-$PGF_2\alpha$, the next product in PG metabolism. Rabbit antibodies against 13,14-dihydro-15-keto-$PGF_2\alpha$ have also been produced. Serum for immunoassay was extracted with a methylal/ethanol mixture and dialyzed, and 78% of added $PGF_2\alpha$ was recovered. Serum $PGF_2\alpha$ levels were 889 pg/ml in 32 men, 272 in 46 women, rose to a peak of 825 after 17-20 weeks of pregnancy (p<.01), fell to 340 at 33-36 weeks, and 360 at 37-41 weeks. During labor, $PGF_2\alpha$ rose significantly during the first stage (5-9 cm cervical dilation), then declined throughout the second stage and in the immediate postpartum period. (LJG)   005661

0790

HAFEZ, E.S.E.
Transport of spermatozoa in the female reproductive tract.
American Journal of Obstetrics and Gynecology 115(5): 703-717. March 1, 1973.

This comprehensive review discusses anatomic, physiologic, and dynamic aspects of sperm transport in the female reproductive tract, drawing information from man, and other mammalian and nonmammalian species. The anatomic introduction considers male and female genitalia and coital behavior. Semen may be deposited in the vagina or uterus, with or without coagulation or a vaginal plug. From the vaginal pool, sperm travel at 20-50 μm per second through cervical mucus, which by its anisotropic structure directs sperm upward. It is thought that the major impetus of sperm movement from vagina, through the uterus and into the tubes is muscular contraction, although how the oviduct transports the eggs and sperm simultaneously in opposite directions is unknown. There are 3 phases of transport: a rapid movement in minutes; a colonization by sperm of cervical crypts, glands, and fluids; and a prolonged release. Factors affecting sperm transport include prostaglandins in semen, genetic, physiologic, endocrine, immunologic factors, and contraceptives. The sperm are lost by phagocytosis, or via the vagina or peritoneal cavity. (LJG)   005621

0791
HAFS, H.D.; LOUIS, T.M.; STELLFLUG, J.N.
Ductus deferens and ejaculated sperm after $PGF_2\alpha$.
Journal of Animal Science 37(1): 313. 1973.

The number of sperm in the deferent duct and in ejaculates after treatment of rabbits with $PGF_2\alpha$ was measured. In 2 tests with 5 mg $PGF_2\alpha$ (Tham salt) sc, both control groups averaged 33 million sperm, treated rabbits had 70 and 26 million at 10 minutes, 53 and 89 million at 30 minutes, and 89 million at 60 minutes. In a third test, 4 controls averaged 74 million, and 4 rabbits given 10 mg $PGF_2\alpha$ sc averaged 203 million sperm in the deferent ducts at 20, 40, and 60 minutes. When 8 rabbits were ejaculated twice every second day, 5 mg $PGF_2\alpha$ sc doubled sperm in the first ejaculate (100 million vs. 50 million in controls), but not in the second (98 vs. 104 million). When 7 rabbits were ejaculated 4 times once weekly, 2.5 mg $PGF_2\alpha$ given 2 and 4 hours before ejaculation only increased sperm in the first ejaculate (369 vs. 142 million, $p < .01$). Thus the number of sperm in the first ejaculate increased after $PGF_2\alpha$ administration, probably because of increased numbers in the deferent duct. (LJG) 005902

0792
HALE, R.W.; PION, R.J.; SCHINER, W.B.
Vaginal administration of prostaglandins to induce early abortion.
Advances in the Biosciences 9: 561-565. 1973.

10 patients pregnant 7-8 weeks were given 50 mg $PGF_2\alpha$ intravaginally every hour for 12 hours. There were 3 complete and 1 incomplete abortions within 48 hours in this group. Of 10 patients pregnant less than 2 weeks, 6 were given the drug in 25-mg doses every 4 hours to a maximum dose of 225 mg and none aborted. Of the 4 patients given 400-600 mg, 3 aborted. 19 of the 20 patients experienced severe uterine cramping and bleeding. (HC)   006798

0793
HARGROVE, J.L.; SEELEY, R.R.; ELLIS, L.G.C.
Contraction of rabbit testes in vitro: permissive role of prostaglandins for the actions of calcium and some smooth-muscle stimulating agents.
Prostaglandins 3(4): 469-480. 1973.

Fresh Tyrode's solution abolished the contractions of rabbit testes in vitro stimulated by Ca^{++}, serotonin or acetylcholine. PGE_2 added to the medium restored the contractions to each 3 agents. A reciprocal dependency was observed between Ca^{++} and PGE_2 in stimulating contractions. PGE_2 potentiated the stimulatory action of epinephrine or histamine. The alpha-blocking agent, ergotamine tartrate, inhibited the action of epinephrine. Isoproternol inhibited testicular contractions evoked by PGE_2. The present data indicates an increase in the effective permeability of smooth muscle cells to calcium in the presence of prostaglandins. (Authors' modified)   005622

0794

HARGROVE, J.L.; SEELEY, R.R.; JOHNSON, J.M.; ELLIS, L.C.
Prostaglandin-like substances: initiation and maintenance of rabbit testicular contraction in vitro.
Proceedings of the Society for Experimental Biology and Medicine 142(1): 205-209. 1973.

Rabbit testes had spontaneous contractions when suspended in a Tyrode's solution bath medium. Changing the medium markedly reduced the activity. Addition of the bath medium of an active preparation to an inactive preparation increased tonus and initiated contractions. PGE_1 at concentrations of 10^{-9} to 10^{-8} M stimulated inactive preparations but inhibited active preparations. Concentrations greater than 10^{-7} were inhibitory. Treatment with PGE_2 gave similar results. Variability in response of smooth muscle to exogenous prostaglandins may be due to endogenous levels of these substances. (HC)   006760

0795

HARPER, M.J.K.; SKARNES, R.C.
The role of prostaglandin in endotoxin-induced abortion and fetal death.
Advances in the Biosciences 9: 789-793. 1973.

Previous results had suggested that fetal death produced by endotoxin was due to release of serotonin (5HT). The present experiments indicate that although 5HT is effective in inducing fetal death in day-16 pregnant mice, it is unlikely to be the causative agent following endotoxin. Doses of cyproheptadine and methysergide (1 mg/kg), which completely block the action of a maximally effective dose of 5HT, do not reduce fetal death postendotoxin. A larger dose of cyproheptadine (5 mg/kg) does reduce fetal death following endotoxin or $PGF_2\alpha$ injection, and this same dose combined with 2 mg/kg of indomethacin (2 hr before challenge) virtually eliminates abortion and fetal death produced by endotoxin. This same pretreatment also greatly reduces the abortifacient and fetal death actions of $PGF_2\alpha$. We believe that indomethacin and cyproheptadine exert their protective effects by preventing synthesis and release respectively of prostaglandins following endotoxin or exogenous prostaglandins. (Authors)   006499

0796

HASPELS, A.A.; NETH, F.
Induction of abortion 1. by intravenous and 2. by intrauterine administration of $PGF_2\alpha$ (extra- and intraamniotic).
Advances in the Biosciences 9: 515-524. 1973.

10 patients pregnant 7-16 weeks were given 25-200 µg/minute $PGF_2\alpha$ intravenously. 7 aborted, 5 were complete, the IAT averaged 12 hours, and the total dose was 75 mg. Vomiting and diarrhea were predominant side effects. Uterine response was an initial rise of tone, then frequent contractions of increasing amplitude. An intrauterine dose of 375-750 µg given repeatedly to 11 women pregnant 7-27 weeks produced abortion in 10; 8 were complete, total dose averaged 10 mg. There were virtually no side effects. A single intrauterine dose of .5 mg $PGF_2\alpha$ was given to 11 patients pregnant 7-13 weeks. 2 patients aborted, several had weak contractions, and all had cervical dilation. Nausea and vomiting occurred in 5 patients. Repeated intraamniotic injections of 5 mg/hour were given to 9 patients pregnant 12-20 weeks. All aborted, 4 were complete, total dose was 25 mg, IAT averaged 13 hours; there were virtually no side effects. Uterine activity was similar to that seen after intravenous administration. (HC)   006758

0797

HAVRANEK, F.; JUNGMANNOVA, C.; HODR, J.; DRASNAR, J.

Use of prostaglandin $F_2\alpha$ for interruption of pregnancy during the first and second trimester. (CZ, summaries in RS, EN)

Ceskoslovenska Gynekologie 38(6): 432-435. 1973.

5 first and 15 second trimester pregnancies were terminated with 250 to 750 μg $PGF_2\alpha$ administered extraamniotically at 1 to 2 hour intervals by Foley catheter. The instillation evacuation interval was 9 to 28 hours, the average being 18 hours. Total dose of prostaglandin was 2.5 to 10 mg, the average being 6.7 mg. Evacuation was complete in 3, incomplete in 9, and failed in 8 cases. Side effects, mainly nausea, vcmiting, and diarrhea, were seen in 7 women. 15 patients had contractions painful enough to have supplemental treatment with analgesics and spasmolytics. (RMS)   005496

0798

HAWK, H.W.

Estrus, uterine motility and sperm transport after prostaglandin.

Journal of Animal Science 37: 314. 1973.

Ewes given $PGF_2\alpha$ im on day 9 to evoke premature estrus were compared with medroxyprogesterone acetate (MAP) treated ewes (by intravaginal sponge) and untreated controls with respect to uterine motility and sperm transport. 2 out of 17 ewes, given 1, 2, 5, or 10 mg PG, were in estrus 2-3 days later, but 43 out of 50 given 10 mg or 15 mg PG in divided doses, came into estrus. 10 PG and 10 MAP treated ewes had 18 and 11 uterine contractions per 10 minutes moving toward the oviducts, compared with 38 contractions in 10 controls ($p<.01$); and MAP treated ewes had 17 contractions moving toward the cervix, compared with 9 per 10 minutes in controls and PG treated ewes ($p<.01$). Numbers of stermatozoa in oviducts 24 hours after mating were 20,700 in controls, 5100 in PG treated ($p<.02$), and 8200 in MAP treated ewes; but numbers in other segments did not differ. 9 controls, 7 PG and 10 MAP treated ewes had cleaving ova, with no significant differences in numbers of accessory sperm cells. The results indicate that PG given intramuscularly induces luteolysis; uterine motility and the efficiency of sperm transport may be somewhat impaired at the ensuing estrus. (LJG)   005977

0799

HAWK, H.W.; CONLEY, H.H.

Induced estrus and myometrial motility in sheep.

Journal of Animal Science 37(1): 314. Abstract 337. 1973.

Myometrial activity was recorded at estrus, 2-4 days after treatment. 6 ewes were treated with 10 mg $PGF_2\alpha$ im on Day 9, 18 ewes with 300 μg melengestrol acetate or 60 mg medroxyprogesterone acetate orally; 11 ewes had medroxyprogesterone acetate sponges inserted intravaginally from Days 8-25; and 21 controls had blank sponges inserted from Day 8 to Day 15. Isotonic contractions were recorded from 5 by 25 mm strips bathed in oxygenated Krebs-Ringer bicarbonate, at 12.4 amplification, giving a mean amplitude of 65 mm. Progestogen shortened relaxation time significantly from 48 seconds in controls to 18 seconds ($p<.01$) and PG to 20 seconds ($p<.05$). Medroxyprogesterone sponges increased length of contractions from 50 to 114 seconds ($p<.01$). All treatments increased the proportion of time under tensions above the base line ($p<.01-.05$). Results indicate that treating ewes with PG or progestogen changes uterine contractility in vitro at the following estrus. (LJG)   006020

0800

HAWK, H.W.

Uterine motility and sperm transport in the estrous ewe after prostaglandin induced regression of corpora lutea.

Journal of Animal Science 37(6): 1380-1385. 1973.

Parous cycling ewes given $PGF_2\alpha$ to evoke premature estrus were compared for uterine motility and sperm transport with untreated and progestagen-treated controls. 8-year-old ewes came into estrus 53-60 hours later if given 15 mg $PGF_2\alpha$ Tham salt or 10 mg in split doses, but not if given 1, 2, 5, or 10 mg. 60 mg medroxyprogesterone acetate was applied intravaginally in a sponge from Day 9 until Day 25. Uterine motility, observed for 10 minutes 6 hours after estrus during laparotomy in the exteriorized uterus was classified as 'contraction moving toward the oviduct,' 'toward the cervix', or 'other'. PG- and progestagen-treated ewes had fewer contractions moving toward the oviduct ($p<.01$) than controls, whereas PG had less and progestagen groups had more contractions toward the cervix. Sperm counts were lower in the oviduct in PG-treated ewes ($p<.02$) when observed 24 hours after natural mating. 3 days after mating, no differences were observed in numbers of ewes with cleaving ova, or numbers of accessory sperm cells attached to the zona pellucida although the sperm numbers varied widely. (LJG)   006042

0801

HEARNSHAW, H.; RESTALL, B.J.; GLEESON, A.R.

Observations on the luteolytic effects of prostaglandin $F_2\alpha$ during the oestrous cycle and early pregnancy in the ewe.

Journal of Reproduction and Fertility 32(2): 322-323. 1973.

$PGF_2\alpha$ was infused into the uterine vein of ewes at a rate of 45 or 67 µg/hour for 7-9 hours on Days 1-8 of estrous or Days 13-16 of pregnancy. The drug had no effect on Days 1 or 2; luteolysis (defined as fall in progesterone, return to estrus, and presence of ovulation) occurred in all animals treated on Day 3, in some of the animals on Days 4, 5, and 6, and in all animals by Day 8. 10 of 12 ewes given $PGF_2\alpha$ on Days 13-16 of pregnancy returned to estrus, showed a fall in progesterone level, and were confirmed nonpregnant. (HC)   006773

0802

HEMPTINNE, D., DE; SCHUDDINCK, L.; THEIRY, M.; MARTENS, G.

Neonatal bilirubinaemia -- effect of oxytocic compounds (oxytocin and prostaglandins) and vacuum extraction.

International Research Communications System. (73-12) 10-14-3. December 1973. 1 p.

Serum bilirubin levels were assayed at birth and on Day 3 in neonates of women in whom labor was induced by oxytocin, $PGF_2\alpha$ 2-20 µg/minute intravenously, or PGE_2.5-3.5 mg orally. There were no consistent differences found between these values and those obtained in the controls. (HC)   006896

0803

HENNAM, J.F.; TYLER, J.

A cost reductive method for the radioimmunoassay of steroids and prostaglandins.

Journal of Reproduction and Fertility 35(3): 603-604. 1973.

This method of radioimmunoassaying PGE_2, $PGF_2\alpha$, testosterone, testosterone glucuronoside, cortisol, 17alpha-hydroxyprogesterone, and androstenedione reduces cost of beta scintillation counting about 65% by using the assay tube for counting and only 1 ml counting fluid. Following incubation with anitsera, add 1 ml ammonium sulphate (65% saturated for 300 μl, or 60% for 200 μl, buffered at the assay pH) containing 40 mg powdered calcium sulfate dihydrate to each tube. Resuspend the resulting precipitate in 200-400 μl distilled water. Add 1 ml scintillation fluid, cap the tube, place it in an uncapped counting vial, and count using a quench curve. The same basic saving could be obtained for progesterone, estradiol, and estrone, which are precipitated at high salt concentrations, by extracting free hormone from the charcoal-dextran precipitate with 300 μl methanol. (LJG)   005912

0804
HENZL, M.R.; TOMLINSON, R.V.
    Prostaglandin antagonists.
    Research in Prostaglandins 3(1): 1-3. July 1973.

Competitive antagonists of PG-induced smooth muscle contractions include 7-oxa-13-prostynoic acid, polyphloretin phosphates, and derivatives of dibenzoxazepine. PG synthesis can be inhibited at its second step by several fatty acid substrates such as linoleic, linolenic and 8c,12t,14c-eicosatrienoic acid which catalyze a destruction of a microsomal dioxygenase. Bicyclo (2:2:1) heptene derivative is believed to inhibit PGE synthesis in its final step by acting on an endoperoxide isomerase. The nonsteroidal antiinflammatory agents, including indomethacin, naproxen, mefenamic acid, aspirin, and sodium salicylate, irreversibly inactivate the dioxygenase complex. A basic application of PG antagonists is in the improvement of the measurement by radioimmunoassay of endogenous levels of PGs in blood, since there is evidence that PG synthesis may take place after samples are collected. Also these antagonists can be used to create PG deprivation conditions so that PG normal function might be better gauged. Additionally, they can be used therapeutically in the treatment of premature labor and general control of uterine activities in which PGs play a role. (MJN)   006577

0805
HENZL, M.R.; ORTEGA, E.; CORTES-GALLEGOS, V.; TOMLINSON, R.V.; SEGRE, E.J.
    Prostaglandin E_2 and the luteal phase of the menstrual cycle: effects on blood progesterone, estradiol, cortisol and growth hormone levels.
    Journal of Clinical Endocrinology and Metabolism 36(4): 784-787. 1973.

PGE_2, 5 to 15 μg per minute, was infused iv in 7 normally cycling women from 9 a.m. to 5 p.m. for 3 consecutive days, and 1500 I.U. human chorionic gonadotropin (HCG) was injected im on 4 alternating days before PG infusion in 1 woman; 3 controls received saline. Blood was sampled before, at 2 hour intervals, and after infusion, for progesterone, estradiol, cortisol, and growth hormone (GH), determined by radioimmunoassay or protein binding methods. Progesterone values fluctuated normally with no definite depression, even when PG infusion took place on the day of ovulation. In the subject treated with HCG, PG failed to prevent a slight rise in progesterone. Estradiol was normal, measured in only 1 subject on Days 1, 2 and 3 after ovulation. Cortisol and GH tended to peak at some time during PG infusion. All women experienced slight increases in temperature and pulse, decreases in blood pressure, mild nausea, headache, sweating and uterine cramps, with labor-like uterine contractions. All menstruated as expected except 1 who was later confirmed pregnant. Since progesterone remained normal, and the pregnant subject infused on Postovulatory Days 3, 4, and 5 was being treated for sterility of 4 years duration, these data cast serious doubt on the possiblity that PGE_2 could be luteolytic or effective for inducing menses. (LJG)   005658

0806

HENZL, M.R.; AREVALO-T., N.; NORIEGA, L.; AZNAR, R.; ORTEGA, E.; SEGRE, E.
Quantitation of uterine activity after vaginal administration of prostaglandins.
Advances in the Biosciences 9: 767-772. 1973.

A solution of prostaglandin $E_2(PGE_2)$ and $F_2\alpha$ $(PGF_2\alpha)$ was administered vaginally to a group of nonpregnant women during the luteal phase (cycle day 20-22). A total of 8-10 mg PGE_2or 40 mg of $PGF_2\alpha$, in divided doses, changed the pattern of luteal phase uterine activity into well-defined cyclic contractions. The uterine responses were graded and compared to the uterine activity after intravenous infusion of PGE_2. PGE_2in a solution and in vaginal suppositories was used to induce uterine contractions in pregnant women with dead fetuses or hydatidiform moles. Preliminary results suggest that a testing system using the nonpregnant uterus can serve as a model for potency estimation of new PGs or novel PG formulations. (Authors)   006495

0807

HERLITZ, H.; ROSBERG, S.
Effects of LH and prostaglandin E_2on cyclic AMP levels in rat corpus luteum of different ages.
Acta Endocrinologica Suppl. 177: 316. 1973.

Cyclic AMP (cAMP) was measured by protein binding in corpora lutea incubated with LH or PGE_2in Krebs bicarbonate. 30-day-old rats were injected with 10 IU of PMS (Gestyl), which induces ovulation between Days 32 and 33. Corpora lutea were isolated from Days 33 to 40 and incubated. The cAMP recovered from corpora lutea of 34-day-old rats was PGE_2, with a 1 μg/ml minimal effective dose for both. As corpora lutea matured, their proportional to dose of LH or cAMP dose response lost sensitivity to LH but not to PGE_2. (LJG)   005892

0808

HERTELENDY, F.
Block of oxytocin-induced parturition and oviposition by prostaglandin inhibitors.
Life Sciences 13(11): 1581-1589. 1973.

Labor induced by oxytocin in rabbits and oviposition induced by oxytocin or PGE_1in Japanese quail (Coturnix coturnix japonica) could be prevented by indomethacin or eicosatetraynoic acid. Control rabbits delivered within 4 minutes after receiving 100 mU oxytocin iv on Pregnancy Day 31. 4 rabbits were treated with 25 mg indomethacin orally 1 hour before 100 mU oxytocin on 2 consecutive evenings; all failed to deliver. 2 of these were treated on the third day with 100 mU oxytocin alone. Parturition began within 3 minutes. Intrauterine pressure recordings showed complete inhibition of response to oxytocin. 500 mU intrauterine oxytocin given between 0900-1100 hours induced premature oviposition within mean 6.8 minutes in 5 out of 8 control quail, which usually laid between 1400-1800 hours. .5 mg oral indomethacin 15 minutes before oxytocin postponed laying for at least 1 hour. Intrauterine PGE_1.025 μg induced laying within 2.8 minutes in 6 quail; this induction was not inhibited by 1 mg, but 2 mg indomethacin prevented laying in 4 of 8 quail. Intrauterine eicosatetraynoic acid 2.5 μg blocked premature laying in all quail induced by 1000 mU oxytocin. Several hypothetical roles for PGs in parturition are discussed. (LJG)   005652

0809

HERTELENDY, F.; WOODS, R.; JAFFE, B.M.
Prostaglandin E levels in peripheral blood during labor.
Prostaglandins 3(2): 223-227. February 1973.

Plasma or serum concentrations of PGE in women pregnant 29-34 weeks did not differ from those seen in controls. Levels rose significantly during early labor, began to drop during the second phase, and approximated control values by 24 hours postpartum. A similar pattern was noted during oxytocin-induced labor. Highest levels measured were in cord blood. (HC)   006566

0810
HICKL, E.J.; MICKAN, H.; WALTHER, D.
  Kombination von Prostaglandin $F_2\alpha$ und Paracervicalblock bei Missed Abortion und intrauterinem Fruchttod. [Combination of prostaglandin $F_2\alpha$ and paracervical block in the therapy of missed abortion and intrauterine fetal death].
  Klinische Wochenschrift 51(3): 140-141. 1973.

  $PGF_2\alpha$ was successful in treating 8 women with missed abortion (16-27 weeks gestation) and 5 with intrauterine death (32-43 weeks gestation) but only after paracervical block in 9 cases. The patients, aged 16-43, all had undilated cervices. They received $PGF_2\alpha$ in 5% glucose, 25 μg per minute for 30 minutes, then 100 μg per minute, totaling 3.8 to 50 mg $PGF_2\alpha$ over 3:15 to 12:45 hours. In all but 2 patients, light contractions appeared within minutes and regular labor within 2 hours after PG, but the first 3 patients remained undilated. After paracervical block with 10 ml .5% bupivacain (Carbostesin), repeated every 2 hours as needed, 9 patients achieved cervical dilatation and explusion of the uterine contents. Only one of them required curettage. Side effects included nausea and vomiting, easily treated with Paspertin, and local inflammation of the vein. (LJG)   005654

0811
HILL, J.R., Jr.; DICKEY, J.F.; HENRICKS, D.M.
  Estrus and ovulation in $PGF_2\alpha$/PMS treated heifers.
  Journal of Animal Science 37: 315. 1973.

  Estrus was induced in 5 of 18 heifers given 2 mg $PGF_2\alpha$ intrauterine on Days 3 to 5 of the estrus cycle, in 16 of 18 heifers given $PGF_2\alpha$ on Days 9 or 10, and in 16 of 18 given $PGF_2\alpha$ on Days 16 or 17. PMS alone or given either 2 days prior to or on the same day as $PGF_2\alpha$ resulted in multiple ovulations in most of the heifers. (HC)   005917

0812
HILLIER, K.; DUTTON, A.; CORKER, C.S.
  The effect of prostaglandin $F_2\alpha$ on plasma steroid and LH levels in the luteal phase of the menstrual cycle.
  Advances in the Biosciences 9: 673-678. 1973.

  Intravenous $PGF_2\alpha$ (12.5-250 μg/min) in four volunteers in the mid-late luteal phase of the menstrual cycle caused a prolonged fall in plasma progesterone levels, a slight decline in estrogens, and a transient elevation in LH. Early luteal phase infusions did not affect progesterone levels. Spontaneous cyclic variation in progesterone levels may contribute to the decline observed in the late luteal phase. The luteal phase was not shortened. (Authors)   006485

0813
HINGORANI, V.; GANESH, K.
  Induction of abortion by extraamniotic administration of prostaglandin $F_2\alpha$.
  Advances in the Biosciences 9: 665-668. 1973.

20 patients, 8 to 22 weeks pregnant, were given PGF$_2\alpha$ by the extraamniotic route, 500 µg initially and 750 µg every 2 hours until abortion or 30 hours. Complete abortion occurred in 6 patients; another 10 had incomplete evacuation, which was easily terminated vaginally. IAT averaged 19 hours; average dose was 6.3 mg. Only 2 patients had mild nausea and vomiting, 1 had hyperthermia. (HC)   006483

0814
HINGORANI, V.; DUA, A.
Present status of prostaglandin in conception control.
Proceedings of the Indian National Science Academy 39(3,Pt.B): 420-428. June 1973.

The authors reviewed the use of PGs for terminating pregnancies. They noted that given intravenously, these compounds are effective at any stage of gestation but side effects often occur. Intravaginal administration has also led to systemic side effects. Intrauterine PG has been useful in first-trimester pregnancies for preoperative dilatation of the cervix. The intraamniotic route is favored for terminating second-trimester pregnancies. (HC)   006948

0815
HINMAN, J.W.
Prostaglandins: perspective in obstetrics and gynecology.
Journal of Reproductive Medicine 10(3): 118-120. March 1973.

Before a clinician uses PGs, he should have a basic knowledge of their biosynthesis, metabolism, and tissue differences. PGs function like the 'second messenger' cAMP rather than like hormones or steroids. PGs are synthesized by insoluble microsomal multienzyme systems from arachidonic acid and bis-homo-gamma-linoleic acid, with molecular oxygen and cofactors. Individual enzymes have not been well characterized, but PGEs and PGFs are derived from the same intermediate, and the system also forms other PGs and open chain hydroxy fatty acids of unknown significance. Inhibitors of PG synthesis, aspirin and indomethacin, have demonstrated that overproduction of PGs occurs in some pathologic states, but the inhibitors can be toxic when they upset normal PG modulation (e.g., gastric ulceration caused by aspirin). Biosynthesis is initiated by hormonal or mechanical stimuli on cell membranes. PGs are degraded rapidly by oxidation of the 15-OH to carbonyl and excreted in urine in man. The physiologic significance of PGAs is controversial; they are shown to be metabolized to PGB and PGC, to occur in human serum at .75-3.0 ng/ml, and to regulate aldosterone secretion. Picogram and nanogram levels of PGs must be detected for physiologic studies by radioimmunoassay or mass spectrometry of deuterium-labeled PGs. Most of our knowledge of PGs comes from sheep, but species differences are so great that for example, PGF$_2\alpha$ regulates luteolysis in sheep and uterine contraction in humans, but not sheep uterine activity or human luteolysis. (LJG)   005660

0816
HIXON, J.E.; NADARAJA, R.; SCHECTER, R.J.; HANSEL, W.
Prostaglandin F$_2\alpha$-induced stimulation of estrone and estradiol-17β secretion in cattle.
Prostaglandins 4(5): 679-687. November 1973.

To examine whether PGs influence estrogen secretion, which synchronizes luteolysis, estrus, and ovulation, PGF$_2\alpha$ was injected into the ovaries of 5 Holstein heifers and PGF, estrone, estradiol-17β, and progesterone were measured by radioimmunoassay. 300 µg PGF$_2\alpha$ was injected into the corpus luteum on Cycle Days 12-14. 4 controls received Tris buffer vehicle. Control PGF$_2\alpha$ levels rose from 78 to 144 pg per ml in the first 15 minutes; levels in injected heifers rose to about 400, fell below control levels at 4 hours, and rose slightly for the next 24 hours. Progesterone dipped sharply, then remained somewhat depressed for the next 72 hours, compared to controls. Estradiol and estrone peaked at variable times in different animals, usually once in the first 24 hours and again at about 64 hours after $PGF_2\alpha$. $PGF_2\alpha$ failed to induce luteolysis or estrus. (LJG)   005637

0817
HORTON, E.W.; JONES, R.L.; MARR, C.G.
  Effects of aspirin on prostaglandin and fructose levels in human semen.
  Journal of Reproduction and Fertility 33: 385-392. 1973.

The effect of aspirin on fructose and prostaglandin levels in human semen was studied in 2 subjects. Each received aspirin at 2 dose levels of 3.6 and 7.2 gm/day for 3 days. PGE and 19-hydroxy PGA and PGB were separated by thin-layer chromatography and quantitized spectrophotometrically. Aspirin was found to have no significant effect on seminal fluid fructose levels. PGE concentrations in the 2 subjects fell 52% and 60% for the 3.6 gm/day dose schedule, and by 78% and 82% for the 7.2 gm/day level. Concentrations of 19-hydroxy PGA and PGB in the semen were also reduced. A significant decrease in $PGF_2\alpha$ levels, as determined by the stimulation of isolated rabbit jejunum muscle, was noted in 1 subject at the higher aspirin dosage. Aspirin administered at a level of 7.2 gm/day was associated with toxic side effects in both subjects and did not completely inhibit prostaglandin production. (JSL)   006881

0818
HORTON, E.W.; POYSER, N.L.
  Elongation of oestrous cycle in the guinea-pig following subcutaneous or intra-uterine administration
    of indomethacin.
  British Journal of Pharmacology 49(1): 98-105. 1973.

Estrous cycle length in guinea pigs was increased by indomethacin, an inhibitor of PG synthesis, given sc or applied locally in the uterus by paraffin implants; the mechanism was investigated by measuring corpora lutea, progesterone levels, PG synthesis by uterine tissue, and PG level in uterine fluid. Cycle Day 1 was defined by maximal influx of leucocytes and vaginal cornification checked in daily smears. Indomethacin sc, 20 mg per day from cycle Day 7, increased cycle length significantly by 3 days (p<.001); this dose killed 2 out of 14, but 4 mg per day was ineffective. Indomethacin, 4 paraffin wax implants, 25 mg each, per uterine horn, with pentobarbitone and procaine anesthesia and tetracycline postoperatively, increased cycle length in 6 guinea pigs beyond 47 days. By measuring indomethacin remaining in the implants, the release rate was calculated as .2 to .6 mg per day. Guinea pigs with implants killed on cycle Day 47 had plasma progesterone 3.8 to 8.9 ng per ml, compared to .27 to .62 in controls bearing pure wax implants killed on Day 14-15; corpora lutea were 4.2 to 6.0 cubic mm compared to .55-.95 (regressed) in controls. PGE_2 and $PGF_2\alpha$ were detected at low levels in homogenized uteri incubated 90 minutes, and appeared higher at the end of the cycle. $PGF_2\alpha$ was almost as high in control uteri as in indomethacin uteri. All uteri were hypertrophied with 1 to 63 ml of alkaline intraluminal fluid, due to uterine ligation. (LJG)   005651

0819
HUMES, J.L.; SIGAL, L.H.; KUEHL, F.A., Jr.
  Effects of hormones and antagonists upon cyclic AMP formation in bovine luteal cells.
  Federation Proceedings 32: 536Abs. 1973.

10 μg/ml of LH stimulated cAMP formation tenfold in bovine luteal cells. Removal of LH with buffer after 5 minutes stopped cAMP formation. 10-200 μg/ml PGE_1 increased cAMP formation 1/6 that

induced by LH. A synergistic relationship between PGE_1 and LH could not be determined, as has been demonstrated in granulosa cells. 7-oxa-13-prostynoic acid inhibited the action of both LH and PGE_1. A prostaglandin synthetase inhibitor, fluoroindomethacin, did not exhibit any inhibitory effect on PGE_1 or LH but did inhibit the arachidonic acid-stimulated PGF increase. (RAS)   006629

0820

IGEL, Von H.; LAU, H.-U.; HENGST, P.; HALLE, H.

Erste Erfahrungen mit dem Prostaglandin $F_2\alpha$ bei der kuenstlichen Abortusausloesung. [Initial experiences with prostaglandin $F_2\alpha$ in induced abortion]

Zentralblatt fuer Gynaekologie 95(11): 353-357. 1973.

Extraamniotic $PGF_2\alpha$ was administered by indwelling polyethylene catheter to 36 primiparas (aged 14-40 years) and to 32 multiparas for therapeutic abortion. $PGF_2\alpha$ doses ranged from 250, 500, or 750 μg to 1 or 2 mg, given at 2-hour intervals. The primiparas received mean total dose of 11.5 mg over 28.5 hours, with 30 (83%) complete abortions and 6 (17%) incomplete abortions (dilatation of cervix without expulsion of uterine contents). The multiparas received mean total dose of 10.68 mg over 24.6 hours with 27 (84%) complete abortions, 4 (13%) incomplete, and 1 (3%) failure. Oxytocin drip infusion was given to 11 women who had no cervical dilatation. There were brief side effects: nausea in 6% vomiting in 13%, sweating in 3%, and fever in 9% of patients. 3 patients lost 500-600 ml blood. All abortions were followed by curettage. During the course of this trial, the authors came to prefer a higher dose schedule because this shortened abortion time without significantly increasing side effects. They suggest 3 doses of 1 mg over a 6-hour period, followed by 1.5 mg and 2 mg, 2 hours apart, as needed. (LJG)   006011

0821

JACKS, F.; SETCHELL, B.P.

A technique for studying the transfer of substances from venous to arterial blood in the spermatic cord of wallabies and rams.

Journal of Physiology 233: 17p-18p. June 1973.

This technique for quantitating substances infused into the venous, and recovered from the arterial, circulation of the testis in wallabies and rams was perfomred with labeled $PGF_2\alpha$, testosterone, and water. After pentobarbitone anesthesia, the animals were cannulated through an 18-gauge needle in a vein on the ventral surface of the wallaby testis, or a vein under the head of the ram epididymis. Cannulas were also inserted in the aorta via the saphenous artery, in the internal spermatic vein, and in the testicular artery. 20 μCi tritiated $PGF_2\alpha$, 20 μCi carbon-14 testosterone, and 200 μCi tritiated waier were mixed with 5 ml homologous plasma and infused at 20 μl/minute into wallabies or at 55 μl/minute into rams. In wallabies, blood was sampled from the aorta every 15 minutes, from the testicular artery every 30 minutes, and from the internal spermatic vein at the end of the infusion. In rams, blood was drawn continuously from these catheters. Recovery of label was stable after 15 minutes, and transfer was computed from the ratio (testicular artery-aorta)/(internal spermatic vein-aorta). The recovery was 11%, 2%, and 9% in 4 rams and 1%, 10%, and 21% in 2 wallabies for $PGF_2\alpha$, testosterone, and water, respectively. (LJG)   005650

0822

JAFFE, B.M.; PARKER, C.W.

Physiologic implications of prostaglandin radioimmunoassay.

Metabolism 22(8): 1129-1137. 1973.

This review describes a technique of assaying PGF, PGA and PGF separately in samples of plasma extracted and chromatographed on silicic acid to improve specificity and then cities results of PG assays in several tissues. The plasma technique uses tritium labeled PGs for recovery and dextran-charcoal to absorb unbound PGs. Antibodies do not cross react with 15-keto-PGs; $PGF_2\alpha$ antibodies are specific except for $PGF_1\alpha$; PGA_1 and PGE_1 antibodies bind unlabeled PG preferentially. Reports of plasma measurements are conflicting: possibly mechanics of sampling blood introduce artifacts. Indomethacin decreased plasma levels of PGA 87% and PGE 70%. Circulating PGA levels are higher than PGE and PGF because lungs do not convert PGAs to 15-keto derivatives. Half-lives of circulating $PGF_2\alpha$ were 1·5 minutes. PGEs were increased in studies on renal ischemia, and on cell growth, particularly carcinomas and mouse fibrosarcoma. PGEs and PGFs have been implicated in control of intraocular pressure and thyroid hormone release. $PGF_2\alpha$ was an abortifacient in women at 2 ng per ml plasma. It increased in midtrimester, labor and delivery, and in midcycle of sheep estrous cycle. PGE_1 was also reported increased in labor and delivery, but not during pregnancy. (LJG)   005924

0823

JAFFE, B.M.; BEHRMAN, H.R.; PARKER, C.W.
Radioimmunoassay measurement of prostaglandins E, A, and F, in human plasma.
Journal of Clinical Investigation 52(2): 398-405. February 1973.

A modified method for radioimmunoassay of PGs in plasma is described. Protein binding is eliminated by use of an organic solvent extraction and cross-reactivity is circumvented by column chromatography. Mean plasma concentrations of PGA in normal donors were 1024, 888 and 1285 picograms per ml in males, females, and females on oral contraceptives, respectively. For PGE the values were 378, 316 and 478 and for PGF 84, 154, and 162 pg/ml. PGE levels were markedly elevated in patients with endocrine-active tumors. PGE levels were greatly reduced in indomethacin-treated rats. (HC)
006557

0824

JANSON, P.O.; AHREN, K.
Effects of prostaglandins ($F_2\alpha$) on ovarian blood flow.
Acta Physiologica Scandinavica S396: 20. 1973.

The vascular effects of a 50 μg/kg iv dose of $PGF_2\alpha$ on the rabbit ovary were recorded. The blood flow of the ovarian vein was measured in 1 set of animals. In another group the ovarian blood flow was measured by means of ytterbium-169- and scandium-46-labeled 35 plus or minus 5 μ microspheres. $PGF_2\alpha$ caused marked bradycardia and arterial hypotonia for 20-30 minutes, which could be modified by iv atropine or vagotomy. Simultaneously, blood flow decreased in the ovarian vein, whole ovary, and corpus luteum. All blood flows returned to normal at the same time as the systemic pressure; iv $PGF_2\alpha$ had no specific effect on ovarian blood flow. (LJG)   006111

0825

JARVINEN, P.A.; PENNANEN, S.; YLOSTALO, P.
Induction of abortion by intra- and extra-amniotic prostaglandin $F_2\alpha$ administration.
Prostaglandins 3(4): 491-504. 1973.

Patients in the eleventh to twentieth week of pregnancy were given $PGF_2\alpha$, 61 by the intraamniotic route and 54 extraamniotically. Average doses were 35 mg (given in 5- to 25-mg doses) for the intraamniotic group and 6358 μg for the extraamniotic group given as 3 to 6 doses at 1- to 2-hour intervals. 56 of the intraamniotic group aborted, 19 completely. 39 of the extraamniotic group aborted, 19 completely. The IAT averaged 20 hours for both groups. Side effects, mainly vomiting, diarrhea, or flush, were noted in 72% of the intraamniotic patients and 54% of the extraamniotic group. The differences between the routes in success rate and frequency of side effects is due to the difference in doses used. The authors state that the intraamniotic route is generally better and easier to perform. (HC)   005627

0826
JOCHLE, W.
Coitus-induced ovulation.
Contraception 7(6): 523-564. June 1973

This review of coitus-induced ovulation covers its occurrence among mammals, its required anatomical and neurohumoral apparatus, the nature of the follicular phase, documented paracyclic and coitus-induced ovulation or conception in women, and women's reproductive psychology. Mammals may be divided into strictly coitus induced reflex ovulators, facultative reflex ovulators, species in which coitus hastens ovulation, and cyclic ovulators. Examples are tabulated from genera from Marsupia to Primates. There are possible 3 pathways linking genital tract, hypophysis and ovaries: a neurohormonal axis used in cyclic ovulation, a nervous pathway activated by coitus resulting in LH and oxytocic release, and possibly a direct autonomic or biochemical stimulation by coitus. The biochemical path may be PGs originating from the female or from semen and acting as a luteolysin either by causing contractions of uterus, tubes, and ovaries or by directly causing follicular rupture. The follicular phase is typically 4 days long in rodents, 8 in horses, 12-16 in humans, and higher multiples of 4 in monkeys, with 4 day cyclic LH surges and follicular maturation. Evidence from autopsy, rape, and short coital exposures documents that 10% to 65% of human conceptions arose during the 'safe period ' or amenorrhea. Indirect evidence from holidays and blackouts suggests high conception rates during the follicular phase. Psychologic stimuli are known to alter normal menstrual cycles and cause pseudocyesis, amenorrhea, and probably paracyclic ovulation. (LJG)   005624

0827
JOCHLE, W.; TOMLINSON, R.V.; ANDERSEN, A.C.
Prostaglandin effects on plasma progesterone levels in the pregnant and cycling dog (beagle).
Prostaglandins 3(2): 209-217. February 1973.

Dogs were given intravenous infusions of $PGF_2\alpha$ 50 μg/hour or $PGE_2$5 μg/hour for 6 hours during metestrus. Pregnant and cycling dogs received .5-2.0 mg doses $PGF_2\alpha$ intravenously. In 3 animals plasma progesterone levels fell approximately 90% after $PGF_2\alpha$ infusions or 40% after PGE_2. Single injections of $PGF_2\alpha$ also decreased progesterone levels in 2 pregnant and 1 cycling dog. $PGF_2\alpha$ 1-4 mg/day had no effect on the gestation period. At all dose levels, $PGF_2\alpha$ induced emesis. (HC)
006886

0828
JOHNSON, M.; JESSUP, R.; RAMWELL, P.
Ultraviolet light modification of the prostaglandin receptor.
Prostaglandins 4(4): 593-605. October 1973.

The effects of ultraviolet light (UV), dithiothreitol (DTT) a disulfide reducing agent, dithionitrobenzoic acid (DTNB) a sulfhydryl oxidant, N-ethyl maleimide (NEM) a sulfhydryl alkylator, and p-hydroxymer-curibenzoate (PHMB) a sulfhydryl complexing agent, were explored on PGE_1receptors in ADP induced human platelet aggregation and isotonic virgin rat uterus contractions. Platelet aggregation was monitored in a recording spectrophotometer at 600 nm, initiated by 500 ng per ml ADP or 253.7 nm UV light. 10 ng per ml PGE_1 inhibited UV-induced platelet aggregation 38.2%, and ADP-induced aggregation 71.7%. Prior UV exposure or .1 mM DTT lessened the inhibitory effect of PGE_1 10-50%, depending on time of exposure. .1 mM DTNB added during UV exposure permitted PGE_1 to inhibit UV aggregation as well as it did ADP-induced platelet aggregation, and prevented the protective effect of prior UV irradiation against PGE_1 inhibition. .1 mM NEM combined with DTNB prevented DTNB from permitting PGE_1 to inhibit UV induced aggregation. 30 minutes of UV exposure increased the amount of PGE_1 or $PGF_2\alpha$ required to elicit half-maximal rat uterine contractions. .2 mM DTNB reversed this inhibition when applied for 30 minutes after UV, and prevented it when applied during UV exposure. (LJG)   005618

0829

JONSSON, H.T.; SHELTON, V.L.; BAGGETT, B.

Stimulation of adenyl cyclase by prostaglandins in rabbit corpus luteum.

In: Kahn, R.H. and Lands, W.E.M., eds. Prostaglandins and cyclic AMP. New York, Academic Press, 1973. p 183-186.

Homogenates of 4 and 10 day corpora lutea from LH-induced pseudopregnant rabbits were incubated for 30 minutes with labeled ATP and the adenyl cyclase activity measured by the amount of cAMP formed. PGE_1 stimulated adenyl cyclase activity more than PGE_2; the stimulation of PG + LH was greater than with PG or LH alone. $PGF_2\alpha$ had a greater effect on adenyl cyclase activity in the 10-day than in the 4-day corpora lutea but did not augment LH-induced stimulation. Prostaglandin concentrations were 6 μg/ml; LH was 10 μg/ml. The authors conclude that prostaglandins regulate the corpus luteum through their action on adenyl cyclase, but are not mediators of the stimulatory effect of LH. (HC)   005546

0830

JUBIZ, W.; FRAILEY, J.

Seminal fluid and plasma prostaglandin responses to aspirin in normal subjects.

Fertility and Sterility 24(12): 977-978. December 1973.

The concentration of PGE and PGF in seminal fluid and plasma of 5 normal males dropped significantly after 5 consecutive daily doses of 3.25 gm of aspirin. The authors suggest that this decrease in seminal fluid PG content could lead to infertility. (HC)   005825

0831

KARIM, S.M.M.; AMY, J.-J.;

Effect of prostaglandin 16,16 dimethyl E_2 methyl ester on the pregnant human uterus.

Prostaglandins 4(4): 581-592. 1973.

22 patients, 12 to 22 weeks pregnant, were given 16,16 dimethyl PGE_2 methyl ester. 8 patients received 1 to 3 25-μg doses intravenously; 6 aborted completely, 5 in 6 to 10 hours and the sixth 20 hours after administration. Shivering and hyperthermia occurred frequently. 4 patients were given 25- or 50-μg injections, 4 to 9 times intramuscularly (maximum 300 μg); 2 aborted but only after 27 (complete) and 59 (incomplete) hours; vomiting and hyperthermia occurred. Doses of 100 to 800 μg 2 to 4 times by the intraamniotic route were given to 5 patients. 1 aborted completely after 14 hours; the only side effect was hyperthermia. 5 patients were given 100 to 500 μg doses 2 to 3 times orally. 1 aborted completely after 14 hours with emesis and diarrhea. A single intravenous dose of 25 μg produced uterine contractions sufficient for abortion; doses of 100 to 400 μg intraamniotically

increased myometrial activity but abortion did not occur. The authors conclude that further investigation of this drug as an abortifacient is not warranted. (HC)   005617

0832

KARIM, S.M.M.

Intrauterine prostaglandins for outpatient termination of very early pregnancy.

Lancet 2: 794. 1973.

6 to 14 days after their last menstrual period, 12 patients were given 1 mg PGE_2 and 4 patients received 4 mg $PGF_2\alpha$ as a single intrauterine dose. Uterine bleeding occurred in 2 to 5 hours, and pregnancy test was negative in all patients within 14 days. 12 patients reported uterine cramps and 6 patients vomited. (HC)   005623

0833

KARIM, S.M.M.; HILLIER, K.

Pharmacology and therapeutic applications of prostaglandins in the human reproductive system. In: Cuthbert, M.F., ed. The prostaglandins: pharmacological and therapeutic advances. Philadelphia, J.B. Lippincott Co., 1973. p. 167-199.

The effect of PGs are enumerated in detail, largely from the authors' work, on the nonpregnant human uterus in vitro and in vivo, menstruation, induced abortion, and labor. PGEs, PGA, PGB, and human seminal plasma inhibit motility of the nonpregnant uterus in vitro, but $PGF_1\alpha$ or $PGF_2\alpha$ increase it. All PGs stimulate the nonpregnant uterus in situ in a noncyclic fashion, except for possibly greater sensitivity before menstruation. The nonpregnant uterus is more sensitive than the pregnant, but less so than the term uterus, having thresholds for PGE_1 of 20-50 μg, and for $PGF_2\alpha$ about 50 μg. PGE_2, $PGF_2\alpha$, $PGF_1\alpha$, and PGE_1, are now considered the menstrual stimulants found in endometrium and in the circulation, perhaps higher during dysmenorrhea. The author brought on menstruation by intravaginal PGE_2 and $PGF_2\alpha$ in 11 of 12 women. Pregnant uterus in vitro was stimulated more strongly than nonpregnant uterus by semen extracts, low doses of PGE_1, and $PGF_2\alpha$. The studies on the response to PGs of pregnant uterus in vivo were all induced abortion trials. The authors believe that the slow infusions at constant rate produce the most clinically complete results, for example, 105 complete abortions and 130 successes with PGE_2, 5 μg per minute, in 139 women. Several small scale trials by others with constant, single injection, or increasing intravenous doses are summarized. Intramuscular and subcutaneous PGs are effective but the injections are very painful. A single intraamniotic injection of 5 μg PGE_2 or 25 μg $PGF_2\alpha$ will often terminate midpregnancy. The author aborted 30 women 7-23 weeks pregnant with 20 mg PGE_2 and 15 women 9-22 weeks pregnant with 50 mg $PGF_2\alpha$ given every 2.5 hours by the intravaginal route, with low incidence of side effects. 2 small trials with the intrauterine extraamniotic route with overall 29% complete abortions are summarized. At 10-fold higher concentrations and with severe vomiting and diarrhea, oral PGs will induce term labor. On term pregnant uterus in vitro, PGEs stimulated the upper segment more strongly than the lower; PGFs stimulated the upper segment, but in the lower uterine segment inhibition followed stimulation. Term deliveries were induced in 93% of 100 women with 5-10 μg per minute $PGF_2\alpha$, and were highly successful in over 500 women with .5-2.0 μg per minute PGE_2. 2 studies have been done so far comparing PGE_2 and or $PGF_2\alpha$ with oxytocin. Labor has been induced by repeated oral or vaginal PGE_2 or $PGF_2\alpha$. At higher doses, PGs are more effective than oxytocin for missed abortion, missed labor, and hydatidiform mole. (LJG)   005907

0834

KARIM, S.M.M.; HILLIER, K.

Prostaglandins and the induction of labour.

Research in Prostaglandins 2(6): 1-2. May 1973.

Intravenously $PGF_2\alpha$, 5 to 20 μg per minute, or PGE_2, .5 to 2 μg per minute, effectively stimulated the term uterus. Results compare favorably with oxytocin. Occasional nausea and vomiting were reported, reports on hypertonus are equivocal. $PGF_2\alpha$, 5 to 25 mg orally every 2 hours, or PGE_2, .5 to 2.5 mg, provide adequate uterine stimulation. Vomiting and diarrhea occur less with PGE_2. (HC)   006764

0835

KARIM, S.M.M.; SHARMA, S.D.; FILSHIE, G.M.; SALMON, J.A.; GANESAN, P.A.

Termination of pregnancy with prostaglandin analogs.

Advances in the Biosciences 9: 811-830. 1973.

3 women in the second trimester of pregnancy were given intravenous doses of PGE_2 and subsequently its methyl ester; another 3 received $PGF_2\alpha$ and subsequently its methyl ester. The uterine stimulant effect of PGE_2 methyl ester was 5 times as potent as PGE_2; $PGF_2\alpha$ methyl ester and $PGF_2\alpha$ were equipotent. Given by the intramuscular, intravaginal, or intraamniotic route, 15-methyl analogues were more potent and had a longer duration of action as uterine stimulants than the parent compounds. 25 to 100 μg of 15-methyl PGE_2 methyl ester given intramuscularly to 6 women 7 to 16 weeks pregnant produced abortion in all (4 complete) within 18 hours; cold and/or shivering occurred in 4 patients. 50 or 100 μg of this drug intravaginally caused complete abortion within 20 hours in 5 of 6 patients 13 to 16 weeks pregnant; cold and/or shivering occurred in 2. In 38 of 42 patients 13 to 22 weeks pregnant, 1 or 2 50-μg doses of this drug intraamniotically was sufficient to produce abortion; 2 patients required 3 doses, 2 others required 4 doses; nausea and vomiting occurred. 18 women successfully aborted after receiving 100 μg doses of this drug intraamniotically, 13 with 1 dose, the others had 2 doses. A single 200 μg dose was successful in all. There were no side effects using this route. Injection of 50 to 75 μg into the placenta, intraperitoneally or intravenously, produced abortion and side effects as described above. Using 15-methyl $PGF_2\alpha$, doses of .25 to 1.5 mg intramuscularly, .5 to 1.5 mg intravaginally, or .5 to 1.5 mg intraamniotically was successful in inducing abortion in 8 to 32 hours in all 17 women tested, all but 2 of whom were in the second trimester. Vomiting and diarrhea occurred in most women given the drug intramuscularly or intravaginally. Of 5 patients 6 to 10 weeks pregnant a dose of 50 μg 15-methyl PGE_2 methyl ester intramuscularly caused abortion and a decrease in progesterone, estrogen, and HCG levels in 4. Abortion failed and no change in hormone levels was seen in the fifth patient. (HC)   006502

0836

KARIM, S.M.M.; HILLIER, K.

The role of prostaglandins in myometrial contraction.

In: Josimovich, J.B., ed. Uterine contraction--side effects of steroidal contraceptives. New York, Wiley-Interscience, 1973. (Volume 1 of Problems of Human Reproduction) p. 141-169.

The bulk of this review contains an explanation of the pharmacologic discrepancies in effects of PGs on motility of uterus, depending on PG, dose, in vitro or in vivo, local or systemic application, pregnant or not, length of gestation, phase of the cycle and portion of the uterus tested. The lipid menstrual stimulant is probably a mixture of PGE_2, $PGF_2\alpha$, $PGF_1\alpha$, and PGE_1, and is probably of physiologic significance because these PGs activate nonpregnant myometrial contraction; .5 μg local $PGF_2\alpha$ will initiate contraction after dilatation and curettage; $PGF_2\alpha$ and PGE_2 will stimulate contractions and bleeding when given at the time of menstruation; and $PGF_2\alpha$ level in menstrual fluid is higher in women with primary dysmenorrhea. Association of PGE_1, E_2, $F_1\alpha$, and $F_2\alpha$ with spontaneous contractions of labor has been observed in amniotic fluid, decidua, and maternal venous blood. These PGs will stimulate the uterus at any stage of pregnancy. PGs may be involved in the action of the IUD. The common fear that coitus during pregnancy will instigate premature labor was disproved with respect to PGs since the amount of PGs in an ejaculate administered vaginally was shown not to stimulate the pregnant uterus. Hormonal milieu affects PG's action on myometrium: PGEs are more effective at ovulation (estrogen influence), but PGFs are more active before menstruation. The uterus may be sensitized to other spasmogens if PGs are given first. (LJG)   006312

0837

KEICHLINE, L.D.; HAGEN, A.A.

A comparison of the effects of prostaglandin E_1 and luteinizing hormone on cAMP levels in the rat testis.

Federation Proceedings 32(3): 298Abs. 1973.

At 15 minutes 10 μg PGE_1 yielded peak cAMP levels of 5.15 pM/mg of testis as measured by protein binding assay. 10 μg of LH produced peak cAMP levels of 6.22 pM/mg of testis at 40 minutes. Concentration-effect studies of PGE_1 and LH had similar maxima for cAMP levels. PGE_1 and LH combined did not increase cAMP levels above either one alone. 7-oxa-13-prostynoic acid inhibited PGE_1 or LH stimulation of cAMP; additionally, PGE_1 or LH could not overcome this inhibition. (RAS) 006728

0838

KEIRSE, M.J.N.C.; TURNBULL, A.C.

E prostaglandins in amniotic fluid during late pregnancy and labour.

Journal of Obstetrics and Gynaecology of the British Commonwealth 80: 970-973. 1973.

Amniotic fluid obtained from 50 women at or near term or during labor was analyzed for PGE content using highly specific column and gas liquid chromatography. No PGE was detected in samples obtained before labor. PGE_2 content during labor ranged from 1.2 to 17 ng per ml, the concentration being proportional to the degree of cervical dilatation. PGE_1 was not detected in any sample. The authors suggest that PGE_2 may be produced as a consequence of labor. (HC)   005822

0839

KEIRSE, M.J.N.C.; TURNBULL, A.C.

Gas chromatographic determination of E prostaglandins in human amniotic fluid.

Prostaglandins 4(2): 263-267. August 1973.

A simple procedure is described for the extraction and purification of prostaglandins from amniotic fluid. Combined gas chromatography with electron-capture detection, it provides a rapid and specific method for the determination of PGE in human amniotic fluid. This procedure enabled the authors to demonstrate the absence of PGE_1 with a lower limit usually between .2 and .4 ng/ml, even in the presence of large amounts (>10 ng/ml) of PGE_2. (Authors' modified)   005625

0840

KEIRSE, M.J.N.C.; PATTEN, P.T.; ANDERSON, A.B.M.; TURNBULL, A.C.

Pregnant sheep mymometrium responds to prostaglandins in vitro but not in vivo.

International Research Communications System (73-4) 8-5-1. April 1973. 1 p.

3 sheep pregnant for 118, 120, and 140 days were infused with $PGF_2\alpha$ via catheter into the fetal vena cava, and the ewe at 140 days had a catheter also into the uterine artery. No uterine contractions were observed after total doses of 60 mg $PGF_2\alpha$ over 59 hours (up to 22.5 μg/minute), 13 mg over 30 hours (up to 16.6 μg/minute), and 1.5 mg at 5 μg/minute intrafetally repeated 2 hours later into the uterine artery. The in vitro recordings were taken from myometrial strips from 6 ewes pregnant 120-145 days, in oxygenated Krebs-Henseleit medium at 37 degrees C, using isometric, isotonic, and sucrose-gap methods. 5 of 6 myometria responded to at least 10 ng/ml PGE_1 or 50 ng/ml $PGF_2\alpha$ by beginning or increasing activity. They exhibited prolonged 'enchancememt' to other spasmogens such as calcium ion. Spontaneous isotonic contractions were inhibited by 10^{-5} gm/ml indomethacin. Isometric contractions to PGE_1 were accompanied by depolarization and action potentials. (LJG) 006344

0841

KELLY, J.; FLYNN, A.M; BERTRAND, P.V.
A comparison of oral prostaglandin E_2 and intravenous Syntocinon in the induction of labour.
Journal of Obstetrics and Gynaecology of the British Commonwealth 80: 923-926. October 1973.

To compare the efficacy of oral PGE_2 and Syntocinon in the induction of labor, 49 patients received PGE_2 and 49 matched controls received Syntocinon. PGE_2 was successful in inducing full dilatation and active labor within 24 hours in 47 patients and Syntocinon in 46. There was no significant difference between the 2 groups for induction to delivery interval, Apgar scores of babies at 1 and 5 minutes, or incidence of postpartum hemorrhage. Resting uterine tone, frequency of contractions, and incidence of incoordinate uterine activity were all significantly greater in the Syntocinon group. Vomiting and diarrhea were more common with PGE_2 patients. Greater incidences of pyrexia and acetonuria with Syntocinon were not statistically significant. (Authors' modified)   005492

0842

KIRTON, K.T.
Biochemical effects of prostaglandins as they might relate to uterine contraction.
In: Josimovich, J.B., ed. Uterine contractions -- side effects of steroidal contraceptives. New York, Wiley-Interscience, 1973. (Vol. 1 of Problems of Human Reproduction.) p. 193-203.

This review goes from smooth muscle physiology and classical mass-action theory of drug receptor interaction to general theories of PG action in an attempt to generate a model of PG action on myometrium. The drug-receptor interaction is dependent on drug concentration, affinity, and intrinsic activity. In smooth muscle contraction, Ca^{++} ions move into the cell from 2 pools, the extracellular and that bound to the fiber. PGs stimilate contraction without being inhibited by atropin, antihistamines, serotonin antagonists, alpha or beta adrenergic blocking agents. PGEs and PGFs can decrease sensitivity to oxytocin, histamine and vasopressin. PGs may have a specific uterine receptor, and may affect contraction directly by a short-lived depolarization and a Ca^{++} independent long-lasting coupling of excitation and contraction. Ion flux mechanisms in other systems like toad bladder are influenced by PGs; some are related to adenyl cyclase or to cholinergic receptors. Ca^{++} may be the initial event of PG action universally. (LJG)   006314

0843

KIRTON, K.T.; KOERING, M.J.
Prostaglandin $F_2\alpha$ and primate corpus luteum: a correlation of structure and function.
Fertility and Sterility 24(12): 926-934. December 1973.

Corpora lutea were removed for light and electron microscopy observation and plasma progesterone was measured in 4 rhesus monkeys infused with 45 mg $PGF_2\alpha$ over a 24-hour-period between Day 22-25 of the menstrual cycle, and in 3 given saline. Peripheral plasma progesterone fell from mean 5.4 ng per ml (range 2.2-7.3) to mean 2.4 (range .7-4.3) after infusion of $PGF_2\alpha$, and from mean 6.9 (range 5.6-8.6) to 6.7 (6.0-7.1) in controls. In control rhesus monkeys the granulosa cells of the corpus luteum constituted a homogeneous population of large spherical cells, but the treated tissues contained patches of smaller cells. In both there was a central nucleus, with abundant endoplasmic reticulum, Golgi appartus, mitochondria, and lipid droplets. Lipid droplets in control granulosa cells were spherical, with either retained or lucent contents, but in PG treated cells the droplets were variable in size and shape, often containing faint 'residual luminal structures.' It was remarked that the changes in PG treated tissues appeared premature, but whether they were physiologic was unknown. (LJG)   005824

0844

KIRTON, K.T.; WYNGARDEN, L.J.; BERGSTROM, K.K.
Prostaglandins and myometrial contractility.
Advances in the Biosciences 9: 651-655. 1973.

Incubation of myometrial strips with varying amounts of unlabeled PGE_1 inhibited the amount of labeled PGE_1 bound to the tissue in a dose-related fashion suggesting the presence of, and a finite binding capacity for, prostaglandin receptors. Addition of nonlabeled $PGF_2\alpha$ decreased the amount of labeled $PGF_2\alpha$ bound to tissue following subcutaneous injection of labeled PG. Uterine concentrations were higher than skeletal muscle but lower than blood, indicating extreme sensitivity of the uterus to PG. PGE_1 and PGE_2, but not $PGF_2\alpha$, injected subcutaneously increased uterine cAMP levels for estrous, but not for pseudopregnant hamsters. These data suggest that stimulation of adenyl cyclase is not a step in the mediation of PG-induced uterine contractility. A probable mechanism may be a change in myometrial smooth muscle cell membrane permeability and/or alteration of (HC)   006768

0845

KIRTON, K.T.
Prostaglandins and steroidogenesis.
Advances in the Biosciences 9: 645-650. 1973.

The author noted that exogenous prostaglandins have significantly different effects on corpus luteum function in primate and subprimate specimens. In most subprimates prostaglandins act as a direct luteolytic factor. Although $PGF_2\alpha$ terminates pregnancy in both human and monkey, an initial attempt to demonstrate a direct lytic effect in nonpregnant monkey by $PGF_2\alpha$ injection in early luteal phase has been successful. However, injection of $PGF_2\alpha$ in late luteal phrase significantly shortened the cycle by 3 days. There remains little evidence to suggest a role of PG in normal primate corpus luteum function. (RAS)   006771

0846

KLOECK, F.K.; JUNG, H.
In vitro release of prostaglandins from the human myometrium under the influence of stretching.
American Journal of Obstetrics and Gynecology 115(8): 1066-1069. April 15, 1973.

PGE and PGF concentrations were estimated by rat stomach and rabbit jejunum bioassays in stretched human myometrium and in fluid flowing through the organ bath while contractions of human and rat uterus were recorded on a kymograph. When human fundal strips, 30-40 mm by 2-3 mm, were stretched by 1-gm and then 9-gm weights, the spontaneous contractions increased in frequency and amplitude. Mean PGE activity in the Ringer's solution increased in 2 experiments from 55.7 and 51.9 to 449.7 and 340.3 ng/l, while PGE in fundal muscle fell from 15.9 to 8.4 ng/gm. $PGF_2\alpha$ activity simultaneously fell in the fluid from 106.6 to 63.7, and in muscle tissue from 30.6 to 15.8 ng/gm. The increase in concentration of PGE and decrease in PGF suggest an enzymatic mechanism rather than an autooxidative conversion of unsaturated fatty acids to prostaglandins. (Authors' modified)   005626

0847
KOERING, M.J.
Normal and prostaglandin $F_2\alpha$-induced luteolysis in the pseudopregnant rabbit.
Anatomical Record 175(2): 361-362. 1973.

Histological examination of corpora lutea and progestin determination from pseudopregnant rabbits during normal and $PGF_2\alpha$ induced luteolysis were examined on Days 9, 16, 18, and 20 of pseudopregnancy. 3 rabbits, killed on Day 9, 1 day after 400 μg per kg $PGF_2\alpha$ sc, had decreasing progestins indicative of luteolysis. Most of their corpus luteum cells examined under light and electron microscopes appeared smaller, and had small secondary lysosomes. Some cells had large lysosomes and subendothelial spaces. All had lipid droplets of medium-density. During normal luteal regression, some cells had lucent lipid droplets on pseudopregnancy Day 16; more cells had coalescing lucent droplets and some had myelin figures on Day 18. (LJG)   006770

0848
KOERING, M.J.; KIRTON, K.T.
The effects of prostaglandin $F_2\alpha$ on the structure and function of the rabbit ovary.
Biology of Reproduction 9(3): 226-245. 1973.

Pregnant rabbits were given $PGF_2\alpha$ 250 μg/kg sc twice daily for 3 days and killed on Day 8, 14, or 21 for light and electron microscope studies of ovarian structure. Upon gross observation ovaries varied in size regardless of stage of pregnancy or treatment. Corpora lutea (CL) in controls contained large round cells with whorled smooth endoplasmic reticulum (ER) surrounding lipid droplets, particularly developed on Day 14. In PG-treated ovaries, CL had smaller cells with more prominent nuclei, and autophagic vacuoles whose location in or next to cells was difficult to define. The luteal cells at Day 14 showed the greatest morphological change. Interstitial cells were typically smaller with empty lipid droplets and were not affected by PG, nor did PG affect developing follicles. Peripheral serum progestins fell to less than .2 ng/ml as a result of $PGF_2\alpha$. (LJG)   005497

0849
KORDA, A.; SHEARMAN, R.P. SMITH, I.D.
Some observations on the use of $PGF_2\alpha$ for termination of pregnancy.
Australian and New Zealand Journal of Obstetrics and Gynaecology 13(1): 44. 1973.

Of 14 women pregnant 12-19 weeks, 7 who were given 9.25-37.5 mg $PGF_2\alpha$ by the intrauterine route aborted within 72 hours, 4 completely. Via the intraamniotic route 30-75 mg $PGF_2\alpha$ induced complete abortions within 48 hours in all of the other 7 women treated. Nausea was noted in 10 women. (HC)
006327

0850

KUNZE, H.; BOHN, E.; KURZ, C.-S.; VOGT, W.

Hemmung der Prostaglandinbiosynthese durch ein Loklanasthetikum, Nupercain. [Inhibition of prostaglandin biosythesis in bovine seminal vesicles by a local anesthetic, nupercaine.]
Naunyn-Schmiedebergs Archives of Pharmacology 279(S): R20. 1973.

Incubation of bovine seminal vesicle homogenates with nupercaine decreases PG formation. Calcium antagonizes and free arachidonic acid blocks the effect of nupercaine. Phospholipase activity is also reduced by nupercaine. The authors conclude that nupercaine acts by inhibiting calcium-dependent phospholipase leading to a decreased concentration of free fatty acids thus inhibiting PG biosynthesis. PG synthetase is not affected by nupercaine. (HC)   006155

0851

KVINNSLAND, S.

Estradiol-17β, cyclic AMP and prostaglandins: in vivo and in vitro studies on the cervicovaginal epithelium from neonatal mice.
Life Sciences Part I 12(8): 373-384. 1973.

The development of an estrogen-induced antigen in the neonatal mouse cervicovaginal epithelium was studied in vivo and in vitro under the influence of estradiol. cAMP, theophylline, PGE_1, PGE_2, $PGF_2\alpha$, propranolol, isoproterenol and cortisone. For studies in vitro, the cervicovaginal tissues from mice less than 24 hours old were cultured on millipore filters with test compounds for 40-48 hours; for studies in vitro, 3-day-old mice were injected sc with varying compounds for 2 days. The antigen was estimated by indirect immunofluorescence, with an antiserum from rabbits immunized against adult mouse vaginal epithelium. In cultured tissues, cAMP and dibutyryl cAMP increased antigen at optimal concentrations of 10^{-4}M. 4 x 10^{-6}M estradiol combined with 10^{-5}M dibutyryl cAMP elicited as high a level of antigen. Propranolol, a beta adrenergic inhibitor, inhibited the combined stimulation, but the beta stimulator isoproterenol did not reverse this inhibition. 5'-AMP, cGMP and dibutyryl cGMP were ineffective. Theophylline alone increased antigen slightly. PGE_1, 7 x 10^{-5}M, but not PGE_2 or $PGF_2\alpha$, induced moderate antigen production. In vivo, 5 μg daily estradiol increased antigen content, and proliferation of the basal zone of the epithelium. Propranolol inhibited antigen in the epithelial zone but not basal hyperplasia. (LJG)   005709

0852

LeMAIRE, W.J.; YANG, N.S.T.; BEHRMAN, H.H.; MARSH, J.M.

Preovulatory changes in the concentration of prostaglandins in rabbit Graafian follicles.
Prostaglandins 3(3): 367-376. March 1973.

Rabbit Graafian follicles were obtained during estrus and at 1, 5, and 9 hours following the initiation of ovulation by the injection of 100 IU of human chorionic gonadotropin (HCG). The mean values for PGF and PGE during estrus, as determined by radioimmunoassay, were 58.2 and 256 pg/follicle, respectively. Levels 1 hour after HCG injection were not significantly different. After 5 hours a nearly tenfold increase in PGF (543 pg/follicle) and a threefold increase in PGE (770 pg/follicle) were noted. 9 hours after HCG injection, PGF levels had risen sixtyfold, and PGE approximately fifteenfold (3700 and 4150 pg/follicle, respectively). This increase in prostaglandin concentrations during induced ovulation may indicate a physiological role for the prostaglandins in the process of ovulation. (JSL) 006775

0853

LeTELLIER, P.R.; SMITH, W.L., Jr.; LANDS, W.E.M.

Effect of metal-complexing agents on the oxygenase activity of sheep vesicular glands.

Prostaglandins 4(6): 837-843. December 1973.

[I]50 values (concentration of inhibitor that gives 50% inhibition) for 12 copper chelators and 1 inhibitor of activition were estimated for sheep vesicular gland oxygenase and soybean lipoxygenase. Sheep oxygenase, from acetone powder, was assayed in unactivated and phenol-activated forms by oxygen monitor. Reaction mixtures contained 4 mg vesicular gland acetone powder or .02 mg soybean lipoxygenase added to 2.9 ml of .1 M tris-HCl (pH 8.5), plus inhibitor 15 seconds later; oxygenation was begun with 23 μM 5,8,11,14- eicosatetraenoic acid. [I]50 values (in mM) for activated oxygenase were as follows: diethyldithiocarbamate, 2; ethylxanthate, 25; bathocuproine sulfonate, 7; toluene-3,4-dithiol, .1; dithizone, .8; NaCN, .5; neocuproine, 1; cuprizone, 6; o-phenanthroline, 2; 2,2'-bipyridone, 8; tiron, 50; and m-phenanthroline, 1. Unactivated oxygenase was inhibited by all these agents, and lipoxygenase gave comparable results except for no inhibition by diethyldithiocarbamate, ethylxanthate, or NaCN. Inhibition of activated oxygenase by diethyldithiocarbamate, NaCN, and o-phenanthroline was reversible. Since 5 of these inhibitors can also complex iron, either copper or iron is required in fatty acid oxygenation in PG synthesis. (LJG)   005676

0854

LABHSETWAR, A.P.

A comparative study of some effects of prostaglandin $F_2\alpha$, $F_1\alpha$, E_2, and E_1 on reproductive processes of rats and hamsters.

Advances in the Biosciences 9: 641-644. 1973.

The minimum PG doses required for 100% abortion when given subcutaneously to hamsters once a day from Days 4 through 6 of pregnancy were .12 mg/kg for $PGF_2\alpha$ and $PGF_1\alpha$, 1.5 mg/kg for PGE_2, and 3.0 mg/kg for PGE_1. When given orally, PGF values were 10 mg/kg and PGE values were greater than 25 mg/kg. In rats, the drugs were given twice a day and the values were 2, 2, and 10 mg/kg subcutaneously for $PGF_2\alpha$, $PGF_1\alpha$, and PGE, respectively. PGE_1 had no antifertility effect at sublethal doses in rats. Administered to rats at the above doses twice daily from Days 1 to 5, the PGE compounds, but not $PGF_2\alpha$, delayed the passage of blastocysts from the fallopian tubes and uterus. Similarly, the PGE compounds, but not $PGF_2\alpha$, markedly reduced the number of spots seen in the uteri of these rats 15-30 minutes after an intravenous injection of Pontamine blue dye. Prostaglandins given subcutaneously to hamsters twice on Day 14 induced littering of all pups by Day 15; control animals delivered on Day 16. Oxytocin failed to induce premature littering. Intratesticular injections of $PGF_2\alpha$ caused atrophy of the testis and accessory sex glands. (HC)   006733

0855

LABHSETWAR, A.P.

Do prostaglandins stimulate LH release and thereby cause luteolysis?

Prostaglandins 3(5): 729-732. May 1973.

The author evaluated a recent paper by Cerini and Christie (Prostaglandins, April 1973) which sets out to test the hypothesis that the luteolytic effect of $PGF_2\alpha$ is mediated through an increased secretion of LH from the pituitary. Several factors may account for their negative results. First, they used a low dose of $PGF_2\alpha$ (5 μg/hour), although it has been shown that infusion of as much as 25 μg/hour of $PGF_2\alpha$ into systemic circulation will not cause luteolysis in sheep. Additionally, they used spayed ewes in which the pituitary may already have been secreting maximal amounts of LH and further stimulation would not be possible. Finally, no one has yet demonstrated a luteolytic effect of LH in

284

sheep, so that even if they had demonstrated increased levels of LH following infusion of $PGF_2\alpha$, no conclusion might have been drawn that this LH increase is the mediator of luteolytic action of $PGF_2\alpha$ in sheep. (MJN)   006756

0856
LABHSETWAR, A.P.
   Effects of prostaglandins E_1, E_2 and $F_2\alpha$ on zygote transport in rats: induction of delayed implantation.
   Prostaglandins 4(1): 115-125. July 1973.

   PGE_1, PGE_2, and $PGF_2\alpha$ injected in pregnant rats on Days 1-3 caused delayed development of morulae on Day 4 and delayed transport and implantation on Day 6. Rats received subcutaneous injections of .5 mg PGE_1, 1 mg PGE_2 (equivalent of Tham salt), .2 mg $PGF_2\alpha$, or .2 ml phosphate buffer twice daily. Tubal flushings on Day 4 yielded 8 or 16 cell morulae from controls and $PGF_2\alpha$ rats, but 2 or 4 cell morulae from rats given PGEs. Only 1 rat given $PGF_2\alpha$ had zygotes in the uterus. On Day 6, controls and rats given $PGF_2\alpha$ produced no blastocysts in tubal or uterine flushings. 71% of PGE_1-treated and 40% of PGE_2-treated rats had blastocysts in uterine flusings, and 64% of PGE_1-treated and 73% of PGE_2-treated rats had tubal blastocysts. It was concluded that PGEs cause tubal retention and delayed implantation of blastocysts. (LJG)   005839

0857
LABHSETWAR, A.P.; ZOLOVICK, A.
   Hypothalamic interaction between prostaglandins and catecholamines in promoting gonadotrophin secretion for ovulation.
   Nature 246(150): 55-56. November 14, 1973.

   Immature rats treated to induce an ovulatory response were given bilateral injections of aspirin into the anterior hypothalamus. Ovulation was markedly inhibited; $PGF_2\alpha$ or dopamine, given at the same time as aspirin, blocked the aspirin effect. Direct intrapituitary injection of aspirin had no effect on ovulation. The authors concluded that aspirin blocks ovulation by inhibiting PG synthesis and that $PGF_2\alpha$ and dopamine interact to potentiate the adrenergic transmission for gonadotrophin secretion. (HC)   006762

0858
LABHSETWAR, A.P.
   Neuroendocrine basis of ovulation in hamsters treated with prostaglandin $F_2\alpha$.
   Endocrinology 92(2): 606-610. 1973.

   The neuroendocrine basis of ovulation in hamsters was studied. A single injection of 250 μg of $PGF_2\alpha$ in phosphate buffer was administered to adult hamsters on specific days of pregnancy; animals were then killed at selected intervals and the number of tubal ova were counted. Injection on Day 3 and autopsy on Day 7 showed that none of the treated hamsters, but all of the controls, were pregnant. Ova were found in all of the treated animals. Ovulation was expected to have occurred on the night of Day 4 since a rise in serum LH was observed on Day 4. $PGF_2\alpha$ was thus thought to be prompt in terminating luteal function, in reducing the secretion of progesterone, and in enhancing the secretion of estrogen. The administration of antiestrogen was seen to interfere with ovulation. Unilateral ovariectomy prior to injection of $PGF_2\alpha$ caused a doubling in the number of ova shed by the remaining ovary; the minimum ovariectomy-ovulation interval was thought to be 48 hours for such compensation. It was suggested that $PGF_2\alpha$ by decreasing progesterone secretion makes the ovaries sensitive to gonadotropins. (MLH)   006530

0859
LABHSETWAR, A.P.
   Prostaglandin E_1: studies on antifertility and luteolytic effects in hamsters and rats.
   Biology of Reproduction 8(1): 103-111. 1973.

PGE$_1$ and PGF$_2\alpha$ were given to hamsters and rats in a study of their antifertility effects on implantation, corpora lutea, plasma progesterone, premature labor, and deciduomata in pseudopregnancy. In hamsters, PGE$_1$ 300 μg/day sc given either on Pregnancy Days 4-6 or on Days 6-8 prevented implantation in all animals, as did 25 μg PGF$_2\alpha$ given on Day 4-6. Orally, 2 mg/day PGE$_1$ prevented implantation in 4 out of 5 hamsters, but 1 mg PGF$_2\alpha$ prevented all implantation. PGE$_1$ 250 μg/day sc on Days 4-6 reduced plasma progesterone from 26.6 ng/ml to 3.0 in hamsters. Progesterone 4 mg sc on Day 4-6 restored pregnancy and maintained progesterone at 23.1 ng/ml. 300 μg PGE$_1$ given on Days 1-3 did not affect number or size of implanted embryos. 66% of hamsters given 500 μg PGE$_1$ twice on Day 14 and once on Day 15 gave birth 2 days early. Normal deciduomata were not visible macroscopically in pseudopregnant hamsters (tubes ligated on Day 1 and uteri traumatized by suture on Day 4) given PGE$_1$ 300 μg on Days 4-6. In rats, .5 mg PGE$_1$ sc twice daily on Days 4-7 was ineffective, .75 mg reduced implants from mean 12.5 to 8.7 and 1 mg was lethal. PGF$_2\alpha$ .2 mg 2 times on Days 4-7 prevented all implantation. When pontamine blue was given to visualize number of implants and corpora lutea, control rats had equal numbers on Day 6, but the implants in PGE$_1$-treated rats were not evident until Day 11. PGE$_1$ .5 mg twice daily on Days 1-4 reduced number and size of implants; this reduction could not be reversed by progesterone. PGE$_1$ is probably luteolytic in hamsters but delays implantation in rats. (LJG)   005670

0860
LABRIE, F.; GAUTHIER, M.; PELLETIER, G.; BORGEAT, P.; LEMAY, A.; GOUGE, J.J.
   Role of microtubules in basal and stimulated release of growth hormone and prolactin in rat adenohypophysis in vitro.
   Endocrinology 93(4): 903-914. 1973.

The role of microtubules in secretion of growth hormone (GH) and prolactin was investigated by incubating rat anterior pituitaries in vitro with the inhibitors deuterium oxide (D20) and vincristine sulfate, the stimulants PGE$_1$, growth hormone releasing factor (GH-RF) and 25 mM potassium, by measuring cAMP release and by electron microscopy. 2 incubations took place for 45 minute periods in Krebs Ringer bicarbonate with 11 mM D-glucose; peptide hormones released were quantitated by polyacrylamide gel electrophoresis and radioimmunoassay. 25-50% D20 inhibited GH 13.2-23% and prolactin release 22.9-24%. Microtubules appeared normal but perhaps longer. D20 inhibited release of both hormones in presence of all stimulants, l mM butyryl cAMP, .01 mM PGF1, GH-RF, and 25 mM K, from 18.1 to 63.8%. Vincristine, 10^{-7} to 10^{-4} M, inhibited prolactin release up to 51%, particularly in the second incubation. .01 mM vincristine inhibited the stimulation of both hormones by 25 mM K, .01 mM PGE$_1$ and GH-RF, from 25.3% to 65.2% in the second incubation period. No inhibition of cAMP release was detected. With vincristine, microtubules disappeared and crystalline structures with a 300 Angstrom periodicity appeared. (LJG)   005935

0861
LAMOND, D.R.
   Rebuttal.
   Prostaglandins 3(5): 702. May 1973.

Methinks John McCracken protests too much. In all innocence, we merely question the claim that if PGF$_2\alpha$ is the luteolysin in the ewe, then counter-current transfer between the conjoined V. ovarica and

A. ovarica is not the only mechanism. Others are even questioning the concept that $PGF_2\alpha$ is the natural luteolysin in the ewe. Rate of blood flow should be measured and related to natural fluctuation of $PGF_2\alpha$ in the uterine vein together with fluctuation in secretion rate of progesterone from the corpus luteum. Until then, the problem of ipsilateral uteroovarian relations does not need to be elevated to the status of a scientific controversy. (Authors' modified)   005933

0862

LAMOND, D.R.; TOMLINSON, R.V.; DROST, M.; HENRICKS, D.M.; JOCHLE, W.
Studies of prostaglandin $F_2\alpha$ in the cow.
Prostaglandins 4(2): 269-284. August 1973.

$PGF_2\alpha$ was injected into the uterine artery, uterine lumen, or jugular vein of pregnant cows and cows in luteal phase, and $PGF_2\alpha$, progestins, and estrogens were measured by radioimmunoassay in a study of luteolysis. In a cycling cow 20 mg $PGF_2\alpha$ iv raised plasma $PGF_2\alpha$ levels to 120 ng/ml in 1 minute, but was cleared in 10 minutes to normal (.5-2 ng/ml). Progesterone rose briefly 5-10 minutes following injection. The cow went into estrus 3 days later, 5-7 days prematurely. When 2-4 mg $PGF_2\alpha$ was injected in the ipsilateral uterine horn in cows whose ovarian artery had been sectioned, luteolysis also resulted. When 2-4 mg $PGF_2\alpha$ was injected into the uterine horn on the corpus luteum side in several cows in the luteal phase, $PGF_2\alpha$ rose in ovarian vein plasma but was normal within 30 minutes; progesterone peaked 15-30 minutes after $PGF_2\alpha$ but measured about .3 ng/ml in 24 hours; estrogens fluctuated independently. Catheterization by laparotomy of the ovarian vein did not affect $PGF_2\alpha$ levels in the luteal phase. Pregnant cows' $PGF_2\alpha$ levels were not affected by cesarean section. Abortion occurred 5 days after a stepwise infusion into the uterine artery totalling 16.8 mg $PGF_2\alpha$ in 7.5 hours, after 28 mg into the uterine artery in 28 hours, and after 10 mg in 10 hours by the intrauterine route. Plasma progesterone decreased 50% within 4 hours and decreased steadily during luteal regression. Since no uterine contractions were observed and luteolysis took place despite sectioning the ovarian artery, luteolysis as such is the probable mechanism of $PGF_2\alpha$ in cows. (LJG)   005934

0863

LAMOND, D.R.; DROST, M.
The counter-current transfer of prostaglandin in the ewe.
Prostaglandins 3(5): 691-695. May 1973.

A section of the ovarian artery or of the uterine branch of this artery distal to where countercurrent transfer of $PGF_2\alpha$ occurs was removed from 6 ewes on Day 8 of the estrous cycle. All animals returned to estrus on Day 15 to 18; all had ovulated; surgical manipulation had no influence on progesterone blood levels. The authors concluded that if $PGF_2\alpha$ is the luteolytic factor, countercurrent transfer between the ovarian vein and artery cannot be the only mechanism. (HC)   006767

0864

LAMPRECHT, S.A.; ZOR, U.; TSAFRIRI, A.; LINDNER, H.R.
Action of prostaglandin E_2 and of luteinizing hormone on ovarian adenylate cyclase, protein kinase and ornithine decarboxylase activity during postnatal development and maturity in the rat.
Journal of Endocrinology 57(2): 217-233. 1973.

Rat ovarian adenylate cyclase, protein, kinase and ornithine decarboxylase activities were determined in vitro in presence of luteinizing hormone (LH) and/or PGE_2 to analyze the pathway of ovarian response to trophic hormones. With PGE_2 10 μg per ml, intact ovaries increased cAMP production maximally 300% in 5 minutes. LH and PGE_2 stimulated cAMP formation and cAMP tissue levels in isolated Graafian follicles, corpora lutea, luteinized ovaries, ovaries from rats sterilized by androgen or constant light, and in 30-day-old juvenile ovaries. When Graafian follicles were cultured 18 hours, LH and PGE_2 increased cAMP synthesis 851% and 1005%; after culture with LH, the increases were 114% and 441%. Oxaprostynoic acid, $1 \cdot 100$ μg per ml, an inhibitor of PGE_2 action, and indomethacin 50 μg per ml, an inhibitor of PG synthesis, failed to prevent cAMP stimulation by LH and PGE_2. Cyclic AMP tissue level increased 582% in prepubertal rat ovaries incubated with LH, 830% with PGE_2, and 1126% with both, slightly less than an additive effect. Rat ovaries were resistant to LH from 1 week before term until 2 weeks of age, but did react to PGE_2 before birth. Cyclic AMP stimulated protein kinase activity at 12, and 28, but not 7 days of age. In 28 day ovaries, LH and PGE_2 activated protein kinase and saturated free cAMP binding sites. PGE_2 and LH stimulated glucose oxidation and ornithine decarboxylase similarly. PGE_2 evokes LH release in intact rats. (LJG)   006504

0865
LANGE, A.P.; SECHER, N.J.
Delivery induced with prostaglandin $F_2\alpha$ in cases of foetal death.
Danish Medical Bulletin 20(2): [11] 1973.

9 consecutive cases of fetal death were delivered, 6 on the first day, and 3 on the second day, of intravenous $PGF_2\alpha$. These cases and others reported suggest that $PGF_2\alpha$ or PGE_2 is preferable to oxytocin for induction of labor in fetal death. (Authors' modified)   006794

0866
LANGE, A.P.; SECHER, N.J.; PEDERSEN, G.T.
Intrauterine extra-amniotic instillation of prostaglandin $F_2\alpha$ in abortion compared with the extra-amniotic hypertonic saline method.
Acta Endocrinologica Suppl. 177: 322. 1973.

80 patients received a single extraamniotic instillation of 20% saline, repeated in 24 hours if necessary, and intravenous oxytocin was then added if necessary. An equal number of women received .2-.75 mg $PGF_2\alpha$ every 2-3 hours. The authors stated that $PGF_2\alpha$ was safer and more effective without presenting specific data on total dosages or outcome. (HC)   005728

0867
LAROS, R.K., Jr.; WITTING, W.C.; WORK, B.A. Jr.
Uterine activity response to constant infusion of prostaglandin $F_2\alpha$ in term human pregnancy.
Clinical Pharmacology and Therapeutics 14(1): 140. 1973.

Prostaglandin $F_2\alpha$ was administered for 4-hours' constant infusion (dosage ranging from .5 to 16 μg/min) to 60 full-term pregnant women (30 having ruptured membranes) to study the uterine response. 2 hours prior to prostaglandin infusion uterine activity was monitored by measuring amniotic fluid pressure. If labor had not begun by the end of the $PGF_2\alpha$ infusion, uterine activity was again measured for 2 hours. Administration of 2 μg/min and less had little effect in increasing uterine activity, but 4 μg/min or more usually increased uterine activity and resulted in labor and delivery. Observations made during the study included: 1) Infusion levels above 4 μg/min sometimes were accompanied by hypersystole and hypertonus; 2) the labor graph often showed an unexplained 1-hour plateau at 5-6 cm of dilation; and 3) effective labor measured by cervical dilation occurred in some patients without an 'adequate' contraction pattern. (IC)   006435

0868

LARSEN, J.F.; PEDERSEN, P.H.; SORENSEN, B.
Prostaglandin $F_2\alpha$ as an oxytocic agent in the second and third trimesters.
Danish Medical Bulletin 20(2): 11. 1973.

$PGF_2\alpha$ was infused for induction of labor in second and third trimesters at 15 μg per ml in isotonic glucose for normal delivery or 25 μg per ml for dead or anencephalic fetuses, in 44 patients. Total dose ranged from 3-27.5 mg (mean 12.9) in normal deliveries, and 5-100 mg (mean 45) for fetal death or anencephaly. Infusion ranged from 3.75-23 hours (mean 10.5) in the total group. 89% of the patients delivered and less than half had side effects. (LJG)   006777

0869

LARSEN, J.W.; HANSON, T.M.; CALDWELL, B.V.; SPEROFF, L.
The effect of estradiol infusion on uterine activity and peripheral levels of prostaglandin F and progesterone.
American Journal of Obstetrics and Gynecology 117(2): 276-279. September 15, 1973.

Estradiol-17β, 400 μg to 200 mg, was infused over 2-4 hours in 5 normal-term pregnant women and in 4 preeclamptic women at term to discover whether prostaglandins would rise and progesterone fall as they do in sheep parturition. The dose schedules were 400 μg estradiol over 4 hours in 2 patients, 4 mg over 4 hours in 3 patients, 40 mg over 2 hours in 1 patients, and 200 mg over 4 hours in another 3 patients. As expected, estradiol concentration rose during infusion, but PGF and progesterone were not changed. Base-line uterine activity assessed for 1 hour before treatment consisted of fleeting increases in tone. 8 of 9 infused patients developed mildly painful contractions 30-40 seconds in length at 3-7 minute intervals after the second hour of infusion. The patient who did not respond had a twin pregnancy. Bishop score rose from 6 to 8 in 1 patient. These results corroborated the observations that progesterone level does not fall unless PG increases, the placenta is disrupted, or the uterus is in active labor. (LJG)   005506

0870

LAUDERDALE, J.W.; CHENAULT, J.R.; SEQUIN, B.E.; THATCHER, W.W.
Fertility of cattle after $PGF_2\alpha$ treatment.
Journal of Animal Science 37(1): 319. Abstract 356. 1973.

Pregnancy rates were compared in cows (housed at Upjohn Co., University of Florida or Michigan State University) inseminated about 12 hours after onset of estrus, 12 hours after onset of estrus within 7 days after injection with 30 mg $PGF_2\alpha$-tham salt, or 72 and 90 hours after injection with 30 mg $PGF_2\alpha$-tham salt. Since pregnancy rates (estimated by palpation 35-60 days after insemination) showed statistically insignificant interaction between treatment groups and location, data were pooled between locations within treatment groups to yield 66% and 58% pregnancy in estrus controls, 51% and 57% in PG-induced estrus, and 60% and 58% in PG treatment without regard to estrus cycle. Thus fertility of cattle inseminated at estrus induced by $PGF_2\alpha$ or at timed intervals after $PGF_2\alpha$ was comparable to controls. (LJG)   006017

0871

LAUERSEN, N.H.; WILSON, K.H.
Continuous prostaglandin-$F_2\alpha$ infusion for middle-trimester abortion.
Lancet 1(7813): 1195. May 26, 1973.

An initial dose of 250 mg PGF$_2\alpha$ was instilled followed by 750 mg in 5 minutes and 1 mg in 30 minutes. 1 mg was then instilled at 6-hour intervals. This dose schedule was unsatisfactory. Increasing the frequency of administration to every 2-3 hours also proved unsatisfactory. Midtrimester abortion was then induced with a continuous extraovular infusion of prostaglandin F$_2\alpha$. The patients complained little of pain and side effects. (AS)   006526

0872
LAUERSON, N.H.; RAGHAVAN, K.S.; WILSON, K.H.; FUCHS, F.; NIEMANN, W.H.
Effects of prostaglandin F$_2\alpha$, oxytocin, and ethanol on the uterus of the pregnant baboon.
American Journal of Obstetrics and Gynecology 115(7): 912-918. April 1, 1973.

The dose response of uterine contractions to oxytocin and PGF$_2\alpha$ was recorded transabdominally in pregnant baboons 2-18 days before term. 7 baboons sedated with phencyclidine HCl received oxytocin or PGF$_2\alpha$ intravenously at constant infusion rate. PGF$_2\alpha$ had a variable effect at 10 μg/minute, increased frequency of contractions in all animals to a mean 128 Montevideo Units at 20 μg, increased activity to 211 M.U. at 50 μg, and was followed by a slow decrease in activity. Oxytocin's threshold dose was 4 mU/minute, mean 82 M.U. in all animals. Activity was 134 M.U. at 8 mU principally because of increased amplitude, 169 M.U. at 15 mU., and activity ceased rapidly after infusion was discontinued. Ethanol 7.5 ml/kg was in effective in preventing increased contraction from PGF$_2\alpha$; however, for low doses of oxytocin, ethanol diminished uterine response. (LJG) 005671

0873
LEE, J.
Prostaglandins as therapeutic agents.
Archives of Internal Medicine 131(2): 294-300. February 1973.

Prostaglandins are mentioned as oxytocic agents which can induce labor or terminate pregnancy. The relationship of prostaglandins to hypertension is the article's major concern. (RAS)   006001

0874
LEHMANN, F.; PETERS, F.; BRECKWOLDT, M.; BETTENDORF, G.
Plasma progestin and plasma estrogen levels during infusion of PGF$_2\alpha$ in the human.
Advances in the Biosciences 9: 679-688. 1973.

32 patients, 10 to 28 weeks pregnant, were given PGF$_2\alpha$ either by intravenous infusion or extraamniotic injection. The rate of infusion was 50 μg/min the first hour, then 75 μg/min, 100 μg/min, and 125 μg/min during subsequent hours. For extraamniotic administration the dose was .25 mg every hour. Blood samples were taken periodically before and during drug administration. Complete abortion occurred in 14 of the 16 women given PGF$_2\alpha$ intravenously; the average dose was 60 mg and the length of application averaged 9 hours. By the extraamniotic route, 11 of 16 patients aborted; the average dose was 5 mg and the time of application averaged 20 hours. In patients that aborted, progestin levels at time of abortion were 60% to 65% of initial levels. Rate of decline was immediate when PGF$_2\alpha$ was given intravenously. Progestins stayed constant for approximately 15 hours and then declined when PG was given extraamniotically. In 5 women in corpus luteum phase given 25 mg PGF$_2\alpha$ by intravenous infusion over a 4-hour period, steroid levels showed no consistent pattern. The authors conclude from the results that PGF$_2\alpha$ is not luteolytic in the human at the given dosage and that steroid fluctuations are due to contraction-induced mechanical alterations of placental hormone genesis. (HC)   006486

0875

LERNER, L.J.; CARMINATI, P.; RUBIN, B.L.

Effects of prostaglandins PGE₂and PGF₂α on the adenyl cyclase activity of various segments of the immature rabbit oviduct and uterus.

Proceedings of the Society for Experimental Biology and Medicine 143: 536-539. 1973.

Sections of the oviductal ampulla, isthmus, uterotubal junction, and uterus of rabbits were incubated with prostaglandin and tritium-labelled adenine for 15 minutes. Tritium-labelled cAMP was isolated by column chromatography and assayed using a liquid scintillation counter. PGE₂produced concentration-related increases in cAMP by the uterotubal junction tissue, minimum concentrations being .001 to .01 μg per ml and maximum being 1.0 to 10 μg per ml. All sections of the oviduct had approximately the same degree of cAMP activity; however, PGE₂at 10 μg per ml increased the synthesis of the uterotubal junction, isthmus, and ampulla cAMP 3.0, 2.5 and 1.7 times, respectively. PGF₂α at 10 μg per ml had virtually no effect on cAMP of any section of the oviduct. Theophylline tripled cAMP activity of each oviduct section and also potentiated the action of PGE₂. PGE₂at 10 μg per ml doubled cAMP activity of the uterus; PGF₂α was ineffective. The results indicated a gradient effect for PGE₂, lowest at the ovarian end of the oviduct and maximum at the segment where ova and blastocysts accumulate and are then ejected into the uterus. (HC)   005708

0876

LERNER, L.J.; CARMINATI, P.

Inhibition of prostaglandin synthetase by steroidal and nonsteroidal estrogens and anti-estrogens and by anti-inflammatory agents.

Acta Endocrinologica Suppl. 177: 313. 1973.

Using bull seminal vesicle as a test tissue, the authors found that the best antiinflammatory drug for inhibition of prostaglandin synthetase was indomethacin. Estradiol was almost inactive, estriol was very weak, and diethylstilbestrol was .2 times as active as indomethacin. There was a correlation between inhibition of prostaglandin synthetase and the antiinflammatory activity of several antiestrogens, suggesting a common biological activity, the inhibition of prostaglandin synthetase. (HC)   005732

0877

LEVINE, N.; RINALDO, J.; SCHULTZ, S.G.

Ion transport in guinea pig seminal vesicles.

Federation Proceedings 32: 309Abs. 1973.

Guinea pig seminal vesicles were suspended in a phosphate buffer containing glucose and glutamate. The transmural potential difference (PD) and the short circuit current (SCC) declined during the first hour but then rose, reaching steady state in 2 1/2-3 hours. The spontaneous rise was caused primarily by chloride ion flux. SCC and PD were increased by PGF₂α, theophylline, and dibutyryl cyclic AMP for reasons as yet unclear. (HC)   005707

0878

LEVITT, M.J.

The chemistry of prostaglandins in relation to the female reproductive tract.

In: Josimovich, J.B., ed. Uterine contraction -- side effects of steroidal contraceptives. New York, Wiley-Intrascience, 1973. (Vol. 1 of Problems of Human Reproduction.) p. 179-191.

The steric structure and chemical activity of natural PGs are described. The cyclopentane ring, C13,14 double bond, and possibly the carboxyl group (which is likely to be ionized at pH 7.4) determine a rigid three dimensional relationship between the oxygen functions which is critical for biological activity. If it is assumed that the oxygen groups are held in a coplanar position, it is possible to find parallel oxygen functions in indomethacin, aspirin, sodium salicylate, and polyphloretin phosphate, all PG antagonists, and in the kinins, which mimic PG action. PGs may be measured by bioassay (1 ng sensitivity), spectrophotometer (1 μg), radioimmunoassay (50 pg), enzymatic analysis (50 ng or higher), gas liquid chromatography of O-methyloximes (10 ng), or possibly with greater sensitivity by scanning selected mass spectral lines. In reproductive tissues, PGs are released by seminal vesicles into semen, by secretory endometrium into menstrual fluid, and are found in amniotic fluid, placenta, and cord blood. PGs are known to be low in infertile men's semen, PGFs high in dysmenorrheic women's menstrual fluid, and associated with uterine contraction, but a unified mechanism of action is unknown. (LJG)   006313

0879
LEWIS, R.B.; SCHULMAN, J.D.
   Influence of acetylsalicylic acid, an inhibitor of prostaglandin synthesis, on the duration of human
      gestation and labour.
   Lancet 2(7839): 1159-1161. November 24, 1973.

A study was made of the effects of long-term, high-dose aspirin use on gestation in 103 patients receiving therapeutic aspirin in excess of 3250 mg (50 grains) daily. 52 patients had similar patient history but were not receiving aspirin; another control group had 50 normal patients. Patients taking aspirin had an average gestation over 1 week longer than control groups which was highly significant (p<.025); 42% of the aspirin group had gestations greater than 42 weeks while only 3% of the combined control groups did. The average length of labor in the aspirin group was 12 hours 7 minutes, about 5 hours longer than combined controls and was highly significant (p<.005). Aspirin patients had an average blood loss of 340 ml, about 100 ml more than controls. The results are compatible with the known capacity of aspirin to inhibit synthesis of PGs and suggest that endogenous PGs are important to the duration of gestation and labor.   005507

0880
LIGGINS, G.C.
   Fetal influences on myometrial contractility.
   Clinical Obstetrics and Gynecology 16: 148-165. 1973.

Current knowledge and speculation about fetal control of parturition in sheep, goat, guinea pig, cow, pig, rat, rabbit, rhesus monkey, and human are presented. In sheep, fetal cortisol responding to ACTH induces placental sulfatases and consequent release of unconjugated estradiol and estrone and decreases placental progesterone production, which instigates $PGF_2\alpha$ formation in maternal placenta and myometrium. Goats control labor similarly except that 80% of their progesterone is made by the corpus luteum, and most of their estrogen is free estradiol; $PGF_2\alpha$ is the probable stimulus for luteolysis needed to initiate labor. The fetal stimulus of parturition in guinea pigs is uncertain. It is known that maternal cortisol, a functioning fetal pituitary axis, and altered progesterone metabolic clearance rate are required. Knowledge about control of parturition in cows, pigs, rats, rabbits, and rhesus monkeys is incomplete. In humans, circumstantial evidence from disorders such as anencephaly and adrenal hypoplasia, which cause prolonged gestation, implicates the fetal hypothalamus and adrenal in control of delivery. Estrogen has a permissive role; the 'progesterone block' theory and the oxytocin hypothesis are still unproven; it is unknown whether the high PGE_1, PGE_2, $PGF_1\alpha$ and $PGF_2\alpha$ levels in amniotic fluid, blood, and decidual tissues are the cause or consequence of labor. An extensive discussion of the PG-lysosomal theory of control of menstruation, abortion, and labor is included. (LJG)   005720

0881
LIGGINS, G.C.
   Fetal influences on uterine contractility.
   In: Josimovich, J.B., ed. Uterine contraction--side effects of steroidal contraceptives. New York, Wiley-Interscience, 1973. (Volume 1 of Problems of Human Reproduction) p. 205-218.

The author documents a proposed hypothesis of parturition in lamb. Cortisol from hypothalmic action inhibits progesterone and stimulates estrogen. A simultaneous rise in $PGF_2\alpha$ leads to uterine contractions. The uncertainity of the role of fetal adrenal activity complicates any hypothesis about human parturition, nonetheless a mechanism dominated by fetal control is emerging. (RAS) 006315

0882
LIGGINS, G.C.
   Hormonal interactions in the mechanism of parturition.
   Memoirs of the Society for Endocrinology 20: 119-139. 1973.

This essay on the hormonal factors in parturition concerns chiefly sheep and develops a model involving $PGF_2\alpha$ in initiation labor. The author proposes that fetal cortisol initiates a fall in progesterone secretion, which allows existing estrogen to stimulate $PGF_2\alpha$ synthesis by the decidua. $PGF_2\alpha$ increases sensitivity to constant oxytocin levels and may also have direct oxytocic effects. Finally reflex release of oxytocin augments labor. This model differs from the 'classic' theory in that uterine contractions are not spontaneous, and estrogen and progesterone act on decidua rather than on myometrium. Some of the evidence cited was: maintenance of pregnancy by progesterone can be overruled by exogenous corticoids; progesterone can inhibit $PGF_2\alpha$ synthesis by decidua in vitro, or PG stimulated by stilbestrol in vivo; the threshold oxytocin dose fell under estrogen or $PGF_2\alpha$ treatment independent of progesterone level. Neither oxytocin level nor uterine volume are influential in initiating labor. (LJG)   006290

0883
LIGGINS, G.C.; FAIRCLOUGH, R.J.; GRIEVES, S.A.; KENDALL, J.Z.; KNOX, B.S.
   The mechanism of initiation of parturition in the ewe.
   Recent Progress in Hormone Research 29: 111-159 1973.

The author reviews studies of the mechanism of parturition in the ewe. Cortisol levels, stimulated by hypothalmic activity, act on the placenta to increase estrogen and decrease progesterone content. An associative rise in unconjugated estrogen occurs, initiating a rise of $PGF_2\alpha$ in maternal cotyledons and myometria. Progesterone inhibits release of $PGF_2\alpha$; thus falling levels of progesterone consequent upon the action of fetal cortisol facilitate synthesis and release of $PGF_2\alpha$. Nevertheless, under experimental conditions when progesterone levels are maintained, estrogen can promote both the synthesis of $PGF_2\alpha$ and parturition. The myometrium responds to $PGF_2\alpha$ with heightened sensitivity to oxytocin, which may permit unchanged levels of circulating oxytocin to induce uterine contractions. It is uncertain in the sheep whether $PGF_2\alpha$ itself has a direct oxytocic action. (Authors' modified.) 006334

0884

LIGGINS, G.C.

The physiological role of prostaglandins in parturition.

Journal of Reproduction and Fertility Suppl. 18: 143-150. 1973.

The role of PGs, in particular $PGF_2\alpha$, in the physiological mechanisms initiating parturition in man, in sheep, and in the goat was discussed. PGs are thought to be synthesized during early labor primarily in the decidua but also in the myometrium. Control of the synthesis and release varies with the species. In man it is thought that progesterone may be responsible for developing the capability for PG synthesis, perhaps by stimulating the transformation of stromal cells of the endometrium to decidual cells. It is also suggested that the fetus may take part in the control of PG release by means of feto-maternal interaction at the membranes. In sheep it is proposed that increased fetal secretion of cortisol is followed by a rise in the production of free estrogen which stimulates synthesis of $PGF_2\alpha$ in maternal placenta and myometrium; the effect of the PG on the myometrium could be a direct oxytocic effect or could be a response to circulating oxytocin. In the goat it is thought that $PGF_2\alpha$ may be the luteolytic messenger between the conceptus and the corpus luteum; the adrenals of the fetal goat may influence parturition mechanisms. A model for $PGF_2\alpha$ action on myometrial contractility is hypothesized. (MLH)   006234

0885

LIGGINS, G.C.; GRIEVES, S.A.; McKENZIE, P.A.

The response of prostaglandin $F_2\alpha$ to ACTH administered to one of twin fetal lambs.

New Zealand Medical Journal 78(501): 365. 1973.

The jugular veins of fetuses in sheep with twin pregnancies were catheterized and ACTH was infused into one. Ewes were sacrificed after 72-96 hours and $PGF_2\alpha$ in the placenta, myometrium, and the blood of each uterine vein was measured by immunoassay. $PGF_2\alpha$ from the infused horn rose from 1 ng/ml of plasma to 40 ng/ml. No change in $PGF_2\alpha$ levels were noted in the blood of the non-infused horn. $PGF_2\alpha$ levels were comparable in both infused and non-infused placental and myometrial tissues. It is concluded that the release of $PGF_2\alpha$ resulting from the action of fetal cortisol is a local effect. (JSL)   005995

0886

LINDMARK, G.; MELANDER, S.; NILSSON, B.A.; ZADOR, G.

Inhibition of prostaglandin induced uterine activity in the second trimester of pregnancy by beta-mimetic adrenergic agents.

Prostaglandins 3(4): 481-489. April 1973.

The beta-adrenergic stimulants orciprenaline (Alupent) and ritodrine were each given intravenously after $PGF_2\alpha$ to 4 women pregnant 16-24 weeks to counteract uterine contraction. Orciprenaline was infused at 10, 15, and 20 µg/minute successively at 15-minute intervals after 1 mg doses of $PGF_2\alpha$ had been given hourly by extraamniotic catheter and contractions had begun. Orciprenaline decreased basal tone, frequency, and intensity of uterine contractions (almost completely in 1 case), and increased maternal heart rate. After Orciprenaline, contractions resumed. Ritodrine was infused at 100, 200, and 300 µg/minute at 30-minute intervals after 25 mg intraamniotic $PGF_2\alpha$ had induced contractions. Ritodrine also depressed intensity and frequency of uterine contractions and resulted in tachycardia with increased systolic and decreased diastolic pressure and increased pulse amplitude. Subjectively the women noted palpitations and tachycardia, but no other side effects. No sedatives or analgesics were administered. (LJG)   005669

0887

LIPPERT, T.H.; MODLY, T.

Induction of abortion by the extra-amniotic administration of prostaglandin gels.

Journal of Obstetrics and Gyneacology of the British Commonwealth 80: 1025-1027. November 1973.

PGE_2 and $PGF_2\alpha$ were administered extra-amniotically to terminate pregnancy in 20 patients, age 21-44 years, their period of amenorrhea ranging from 13-17 weeks. PGs were incorporated in a gel base, the final concentration being 2.5 mg PG/ml gel. Initial doses were .5 ml PGE_2 in 14 women and 1-1.5 ml $PGF_2\alpha$ in 6 women; subsequent injections were given at 2-3 hourly intervals. All cases were successful, the mean induction to abortion interval being 11 hours 35 minutes for PGE_2 and 17 hours 38 minutes for $PGF_2\alpha$. The mean amount of PG needed for abortion was 4.6 mg PGE_2 and 17.7 mg $PGF_2\alpha$. The placenta was expelled spontaneously in all patients treated with PGE_2 and in 2 of the 6 patients treated with $PGF_2\alpha$. Painful uterine contractions were reported as the most distressing side effect; vomiting and fever were reported in selective cases among PGE_2-treated patients. The extra-amniotic administration of the PG gel was considered favorable due to the relatively short induction-abortion interval, the possibility of administering high doses without undesirable side effects, and the stability of the PGs in the gel base. (MLH)   005823

0888

LIPPERT, T.H.; BARTSCHI, R.; LUTHI, A.

Therapeutic abortion by intrauterine infusion of prostaglandin $F_2\alpha$.

Archiv fur Gynaekologie 213(3): 197-201. 1973.

A continuous intrauterine infusion of $PGF_2\alpha$, 12.5 μg/minute the first hour increased hourly to 100 μg/minute, was given to 12 patients, 12-16 weeks pregnant. The mean total dose used was 45 mg. The induction-abortion interval averaged 22 hours 35 minutes for primiparas and 11 hours 50 minutes for multiparas; abortion was complete in 2 and incomplete in 10 patients. All patients complained of very painful uterine contractions. (HC)   005750

0889

LIPSETT, M.B.

The prostaglandins: progress in research. [Book review]

Archives of Internal Medicine 132(2): 300. August 1973.

This collection of review articles aims to combine hypotheses and data useful to researchers who want to learn about other fields of PG research. The selection includes items on reproduction by Karim, termed optomistic by the reviewer, and on pharmacology of the cardiovascular, renal, gastrointestinal, and pulmonary systems. The reviewer found fault with an inappropriate chemical chapter, a nonproductive chapter on platelet aggregation, and a methodologic chapter which omitted radioimmunoassay. (LJG)   006383

0890

LOUIS, T.M.; STELLFLUG, J.N.; SEGUIN, B.E.; HAFS, H.D.

Disappearance of injected $PGF_2\alpha$ in heifers.

Journal of Animal Science 37(1): 319-320. Abstract 359. 1973

Plasma PGF and progesterone were measured by radioimmunoassay in 6 heifers after 30 mg $PGF_2\alpha$ Tham salt im, in 4 heifers after 2 injections of 15 mg $PGF_2\alpha$ at 6-hour intervals, and in 6 heifers after 60 mg $PGF_2\alpha$. Plasma samples were taken from the jugular at 10-minute intervals, then at 1, 1.5, 2, 4,

6, and 12 hours. 16 controls averaged .8 ng/ml PGF. After 30 mg $PGF_2\alpha$, mean PGF was 4.9 ng/ml 10-30 minutes later and reached .9 ng in 1.5 hours. After 15 mg $PGF_2\alpha$, PGF averaged 4.6 ng/ml 10-30 minutes later and 2.0 ng/ml 6 hours after the second $PGF_2\alpha$ in injection. Then PGF ranged from 4.7-5.9 ng/ml for 1 hour and reached .8 ng/ml 12 hours after the first injection. After 60 mg $PGF_2\alpha$, PGF averaged 4.4 ng/ml 10-30 minutes later, 1.6 at 2 hours but rose to 4.6 at 12 hours. Progesterone decreased from mean 3.9 ng/ml to 2.1 at 6 hours and 1.7 at 12 hours, with no significant differences among the 3 groups. (LJG)   006025

0891

LOUIS, T.M.; HAFS, H.D.; SEGUIN, B.E.
Progesterone, LH, estrus and ovulation after prostaglandin $F_2\alpha$ in heifers.
Proceedings of the Society for Experimental Biology and Medicine 143(1): 152-155. 1973.

30 mg $PGF_2\alpha$ was given intramuscularly to 5 heifers during diestrus (8-14 days after estrus) and to 6 heifers during metestrus (Day 3). Another 6 heifers received $PGF_2\alpha$ intravaginally during diestrus. Signs of behavioral estrus and corpus luteum diameter were monitored. Serum levels of LH and progesterone were estimated by radioimmunoassay. In those animals given $PGF_2\alpha$ intramuscularly of intravaginally during diestrus, serum progesterone levels fell from 4-5 ng/ml to 1 ng/ml 48 hours later. $PGF_2\alpha$ given during metestrus had no effect on corpus luteum function as progesterone levels rose continuously. Changes in corpus luteum diameter were similar to those noted for progesterone. Intervals to onset of estrus, peak LH, and ovulation were longer after intravaginal than after intramuscular administration of $PGF_2\alpha$. The authors concluded that $PGF_2\alpha$ is luteolytic when given intramuscularly during diestrus and that this route is more effective than the intravaginal route. (HC)
005702

0892

LUBICZ-NAWROCKI, C.M.; SAKSENA, S.K.; CHANG, M.C.
The effect of prostaglandins E_1 and $F_2\alpha$ on the fertilizing ability of hamster spermatozoa.
Journal of Reproduction and Fertility 35(3): 557-559. 1973.

Male hamsters were given 50 μg of PGE_1 or $PGF_2\alpha$ subcutaneously daily in 2 divided doses for 10 days. Animals were then paired overnight with females in estrous. The degree of fertilization was determined by sacrificing the females 40 hours later and counting the number of eggs containing sperm. Plasma testosterone levels were estimated in the males 12 hours after the last PG injection. Fertilizing ability was 100% in all animals, but PGE_1 caused a significant decrease in plasma testosterone suggesting that fertilizing ability is not androgen-dependent. (HC)   005762

0893

LUNDQUIST, L.E.; DICKEY, J.F.; HENRICKS, D.M.; HILL, J.R.
Prostaglandin treated bovine corpora lutea cells in culture.
Journal of Dairy Science 56(2): 306-307. 1973.

Cells from corpora lutea obtained from cows 8 days postestrous were incubated with stearic acid, LH, and/or $PGF_2\alpha$. Histochemical assays revealed more cholesterol and acidic lipids in cells cultured with $PGF_2\alpha$. Electron microscopic evaluation revealed a marked vacuolization of the cytoplasm of cells incubated with $PGF_2\alpha$. (HC)   006372

0894

LYNEHAM, R.C.; McLEOD, J.G.; SMITH, I.D.; LOW, P.A.; SHEARMAN, R.P.; KORDA, A.R.
Convulsions and electroencephalogram abnormalities after intra-amniotic prostaglandin $F_2\alpha$.
Lancet 2(7836): 1003-1005. November 3, 1973.

5 out of 320 women having midtrimester abortion and 2 of 7 patients with epileptic history had convulsions during intraamniotic $PGF_2\alpha$. 4 out of 8 had abnormalities in EEG after $PGF_2\alpha$ administration. All patients were pregnant 14-20 weeks and had been given 30 mg $PGF_2\alpha$ transabdominally, with 15 mg 24 and 36 hours later if needed. The seizures occurred 7-25 hours after PG and from 11 hours before abortion to 1 hour after. In the EEG study, 7 out of 8 outpatient pretreatment recordings were normal. 4 were normal 4 hours after PG: 1 41-year-old woman had generalized 4-7 Hz activity occasionally slowing to 2-3 Hz over both fronto-temporal areas and frequent bilateral spikes. 2 women had 1-sided abnormalities and 1 had generalized asymetrical theta waves. (LJG)   005931

0895

MacKENZIE, I.Z.; HILLIER, K.; EMBREY, M.P.
Convulsions and prostaglandin induced abortion.
Lancet 2(7841): 1323. 1973.

In Oxford, England (1970-73) 615 pregnancies were terminated using PGs. Intravenous infusion of oxytocin was used in 50% of cases. The routes of administration and dose regimen were the following: 1) intraamniotic $PGF_2\alpha$, 19 cases, 25 mg plus 25 mg at 24 hours; PGE_2, 53 cases, 10 mg plus 10 mg at 6 hours; 2) extraamniotic $PGF_2\alpha$, 111 cases, 750 µg 2-hourly; PGE_2, 374 cases, 200 µg 2-hourly; 3) intravenous $PGF_2\alpha$, 10 cases, 25-200 µg per minute; PGE_2, 48 cases, 2.5-20 µg per minute. 7 patients had a history of epileptic seizures but none of the 615 had convulsions during the administration of PGs. Abortion occurred in 7-35.5 hours. Reference is made to other reports of PG induced abortion where convulsions did occur, and it was suggested that this may be caused by circulating PG metabolites. Further investigation is suggested. (IC)   005508

0896

MacVICAR, J.
Acceleration and augmentation of labour.
Scottish Medical Journal 18(6): 201-214. 1973.

This review gives historical perspectives and detailed directions for accelerating labor, meaning amniotomy to reduce induction to delivery time, and for augmenting labor, meaning giving oxytocins to shorten the length of labor already begun spontaneously. Factors affecting outcome of amniotomy include ripeness of the cervix, parity, age, timing, and method of amniotomy. It was once hoped that PGs could take the place of amniotomy, but induced labor can be enhanced by forewater rupture followed by PGs and oxytocin. With proper timing, dosage, iv route, and use of a Cardiff pump, labor can be completed the same day. Augmentation is appropriate for primigravidae or women having second or third babies, and for many obstetric problems often thought to contraindicate oxytocics. Cephalopelvic disproportion, cesarean section, and hypertonic uterine activity are compatible with augmentation, if carefully managed. The trend is toward increasing use of induced labor. (LJG)
005751

0897
McCRACKEN, J.A.
Comment.
Prostaglandins 3(5): 696-701. May 1973.

Evidence is cited which supports the transport of a luteolytic substance from the uterine vein to the ovarian artery by means of a countercurrent mechanism. Experiments mentioned included 1) the use of labeled $PGF_2\alpha$, 2) injection of cold $PGF_2\alpha$, and 3) infusion into autotransplanted ovary. The author suggested that the recent research by Lammond and Drost, which negated the countercurrent concept, 1) lacked adequate controls, 2) should have investigated ligation of venous connnections, and 3) should have considered variations in the anatomy of animals tested. (RAS)   005981

0898
McCRACKEN, J.A.; BARCIKOWSKI, B.; CARLSON, J.C.; GREEN, K.; SAMUELSSON, B.
The physiological role of prostaglandin $F_2\alpha$ in corpus luteum regression.
Advances in the Biosciences 9: 599-624. 1973.

The role of PGs in the initiation of corpus luteum regression in sheep was studied with either the ovary alone or both the ovary and uterus autotransplanted with vascular anastomoses to jugulocarotid skin loops in the neck. The advantage of this model is easy access to the arterial and venous circulation in the transplanted organs of the conscious undisturbed sheep for long periods. The ovary and uterus in the sheep had to be contiguous for normal cyclic luteolysis. By cross-circulating blood, a luteolytic factor was found in uterine vein blood at the time of luteal regression which was mimicked by infusion of $PGF_2\alpha$. The transport of the luteolytic factor took place via the circulation and probably by countercurrent transfer. Confirmation of $PGF_2\alpha$ as a uterine luteolysin in a sheep came from identification and measurement of $PGF_2\alpha$ in uterine vein blood by gas-liquid chromatography and mass spectrometry. By introducing 25 μg/hour $PGF_2\alpha$ into the uterine vein, premature luteal regression was induced consistently, confirming not only the local transport of $PGF_2\alpha$ from uterus to ovary, but also the role of $PGF_2\alpha$ as a luteolytic hormone in sheep. The application of PGs as luteolytic agents in fertility control in sheep, the hormonal control of the endogenous release of PGs from the ovine uterus, an apparent antagonism between gonadotropins and $PGF_2\alpha$ at the ovarian level, and presumptive evidence for impaired microcirculation as the mechanism of luteolysis are discussed.
(Authors' modified)   006732

0899
McCRACKEN, J.A.; BAIRD, D.T.; CARLSON, J.C.; GODING, J.R.; BARCIKOWSKI, B.
The role of prostaglandins in luteal regression.
Journal of Reproduction and Fertility Suppl. 18: 133-142. 1973.

The role of the uterus in luteal regression, the route taken by $PGF_2\alpha$, the role of hormones, and the mechanism of luteolysis are reviewed. Hysterectomy shortens the life of the corpus luteum (CL) in guinea pig, sheep, cow, pig, pseudopregnant rabbit, rat, and hamster, indicating the existence of a uterine luteolysin, defined as a substance that diminishes progesterone secretion, CL life span, or both. The route of this luteolytic agent is via the uterine vein directly to the ovarian artery, not via systemic circulation (as shown by selective vascular ligation and by transplantation of ovaries in sheep). $PGF_2\alpha$ infused into the ovary at 2.5 μg/hour mimicks luteal regression; tritiated $PGF_2\alpha$ was detected in ovarian arterial blood when infused into the underlying uterine vein, suggesting that countercurrent flow is the means of transport. $PGF_2\alpha$ was then indentified and measured in uterine vein blood of individual sheep by gas liquid chromatography and mass spectrometry. $PGF_2\alpha$ rose to 25 ng/ml on Cycle Day 15 from less than 2 ng/ml and was associated with with a fall in progesterone, the

best indicator of luteal regression. Premature luteal regression could be induced by infusing $PGF_2\alpha$ at 25 μg/hour into the uterine vein. The hormonal milieu required for luteal regression is progesterone priming followed by estrogen surges. Progesterone can increase uterine $PGF_2\alpha$ content, and estradiol infusion into the uterine artery increases $PGF_2\alpha$ release tenfold into the uterine vein. The means by which pregnancy overrules luteolysis is unknown, although prolactin can overcome luteolysis by $PGF_2\alpha$. The mode of action of $PGF_2\alpha$ as a luteolysin is hypothetical: local redistribution of blood flow within the ovary and inhibition of cholesterol synthetase have been suggested. The fact that $PGF_2\alpha$ is luteotrophic in some in vitro systems makes the phenomenon difficult to completely understand. (LJG)
006538

0900
MAGEE, W.E.; ARMOUR, S.B.; MILLER, O.V.
Absorption of prostaglandins by the intestine and vagina of the rat.
Biochimica et Biophysica Acta 306: 270-282. 1973.

In rats, 1 cannula was inserted through the stomach into the duodenum, the other above the ileocecal valve. Labelled PG solutions were placed in the upper cannula, samples were periodically removed from the lower. Blood, urine and tissue samples were obtained at the end of the experiment. Separation and identification of PG's was done by extraction, thin-layer chromatography and radioactivity counting. PGE_2 was absorbed rapidly with a half absorption time fo 40 minutes. 2.5 to 3.0% of the administered dose appeared in the serum in 30 to 60 minutes, nearly all of which was in the form of a 15-keto metabolite. The half absorption time value for $PGF_2\alpha$ was 65 minutes. Serum levels were .04 to .2% of the dose for $PGF_2\alpha$ and too low to be processed for the 15-methyl analog. The urine contained less than 1% of the dose, 1.5% was collected from the bile duct. Absorption of 15-methyl $PGF_2\alpha$ methyl ester was biphasic, the rapid initial phase ceasing as the ester became cleaved to the free acid. Up to 7% of the dose appeared in the serum in 30 minutes, half being 15-methyl $PGF_2\alpha$, none was the methyl ester. Absorption from the large intestine was more rapid for the 15-methyl ester than for $PGF_2\alpha$; nearly all material in the serum was, however, the free acid. Rate of absorption of the methyl ester slowed as the ester group was hydrolyzed. Placed into the vagina, over 1.5% of the dose of 15-methyl $PGF_2\alpha$ methyl ester appeared in the serum and 7.5% in the urine in 30 minutes. 20% of the material in the serum was 15-methyl $PGF_2\alpha$; there was no methyl ester present. In a liver perfusion experiment, 15-methyl $PGF_2\alpha$ methyl ester was rapidly acted upon by esterase and then converted to polar compounds. The authors conclude that although the primary step in metabolism of PG is via 15-dehydrogenation, the liver extensively degrades the compounds to polar metabolites. Barriers to successful oral dosage forms are not in their rate of uptake, but rather in the extensive metabolism, especially 15-dehydrogenation, occurring during absorption. (HC)   005705

0901
MAGLIULO, A.
Prostaglandins.
Journal of Chemical Education 50(9): 602-603. September 1973.

Prostaglandin structure, synthesis, metabolism, extraction, isolation, and physiological activity are summarized. Intravenous prostaglandins can induce labor or terminate early or late pregnancies. Prostaglandins can modify the action of cAMP and are therefore implicated in lipid metabolism and tissue innervation. (RAS)   005985

0902

MAKINO, T.

Study on the intracellular mechanism of LH release in the anterior pituitary.
American Journal of Obstetrics and Gynecology 115: 606-614. March 1973.

Radioimmunoassay of cAMP in rat anterior pituitary extract was carried out to investigate hormone release and alteration of intracellular cAMP content in response to the synthetic gonadotrophin-releasing factor Gn-Rf and other releasing or inhibiting factors, including PGE_1 and $PGF_2\alpha$. .5 and 20 μg PGE_1 and 30 and 300 μg $PGF_2\alpha$ markedly increased cAMP but not LH and FSH release at the high concentrations. Reduced cAMP but enhanced LH and FSH release were seen at the low concentrations. Indomethacin and 7-oxa-13 prostynoic acid did not change basal LH and FSH secretion. Gn-Rf showed an additive effect on LH and FSH release and cAMP formation stimulated by PGE_1. (RMS)   006436

0903

MALOFIEJEW, M.; LAUDANSKI, T.; KOSTRUZEWSKA, A.

Wplyw prostaglandyna $A_2(PGA_2)$ na zjawiska elektrobiologiczne W izolowanym myometrium szczura.
[The effect of A_2prostaglandin (PGA_2) upon the electrobiological phenomenons in isolated myometrium of rat]
Ginekologia Polska 44(3): 255-229. March 1973.

PGA_2 in increasing doses stimulated or increased the existing electrobiological and mechanical activity of isolated myometrium in rat. (Authors' modified)   006373

0904

MARION, G.B.; RACHOW, T.E.

The histological effects of prostaglandin $F_2\alpha$ on bovine ovarian structures.
Biology of Reproduction 9: 105-106. 1973.

5 normally cycling heifers were given 6 mg $PGF_2\alpha$ directly into the uterus by catheter on Day 9 of the estrous cycle to determine its effect on ovarian vesicular follicles. The heifers and 5 untreated controls were killed on the day of estrus, an average of 2.8 days after treatment. The total number of follicles and distribution of vesicular follicles by size were similar in the 2 groups. The percent of normal follicles was significantly higher in heifers treated with $PGF_2\alpha$ ($p < .01$). $PGF_2\alpha$ destroyed pretreatment luteal tissue, which appeared smaller and contained vacuolated cells and pyknotic nuclei. (LJG)
005684

0905

MEMON, G.N.

Effects of intratesticular injections of prostaglandins on the testes and accessory sex glands of rats.
Contraception 8(4): 361-370. October 1973.

Adult male rats were given 1.3 mg $PGF_2\alpha$ or 1.0 mg PGE_2 by injection into the testes either bilaterally for 1 or 3 days or unilaterally for 1 day. Bilaterally for 3 days, either type of PG significantly reduced the weight of testes, seminal vesicles, and ventral prostate. Single bilateral (or unilateral) injections of the PGs also reduced testicular weight significantly (in the treated organ only). Plasma levels of testosterone were significantly lower while LH levels were significantly higher following 3 days of bilateral injections of either $PGF_2\alpha$ or PGE_2. The author concluded that the PG effect was local rather than systemic, perhaps an alteration in testicular hemodynamics which could cause subsequent atrophy of the testes. (HC)   006004

0906
MICHAEL, C.M.
Prostaglandins in swine testes.
Lipids 8(2): 92-93. 1973.

Homogenates of testicular tissue from 3 swine were extracted and prostaglandin content assayed by column and thin layer chromatography followed by spectrofluorometry. PGE's average was 1.5 μg/g tissue while PGF's averaged .7 μg/g. (HC)   005734

0907
MIDAK, E.
The action of prostaglandin E_1(PGE$_1$) on human pregnant (at the time of labour) and non-pregnant uterus in vitro and modification of this action by adrenergic receptor blocking agents.
Acta Physiologica Polonica 24(4): 537-550. 1973.

Isotonic contractions of myometrium from 10 term cesarean sections and 26 hysterectomies were recorded in response to 10-100 ng per ml PGE$_2$, phentolamine (alpha adrenergic blocker) and LB-46 (beta adrenergic blocker). 124 fragments were tested usually within 1 day, but sometimes after 2 days of refrigeration. 20 x 3 x 2 mm pieces were bathed in 20 ml Krebs solution at 37 degrees C, weighted with 3-4 gm for 3-4 hours, and .5-1 gm after optimal spontaneous contractions were obtained. Pregnant transverse myometrial strips contracted with a mean frequency of 7.5 minutes; only 1 of 4 was inhibited by 50 ng PGE$_1$per ml, and 3 of 4 by 100 ng per ml. In proliferative phase, longitudinal strips of non-pregnant uterus were inhibited 36% at 25 ng, 59% at 50 ng, and 82% at 100 ng PGE$_1$per ml. At ovulation activity was decreased 25% by 10 ng, 38% by 25 ng, a very marked decrease at 50 ng, and completely inhibited at 100 ng PGE$_1$per ml. Secretory myometrium was inhibited 20% by 10 ng, 50% by 25 ng, 80% by 50 ng, 100% by 100 ng PGE$_1$per ml. After menopause, 25 ng decreased activity 40%, 50 ng 68%, and 100 ng inhibited activity 97% (% mean amplitude). Inhibition by PGE$_1$ was lessened after 10 minutes incubation with 10^{-7}or 10^{-6}gm per ml phentolamine in 4 of 7 experiments. LB-46 (4[-2-hydroxy-3-isopropylaminopropoxy]-indole, Sandoz) had no effect. (LJG) 006962

0908
MIDWINTER, A.; SHEPHERD, A.; BOWEN, M.
Continuous extra-amniotic prostaglandin $E_2$for therapeutic termination and the effectiveness of various infusion rates and dosages.
Journal of Obstetrics and Gynaecology of the British Commonwealth 80: 371-373. April 1973.

PGE$_2$was infused continuously in the extraamniotic space at 46.5, 66.5, 93, 133.5 or 333.5 μg per hour to induce abortion in 91 women 10-20 weeks pregnant for a dose-effect study. The study group included all healthy patients aged 15-40 years who had no more than 4 term pregnancies or 1 miscarriage. PGE$_2$was infused at 25 or 50 μg per ml with a 50 ml syringe using Palmer or Meltec infusion pumps. Patients received nitrous oxide and pethidine; pulse and respiration rates were recorded hourly and blood pressure and temperature measured every 4 hours. At 46.5 μg per hour 13 out of 20 aborted within 36 hours. Of 68 patients given 66.5 to 133.5 μg per hour, 61 aborted within 36 hours; no significant relationships appeared between induction interval and age, parity, or gestational age in each of the 3 dose groups. The 333.5 μg per hour dose produced persistent diarrhea, vomiting, and continuous pain in 2 patients, and increased blood pressure from 120/70 to 170/100 in 1. Thus 66.5 μg per hour was judged the optimum dose because only 1 syringe was needed and the least PGE$_1$was used. (LJG)   005685

0909
MILLER, A.W.F.
The use of prostaglandins for missed abortion, foetal death in utero and hydatidiform mole.
In: Jacomb, R.G., ed. The use of prostaglandin E$_2$ and F$_2\alpha$ in obstetrics and gynecology. (Proceedings of the Upjohn Prostaglandin Symposium, London, September 21, 1972.) Miami, Florida, Symposia Specialists, 1973. p. 63-67.

Other authors have reported that infusing PGE$_2$5 μg/minute intravenously for 8.5-11.5 hours was successful in all 17 cases of 'missed abortions,' i.e., fetal death in utero before 28 weeks. Similarly, infused at a rate of .5-2.5 μg/minute for 8.5-11.5 hours, the drug was successful in 43 of 48 cases of intrauterine death after 28 weeks. The author, using PGE$_2$2.5-5 μg/minute, successfully terminated 7 of 10 cases of missed abortions and all 7 cases of intrauterine death. Average dose used was 3.4-5 mg, infusion time was 12.5-14.7 hours. Pyrexia, nausea, vomiting, and diarrhea were noted. The author concluded that this method compares favorably with alternative forms of treatment. (HC) 006955

0910
MOCSARY, P.; CSAPO, A.I.
'Delayed menstruation' induced by prostaglandin in pregnant patients.
Lancet 2(7830): 683. September 1973.

After complete abortion was induced in the form of 'delayed menstruation' in 20 of 22 early pregnant (11 ± days missed period) sedated patients with 5 mg PGF$_2\alpha$ an additional study using 65 early pregnant (11 ± days) volunteers was made. 50 received 5 mg PGF$_2\alpha$ and 15 received 1 mg PGE$_2$. PGs were delivered by catheter near the fundus. Patients were sedated with diazepam 20 mg, pethidine 100 mg and atropine .4 mg. PGF$_2\alpha$ was injected at 250 μg per minute for 20 minutes, PGE$_2$, at 50 μg per minute for 20 minutes. A single dose of 5 mg PGF$_2\alpha$ successfully provoked abortion in all instances with bleeding beginning 4.6 ± .5 hours after PG impact. By Day 10 Pregnosticon test was negative in all 50. Results with PGE$_2$were similar. Bleeding began at 6.2 ± .4 hours after PG impact and by Day 10 pregnancy tests were negative in all 15. The only side effect was vomiting in 13 of the 50 PGF$_2\alpha$ treated and in 5 of the 15 PGE$_2$patients. There were no complications during follow-up. Patients were hospitalized over night but it was suggested that the procedure could be managed on an out-patient basis as bleeding resembles menstruation. (IC)   006451

0911
MOREWOOD, G.A.
Therapeutic abortion employing the synergistic action of extraamniotic prostaglandin E$_2$and an intravenous infusion of oxytocin.
Journal of Obstetrics and Gynaecology of the British Commonwealth 80: 473-475. May 1973.

Termination of second trimester pregnancy was performed in 38 patients by extraamniotic administration of PGE$_2$intravenous oxytocin. 36 pregnancies were terminated within 24 hours and all within 32 hours. In no case were membranes ruptured inadvertently during introduction of the Foley catheter. In 23 patients the abortion was incomplete, and surgical evacuation of the uterus was performed. At 6 weeks postabortal follow-up no patients had experienced heavy vaginal bleeding after discharge from the hospital. Side effects included vomiting in 15 patients (40%) and diarrhea in 6. Blood loss was less than 300 ml in 33 patients, between 300-600 ml in 4, and over 1000 ml in 1. A brief review of other extraamniotic PGE_2 studies was referred to. The 40% vomiting incidence in this study was higher than others, but blood loss was similar. (IC)   006431

0912
MORIN, R.J.
Effects of prostaglandins E_1 and $F_2\alpha$ and of polyphloretin phosphate on cholesterol esterification by rabbit ovarian subcellular fractions.
Research Communications in Chemical Pathology and Pharmacology 6(1): 195-206. July 1973.

PGE_1, $PGF_2\alpha$, and polyphloretin phosphate (PPP) were included in assays of microsomal ATP-CoA dependent and mitochondrial cofactor independent acyltransferase from corpus luteum and interstitial portions of 8-day pregnant New Zealand rabbits. The mitochondrial enzyme incubation contained mitochondria precipitate and .5 ml citrate-phosphate buffer pH 4.4, with .05 μc/10 nmoles cholesterol-4-14C or palmitic acid-1-14C; microsomal enzyme incubations contained .5 ml microsomes, 12 μmoles ATP, .6 μmoles CoA, 10 μmoles MgCl2, and substrates as above in .5 ml .1 M phosphate buffer pH 7.1. Products were determined by thin layer chromatography and scintillation counting, or fluorometrically in cholesterol ester hydrolase assays. PGE_1 10^{-5} inhibited both mitochondrial enzymes 70-75%, and $PGF_2\alpha$ less so, at .01 mM. Neither PG inhibited at .001 mM or less. PPP .1 μg per ml decreased esterification to 80% of control, at 10 μg per ml to less than 50%, and to 100 μg per ml inhibited completely. Microsomal esterification was inhibited slightly by PGE_1, and negligibly by $PGF_2\alpha$, both at .01 mM. PPP at 10 mg per ml lowered esterification to about 60% of controls. In both tissues PGs and PPP combined produced additive suppression of both enzymes. Cholesteryl ester hydrolase activity was not inhibited by PGs or PPP. (LJG)   005712

0913
MOSLER, K.H.; CZEKANOWSKI, R.; DORNHOFER, W.
Comparative studies of the effect of oxytocin and prostaglandin $F_2\alpha$ in the uterus.
Advances in the Biosciences 9: 751-760. 1973.

A comparative study of prostaglandin $F_2\alpha$ and oxytocin in the uterus, including side effects on the heart, circulation, and intestine, was performed. In the nonpregnant uterus, prostaglandin $F_2\alpha$ and oxytocin only had a tocergic effect when highly concentrated. The side effects of prostaglandin on the heart frequency, changes of ECG, blood pressure, and intestinal motility were stronger than in similar uterine effects with oxytocin. When applying lower effective doses in the term uterus, the side effects decrease intravenous injections only resulted in vessel reactions (thrombophlebitis) when high concentrations were used. Special safety measures for its prevention were mentioned. The comparative in vivo and in vitro study in man and animal included the total spectrum of effects and side effects of prostaglandin $F_2\alpha$ fraction and its comparison to the effect of oxytocin, in concentrations increasing uterine activity. (Authors' modified)   006493

0914
MOSLER, K.H.; DORNHOFER, W.
The inhibition of premature labor by beta-adrenergic sympathomimetics and prostaglandin-antagonists in managing premature onset of labor.
Naunyn-Schmiedebergs Archives of Pharmacology Suppl. 277: R48. 1973.

A beta-adrenergic sympathomimetic and a spasmolytic in the acute phase supplemented for long term treatment with acetylsalicylic acid to decrease excessive production of PG in the uterus were utilized for inhibition of premature labor. (HC)   005718

0915
MOSLER, K.H.
Vergleichende Untersuchungen ueber die Wirkung von Oxytocin und Prostaglandin $F_2\alpha$ auf den Uterus. [Comparative studies on the effect of oxytocin and prostaglandin $F_2\alpha$ on the uterus.] Archiv fuer Gynaekologie 214(1-4): 304-307. 1973.

Several studies comparing effects of $PGF_2\alpha$ and oxytocin on rat and human uteri are presented. There was no difference between the effects of oxytocin and $PGF_2\alpha$ on electric activity or contractility of rat uterus, but only $PGF_2\alpha$ increased the heart rate. In an experiment recording contractions in rat uterus and small intestine, $PGF_2\alpha$ doses ineffective on the uterus did increase intestinal motility. In perfused nongravid human uteri, $PGF_2\alpha$ at 10^{-6}gm/ml and oxytocin at .003 I.E./ml each increased uterine contraction, and $PGF_2\alpha$ also increased perfusion pressure and contraction of the uterine artery. In cases of fetal death, both have a slight effect on uterine activity, and the uterine sensitivity to oxytocin increases with the PG level. In 1 case of a term pregnancy, oxytocin caused deceleration of the fetal heartbeat, and $PGF_2\alpha$, 50-100 μg/ml for 10 minutes brought on bradycardia. Th1165a, a beta-adrenergic stimulator, was an effective antidote at 40 μg iv. PGs should only be given under intensive care. (LJG)   006078

0916
NAFTOLIN, F.
Prostaglandins: basic reproductive consideration and new clinical applications.
In: 'Prostaglandins and Cyclic AMP,' eds. R.H. Kahn and W.E.M. Lands, pp. 157-182, Academic Press, New York. 1973.

Although a correlation has been made between low levels of PGE and male infertility, additional research is needed. Prostaglandins are effective oxytocic agents when an early delivery is required, i.e., dead fetus, trophoblastic disease. Various routes of administration for abortion were compared; intravenous and intravaginal have a poor success rate with troublesome side effects; extraovular and intrauterine have a 80% success rate with minimal side effects. (RAS)   005545

0917
NAISMITH, W.C.M.K.; BARR, W.; MacVICAR, J.
Comparison of intravenous prostaglandins $F_2\alpha$ and E_2 with intravenous oxytocin in the induction of labour.
Journal of Obstetrics and Gynaecology of the British Commonwealth 80: 531-535. June 1973.

Intravenous $PGF_2\alpha$, PGE_2, and oxytocin (Syntocinon) were compared as inducers of labor in 40 term pregnant primigravidae matched in age, height, weight, Bishop inducibility score. All were 7-15 days past expected delivery. Initial doses were 2.66 mU oxytocin per minute, 2.5 μg $PGF_2\alpha$ per minute, or .25 μg PGE_2 per minute; oxytocin dose was doubled every 15 minutes or PGs every 30 minutes, to maxium of 340 mU oxytocin, 80 μg $PGF_2\alpha$ or 8 μg PGE_2 per minute. Amnitomy was postponed until labor was established. 19 of 20 given oxytocin delivered within a mean interval of 10 hours; 1 required cesarean section. 2 of 10 $PGF_2\alpha$ patients and 3 of 10 PGE_2 patients required oxytocin for undilated cervix after 12 hours, and mean interval for the remaining 15 was 13 1/2 hours. 9 of 20 oxytocin patients, 2 of 10 $PGF_2\alpha$ patients and 5 of 10 PGE_2 patients delivered spontaneously; 23 who were assisted had epidural block. No significant differences in side effects or Apgar scores were observed. (LJG)　005465

0918

NAKANO, J.; KOSS, M.C.
Pathophysiologic roles of prostaglandins and the action of aspirin-like drugs.
Southern Medical Journal 66(6): 709-723. June 1973.

The biochemistry of PGs, pharmacology of PGs, the nonsteroidal antiinflammatory drugs, and the role of PGs in clinical disorders are extensively reviewed. The biochemistry section includes the relationship of PGs to cyclic AMP and to autonomic transmission. The pathologic conditions covered are inflammation, trauma, endotoxin shock, pulmonary embolism, diarrhea associated with cholera and medullary carcinoma, dysmenorrhea and other disorders of uterine motility, IUD contraception, luteolysis, male sterility, renal and essential hypertension, asthma, neoplasm, platelet aggregation, polycythemia, hemolytic and sickle cell anemia, essential fatty acid deficiency and peptic ulcer. (LJG)
006272

0919

NANCARROW, C.D.; BUCKMASTER, J.; CHAMLEY, W.; COX, R.I.; CUMMING, I.A.; CUMMINS, L.; DRINAN, J.P.; FINDLAY, J.K.; GODING, J.R.; RESTALL, B.J.; SCHNEIDER, W.; THORBURN, G.D.
Hormonal changes around oestrus in the cow.
Journal of Reproduction and Fertility 32: 320-321. 1973.

Blood was sampled from cows by catheter every 2 hours for 9 days from the jugular vein for luteinizing hormone (LH), and from the uteroovarian vein for progesterone, estradiol-17β, and PGF activity. There were active estrogen secretion, PGF surges and a decline in progesterone before Day 0 (day of LH peak). A PGF peak was detected after the LH peak. Then estrone was secreted at 400-900 pg per ml on Day 2-3, while estradiol was negligible (Day 0-3). On Day 3 PGF, estrogen, and progesterone began to rise; progesterone rose from 1 ng per ml to 5 ng on Day 7 and lasted until Day 16. (LJG)
005648

0920

NATHANIELSZ, P.W.; ABEL, M.
Initiation of parturition in the rabbit by maternal and foetal administration of cortisol: effect of rate and duration of administration: suppression of delivery by progesterone.
Journal of Endocrinology 57: 47-54. 1973.

Continuous infusion of cortisol at rates from .31 to 1.24 mg/hour into pregnant rabbits from Day 21 of pregnancy onward induced delivery in approximately 72 hours. Similar rates of infusion into the fetal amniotic sac induced delivery with a shorter latency. The authors noted Liggins' work, which showed prostaglandin production to be a secondary effect of cortisol infusion in sheep. Maternal infusion of 1.24 mg/hour cortisol produced parturition providing it was continued for at least 24 hours. If after this period the infusion was terminated, delivery still occurred after a further 48 hours (i.e., 72 hours in all). 10 mg progesterone was administered im on Day 21 and on subsequent Days 22-27. On Day 28, 5 mg progesterone and on Day 29, 2.5 mg progesterone were injected. Progesterone blocked the effect of .31 mg/hour cortisol from operation until delivery. (Authors' modified)　006091

0921
NEGRI, L.; ERSPAMER, G.F.; PICCINELLI, D.
    The action of polypeptides, amines and prostaglandins on isolated smooth muscle preparations of
        seminal vesicles and deferent ducts.
    Archives Internationales de Pharmacodynamie et de Therapie 205(1): 23-28. 1973.

Acetylcholine at threshold concentrations of .1 to .5 μg per ml produced contractions of the isolated
hamster seminal vesicle; all other compounds tested were inactive in concentrations up to 1 μg per ml.
Rat and guinea pig seminal vesicles were virtually insensitive to vasopressin, oxytocin and $PGF_1\alpha$.
Bradykinin and acetylcholine were the most active on rat tissue, histamine and acetylcholine were the
most potent on guinea pig tissue. PGE_1 had a stimulant effect equal in degree on both tissues. Using
the vas deferens, vasopressin, oxytocin and $PGF_1\alpha$ were ineffective; the other compounds tested had
an intense stimulant effect. Dose-response relationships were usually unsatisfactory and tachyphylaxis was not infrequent. The authors, therefore, conclude that smooth muscle of the male genital tract is
not first-choice for bioassay of these compounds. (HC)   005710

0922
NEWTON, J; COLLINS, W.
    Plasma hormone changes during infusions of prostaglandins $F_2\alpha$ and E_2 for therapeutic abortion.
    Advances in the Biosciences 9: 689-699. 1973.

9 patients, 10 to 20 weeks pregnant, were given an intravenous infusion of $PGF_2\alpha$ at 25 μg/minute for
30 minutes and then 50 μg/minute. 5 aborted, the IAT averaging 15 1/2 hours. Infusion was
terminated in the other 4 patients due to undesirable side effects. 2 patients, 8 weeks pregnant, were
given PGE_2 rather than $PGF_2\alpha$. Neither aborted, and infusions were terminated due to side effects. In
patients that aborted, steroid hormone levels decreased 40% to 50%. In those patients that did not
abort, progesterone levels fell, 17 alpha-hydroxyprogesterone rose intermittently, estrogens increased
or decreased 10% to 50%. The authors conclude that the mode of action of prostaglandins in abortion
may be a vasoconstrictor effect on the arterioles supplying the steroid-producing tissues; a direct
action on the corpus luteum is probably excluded. (HC)   006487

0923
NIKOLASEV, V.; RESCH, B.A.; MESZAROS, J.; SZONTAGH, F.E.; KARADY, I.
    Phospholipid content of early human placenta and fatty acid composition of the individual
        phospholipids.
    Steroids and Lipids Research 4: 76-85. 1973.

Phospholipids were extracted from 5 to 12 week human placentae, separated by thin-layer
chromatography and their fatty acid composition determined using gas liquid chromatography.
Phospholipid content does not change during this period; phosphatidylcholine and phosphatidyletha-
nolamine together comprise more than 50% of the total amount. Saturated and unsaturated fatty
acids having 14 to 22 carbon atoms were found in the phospholipids, the saturated acids
predominating. The authors conclude that the fatty acids may be precursors of PG. (HC)   006715

0924
NIKOLASEV, V.; RESCH, B.A.; MESZAROS, J.; SZONTAGH, F.E.; KARADY, I.
    The content of possible fatty acid precursors of prostaglandins in the early human placenta.
    International Research Communications System. (73-9) 3-7-16. September 1973. 1 p.

Lipids from 5- and 12-week-old human placenta were analyzed by thin-layer and gas chromatography. Significant quantities of arachidonic acid, the precursor for PGE_2 and $PGF_2\alpha$, were found in placental phospholipids (phosphatidyl choline, inositol, and ethanolamine) and triglycerides. The free fatty acid fraction of the lipids contained significant quantities of eicosatrienoic acid, the precursor for PGE_1 and $PGF_1\alpha$. (JSL)   006894

0925

NILSSON, L.; ROSBERG; S.
Effects of gonadotrophins and prostaglandins (PG) on the metabolism of isolated rat ovarian follicles.
Acta Endocrinologica Suppl. 177: 315. 1973.

Premature growth of follicles and LH release were induced in 30-day-old rats. 2-3 days later, just before and after ovulation, the follicles were isolated and incubated with hormones or amino acids. LH or FSH stimulated glycolysis; PGE_1 and PGE_2 had little effect. Amino acid uptake and incorporation into protein were not influenced by any of the above substances. LH and PG elevated cAMP levels in the follicles. (HC)   005716

0926

NODEN, P.A.; OXENDER, W.D.; HAFS, H.D.
LH after $PGF_2\alpha$ in mares.
Journal of Animal Science 37(1): 323. Abstract 373. 1973.

Jugular plasma LH was determined by radioimmunnoassay in 6 mares after 10 mg intrauterine $PGF_2\alpha$ given on Day 7-9 after ovulation, and during the preceeding and following estrous cycles. In both control cycles LH averaged 57 ng per ml in diestrus, 323 on estrus Day 1, 1030 after ovulation, remained high for 42 hours, then declined to 123 7-9 days after ovulation. After $PGF_2\alpha$, LH rose to 317 ng per ml on estrus Day 1, 2.2 days later, peaked at 1063 18 hours after ovulation, and remained high for 42 hours, then declined to 90 ng per ml 10 days after ovulation. Estrus usually began when progesterone dipped below 1 ng per ml, mean 2.3 days after the first definite LH peak. Thus LH changes after $PGF_2\alpha$ resembled those which began about 2 days before normal estrus. (LJG) 006043

0927

NODEN, P.A.; HAFS, H.D.; OXENDER, W.D.
Progesterone, estrus, and ovulation after prostaglandin $F_2\alpha$ in horses.
Federation Proceedings 32(3): 229Abs. 1973.

$PGF_2\alpha$ Tham salt was given to mares 7-9 days after ovulation, 10 mg intrauterine (iu) to 6 mares and 15 mg sc to 8 mares, and plasma progesterone was measured by radioimmunoassay to determine whether PG was luteolytic. In 8 controls the interestrual interval was 15.0 days. After iu PG, mean progesterone fell from 13.6 to 5.8 to 2.6 to .9 ng/ml in 12, 24, and 48 hours; estrus occurred 2.2 days later; ovulation occurred 5.8 days after estrus and estrus lasted 7.5 days. After sc PG, progesterone fell from 10.4 to 4.6 to 2.0 to .6 ng/ml in 12, 24, and 48 hours; estrus occurred 2.5 days later; ovulation occurred 5.8 days after estrus and estrus lasted 7.5 days. In controls, ovulation took place 4.1 days after estrus and estrus lasted 5.7 days; progesterone was .7 ng/ml at ovulation and 14-16 ng in diestrus. Progesterone was .4 ng/ml at ovulation in treated mares and 12-13 ng/ml in diestrus. Thus $PGF_2\alpha$ was luteolytic in mares, producing normal estrus. (LJG)   005423

0928
NOVY, M.J.; COOK, M.J.
Redistribution of blood flow by prostaglandin F$_2\alpha$ in the rabbit ovary.
American Journal of Obstetrics and Gynecology 117(3): 381-385. October 1, 1973.

Hemodynamic measurements were made on 7 rabbits during estrus and on 11 rabbits on Days 8 to 10 of pseudopregnancy. Ovarian blood flow measured using radioactive microspheres was 5-fold higher on Day 9 than before ovulation and was 2% of the total cardiac output. There was a direct relationship between corpus luteum weight and degree of ovarian blood flow. After PGF$_2\alpha$, 500 or 1000 μg intravenously, corpus luteum blood flow decreased as did the percentage of cardiac output going to the corpora lutea. However, blood flow to interstitial and follicular tissue increased. The only systemic cardiovascular effect of PGF$_2\alpha$ was a decrease in arterial pressure. The authors suggest that PGs are locally involved in the regulation of ovarian circulation, more on the afferent blood vessels than on the uteroovarian vein. (HC)   005828

0929
NYBERG, R.
Therapeutic abortion by intraamniotic administration of prostaglandin F$_2\alpha$.
Advances in the Biosciences 9: 533-537. 1973.

Prostaglandin F$_2\alpha$ was instilled intramniotically in 101 patients with a gestational age of 10-26 weeks. A single dose varying from 15-35 mg was given. Nineteen percent of the patients needed a second injection. Abortion was induced in 95% of the patients after 1 or 2 injections. Curettage was necessary in 62%. The incidence of vomiting and/or diarrhea was 0.06 episodes per hr. (Author)   006471

0930
O'SHEA, J.D.; LEE, C.S.
Effects of section of utero-ovarian vascular connections on the duration of pseudopregnancy in the rat.
Journal of Reproduction and Fertility 33(2): 245-253. 1973.

2 experiments were performed on female hooded Wistar rats to determine the effects of section of uteroovarian connections on the duration of pseudopregnancy. Ligation and section of the anterior uterine blood vessels on Day 6 of pseudopregnancy caused a prolongation of the pseudopregnancy during which the operation was performed, and also of the subsequent pseudopregnancy. Neither section of the fallopian tube nor that of the mesosalpinx and broad ligament significantly affected the duration of pseudopregnancy. When the anterior uterine vessels were sectioned separately, it was shown that severing the artery led to a prolongation of pseudopregnancy. Section of the vein did not significantly affect the duration of pseudopregnancy. This last observation casts doubt on the hypothesis by Pharriss that PGF$_2\alpha$ released by the uterus can pass down the uteroovarian vein and exert a venoconstrictor action, and indirectly initiate luteolysis. McCracken's hypothesis that PGF$_2\alpha$ may pass down the uterine vein and be transferred by countercurrent mechanism to the adjacent ovarian artery is also unsupported by the experiment. These results suggest that the effects on pseudopregnancy of section of uteroovarian connections depend primarily on interruption of the anterior uterine artery. (Authors' modified)   005649

308

0931
OAKES, G.; MOFID, M.; BRINKMAN, C.R., 3rd; ASSALI, N.S.
Insensitivity of the sheep to prostaglandins.
Proceedings of the Society for Experimental Biology and Medicine 142(1): 194-197. 1973.

Single intravenous progressively increasing doses of .4-3.0 $\mu g/kg$ PGE_2 or .4-10.0 $\mu g/kg$ $PGF_2\alpha$ had no effect on the cardiovascular system or intraamniotic pressure of pregnant ewes. Continuous infusion of PGE_2 at 10 μg/minute for 20 minutes in 1 animal also had no effect. Doses of .5-100 mg/kg $PGF_2\alpha$ given intravenously to the fetus did not affect its heart rate or arterial pressure. The authors concluded that their data do not support the hypothesis that prostaglandins play a role in initiating labor in sheep. (HC)   006757

0932
OSHIMA, K. MATSUMOTO, K.
Absorption of prostaglandin E_2 and uterine sensitivity of the non-pregnant and pregnant monkey in vivo.
Prostaglandins 3(4): 447-455. April 1973.

Plasma levels of PGE_2 and $PGF_2\alpha$, as well as uterine contractility, were measured in pregnant and nonpregnant monkeys after oral or iv PG. Plasma levels were determined by radioactive label in 4 nonpregnant Rhesus monkeys given 1 mg per kg PGE_2-CD; or by radioimmunoassay in 3 nonpregnant Japanese monkeys given 143 μCi PGE_2-CD; uterine contractions were recorded after iv PGE_2 and $PGF_2\alpha$ in 3 stump-tailed monkeys, in 3 Japanese monkeys in third trimester given PGE_2 orally and $PGF_2\alpha$ intramniotically. Plasma PGE_2 levels by radioimmunoassay rose from 84 to a maximum of 303 ng per ml at 90 minutes after oral PGE_2, then fell to 125 at 4 hours and 26 ng per ml at 24 hours. Peak tritium label was also detected at 90 minutes, but fell more gradually for the first 4 hours. 5 μg per kg PGE_2 iv was recovered at a peak level of about 1500 ng per ml in 2 minutes, and was negligible in 5 minutes. Uterine mobility varied with season and phase of sexual cycle as follows: in nonpregnant monkeys, uterus was highly sensitive to PGE_2 in spring, but less sensitive in summer (post-breeding period.) Third trimester pregnant monkeys were more sensitive to PGE_2 than $PGF_2\alpha$. In lactating monkeys with spontaneous rhythmic contractions, $PGF_2\alpha$ slightly stimulated and PGE_2 inhibited uterine activity. Oral PGE_2 induced labor in 2 out of 3 Japanese monkeys at term when given in 2 oral doses of 1 mg per kg, 8 hours apart. In the third monkey, 2 mg intraamniotic $PGF_2\alpha$ induced delivery 12 hours later. Thus to maintain high plasma PGE_2 levels, doses should be repeated every 1 1/2 to 4 hours. (LJG)   005687

0933
OSTERGARD, D.R.
The cervical relaxant properties of prostaglandin E_2 in non pregnant subjects.
Prostaglandins 4(5): 701-702. November 1973.

5 non-pregnant women were administered a 20 mg suppository of PGE_2 intravaginally to determine its potential use as a precurettage cervical dilator. All patients demonstrated severe side effects, the most common being nausea, chills, pyrexia, and cramping. 1 patient developed hypotension lasting 1.5 hours. None of the 5 showed any marked dilation. (RAS)   005640

0934

OXENDER, W.D.; HAFS, H.D.; NODEN, P.A.

Controlled ovulation in mares and cows with prostaglandins.

Journal of the American Veterinary Association 163(10): 1182. 1973.

In mares, progesterone levels decreased from 13.5 ng/ml to .6 ng/ml when estrous behavior began 48 hours after $PGF_2\alpha$ administration. Ovulation occured 8 days after treatment. In cows, progesterone levels fell from 4.2 ng/ml to 1 ng/ml in 24 hours, and ovulation occurred 4 days after treatment with $PGF_2\alpha$. (HC)   005713

0935

PARK, I.J.; WENTZ, A.C.; JONES, H.W.

The viability of fetal skin of abortuses induced by saline or prostaglandin.

American Journal of Obstetrics and Gynecology 115(2): 274-275. January 15, 1973.

Fetal skin from abortuses induced by saline or $PGF_2\alpha$ were cultured to determine their suitability for biochemical and cytogenetic study. 2/3 of the prostaglandin group grew and none of the hypertonic saline group cultured. The mechanism by which $PGF_2\alpha$ induces abortion apparently does not cause immediate death of the cells of the fetal skin as does hypertonic saline. (RAS)   006765

0936

PATRONO, C.

Radioimmunoassay of prostaglandin $F_2\alpha$ in unextracted human plasma.

Journal of Nuclear Biology and Medicine 17: 25-29. 1973.

Antibody-bound $PGF_2\alpha$ was prepared by incubating guinea pig antiserum to $PGF_2\alpha$-rabbit serum albumin conjugate with tritium-labeled $PGF_2\alpha$. Sensitivity of the assay is 5 pg/ml, and the specificity is nearly exclusive with only $PGF_1\alpha$ cross-reacting approximately 20%. $PGF_2\alpha$ plasma levels in men and in pregnant as well as nonpregnant women were less than 50 pg/ml. During labor at the peak of uterine contractions values were 180-400 pg/ml. During spontaneous abortion the levels recorded were 350-400 pg/ml. In cirrhotic patients plasma $PGF_2\alpha$ levels of 120-360 pg/ml were noted. The authors concluded that endogenous plasma $PGF_2\alpha$ levels are considerably lower than those estimated by others using mass spectrometry, and methods with greater sensitivity must be developed to monitor physiological changes. (HC)   006179

0937

PERSAUD, T.V.N.

Pregnancy and progeny in rats treated with prostaglandin E_2.

Prostaglandins 3(3): 299-305. March 1973.

The teratogenic and embryotoxic effects of PGE_2given to rats ip on Gestation Days 12-15, and intraamniotically on Day 15, were evaluated. $PGE_2$was given intraamniotically to fetuses in the right uterine horn and vehicle on the left during laparotomy under pentobarbital anesthesia, and all fetuses were evaluated on Gestation Day 20 after cesarean section. Intraperitoneal PGE_2, 300 μg daily from Days 12-15, resulted in 93 implantations, 5(5.4%) resorptions, and 16 fetuses (18.2%) with hemorrhagic lesions and edema among the 88 liveborn fetuses of 12 mother rats. 8 control rats produced 86 implantations, also all liveborn, and 3 (3.5%) resorptions. Intraamniotic PGE_2, 100 μg, caused resorption of all 62 fetuses on the right (treated) side, compared to 11 (16.4%) of 67 control fetuses given vehicle. No vaginal bleeding or premature labor occurred in mothers, nor were any structural defects noted in any treated or control fetuses. (LJG)   005688

0938
PERSAUD, T.V.N.
  Prostaglandin E_2 and fetal development in rats: a preliminary report.
  Teratology 7: A-25. 1973.

Intraamniotic injection of 100 μg PGE_2 on day 15 of gestation did not induce labor in rats; all fetuses were dead when the mothers were sacrificed 5 days later. 300 μg injections intraperitoneally from day 12 to 15 did not interrupt pregnancy; all fetuses were alive on day 20 but extensive edema and hemorrhagic lesions were noted in 20% of the offspring. Average weight of the fetuses was not significantly altered, there was no definitive evidence of a teratogenic effect. The authors conclude that either PGE_2 is not a uterine stimulant in the rat or that ineffective concentrations reached the uterus due to rapid metabolism of PG. (HC)   005715

0939
PERSAUD, T.V.N.
  Prostaglandin E_2 effects on placenta and fetus following intraamniotic administration in rats.
  International Research Communications System (73-8) 8-5-4. September 1973. 1 p.

A dose of 100 μg of PGE_2 was given intraamniotically on Day 15 to pregnant rats; others were given in addition, 5 mg progesterone daily from Days 15 through 18. There was an 80-90% incidence of fetal resorptions in each group, but there were no congenital malformations. There was no luteolysis but placental changes were extensive. The authors suggested that PGE_2 terminates pregnancy in the rat by a direct action on the fetoplacental unit rather than through luteolysis. (HC)   006930

0940
PERSAUD, T.V.N.; MANN, R.A.; MOORE, K.L.
  Teratological studies with prostaglandin E_2 in chick embryos.
  Prostaglandins 4(3): 343-350. 1973.

PGE_2 (20-100 μg) was administered to 48 hour and 72 hour chick embryos. The developing chick embryo was most sensitive to PG treatment at 48 hours incubation. The incidence of embryonic mortality and malformations was higher at dose levels of 50 and 100 μg. No lethal effect was noted at 72 hours but there was a high percentage of lethality at 48 hours. A high incidence of abnormal embryos increased with dose levels of PG and was induced in both 48 and 72 hour incubations, but this was significant only in 48 hour embryos treated with 100 μg PG. Embryos showed no sign of growth retardation. (Authors' modified)   005437

0941
PERSIANINOV, L.S.; MANUILOVA, I.A.; CHERNUKHA, E.A.
  The results of using prostaglandin $F_2\alpha$ for induction and stimulation of labor.
  Advances in the Biosciences 9: 585-592. 1973.

We have used $PGF_2\alpha$ in 21 women, 11 for induction and 10 for stimulation of labor. Indications for induction of labor were: serotinus labor in 5, rhesus-negative blood with sensibilization in 3, and premature rupture membranes in 3. Primary and secondary inertia of the uterus were indications for labor stimulation. In induction of labor we paid great attention to the state of cervix uteri; in 8 out of 11 women cervix was mature. We used a solution of prostaglandins which contained 5 mg of $PGF_2\alpha$ in 1000.0 ml of 5% glucose at 20 drops/min gradually increased to 40 drops/min. We used radiotelemetry and cardiotocography during labor, which gave us the possibility of registering uterine contractility and fetal heart rate before and during the $PGF_2\alpha$ administration. Using radiotelemetry we found that the character of the curve began to change immediately after the PGF2 administration, and in 15-30 min uterine contractions appeared as normal labor. The duration and intensity of uterine contractions increased, but the interval between contractions was not actually changed. The fetal heart rate before and during stimulation with $PGF_2\alpha$ remained unchanged. Thus we found that the administration of $PGF_2\alpha$ is very effective for induction and stimulation of labor. The greatest advantage of $PGF_2\alpha$ is its lack of effect on the fetus and its ability to decrease postpartum bleeding. (Authors' modified)   006477

0942

PEXTON, J.E.; FORD, S.P.; WILSON, L., Jr.; BUTCHER, R.L.; INSKEEP, E.K.
Prostaglandins F (PGF) in IUD-treated ewes.
Journal of Animal Science 37: 324. 1973.

The presence of either unilateral or bilateral IUD in ewes from Day 2 to Day 5 increased mean uterine venous plasma levels of PGF from control levels of .5-8.0 ng/ml. (HC)   005748

0943

PHARRISS, B.B.; BEHRMAN, H.R.
Gonadal function.
In: Ramwell, P.W., ed. The Prostaglandins, Vol. 1. New York, Plenum Press, 1973. p. 347-363.

The effects of PGs on gonads are discussed in the format of gonadotrophic, luteolytic and abortifacient effects on ovaries, and their known effects on testis. PGs act like LH on rat, rabbit, and cow ovary in vitro; for example they increase progesterone production, ascorbic acid depletion, 20alpha-hydroxypregn-4-en-3-one release, and their action involves cAMP. In vivo, PGs are usually luteolytic. The mechanism is independent of hypothalamus and uterus, as shown by ablation experiments, but may entail decreased ovarian blood flow, response to gonadotropins, or inhibition of enzymes of progesterone biosynthesis. Much evidence documents that $PGF_2\alpha$ is the physiologic uterine luteolysin in bicornate species. Since pregnancy can be terminated only on Days 4-6 or 9-12 in rats, progestins can reverse the effect of PG, and abortion begins with falling progesterone levels, luteolysis is sufficient to account for abortion. In primates, the abortive mechanism is the oxytocic action of PGs. In testis, PGs can affect testosterone secretion in vitro, testis capsule contraction, and blood flow in vivo. (LJG)   006912

0944

PHILLIPS, L.L.
Effect of prostaglandins on the coagulation mechanism of the pregnant rat.
American Journal of Obstetrics and Gynecology 115(2): 227-232. January 15, 1973.

Pregnant rats were given prostaglandin E_2 or $F_2\alpha$ to determine the effect on the coagulation system and on gestation. Intramuscular administration of $PGF_2\alpha$ (1,000 mg per day for 3 days) beginning on Day 6 of pregnancy induced resorption or abortion. PGE_2 (1,000 to 1,500 mg intravenously in divided doses at 24 hour intervals) beginning on Day 10 or 12 of gestation and intravenous $PGF_2\alpha$ (225 mg per day for 5 days) beginning on Day 4 or 5 did not cause abortion or resorption. Prothrombin and activated partial thromboplastin times were not affected, but there was a slight prolongation of plastic partial thromboplastin time. Fibrinogen levels remained within normal. Platelet counts were also normal for rats in most cases but were slightly depressed (800,000) 3 and 24 hours after the first intravenous administration of $PGF_2\alpha$. The prostaglandins do not appear to cause significant intravascular coagulation in pregnant rats. (Author)   006769

0945
PICKLES, V.R.
   Is your pain really necessary?
   British Medical Journal 3(5875): 349. August 11, 1973.

   The author suggested that indomethacin, an inhibitor of prostaglandin synthesis, may be helpful in the treatment of dysmenorrhea. Dysmenorrhea may involve an abnormal production of the 'menstrual stimulant' prostaglandin. (RAS)   005810

0946
PIJNENBORG, R.; BROSENS, I.; ROBERTSON, W.B.
   Prostaglandin ($F_2\alpha$) induced histological changes in conceptuses and deciduo mata of pregnant golden hamsters.
   International Research Communications System (73-12) 8-5-9. 1973.

   $PGF_2\alpha$, 200 or 300 μg as a single injection or 100 μg given on 2 or 3 consecutive days, caused blighted pregnancies and necrosis in the deciduomata. Histological examination revealed that necrosis started in the antimesometrial, the older decidual cells, and then spread circumferentially to the mesometrial basalis decidua. The authors conclude that PG causes corpus luteum necrosis followed by decreased progesterone, decidual necrosis, and then fetal death. (HC)   006891

0947
POULOS, A.; VOGLMAYR, J.K.; WHITE, I.G.
   Phospholipid changes in spermatozoa during passage through the genital tract of the bull.
   Biochimica et Biophysica Acta 306: 194-202. 1973.

   The phospholipids and phospholipid-linked fatty acids and aldehydes of testicular, epididymal, and ejaculated bull spermatozoa were analyzed. Spermatozoa were collected by cannulating the rete testis and ductus deferens, and the lipids, were quantitated by extraction and thin-layer and gas-liquid chromatography of lipids or derivatives. Testicular spermatozoa had about twice the phospholipid content of cauda epididymal or ejaculated spermatozoa, which were similar; all phospholipids decreased during passage through the epididymus except choline plasmologen. Amounts of most phospholipid-bound fatty acids, particularly palmitic acid, declined, but docosahexenoic acid, the principal fatty acid in ejaculated spermatozoa, decreased the least. Arachidonic acid, a precursor of PGE_2 and $PGF_2\alpha$, decreased from 5.5% by weight in testicular to 2.3% in epididymal and 3.2% in ejaculated spermatozoa (p<.01). If arachidonic acid conversion to PGs regulates lipolysis, a means of controlling degradation of spermatozoal phospholipids is possible. (LJG)   005717

0948
POULOS, A.; DARIN-BENNETT, A.; WHITE, I.G.
   The phospholipid-bound fatty acids and aldehydes of mammalian spermatozoa.
   Comparative Biochemistry and Physiology 46B: 541-549. 1973.

The phospholipid-bound fatty acids and aldehydes of ram, bull, boar, rabbit, and human spermatozoa were analyzed by thin-layer and gas-liquid chromatography. Docosahexenoic acid was the predominant fatty acid in all species except rabbit, from 37.7% of total fatty acids by weight in boar to 61.4% in ram. Rabbit spermatozoa had 39% docosapentanoic acid. The major saturated fatty acid was palmitic acid. The 5 species fell into 2 groups: bull, boar, and ram sperm had lower stearic acid and oleic acid but higher polyunsaturated/saturated ratios than man and rabbit. Arachidonic acid, the precursor of PGE_2 and $PGF_2\alpha$, made up 3.2% of fatty acid weight in boar, 3.5% in bull, 4.5% in ram, and 5.1% in human spermatozoa, but less than 1% in rabbit. The major fatty aldehyde, was palmitaldehyde, ranging from 52.1% in human to 91.2% in ram spermatozoa. (LJG)   005714

0949
POYSER, N.L.
The formation of prostaglandins by the guinea-pig uterus and the effect of indomethacin.
Advances in the Biosciences 9: 631-634. 1973.

Prostaglandin $F_2\alpha$ was detectable in the guinea pig uterus in small amounts on Days 13, 14, and 15 only of the estrous cycle. Prostaglandin E_2 was detectable in smaller amounts on Days 14 and 15. On incubation the homogenized uterus synthesized $PGF_2\alpha$ and, to a lesser extent, PGE_2 from endogenous precursors on every day of the cycle studied. The amounts of each prostaglandin produced were greater toward the end of the cycle. Indomethacin (5 μg per ml) inhibited the biosynthesis of $PGF_2\alpha$ and PGE_2 by 62% and 77%, respectively. In vivo indomethacin (20 mg daily) injected from Day 10 significantly increased estrous cycle length in guinea pigs from 16.3 ± .2 to 19.3 ± .5 days (12 animals, $p<.01$). It is evident that toward the end of the estrous cycle the uterus is able to synthesize larger amounts of prostaglandins, especially $PGF_2\alpha$. This observation supports the view that $PGF_2\alpha$, a potent luteolytic substance, is the uterine luteolytic hormone. In the inhibition of uterine $PGF_2\alpha$ biosynthesis by indomethacin observed in vitro occurred in the vivo experiments, prevention of $PGF_2\alpha$ synthesis may account for the lengthening of the estrous cycle. (Author's modified)   006480

0950
PURI, C.P.; HINGORANI, V.; LAUMAS, K.R.
Effect of prostaglandin $F_2\alpha$ on progestin synthesis in the human and rabbit ovary.
Advances in the Biosciences 9: 657-663. 1973.

Prostaglandin $F_2\alpha$ ($PGF_2\alpha$) has been demonstrated to stimulate progesterone (P) and 20a-hydroxyprogesterone (20a-OH P) synthesis in vitro in the corpus luteum of the human menstrual cycle and pregnancy and also in the pseudopregnant rabbit ovary. Interestingly, a decrease in P synthesis in vitro was observed in a corpus luteum from a pregnant woman who had been administered extraamniotic $PGF_2\alpha$ for 30 hr. This preliminary observation indicated that pretreatment of $PGF_2\alpha$ for a prolonged period of time also decreased P synthesis in vitro. This single observation on the human received support from experiments on pseudopregnant rabbits which were intravenously infused with $PGF_2\alpha$ for 6 hr. The in vitro capacity of the ovary to synthesize P was markedly decreased under these conditions. The results led to the conclusion that short-term incubation may produce only a stimulatory effect on P synthesis; however, exposure of the ovary to $PGF_2\alpha$ for a comparatively longer period of time may produce a decrease in P synthesis in vitro. (Authors)   006482

0951
RAJAN, R.
Amniotic fluid assays in high-risk pregnancy.
Clinical Obstetrics and Gynecology 16: 313-328. 1973.

Current knowledge and its clinical application of amniotic fluid volume, osmolarity, hormones, cytology, phospholipids, prostaglandins, creatinine, protein, hydroxyproline, bilirubin and meconium are summarized. Parameters useful for assessing fetal maturity include lecithin sphingomyelin ratio, creatinine and bilirubin levels, and proportion of cornified cells. Measurements of bilirubin are valuable for Rh disease. Fluid volume is a good indicator of hypertension, toxemia, diabetes, and postmaturity. Current research points to relationships between adrenal steroids and congenital adrenal disease or prematurity, between pressor amines and anoxia, between estrogens and anencephaly. Techniques for studying amniotic fluid include ultrasound to locate the placenta preparatory to amniocentesis, amnioscopy to visualize pigments via the cervix, dye markers to quantitate fluid volume, and cell culture or karyotyping for diagnosis of genetic disorders. PGEs inhibit myometrial activity; PGFs stimulate, and are known to increase to a peak before onset of labor. (LJG) 005719

0952
RAO, C.V.
Receptors for prostaglandins and gonadotropins in the cell membranes of bovine corpus luteum.
Prostaglandins 4(4): 567-576. October 1973.

Bovine corpus luteum cell membrane preparations were incubated with $3H$-PGE_1 and 125-I-HCG, and either unlabeled PGE_1, PGE_2, $PGF_1\alpha$, $PGF_2\alpha$ or 7-oxa-13-prostynoic acid, a PG antagonist, to define the characteristics of PG and gonadotropin receptors. Cell membranes were prepared from the 2000 x g x 10 minutes pellet of filtered corpus luteum homogenate. Incubation was for 1 hour with PG and 4 hours with HCG in .01M Tris-HCl pH containing .001 M dithiothreitol, .001 M CaCl2 and .1% gelatin $3H$-PGE_1 binding was reduced 80-90% by unlabeled PGE_1 or PGE_2 at 1.4×10^{-7}M, by $PGF_1\alpha$ and $PGF_2\alpha$ at 1.4×10^{-5}M. 125-I-HCG binding was completely inhibited by unlabeled HCG at 5×10^{-8}M. Neither labeled substance was displaced by the opposite unlabeled one, suggesting that they bind to different receptor sites. Prostynoic acid inhibited $3H$-PGE_1 binding only. (LJG) 005615

0953
RAO, V.S.N.; SHARMA, P.L.
Effect of intraperitoneal administration of (+)-INPEA on oxytocin and prostaglandin evoked responses of the isolated rat uterus.
British Journal of Pharmacology 48: 344P-345P. 1973.

In a study of the effects of optical isomers of INPEA (N-isopropyl-p-nitrophenyl ethanolamine hydrochloride) on oxytocin- and PG-evoked responses of isolated rat uterus it was found that (-)-INPEA showed weak antioxytocin activity and no effect on PGs, but (+)-INPEA potentiated the action of both when added in concentration of 1×10^{-5}g per ml. The present investigation was undertaken to discover the effects of (+)-INPEA on PGE_1, PGE_2, $PGF_2\alpha$ and oxytocin-evoked response upon intraperitoneal administration. Uterine strips from (+)-INPEA treated rats (10 mg/kg body weight intraperitoneal 1 hour before experiment) were suspended in a bath of aerated de Jalon's solution at 29 degrees C and equilibrated for 30 minutes. Responses to graded doses of antagonists were recorded on a potentiometric recorder. In all 7 experiments there was marked increase in amplitude of oxytocin- and PG-evoked response compared to controls. Results showed that intraperitoneal (+)-INPEA sensitizes the uterine tissue to the action of oxytocin and PGs. This potentiating effect of (+)-INPEA may allow the use of smaller doses of PGs and oxytocin in induction of labor and abortion thus decreasing their side effects. (Authors' modified) 006457

0954

RAZ, A.; STERN, H.; KENIG-WAKSHAL, R.

Indomethacin and aspirin inhibition of prostaglandin E_2synthesis by sheep seminal vesicles microsome powder and seminal vesicles slices.

Prostaglandins 3(3): 337-352. March 1973.

The site of inhibition of PG synthesis by aspirin and indomethacin was studied by comparing PG synthesis in vitro by acetone-pentane powder of sheep seminal vesicle microsomes and microsome prepared from slices which were previously incubated with the inhibitors. The incubation and assay medium contained, per 50 mg microsome powder, 65 μmoles Na-EDTA, 5 μmoles reduced gluthathionel, .4 μmoles hydroquinone in 1.5 ml pH 8.1, with .1 ml Bucher medium and .75 mg [3H]-arachidonic acid in .5 ml added. $PGE_2$and $PGF_2\alpha$ were estimated by thin layer chromatography and scintillation counting. Indomethacin inhibited PG synthetase in enzyme powder from 5.1% at .26 μM to 91.1% at 4.63 μM; aspirin exhibited comparable inhibition at 300-500 μM, i.e. was 50-100 times less effective. In microsomes, 95.8 μM indomethacin was required to inhibit 77.1%, and 1920 μM aspirin to inhibit 49.5%. When the enzyme powder was incubated with indomethacin, in assay medium without substrate, and washed or dialyzed up to 42 hours, then assayed, inhibition was irreversible. A 42-hour dialysis against 2 changes of .15 M potassium phosphate buffer pH 8.0 decreased inhibition from 66.5% to 54.7%. Finally vesicle slices (1 x 1 cm x .5 mm) were preincubated with 27.5 μM indomethacin or 550 μM aspirin for .5 or 4 hours at 37 degrees, washed and assayed as microsomes or acetonepentane powder. Aspirin inhibited PGE_2synthesis 92.4% to 96.4% in the 2 preparations and 2 incubation times. Indomethacin had no effect. It was concluded that indomethacin does not penetrate cell membranes in the slices, and probably binds to or inactivates PG synthetase in cell free preparations. (LJG)   006579

0955

RAZ, A.; STERN, H.; KENIG-WAKSHAL, R.

Inhibition of prostaglandin synthesis by anti-inflammatory drugs; differential effects of indomethacin, aspirin and flufenamic acid on sheep seminal vesicle microsomal prostaglandin synthetase and on seminal vesicle slices.

Israel Journal of Medical Science 9:556. 1973.

Indomethacin, flufenamic acid, and aspirin inhibited PGE_2synthesis in sheep microsomal preparations used as a source of PG synthetase. Prostaglandin synthetase activity in seminal vesicle slices after incubation for .5 or 4 hours with aspirin or flufenamic acid was inhibited; indomethacin did not affect this enzyme's activity. The authors suggest that in vivo inhibition by indomethacin may not be mediated via direct inhibition of prostaglandin synthesis. (HC)   005409

0956

REMBIESA, R.; WARCHOL, A.

The influence of pH on the conversion of arachidonic acid to $PGF_2\alpha$ by bovine seminal vesicles microsome (BSVM).

Prostaglandins 4(3): 441-442. September 1973.

This letter reports that production of $PGE_2$and $PGF_2\alpha$ from 14C- arachidonic acid by bovine seminal vesicle microsomes is dependent on pH rather than addition of L-epinephrine or hydroquinone. The reaction mixture included 11 μg reduced glutathione, .92 μg arachidonic, 20 mg microsomal protein, 5 μg L-epinephrine or hydroquinone, in .05 M tris HC1, in 2.0 ml volume. $PGF_2\alpha$ production increased from pH 7 to 8 and leveled off at pH 8.5 with epinephrine or hydroquinone, but increased at pH 8.5 without cofactors. PGE_2peaked at pH 7.5 and fell at higher pH, although the peak was broader with epinephrine added. In these 60 minute incubations, PG production rate was comparable whether hydroquinone, epinephrine, or neither, were included. (LJG)   005447

0957

RESTALL, B.J.; HEARNSHAW, H.R.; GLEESON, A.R.; THORBURN, G.D.
Observations on the luteolytic action of prostaglandin $F_2\alpha$ in the ewe.
Journal of Reproduction and Fertility 32(2): 325-326. 1973.

Infusion of $PGF_2\alpha$ into the uterine vein of ewes at a rate of 80 μg/hour for 6 hours resulted in luteolysis (deduced from a fall in plasma progesterone levels and occurrence of estrus). Infusing the drug into the jugular vein failed to elicit this response. Only 1 of 4 ewes given 8 mg/hour guanethidine at the same time as the $PGF_2\alpha$ underwent luteolysis. This would suggest involvement of the sympathetic nervous system. (HC)   006772

0958

RING, A.; KRESS, D.; SEMM, K.
Weheninduktion mit Prostaglandin $F_2\alpha$. [Induction of labor with prostaglandin $F_2\alpha$].
Archiv fuer Gynaekologie 214(1-4): 459-461. 1973.

71 cases of labor induced by intravenous $PGF_2\alpha$ and 6 cases of intrauterine death induced by intraamniotic $PGF_2\alpha$ are reported. 10 women given an average of 60 mg $PGF_2\alpha$ aborted in a mean 12 hours and 45 minutes, 7 completely, 1 incompletely. Women with missed abortion received mean 65.1 mg and took a mean 15 hours and 20 minutes to abort completely in 66% and incompletely in 12.5%. 13 cases of intrauterine death were delivered in a mean 7 hours after mean 25.2 mg $PGF_2\alpha$. Of 42 women with labor induced by a mean 2.5 mg $PGF_2\alpha$, 38 (90%) delivered in mean 5 hours and 33 minutes. Side effects included nausea, vomiting, diarrhea, phlebitis and 1 case of allergic exanthema. 6 cases of intrauterine death (patients were from 19 to 36 years of age and 20 to 36 weeks of gestation) received 25 mg $PGF_2\alpha$ intraamniotically, except for 1 given 37.5 mg. Uterine contractions began in 3 to 25 minutes, successful labor lasted 3 to 22 hours, without side effects. After a total of 80 cases treated with $PGF_2\alpha$, the authors believe that intravenous PG is indicated for inducing all deliveries except term deliveries, for which they prefer oxytocin; intraamniotic $PGF_2\alpha$ is indicated for intrauterine death as long as the lengthy duration of uterine activity can be tolerated. (LJG)   005795

0959

ROBERTS, G.; GOMERSALL, R.; ADAMS, M.; TURNBULL, A.C.
Therapeutic abortion by intraamniotic injection of prostaglandins.
Advances in the Biosciences 9: 555-560. 1973.

Intraamniotic injection of either prostaglandin $F_2\alpha$ or E_2 was used in an attempt to induce therapeutic abortion in 27 midpregnancy patients. Termination of pregnancy was successful in 11 of 13 cases using prostaglandin E_2 alone, but in only 6 of 14 cases using prostaglandin $F_2\alpha$; a further 8 patients aborted following intravenous oxytocin stimulation, but the combined procedures failed altogether in 2 patients initially given prostaglandin $F_2\alpha$. The technique was simple, free from serious side effects, and reasonably effective when prostaglandin E_2 was used. In a study of 10 patients to whom prostaglandin $F_2\alpha$ was given, no consistent change was seen in plasma steroid concentrations until after delivery of the fetus and placenta. (Authors)   006473

0960

ROBINSON, I.C.A.F.; WALKER, J.M.
  Assay of ocytocin on the superfused mouse mammary gland using an automatic apparatus.
  Journal of Physiology (London) 234: 6P-7P. 1973.

  2-100 μU oxytocin in .5 ml could be bioassayed with a mean error of 2.6% on mouse mammary gland strips. 4 cm strips from lactating mice were superfused with modified Tyrode solution, pumped at .5 ml per minute, while contractions were recorded with a strain-gauge transducer, amplifier and high-frequency filter on a potentiometric recorder. Serotonin, histamine, bradykinin and $PGF_2\alpha$ in less than 1 μg doses failed to produce contractions. (LJG)   005730

0961

RUDEL, H.W.; KINCL, F.A.; HENZL, M.R.
  Prostaglandins.
  In: Rudel, H.W., Kincl, F.A., and Henzl, M.R. eds. Birth control, contraception and abortion. New York, MacMillan, 1973. p. 197-201.

  PG structure, nomenclature, history, and animal and human reproductive physiology are summarized, then their application to early and midtrimester abortion and missed menstruation are evaluated. In human uterus in vitro, PGEs decrease motility of nonpregnant uterus, but increase activity of pregnant myometrial strips. PGFs stimulate contractions, weakly in midcycle, but strongly in luteal or pregnant uterus. In vivo PGs are ineffective in the proliferative or luteal phase, but increase motility at ovulation. Unlike rodents and domestic animals, women do not respond to PGs by a direct luteolytic effect, but 90% will abort an early (less than 8 weeks) pregnancy upon receiving .3-1.2 μg per minute PGE_1 or 2.5-10 μg per minute $PGF_2\alpha$ iv. In midpregnancy (9-20 weeks) a longer infusion of PG, 13-14 hours instead of 7-8 will induce abortion, but complete abortion in only 20% of patients. It is possible to bring on menstruation with 2-3 mg $PGF_2\alpha$ given by intrauterine administration in 200-500 μg doses over several hours in hospitalized women. Side effects and lack of extensive studies limit the possibility of self-administered vaginal PG for missed menstruation. (LJG)   006903

0962

RUSSELL, P.T.; ALAM, N.; CLARY, P.
  Impaired placental conversion of prostaglandin E_1 to A_1 in toxemia of pregnancy.
  Federation Proceedings 32: 804Abs. 1973.

  Term placentas from normal and eclamptic pregnancies were compared in their ability to convert [3H]PGE_1 to [3H]PGA_1. 10 ml of 15,000 x g supernatant fluid, prepared from 30% homogenate in Tris buffer pH 7.4, was incubated with .2 μg [3H]PGE_1 without cofactors, and products determined by thin-layer chromatography. Normal placentas converted 100% of substrate in 5 minutes, but toxemic placentas converted as little as 10% of the substrate in 10 minutes. Preliminary experiments have ruled out presence of inhibitors in toxemic placenta. (LJG)   006729

0963

RUSSELL, P.T.; STANDER, R.W.
  Prostaglandin defect in toxemia of pregnancy.
  Obstetrics and Gynecology 41(4): 635. April 1973.

  Placentas from normal human subjects and patients with toxemia of pregnancy were studied with respect to their abilities to convert tritiated prostaglandin E (PGE_1) to tritiated prostaglandin A_1(PGA_1).

Incubations of the 15,000g supernatant fraction of placentas from normal subjects quantitatively converted PGE_1 to PGA_1 within 5 minutes. Comparable tissue preparations of placentas from patients with toxemia of pregnancy exhibited significantly less conversion of PGE_1 to PGA_1 during the 5-minute incubation. Initial studies suggest that conversion decrement is proportional to the severity of toxemia as judged by clinical manifestations. Initial combination experiments indicate that the conversion defect is related to an abnormal enzyme system rather than to the presence of an inhibitor. (Authors' modified)   006525

0964

RYAN, M.J.; CLARK, K.E.; VAN ORDEN, D.E.; FARLEY, D.; EDVINSSON, L.; SJOBERG, N.O.; VAN ORDEN, L.S., 3RD; BRODY, M.J.
Role of prostaglandins in estrogen-induced uterine hyperemia.
Prostaglandins 4(5): 629-640. November 1973.

The effect of indomethacin and meclofenamic acid on PGE and PGF levels and uterine hyperemia-produced by estrogen in ovariectomized rats was measured with radioiodinated human serum albumin (RISA). 7 days after ovariectomy rats were treated with estradiol for 5 days. On Day 12 and Day 13 rats recieved prostaglandin inhibitor or vehicle. On Day 14 half received estrogen and RISA and the other half received vehicle and RISA. 30 mg/kg indomethacin reduced the increase in uterine blood volume due to estrogen about 60% (estimated from bar graphs) to about 35% increase ($p < .01$), but this dose of indomethacin was toxic. A dose response test showed no visible intestinal toxicity at 5 mg/kg/day indomethacin and 10 mg/kg/day meclofenamic acid. Uterine blood volume tests (similar protocol) were reduced from about 100% increase in untreated rats to about 65% ($p < .05$) with 5 mg indomethacin, and to about 75% with 10 mg meclofenamic acid (n.s.). PGF levels were about 14 and 25 ng per horn in control and estrogen-treated rats, about 8 and 17 in comparable rats given 10 mg meclofenamic acid, and about 3 and 5 in comparable rats given 4 mg indomethacin. In another test the uterine blood volume increase was reduced to about 70% ($p < .05$) with 20 mg meclofenamic acid, compared with about 115% with estrogen. Increases in PGE and PGF were eliminated by 20 mg meclofenamic acid. Thus the PG levels and uterine hyperemia were correlated. (LJG)   005632

0965

SADOVSKY, E.; DORA, A.; PFEIFER, Y.; POLISHUK, W.Z.; SULMAN, F.G.
Attempts to delay labor in rats by a serotonin antagonist, cyproheptadine.
Israel Journal of Medical Sciences 9(11-12): 1590-1591. November-December 1973.

12 pregnant rats received .5 mg and 12 received 1.25 mg cyproheptadine, a serotonin antagonist, sc twice daily for the last week of gestation. No inhibition of deliveries compared to 12 controls occurred, indicationg that serotonin probably plays no role in initiating labor in rats. (LJG)   006051

0966

SAKSENA, S.K.; LAU, I.F.
Effect of exogenous estradiol and progesterone on the uterine tissue levels of prostaglandin $F_2\alpha$ ($PGF_2\alpha$) in ovariectomized mice.
Prostaglandins 3(3): 317-322. March 1973.

Ovariectomized mice were treated 2 weeks later with .05 μg estradiol and/or 1 mg progesterone subcutaneously for 6 consecutive days. Uterine content of $PGF_2\alpha$ in control animals averaged 2 ng but was 7 ng in those animals given progesterone or estradiol for 6 days and in those given estradiol for 3 days followed by progesterone for 3 days. The mean value, however, in those animals given progesterone for 3 days and then estrogen for 3 days was 11 ng. (HC)   006766

0967
SAKSENA, S.K.; STEELE, R.; HARPER, M.J.K.
   Effects of exogenous oestradiol and progesterone on serum levels of prostaglandin F and luteinizing hormone in chronically ovariectomized rats.
   Journal of Reproduction and Fertility 32(3): 495-499. 1973.

Ovariectomized rats received daily subcutaneous injection of 5 μg 17β-estradiol or 2 mg progesterone beginning 35 days after the operation. Estradiol for 6 days or estradiol for 3 days followed by progesterone for 3 days virtually eliminated the postcastration rise in serum LH levels. Progesterone for 6 days or progesterone for 3 days followed by estradiol for 3 days did not interfere with the LH rise. None of the treatments significantly altered serum PGF levels. The authors suggested that the time of sampling after estradiol injection rather than the sequence in which the steroids are administered determines the effect on LH levels. They further suggested that the failure of estrogen to elevate PGF levels, contrary to what had been noted in other species, may be explained by a small change being masked by the high base-line values. (HC)   006932

0968
SAKSENA, S.K.; El-SAFOURY, S.; BARTKE, A.
   Prostaglandins E$_2$ and F$_2\alpha$ decrease plasma testosterone levels in male rats.
   Prostaglandins 4(2): 235-242. August 1973.

PGE$_2$ and PGF$_2\alpha$, but not PGE$_1$ or PGF$_1\alpha$, decreased radioimmuoassayed plasma testosterone, LH, and testis weight in mature male rats. In the first trial, 500 μg PGF$_2\alpha$ in saline injected sc at 0900 and 2100 for 3 days into fertile 350-400 g rats decreased testosterone from 2.79 to .73 ng per ml plasma and LH from 15.4 to 6.13 ng per ml. PGE$_2$ (same dose and schedule) reduced testosterone to .37 ng per ml, LH to 9.7 ng per ml and testis weight from 926.5 to 837.2 mg per 100 g body weight. In the second trial, 250-300 g rats given 250 μg PG in sesame oil every 12 hours for 4 days had 1.03 ng per ml plasma testosterone after PGE$_2$ and 1.39 ng per ml after PGF$_2\alpha$. Control values were 2.34 ng per ml. 250 μg PGE$_1$ or PGF$_1\alpha$ produced no significant changes. (LJG)   005690

0969
SAKSENA, S.K.; SHAIKH, S.A.; SHAIKH, A.A.
   Uterine and peripheral plasma F-prostaglandins correlated with peripheral progesterone in cyclic rats.
   Prostaglandins 4(2): 243-249. August 1973.

Uterine and peripheral venous blood samples were collected from rats each day of the estrous cycle. PGF and progesterone content were estimated using a double antibody radioimmunoassay. Uterine plasma PGF values were lowest (6 ng/ml) at 1000 to 1200 on day of estrous and highest (12.5 ng/ml) on day of metestrus. Small peaks occurred on the day of diestrus and at 1400 to 1600 hours day of proestrus. Peripheral plasma PGF also peaked (6.5 ng/ml) between 1400 hours and 1600 hours on day of proestrous and dropped to a low of 3.3 ng/ml 5 hours later. Simultaneously, plasma progesterone levels rose from a low of 19 ng/ml at 1400 hours to 1600 hours on the day of proestrous to a high of 46 ng/ml 5 hours later. The authors suggest that the uterine plasma PGF peak on the day of metestrus might be responsible for the subsequent fall in progesterone and that the peak on the day of proestrus could be involved in the process of ovulation. (HC)   005701

0970

SAKSENA, S.K.; LAU, I.F.

Uterine tissue content of F prostaglandins in the cyclic mouse, their role in ovulation and changes in content caused by estradiol, progesterone arachidonic acid and indomethacin.

Biology of Reproduction 9: 69. 1973.

Uterine PGF levels in mice ranged from 9.8 ng on the day of metestrus to 22.2 ng late on the day of proestrus. Indomethacin markedly reduced the peak, but estradiol and progesterone increased PGF levels. Indomethacin blocked ovulation; this was partially reversed by PGE_2 or $PGF_2\alpha$ and completely reversed by LH and HCG. The authors conclude that PGE and PGF may be involved in ovulation, that estradiol is more effective than progesterone in stimulating synthesis/release of PG by the uterus, and that indomethacin prevents the ovulatory effect of LH on the ovary. (HC)   005699

0971

SALDANA, L.; SCHULMAN, H.; YANG, W.-H.

On the mechanism of midtrimester abortions induced by the prostaglandin 'impact.'

Prostaglandins 3(6): 847-858. June 1973.

An intraamniotic $PGF_2\alpha$ study was designed to examine its mechanism of action. Using Csapo's technique, a single dose of 24.3 ± 1.1 mg $PGF_2\alpha$ was given intraamniotically to 20 sedated, 16-week pregnant patients to provoke a PG impact (PGI) caused by consequent progesterone withdrawal and conversion of the normal pregnant uterus into a reactive organ. Side effects were slight and further $PGF_2\alpha$ treatment was necessary in only 4 cases. When the uterus was reactive, 50 mU per minute of oxytocin was infused intravenously to ease the evolution of IUP to 93 ± 3 mm Hg, promoting clinical progress. All 20 patients aborted both the fetus and placenta in 16.5 ± 2.1 hours. Plasma progesterone levels decreased significantly after 3 hours ($p<.05$). The mean progesterone (P) levels were significantly higher in patients aborting rapidly than those of patients taking longer. As progesterone levels decreased, the cyclic IUP and reactivity of the uterus to oxytocin increased and the rate of P withdrawal significantly affected the IAT. This study substantiates the conclusion that the effect of PGI is to provoke P withdrawal, thus converting the refractory pregnant uterus into a reactive organ. (IC)   006466

0972

SALMON, J.A.; AMY, J.J.

Levels of prostaglandin $F_2\alpha$ in amniotic fluid during pregnancy and labor.

Prostaglandins 4(4): 523-533. October 1973.

A highly specific anti-serum for $PGF_2\alpha$ was prepared by injecting rabbits with $PGF_2\alpha$ conjugated to bovine serum albumin. This anti-serum and a tracer amount of tritiated $PGF_2\alpha$ were then added to samples of amniotic fluid obtained from women during pregnancy and labor; $PGF_2\alpha$ content was assayed in a scintillation counter. Levels of $PGF_2\alpha$ were less than 1.0 ng per ml from Weeks 15 to 35, rose to 2 ng per ml Weeks 40-43 and then to 6 ng per ml during labor. $PGF_2\alpha$ levels during labor paralleled the degree of cervical dilatation. The technique utilized eliminates the need for extraction and chromatography and thereby increases the sensitivity of the assay. The decidua is suggested as the source of PG detected in amniotic fluid. The increased PG could play a role in initiation of labor or could have been released by uterine distension. (HC)   005609

0973
SALOMY, M.; HALBRECHT, I.
Immediate and late effect of PGE$_1$ on the rabbit oviduct (in vivo studies).
Advances in the Biosciences 9: 795-803. 1973.

The effect of PGE$_1$ on the contractions of the rabbit oviduct was assessed by local injections into the oviduct. Each injection contained .4 µg PGE$_1$ solved in .2 cc saline solution. Pressure changes from the oviducts and uterus performed simultaneously on the same rabbits were recorded daily. The following results were observed: 1. An immediate paralyzing effect on the contractions of the injected oviduct for 3-5 minutes. The contractions return to the starting point gradually in 5-20 minutes. The uterus of the same side starts contracting strongly and more frequently in 3-5 minutes following the injection. The injections were repeated 6 times in intervals of 30 minutes with the same results. 2. The contractions of the injected oviduct increased gradually 5-7 hours following the injection reaching the maximum level after 9-14 hours up to 30 hours. The contralateral side started with strong contractions 12-16 hours after the first injection. The pattern of the contractions returned to its starting point after 48 hours. Using iv injections of PGE$_1$ 200 µg each time we observed the same results on both sides. These changes in the pattern of the contractions of the oviduct are similar to the changes observed following induction of ovulation in the rabbit with HCG given iv. (Authors' modified)
006500

0974
SAMUELSSON, B.
Methods for determination of prostaglandin formation in man.
In: Bergstrom, S., ed. Report from meetings of the prostaglandin task force steering committee, Chapel Hill, June 8-10, 1972, Stockholm, October 2-3, 1972, Geneva, February 26-28, 1972. Stockholm, WHO Research and Training Centre on Human Reproduction, Karolinska Institutet, 1973. (Prostaglandins in Fertility Control 3) p. 37-42.

This review describes how the major urine metabolites of PGEs and PGFs were quantitated by gas liquid and mass spectommetry techniques, measured to calculate amount of PGEs and PGFs synthesized, and measured after intake of indomethacin or salicylates to estimate inhibition of PG synthesis. Plasma metabolites of PGEs and PGFs were measured to calculate plasma PGE and PGF levels. The major urine metabolite of PGE$_1$ and PGE$_2$ is a 16 carbon dicarboxylic acid, 7alpha-hydroxy-5,11-diketo-tetranor-prostane-1,16-dioic acid, and that of PGF$_1\alpha$ and PGF$_2\alpha$ is 3alpha,7alpha-dihydroxy-11-ketotetraner-prostane-1,16-dioic acid. 24-hour urine contained 7-27 µg of the PGE metabolite, higher in men than in women. PG turnover was estimated by quantitating urine metabolites to be 18-38 µg PGE and 36-61 µg PGF synthesized per 24 hours in women, and 46-333 µg PGE and 42-120 µg PGF per 24 hours in men. 4 doses of 50 mg indomethacin suppressed the urine PGE metabolite 77-98%; 4 doses of .75 gm aspirin or salicylic acid were comparable. The fate of PGE$_2$ in plasma after iv administration showed that within 1.5 minutes less than 3% of PGE$_2$ remained; 40% was recovered as 15-keto-dihydro-PGE$_2$ (product of 15-dehydrogenase and 13-reductase). After iv PGF$_2\alpha$ infusion in women, the 15-keto-dihydro-PGF$_2\alpha$ was 10-70 times higher than PGF$_2\alpha$; implying that the normal PGF$_2\alpha$ level should be less than 2 pg per ml and that of 15-keto-dihydro-PGF$_2\alpha$ is up to 50 pg per ml. Thus plasma PG levels often reported are 100-1000 fold too high; the 15-keto-dihydro-PGF$_2\alpha$ would be a more accurate measure of plasma PG. Radioimmunoassay and mass spectrometry methods have been developed for 15-keto-dihydro-PGF$_2\alpha$ in the author's laboratory. (LJG)   006736

322

0975

SANTOS, A.A.; HERMIER, C.; NETTER, A.

Etude in vitro de la synthese de la progesterone dans le corps jaune cyclique humain: role de la prostaglandine $F_2\alpha$. [In vitro study of synthesis of progesterone in the human cyclic corpus luteum: role of prostaglandin $F_2\alpha$].

FEBS Letter 34(2): 179-184. August 1973.

The effects of $PGF_2\alpha$, HCG, imidazole, theophylline, and acetylsalicylic acid on progesterone synthesis by human cyclic corpus luteum are quantitated. Corpora lutea were removed from patients without endocrine disorders during luteal phase, minced into 10 mg pieces and preincubated within 1 hour in 5 ml Krebs-Ringer-bicarbonate-glucose with .5% bovine serum albumin. In 3 hour incubations mean progesterone increased from 42.4 μg per g after preincubation to 58.6 in presence of 2.5 μg per ml $PGF_2\alpha$, 61.8 in 5 μg PG, 91.5 in 7.5 μg PG, 93.9 in 10 μg PG, 64.7 in 12.5 μg PG and 75.9 in μg PG. Mean progesterone released was 109.9 μg per g incubated with 20 IU HCG per ml, 221.6 with 30 IU and 85.6 μg with 50 IU HCG. Incubations with 20 IU HCG and 10 μg $PGF_2\alpha$ per ml, and with 30 IU HCG and 7.5 μg $PGF_2\alpha$ per ml were additive. Theophylline 1.5 μg per ml stimulated progesterone release, and when combined with 7.5 μg per ml $PGF_2\alpha$, was additive. Imidazole 8 μmol per ml inhibited the expected stimulation by PG. Acetylsalicylic acid from $10^{-7}M$ to $10^{-2}M$ had no effect. These results (except for the aspirin experiments) suggest a role for adenyl cyclase and separate receptors for $PGF_2\alpha$ and HCG in progesterone synthesis by human corpus luteum. (LJG)   006067

0976

SATO, T.; AMI, K.; SHINADA, T.; IGARASHI, M.

Plasma progesterone and prostaglandin $F_2\alpha$ levels during the insertion of prostaglandin $F_2\alpha$ vaginal tablet.

Prostaglandins 4(1): 107-113. July 1973.

$PGF_2\alpha$ in a 50 mg vaginal tablet was given twice at 1 hour intervals to 3 women in the luteal phase and 2 times in 1 hour to 3 women 6-10 weeks pregnant; plasma $PGF_2\alpha$ and progesterone were measured for up to 24 hours. The pregnant women experienced vaginal bleeding 2.5-3 hours after the first tablet, 1 had a complete abortion 4 hours later, and 1 had an incomplete abortion. PG level rose and progesterone level decreased slightly. The nonpregnant women began a menstrual-like bleeding 2.5-3 hours after $PGF_2\alpha$ but plasma progesterone did not change significantly. (LJG)   005838

0977

SATO, T.; AMI, K.; MATSUMOTO, S.

The induction of abortion and menstruation by the intravaginal administration of prostaglandin $F_2\alpha$.

American Journal of Obstetrics and Gynecology 116(2): 287-289. May 15, 1973.

16 patients pregnant 6-11 weeks were given 50 mg $PGF_2\alpha$ tablets intravaginally at 1- to 2-hour intervals. 7 aborted completely, 9 incompletely; total dose was 100-250 mg. All patients had nausea, vomiting, and diarrhea. For induction of menstruation, 10 women were given 25-100 mg intravaginally. Menstruallike bleeding occurred in those women given the drug 2-3 days before the expected start of menstruation but did not occur if the drug was given 5-13 days before menstruation. Nausea, vomiting, and diarrhea were observed in 6 of these patients. (HC)   006774

0978

SCARAMUZZI, R.J.; BAIRD, D.T.; WHEELER, A.G.; LAND, R.B.
The oestrous cycle of the ewe following active immunisation against prostaglandin $F_2\alpha$.
Acta Endocrinologica Suppl. 177: 318. 1973.

6 ewes were actively immunized against a $PGF_2\alpha$-protein conjugate of bovine serum albumin or keyhole limpet hemocyanin. Antibody titers rose steadily, reaching a plateau in 90 days. Plasma $PGF_2\alpha$ levels at this time were .8-7.1 ng/ml. Estrous activity stopped in 4 of the animals within 1 month after immunization. At least 1 marked corpus luteum was observed in 3 of 6 experimental ewes 28 days after laparoscopy, 62 days after immunization. Surgical enucleation of the persistent corpora lutea was followed by mating within 3 days. No immunized animal showed estrus during the subsequent 42 days; all developed normal corpora lutea, but none became pregnant. (HC)   005727

0979

SCHER, J.; BAILLIE, P.; JESSOP, S.; HENDRIE, B.
A comparison between the effects of prostaglandin $F_2\alpha$ and oxytocin on fluid balance during induction of labour in patients suffering from pre-eclampsia.
South African Medical Journal 47: 1291-1292. July 23, 1973.

Water balance, serum and urinary osmolality, creatinine clearance, and renal function were compared in 8 pre-eclamptic primigravidae given oxytocin and 9 given $PGF_2\alpha$ to induce labor. Patients were selected with these criteria: blood pressure over 140/90; comparable proteinuria; symptoms escalating after 28 weeks of pregnancy; no history of renal disease or hypertension. Drugs were infused in 5% dextrose by standard methods and no oral fluids were permitted. Urine volume and osmolality were measured by catheter every 30 minutes and plasma osmolality and creatinine every 3 hours. Fluid intake was 43.5 ml per hour, greater in patients receiving oxytocin ($p < .05$) but no significant differences appeared in fluid output, urine osmolality or volume. This data suggests an antidiuretic effect of oxytocin and an indication for the use of prostaglandin rather than oxytocin in cases of pre-eclampsia. (LJG)   005454

0980

SCHER, J.; BAILLIE, P.
The effect of beta-adrenergic agents on prostaglandins stimulated labor.
Advances in the Biosciences 9: 743-749. 1973.

The effect of beta-adrenergic administration (orciprenaline) on oxytocin- and prostaglandin-induced uterine activity at term in the same subject was investigated in 10 cases. The orciprenaline has less effect on the prostaglandin-induced activity. This is consistent with the hypothesis that prostaglandins may be implicated in the stage of rapid cervical dilatation and suggests the use of antiprostaglandins in advanced premature labor. (Authors)   006492

0981

SCHLEGEL, W.; DEMERS, L.M.
Human placental prostaglandin dehydrogenase: partial purification and characterization.
Federation Proceedings 32: 803Abs. 1973.

The authors describe the chromatographic procedures used for the 100-fold purification of prostaglandin dehydrogenase from term human placenta. There is more enzyme in this tissue than in immature (10 to 20 weeks) placenta and estrogen increases its activity 10-fold. (HC)   006704

0982

SCHNEIDER, G.; MULROW, P.J.

Regulation of aldosterone production during pregnancy in the rat.

Endocrinology 92(4): 1208-1215. April 1973.

Plasma renin activity (PRA), electrolyte, PGA, maternal and fetal adrenal corticosterone and aldosterone release in vitro were determined in 2 strains of rats (dark agouti and dark agouti x Lewis F1) given normal (102 mEq/kg Na) and low sodium (negligible Na) diets from Day 14 of pregnancy. The effects of sodium deprivation on the fetal renin-aldosterone system were also examined. In dark agouti rats, PRA was not increased during pregnancy, but was doubled during sodium depletion. Aldosterone production increased during pregnancy (p<.0075), but changed no further during sodium depletion. In F1 rats, both PRA and aldosterone production increased during pregnancy and was not further stimulated by sodium depletion. Fetal aldosterone production was doubled by maternal sodium depletion in both strains (p<.01), but renal renin content remained unchanged. Plasma potassium in the mother was increased in pregnancy (p<.025) and increased further with sodium depletion (p<.001). Plasma PGA was not affected by pregnancy or sodium restriction. It was suggested that increased plasma potassium is associated with the increased aldosterone production observed during pregnancy in rats and may also be a controlling factor in fetal aldosterone production. (Authors' modified)   006345

0983

SEELEY, R.R.; HARGROVE, J.L.; SANDERS, R.T.; ELLIS, L.C.

Response of rabbit testicular capsular contractions to testosterone, prostaglandin E_1, and isoproterenol in vitro and in vivo.

Proceedings of the Society for Experimental Biology and Medicine 144(1): 329-332. 1973.

Rabbit testicular contractions initiated by PGE_2 in vitro were inhibited by 5×10^{-3} mg/ml hydroxyprogesterone, testosterone, and several other steroids. Spontaneous contractions were not eliminated by repeatedly changing the bathing medium of the in vivo preparations nor were they affected by testosterone in the bath medium (.12 mg/ml) or injected intravenously (18 mg/kg). The stimulatory effect of PGE_1 was partially inhibited by testosterone at 2.5×10^{-2} mg/ml. PGE_1 stimulated and at higher concentrations inhibited contractions with equal potency in vitro and in vivo. Isoproterenol inhibited motility when added to the bath medium (5×10^{-3} mg/ml) or injected intravenously (1.7 mg/kg). The authors concluded that the in vivo preparation is more stable and that endogenous steroids do not play a role in regulating testicular contractions, whereas nervous mechanisms and prostaglandins are important regulators. (HC)   005729

0984

SELLER, M.J.; CAMPBELL, S.; COLTART, T.M.; SINGER, J.D.

Early termination of anencephalic pregnancy after detection by raised alpha-fetoprotein levels.

Lancet 2: 73. July 14, 1973.

24 mg of $PGF_2\alpha$ was instilled as an abortifacient into the amniotic cavity of a women with an anencephalic pregnancy. The authors suggested that anencephaly can be determined in utero by measuring the alpha-fetoprotein content of the amniotic fluid. (RAS)   006046

0985

SEPPALA, M.; RENKONEN, O.V.; VARA, P.

Extra-amniotic prostaglandin-oxytocin abortion: comparison with the intra-amniotic method.

Prostaglandins 3(1): 17-28. January 1973.

23 midtrimester patients received .5-1.0 mg $PGF_2\alpha$ by the intrauterine route at 1-hour intervals. 35 women were also given oxytocin simultaneously by the intrauterine route. 10 women received .5 mg $PGF_2\alpha$ every 2 hours with intravenous oxytocin begun 3 hours before. 58 women were given a single 25 mg dose intraamniotically (supplemented 24 hours later if necessary) along with intravenous oxytocin begun before the $PGF_2\alpha$ injection. 78% of the women given only intrauterine $PGF_2\alpha$ aborted; 35% were complete, the IAT averaged 27 hours, the average dose was 20 mg. Nausea, vomiting, decrease in hemoglobin, and signs of infection were frequent. 83% of the patients given intrauterine $PGF_2\alpha$ and oxytocin simultaneously aborted; 45% were complete, the IAT averaged 24 hours, and the average dose was 13 mg. Nausea, vomiting, decreased hemoglobin, and signs of infection occurred frequently. 90% of the women given intrauterine $PGF_2\alpha$ and intravenous oxytocin aborted; 50% were complete, the mean IAT was 22 hours, and the average dose was 6.25 mg. Headache, neausea, and vomiting were frequent. 95% of the patients that received intraamniotic $PGF_2\alpha$ and intravenous oxytocin aborted; 45% were complete, the IAT average was 20 hours, and the mean $PGF_2\alpha$ dose was 27.5 mg. Nausea, vomiting, and signs of infection occurred frequently. (RAS)   006759

0986

SHAIKH, A.A.; SAKSENA, S.K.

Cyclic changes in uterine venous and peripheral plasma levels of F prostaglandins correlated with peripheral progesterone levels in the golden hamster.

Advances in the Biosciences 9: 635-639. 1973.

The concentration of F series prostaglandins (PGF) were measured in the uterine venous plasma and peripheral plasma of hamsters during the estrous cycle by radioimmunoassay. Radioimmunoassay measurements of total progestins were also made in the peripheral plasma. Uterine venous blood and peripheral blood were collected throughout the estrous cycle and at short time intervals on day 4 (day 4 = day before ovulation and day 1 = day of ovulation). The lowest concentrations of PGF in the uterine venous plasma were recorded between 1000-1200 h of day 4 (0.98 ng/ml), and the peak concentrations (7.1 ng/ml) were recorded between 1900-2100 h of the same day. These values showed a significant drop before ovulation. Another peak of PGF was recorded on day 3. The peripheral PGF values showed a pattern similar to that of uterine PGF. Total progestin values dropped steadily from day 1 onwards reaching their lowest levels between 100-1200 h on day 4. Thereafter, these levels started to rise reaching a peak just after ovulation. Estrone, estradiol, and progesterone levels have been determined in hamster ovarian venous plasma using the same time schedule. From these studies, it appears that the secretion of estrogens, progestins, and PGFs are interdependent. (Authors)   006481

0987

SHAIKH, A.A.; BIRCHALL, K.; SAKSENA, S.K.

Steroids in the ovarian venous plasma and F prostaglandins in the peripheral plasma during pseudopregnancy and days 1-9 of pregnancy in the golden hamster.

Prostaglandins 4(1): 17-30. July 1973.

Estrone, estradiol, progesterone, and 17alpha-hydroxy-progesterone were determined daily in ovarian venous blood and PGF in peripheral plasma until Day 9 of pseudopregnancy or pregnancy, in mature 4-day cycling golden hamsters. Ovarian blood was collected under pentobarbital anesthesia from hamsters, confirmed to be pseudopregnant or pregnant, at 1000-1200 daily and also at 1900-2100 and 2300-0100 on Days 4 and 8. Steroids and PGF were extracted, chromatographed, and radioimmunoassayed. In pseudopregnancy, estradiol rose to a plateau on Day 3-4, dropped, and peaked on Day 8. Estrone resembled estradiol, with less magnitude. Progesterone and OH-progesterone decreased, then peaked on pseudopregnancy Day 6. (PGF rose slightly from .83 to 1.1 ng per ml, peaked abruptly to 2.59 on Day 4, fell and peaked on Day 9 at 2.53 in ovulating hamsters and 3.03 in hamsters that did not ovulate.) In pregnant hamsters estradiol and estrone also increased up to Day 4, fell, and rose again by Day 9. Progesterone showed peaks on Days 4 and 9, but the hydroxy-progesterone level increased on Day 9 only. PGF fell from 2.47 to .48 by 1900-2100 of Day 4, rose to 2.67 by 2300-0100. PGF ranged from 1.21 to 1.97 on Days 5-9. These results differed from rat data in that hamsters secreted more estrogen and less progestin and hamsters showed increased estrogen, progestin, and PGF before implantation on Day 5. (LJG)   005831

0988

SHARMA, S.C.; FITZPATRICK, R.J.; LANHAM, J.

Development of a potent antiserum to prostaglandin F and measurement of prostaglandin F in the peripheral circulation of parturient sheep.

Journal of Reproduction and Fertility 35(3): 598-599. 1973.

Specific highly-potent antisera were developed in rabbits against a conjugate of $PGF_2\alpha$ with human serum albumin. This antiserum was used as the basis of a radioimmunoassay for $PGF_2\alpha$ in plasma samples taken from 8 ewes during the last week before expected day of delivery. PGF levels began to rise about 40 hours before parturition while progesterone levels started to fall and total estrogens began to rise 30 to 33 hours before parturition. (HC)   005761

0989

SHARMA, S.C.; HIBBARD, B.M.; HAMLETT, J.D.; FITZPATRICK, R.J.

Prostaglandin $F_2\alpha$ concentrations in peripheral blood during the first stage of normal labour.

British Medical Journal 1(5855): 709-711. March 24, 1973.

Blood samples were taken periodically from the antecubital vein of women in the first stage of labor. The samples were assayed with tritiated $PGF_2\alpha$ and rabbit antiserum raised against $PGF_2\alpha$-albumin conjugate. Using intermittent sampling in 4 patients, $PGF_2\alpha$ concentrations ranged from less than 75 to 427 pg/ml, and uterine contractions were associated with, but not correlated with, elevated $PGF_2\alpha$ concentrations. Using a more detailed sampling in 6 patients, maximum $PGF_2\alpha$ concentrations (300 pg/ml) were obtained 15-45 seconds after uterine contractions. Assuming circulation time from uterine to antecubital vein to be less than 60 seconds, the authors suggest that prostaglandins are released by, rather than initiate, uterine contractions. (HC)   005698

0990

SHARMA, S.D.; FILSHIE, S.E.; FILSHIE, M.; REICH, L.; KARIM, S.M.M.

Prostaglandins - an alternative to menstrual extraction.

Menstrual Regulation Conference, Honolulu, Hawaii, 9 p. 1973.

Women who had had a delay of 1 to 15 days in their menses were given 15-methyl PGE_2methyl ester 30 or 40 μg vaginally 1, 2 or 3 times per day for 3 to 4 days. Of the 11 women given the drug 1 time per day, menstruation was induced in 2 of the 9 that were pregnant and in both non-pregnant patients. 6 patients had diarrhea, 3 had abdominal cramps. Giving the drug 2 times per day induced menstruation in 4 of 8 pregnant women and in all 3 non-pregnant. 5 patients had diarrhea.

Menstruation was induced in 8 of 12 pregnant and in 7 of 8 non-pregnant women given the drug 3 times a day. Side effects included diarrhea, abdominal cramps, shivering. Total dose in all 3 protocols averaged 140 to 165 μg. Problems associated with the method include inaccurate dating of the last menstrual period, unreliable diagnosis of pregnancy, troublesome side effects, and vaginal bleeding but intact pregnancy. The authors discuss mechanisms of action for PG, namely luteolytic, mechanical, and vascular. (HC)   006938

0991

SHARP, D.S.; BURSLEM, R.W.
  The termination of pregnancy by intravenous infusion of a synthetic prostaglandin $F_2\alpha$ analoque.
  Journal of Obstetrics and Gynaecology of the British Commonwealth 80(2): 138-141. February 1973.

ICI 74205, a synthetic form (that is a mixture of optical isomers) of $PGF_2\alpha$, with an alkyl side chain extended by an ethyl group, was given intravenously to 5 women pregnant 16-17 weeks and to 1 woman pregnant 9 weeks at a rate of 5-10 μg per minute. Abortion was complete in 5, incomplete in 1 patient; the IAT averaged 25 hours 15 minutes; the total dose given averaged 150 mg. All patients had diarrhea and vomiting. All patients had a fall in plasma estradiol and progesterone levels beginning 2 hours after the start of the infusion. (HC)   005697

0992

SHEARMAN, R.; SMITH, I.; KORDA, A.
  Intrauterine administration of $PGF_2\alpha$ during the second trimester of pregnancy; clinical and hormonal results.
  Journal of Reproduction and Fertility 32(2): 321-322. 1973.

The clinical and hormonal response of 41 patients (12-20 weeks gestation) undergoing $PGF_2\alpha$-induced abortion was studied. 34 patients in whom the uterus was palpable abdominally were given $PGF_2\alpha$ by transabdominal intraamniotic catheter in an initial dose of 30 mg followed by 15 mg at 24 hours and 42 hours. 7 patients received $PGF_2\alpha$ by transcervical route into the extraovular space in doses of 375-750 μg every 1-2 hours until abortion occurred. Pregnancy was terminated in all patients. 2 of the 34 receiving intraamniotic $PGF_2\alpha$ failed to abort within 72 hours; 1 aborted after transcervical $PGF_2\alpha$ and the other required curettage. The mean induction-abortion interval was 28 hours in the remaining 32 women. 1 patient aborted through a spontaneous cervicovaginal fistula, a complication known with saline but not previously with PGs. 3 patients having transcervical terminations developed genital tract infection; the mean induction-abortion interval was 36 hours. 5 patients experienced vomiting after each injection. The high incidence of induced lactation and the probable importance of chronoperiodicity raise fundamental questions concerning hypothalamic stimulation and endogenous biological rhythms. (IC)   006437

0993

SHELTON, J.N.
  Prostaglandin $F_2\alpha$ for synchronization of oestrus in beef cattle.
  Australian Veterinary Journal 49(9): 442-444. September 1973.

Estrus was induced in nonpregnant cows 2-6 days after treatment with intrauterine $PGF_2\alpha$. The author was unable to establish a dose-response relationship, the explanation offered being that the biological activity of PG deteriorated during the 3-month period of this study. (HC)   005752

0994

SHERWOOD, O.D.; BIRKHIMER, M.L.; PARKES, D.G.

The gonadotropin-primed rat: A model for testing agents that influence steroid biosynthesis.

Endocrinology 93(3): 723-728. 1973.

To study effects on steroidogenesis, 26 immature female rats were given pregnant mare serum gonadotropin (PMSG) and human chorionic gonadotropin (HCG), and correlations were made of PMSG and HCG administration on ovarian, adrenal, and pituitary weights, plasma progestin and LH concentrations, and changes produced by injection of $PGF_2\alpha$, aminoglutethimide phosphate, metyrapone, LH-releasing factor, and ergocornine methanesulfonate. Day 5 gonadotropin-primed rats (D5GPR) were injected with 177 μg $PGF_2\alpha$. Plasma samples were analyzed at 6 time intervals (10-240 minutes) after administration. Following injection, plasma progestin concentrations dropped to 50% of control concentrations after 20-30 minutes, then increased. In a second $PGF_2\alpha$ experiment D5GPR were injected with doses ranging from 1 ng-88 μg. The minimum amount of $PGF_2\alpha$ needed to obtain a maximal reduction of plasma progestin concentration 30 minutes after treatment (50% control levels) was 3-5 μg. The D5GPR is a sensitive biological model to guide synthesis toward an effective luteolytic PG analogue. Although the immediate decrease in plasma progestin concentrations brought about by natural PGs and their analogues may not accurately reflect their ability to produce long-term luteolysis, the progestin concentrations remain high for several days in the PMSG-HCG primed rat; therefore this model could be used to compare long-range effects of repeated doses of various PG analogues on plasma progestin concentrations. Since the D5GPR model is an in vivo model, it may be useful in search for orally active PG analogues. (IC)   006453

0995

SHINE, I.; LAL, S.

Prostaglandin E_2 and sickling.

New England Journal of Medicine 289(19): 1040. November 8, 1973.

Microscopic examination of erythrocytes treated with PGE_2 10 to 200 ng per ml revealed no enhancement of sickling. Similarly, PGE_2 did not reduce erythrocyte sedimentation rate or change blood viscosity. The study was undertaken in reaction to speculation that PGE_2 used to induce labor or abortion may initiate sickling. (HC)   006266

0996

SHUTT, D.A.; SMITH, I.D.; SHEARMAN, R.P.

Changes in estrogen levels in maternal venous plasma after intra-amniotic infusion of prostaglandin $F_2\alpha$ for therapeutic abortion.

Prostaglandins 4(2): 291-299. August 1973.

10 women pregnant 14-19 weeks were given $PGF_2\alpha$ intraamniotically by transamniotic injection, 30 mg initially with additional 15 mg doses given at 24 and 42 hours if necessary. All women aborted completely, the mean IAT being 24 ± 12 hours. Peripheral plasma levels of estrone, estradiol-17β, and estriol decreased 80%, the rate being more rapid during the first and last quarters of the IAT. There was no relationship between mean preinfusion level or pattern of change in estrogen levels and the IAT. The authors suggest that the decrease in estrogen level following $PGF_2\alpha$ administration could be the result of direct effect on the fetal-placental unit. (HC)   005700

0997
SMITH, A.P.
The effects of intravenous infusion of graded doses of prostaglandins $F_2\alpha$ and E_2on lung resistance in
patients undergoing termination of pregnancy.
Clinical Science 44(1): 17-25. 1973.

7 patients pregnant 11-18 weeks given intravenous infusions of PGE_2at a rate of 2.5 μg/minute the
first 30 minutes, increased to 20 μg/minute over 8 hours. Another 8 patients received $PGF_2\alpha$ 5
μg/minute increased gradually to 200 μg/minute. Of the women given PGE_2, lung resistance
increased in 5, the mean change for the group being a small but significant increase at the 10-20
μg/minute rate. 5 of the patients given $PGF_2\alpha$ had increased lung resistance, the mean change being
small but significant at the 200 μg/minute rate. The authors conclude that PG infusions for
terminating pregnancy should be contraindicated in patients with preexisting bronchoconstricting
conditions such as asthma. Since PGE_2relaxes bronchial smooth muscle, the authors hypothesize that
the increased lung resistance noted following its use in this study may have been due to a metabolite,
a situation that might also account for its uterine-constricting effect. (HC)   005499

0998
SMITH, G.W.; ABEL, M.; NATHANIELSZ, P.W.
Uterine sensitivity to prostaglandins in the pregnant rabbit: a method for testing the effect of previous
administration of small doses of $PGF_2\alpha$ for short periods.
Prostaglandins 3(4): 525-530. April 1973.

On Day 21 of pregnancy, a catheter was inserted via the femoral artery of rabbits to the origin of the
uterine and ovarian arteries and a pressure transducer was placed in one horn of the uterus.
Spontaneous contractions were absent or minimal; $PGF_2\alpha$ infused at 1-10 μg/minute for 10 minutes
had no effect. However, following an infusion of 1 μg/hour for 11 1/2 to 16 hours, regular
spontaneous contractions were elicited and sensitivity to $PGF_2\alpha$ became marked. Absence of
contractions in controls and prior to continuous infusion of $PGF_2\alpha$ demonstrates that the development
of sensitivity is not related to anesthesia or surgery. (HC)   005724

0999
SMITH, I.D.; SHEARMAN, R.P.; KORDA, A.R.
Chronoperiodicity in the response to the intra-amniotic injection of prostaglandin $F_2\alpha$ in the human.
Nature 241: 279-280. January 26, 1973.

$PGF_2\alpha$ was injected transabdominally into the amniotic sac of pregnant human females in a dose of 30
mg. The mean induction-abortion interval (in hours) ± the standard error for parous women, 19.7 ±
1.7, was less than that for nulliparous women 25.6 ± 1.6 (p<.05). The induction-abortion interval
was shortest when the initial injection of $PGF_2\alpha$ was administered at 6 p.m. In the nulliparous women,
the mean induction-abortion interval was significantly less following $PGF_2\alpha$ administration at 5-7 p.m.
than at any other time. In terms of the total dose of $PGF_2\alpha$ required to terminate the pregnancy, there
was a highly significant difference (p<.002) between the group treated at 5-7 p.m. and the group
treated at other times. (AS)   006429

1000

SMITH, I.D.; TEMPLE, D.M.

The influence of analgesic drugs on the action of prostaglandin $F_2\alpha$ on the human uterus in vivo and human and rabbit myometrial strips in vitro.

Prostaglandins 4(4): 469-477. 1973.

To investigate the possibility of antagonistic or synergistic effects of concomitant therapy during the termination of pregnancy or induction of labor with PGs, patients 12-20 weeks pregnant undergoing $PGF_2\alpha$-induced abortions were treated with 2 types of analgesic drug. The group receiving a mixture of dextropropoxyphene hydrochloride and acetaminophen (Di-gesic) showed a significantly prolonged induction-abortion interval compared with the group treated with meperidine. Each analgesic drug was tested by cumulative addition to isolated preparations of myometrial tissue, which was obtained either from humans during hysterectomy or freshly killed rabbits and stimulated to contract regularly by preaddition of $PGF_2\alpha$ to the organ bath. Meperidine and acetaminophen did not influence myometrial contractions in vitro except at concentrations much higher than plasma concentrations produced by analgesic doses of these drugs, but dextropropoxyphene inhibited the contractions at concentrations comparable to the patients' plasma concentrations. It is suggested that the stimulant effect of $PGF_2\alpha$ on the uterus in vivo is inhibited by dextropropoxyphene, and that this accounts for the prolongation of induction-abortion in these patients. (Authors' modified)   005529

1001

SOLOFF, M.; SWARTZ, T.; MORRISON, M.; SAFFRAN, M.

Oxytocin receptors: oxytocin analogs, but not prostaglandins, compete with 3H-oxytocin for uptake by rat uterus.

Endocrinology 92(1): 104-107. 1973.

Uterine pieces from adult female rats pretreated with estrogen were incubated with tritiated oxytocin and nonradioactive oxytocin or its analogues [4-threonine]-oxytocin, [8-valine]-oxytocin and [8-lysine]-vasopressin), PGA_1, PGE_1, PGE_2, and $PGF_2\alpha$. The incubation included 20-30 mg uterine tissue and 8 x 10^{-7}Ci (about 40 ng) tritiated oxytocin in 1 ml Tyrodes solution (pH 7.6) at 20 degrees C for 1 hour. Radioactivity was determined by combustion. Uptake of tritiated oxytocin was 2.5 times greater than that of carbon-14-labeled insulin, interpreted as active binding rather than diffusion. Tritium uptake was reduced competitively with parallel log-concentration curves by [4-threonine]-oxytocin, [8-valine]-oxytocin, and [8-lysine]-vasopressin, in order of their oxytocic potency. [4-proline]-oxytocin and the PGs were inactive in competing with binding of tritiated oxytocin. The specific activity of tritiated oxytocin (20 Ci/mM) was too low to estimate concentration needed for half-maximal biologic response, as has been done with glucagon, insulin, and oxytocin in other tissues. Prostaglandins E_1, E_2, A_1 and the tromethamine salt of $F_2\alpha$ had no effect on the uptake of tritiated oxytocin. It was suggested that oxytocin receptors exist in the uterus and are different from prostaglandin receptor sites. (LJG)   006742

1002

SOLOFF, M.S.; MORRISON, M.J.; SWARTZ, T.L.

The specific uptake of radioactivity from [3H] prostaglandin E_1 by rat uterus.

Prostaglandins 4(6): 853-861. December 1973.

Uterine pieces from rats given 5 μg diethylstilbestrol propionate sc for 2 days were incubated for 1 hour at 20 degrees with $3H$-PGE_1 and various nonradioactive PGs to locate specific PG binding sites. Tissues and 20,000 x g pellets were oxidized to tritiated water and 20,000 x g supernatant fluid was filtered through Sephadex G-25 for determination of radioactivity in fractions, and in samples of incubation medium. Results, expressed as linear reduction of recovered tritium in proportion to log concentration of competing PG, showed that PGE_1 and PGE_2 were comparable in their ability to compete for uptake of $3H\text{-}PGE_1$ from medium, but $PGF_2\alpha$, PGA_1, PGB_1 were 2.5% or less as effective as competitors. All regressions were parallel, indicating a common set of uptake sites. About 9% of the radioactivity was associated with the 20,000 x g pellet, the rest in the supernatant fluid. The radioactivity in the pellet was reduced 70% by 1 μg PGE_1; that in the fluid was reduced 20% by 1 μg PGE_1. 2% of radioactivity was eluted from the Sephadex column, indicating no binding to macromolecules. Uptake sites probably are located in cellular particles, but not in those sedimented at 20,000 x g for 10 minutes. (LJG)   005678

1003
SPARKS, R.M.; LEE, C.M.
Clinical use of PG's in fertility control. SDT: Population Report: Prostaglandins G(1): 1-15. April 1973.

PG's, found in species as different as man and coral, can apparently modify the activity of the tissue in which they are located by serving as mediators in the formation of cyclic 3',5'-adenosinemonophosphate. Their clinical application in the management of human fertility has attracted much public attention and intensive study. PG's in secretions and extracts of human prostate glands and seminal vesicles were found by several researchers independently to greatly lower blood pressure and stimulate the uterus. By the late 1950's and early 1960's, natural PG's were being isolated and by the late 1960's they were being synthesized chemically. Since then research has been underway to study their role in the induction of labor and abortion. The data from many different studies are reported but it is difficult to compare success rates in disparate and uncoordinated research. It is also unsatisfactory to try to draw conclusions about side effects and dosage. Of the 99 articles from which data were drawn, 27 involve PG's used in the first trimester of pregnancy, 45 in the second, and 21 in the third. Only $PGF_2\alpha$ and PGE_2 were discussed extensively. The different routes of administration reported, in decreasing order of usage, were intravenous, oral, intraamniotic, intravaginal, intrauterine, and subcutaneous. Continuous or repeated application was necessary in nearly every situation where high success rates were achieved. The time required for induction of abortion usually ranged from 20-30 hours and for induction of labor about half as long. Success in inducing labor was clearly more easily achieved than success in interrupting an early or mid-term pregnancy. Since the first investigations were initiated, clinical research networks have been established so that standard protocols can be followed and worldwide data can be pooled for more precise evaluation. Data from WHO studies presented in 1972 confirmed earlier indications that intrauterine and intraamniotic administration were the preferred routes and the late first or early second trimester was the preferred time for PG use. Single dose intraamniotic injection has been reported more successful than multiple dosage schedules. PG's are now commercially available for clinical use in certain facilities in the United Kingdom. The use of synthesized PG analogues is considered promising; they are proving more potent than natural compounds and side effects seem less intense. (MLH)   006977

1004
SPARKS, R.M.; GAIL, L.J.
PG fertility control research -- maps and directory.
Population Report: Prostaglandins G(2): 17-24. June 1973.

Research on PGs has increased steadily over the last decade, and the greatest interest has been the application of PG's in human fertility control and management. Work is now underway in at least 22 countries with the number of both clinical and laboratory reports approximately doubling every year. The main centers of PG research include the Karolinska Institute in Stockholm, Sweden; the Upjohn

Company; Makerere University in Uganda; Harvard University and Oxford University. A worldwide directory of prostaglandin research is included in the report. (MLH)　006976

1005

SPELLACY, W.N.; GALL, S.A.; SHEVACH, A.B.; HOLSINGER, K.K.
　The induction of labor at term.
　Obstetrics and Gynecology 41(1): 14-21. January 1973.

Using a randomized double-blind design, 222 women, 36 to 43 weeks pregnant, received either oxytocin (.5 mU per minute) or $PGF_2\alpha$ (2.5 μg per minute) by intravenous infusion. The rate of infusion was doubled at 1/2 to 1 hour intervals. Amniotomy was performed 1 1/2 hours after the infusion was begun. Induction was successful in 74% of the $PGF_2\alpha$ group and in 66% of the oxytocin group. The IDT means were 6 1/2 hours and 7 hours for $PGF_2\alpha$ and oxytocin groups, respectively. Cesarean sections were necessary in 9 women given $PGF_2\alpha$, none given oxytocin. Uterine hypertonus, hot flashes and phlebitis were noted significantly more often in the $PGF_2\alpha$ group; the incidence of nausea, vomiting, diarrhea, and fetal bradycardia was similar in both groups. Maternal blood chemistry data, mean weight and Apgar scores of the infants were similar for both groups. The authors note that their success rate was lower than those reported by others and suggest that the difference may be the higher initial Bishop inducibility score in the studies with higher success rates. The authors conclude that with the lack of increased efficacy and with the increased number of complication, the use of $PGF_2\alpha$ is not a panacea for labor induction. (This paper also summarizes 12 studies performed between 1968 and 1972 in which $PGF_2\alpha$ was used intravenously to induce labor at term.) (HC) 006779

1006

SPEROFF, L.
　An essay: prostaglandins and toxemia of pregnancy.
　Prostaglandins 3(5): 721-728. May 1973.

The author reviewed historical understanding of toxemia, renin physiology in pregnancy and experimental attempts to stimulate uterine ischemia and suggested that inadequate PG production or response to PG may be a cause of toxemia. Toxemia is a syndrome of edema, hypertension, proteinuria with fibrin deposition under the kidney glomerular capillary endothelium, sometimes accompanied by convulsions, and often resulting in neonatal morbidity, mortality, or central nervous system deficiency. The existence of a humoral toxin or pressor has long been postulated, and in 1958 was demonstrated by parabiosis experiments. Renin is known to be produced by trophoblast as well as kidney. Renin acts independently of renal control during pregnancy and is high in normotensive pregnany women but low in toxemic patients. In pregnant dogs with constricted uterine arteries, a toxemialike syndrome developed with hypertension, proteinuria, hypernatremia, low peripheral renin, but high renin remained in the uterine vein. By calculating uptake of dehydroepiandrosterone sulfate by the fetoplacental unit, impaired uteroplacental blood flow was shown to develop gradually into toxemia. The only evidence that PG is low in toxemia is that toxemic placentas contained less PGE activity by bioassay than normal placenta. The author suggested that in cases of reduced intrauterine blood flow, the placenta may release renin, resulting in increased angiotensin. The resulting increased peripheral resistance may prompt the placenta to secrete PGs to normalize blood pressure, but in toxemia PG production or response is inadequate. (LJG)　006755

1007
SPEROFF, L.
Physiologic and pharmacologic roles for prostaglandins in obstetrics.
Clinical Obstetrics and Gynecology 16: 109-129. 1973.

This review is a critical discussion of the role of PGs in mammalian parturition, their use in inducing human labor, and their potential involvement in pathology of toxemic pregnancy, which follows an introduction covering PG biochemistry and nomenclature. Parturition is partly controlled by the fetus, which release ACTH, cortisol, and pulmonary surfactants; by progesterone and estrogens, which alter myometrial activity; by $PGF_2\alpha$ which rises during labor and is oxytocic in some species; and perhaps by oxytocin. Current understanding is hampered by species differences, by difficulty in sampling substances, particularly PGs, from the local site at the proper time, and by the complexity of the parturition process. In obstetrics $PGF_2\alpha$ is comparable to oxytocin, except it can cause uterine hypertension and consequent fetal damage without warning. PGE_2or its 15-methyl derivative given orally is promising for inducing labor, $PGF_2\alpha$ is appropriate for intrauterine death, missed abortion, anencephaly, molar pregnancy, and perhaps prematurely ruptured membranes or 'unripe' cervix in preterm patients. The author suggests that PGA modulates vasopressor effects of the renin-angiotensin-aldosterone system. He hypothesizes, that in conditions of reduced blood flow through the uteroplacental bed this homeostatic system is overtaxed to the extent that PGA production by the trophoblast is defective or response to PGA is lost. It may be feasible to administer PGA to relieve toxemia. (LJG)   005738

1008
SPIES, H.G.; NORMAN, R.L.
Luteinizing hormone release and ovulation induced by the intraventricular infusion of prostaglandin E_1
    into pentobarbital-blocked rats.
Prostaglandins 4(1): 131-141. July 1973.

The effect of $PGE_1$and $PGF_2\alpha$ infused for 45-120 minutes into the third ventricle during proestrus in rats anesthetized with pentobarbital before the critical period of LH release at 1400 hours on ovulation and on LH levels was examined to study whether PGs affect LH release by stimulating the central nervous system. Rats were operated on proestrus morning under pentobarbital anesthesia (2 mg per 100 gm ip). A 24 gauge stainless steel cannula was placed stereotaxically and anchored in the third ventricle and a polyvinyl catheter was inserted in the atrium through the jugular vein and sutured in place. 88% of these rats ovulated (mean 10.5 ova). Pentobarbital injected again at 1300 hours blocked ovulation in 80%. 20 μg PGE_1reversed the blockade in 64%; 10 μg in 70% (mean 10.7 ova); 5 μg in 50%. 10 μg PGE_1did not permit ovulation if given sc or infused into the pituitary nor did 10-30 μg $PGF_2\alpha$ infused into the third ventricle. 86% of PG controls given 5 μl ethanol failed to ovulate. LH rose to 2400 ng for 2.5 hours in controls, to 300 ng in rats given ethanol, to 900 ng for 3 hours in rats given 10 μg PGE_1. In rats treated on diestrus Day 3, PGE_1caused elevated LH and ovulation 24 hours early, but had no effect in rats not given pentobarbital. (LJG)   005841

1009
SPIES, H.G.; NORMAN, R.L.; CAMPBELL, E.C.
LH release and ovulation in rats after intraventricular infusion of prostaglandin E_1.
Federation Proceedings 32(3): 239. 155Abs. 1973.

$PGE_1$was infused into the third ventricle or pituitary and peripheral LH determined in pentobarbital-blocked female rats. A stainless cannula was placed stereotaxically at 0900-1100 hours of proestrus, in PGE_1in 5 μl ethanol was infused from 1300-1600 hours and 100-150 μl atrial blood samples, taken

334

every 30 minutes for 6 hours, were radioimmunoassayed for LH. 13 out or 18 rats infused with 10 μg PGE$_1$ intraventricularly and all 4 rats given 20 μg PGE$_1$ ovulated. 2 of 4 rats given 5 μg PGE$_1$ ovulated partially. 1 of 7 rats given 5 μl ethanol vehicle ovulated partially. 2 of 7 rats infused with PGE$_1$ into the pituitary ovulated. 10 μg PGE$_1$ sc failed to induce ovulation in 6 pentobarbital treated rats. LH increased 30-60 minutes earlier and returned to normal 2 hours earlier in PGE$_1$ treated rats than in controls, but LH levels were 2-5 times higher in controls than in PGE$_1$ treated rats. PGE$_1$ appears to stimulate LH release at a site in the central nervous system above the pituitary. (LJG)   005458

1010
SPILMAN, C.H.; HARPER, M.J.K.
Effect of prostaglandins on oviduct motility in estrous rabbits.
Biology of Reproduction 9: 36-45. 1973.

Balloon-tipped fluid-filled catheters were placed in the proximal isthmus of both oviducts of unmated, unanesthetized rabbits. PGE$_1$ and PGE$_2$ at doses of 5 to 20 μg per kg iv initially suppressed activity and then markedly decreased basic tone and frequency of spontaneous contractions. PGF$_1\alpha$ and PGF$_2\alpha$ at 10 to 40 μg per kg iv caused an initial spasmodic contraction followed by an increase in amplitude and frequency of contractions. Duration of effect increased with dose with the PGEs but was constant with the PGFs. The E and F PGs were mutually antagonistic, the effect being greater with longer intervals between injections. In some animals refractoriness to the effect of PGE or PGF was noted. Administering the PGE compounds sc resulted in a response qualitatively similar but longer in duration than after iv administration. However, when the PGF compounds were given sc there was no initial spasmotic contraction and duration was similar to that following iv injections. Other investigators have reported that PGE$_1$ and PGE$_2$ increased tone and amplitude of contractions of the oviduct isthmus in vitro. The explanation of the difference between these data and the results of the present study is that PGE$_1$ stimulates longitudinal muscle, movements of which are more readily detected in vitro, but it relaxes circular muscle, changes in which are more readily detected in vivo. Variation among animals in sensitivity to PG may be associated with variation in degree of spontaneous activity. The mutually antagonistic effect of PGE and PGF suggests an involvement of these substances in the normal passage of embryos through the oviduct. (HC)   005741

1011
SPILMAN, C.H.; FINN, A.E.; NORLAND, J.F.
Effect of prostaglandins on sperm transport and fertilization in the rabbit.
Prostaglandins 4(1): 57-64. July 1973.

The effect of PGE$_1$, PGE$_2$, and PGF$_2\alpha$, sc and suspended in semen, on sperm transport and fertilization was studied in rabbits. Females were inseminated with 50 million sperm in dilute semen, injected immediately with 75-100 I.U. human chorionic gonadotropin or 2 mg luteinizing hormone. 1 uterotubal junction was ligated with a 2-0 silk suture 2.5-3 hours later. PGs were injected sc, 1 mg per kg before, 1, 2 and 2.75 hours after insemination, or 100 μg PG was mixed with semen. 28-30 hours later oviducts were flushed and 2-4 cell zygotes or unfertilized ova counted. PGF$_2\alpha$ sc produced 50% or more fertilized eggs in 6 out of 10 ligated oviducts (p<.05), but only 2 out of 6 PGE$_2$ treated oviducts and 1 out of 7 untreated controls had any fertilized eggs. Intravaginal PGF$_2\alpha$ resulted in 50% or more fertilized eggs in 6 of 8 ligated oviducts (p<.05). Only 4 out of 9 PGE$_2$ treated, 1 out of 6 PGE$_1$ treated, and 1 out of 6 controls had any fertilized eggs. Fertilization in nonligated oviducts was normal in all rabbits. The results suggested an all-or-none effect by PGF$_2\alpha$ on motility of female reproductive tract, since only 1 ligated oviduct from a rabbit given PGF$_2\alpha$ sc yielded less than 50% eggs fertilized if any fertilized ova were recovered. (LJG)   005833

1012

SPILMAN, C.H.; FORBES, A.D.; NORLAND, J.F.

Oviduct motility during the rhesus monkey menstrual cycle: effect of prostaglandins.

Biology of Reproduction 9: 68. 1973.

PGE_1, PGE_2, and $PGF_2\alpha$ intravenously at 5-20 μg/kg had little or no effect on spontaneous activity of the oviduct during the follicular phase of the menstrual cycle in monkeys. After ovulation, activity was suppressed by PGE_1 and PGE_2, but tone and amplitude were increased by $PGF_2\alpha$. (HC)   005742

1013

SRIVASTAVA, K.C.; CLAUSEN, J.

Extraction of prostaglandins E_1, E_2, and E_3 from human seminal plasma.

Lipids 8(7): 431-433. 1973.

A modified procedure for isolating microgram quantities of pure PGE compounds is described. Preliminary purification from other lipids by extraction procedures is followed by separation of PGs into groups using column chromatography. The final step, separating PGs of each group, is done by thin layer chomatography (TLC) on silica gel and then argentation rather than the time-consuming reversed phase partition chromatography. A total of 9.87 mg PGE was extracted from 300 ml human seminal fluid. This was dissolved in 10 ml of chloroform and 200 μl quantities were spotted. Impurities were removed by the silica gel TLC, only 3 zones (E_1, E_2, and E_3) being present in the argentation TLC. The average recovery of PGE was 55% from the argentation TLC (24% E_1, 21% E_2, and 10% E_3). An additional 20%-25% PGE was left behind on the plate. (HC)   005734

1014

STAGE, A.H.

Severe burns in the pregnant patient.

Obstetrics and Gynecology 42(2): 259-261. August 1973.

3 cases of burns in pregnant women are presented with a review of findings of other authors. Premature labor may be a hazard in second and third degree burns involving over 30% of the body surface. Thermal injuries to the skin have been shown to result in synthesis and release of PGE_2, perhaps by release of lysosomal phospholipases that act upon the phospholipids of the cell membrane to cause a local concentration of polyunsaturated fatty acid PG precursors. PGE_2 causes uterine contraction and it is postulated that its production by burn tissue may cause premature labor. (IC) 006430

1015

STALLWORTHY, J.

A review of procedures in therapeutic termination of pregnancy.

In: Jacomb, R.G., ed. The use of prostaglandins E_2 and $F_2\alpha$ in obstetrics and gynecology. (Proceedings of the Upjohn Prostaglandins Symposium, London, September 21, 1972.) Miami, Florida, Symposia Specialists, 1973. p. 59-61.

The author emphasized that no technique of abortion is free of complications, and avoidance of pregnancy is more logical and safer than its termination. The easiest and safest time for abortion is the first trimester, and the method of choice is suction curettage. For midtrimester abortions the use of prostaglandins is advocated. (HC)   006957

1016

STELLFLUG, J.N.; LOUIS, T.M.; SEQUIN, B.E.; HAFS, H.D.
Luteolysis after 30 or 60 mg PGF$_2\alpha$ in heifers.
Journal of Animal Science 37: 330. 1973.

30 mg PGF$_2\alpha$ given intramuscularly to heifers in diestrus either as a single dose or as 2 doses, 15-mg, at 6 hour intervals decreased serum progesterone levels to 1 ng/ml or less within 24 hours after administration. Estrus began at 55 to 60 hours, LH peaked at 65 hours, and ovulation occurred at 90 hours. Following a 60 mg dose, serum progesterone levels fell to less than 1 ng/ml within 12 hours, estrus was at 50 hours, LH peaked at 57 hours, and ovulation occurred at 78 hours. (HC)   005792

1017

STJARNE, L.
Alpha-adrenoceptor mediated feed-back control of sympathetic neurotransmitter secretion in guinea-pig vas deferens.
Nature New Biology 241(110): 190-191. February 7, 1973.

Evidence from several tissues and species has accumulated for 2 separate mechanisms of modulating secretion of norepinephrine upon sympathetic stimulation; negative feedback control by PGE, and a possibly presynaptic negative feedback by alpha adrenoceptors. In the author's system using field-stimulated isolated guinea pig vas deferens, release of tritiated norepinephrine can be measured because inhibitors desipramine and normetanephrine prevent rebinding of norepinephrine. The addition of eicosatetraynoic acid to prevent PGE synthesis increased release of norepinephrine upon stimulation, providing evidence for a 'PGE-mediated braking system.' Addition of phentolamine, an alpha blocker, further enhanced norepinephrine release, suggesting removal of a second independent braking system. With PGE synthesis blocked, addition of methoxamine, and alpha stimulator, depressed norepinephrine efflux in proportion to dose. Very low doses of PGE$_2$ would also reverse the inhibition by methoxamine. Probably, the PGE control is proximal in the chain of events from nerve stimulation to norepinephrine release. (LJG)   005457

1018

STJARNE, L.
Prostaglandin versus alpha-adrenoceptor-mediated control of sympathetic neurotransmitter secretion in guinea-pig isolated vas deferens.
European Journal of Pharmacology 22: 233-238. 1973.

To investigate a possible modulation of norepinephrine secretion by PGE or by alpha adrenoceptors, total efflux of norepinephrine was quantitated in isolated guinea pig vas deferens stimulated by low frequency electrical field stimulation. By first labeling norepinephrine stores with tritiated norepineph-rine, showing that over 90% of collected radioactivity was norepinephrine, then adding desmethylimi-pramine 6x10^{-7}M to block extraneuronal binding, the experiment was designed to measure secretion of norepinephrine due to nerve stimulation only. Desmethyl imipramine and normetanephrine increased radioactivity in the perfusate by a mean 182.2% in 5 tests. Eicosatetraynoic acid, an inhibitor of PG synthesis, 9 μg per ml enhanced recovery of tritium 146.9% (mean of 15 experiments) at stimulation rate of 5 per second only. Indomethacin 4 μg per ml also increased tritium release 141.8% in 5 tests. Phenoxybenzamine, an alpha blocker, added to the above inhibitors at .4 μg per ml further enhanced norepinephrine release 253.7% (mean of 6 experiments) above the level achieved after inhibition of PGE synthesis. The addition of PGE$_2$ 4 ng per ml reversed the increase due to phenoxybenzamine. The author states that this is first direct evidence for a local feedback control of sympathetic neurotransmission at low frequency stimulation by PGE. (LJG)   006503

1019

STJARNE, L.

Uncompetitive character of inhibition by prostaglandin E_2 of the enhancing effect of alpha-adrenoceptor blocking drug on noradrenaline secretion in isolated guinea-pig vas deferens.

Acta Physiologica Scandinavica 89(2): 278-282. 1973.

Noradrenaline secretion and muscular contractions were induced by nerve stimulation in isolated, superfused guinea pig vas deferens preparations. Phentolamine at concentrations of 7.5×10^{-9}M to 10^{-6}M induced secretion of noradrenaline in a dose-dependent manner. PGE_2 at concentrations of 1.0- 25×10^{-9}M antagonized varying concentrations of phentolamine to about the same extent, Phentolamine shortened the latency between electrical stimulation and muscular contraction and also increased the amplitude in a dose-related manner. PGE_2 had little, if any, effect on this response to nerve stimulation. The authors concluded that the antagonism between PGE_2 and phentolamine is not competitive; different sets of alpha-adrenergic receptors are involved, one dependent and one independent of PGE. (HC)   005739

1020

STRAUSS, J.F., III.; STAMBAUGH, R.L.

Prostaglandin-induced luteolysis in the rat.

Federation Proceedings 32(3): 242Abs. 1973.

20alpha-hydroxysteroid dehydrogenase (20a-SDH) in rat corpora lutea (CL) is a marker of luteolysis. Appearance of 20a-SDH is inhibited by prolactin and stimulated by LH or HCG. Rats treated with prostaglandin $F_2\alpha$ (PGF) on Days 8 and 9 of pregnancy begin resorbing implants by Day 10, and the CL show high 20a-SDH (113 mU/mg vs. .8 mU/mg protein control CL). Exogenous progesterone maintained implants in PGF-treated rats, but 20a-SDH appeared in the CL, marking luteolysis. Estradiol benzoate given concomitantly with PGF did not reverse PGF action, but HCG in large doses maintained implants and partially blocked appearance of 20a-SDH. Prolactin reduced levels of 20a-SDH induced by PGF. HCG given alone on Days 8 and 9 of pregnancy could not mimic the effect of PGF. Given on Days 14 and 15, PGF induced 20a-SDH by Day 16 of pregnancy in CL of intact and hypophysectomized rats, but embryos appeared normal. Concomitant HCG, prolactin, progesterone, or estradiol benzoate did not prevent appearance of 20a-SDH in CL. The data indicate that PGF antagonizes the action of prolactin and possibly LH on the rat CL of early pregnancy. In late pregnancy, antagonism to prolactin may be accompanied by an LH-like luteolytic action. (Authors) 005471

1021

STUDY GROUP ON PROSTAGLANDINS

Report of Study Group on Prostaglandins.

In: Bergstrom, S., ed. Report from meetings of the prostaglandin task force steering committee, Chapel Hill, June 8-10, 1972, Stockholm, October 2-3, 1972, Geneva, February 26-28, 1972. Stockholm, WHO Research and Training Centre on Human Reproduction, Karolinska Institutet, 1973. (Prostaglandins in Fertility Control 3) p. 5-7.

Induction of abortion in 179 cases using extraamniotic $PGF_2\alpha$ had a 57.5%, 79.2%, and 85.5% success rate in 24, 36, and 48 hours, respectively. Extraamniotic PGE_2 had 78%, 82%, and 98% success rates at 24, 36, and 48 hours, respectively. A repeated dose of 25 mg $PGF_2\alpha$ in 24 hours intraamniotically had 90% success. A 95% abortion success rate was reached with repeated doses at 6, 24, and 30 hours. The group strongly recommended additional studies with 15-methyl $PGF_2\alpha$

because preliminary indications suggest a favorable abortion rate. Intravaginal administration of PGE and $PGF_2\alpha$ met with clinically unacceptable side effects. (RAS)   006581

1022

STYLOS, W.A.; BURSTEIN, S.; ROSENFELD, J.; RITZI, E.M.; WATSON, D.J.
A radioimmunoassay for the initial metabolites of the F prostaglandins.
Prostaglandins 4(4): 553-565. October 1973.

Antibodies to the PGF metabolite, 9alpha,11alpha-dihydro-15-keto-prostanoic acid, were produced with bovine serum albumin conjugates in rabbits, and standard curves, cross-reaction studies, and double antibody radioimmunoassay of human plasma, urine, and seminal plasma were performed. The antibody elicited in a rabbit given 4 1 mg doses sc and bled 5 months later did not cross react with primary PGs, but there was 50% cross-reaction with 9alpha,11alpha-dihydroxy-15-keto-prost-5-enoic acid, and a 23% cross-reaction with 9alpha,11alpha,15-trihydroxy prostanoic acid (FOa). Biologic fluids were centrifuged, extracted, and chromatographed on silicic acid or assayed directly. Mean metabolite levels in blood plasma were 63.3 and 67.0 pg/ml, in urine 176.8 and 450.0 pg/ml, and in seminal plasma 24,500 and 27,100 in extracted and untreated fluids, respectively. (LJG)   005614

1023

SWEETMAN, B.J.; WATSON, J.T.; CARR, K.; OATES, J.A.; FROLICH, J.C.
Quantitative vapor-phase analysis of prostaglandin $F_2\alpha$ in female human urine.
Prostaglandins 3(3): 385-387. March 1973.

The methyl ester-tri-trimethylsilyl derivative of $PGF_2\alpha$ successfully fulfilled the criteria of high yield, chemical and thermal stability, good vapor phase properties, and several high-mass ions on mass spectrometry, for quantitation in female human urine, with a 50 picogram sensitivity. 200 ml urine samples were mixed with 1-4 μg 3,3,4,4-(2H)-$PGF_2\alpha$, extracted, chromatographed on an Amberlite XAD-2 column, partitioned, and evaporated. The pure extract was treated with diazomethane, then pyridine-bis-(trimethylsilyl)-acetamide. The derivative was chromatographed on a .5 m 1% Dexsil 300 gas liquid column at 215 degrees C and monitored at m/e 423/427, m/e 513/517, or m/e 494/498. Preliminary results gave a range of .5-5 ng per ml $PGF_2\alpha$ in female human urine. (LJG)   006081

1024

SWERDLOFF, R.S.; GROVER, P.K.; JACOBS, H.S.; BAIN, J.
Search for a substance which selectively inhibits FSH -- effects of steroids and prostaglandins on serum FSH and LH levels.
Steroids 27(5): 703-722. 1973.

In an attempt to find a preferential inhibitor of serum FSH, a number of C18, C19, and C21 steroids were evaluated for suppressive effects on LH and FSH and stimulatory effects upon ventral prostate weight in the acutely castrate adult male rat. C21 steroids had no effect. The C19 steroids were listed in order of potency with regard to suppressing LH and FSH, and stimulating ventral prostate weight. All C19 steroids that had an inhibitory effect on gonadotropins also stimulated ventral prostate weight, and all demonstrated preferential inhibition of LH over FSH. 2 C18 steroids were tested, estradiol-17β and estrone; estradiol-17β suppressed LH and FSH in a parallel fashion. Estrone was less potent than estradiol-17β. 300 μg/day of $PGF_2\alpha$, PGE_1, PGE_2, and PGA_2 were evaluated with regard to effect on gonadotropins and ventral prostate weight; none demonstrated effect. (Authors' modified)   005476

1025

SWERDLOFF, R.S.; GROVER, P.K.
Search for inhibin-effect of steroids and prostaglandins on serum FSH and LH levels.
Clinical Research 21: 256. 1973.

The effects of steroids and prostaglandins on serum LH and FSH concentrations in castrated male rats were evaluated. C-19 compounds inhibited serum LH and was correlated with an increase in ventral prostate weight. Estradiol and estrone suppressed LH and FSH. All prostaglandins and C-21 compounds tested were without effect. No compound had a preferential effect on lowering FSH as compared with LH. (RAS)   005787

1026

SYMONDS, M.; FAHMY, D.; MORGAN, C.; ROBERTS, G.; GOMERSALL, C.R.
Steroid profiles in prostaglandin induced abortion.
Australian and New Zealand Journal of Obstetrics and Gynaecology 13(1): 44-45. 1973.

In 7 patients 16-20 weeks pregnant plasma progesterone and estrogen levels showed considerable fluctuation before and after intraamniotic injection of 25 mg of $PGF_2\alpha$. The authors suggest that interference with steroidogenesis is not the vital factor in the abortion process. (HC)   005753

1027

TERAKI, Y.; MIYASAKA, M.; HORISAKA, K.
Effects of prostaglandin, 5-hydroxytryptamine and polypeptides on circulation and uterine contraction in rodents.
Japanese Journal of Pharmacology Suppl. 23: 119. 1973.

PGE_1, PGE_2, $PGF_2\alpha$, serotonin, oxytocin, vasopressin, hypertensin, bradykinin, and octapressin were compared in the uterus and circulatory system of the rat and rabbit by several methods. Uterine contractions in vivo were greater in response to $PGF_2\alpha$ than to PGE_1 or PGE_2, and were inconsistent to oxytocin. All PGs decreased blood pressure and stimulated respiration. Serotonin decreased blood pressure; oxytocin was without effect. Pelvic angiography in rabbits showed that serotonin caused vascular contraction, PG less contraction, and oxytocin had no effect. Serotonin and PG, but not oxytocin, depressed pCO2 and pH of blood gases. (LJG)   006951

1028

TERVIT, H.R.; ROWSON, L.E.A.; BRAND, A.
Synchronization of oestrus in cattle using a prostaglandin $F_2\alpha$ analogue (ICI 79939).
Journal of Reproduction and Fertility 34: 179-181. 1973.

The successful induction and synchronization of estrus in cattle treated with a potent $PGF_2\alpha$ analogue by either the intrauterine or intramuscular route is reported. 251 heifers were given $PGF_2\alpha$ analogue, ICI 79959, between days 5-16 of the estrus cycle. The cattle were used as donors or as recipients for egg transfer experiments. The donors received 2000 I.U. PMSG intramuscularly 1-2 days before $PGF_2\alpha$. The analogue (200 μg per ml water) was deposited in the ipsilateral uterine horn or injected intramuscularly into the gluteal region. The treated cattle were placed with vasectomized bulls until the onset of estrus. Satisfactory synchronization of estrus was achieved when test animals (50%) began estrus within 3 days. The intrauterine administration of 300 μg $PGF_2\alpha$ per day on 2 consecutive days gave precise synchronization of estrus. 200 μg per day of $PGF_2\alpha$ also gave satisfactory synchronization of estrus whether or not animals were pretreated with PMSG, although those having

PMSG showed estrus 1 day earlier. A total dose of 350 μg PGF$_2$α gave satisfactory synchronization of estrus when PMSG was also used. In cattle not treated with PMSG, more precise synchronization may have been achieved by increasing the total dose of intramuscularly administered analogue. (IC) 006448

1029

THIERY, M.; KETS, H. VAN; YO LE SIAN, A.; HEMPTINNE, D. de; VRIJENS, M.; CHEF, R.
Early diagnosis of anencephaly.
Lancet 1: 599-600. March 17, 1973.

A case of anencephalous gestation took 46 hours to abort with intraamniotic PGF$_2$α. The patient was a 26 year old mother of 2 normal children, 21 weeks pregnant, examined in a routine ultrasound screening program at Maternite Reine Astrid, Charleroi, Belgium. Several ultrasound exams with different equipment and observers showed no normal cephalic outlines in a fetus with a 5 cm thoracic diameter. The woman received epidural anesthesia, and 25 mg PGF$_2$α intraamniotically at 2 p.m. on January 18, 25 mg more at 10 a.m. on January 19, oxytocin from 1 to 5 p.m. and 35 mg PGF$_2$α at 11 a.m. on January 20. A macerated female fetus of 350 gm was delivered 12:30 a.m. that day. The authors thought that the continued myometrial activity, like that of first stage labor, with uneffaced and intact cervix, suggested that increased and more frequent doses of PGF$_2$α, or use of a PG analog, might have shortened the induction-to-abortion interval. (LJG)   005945

1030

THIERY, M.; HEMPTINNE, D., DE; VANDERHEYDEN, K.; YO LE SIAN, A.; DEROM, R.; KETS, H., VAN,; MARTENS, G.
Elective induction of term labor with amniotomy and oral prostaglandin E$_2$.
European Journal of Obstetrics, Gynecology and Reproductive Biology 3(5): 159-166. 1973.

50 women at term were given after amniotomy an initial dose of .5 mg PGE$_2$, a second dose of .5-1.0 mg 60 minutes later, and subsequent doses at 2-hour intervals if necessary. Induction was successful in 48 (46 complete); the mean IDT was 7 1/2 hours for primiparae and 4 1/2 hours for multiparae. Mean total dose was 2.3 mg for primiparae and 1.6 mg for multiparae. Vomiting occurred in 7 primiparae and 2 multiparae. Transient uterine hypertonus occurred in 2 women. Fetal heart rate patterns were normal in the multiparae, but 10 cases of bradycardia were found in the primiparae. The 1-minute Apgar score was low in 6 primiparae fetuses. pH values deviated from the norm in 5 infants in the primiparae group. The authors found the oral route to be slightly less favorable than intravenous infusion and the fetal outcome to be slightly worse than that with oxytocin-induced or spontaneous labor. The authors concluded that the slight but sigificant tendency of the fetus to develop acidosis and hypoxia during the second stage of labor warrants careful constant electronic surveillance of the unborn. (HC)   006281

1031

THIERY, M.
Fetal aspects of the elective induction of term labor with prostaglandins.
In: Abstracts of the [7th] World Congress of Obstetrics and Gynaecology, Moscow, U.S.S.R., August 12-18, 1973. Amsterdam, Excerpta Medica, 1972. (International Congress Series No. 279) Abstract 279.

The author studied fetal and neonatal well-being following the administration of oxytocin and PGF$_2$α intravenously and PGE$_2$orally for the induction of labor. He reported that there is no danger associated with any of these procedures if uterine hyperstimulation, most often associated with intravenous $PGF_2\alpha$, is avoided. (HC)   006721

1032
THIERY, M.; WILLIGHAGEN, R.G.J.
  Prostaglandins -- effect on the enzyme content of the human placenta.
  International Research Communications System (73-9) 10-26-2. September 1973. 1 p.

The distribution of 18 enzymes, glycogen, and lipids were compared histochemically in 96 term human placentae of 13 women given PGE_2.5-3.5 mg orally, 35 given $PGF_2\alpha$ 2-40 μg/minute, 28 given $PGF_2\alpha$ 2-20 μg/minute totaling 36-62.5 mg; 20 control placentae included 2 delivered by cesarean section, 7 induced by oxytocin, and 11 delivered spontaneously. The following enzymes were detected with comparable individual and local variations between groups: acid and alkaline phosphatase, alpha-naphthyl esterase, succinic acid dehydrogenase, lactic acid dehydrogenase, beta-OH-butyric acid dehydrogenase, isocitric dehydrogenase, glucose-6-phosphate dehydrogenase, 5-nucleotidase, adenosine triphosphatase, indoxyl esterase, aminopeptidase, alphaglycerophosphate, and 3beta-ol-steroid dehydrogenase. These enzymes showed little or no activity: naphthyl esterase, primary and secondary alcohol dehydrogenase, and 17beta-ol-steriod dehydrogenase. (LJG)   006400

1033
THIERY, M.; VROMAN, S.; DEROM, R.; KETS, H., VAN
  The fetal influence of prostaglandin $F_2\alpha$.
  In: Bossart, H., Cruz, J.M., and Huber, A., eds. Perinatal Medicine. (Third European Congress on Perinatal Medicine, Lausanne, 1972.) Bern, Hans Huber, 1973. p. 368.

Labor was induced in 25 women at term using intravenous $PGF_2\alpha$ infusions. 2 women had uterine hypertonus; there was 1 case of fetal bradycardia. Fetuses were biochemically normal at birth, but with the acid-base parameters pH and BE, significant differences were found in blood from the umbilical artery. (HC)   006708

1034
THOMPSON, I.E.
  The IUD and prostaglandins: a review of the evidence.
  Obstetrics and Gynecology 42(4): 617-620. October 1973.

The hypothesis that the contraceptive action of the IUD is mediated by prostaglandins is based on the luteolytic effect of $PGF_2\alpha$ in some mammals and on presumptive evidence in primates. In guinea pigs $PGF_2\alpha$ is secreted in the uterus on cycle Days 13, 14 and 15 and is believed necessary for luteolysis. $PGF_2\alpha$ levels measured in vivo and in vitro increase in the guinea pig uterus when a foreign body is inserted. The quantitative details on how $PGF_2\alpha$ evokes luteolysis have been estimated in sheep, and studies are also reported on rats, rabbits, and hamsters. In primates the IUD could not work by causing luteolysis, since there is no anatomical route for a uterine factor to reach the ovary nor do hysterectomy or IUDs affect progesterone levels. In monkeys IUDs alter motility of myometrium and oviduct, but do not influence ovulation, sperm transport, fertilization or cleavage. IUDs also affect intrauterine metals, leucocytes, macrophages, histamine and biochemical composition, all of which are known to be affected by PGs. In humans $PGF_2\alpha$ increases during the luteal phase, and $PGF_2\alpha$ and PGE_2 have been observed in menstrual fluid. These PGs and IUDs alter uterine and tubal motility. It is postulated that the endometrium, stimulated by the IUD, releases PGs locally. (LJG)   005468

1035

THORBURN, G.D.; COX, R.I.; CURRIE, W.B.; RESTALL, B.J.; SCHNEIDER, W.
  Prostaglandin F and progesterone concentrations in the uteroovarian venous plasma of the ewe during
    the oestrous cycle and early pregnancy.
  Journal of Reproduction and Fertility Suppl. 18: 151-158. 1973.

Uteroovarian and jugular venous blood samples were collected from 4 ewes every 2-3 hours for
varying periods of time from estrous-cycle Day 10. Progesterone concentrations in peripheral plasma
peaked at 5-22 ng/ml on Days 13-16. In the uteroovarian blood, transient but marked progesterone
decreases followed each PGF peak. On Days 15 and 16 PGF peaks coincided with very low
progesterone levels. In pregnant ewes, the PGF peaks were absent and there was no correlation
between progesterone and PGF changes. The authors theorized that a series of PGF peaks are
necessary for luteolysis and that the limited information obtained thus far strongly promotes $PGF_2\alpha$ as
the ovine luteolytic factor. Decreased effectiveness of PGF in this capacity in pregnant animals
suggests the conceptus may suppress PGF by direct or indirect action. (HC)   005736

1036

TOPPOZADA, M.; BYGDEMAN, M.; PAPAGEORGIOU, C.; WIQVIST, N.
  Administration of 15-methyl-prostaglandin $F_2\alpha$ as a pre-operative mean of cervical dilatation.
  Prostaglandins 4(3): 371-379. September 1973.

15-methyl-$PGF_2\alpha$ was given as 1 extraamniotic injection (200-500 μg) or 2 im injections (300-800 μg)
6 hours apart to facilitate cervical dilatation in 67 women 9-13 weeks pregnant undergoing vacuum
aspiration. 21 (47%) of those given extraamniotic PG (45) aborted spontaneously in a mean interval of
14.9 hours. 17 of these were incomplete abortions; 17 achieved dilatation of 10 mm or more; one
third vomited and 1 had diarrhea. Pain requiring a mean of 1.45 analgesic injections occurred just
after PG injection. The patients given PG im (22) had variable results with only a hint of better
outcome or more side effects at higher doses. 9 (41%) aborted incompletely before aspiration. These
patients had less uterine pain but a mean of 3 episodes of diarrhea each and more vomiting than the
extraamniotic group. In patients who did not abort spontaneously the cervix was dilated sufficiently, or
easily dilated, for vacuum aspiration. The authors recommend use of 15-methyl-$PGF_2\alpha$ for those late
in the first trimester, with large uteri, or nulliparae. (LJG)   005440

1037

TOPPOZADA, M.; BEGUIN, F.; BYGDEMAN, M.; WIDE, L.; WIQVIST, N.
  Postconceptional fertility control by prostaglandins.
  Advances in the Biosciences 9: 567-573. 1973.

80 patients less than 8 days after their missed menstrual period were given $PGF_2\alpha$; 1) by intravenous
infusion at a rate of 68 or 78 μg/minute for 5 or 8 hours, 2) by the intravaginal route as a tablet or
solution at a dose of 50 mg 1 to 4 times at 2- to 4-hour intervals, or 3) as a single intrauterine injection
of 500 μg. PGE_2(10-20 mg 1-4 times at 2- to 4-hour intervals, intravaginally) and 15-methyl $PGF_2\alpha$ (25
to 400 μg as a single intrauterine dose) were also used. Uterine contractility varied greatly among the
patients, but in general there was rapid tone elevation, frequent small-amplitude contractions, never
very high pressure, and direct correlation between uterine activity and abortion. Uterine bleeding was
induced in 41% of the 29 nonpregnant and 70% of the 51 pregnant patients, but this response did not
correlate with the abortion rate. $PGF_2\alpha$ infused for 5 hours produced abortions in 3 of 7 patients; given
for 8 hours 10 of 15 patients aborted. The intravaginal route was effective in 3 of 8 cases, the tablets
in 0 of 2. 2 of 4 patients given intrauterine $PGF_2\alpha$ aborted. PGE_2tablets were ineffective in all 4
patients tested, 15-methyl intrauterine $PGF_2\alpha$ was effective in 6 of 11 cases. Using the vaginal or

intravenous route, vomiting and diarrhea occurred. Intense uterine pain followed intrauterine administration. The conclusion is that use of prostaglandins is not satisfactory for postconceptional fertility control. (HC)   006474

1038

TOPPOZADA, M.; BYGDEMAN, M.; WIQVIST, N.

Systemic and local administration of prostaglandins for postconceptional fertility control.

In: Bergstrom, S., ed. Report from meetings of the prostaglandin task force steering committee, Chapel Hill, June 8-10, 1972, Stockholm, October 2-3, 1972, Geneva, February 26-28, 1972. Stockholm, WHO Research and Training Centre on Human Reproduction, Karolinska Institutet, 1973. (Prostaglandins in Fertility Control 3) p. 108-115.

90 women with missed menstruation, of whom 59 proved pregnant of less than 2 weeks duration, received PGE_2, $PGF_2\alpha$, or 15-methyl-$PGF_2\alpha$ by intrauterine, intravaginal, intravenous, or intramuscular routes, and were tested 2 weeks later for pregnancy by radioimmunoassay of plasma for LH and urine for HCG. Intravenous $PGF_2\alpha$, 68 μg per minute for 5 hours, aborted 3 of 7 pregnancies; $PGF_2\alpha$, 78 μg per minute for 8 hours, aborted 10 of 15. 50 mg intravaginal $PGF_2\alpha$ solution repeated 1-4 times aborted 3 of 10 pregnancies. 800 μg intramuscular 15-methyl-$PGF_2\alpha$ aborted 4 of 6, with mean 5.8 episodes of side effects. 500 μg $PGF_2\alpha$ injected by catheter via intrauterine route aborted 2 of 4, and 25-400 μg 15-methyl-$PGF_2\alpha$ aborted 7 of 13. These trials were disappointing because side effects were high, vaginal route was difficult, intrauterine PG was expelled, iv route was inconvenient. Most important, 12 of 31 nonpregnant women had vaginal bleeding, 40 of 59 pregnant women bled, but only 22 aborted, and 19 of 59 pregnant women had no bleeding but 7 of these 19 did abort. It was hoped that vaginal 15-methyl-$PGF_2\alpha$, which had not been tried, would be more successful. (LJG) 006725

1039

TREDWAY, D.R.; MISHELL, D.R., Jr.

Therapeutic abortion of early human gestation with vaginal suppositories of prostaglandin $F_2\alpha$.

American Journal of Obstetrics and Gynecology 116: 795-798. July 1973.

The effectiveness of intravaginal suppositories of $PGF_2\alpha$ for therapeutic abortion was studied in 10 women having gestations less than 6 weeks. A total dosage of 200-600 mg administered in suppositories of 50 mg at 2-4 hour intervals for a 24-hour period were given. Serum human chorionic gonadotropin (HCG) and progesterone levels were measured by radioimmunoassay prior, during, and 2 weeks after therapy. All patients had significant side effects, mainly diarrhea and vomiting, indicating that systemic absorption took place. 7 patients aborted completely, 1 had an incomplete abortion requiring curettage. 2 failed to abort and had dilatations and curettages. The 7 patients who completely aborted had a fall in HCG levels, while the HCG continued to rise in the 3 D & C patients. (Authors' modified)   006433

1040

TSAFRIRI, A.; KOCH, Y.; LINDER, H.R.

Ovulation rate and serum LH levels in rats treated with indomethacin or prostaglandin E_2.

Prostaglandins 3(4): 461-467. April 1973.

To ascertain whether indomethacin, an inhibitor of PG synthesis, prevents ovulation by inhibiting LH release, serum LH levels were radioimmunoassayed at 1730-1800 hours of proestrus, and ovulation was examined on estrus morning in rats treated with indomethacin. When indomethacin 10 mg per rat was administered ip at 1430 on proestrus, 21% of the rats ovulated and the number of ova shed was reduced to 4% of the controls, but there was no significant change in peak serum LH level: 1122 ng/ml in treated rats compared with 975 in controls. PGE_2given sc 25-750 μg per rat at 2400 of proestrus overcame antiovulatory action of indomethacin: 71%-90% of the rats ovulated, though the number of eggs shed was 24%-55% of controls. Indomethacin still blocked ovulation when given at 2000 after the proestrus LH surge, but not at 2400. 1500 μg PGE_2in early afternoon of proestrus increased serum LH levels in rats in which the cyclic LH surge had been blocked with pentobarbital ip 30 mg/kg, and induced ovulation in 67% of these animals. Direct measurements confirm that indomethacin does not block LH release but interferes with a late phase of the ovulatory process. PGE_2reverses the action of indomethacin and has a effect causing LH release. (Authors' modified) 005694

1041

TSAI, T.H.; LEIGHTON, J.

Potentiating effects of prostaglandin $E_2(PGE_2)$ on the response of the isolated guinea-pig vas deferens to 1-norepinephrine (NE) and acetylcholine(Ach).

Pharmacologist 15: 209. 1973.

Stimulation of the isolated guinea-pig vas deferens by 10^{-8}M and 10^{-6}M PGE_2in a Tyrode solution caused a parallel shift in the dose-response curve of 1-norepinephrine to the left (2.7 and 9.7 fold) with an increase in maximum (72 and 48%), but shifted the dose-response curve of acetylcholine to the left (3 and 4.7 fold) without affecting the maximum. In Krebs, $PGE_2(10^{-6}$M) had similar results but the shift was significantly smaller. In Krebs, decreasing (K^+increased the potentiation by PGE_2while decreasing (Ca^{++}) did not alter the potentiation. In Tyrode, increasing (K^+) decreased the potentiation by PGE_2while decreasing (K^+) or increasing (Ca^{++}) did not alter the effects. The results indicate that in the isolated guinea-pig vas deferens, the potentiating effects of $PGE_2$are 1) relatively non-specific, 2) characterized by shifting the dose-response curve of NE to the left with an increase in maximum and 3) influenced by the potassium ion concentration but not calcium ion concentration. (Authors' modified)   005596

1042

TURNBULL, A.C.

Dinoprostone (prostin E_2): a prostaglandin for clinical use in obstetrics and gynaecology.

Prescribers' Journal 13(2): 25-31. April 1973.

The history, occurrence, and physiology of PGs are briefly summarized as an introduction to a description of the use of PGE_2marketed as 'Dinoprostone' for induction of term labor and for midtrimester abortion. For term labor, .75 ml ampoules containing 1 mg PGE_2per ml ethanol are provided, to be diluted in 500 ml saline for a final concentration of 1.5 μg/ml. This solution may be administered intravenously by constant infusion pump or pediatric drip set (60 drops/ml), starting at .25 μg/minute, increased stepwise to a maximum of 2 μg/minute as needed. If uterine hypertonus or fetal distress occur, infusion is stopped and resumed at 1/2 the last dose. $PGE_2$is useful for difficult inductions, may be supplemented with oxytocin, and is suggested for intrauterine death at twice normal dosage. For midtrimester abortion, 2 ml ampoules are mixed with 18 ml dilutent for intraovular admininstration by Foley catheter at 2.5-40 ml, depending on gestation. 5 ml ampoules with 5 mg PGE_2in ethanol are supplied, to be diluted with 1000 ml saline for a final concentration of 5 μg/ml for intravenous use. Recommended dose rates are 2.5 μg/minute for 30 minutes, increased to 10 μg/minute after 4 hours if needed. Side effects of vomiting, diarrhea, phlebitis, and vasodilation are minor with the intrauterine route but are more severe with the intravenous midtrimester abortion.

The widely publicized vaginal route for inducing abortion or menstruation is unreliable and evokes severe side effects. (LJG)   005791

1043
TUVEMO, T.; WIDE, L.
Prostaglandin release from the human umbilical artery in vitro.
Prostaglandins 4(5): 689-694. November 1973.

$PGF_2\alpha$ released into Krebs bicarbonate glucose solution bathing spiral strips of human umbilical arteries was measured by radioimmunoassay. In 4 experiments, fluid collected after 7 hours of incubation yielded about 7.5 ng $PGF_2\alpha$ per g wet weight of artery per hour. Indomethacin, 50 μg per ml, an inhibitor of PG synthesis, reduced recovery of $PGF_2\alpha$ to less than .5 ng per ml sample fluid. When intrinsic tone of the spiral strip was recorded on a smooth muscle transducer, a constant spontaneous tone was maintained for 2.5 hours, but the strips relaxed when bathed in indomethacin 40 μg per ml. (LJG)   005638

1044
VANDERHOEK, J.Y.; LANDS, W.E.M.
Acetylenic inhibitors of sheep vesicular gland oxygenase.
Biochimica et Biophysica Acta 296(2): 374-381. 1973.

Eicosa-5,8,11,14-tetraynoic acid and 3 analogs inhibit fatty acid oxygenase from sheep vesicular gland acetone powder in 2 patterns: time-dependent destruction and instant concentration-dependent inhibition. The time-dependent inhibition was analyzed by preincubating 2 mg enzyme powder, .66 mM phenol added for activation, with 6 mM eicosatetraynoic acid and adding substrate to show decreased rate. Diluting inhibitor to 5 μM did not restore activity. Time-dependent inhibition did not occur in anaerobic conditions. Inhibition caused by adding glutathione and glutathione peroxidase was reversed by adding N-ethylmaleimide to remove glutathione, to the preincubation mixture. Time-dependent inhibition was also prevented by 9.1 mM diethyldithiocarbamic acid. The ynoic acids delta 10a-18:1, delta 13a-18:1 and delta 9a-18:2 also inhibited oxygenase. Inhibition constants (KIi) and first order rate constants (k2') of inactivation are shown. Together these properties suggest that eicosatetraynoic acid inhibits similarly to substrate-catalyzed enzyme destruction, probably through an unstable intermediate, although eicosatetraynoic did not react as substrate. (LJG)   005478

1045
VANDERHOEK, J.Y.; LANDS, W.E.M.
The inhibition of the fatty acid oxygenase of sheep vesicular gland by antioxidants.
Biochimica et Biophysica Acta 296(2): 382-385. 1973.

Antioxidants inhibit the fatty acid oxygenase of sheep vesicular gland and soybean lipoxygenase in an instantaneous, reversible manner. Their inhibitory effectiveness was not related to their traditional antioxidant potencies but seemed to depend on the nature of the enzyme. The destructive effect of eicosa-5,8,11,14-tetraynoic acid, a substrate analog, on vesicular gland oxygenase could be prevented in the presence of either alpha-naphthol or 2,2,4-trimethyl-6-ethoxy-1,2-dihydroquinoline (Santoquin). (Authors)   005479

1046

VANE. J.R.; WILLIAMS, K.I.

The contribution of prostaglandin production to contractions of the isolated uterus of the rat.

British Journal of Pharmacology 48: 629-639. 1973.

Isotonic contractions of isolated nonpregnant rat uteri induced by $PGF_2\alpha$, 10 to 50 ng per ml, or acetylcholine were not antagonized by indomethacin. The stimulatory effect of oxytocin on nonpregnant uteri was antagonized by indomethacin but contractions of pregnant uteri were not altered. Spontaneous intermittent contractions of uteri from 17 to 22 day pregnant rats were abolished by indomethacin. Bioassay, thin-layer and gas chromatography revealed the presence of mainly $PGF_2\alpha$ and some PGE_2 in the fluid bathing these uteri. Levels of $PGF_2\alpha$ were much higher on Day 22 (day of delivery) than on Days 19 to 21. Indomethacin reduced both contractile activity and output by PG of pregnant rat uteri; addition of PG to the bath restored contractions to uteri rendered quiescent by indomethacin. The authors suggest that production of PG by the uterus plays a role in the initiation of parturition. (HC)   005743

1047

VENTURA, W.P.; FREUND, M.

Evidence for a new class of uterine stimulants in rat semen and male accessory gland secretions.

Journal of Reproduction and Fertility 33: 507-511. 1973.

PG content was measured in rat semen and prostate gland fluids, and the nature of the uterine spasmogenic agent was analyzed. Each ml of semen was equivalent to 200 ng PGE_1 in ovariectomized rat uterus bioassay in vitro, but after extraction of 91% of PGs with ethyl acetate at pH 3, each ml semen contained only 1 ng PGE_1 and .5 ng PGF1 per ml; seminal vesicle fluid contained 1 ng PGE_1 and 1 ng PGF1 per ml; anterior prostate 13 ng PGE_1 and 10 ng PGF1 per ml; lateral prostate 13 ng PGE_1 and 13 ng PGF1 per ml; posterior prostate 10 ng PGE_1 and 3 ng PGF1 per ml. The aqueous fraction contained virtually all the spasmogenic activity. This material, when extracted by the Folch and the Svennerholm methods, was soluble in water or chloroform:methanol, was non-dialysable through cellophane, and was similar to bovine brain ganglioside in hexose sialic acid, hexose hexosamine, sialic acid hexosamine, and sialic acid sphingosine content, and Rf values. Rat semem spasmogens, rat prostatic fluid spasmogens, and bovine brain gangliosides were equally potent uterine stimulants in the rat bioassay. (LJG)   005695

1048

VENTURA, W.P.; FREUND, M.

In-vitro effects of PGE_1 on the motility of the guinea pig and rat female tracts.

Federation Proceedings 32: 787Abs. 1973.

43 guinea pigs and 26 rats pretreated with estrogen or progesterone were used to record the in vitro motility and response to PGE_1 of vagina, uterus, and horns. PGE_1 .01-160 ng/ml increased force, work, duration, and frequency of contraction in both species. .02 ng/ml PGE_1 for guinea pigs and 40 ng/ml PGE_1 for rats were found to be the minimum stimulatory dose. Further investigation indicated that doubling drug doses in successive treatment periods could give misleading results and therefore should be compared with the use of constant drug doses. (RAS)   006686

1049
VIRUTAMASEN, P.; WRIGHT, K.H.; WALLACH, E.E.
    Monkey ovarian contractility--its relationship to ovulation.
    Fertility and Sterility 24(10): 763-771. October 1973.

Isometric contractions of ovarian smooth muscle in 5 rhesus monkeys in vivo and 20 ovaries in vitro were recorded in physiologic and pharmacologic studies with oxytocin, PGE_2, $PGF_2\alpha$, and adrenergic drugs. Monkeys were given .5-1 mg/kg phencyclidine HCl im and 5-10 mg/kg Na pentobarbital iv. For in vivo experiments, the ovary was dissected free and isolated by a cylinder. For in vitro studies, ovaries were removed, placed in oxygenated modified Krebs buffer, the hilus was removed, and the remaining tissue bisected. In vivo, 10 ovaries contracted spontaneously and were stimulated by oxytocin. In vitro, all 6 follicular ovaries contracted spontaneously, 4 in a regular pattern 2-3 times per minute. 3 of 9 postovulatory ovaries contracted spontaneously. 20 ovaries without known cycle dates were tested pharmacologically. $PGE_2$4-1000 ng/ml had no effect on 2; $PGF_2\alpha$ 1-1000 ng/ml increased amplitude, frequency, and/or tone in 27 experiments. PGE_2inhibited ovaries already stimulated with $PGF_2\alpha$. Norepinephrine stimulated 17 out of 23 ovaries. Propranolol (beta-adrenergic inhibitor) did not prevent PGE_2inhibition. Isoproterenol (beta-adrenergic stimulator) decreased response to $PGF_2\alpha$. Phenoxybenzamine (alpha-adrenergic inhibitor) increased contractions due to $PGF_2\alpha$. Oxytocin initiated or immediately increased contraction and acted synergistically with $PGF_2\alpha$ and norepinephrine. The ovary with a corpus luteum from a pregnant monkey was inert, but the other ovary responded characteristically to drugs. (LJG)   005937

1050
VOGLMAYR, J.K.
    Prostaglandin $F_2\alpha$ concentration in the genital tract secretions of dairy bulls.
    Prostaglandins 4(5): 673-678. November 1973.

$PGF_2\alpha$ concentrations were determined by radioimmunoassay in rete testis fluid, cauda epididymal plasma, seminal plasma, and coccygeal venous blood of dairy bulls, and the effect of $PGF_2\alpha$ on oxidation, glucose uptake, and lipid synthesis by spermatozoa was assessed by Warburg techniques. Mean $PGF_2\alpha$ concentrations in 4 bulls were .17 ng per ml in rete testis fluid and seminal plasma, .14 in coccygeal venous blood plasma, and 1.61 in cauda epididymal plasma. $PGF_2\alpha$ 100 ng per ml did not effect oxygen uptake, glucose uptake, glucose oxidation, or incorporation of glucose carbon into lipid by ejaculated spermatozoa in calcium-free Krebs-Ringer phosphate with or without glucose. These parameters were increased significantly by testosterone, 50 μg per ml, and phosphatidylinositol, 250 μg per ml. (LJG)   005636

1051
VULLIEMOZ, Y.; VEROSKY, M.; FINSTER, M.; TRINER, L.
    The effect of $PGE_1$and catecholamines on the cAMP system in rat and human uterine smooth muscle.
    Pharmacologist 15: 230. 1973.

PGE_1at concentrations of .01 μg/ml to 50 μg/ml stimulated adenyl cyclase activity in rat uterus and inhibited isoproterenol-induced cAMP formation in a dose-related manner with homogenates and intact tissue. Isoproterenol, epinephrine, norepinephrine, and PGE_1all increased cAMP formation in human myometrium. PGE_1also decreased the effect of isoproterenol. The authors suggest that this interaction may represent a basic mechanism in controlling uterine tone. (HC)   005735

348

1052

WAKELING, A.E.; KIRTON, K.T.; WYNGARDEN, L.J.
 Prostaglandin receptors in the hamster uterus during the estrous cycle.
 Prostaglandins 4(1): 1-8. July 1973.

Specific binding of tritiated PGE_1 and of tritiated $PGF_2\alpha$ by hamster uterus was measured on each day of the estrous cycle. About 100 mg of uterus slices were incubated in 1 ml .25 M sucrose in .02 M potassium phosphate with .5 mM calcium and 10 μg/ml indomethacin for 1 hour at 37 degrees C. .05 μCi/ml of tritiated PGE_1, 48 Ci/mM or of tritiated $PGF_2\alpha$ 7.5 Ci/mM with or without 100-fold excess unlabeled PG were added. The difference between the tritium bound in the presence and absence of excess unlabeled PG represents the specifically bound PG. PGE_1 binding was detected on each cycle day (p<.001) between labeled and excess unlabeled incubation, with binding the highest on proestrus (Day 3), and on the day after ovulation values were the lowest. $PGF_2\alpha$ binding was greater than binding on PGE_1, considering lower specific activity of tritiated $PGF_2\alpha$, but day-to-day differences were not significant. (LJG)   005829

1053

WAKELING, A.E.; SPILMAN, C.H.
 Prostaglandin specific binding in the rabbit oviduct.
 Prostaglandins 4(3): 405-414. September 1973.

Apulla, distal, and proximal isthmus of oviducts from estrus and pregnant rabbits were incubated with 3H-prostaglandin E_1 or 3H-$PGF_2\alpha$ in the presence or absence of nonradioactive PG. Specific binding of 3H-PG was assessed as the difference between 3H-PG bound in the presence or absence of nonradioactive PG. Significant specific binding of PGE_1 and $PGF_2\alpha$ was detected in all sections of the oviduct for both estrus and pregnant rabbits, except $PGF_2\alpha$ in the proximal isthmus of pregnant ones. The mass of specifically bound $PGF_2\alpha$ was almost 5 times that of PGE_1 indicating that $PGF_2\alpha$ may be more important in the control of oviduct function. Changes in specific binding of PGE_1 or $PGF_2\alpha$ in different sections of the oviduct for estrus and pregnant rabbits correlated with the differential effects of E- and F-series PGs on oviduct motility and functional changes during the time of ovum transport. The correlation of specific binding with oviduct function suggests the presence of physiological receptors for PGs in oviduct tissue. (Authors' modified)   005443

1054

WALKER, F.M.M.; POYSER, N.L.
 Production of prostaglandins by the early pregnant guinea pig uterus.
 Acta Endocrinologica Suppl. 177: 312. 1973.

The amount of $PGF_2\alpha$ produced by homogenates of Day 15 non-pregnant guinea pig uterine tissue averaged 110 ng/100 mg tissue as compared to 13 ng/100 mg tissue for Day 15 pregnant animals. The authors suggest that the conceptus decreases $PGF_2\alpha$ levels either by inhibiting its synthesis or increasing its metabolism. (HC)   005733

1055

WALTMAN, R.; TRICOMI, V.; PALAV, A.
 Aspirin and indomethacin: effect on instillation/abortion time of mid-trimester hypertonic saline
  induced abortion.
 Prostaglandins 3(1): 47-58. January 1973.

The effect of aspirin or indomethacin on the outcome of midtrimester abortion by hypertonic saline was explored. 36 patients recieved 650 mg aspirin orally every 6 hours (up to 10 doses), 36 received indomethacin 25 mg every 6 hours (up to 8 doses), and 50 patients were untreated controls. All 16-21 weeks pregnant subjects were given 200 ml 20% saline transabdominally after aspiration of 200 ml amniotic fluid. Induction-to-abortion intervals were mean 36 hours (range 11-109) in controls, 45 hours (14-102) in patients given aspirin, and 68 hours (29-117) in those given indomethacin. The proportion of patients with intervals over 50 hours were 22%, 33% and 73% in these groups respectively. 3 controls, 9 aspirin patients and 1 indomethacin patient were also treated with oxytocin according to predetermined criteria, and 8 indomethacin treated patients were given a second saline injection. (LJG)   006776

1056

WALTMAN, R.; TRICOMI, V.; SHABANAH, E.H.; ARENAS, R.
The effect of anti-inflammatory drugs on parturition parameters in the rat.
Prostaglandins 4(1): 93-106. July 1973.

3 salicylates, cortisone acetate, and phenobarbital were given orally to pregnant rats from Days 19-21 or on Days 19-21, to test their effect on length of gestation, and duration and outcome of labor. Drugs were given at a dose of 10 mg per kg twice daily, except cortisone acetate which was given at 5 and 125 mg per kg. Pregnancy was calculated from 0200 on the morning following mating from 1600-2200. Gestation was 520.3 hours in untreated controls, 521.8 hours in controls gavaged with distilled water. There was no significant variation from controls for sodium salicylate, 5 mg cortisone acetate, and phenobarbitol, but gestation was significantly longer after acetylsalicylic acid (2 and 3 days treatment), salicylic acid, and 125 mg cortisone acetate (537.1 hours to 539.4 hours). The salicylates prolonged parturition and incidence of increased bleeding but steroid did not. Fetal deaths occurred with sodium salicylate and 3 days of acetylsalicylic acid. The authors hypothesized that acidic nonsteroid antiinflammatory drugs stabilize lysosomal membranes and suppress prostaglandin synthesis directly, while the glucosteroids simply stabilize lysosomes to evoke these effects on parturition. (LJG)   005837

1057

WATSON, J.; LEASK, J.T.S.; ALAM, M.; ADAMS, P.M.
Steroid secretion by superfused porcine corpus luteum tissue.
Acta Endocrinologica Suppl. 177: 338. 1973.

Using a superfusion system in which the incubation medium is continuously being changed, a steady state for estrogen and progesterone secretion by porcine corpus lutem tissue was reached in 2 to 3 hours. Addition of LH, PGE_2, or $PGF_2\alpha$ stimulated steroid secretion. (HC)   005745

1058

WEEKS, J.R.
Tachyphylactic response of the rat uterus in vivo to prostaglandins E_2 and $F_2\alpha$.
Advances in the Biosciences 9: 773-777. 1973

The effect of PGE_2 and $PGF_2\alpha$ on uterine motility was studied in unanesthetized rats with chronic uterine and venous cannulas. Single intravenous injections increased motility. An intravenous infusion of either prostaglandin first increased uterine motility, but after 3 hr the uterus became quiescent and no longer responded to another injection. Becasue of this tachyphylaxis, such a preparation appears unsuitable for evaluating long-acting prostaglandin preparations. (Author)   006496

1059

WEEKS, J.R.; DUCHARME, D.W.; MAGEE, W.E.; MILLER, W.
   The biological activity of the (15S)-15-methyl analogs of prostaglandins E_2 and $F_2\alpha$.
   Journal of Pharmacology and Experimental Therapuetics 186(1): 67-74. 1973.

The biological activity of the (15S)-15-methyl analogs of prostaglandin (PG)E_2 methyl ester and PGF$_2\alpha$ methyl ester were compared in vitro on the isolated gerbil colon and rat uterus and in vivo for their antifertility effects in the hamster and cardiovascular effects in anesthetized dogs and rats. The initial step in the metabolism of PGE$_2$ and PGF$_2\alpha$ is an oxidation at carbon 15 by 15-hydroxy prostaglandin dehydrogenase. These analogs are not substrates for this enzyme. 15 methylation caused only slight changes in the activity on isolated tissues and in the magnitude and duration of the cardiovascular responses. These data are consistent with efficient means of inactivation in vivo other than 15-dehydrogenation. However, at high doses 15-methylation of PGE$_2$ methyl ester quantitatively changed the mechanism of its depressor action in the dog. PGE$_2$ methyl ester, at 1 and 3.2 ng/kg, lowered blood pressure by a decrease in total peripheral resistance, whereas the 15-methyl analog did so by depressing myocardial contractility and cardiac output. In striking contrast, the antifertility activity of the 15-methyl analogs was increased 50- to over 100-fold over the PGE$_2$ and PGF$_2\alpha$ parent compounds, respectively. Female hamsters were injected with prostaglandins 4 days after mating. On the eighth day, animals were sacrificed and considered pregnant if one or more implantation sites were found in the uterus. 500 μg PGE$_2$ methyl ester and 12.5 μg 15-methyl PGE$_2$ methyl ester terminated 7 of 8 and 8 of 8 pregnancies, respectively. 160 μg PGF$_2\alpha$ methyl ester and 2 μg 15-methyl PGF$_2\alpha$ methyl ester both terminated 8 of 8 pregnancies. (Authors' modified)   006248

1060

WENTZ, A.C.; AUSTIN, K.; KING, T.M.
   Abortifacient efficacy of intravaginal prostaglandin $F_2\alpha$.
   American Journal of Obstetrics and Gynecology 115(1): 27-32. January 1, 1973.

Intravaginal PGF$_2\alpha$ was given to 20 early midtrimester patients, 7 of whom had intrauterine catheters for monitoring contractility, to evaluate this route for potential self-administration. The women were aged 16-30, second to seventh pregnancy and ninth to sixteenth gestation. Lactose tablets impregnated with 50 mg of PGF$_2\alpha$ were administered to maintain a frequent and intense contraction. Antiemetic, antidiarrheal and analgesic drugs were administered when needed. 19 aborted in mean 17 hours and 50 minutes (range 10 hours and 30 minutes to 29 hours), 7 completely, after mean 590 mg PG (range 350-900). The woman whose dilatation ceased to progress after 23 hours was given hypertonic saline and aborted 24 hours later. Blood loss averaged 111 ml and the only significant change in laboratory tests was doubling in white blood cells. Diarrhea occurred in 13, vomiting in 18, fever over 100.4 degrees F. in 11, but of 15 interviewed 4 weeks later, only 3 listed side effects as the worst aspect of the procedure. (LJG)   006738

1061

WENTZ, A.C.; JONES, G.S.; BLEDSOE, T.; ROCCO, L.
   Effects of PGF$_2\alpha$ infusion on human cortisol biosynthesis.
   Prostaglandins 3(2): 155-172. February 1973.

To study the effect of PG on cortisol release, PGF$_2\alpha$ was infused intravenously in 17 women, 7 of whom received dexamethasone to suppress ACTH and 3 of whom were given 2500 ml saline. 25 mg PGF$_2\alpha$ was infused over 8 hours and 20 minutes at 50 μg/minute. Cortisol was measured by radioimmunoassay, either in blood samples taken every 30 minutes or in urine samples taken every 24 hours. Mean daily cortisol excretion was 31 μg per 24 hours before, 92 μg during, and 28 μg after the experiment. In the women sampled every 20 minutes, cortisol rose to a plateau during the infusion and remained high regardless of time of day. Urine volume increased approximately twofold on the study day and about threefold in those given 2500 ml saline. This water loading did not blunt the increased cortisol secretion. 7 volunteers received 1 mg dexamethasone at midnight for 4 days; their cortisol excretion was lower than average on the pre- and postinfusion days and was not increased during $PGF_2\alpha$ infusion. PG side effects such as vomiting, diarrhea, fever, uterine cramps, phlegm, and coughing were generally associated with high cortisol response, since 5 of 7 women without side effects had little or no increase in cortisol. (LJG)   006741

1062
WENTZ, A.C.; BURNETT, L.S.; ATIENZA, M.F.; KING, T.M.
Experience with intra-amniotic prostaglandin $F_2\alpha$ for abortion.
American Journal of Obstetrics and Gynecology 117(4): 513-521. October 15, 1973.

$PGF_2\alpha$ was given intraamniotically to 132 women pregnant 12-21 weeks, who were divided into 4 groups according to dosage regimen. Of 48 women given an initial dose of 25 mg and subsequent 5-25 mg doses if needed, 45 aborted, 29 completely. The IAT averaged 14 1/4 hours and the mean dose was 38 mg. 43 women received 30 mg initially, 25 mg 6-8 hours later, and another 25 mg dose at 24 hours if needed. 40 aborted, 32 completely. Average dose was 58 mg and the IAT averaged 17 1/4 hours. 16 of 17 women given 40 mg initially and 24 hours later if necessary aborted, 9 completely. Mean total dose was 59 mg and IAT averaged 23 hours. Another 24 women were given 40 mg initially followed by 20 mg doses 8 hours and 24 hours later if necessary. All aborted, 19 completely. The IAT averaged 18 hours and the average dose was 63 mg. 70% of the patients in this study vomited and 20% had fever. Sepsis, systemic reaction, and cervical laceration were serious complications noted in 5 patients. The authors commented that the efficacy rate of intraamniotic $PGF_2\alpha$ approaches that obtained with hypertonic saline but the necessity of multiple doses and close monitoring of the patients are disadvantages. There was no significant advantage of any initial dose used, and repeated doses at 6- to 8-hour intervals maximized efficiency. The intraamniotic route minimized all the side effects associated with $PGF_2\alpha$ except emesis, which was dose-dependent. Advantages of this method over saline are that 1) it can be used in patients with smaller gestations, 2) there is no removal or exchange of large volumes of fluid, and 3) induction of labor is rapid and the IAT is short. (HC)   006917

1063
WENTZ, A.C.; JONES, G.S.
Intravenous prostaglandin $F_2\alpha$ for induction on menses.
Fertility and Sterility 24(8): 569-577. August 1973.

13 women who were approximately 10 days past expected menses were given 25 mg $PGF_2\alpha$ intravenously at a rate of 50 μg/minute. 12 of the women bled, the degree of bleeding comparable to a normal menstrual period. 5 of the 9 women who were pregnant had negative pregnancy tests after $PGF_2\alpha$ treatment but in 4 pregnancy continued. Histological abnormalities following the infusion were noted in the women pregnant at the beginning of the study. Side effects included vomiting, diarrhea, coughing, and abdominal cramping and pain. Progesterone levels fell in the pregnant women; estradiol and estrone values were not altered. The authors conclude that pregnancy was not reliably terminated by this procedure and gestations that continued appeared to be damaged. (HC)   006586

1064

WENTZ, A.C.; THOMPSON, B.H.; KING, T.M.
Posterior cervical rupture following prostaglandin-induced mid-trimester abortion.
American Journal of Obstetrics and Gynecology 115(8): 1107-1110. April 15, 1973.

2 cases of transverse posterior cervical rupture during midtrimester abortion induced by $PGF_2\alpha$ are reported. Both were healthy nulligravidae with normal pregnancy. The first, 16 years of age, 21 weeks gestation, received 5 mg $PGF_2\alpha$ transabdominally, then 25 mg repeated at 10 minutes and 8 hours. Her cervix was 80% effaced but undilated at 12 hours, and dilated at 22.5 hours. Complete abortion followed at 23 hours. 10 minutes later there was vaginal bleeding, and she was given 10 I.U. oxytocin im and .2 mg methylergonovine orally. The tear was then detected by elevating the cervix anteriorly, 7-8 cm from midline to the right. Upon laparotomy the apex of the defect was felt below the uterine vessels. It was repaired vaginally. The second case was 19 years, 20 weeks gestation, and was given 105 mg $PGF_2\alpha$ in 34 hours. At 27 hours her cervix was 80% effaced, 1 cm dilated, and amniotomy was performed. At 39 hours the 2 cm rigid external cervical os was visible at the introitus during contractions. At 40 hours 2 fetuses and a placenta were delivered. Severe bleeding then occurred and a transverse posterior cervical laceration not extending through the internal os was repaired vaginally. In this series of 102 midtrimester abortions, the cervix often appeared more rigid than in hypertonic saline abortions. Definitely in the second, and probably in the first case, the rupture began near the intact external os in the posterior endocervical canal. (LJG)   005696

1065

WENTZ, A.C.; BELL, W.R.
The coagulopathy of induced abortion.
In: Abstracts, Fourth International Congress on Thrombosis and Haemostasis, June 19-22, 1973, Vienna, Austria. Abstract 407. p. 442.

The method most commonly used to interrupt second-trimester pregnancy is intraamniotic hypertonic saline. Several complications with this procedure, including severe alteration of the blood coagulation mechanism, have been observed. Serial coagulation studies were performed on 15 patients undergoing abortion by intraamniotic injection of 20% sodium chloride, 20 patients by intraamniotic injection of $PGF_2\alpha$, and 10 normal patients with spontaneous delivery. In the group receiving 20% saline significant decreases in fibrinogen and platelets occurred and fibrinogen-fibrin degradation products became elevated. 1 patient experienced massive generalized exsanguinating hemorrhage. In those receiving $PGF_2\alpha$ and in those having normal delivery these changes were not observed and no problems were encountered. Studies were carried out to define the mechanism of the hypertonic-saline-induced coagulopathy. (Authors' modified)   006935

1066

WENTZ, A.C.; JONES, G.S.
Transient luteolytic effect of prostaglandin $F_2\alpha$ in the human.
Obstetrics and Gynecology 42(2): 172-181. August 1973.

$PGF_2\alpha$ was infused into 13 nonpregnant volunteers at a rate of 50 μg per minute for 8 hours to describe and document detailed hormonal patterns of the luteal phase. 10 patients had a shortened luteal phase. 9 had decreased progesterone values at 8 hours, ranging from 23.5-69.8% of baseline values. No consistent changes in plasma estradiol or estrone were noted. Results indicated a transient effect of $PGF_2\alpha$ on ovarian steroidogenic mechanisms. Plasma progesterone was decreased during PG infusion in every subject. It was postulated that for PG to affect progesterone but not estrogen, biosynthesis and release, there must be a direct effect upon the granulosa cell or its blood supply. A discussion is included with several references to other studies. (Authors' modified)   006723

1067

WHITE, I.G.; KAR, A.B.
Aspects of the physiology of sperm in the female genital tract.
Contraception 3(3): 183-194. March 1973.

Energy metabolism, migration, and fate of spermatozoa in the vagina, cervix, and fallopian tubes are summarized, and data from humans are provided when available on the composition of female genital tract secretions. Sperm leave the acidic vaginal environment within minutes, using seminal plasma substrates fructose, sorbitol, and lactic acid for energy. The high concentration of PGs in human semen may effect sperm transport just after coitus. Sperm can penetrate cervical mucus near ovulation mainly by their own motility. Sperm migration through the uterus and tubes is dependent on uterine and tubal contractions and probably on ciliary motion, but the uterotubal junction prevents millions of sperm from entering the tubes. Although much is known about the composition of vaginal, uterine, and tubal fluid the details of sperm metabolism and capacitation are presumptive. Probably enough oxygen is available for sperm to metabolize aerobically the glucose which is present in uterine and tubal fluids. (LJG)   005692

1068

WHO PROSTAGLANDIN TASK FORCE
Report from the prostaglandin task force meeting, Stockholm, October 2-3 1972.
In: Bergstrom, S., ed. Report from meetings of the prostaglandin task force steering committee, Chapel Hill, June 8-10, 1972, Stockholm, October 2-3, 1972, Geneva, February 26-28, 1972. Stockholm, Who Research and Training Centre on Human Reproduction, Karolinska Institutet, 1973, (Prostaglandins in Fertility Control 3) p. 9-11.

Results of clinical trials of intraamniotic $PGF_2\alpha$ and PGE_2(4 preestablished protocols), extraamniotic $PGF_2\alpha$ and PGE_2(1 protocol, and intrauterine 15(S)-15-methyl-$PGF_2\alpha$ (3 protocols) given for abortion in a collaborative study are summarized. Intraamniotic $PGF_2\alpha$ 25 mg, repeated 6 hours later if necessary, produced 92% success within 48 hours in 89 patients, with higher side effects (3.9-6.9 episodes per abortion), than with a 24 hour interval between doses (2.0-4.5 episodes). 40 mg $PGF_2\alpha$ or 10 mg $PGE_2$was successful within 48 hours in 97-100% of 35 patients. When 10 mg PGE_2did not cause abortion in 6 hours, iv oxytocin 80 mU per minute was given until abortion: this resulted in mean abortion interval of 14.6 hours, compared to 23.8 hours without oxytocin, in 5 patients. 50 mg $PGF_2\alpha$ yielded 94% successful abortion in 18 patients confirmed previous results in 179 patients. 15(S)-15-Me-$PGF_2\alpha$ was applied by an extraamniotic rubber catheter, in 1 or 2 injections of 200-800 μg, in saline mixed with a viscous polysaccharide. There were 28 abortions within 36 hours in 31 patients, within 13.1-15 hours mean abortion interval, with from .4-1.4 episodes of vomiting and very few of diarrhea per trail. 5 mg intraamniotic 15-Me-$PGF_2\alpha$ aborted 37 of 38 patients within mean 19.1 hours with mean 1.5 episodes of vomiting and .4 of diarrhea. 200-500 μg extraamniotic 15-Me-$PGF_2\alpha$ given to 32 late first trimester subjects to facilitate dilatation for aspiration resulted in mean dilatation of 11.4 mm, satisfactory in all but 4 patients, and abortion in some patients. (LJG)   006582

1069

WILKS, J.W.; FORBES, K.K.
Luteotropic action of 15-keto prostaglandin $F_2\alpha$ in the nonpregnant rat.
Biology of Reproduction 9: 95-96. 1973.

Rats were given PG injections subcutaneously daily for 3 to 5 estrous cycles. $PGF_2\alpha$ (15 mg/kg) prolonged cycle length from 4.1 to 6.7 days; 15-keto $PGF_2\alpha$ had no effect on cycle length at either 15 or 150 mg/kg. $PGF_2\alpha$ decreased ovarian progesterone concentration. The high dose of 15-keto $PGF_2\alpha$ increased progesterone content of the ovary and the serum and increased ovarian weight. Corpora lutea did not regress after treatment with the high dose of 15-keto $PGF_2\alpha$; corpora lutea of animals given the low dose of 15-keto or $PGF_2\alpha$ did not differ from controls. It was concluded that doses of 15-keto $PGF_2\alpha$ are luteotropic in the rat. (HC)   005524

1070

WILKS, J.W.; WENTZ, A.C.; JONES, G.S.

Prostaglandin $F_2\alpha$ concentrations in the blood of women during normal menstrual cycles and dysmenorrhea.

Journal of Clinical Endocrinology and Metabolism 37: 469-471. 1973.

Blood samples were collected at 0700, 1300, 1900, and 0100 hours on Days 1, 7, and 20 of the menstrual cycle of 16 women. $PGF_2\alpha$ content of either plasma or serum was quantitated by radioimmunoassay. Blood samples were also collected from 12 women with severe dysmenorrhea, starting 9-12 days before and continuing for 2-4 days after menses. Neither stage of the cycle nor time of sample collection had an effect on $PGF_2\alpha$ concentration in the normal women, mean values being .6-.8 ng/ml. The mean values for the women with dysmenorrhea ranged from .44 to .67 ng/ml. The authors conclude that blood levels of PG are not related to fluctuations in steroid hormone levels and are of little diagnostic value in establishing the cause of dysmenorrhea. (HC)   005737

1071

WILKS, J.W.; FORBES, K.K.; NORLAND, J.F.

Prostaglandins and in vitro ovarian progestin biosynthesis.

Prostaglandins 3(4): 427-437. April 1973.

Rat ovaries and rabbit corpus luteum slices were incubated with PGs and LH to investigate whether PGs stimulate luteal function in vitro. 'Fully luteinized' rat ovaries were prepared by giving 50 I.U. pregnant mare serum (PMS) gonadotropin at 29 days of age, 25 I.U. human chorionic gonadotropin 2.5 days later, and incubating ovaries at 39 days of age. Pseudopregnancy was simulated by giving 4 I.U. PMS at 30 days of age and 6 sc injections of 200 μg prolactin beginning Day 33. Ovaries were incubated Day 36. 1 to 1000 μg per ml $PGF_2\alpha$ did not affect progesterone, specific activity, or total progestins released, and $PGF_2\alpha$ combined with LH resulted in less progesterone than LH alone. $PGF_2\alpha$ also did not affect progesterone synthesis in rabbit tissue, as did LH. $PGF_1\alpha$, PGE_1 and 15keto-$PGF_2\alpha$ had no effect on rat ovaries, but 10 μg/ml PGE_2 inhibited progesterone synthesis ($p < .05$). $PGF_2\alpha$ at 10 μg per ml significantly increased progesterone release by pseudopregnant rat ovaries, but not at 1 or 100 μg per ml, suggesting a steroidogenic action on ovarian follicular or interstitial cells. (LJG) 005693

1072

WILLIAMS, K.I.

Prostaglandin synthesis by the pregnant rat uterus at term and its possible relevance in parturition.

British Journal of Pharmacology 47(3): 628P-629P. 1973.

Spontaneous contractions of uteri from rats 17-20 days pregnant were abolished by 2.8-5.6 μM of indomethacin added to the bath medium. 5.6-11.2 μM were required for suppression of activity of uteri from animals 21-22 days pregnant. The inhibitory effect of indomethacin was reversed by PGE_2

or PGF$_2\alpha$ at 4-16 ng/ml. 66-156 ng/gm/hour of PGF-type material was released into the bath by uteri from rats 19-21 days pregnant, 250-680 ng/gm/hour came from those taken from rats on Day 22. Production of PG by endometrial homogenates from rats 19-21 days pregnant averaged 1790 ng/gm and rose to 24,220 ng/gm on Day 22. The myometrial fractions synthesized a lower amount of PG, averaging 414 ng/gm in rats 19-21 days pregnant. The results suggest that prostaglandin production is important in parturition. (HC)   005482

1073

WIQVIST, N.; BYGDEMAN, M.; TOPPOZADA, M.

Intra-amniotic prostaglandin administration -- a challenge to the currently used methods for induction of midtrimester abortion.

Contraception 8(2): 113-131. August 1973.

Abortifacient efficacy of and uterine response to 40 mg PGF$_2\alpha$ and 1, 2.5, and 5 mg 15-methyl-PGF$_2\alpha$ as a single intraamniotic injection for midtrimester abortion are presented and discussed in comparison with hypertonic saline induction. The 148 healthy volunteers were pregnant 14-24 weeks. PGs were injected for 10 minutes in concentrations of 5 mg/ml for PGF$_2\alpha$ and 1 mg/ml for 15-me-PGF$_2\alpha$, with intravenous analgesics as needed. 54 had uterine pressure recorded through transabdominal polyethylene catheter. Uterine responses were rapid, with peak activity in 2-3 hours after PGF$_2\alpha$ administration. 15-me-PGF$_2\alpha$ had slower but sustained peaks after 9 hours for 1 mg, 6 hours for 2.5 mg and 4 hours for 5 mg. Outcome for 40 mg PGF$_2\alpha$ was as follows: 76% aborted, 44% complete, mean interval 18.5 hours; for 1 mg 15-me-PGF$_2\alpha$: 46% aborted, 67% complete, 20.1 hours; 25 mg: 98% aborted, 54% complete, 18.8 hours; 5 mg: 95% aborted, 42% complete, 18.6 hours. Vomiting was the most common side effect, but incidents and number of analgesic injections were significantly fewer ($p < .05$) with 15-me-PGF$_2\alpha$. 2.5 mg 15-me-PGF$_2\alpha$ was considered the optimal dose. It was concluded that the PGs are superior to saline in their shorter induction time, fewer severe complications, and probably more direct mode of action. (LJG)   005464

1074

WIQVIST, N.; BYGDEMAN, M.

Prostaglandins as early abortifacients.

In: Segal, S.J. Crozier, R., Corfman, P.A. and Condliffe, P.G.,eds. The regulation of mammalian reproduction. Springfield, Ill., Thomas, 1973 p. 484-487.

PGF$_2\alpha$ was administered by intravenous infusion at a rate of 25 to 100 μg per minute to women at various stages of pregnancy. 20 of 22 women less than 8 weeks pregnant aborted; total dose was approximately 30 mg. 10 of 42 women 9 to 20 weeks pregnant aborted; total dose averaged 66.5 mg. Frequency of side effects increased with rate of infusion, uterine pain occurring more often than nausea and diarrhea. Intrauterine administration of repeated doses of PGF$_2\alpha$ in 9 women 9 to 15 weeks pregnant resulted in all patients aborting with the total dose being approximately 1/10 that given intravenously; no gastrointestinal side effects were observed. The authors suggest that the mechanism of action of PG in inducing abortion at early stages of gestation is not suppression of plasma progesterone levels but rather mechanical, i.e. extremely high intrauterine pressure. (HC) 006905

1075

WIQVIST, N.; BYGDEMAN, M.; TOPPOZADA, M.

Prostaglandins in fertility regulation.

In: Bergstrom, S., ed. Report from meetings of the prostaglandin task force steering committee, Chapel Hill, June 8-10, 1972, Stockholm, October 2-3, 1972, Geneva, February 26-28, 1972. Stockholm, WHO Research and Training Centre on Human Reproduction, Karolinska Institutet, 1973. (Prostaglandins in Fertility Control 3) p. 80-107.

This is a comprehensive review of PGs in female reproductive physiology and fertility regulation ranging from luteolysis and ovulation in subprimates, primates, and humans to in vitro and in vivo motility studies with human oviduct and uterus, and concluding with all the applications of PGs for pregnancy termination. Although local PG effects on blood flow are still being considered in some species, $PGF_2\alpha$ is said to be the humoral luteolysin in sheep. In primates a partial, dose-dependent luteolysis can be evoked, but in women PGs do not impair progesterone synthesis. In oviducts and nonpregnant uterus in vivo, and in pregnant uterus in vitro, the primary PGs have dose-dependent qualitatively different effects. PGEs and PGFs all stimulate pregnant uterus in vivo, with PGE_1 about 8 times more potent than $PGF_2\alpha$ and about 30-40 times more so than PGF1 iv. Potency and side effects have unpredictable individual variations. Intrauterine and iv injections or infusions act by first increasing tone, which decreases before cyclic contractions begin. About 2000-3000 PG abortion trials suggest that systemic routes (iv, oral, vaginal) present too many side effects for routine clinic use. Local routes permit much lower doses; the intraamniotic has been more efficacious than the extraamniotic route. The 15-methyl-$PGF_2\alpha$ or 15-methyl-PGE_2-ester analogues may permit 1 dose application, require even lower doses, and elicit more gradual long-lasting uterine activity. None of the natural PGs have been satisfactory for inducing menstruation. For first trimester abortion, PGs may aid late instrumental aspiration by preoperatively dilating the cervix. (LJG)   006735

1076

WIQVIST, N.; BEGUIN, F.; BYGDEMAN, M.; TOPPOZADA, M.

15(S)-15-methyl-prostaglandin $F_2\alpha$: myometrial response and abortifacient efficacy.

Advances in the Biosciences 9: 831-842. 1973.

In midpregnant women a single 10 μg intravenous dose of 15-methyl-$PGF_2\alpha$ was the threshold dose for stimulating uterine contractility. The analog was approximately 10 times more potent than $PGF_2\alpha$, and the duration of effect was nearly double that of the parent compound. 15-methyl-$PGF_2\alpha$ 5 μg/minute intravenously produced the same degree of uterine contractility as 75 μg/minute $PGF_2\alpha$. 8 of 10 women given the analog aborted; 2 of 16 given $PGF_2\alpha$ aborted. Incidence of vomiting and diarrhea was the same in both groups. A single 350 μg dose of 15-methyl-$PGF_2\alpha$ by the intrauterine route produced the same intensity of contractions as did 5 doses of 500 μg of $PGF_2\alpha$. 2 doses 200-400 μg or 1 dose 500-850 μg of 15-methyl-$PGF_2\alpha$ produced abortion in 90% of the women treated, the IAT averaging approximately 14 hours. 9 doses $PGF_2\alpha$ between 250-750 μg were required for the same success rate, and the IAT averaged 21 hours. Vomiting and uterine pain were the most evident side effects. A single 5 mg intraamniotic injection of 15-methyl-$PGF_2\alpha$ or 1 to 2 injections $PGF_2\alpha$ 25 mg produced abortions in 97% of the women treated. The IAT averaged 19 hours for the methyl derivative and 28 hours for the parent compound. The authors concluded that 15-methyl-$PGF_2\alpha$ is superior to $PGF_2\alpha$, primarily because of the effectiveness of a single dose. (HC)   006731

1077

WITTING, W.C.; LAROS, R.K., Jr.; WORK, B.A.

Uterine response to prostaglandin $F_2\alpha$ infusion in term human pregnancy.

Obstetrics and Gynecology 42(4): 581-588. October 1973.

The dose response of uterine activity to $PGF_2\alpha$ infused iv for 4 hours at constant dose was recorded in 60 patients at term pregnancy, including 30 with intact and 2 with prematurely ruptured membranes. 5 women from each group each received .5, 1, 2, 4, 8 or 16 μg $PGF_2\alpha$ per minute. Intraamniotic pressure was recorded for 2 hours before and after, and during the 4 hours of PG infusion. Uterine activity increased throughout infusion in all doses above 2 μg per minute. Patients with ruptured membranes achieved higher activity and success rate (80%), than those with intact membranes (13% delivered) regardless of PG dose or pelvic score. 54 were delivered by vaginal route, 4 had cesarean sections, (2 for failed induction and 2 for fetal distress), 2 failed even after oxytocin. There were no severely depressed infants, but 2 had Apgar scores of 7, at 1 or 5 minutes. In 6 patients cervical dilation ceased for an hour at 5-6 cm; in many cervical dilation progressed faster than expected from observed uterine activity. 8 patients had hypertonus and or hypersystole, 5 of these received 8 or 16 μg PG. A dose of 4 μg per minute was considered safe and effective. According to the authors, the assumption that uterine dose response to PG is analagous to oxytocin has lead to use of exponentially increasing doses of PG and thus severe side effects. (LJG)　006455

1078

YANG, N.S.T.; MARSH, J.M.; LeMARIE, W.J.

Prostaglandin changes induced by ovulatory stimuli in rabbit graafian follicles: the effect of indomethacin.

Prostaglandins 4(3): 395-404. 1973.

Radioimmunoassay was used to determine levels of PGF and PGE in rabbit follicles obtained at estrus and at 5 or 9 hours after ovulatory injections of HCG, LH, or mating. The ovulatory stimuli produced a marked increase in follicular PG levels; Increases produced by HCG and LH, or mating were completely abolished by the intravenous injection of indomethacin 30 minutes prior to gonadotropin treatment or at the time of mating. These results support the concept that PGF and PGE play an obligatory role in the process of ovulation. (Authors' modified)　005442

1079

YIP, S.K.; MA, H.K.; NG, K.H.

Induction of labour with oral prostaglandin E_2.

Journal of Obstetrics and Gynaecology of the British Commonwealth 80: 442-445. 1973.

57 women 35 to 44 weeks pregnant were given .5 mg PGE_2 capsules orally every hour for 4 to 8 hours and then every 2 hours for induction of labor. The success rate was 80%, most of the failures being primiparae. Average total dose was approximately 3.5 mg. Caesarian section was performed on 13 patients. Fetal heart rate was decreased in 3 cases, increased in 2. No uterine hypertonus was noted; 4 women had nausea and vomiting. The authors note that these results compare favorably with those they obtained previously with oxytocin. (HC)　006761

1080

YLIKORKALA, O.; PENNANEN, S.

Human placental lactogen (HPL) levels in maternal serum during abortion induced by intra- and extra-amniotic injection of prostaglandin $F_2\alpha$.

Journal of Obstetrics and Gynaecology of the British Commonwealth 80(10): 927-931. 1973.

HPL concentrations in maternal serum were measured by double antibody radioimmunoassay in 25 patients who were undergoing abortion by intraamniotic (9) or extraamniotic (9) injections of $PGF_2\alpha$, or by extraamniotic hypertonic saline (7). Mean gestational age was 12 weeks. With injection of 150 ml

20% NaCl no change in HPL level occurred, but 5-6 mg extraamniotic $PGF_2\alpha$ and 20-25 mg intraamniotic $PGF_2\alpha$ decreased serum HPL significantly ($p<.01$) 2 hours after injection. $PGF_2\alpha$ administered by intrauterine routes appeared to have direct effect on the placental synthesis or secretion of HPL in early pregnancy. (Authors' modified)   005493

1081

ZADOR, G.; NILSSON, B.A.
Induction of labour with intravenous administration of prostaglandin $F_2\alpha$.
In: Bossart, H., Cruz, J.M., and Huber, A., eds. Perinatal medicine. (Third European Congress on Perinatal Medicine, Lausanne, 1972.) Bern, Hans Huber, 1973. p. 366.

In 22 term pregnant women labor was induced by $PGF_2\alpha$ starting at 3 μg/minute and increasing stepwise. First contractions appeared in about 40 minutes at a $PGF_2\alpha$ rate of 6 μg/minute. 20 of 22 (90.9%) inductions were successful at a mean dose rate of 9 μg/minute, mean total dose of 4.14 mg, and mean time of 6 hours and 58 minutes. There were no abnormalities in contraction pattern (except at dose rates over 12 μg/minute); fetal heart rate; fetal or cord blood pH, pO2 or pCO2, or Apgar scores. (LJG)   005092

1082

ZEROBIN, K.; JOCHLE, W.; STEINGRUBER, C.
Termination of pregnancy with prostaglandin E_2(PGE_2) and $F_2\alpha$ ($PGF_2\alpha$) in cattle.
Prostaglandins 4(6): 891-901. December 1973.

$PGE_2$was given to 16 pregnant Brown Swiss cattle and $PGF_2\alpha$ to 23 to induce parturition. Doses ranged from 5-50 mg; routes included iv, im, and intrauterine; and gestation length ranged from 1 first trimester, 1 second trimester, 21 third trimester; and 16 with delayed gestation. Immediately after iv PGs cows become restless, salivated, defecated, increased regurgitation and rumination, and some had diarrhea; intrauterine PG evoked salivation and defecation only. Uterine activity, recorded with 3 indwelling catheters, resembled the contractions of parturition and lasted from 20-45 minutes after PG until a few hous later. Parturition occurred 1-7 days following PG administration. Most external signs of impending labor, such as relaxation of pelvic ligaments, vulvar edema, and altered appetite and behavior, were absent, but deliveries were rapid and normal in 38 of 39 animals, with maximal relaxation of the birth canal, distension of the cervix, and development of mammary glands. 10 (27%) animals retained their placentas, and after being given antibiotics, delivered the placenta within 10 days. After 260 days of gestation 31 calves survived, 2 were stillborn, 1 died during birth, and 1 died 3 days later of pneumonia. (LJG)   005681

1083

ZOR, U.; KOCH, Y.; LAMPRECHT, S.A.; AUSHER, J.; LINDNER, H.R.
Mechanism of oestradiol action on the rat uterus: independence of cyclic AMP, prostaglandin $E_2$and beta-adrenergic mediation.
Journal of Endocrinology 58(3): 525-533. 1973.

Levels of cAMP in uteri from immature rats were not markedly increased by PGE_2but were altered after 20 minutes incubation with estradiol-17β (10-25 μg/ml) or adrenaline. Propranolol virtually abolished the effect of adrenaline but reduced only slightly the effect of PGE_2. Intraperitoneal or intravenous injections of estradiol had no effect on the cAMP content of uteri from intact or ovariectomized rats. $PGE_2$250 μg per rat or isoprenaline significantly increased the cAMP content. This effect of isoprenaline, but not that of PGE_2, was blocked by propranolol. The authors concluded that 1) stimulation of cAMP production cannot be the mechanism of action of estradiol alone, 2) estradiol does not stimulate PGE$_2$synthesis or activation of beta-adrenergic receptors, and 3) stimulation of cAMP production by PGE$_2$is not mediated by adrenergic receptors. (HC)

BIBLIOGRAPHY OF NON-ABSTRACTED
PROSTAGLANDIN ARTICLES
1971-1973

The following appendix has been prepared by the Upjohn Company. It contains all the articles listed in the Upjohn Prostaglandin Bibliography© for the years 1971, 1972 and 1973 which have not been included as abstracts in the main part of this book.

To aid in the use of this document, some explanations may be helpful:

1. Abbreviations used are in accordance with the Word-Abbreviation List of the American National Standards Institute.

2. Articles are listed in alphabetical order by author. The primary sort has been made by the first three authors; secondary sort is by year of article.

3. Names of authors containing more than 17 letters have been shortened with a slash (/) to indicate the cut-off point. Thus:

   K. Herbaczynska-Cedro   appears as   HERBACZYNSKA-CE/ K

4. Articles for which no author is given (editorials, etc.) are listed under ANONYMOUS.

5. Books are listed with the following order of citation: IN: "Title," names of editors, publisher, city, volume and page numbers, year.

6. Foreign titles have been translated to English; the language is given at the end of the citation, in parentheses.

7. Since the Upjohn computer does not print Greek letter or subscripts, α and β used as part of prostaglandin names are shown as A and B, and subscripts are printed on the line. Thus, $PGF_2\alpha$ appears as PGF2A. When Greek letters appear in other contexts (e.g. α-receptor), they are spelled out (e.g. ALPHA-RECEPTOR).

ABBA GC
PROSTAGLANDINS IN NEONATAL PHYSIOPATHOLOGY
ACTA PAEDIATR LAT 26:160-167(1973);(ITALIAN)

ABDULLA YH   MC FARLANE E
CONTROL OF PROSTAGLANDIN BIOSYNTHESIS IN RAT BRAIN
HOMOGENATES BY ADENINE NUCLEOTIDES
BIOCHEM PHARMACOL 21:2841-2847(1972)

ABDULLA YH   MCFARLANE E
CONTROL OF ADENYLATE KINASE BY PROSTAGLANDINS E2 AND E3
BIOCHEM PHARMACOL 20:1726-1730(1971)

ABDULLA YH   MCFARLANE E
ACTION OF DOPAMINE AND NORADRENALINE ON SYNAPTIC
TRANSMISSION IN SYMPATHETIC GANGLIA OF BROWN FAT
BIOCHEM PHARMACOL 21:592-594(1972)

ABE K
THE EFFECTS OF PROSTAGLANDINS ON THE RENAL FUNCTION
PROC SOC GERONTOL STUDY, OKAYAMA CITY, ONO PHARMACEUTICAL
CO, OSAKA:12-15(1972);(JAPANESE)

ABELL CW   MONAHAN TM
THE ROLE OF ADENOSINE 3',5'-CYCLIC MONOPHOSPHATE IN THE
REGULATION OF MAMMALIAN CELL DIVISION
J CELL BIOL 59:549-558(1973)

ABERG G
INTERACTIONS BETWEEN SALICYLATES AND PROSTAGLANDINS ON THE
TEMPERATURE OF INFLAMED RAT PAWS
INT RES COMMUN SYST (73-4):9-3-3(1973)

ABERG G
INHIBITION OF FLUSH INDUCED BY NICOTINIC ACID
INT RES COMMUN SYS (73-12)8-4-27(1973)

ABRAHAM NA
PROSTAGLANDIN VI - AN EFFICIENT SYNTHESIS OF
11-DEOXYPROSTAGLANDINS
TETRAHEDRON LETT 451-452(1973)

ABRAMOWITZ J   CHAVIN W
THE MELANOSOME DISPERSING ACTIVITY OF PROSTAGLANDINS IN
BLACK GOLDFISH
AM ZOOL 12:673-674(1972)

ABRAMOWITZ J   CHAVIN W
IN VITRO EFFECTS OF PROSTAGLANDINS UPON MELANOSOME
DISPERSION IN THE SKIN OF BLACK GOLDFISH, CARASSIUS
AURATUS L
PROSTAGLANDINS 4:805-818(1973)

ADOLPHE M   GIROUD JP   TIMSIT J
LECHAT P
COMPARATIVE STUDY OF THE EFFECTS OF THE PROSTAGLANDINS E1,
E2, A2, F1A, AND F2A ON MITOSIS OF HELA CELLS IN CULTURE
C R ACAD SCI (D) (PARIS) 277:537-540 (1973);(FRENCH)

ADOLPHSON RL   TOWNLEY RG
COMPARISON OF THE BRONCHODILATOR ACTIVITIES OF ISOPROTERENOL
AND OF PROSTAGLANDIN E1 AEROSOLS
CHEST (SUPPL) 63:5S-6S(1973)

ADVANI AT   PETTIT LD
THE FORMATION CONSTANTS OF THE PROTON AND CALCIUM COMPLEXES
OF PROSTAGLANDINS PGE1 AND PGF1B
CHEM -BIOL INTERACTIONS 7:181-184(1973)

AHERN DG
INHIBITION OF SOYBEAN LIPOXIDASE AND PROSTAGLANDIN
SYNTHETASE AND STUDIES OF THE COMPOSITION OF EPIDERMAL
LIPIDS OF SNAKE SKIN
DISS, BOSTON UNIV(1972)

AHN CS
GLYCOGEN METABOLISM OF THE THYROID.
ENDOCRINOLOGY 88:1341-1348(1971)

AHN CS   ROSENBERG IN
OXIDATION OF 14C-FORMATE IN THYROID SLICES: EFFECTS OF TSH,
DIBUTYRYL CYCLIC 3',5'-AMP (DBCAMP) AND PROSTAGLANDIN E1
(PGE1)
IN:FURTHER ADV THYROID RES TRANS 6TH INT THYROID CONF
VIENNA 1970 2:825-837(1971)

AHREN K   HAMBERGER L   HERLITZ H
HILLENSJO T   NILSSON L   PERKLEV T
SELSTAM G
ASPECTS OF THE MECHANISM OF ACTION OF GONADOTROPHINS
IN: "THE ENDOCRINE FUNCTION OF THE HUMAN TESTIS," VHT JAMES,
M SERIO, L MARTINI, ACADEMIC PRESS, NY/LOND 1:251-272(1973)

AHREN K   PERKLEV T
EFFECTS OF PGE1 AND 7-OXA-13-PROSTYNOIC ACID ON THE
ISOLATED PREPUBERTAL RAT OVARY
ADV BIOSCI (SUPPL) 9:118(1972)

AIKEN JW
ASPIRIN AND INDOMETHACIN PROLONG PARTURITION IN RATS:
EVIDENCE THAT PROSTAGLANDINS CONTRIBUTE TO EXPULSION OF
FOETUS
NATURE (LOND) 240:21-25(1972)

AIKEN JW   VANE JR
BLOCKADE OF ANGIOTENSIN-INDUCED PROSTAGLANDIN RELEASE FROM
DOG KIDNEY BY INDOMETHACIN.
PHARMACOLOGIST 13:293(1971)

AIKEN JW   VANE JR
INHIBITION OF CONVERTING ENZYME OF THE RENIN-ANGIOTENSIN
SYSTEM IN KIDNEYS AND HINDLEGS OF DOGS
CIRC RES 30:263-273(1972)

AIKEN JW   VANE JR
INTRARENAL PROSTAGLANDIN RELEASE ATTENUATES THE RENAL
VASOCONSTRICTOR ACTIVITY OF ANGIOTENSIN
J PHARMACOL EXP THER 184:678-687(1973)

AKISHI Y   IDE H
SECRETION OF CORTISOL AND GROWTH HORMONE BY PGE1
FOLIA ENDOCRINOL JAP 46:35(1971)

AKISHI Y   IDE H
STIMULATORY EFFECT OF PROSTAGLANDIN E1 ON SECRETION OF
GROWTH HORMONE AND CORTISOL
FOLIA ENDOCRINOL JAP 47:360(1971);(JAPANESE)

AKISHI Y   IDE H   MURAO M
EFFECTS OF PROSTAGLANDIN ON METABOLISM OF SUGAR AND LIPID IN
DIABETIC PATIENTS
J JAP DIABETIC SOC 14:97(1971);(JAPANESE)

AL AWQATI Q   GREENOUGH WB III
PROSTAGLANDINS INHIBIT INTESTINAL SODIUM TRANSPORT
NATURE (NEW BIOL) 238:26-27(1972)

AL TAI SA   GRAHAM JPD
THE ACTIONS OF PROSTAGLANDINS E1 AND F2A ON THE PERFUSED
VESSELS OF THE ISOLATED RABBIT EAR
BR J PHARMACOL 44:699-710(1972)

ALABASTER VA
METABOLISM OF VASOACTIVE SUBSTANCES BY THE LUNG
DISS, UNIV LOND (1971)

ALABASTER VA   BAKHLE YS
THE INACTIVATION OF BRADYKININ IN THE PULMONARY CIRCULATION
OF ISOLATED LUNGS
BR J PHARMACOL 45:299-310(1972)

ALANKO K   POPPIUS H
ANTICHOLINERGIC BLOCKING OF PROSTAGLANDIN-INDUCED
BRONCHOCONSTRICTION
BR MED J 1:294(1973)

ALBRO P   THOMAS R   FISHBEIN L
PROSTAGLANDINS: ACTION ON MAST CELLS IN VITRO
PROSTAGLANDINS 1:133-146(1972)

ALBRO PW   FISHBEIN L
ISOLATION AND CHARACTERIZATION OF
5,8,12-TRIHYDROXY-TRANS-9-OCTADECENOIC ACID FROM WHEAT BRAN
PHYTOCHEM 10:631-636(1971)

ALBURN HE   FENICHEL RL
EFFECTS OF PROSTAGLANDINS ON BLOOD PLATELET AGGREGATION
ADV BIOSCI (SUPPL) 9:22(1972)

ALEJANDRO MENDO/ J
PHYSIOLOGY NEWS: THE PROSTAGLANDINS
ANTIOQUIA MED 21:407-417(1971);(SPANISH)

ALEXANDER RW   KENT KM   PISANO JJ
KEISER HR   COOPER T
REGULATION OF CANINE CORONARY BLOOD FLOW BY ENDOGENOUSLY
SYNTHESIZED PROSTAGLANDINS
CIRCULATION (SUPPL IV) 8:IV-107 (1973)

ALLEN DO   ASHMORE J
ASSAY OF GLYCOGEN PHOSPHORYLASE IN ISOLATED FAT CELLS OF THE
RAT
BIOCHEM PHARMACOL 21:1441-1447(1972)

ALLEN JE
LETTER TO THE EDITOR
PROSTAGLANDINS 4:539-542(1973)

ALLEN JE   RASMUSSEN H
PROSTAGLANDIN- AND ISOPROTERENOL INDUCED CHANGES IN THE RBC
MEMBRANE
CLIN RES 19:559(1971)

ALLEN JE   RASMUSSEN H
HUMAN RED BLOOD CELLS: PROSTAGLANDIN E2, EPINEPHRINE, AND
ISOPROTERENOL ALTER DEFORMABILITY
SCIENCE 174:512-514(1971)

ALLEN JE   RASMUSSEN H
SOME EFFECTS OF VASOACTIVE HORMONES ON THE MAMMALIAN RED
BLOOD CELL
IN: "PROSTAGLANDINS IN CELLULAR BIOLOGY," PW RAMWELL AND
BB PHARRISS, PLENUM PRESS, NY/LOND:27-60(1972)

ALPERT JS   HAYNES FW   KNUTSON PA
DALEN JE   DEXTER L
PROSTAGLANDINS AND THE PULMONARY CIRCULATION
CLIN RES 20:360(1972)

ALPERT JS   HAYNES FW   KNUTSON PA
DALEN JE   DEXTER L
PROSTAGLANDINS AND THE PULMONARY CIRCULATION
PHYSIOLOGIST 15:71(1972)

ALPERT JS   HAYNES FW   KNUTSON PA
DALEN JE   DEXTER L
PROSTAGLANDINS AND THE PULMONARY CIRCULATION
PROSTAGLANDINS 3:759-765(1973)

ALPERT JS   HICKLER RB
CARDIOVASCULAR AND RENAL EFFECTS OF RENOMEDULLARY
PROSTAGLANDINS
IN "KIDNEY HORMONES," JW FISHER, ACADEMIC PRESS, LOND:
525-561(1971)

ALSAT E   CEDARD L
STIMULATING ACTION OF PROSTAGLANDINS ON THE PRODUCTION OF
ESTROGENS BY HUMAN PLACENTA PERFUSED IN VITRO
C R ACAD SCI (D) (PARIS) 275:1803-1806(1972):(FRENCH)

ALVAREZ FS   WREN D
SYNTHESIS OF (+-) 11-DEOXYPROSTAGLANDIN E1, F1A AND F1B AND
ITS 15B-EPIMERS BY CONJUGATE ADDITION OF NITROMETHANE TO
2-(6+-CARBOMETHOXYHEXYL)-2-CYCLOPENTEN-1-ONE
TETRAHEDRON LETT 569-572(1973)

ALVAREZ FS   WREN D   PRINCE A
SYNTHESIS OF (+-)-PROSTAGLANDIN E1,
(+-)-11-DEOXYPROSTAGLANDINS E1, F1A, AND F1B, AND
(+-)-9-OXO-13-CIS-PROSTENOIC ACID BY CONJUGATE ADDITION OF
VINYLCOPPER REAGENTS
J AM CHEM SOC 94:7823-7827(1972)

AMBACHE N   ZAR MA
EVIDENCE AGAINST ADRENERGIC MOTOR TRANSMISSION IN THE GUINEA
-PIG VAS DEFERENS
J PHYSIOL (LOND) 216:359-389(1971)

AMBRUS G
PRESENT AND PERSPECTIVE IN RESEARCH ON PROSTAGLANDINS
MAGY KEM LAPFA 26:581-586(1971):(HUNGARIAN)

AMER MS   MARQUIS NR
THE EFFECT OF PROSTAGLANDINS, EPINEPHRINE AND ASPIRIN ON
CYCLIC AMP PHOSPHODIESTERASE ACTIVITY OF HUMAN BLOOD
PLATELETS AND THEIR AGGREGATION
IN: "PROSTAGLANDINS IN CELLULAR BIOLOGY," PW RAMWELL AND
BB PHARRISS, PLENUM PRESS, NY/LOND:93-110(1972)

AMER MS   MCKINNEY GR
POSSIBILITIES FOR DRUG DEVELOPMENT BASED ON THE CYCLIC AMP
SYSTEM
LIFE SCI 13:753-767(1973)

AMOSS M   BLACKWELL R   VALE W
BURGUS R   GUILLEMIN R
STIMULATION OF CONCOMITANT SECRETION IN VITRO OF LH AND
FSH BY HIGHLY PURIFIED HYPOTHALAMIC LRF; EVIDENCE FOR A
PROSTAGLANDIN RECEPTOR FOR THE RELEASE OF LH
PROC INT UNION PHYSIOL SCI, 25TH INT CONGR, MUNICH,
9:17(1971)

AMY JJ   JACKSON DM   GANESAN PA
KARIM SMM
PROSTAGLANDIN 15 (R) 15-METHYL-E2 METHYL ESTER FOR
SUPPRESSION OF GASTRIC ACIDITY IN GRAVIDA AT TERM
BR MED J 4:208-211(1973)

ANDERSEN N
PROGRAM NOTES ON STRUCTURES AND NOMENCLATURE
ANN NY ACAD SCI 180:14-23(1971)

ANDERSEN N
DISCUSSION. (PART I. CHEMISTRY OF PROSTAGLANDINS).
ANN NY ACAD SCI 180:104-106(1971)

ANDERSON ABM   TURNBULL AC
COMPARATIVE ASPECTS OF FACTORS INVOLVED IN THE ONSET OF
LABOUR IN OVINE AND HUMAN PREGNANCY
MEM SOC ENDOCRINOL 20:141-162(1973)

ANDERSON AJ   BROCKLEHURST WE   WILLIS AL
EVIDENCE FOR THE ROLE OF LYSOSOMES IN THE FORMATION OF
PROSTAGLANDINS DURING CARRAGEENIN INDUCED INFLAMMATION IN
THE RAT
PHARMACOL RES COMMUN 3:13-19(1971)

ANDERSON FL   JUBIZ W   KRALIOS A
TSAGARIS TJ
PROSTAGLANDIN (PG) LEVELS DURING HYPOXIA INDUCED PULMONARY
HYPERTENSION IN THE BOVINE
CLIN RES 20:203(1972)

ANDERSON FL   JUBIZ W   KRALIOS AC
TSAGARIS TJ   KUIDA H
PLASMA PROSTAGLANDIN LEVELS DURING ENDOTOXIN SHOCK IN DOGS
CIRCULATION (SUPPL II) 46:124(1972)

ANDERSON FL   JUBIZ W   KRALIOS AC
TSAGARIS TJ
PLASMA PROSTAGLANDIN E LEVELS DURING ENDOTOXIN SHOCK IN DOGS
CLIN RES 21:194(1973)

ANDERSON FL   JUBIZ W   TSAGARIS TJ
KUIDA H
PROSTAGLANDIN F LEVELS DURING ENDOTOXIN INDUCED PULMONARY
HYPERTENSION IN CALVES
CIRCULATION (SUPPL IV) 8:IV-133 (1973)

ANDERSON FL   KRALIOS AC   TSAGARIS TJ
KUIDA H
EFFECTS OF PROSTAGLANDINS F2A AND.E2 ON THE BOVINE
CIRCULATION
PROC SOC EXP BIOL MED 140:1049-1053(1972)

ANDERSON FL   TSAGARIS TJ   KUIDA H
EFFECT OF PROSTAGLANDINS (PG) ON BOVINE PULMONARY
CIRCULATION
FED PROC 30:380(1971)

ANDERSON GG   HOBBINS JC   CALDWELL B
SPEROFF L
PERIPHERAL LEVELS OF F PROSTAGLANDINS IN PATIENTS
RECEIVING PGF2A FOR INDUCTION OF THERAPEUTIC ABORTION
ABSTR SOC GYNECOL INVEST, 19TH ANNU MEET, PHOENIX:20(1972)

ANDERSON GG   HOBBINS JC   RAJKOVIC V
SPEROFF L   CALDWELL BV
MIDTRIMESTER ABORTION USING INTRA-AMNIOTIC PROSTAGLANDIN F2A
WITH INTRAVENOUS SYNTOCINON
J REPROD MED 9:434-436(1972)

ANDERSON GG   HOBBINS JC   SPEROFF L
CALDWELL BV
INTRAVENOUS PROSTAGLANDINS E2 AND F2A AND SYNTOCINON FOR THE
INDUCTION OF TERM LABOR
J REPROD MED 9:287-291(1972)

ANDERSON GG   SPEROFF L
PROSTAGLANDINS AND ABORTION
CLIN OBSTET GYNECOL 14:245-257(1971)

ANDERSON GG   SPEROFF L
PROSTAGLANDINS
SCIENCE 171:502-504(1971)

ANDERSON GG   SPEROFF L   HOBBINS J
PROSTAGLANDIN INFUSIONS FOR INDUCTION OF LABOR AT TERM AND
THERAPEUTIC ABORTION
ABSTR 18TH ANNU MEET SOC GYNECOL INVEST, PHOENIX:15(1971)

ANDERSON TE   SPACKMAN TJ   SCHWARTZ SS
ROENTGEN FINDINGS IN INTESTINAL GANGLIONEUROMATOSIS. ITS
ASSOCIATION WITH MEDULLARY THYROID CARCINOMA AND
PHEOCHROMOCYTOMA.
RADIOLOGY 101:93-96(1971)

ANDERSSON KE   ANDERSSON R   HEDNER P
PERSSON CGA
DUAL EFFECTS ON GALL-BLADDER AND SPHINCTER OF ODDI INDUCED
BY CHOLECYSTOKININ AND PROSTAGLANDINS
ACTA PHARMACOL TOXICOL 31(SUPPL):44(1972)

ANDERSSON KE   ANDERSSON R   HEDNER P
PERSSON CGA
ANALOGOUS EFFECTS OF CHOLECYSTOKININ AND PROSTAGLANDIN E2
ON MECHANICAL ACTIVITY AND TISSUE LEVELS OF CAMP IN BILIARY
SMOOTH MUSCLE
ACTA PHYSIOL SCAND 87:41A-42A(1973)

ANDERSSON KE   ANDERSSON R   HEDNER P
PERSSON CGA
PARALLELISM BETWEEN MECHANICAL AND METABOLIC RESPONSES
TO CHOLECYSTOKININ AND PROSTAGLANDIN E2 IN EXTRAHEPATIC
BILIARY TRACT
ACTA PHYSIOL SCAND 89:571-579(1973)

ANDERSSON KE   HEDNER P   PERSSON CGA
EFFECTS OF PROSTAGLANDIN INHIBITORS ON CONTRACTIONS INDUCED
BY CHOLECYSTOKININ AND PROSTAGLANDIN E2 IN GUINEA-PIG
GALLBLADDER IN VITRO.
ACTA ENDOCRINOL (SUPPL)(KBH)177:355(1973)

ANDERSSON RGG
CYCLIC AMP, A REGULATOR SUBSTANCE IN SMOOTH MUSCLE FUNCTION
AND METABOLISM
ACTA PHYSIOL SCAND (SUPPL 396) : 7(1973)

ANDRESEN O
PROSTAGLANDINS, A REVIEW
NORD VET MED 25:591-598(1973):(NORWEGIAN)

ANGGARD E
STUDIES OF THE ANALYSIS AND METABOLISM OF THE
PROSTAGLANDINS
ANN NY ACAD SCI 180:200-217(1971)

ANGGARD E   BOHMAN SO   GRIFFIN JE III
LARSSON C   MAUNSBACH AB
SUBCELLULAR LOCALIZATION OF THE PROSTAGLANDIN SYSTEM IN THE
RABBIT RENAL PAPILLA
ACTA PHYSIOL SCAND 84:231-246(1972)

ANGGARD E   JONSSON CE
EFFLUX OF PROSTAGLANDINS IN LYMPH FROM SCALDED TISSUE.
ACTA PHYSIOL SCAND 81:440-447(1971)

ANGGARD E   JONSSON CE
FORMATION OF PROSTAGLANDINS IN THE SKIN FOLLOWING A BURN
INJURY
IN: "PROSTAGLANDINS IN CELLULAR BIOLOGY," PW RAMWELL AND
BB PHARRISS, PLENUM PRESS, NY/LOND:269-291(1972)

ANGGARD E   LARSSON C
THE SEQUENCE OF THE EARLY STEPS IN THE METABOLISM OF
PROSTAGLANDIN E1
EUR J PHARMACOL 14:66-70(1971)

ANGGARD E   LARSSON C
PROSTAGLANDIN MEDIATED HYPOTENSIVE EFFECTS OF ARACHIDONIC
ACID IN THE RABBIT
ACTA PHYSIOL SCAND (SUPPL 396) : 18(1973)

ANGGARD E   LARSSON C   SAMUELSSON B
THE DISTRIBUTION OF 15-HYDROXY PROSTAGLANDIN DEHYDROGENASE
AND PROSTAGLANDIN-13-REDUCTASE IN TISSUES OF THE SWINE
ACTA PHYSIOL SCAND 81:396-404(1971)

ANGGARD E   STRANDBERG K
EFFLUX OF PROSTAGLANDIN E2 FROM CAT PAWS PERFUSED WITH
COMPOUND 48/80
ACTA PHYSIOL SCAND 82:333-344(1971)

ANONYMOUS -
PROSTAGLANDINS AND LIPOLYSIS IN RAT ADIPOSE TISSUE
NUTR REV 29:170-173(1971)

ANONYMOUS -
NEW LIGHT ON INFLAMMATION
BR MED J 3:61-62(1971)

ANONYMOUS -
SCIENTISTS SEARCH FOR DRUGS FROM THE SEA
CHEM ENG NEWS 49(23):24-25(1971)

ANONYMOUS -
NONPROPRIETARY NAMES
J AM MED ASSOC 215:2100-2101(1971)

ANONYMOUS -
MANAGEMENT OF THROMBOSIS
LANCET 2:361-362(1971)

ANONYMOUS -
ASPIRIN-LIKE DRUGS AND PROSTAGLANDINS
LANCET 2:363-364(1971)

ANONYMOUS -
HOW ASPIRIN FUNCTIONS
NATURE (LOND) 231:417-418(1971)

ANONYMOUS -
PROSTAGLANDINS: STATEMENT BY COMMITTEE ON SAFETY OF DRUGS
PHARM J 206:186(1971)

ANONYMOUS -
THE ASPIRIN MYSTERY
PHARM J 207:45-46(1971)

ANONYMOUS -
PROSTAGLANDINS
SOUTH MED J 64:374-375(1971)

ANONYMOUS -
PHARMACOLOGIST LEADS PROSTAGLANDIN STUDY
HOSP TRIB 19(1971)

ANONYMOUS -
PROSTAGLANDINS AND ARTHRITIS
SCI NEWS 100:423(1971)

ANONYMOUS -
PROSTAGLANDINS SEEN USEFUL IN EASING RISK FROM LYSOSOME
ENZYME RELEASE
HOSP TRIB:(MARCH 22,1971)

ANONYMOUS -
PROSTAGLANDIN THERAPY FOUND TO ABBREVIATE LABOR AT TERM
HOSP TRIB 5(11):1,20(1971)

ANONYMOUS -
ASPIRIN IS AN ANTI-PROSTAGLANDIN
MED WORLD NEWS 12(29):13-14(1971)

ANONYMOUS -
CYCLIC AMP TURNS OFF TUMOR CELLS IN MICE
MED WORLD NEWS 12(31):16-18(1971)

ANONYMOUS -
ASPIRIN WORKS BY ENHANCING CYCLIC AMP-OR DOES IT?
MED WORLD NEWS 12(31):17(1971)

ANONYMOUS -
CYCLIC AMP. FOR DR. EARL W. SUTHERLAND, A NOBEL PRIZE: FOR
BIOMEDICAL RESEARCH, A NEW DIRECTION
MED WORLD NEWS 12(46):48-58(1971)

ANONYMOUS -
ASPIRIN AND PROSTAGLANDINS
RES PROSTAGLANDINS 1(2):3(1971)

ANONYMOUS -
HOW ASPIRIN WORKS
SCI NEWS 100(3):38(1971)

ANONYMOUS -
PROSTAGLANDINS FORM FASTER AFTER A BURN
HOSP TRIB SEPT. 4:18(1972)

ANONYMOUS -
CYCLIC AMP: 15 YEARS LATER
SCI NEWS 102:148(1972)

ANONYMOUS -
PROSTAGLANDINS: MEDIATORS OF INFLAMMATION
SCIENCE 177:780-781(1972)

ANONYMOUS -
PROSTAGLANDINS FOR THE BODY THAT HAS EVERYTHING.
(ADVERTISEMENT)
NY TIMES JUNE 5:28M(1972)

ANONYMOUS -
NEW THEORY ON HOW ASPIRIN WORKS
WASH POST (JULY 31) : A3(1972)

ANONYMOUS -
PROSTAGLANDIN COMPLICITY IN RHEUMATOID ARTHRITIS
SCI NEWS 102:181-182(1972)

ANONYMOUS -
NAVY STUDY FINDS PROSTAGLANDINS AFFECT RED BLOOD CELLS OF
SEVERELY INJURED
ONR NEWS RELEASE #8-72(1972)

ANONYMOUS -
PROSTAGLANDINS: INVOLVED IN DENTAL DISEASE
SCI NEWS 101:215-216(1972)

ANONYMOUS -
BIRTH CONTROL METHOD TRIED "AFTER-THE-FACT"
HSMHA HEALTH REP 87:41-42(1972)

ANONYMOUS -
PROSTAGLANDINS AND PLATELETS
PROSTAGLANDINS 1 : 119-120(1972)

ANONYMOUS -
SOME SIGNS THE BIG E MAY REDUCE GUM DISEASE
SCI NEWS 102:24(1972)

ANONYMOUS -
THE COLLECTION AND HANDLING OF BIOLOGICAL SAMPLES FOR
PROSTAGLANDIN ANALYSIS
PROSTAGLANDINS 1:279-280(1972)

ANONYMOUS -
ASPIRIN AND PROSTAGLANDINS
MEDICINA (BAIRES)22:198-199(1972):(SPANISH)

ANONYMOUS -
PROMISE BUT NO MIRACLES
NATURE (LOND) 237:482-483(1972)

ANONYMOUS -
PROSTAGLANDINS-SYNTHESIS OR FERMENTATION
PROCESS BIOCHEM 7:3(1972)

ANONYMOUS -
REPORT ON WORLD HEALTH ORGANIZATION MEETING ON
PROSTAGLANDINS
PROSTAGLANDINS 1:171-172(1972)

ANONYMOUS -
TAKING STOCK OF THE CARCINOID SYNDROME
LANCET 2:711-712(1973)

ANONYMOUS -
PROSTAGLANDINS
W VA MED J 69:182(1973)

ANONYMOUS -
HYPERCALCAEMIA, CANCER, AND PROSTAGLANDINS
LANCET 2:1246-1247(1973)

ANONYMOUS -
HYPOPHYSECTOMY IN YOUNG RATS AND PROSTAGLANDIN SYNTHESIS
NUTR REV 31:284-286(1973)

ANONYMOUS -
REPORTS ON THE DISCUSSIONS
IN PHARMACOL THERMOREGUL PROC SATELL SYMP
463-468(1972) (KARGER BASEL) (1973)

ANONYMOUS -
ANALGESICS AND ASTHMA
BR MED J 3:419-420(1973)

ANONYMOUS -
PROSTAGLANDINS FOR INDUCING LABOUR AND ABORTION
DRUG THER BULL 11:41-43(1973)

ANONYMOUS -
OESTROGEN LEVELS IN INFERTILITY AND ABORTION
MED J AUST 2:1107-1108(1973)

APPLEZWEIG N
THE DRUG RESEARCH REVOLUTION. INSIGHTS INTO THE BODY'S
DEEPEST SECRETS ARE OPENING EXCITING NEW AVENUES OF ATTACK
AGAINST MOST MAJOR DISEASES. HERE'S A PREVIEW OF THE DRUGS
OF TOMORROW
CHEM WEEK:25-34(MARCH1,1972)

APPLEZWEIG N
THE DEVELOPMENT OF SYSTEMIC CHEMICAL CONTRACEPTIVES
IN: "MAJOR PROBLEMS IN OBSTETRICS AND GYNECOLOGY," EA
FRIEDMAN, WB SAUNDERS, PHILA/LOND/TORONTO 5:17-37(1973)

APPS MCP   CATER DB
PRODUCTION OF HISTAMINE-LIKE AND PROSTAGLANDIN-LIKE
SUBSTANCES FROM SERUM INCUBATED WITH RAT, DOG, MOUSE OR
HUMAN TUMOURS
BR J EXP PATHOL 54:203-221(1973)

APPS MCP   CATER DB
PRODUCTION OF HISTAMINE-LIKE AND PROSTAGLANDIN-LIKE
SUBSTANCES FROM SERUM INCUBATED WITH RAT, DOG, MOUSE OR
HUMAN TUMOURS
J PATHOL 109:PV(1973)

ARCHER DF
STUDIES ON THE CORPUS LUTEUM OF MIDGESTATION: PROSTAGLANDIN
F2A EFFECT ON CHOLESTEROL
ABSTR SOC GYNECOL INVEST, 20TH ANNU MEET, ATLANTA:47(1973)

ARCHER DF   PETRELLI ES
ALTERATIONS IN BIOSYNTHESIS OF PROGESTERONE IN HUMAN
CORPORA LUTEA OF PREGNANCY BY PROSTAGLANDINS
ADV BIOSCI (SUPPL)9:110(1972)

ARDLIE NG
INFLUENCE OF CA++ AND MG++ AND CAMP ON THE INITIAL CHANGE
IN PLATELET SHAPE INDUCED BY ADP
FED PROC 30:201(1971)

AREHART JL
CORAL: UNEXPECTED BOON IN WORLD WIDE DRUG RESEARCH
OCEANS 4(5):70-72(1971)

ARMSTRONG DT   GRINWICH DL
POSSIBLE ROLE OF PROSTAGLANDINS IN OVULATION
ADV BIOSCI (SUPPL)9:117(1972)

ARNESJO B   IHSE I   OVIST I
INHIBITION OF DUODENAL PANCREATIC ENZYMIC ACTIVITIES BY
POLYPHLORETIN PHOSPHATE WITH SPECIAL REFERENCE TO
PHOSPHOLIPASE A2
ACTA CHEM SCAND 27:2225-2227(1973)

ARRIGO L   MACRI I   RIGON-MACRI P
EFFECTS OF HORMONES AND CHEMICAL MEDIATORS ON SOME
FUNCTIONAL PARAMETERS OF RABBIT HEART
BOLL SOC ITAL BIOL SPER 49:273(1973);(ITALIAN)

ARRIGONI-MARTEL/ E RESTELLI A
RELEASE OF LYSOSOMAL ENZYMES IN EXPERIMENTAL INFLAMMATIONS:
EFFECTS OF ANTI-INFLAMMATORY DRUGS
EUR J PHARMACOL 19:191-198(1972)

ARRIGONI-MARTEL/ E SELVA D   SCHIATTI P
DIFFERENT EFFICACY OF ANTI-INFLAMMATORY DRUGS IN
INHIBITING THE REACTIONS TO INTRADERMAL PHOSPHOLIPASE A AND
PGE2
J INTERN MED RES 1:120-126(1973)

ARTURSON G
THE ACTIVATION OF ENDOGENOUS SUBSTANCE INDUCING PATHOLOGICAL
INCREASES OF MICROVASCULAR PERMEABILITY IN THERMAL INJURY
EUR SURG RES 3:169-170(1971)

ARTURSON G   HAMBERG M   JONSSON CE
PROSTAGLANDINS IN HUMAN BURN BLISTER FLUID
ACTA PHYSIOL SCAND 87:270-276(1973)

ARTURSON G   JONSSON CE
EFFECTS OF INDOMETHACIN ON THE TRANSCAPILLARY LEAKAGE OF
MACROMOLECULES AND THE EFFLUX OF PROSTAGLANDINS IN THE PAW
LYMPH FOLLOWING EXPERIMENTAL SCALDING INJURY
UPSALA J MED SCI 78:181-188(1973)

ARTURSON G   JONSSON CE
EFFECTS OF O-(BETA-HYDROXYETHYL)-RUTOSIDES (HR) AND
INDOMETHACIN ON TRANSCAPILLARY MACROMOLECULAR TRANSPORT
AND PROSTAGLANDINS FOLLOWING SCALDING INJURY
BIBL ANAT (12) : 465-470(1973)

ASAKAWA T   YOSHIDA H
STUDIES ON THE FUNCTIONAL ROLE OF ADENOSINE 3', 5'
MONOPHOSPHATE, HISTAMINE AND PROSTAGLANDIN E1 IN THE CENTRAL
NERVOUS SYSTEM.
JAP J PHARMACOL 21:569-583(1971)

ASAKAWA T   YOSHIDA H
ON THE REGULATION OF FUNCTION OF THE CENTRAL NERVOUS
SYSTEM BY PROSTAGLANDIN
PROC 44TH GEN MEET JAP PHARMACOL SOC, TOKYO:147-148(1971)

ASH RW   CHALLIS JRG   HARRISON FA
HEAP RB   ILLINGWORTH DV   PERRY JS
POYSER NL
HORMONAL CONTROL OF PREGNANCY AND PARTURITION: A COMPARATIVE
ANALYSIS
IN: "FOETAL AND NEONATAL PHYSIOLOGY," CAMBRIDGE UNIV PRESS,
LOND : 551-561(1973)

ASO K   SAKAMOTO N   FARBER E
DENEAU D   KRULIG L   WILKINSON D
THE SYNTHESIS OF PROSTAGLANDINS E2 AND F2A IN PSORIATIC SKIN
J INVEST DERMATOL 60:111(1973)

ASO K   SAKAMOTO N   FARBER E
DENEAU D   KRULIG L   WILKINSON D
THE SYNTHESIS OF PROSTAGLANDINS E2 AND F2A IN PSORIATIC SKIN
CLIN RES 21:739(1973)

ASPINALL RL   CAMMARATA PS   JIU J
BAKER DE   PAUTSCH WF
EFFECT OF VARIOUS PROSTAGLANDINS ON TWO LABORATORY MODELS OF
CHRONIC ARTHRITIC INFLAMMATION
ADV BIOSCI (SUPPL)9:65(1972)

ASPINALL RL   CAMMARATA PS   NAKAO A
JIU J   MIYANO M   BAKER DE
PAUTSCH WF
EFFECT OF VARIOUS PROSTAGLANDINS ON TWO LABORATORY MODELS OF
CHRONIC ARTHRITIC INFLAMMATION
ADV BIOSCI 9:419-425(1973)

ATHERTON A   BORN GVR
IN VIVO MEASUREMENT OF THE ADHESIVENESS OF GRANULOCYTES TO
BLOOD VESSEL WALLS
BIBL ANAT 12:138-145(1973)

ATKINS E   BODEL P
FEVER
N ENGL J MED 286:27-34(1972)

ATTALLAH AA   LEE JB
SPECIFIC BINDING SITES IN THE RABBIT KIDNEY FOR
PROSTAGLANDIN A
PROSTAGLANDINS 4:703-709(1973)

ATTALLAH AA   LEE JB
RADIOIMMUNOASSAY OF PROSTAGLANDIN A--INTRARENAL PGA2 AS A
FACTOR MEDIATING SALINE-INDUCED NATRIURESIS
CIRC RES 33:696-703(1973)

ATTALLAH AA   SCHUSSLER GC
PROSTAGLANDIN BINDING IN HUMAN SERUM FOLLOWS THE POLARITY
RULE
PROSTAGLANDINS 4:479-484(1973)

ATTANASI O   BACCOLINI G   CAGLIOTI L
ROSINI G
AN APPROACH TO THE PROSTAGLANDINE SKELETON THROUGH
ORGANOBORANES REACTIONS
GAZZ CHIM ITAL 103:31-36(1973)

ATTREP KA   MARIANI JM JR   ATTREP M JR
SEARCH FOR PROSTAGLANDIN A1 IN ONION
LIPIDS 8:484-486(1973)

AUGSTEIN J   FARMER JB   LEE TB
SHEARD P   TATTERSALL ML
SELECTIVE INHIBITOR OF SLOW REACTING SUBSTANCE OF
ANAPHYLAXIS
NATURE(NEW BIOL) 245:215-217(1973)

AURBACH GD   MARCUS R   HEERSCHE J
MARX S   NIALL H   TREGEAR GW
KEUTMANN HT   POTTS JT JR
HORMONES AND OTHER FACTORS REGULATING CALCIUM METABOLISM
ANN NY ACAD SCI 185:386-394(1971)

AUSTEN KF
A REVIEW OF IMMUNOLOGICAL, BIOCHEMICAL, AND PHARMACOLOGICAL
FACTORS IN THE RELEASE OF CHEMICAL MEDIATORS FROM HUMAN LUNG
IN: "ASTHMA," KF AUSTEN, LM LICHTENSTEIN, ACADEMIC PRESS,NY:
109-122(1973)

AUSTEN KF
CONTROL MECHANISMS OF THE IMMUNOLOGICAL RELEASE OF CHEMICAL
MEDIATORS
ABSTR, 9TH INT CONG BIOCHEM, STOCKH:306(1973)

AXEN U   BACZYNSKYJ L   DUCHAMP DJ
KIRTON KT   ZIESERL JF JR
DIFFERENTIATION BETWEEN ENDOGENOUS AND EXOGENOUS
(ADMINISTERED) PROSTAGLANDINS IN BIOLOGICAL FLUIDS
ADV BIOSCI (SUPPL)9:15(1972)

AXEN U   BACZYNSKYJ L   DUCHAMP DJ
ZIESERL JF JR
GAS CHROMATOGRAPHY-MASS SPECTROMETRY ASSAY FOR
PROSTAGLANDINS
J REPROD MED 9:372-375(1972)

AXEN U   BACZYNSKYJ L   DUCHAMP DJ
KIRTON KT   ZIESERL JF JR
DIFFERENTIATION BETWEEN ENDOGENOUS AND EXOGENOUS
(ADMINISTERED) PROSTAGLANDINS IN BIOLOGICAL FLUIDS
ADV BIOSCI 9:109-116(1973)

AXEN U   GREEN K   HORLIN D
SAMUELSSON B
MASS SPECTROMETRIC DETERMINATION OF PICOMOLE AMOUNTS OF
PROSTAGLANDINS E2 AND F2A USING SYNTHETIC DEUTERIUM LABELED
CARRIERS
BIOCHEM BIOPHYS RES COMMUN 45(2):519-525(1971)

AXEN U   PIKE JE   SCHNEIDER WP
THE TOTAL SYNTHESIS OF PROSTAGLANDINS
IN: "THE TOTAL SYNTHESIS OF NATURAL PRODUCTS," J APSIMON,
WILEY-INTERSCIENCE, NY 1:81-142(1973)

AZAR S   TOBIAN L
ESSENTIAL FATTY ACIDS AND PAPILLARY INTERSTITIAL CELLS
ABSTR 5TH INT CONGR NEPHROL, MEX CITY:112(1972)

AZAR S   TOBIAN L   ISHII M
PROLONGED WATER DIURESIS AFFECTING SOLUTES AND INTERSTITIAL
CELLS OF RENAL PAPILLA
AM J PHYSIOL 221:75-79(1971)

BACALAO J   RIEBER M
MODIFIED DISTRIBUTION OF SOME CYCLIC AMP PHOSPHODIESTERASE
ACTIVITIES FOLLOWING GROWTH OF L CELL FIBROBLASTS IN
PRESENCE OF PROSTAGLANDIN E1
EXP CELL RES 75:518-521(1972)

BACH GL
ADVERSE REACTIONS OF ANTIRHEUMATIC DRUGS
INT J CLIN PHARMACOL THER TOXICOL 7:198-205(1973)

BACH MA   BACH JF
STUDIES ON THYMUS PRODUCTS. VI. THE EFFECTS OF CYCLIC
NUCLEOTIDES AND PROSTAGLANDINS ON ROSETTE-FORMING CELLS.
INTERACTIONS WITH THYMIC FACTOR
EUR J IMMUNOL 3:778-783(1973)

BACZYNSKYJ L   DUCHAMP DJ   ZIESERL JF JR
AXEN U
COMPUTERIZED QUANTITATION OF DRUGS BY GAS
CHROMATOGRAPHY-MASS SPECTROMETRY
ANAL CHEM 45:479-482(1973)

BAER PG   NAVAR LG
RENAL VASODILATION AND UNCOUPLING OF BLOOD FLOW AND
FILTRATION RATE AUTOREGULATION
KIDNEY INT 4:12-21(1973)

BAGLI J   BOGRI T
PROSTAGLANDINS V -- UTILITY OF THE NEF REACTION IN THE
SYNTHESIS OF PROSTANOIC ACIDS. A TOTAL SYNTHESIS OF
(+-)-11-DEOXY-PGE1, -PGE2, AND THEIR C-15 EPIMERS
TETRAHEDRON LETT 3815-3817(1972)

BAGLI JF   BOGRI T
PROSTAGLANDINS. IV. TOTAL SYNTHESES OF DL-11-DEOXY PGE1 AND
13,14-DIHYDRO DERIVATIVES OF 11-DEOXY PGE1, PGF1A, AND PGF1B
J ORG CHEM 37:2132-2138(1972)

BAILE CA
CONTROL OF FEED INTAKE AND THE FAT DEPOTS.
J DAIRY SCI 54(4): 564-582(1971)

BAILE CA   BEAN SM
FEEDING EFFECTS OF HYPOTHALAMIC INJECTIONS OF
PROSTAGLANDINS
FED PROC 30:375(1971)

BAILE CA   BEAN SM
FEEDING EFFECTS OF HYPOTHALAMIC INJECTIONS OF PROSTAGLANDINS
PROC INT UNION PHYSIOL SCI, 25TH INT CONGR, MUNICH,
9:33(1971)

BAILE CA   MARTIN FH
RELATIONSHIP BETWEEN PROSTAGANDIN E1, POLYPHLORETIN
PHOSPHATE AND A AND B ADRENOCEPTOR-BOUND FEEDING LOCI IN THE
HYPOTHALAMUS OF SHEEP
PHARMACOL BIOCHEM BEHAV 1:539-545(1973)

BAILE CA   MARTIN FH   SIMPSON CW
FEEDING RESPONSE TO PROCAINE, NOREPINEPHRINE, ISOPROTERENOL,
AND PROSTAGLANDIN INJECTIONS INTO THE MEDIAL HYPOTHALAMUS OF
SHEEP
FED PROC 31:397(1972)

BAILE CA   SIMPSON CW   BEAN SM
MCLAUGHLIN CL JACOBS HL
PROSTAGLANDINS AND FOOD INTAKE OF RATS: A COMPONENT OF
ENERGY BALANCE REGULATION?
PHYSIOL BEHAV 10:1077-1085(1973)

BAILLIE P   MILTON PJD
PREMATURE LABOUR AT GROOTE SCHUUR HOSPITAL
S AFR MED J 47:1299-1301(1973)

BAILLIE P   SCHER J
THE EFFECT OF BETA-ADRENERGIC AGENTS ON PROSTAGLANDIN
STIMULATED LABOUR
ADV BIOSCI (SUPPL)9:121(1972)

BAKER AA
OVUM TRANSFER IN THE COW
AUST VET J 49:424-426(1973)

BALDUCCI R   FAVUZZI E   LA STELLA S
SYNTHESIS OF PROSTAGLANDINS
G MED MIL 123:316-331(1973):(ITALIAN)

BALLANTYNE A   JONES RL   UNGAR A
THE ACTIONS OF PROSTAGLANDINS A1 AND A2 ON AIRWAY RESISTANCE
AND COMPLIANCE IN THE CAT
BR J PHARMACOL 47:630P-631P(1973)

BALLARD CA   QUILLIGAN EJ
INTRA-AMNIOTIC PROSTAGLANDIN F2 FOR MID-TRIMESTER ABORTION
ADV BIOSCI (SUPPL)9:90(1972)

BALLARD CA   QUILLIGAN EJ
THERAPEUTIC ABORTION IN THE SECOND TRIMESTER BY
INTRA-AMNIOTIC PROSTAGLANDIN F2A
J REPROD MED 9:397-399(1972)

BALOURDAS TA
EFFECTS OF PROSTAGLANDIN E1 ON THE MICROCIRCULATION AND ON
THE BLOOD PRESSURE. SYNERGISTIC ACTION OF PGE1 AND
BRADYKININ
FED PROC 30:661(1971)

BANERJEE AK   PHILLIPS J   WINNING WW
E-TYPE PROSTAGLANDINS AND GASTRIC ACID SECRETION IN THE RAT
NATURE (NEW BIOL) 238:177-179(1972)

BANERJEE PK
PROSTAGLANDINS
CALCUTTA MED J 70:95-98(1973)

BANWELL JG   SHERR H
EFFECT OF BACTERIAL ENTEROTOXINS ON THE GASTROINTESTINAL
TRACT
GASTROENTEROLOGY 65: 467-497(1973)

BAR HP
EPINEPHRINE- AND PROSTAGLANDIN-SENSITIVE ADENYL CYCLASE
IN MAMMARY GLAND
BIOCHIM BIOPHYS ACTA 321:397-406(1973)

BAR HP   HENDERSON JF
PROPERTIES OF ADENYL CYCLASE FROM EHRLICH ASCITES CELLS
CAN J BIOCHEM 50:1003-1009(1972)

BARAC G
EFFECT OF PROSTAGLANDIN F2A ON DIURESIS AND ON RENAL BLOOD
FLOW IN THE DOG
C R SOC BIOL 165:970-973(1971):(FRENCH)

BARAC G
ACTION OF PROSTAGLANDIN PGE2 (E2) ON URINE OUTPUT AND RENAL
BLOOD FLOW IN THE DOG
C R SOC BIOL (PARIS) 166:221-223(1972):(FRENCH)

BARAC G
ON THE RENAL VASCULAR RECEPTORS OF FOUR PROSTAGLANDINS IN
THE DOG
J PHYSIOL (PARIS) 65:193(1972):(FRENCH)

BARAC G
ON THE CIRCULATORY ACTION OF THE PROSTAGLANDIN PGA1 (A1) IN
THE DOG
J PHYSIOL (PARIS) 67:239A(1973):(FRENCH)

BARAC G   DEBY C   JEUNIAUX F
ROBAYE B   NEURAY J
IS PLASMA CALCIUM CHELATED BY EXOGENOUS PROSTAGLANDIN E1?
ARCH INT PHYSIOL BIOCHIM 81:579(1973); (FRENCH)

BARAC G   DEBY C   NEURAY J
TESTS ON THE INTESTINAL ABSORPTION OF PROSTAGLANDIN PGE1 IN
THE DOG
J PHYSIOL (PARIS) 65:194(1972):(FRENCH)

BARBARASH NA   DAVYDOVA TM   PROKINA NS
DAVIDENKO OI   BUYANOVA ON
THE HYPOTENSIVE ACTIVITY OF PROSTAGLANDINOID SUBSTANCES OF
THE RENAL MEDULLA OF RABBITS WITH DOCA-SALT ACTION
KARDIOLOGIIA 13:81-84(1973):(RUSSIAN)

BARBU N   GLUHOVSCHI G
PROSTAGLANDINS
VIATA MED 20:81-87(1973):(ROMANIAN)

BARDEN TP
INDUCTION OF PRETERM LABOR WITH PROSTAGLANDIN F2A IN
PATIENTS WITH PREMATURE RUPTURE OF MEMBRANES
J REPROD MED 9:339-345(1972)

BARNER HB   KAISER GC   HAHN JW
JELLINEK M   AMAKO H   LEE JB
WILLMAN VL
EFFECTS OF PROSTAGLANDIN A1 ON CARDIOVASCULAR DYNAMICS AND
MYOCARDIAL METABOLISM
J SURG RES 12:168-172(1972)

BARNER HB   KAISER GC   JELLINEK M
LEE JB
EFFECT OF PROSTAGLANDIN A1 ON SEVERAL VASCULAR BEDS IN MAN
AM HEART J 85:584-592(1973)

BARNER HB   KAISER GC   LEE JB
EFFECT OF PROSTAGLANDIN A1 ON SEVERAL VASCULAR BEDS IN MAN
CLIN RES 19: 304(1971) 19: 110(1971)

BARNES AB
THE EFFECT OF OXYTOCIN ON RENAL BLOOD FLOW AND ITS
DISTRIBUTION IN THE DOG
NEPHRON 11:40-57(1973)

BARR W
INDUCTION OF LABOR WITH PROSTAGLANDIN E2
J REPROD MED 9:353-354(1972)

BARR W
ADVANCES IN GYNAECOLOGY
PRACTITIONER 211:459-464(1973)

BARTH LG   BARTH LJ
22SODIUM AND 45CALCIUM UPTAKE DURING EMBRYONIC INDUCTION IN
RANA PIPIENS
DEV BIOL 28:18-34(1972)

BASLOW MH
MARINE TOXINS
ANNU REV PHARMACOL:447-454(1971)

BATTA SK   MARTINI L
ANTI-IMPLANTATION EFFECT OF PROSTAGLANDINS
ABSTR 5TH INT CONGR PHARMACOL, SAN FRANC:16(1972)

BAUDUIN H   CANTRAINE F
"PHOSPHOLIPID EFFECT" AND SECRETION IN THE RAT PANCREAS
BIOCHIM BIOPHYS ACTA 270:248-253(1972)

BAUM H   KIRTLAND SJ   CORVETTI F
CARAFOLI E
PROSTAGLANDIN E1, ANTI-INFLAMMATORY AGENTS AND CALCIUM:
INTERACTIONS AT THE MITOCHONDRIAL MEMBRANE
IN: "MECHANISMS IN BIOENERGETICS, PROC INT CONF ON
MECHANISMS IN BIOENERGETICS," GF AZZONE, L ERNSTER, S PAPA,
E QUAGLIARIELLO, N SILIPRANDI, ACADEMIC PRESS, NY:365-374
(1973)

BAUM T   SHROPSHIRE AT
INFLUENCE OF PROSTAGLANDINS ON AUTONOMIC RESPONSES.
AM J PHYSIOL 221:1470-1475(1971)

BAUMINGER S   ZOR U   LINDNER HR
RADIOIMMUNOLOGICAL ASSAY OF PROSTAGLANDIN SYNTHETASE
ACTIVITY
PROSTAGLANDINS 4:313-324(1973)

BAUMINGER S   ZOR U   LINDNER HR
A RADIOIMMUNOLOGICAL ASSAY OF PROSTAGLANDIN SYNTHETASE
ACTIVITY
ABSTR, 9TH INT CONG BIOCHEM, STOCKH:398(1973)

BAXTER JH
HISTAMINE RELEASE FROM RAT MAST CELLS BY DEXTRAN: EFFECTS OF
ADRENERGIC AGENTS, THEOPHYLLINE AND OTHER DRUGS
PROC SOC EXP BIOL MED 141:576-582(1972)

BAYSAL F
THE EFFECTS OF SOME PROSTAGLANDINS ON THE BLOOD PRESSURE OF
GUINEA-PIG
TURK HIG TECR BIYOL DERG 31:183-204(1971)::(TURKISH)

BAYSAL F   GEMALMAZ A
THE EFFECT OF THE PROSTAGLANDIN E1 ON THE ISOLATED FROG
STOMACH MUSCLE
TURK HIJ TECR BIYOL DERG 32:117-121(1972)::(TURKISH)

BAZAN NG
CHANGES IN FREE FATTY ACIDS OF BRAIN BY DRUG-INDUCED
CONVULSIONS, ELECTROSHOCK AND ANAESTHESIA
J NEUROCHEM 18:1379-1385(1971)

BEATTY CH   BOCEK RM   YOUNG MK
EFFECT OF PROSTAGLANDINS AND EPINEPHRINE ON THE LEVEL OF
CYCLIC AMP-14C OF SMOOTH MUSCLE FROM THE RHESUS MONKEY
PROSTAGLANDINS 4:661-671(1973)

BEAZLEY JM
PROSTAGLANDINS IN HUMAN REPRODUCTION
BR J HOSP MED 5:535-552(1971)

BECHETOILLE A   LEPRETRE C   GUILLAUMAT L
THE EFFECT OF INDOMETHACIN ON OCULAR PRESSURE. PRELIMINARY
STUDY
BULL SOC OPHTHALMOL FR 71:941-945(1971)::(FRENCH)

BECHETOILLE A   LEPRETRE C   GUILLAUMAT L
DO ANTIPROSTAGLANDINS GIVEN SYSTEMICALLY EXERT AN ACTION ON
OCULAR TENSION?
BULL SOC OPHTHALMOL FR 72:295-300(1972)::(FRENCH)

BECK N   EICHENHOLZ AE   REED S
DAVIS B
PARATHYROID HORMONE (PGH) BINDING TO RAT RENAL CORTEX (RC)
AND ITS INHIBITION BY PROSTAGLANDIN E1 (PGE1)
CLIN RES 20:586(1972)

BECK NP   DERUBERTIS FR   MICHELIS MF
FUSCO RO   FIELD JB   DAVIS BB
PROSTAGLANDIN E1 (PGE1) INHIBITION OF RENAL CORTICAL EFFECTS
OF PARATHYROID HORMONE (PTH)
CLIN RES 19:471(1971)

BECK NP   DERUBERTIS FR   MICHELIS MF
FUSCI RO   FIELD JB   DAVIS BB
EFFECT OF PROSTAGLANDIN E1 ON CERTAIN RENAL ACTIONS OF
PARATHYROID HORMONE
J CLIN INVEST 51:2352-2358(1972)

BECK NP   KANEKO T   ZOR U
FIELD JB   DAVIS BB
EFFECTS OF VASOPRESSIN AND PROSTAGLANDIN E1 ON THE ADENYL
CYCLASE-CYCLIC 3',5'-ADENOSINE MONOPHOSPHATE SYSTEM OF THE
RENAL MEDULLA OF THE RAT
J CLIN INVEST 50:2461-2465(1971)

BECK NP   REED SW   DAVIS BB
MURDAUGH HV
EPINEPHRINE (EP) EFFECT ON CYCLIC 3',5'-AMP (CAMP) IN KIDNEY
CLIN RES 19:526(1971)

BECK NP   REED SW   MURDAUGH HV
DAVIS BB
EFFECTS OF CATECHOLAMINES AND THEIR INTERACTION WITH OTHER
HORMONES ON CYCLIC 3',5'-ADENOSINE MONOPHOSPHATE IN THE
KIDNEY
J CLIN INVEST 51:939-944(1972)

BECKER EL
THE RELATION OF ENZYME ACTIVATION TO MEDIATOR SECRETION AND
CHEMOTACTIC RESPONSE
IN: "MECHANISMS IN ALLERGY. REAGIN-MEDIATED
HYPERSENSITIVITY," L GOODFRIEND, AH SEHON, RP ORANGE,
MARCEL DEKKER, NY:339-352(1973)

BECKER GA   ASTER RH
SHORT TERM PLATELET PRESERVATION
CLIN RES 20:480(1972)

BECKER GA   CHALOS MK   TUCCELLI M
ASTER RH
USE OF PGE(1) IN PREPARATION AND STORAGE OF PLATELET
CONCENTRATES
IN: "PROSTAGLANDINS IN CELLULAR BIOLOGY," PW RAMWELL AND
BB PHARRISS, PLENUM PRESS, NY/LOND:61-76(1972)

BECKER GA   CHALOS MK   TUCCELLI M
ASTER RH
PROSTAGLANDIN E1 IN PREPARATION AND STORAGE OF PLATELET
CONCENTRATES
SCIENCE 175:538-539(1972)

BECKER GA   KUNICKI T   ASTER RH
EFFECT OF PROSTAGLANDIN E-1 ON HARVESTING OF PLATELETS
FROM REFRIGERATED WHOLE BLOOD
TRANSFUSION 13:351(1973)

BECKER HD   REEDER DD   THOMPSON JC
THE EFFECT OF PROSTAGLANDIN E1 ON THE RELEASE OF GASTRIN
AND GASTRIC SECRETION ON DOGS
ENDOCRINOLOGY 93:1148-1151(1973)

BECKER KL
PROSTAGLANDINS
POSTGRAD MED 50:264-265(1971)

BEDERKA J   TAKEMORI AE   MILLER JW
ABSORPTION RATES OF VARIOUS SUBSTANCES ADMINISTERED
INTRAMUSCULARLY
EUR J PHARMACOL 15:132-136(1971)

BEDFORD JM
THE RATE OF SPERM PASSAGE INTO THE CERVIX AFTER COITUS IN
THE RABBIT
J REPROD FERTIL 25:211-218(1971)

BEERTHUIS RK   NUGTEREN DH   PABON HJJ
STEENHOEK A   VAN DORP DA
SYNTHESIS OF A SERIES OF POLYUNSATURATED FATTY ACIDS. THEIR
POTENCIES AS ESSENTIAL FATTY ACIDS AND AS PRECURSORS OF
PROSTAGLANDINS
RECL TRAV CHIM PAYS-BASBELG 90:943-960(1971)

BEHRMAN HR
RADIOIMMUNOASSAY OF PROSTAGLANDINS.
PHYSIOLOGIST 14:110(1971).

BEHRMAN HR
REGULATION OF OVARIAN STEROID SECRETION
IN: "ENDOCRINOLOGY. PROC 4TH INT CONG ENDOCRINOL," RO SCOW,
FJG EBLING, IW HENDERSON, EXCERPTA MEDICA, AMST/NY:870-874
(1973)

BEITCH BR   BEITCH I   ZADUNAISKY JA
CHLORIDE TRANSPORT ACTIVATION BY PROSTAGLANDINS IN THE FROG
CORNEA
FED PROC 32:245(1973)

BELAMARICH FA   SIMONEIT LW
AGGREGATION OF DUCK THROMBOCYTES BY 5-HYDROXYTRYPTAMINE
MICROVASC RES 6:229-234(1973)

BELESLIN DB
THE NEUROBIOLOGICAL ASSAY
IN: "METHODS IN PSYCHOBIOLOGY," RD MYERS, ACADEMIC PRESS,
LOND/NY 2:213-250(1972)

BELESLIN DB   MALOBABIC ZS
CATALEPSY PRODUCED BY INTRAVENTRICULAR INJECTION OF NICOTINE
EXPERIENTIA 28 : 427-428(1972)

BELESLIN DB   MYERS RD
RELEASE OF AN UNKNOWN SUBSTANCE FROM BRAIN STRUCTURES OF
UNANESTHETIZED MONKEYS AND CATS
NEUROPHARMACOLOGY 10:121-124(1971)

BELESLIN DB   RADMANOVIC BZ   RAKIC MM
RELEASE DURING CONVULSIONS OF AN UNKNOWN SUBSTANCE INTO THE
CEREBRAL VENTRIGLES OF THE CATS.
BRAIN 35:625-627(1971)

BELIEL OM   SINGER FR   COBURN JW
PROSTAGLANDINS: EFFECT ON PLASMA CALCIUM CONCENTRATION
PROSTAGLANDINS 3:237-241(1973)

BELL C
VASOACTIVE SUBSTANCES IN THE CIRCULATION OF THE PREGNANT
DOG DURING ACUTE FETAL ISCHEMIA
AM J OBSTET GYNECOL 117:1088-1092(1973)

BELL NH
ACTIVATION OF NEUROHYPOPHYSEAL ADENYL CYCLASE BY
NEUROTRANSMITTERS
ABSTR 4TH INT CONGR ENDOCRINOL, INT CONGR SER 256, EXCERPTA
MEDICA, AMST:48(1972)

BELMAN S   TROLL W
THE EFFECT OF 12-O-TETRADECANOYL-PHORBOL-13-ACETATE ON
CYCLIC AMP LEVELS IN MOUSE SKIN.
PROC AM ASSOC CANCER RES 14:21 (1973)

BENGTSSON LP
HORMONAL EFFECTS ON HUMAN MYOMETRIAL ACTIVITY
VITAM HORM 31:257-303(1973)

BENNETT A
CHOLERA AND PROSTAGLANDINS
NATURE (LOND) 231:536(1971)

BENNETT A
EFFECTS OF KININS AND PROSTAGLANDINS ON THE GUT
PROC R SOC MED 64:12-13(1971)

BENNETT A
EFFECTS OF PROSTAGLANDINS ON THE GASTROINTESTINAL TRACT
IN: "THE PROSTAGLANDINS: PROGRESS IN RESEARCH," SMM KARIM,
MTP MED AND TECH PUBL CO, OXFORD:205-221(1972)

BENNETT A
PROSTAGLANDINS AND THE GASTROINTESTINAL TRACT
ACTA HEPATO GASTROENTEROL 20:93-97(1973)

BENNETT A
GASTROINTESTINAL DISORDERS INVOLVING PROSTAGLANDINS
MED TODAY 7:21-23(1973)

BENNETT A
PROSTAGLANDINS IN DISEASE
IN NINTH SYMPOSIUM ON ADVANCED MEDICINE G WALKER NEW YORK
PITMAN PUBLISHING CORPORATION:182-191(1973)

BENNETT A
PROSTAGLANDINS AND THE GUT
IN TOPICS IN GASTROENTEROLOGY SC TRUELOVE AND DP JEWELL
OXFORD BLACKWELL SCIENTIFIC PUBLICATIONS 281-293(1973)

BENNETT A
THE PHARMACOLOGY OF ISOLATED GASTROINTESTINAL MUSCLE
IN:INT ENCYCL PHARMACOL THER,G PETERS,PERGAMON PRESS,NY
2:399-432(1973)

BENNETT A   EAKINS KE   POSNER J
DRUGS WHICH INHIBIT PROSTAGLANDINS ACTION OR SYNTHESIS
HORMONES 3:260(1972)

BENNETT A   ELEY KJ   SCHOLES GB
A POSSIBLE ROLE FOR PROSTAGLANDINS IN GASTROINTESTINAL
MOTILITY
AM J DIG DIS 16:552(1971)

BENNETT A   FLESHLER B
A NONCHOLINERGIC EXCITATORY NERVE PATHWAY IN GUINEA PIG
COLON.
AM J DIG DIS 16:550-551(1971)

BENNETT A   FOX CF   STAMFORD IF
INHIBITION OF PROSTAGLANDIN SYNTHESIS BY BENORYLATE
RHEUMATOL REHABIL (SUPP)12:101-105(1973)

BENNETT A   HARRIS M   JENKINS MV
WILLS MR
THE ROLE OF PROSTAGLANDINS IN BONE RESORPTION BY DENTAL
CYSTS
FED PROC 32:804(1973)

BENNETT A   MISIEWICZ JJ
DRUGS USED IN TREATING DISORDERED MOTILITY OF THE
ALIMENTARY TRACT
IN:INT ENCYCL PHARMACOL THER,G PETERS,PERGAMON PRESS,NY
2:433-455(1973)

BENNETT A   POSNER J
STUDIES ON PROSTAGLANDIN ANTAGONISTS
BR J PHARMACOL 42:584-594(1971)

BENNETT A   STAMFORD IF   UNGER WG
PROSTAGLANDINS AND GASTRIC SECRETION IN MAN
ADV BIOSCI (SUPPL)9:43(1972)

BENNETT A   STAMFORD IF   UNGER WG
PROSTAGLANDIN E2 AND GASTRIC ACID SECRETION IN MAN
J PHYSIOL (LOND) 226:96P-98P(1972)

BENNETT A   STAMFORD IF   UNGER WG
PGE2 AND GASTRIC ACID SECRETION IN MAN
ADV BIOSCI 9:265-269(1973)

BENNETT A   STAMFORD IF   UNGER WG
PROSTAGLANDIN E2 AND GASTRIC ACID SECRETION IN MAN
J PHYSIOL (LOND) 229:349-360(1973)

BENNETT A   STOCKLEY HL
ELECTRICALLY INDUCED CONTRACTIONS OF GUINEA-PIG ISOLATED
ILEUM RESISTANT TO TETRODOTOXIN
BR J PHARMACOL 48:359P-360P(1973)

BENSON MJ   VEALE WL
THERMOREGULATORY RESPONSES TO PERFUSING LOCAL REGIONS OF
BRAIN TISSUE WITH VARIOUS IONS IN THE CONSCIOUS RABBIT
PROC CAN FED BIOL SOC 15:331(1972)

BENTLEY PH
TOTAL SYNTHESES OF PROSTANOIDS
CHEM SOC REV 2:29-48(1973)

BERGSTROM S
THE PROSTAGLANDINS AND THEIR METABOLISM
ACTA PHYSIOL SCAND (SUPPL 396) : 18(1973)

BERGSTROM S
PROSTAGLANDINS AND HUMAN REPRODUCTION
PROC WORLD CONGR FERTIL STERIL 7:40-44(1973)

BERGSTROM S   FARNEBO LO   FUXE K
EFFECT OF PROSTAGLANDIN E2 ON CENTRAL AND PERIPHERAL
CATECHOLAMINE NEURONS
EUR J PHARMACOL 21:362-368(1973)

BERL T   HARBOTTLE JA   SCHRIER RW
EFFECT OF PROSTAGLANDIN (PGE1) ON RENAL WATER EXCRETION
ABSTR 5TH INT CONGR NEPHROL, MEX CITY:113(1972)

BERL T   SCHRIER RW
MECHANISM OF EFFECT OF PROSTAGLANDIN E1 ON RENAL WATER
EXCRETION
J CLIN INVEST 52:463-471(1973)

BERMUDEZ EL   LOBO S
EFFECT OF PROSTAGLANDIN F2A ON THE ISOLATED SPLEEN OF THE
RAT
ACTA CIENT VENEZ 24(SUPPL1):24(1973):(SPANISH)

BERNADY KF   WEISS MJ
PROSTAGLANDINS AND CONGENERS. I. THE SYNTHESIS OF
11,15-BISDEOXY-PROSTAGLANDIN E1, E2, AND F1A. THE
STEREOSPECIFIC CONJUGATE ADDITION OF A LITHIUM
TRANS-L-ALKENYLALANATE
TETRAHEDRON LETT 4083-4086(1972)

BERNADY KF   WEISS MJ
PROSTAGLANDINS AND CONGENERS. V. (1) SYNTHESIS OF
DL-PROSTAGLANDIN E1, DL-11-DEOXYPROSTAGLANDIN E1, AND
DL-15-HYDROXY-9-OXO-13-CISPROSTENOIC ACID VIA CONJUGATE
ADDITION OF 3-TRITYLOXY-L-OCTENYLMAGNESIUM BROMIDES
PROSTAGLANDINS 3:505-508(1973)

BERRY EM   EDMONDS JF   WYLLIE JH
RELEASE OF PROSTAGLANDIN E2 AND UNIDENTIFIED FACTORS FROM
VENTILATED LUNGS
BR J SURG 58:189-192(1971)

BERRY H   FLOWER RJ
THE ASSAY OF ENDOGENOUS CHOLECYSTOKININ AND FACTORS
INFLUENCING ITS RELEASE IN THE DOG AND CAT
GASTROENTEROLOGY 60:409-420(1971)

BERTACCINI G   IMPICCIATORE M   DE CARO G
ACTION OF CAERULEIN AND RELATED SUBSTANCES ON THE PYLORIC
SPHINCTER OF THE ANAESTHETIZED RAT
EUR J PHARMACOL 22 : 320-324(1973)

BERTE F   BUSSI G   SAVIO E
MASCHERPA P
EFFECT OF PROSTAGLANDIN E1 ON MYCOBACTERIUM IN VITRO
BOLL SOC ITAL BIOL SPER 49:19(1973):(ITALIAN)

BERTELLI A
ENZYMES AND ANTIENZYMES IN INFLAMMATORY PROCESSES.
HELV ODONTOL ACTA 15:88-89(1971)

BERTI F   PUGLISI L
EFFECTS OF CYCLIC AMP AND CYCLIC GMP ON SMOOTH MUSCLE
ACTIVITY
ADV CYCLIC NUCL RES 1:567(1972)

BERTI F   TRABUCCHI M   BERNAREGGI V
FUMAGALLI R
PROSTAGLANDINS AND CYCLIC-AMP FORMATION IN CEREBRAL CORTEX
SLICES OF RAT AND RABBIT
ADV BIOSCI (SUPPL)9:76(1972)

BERTI F   TRABUCCHI M   BERNAREGGI V
FUMAGALLI R
THE EFFECTS OF PROSTAGLANDINS ON CYCLIC-AMP FORMATION IN
CEREBRAL CORTEX OF DIFFERENT MAMMALIAN SPECIES
PHARMACOL RES COMMUN 4:253-259(1972)

BERTI F   TRABUCCHI M   BERNAREGGI V
FUMAGALLI R
PROSTAGLANDINS ON CYCLIC-AMP FORMATION IN CEREBRAL CORTEX OF
DIFFERENT MAMMALIAN SPECIES
ADV BIOSCI 9:475-480(1973)

BERTOLINI A
INHIBITION OF THE INTRAVENTRICULAR ACTH BEHAVIORAL EFFECT IN
THE RABBIT BY MEANS OF PROSTAGLANDIN E1 ADMINISTERED BY THE
SAME ROUTE
RIV FARMACOL TER 2:XVII-XX(1971):(ITALIAN)

BERTRAND M   GROSMOND G
THE PROSTAGLANDINS
REV MED VET 124:621-634(1973):(FRENCH)

BESKID M   LORENC R   ROSCISZEWSKA A
(THYROID C CELLULAR ADENOMA; A HISTOCHEMICAL AND BIOCHEMICAL
STUDY)
PRESSE MED 79:2113-2116(1971);(FRENCH)

BESLEY GTN   FRITH DA   SNART RS
INHIBITION OF VASOPRESSIN ACTION IN TOAD BLADDER
ADV BIOSCI (SUPPL)9:31(1972)

BESLEY GTN   FRITH DA   SNART RS
INHIBITION OF VASOPRESSIN ACTION IN TOAD BLADDER
ADV BIOSCI 9:195-200(1973)

BESLEY GTN   SNART RS
EFFECT OF PROSTAGLANDINS ON CYCLIC AMP CONCENTRATIONS
IN TOAD BLADDER AND RAT KIDNEY
FEBS LETT 31:269-272(1973)

BETHEL RA   EAKINS KE
ANTAGONISM BY POLYPHLORETIN PHOSPHATE OF THE INTRAOCULAR
PRESSURE RISE INDUCED BY PROSTAGLANDIN AND FORMALDEHYDE IN
THE RABBIT EYE
FED PROC 30:626(1971)

BETHEL RA   EAKINS KE
THE MECHANISM OF THE ANTAGONISM OF EXPERIMENTALLY INDUCED
OCULAR HYPERTENSION BY POLYPHLORETIN PHOSPHATE.
EXP EYE RES 13:83-91(1971)

BETTINI V   CESSI C   MARTECCHINI M
EFFECT OF PROSTAGLANDIN E1 ON THE RESPONSE OF A PREPARATION
" NERVOUS PHRENICUS - DIAPHRAGMATIC MUSCLE ISOLATE" OF THE
GUINEA PIG
BOLL SOC ITAL BIOL SPER 47:422-425(1971);(ITALIAN)

BETTINI V   CESSI C   MARTECCHINI M
FURTHER OBSERVATIONS ON THE EFFECT OF PROSTAGLANDIN E1 ON
THE NEUROMUSCULAR TRANSMISSION IN MAMMALS
BOLL SOC ITAL BIOL SPER 47:499-502(1971);(ITALIAN)

BEVAN DR
THE SODIUM STORY: EFFECTS OF ANAESTHESIA AND SURGERY ON
INTRARENAL MECHANISMS CONCERNED WITH SODIUM HOMEOSTASIS
PROC R SOC MED 66:1215-1220(1973)

BHAGAT BD   RANA MW   GINN D
EFFECT OF PROSTAGLANDIN E2 (PGE2) ON THE CONTRACTILITY OF
THE SMOOTH MUSCLE.
PHARMACOLOGIST 13:293(1971)

BHANA D   HILLIER K   KARIM SMN
VASOACTIVE SUBSTANCES IN KAPOSI'S SARCOMA
CANCER 27:233-237(1971)

BHANA D   KARIM SMM   CARTER DC
GANESAN PA
THE EFFECT OF ORALLY ADMINISTERED PROSTAGLANDINS A1, A2 AND
15 EPI-A2 ON HUMAN GASTRIC ACID SECRETION
PROSTAGLANDINS 3:307-316(1973)

BHATTACHERJEE P   EAKINS KE
INHIBITION OF THE PG-SYNTHETASE SYSTEMS IN OCULAR TISSUES BY
INDOMETHACIN
PHARMACOLOGIST 15:209(1973)

BIANCHI CP
ASPECTS OF CALCIUM METABOLISM IN SKELETAL MUSCLE
IN PHARMACOL THERMOREGUL PROC SATELL SYMP 469-471(1972)
(KARGER BASEL) (1973)

BIECK PR
ROLE OF CYCLIC AMP IN THE REGULATION OF GASTRIC SECRETION IN
DOGS AND HUMANS
ADV CYCLIC NUCL RES 1:149-161(1972)

BIECK PR   OATES JA   ADKINS RB
INHIBITION OF GASTRIC SECRETION BY ARACHIDONIC ACID IN THE
DOG
CLIN RES 19:387(1971)

BIECK PR   OATES JA   ROBISON GA
ADKINS RB
CYCLIC AMP IN THE REGULATION OF GASTRIC SECRETION IN DOGS
AND HUMANS
AM J PHYSIOL 224(1):158-164(1973)

BIELANSKI A
ROLE OF PROSTAGLANDINS IN REPRODUCTION
MED WETER 29:622-624(1973);(POLISH)

BIGAZZI M   CASCIANO S   CARINI L
MARZOCCA U   ZURLI A
MEDULLARY CARCINOMA: CONSIDERATIONS ABOUT A CASE OF FAMILIAL
PATHOGENESIS
G GERONTOL 21:1049-1053(1973);(ITALIAN)

BILLINGHAM MEJ   MORLEY J   HANSON JM
SHIPOLINI RA   VERNON CA
AN ANTI-INFLAMMATORY PEPTIDE FROM BEE VENOM
NATURE 245:163-164(1973)

BINDRA JS   BINDRA R
PROSTAGLANDINS
PROG DRUG RES 17 : 410-487 (1974)

BINDRA JS   GRODSKI A   SCHAAF TK
COREY EJ
NEW EXTENSIONS OF THE BICYCLO (2-2-1) HEPTANE ROUTE TO
PROSTAGLANDINS
J AM CHEM SOC 95:7522-7523(1973)

BIRNBAUMER L
HORMONE-SENSITIVE ADENYLYL CYCLASES USEFUL MODELS FOR
STUDYING HORMONE RECEPTOR FUNCTIONS IN CELL-FREE SYSTEMS
BIOCHIM BIOPHYS ACTA 300:129-158(1973)

BIRNBAUMER L
POSITIVE AND NEGATIVE CONTROL OF PROSTAGLANDIN AND HORMONE
ACTION ON ADENYLYL CYCLASES BY NUCLEOTIDES AND DIVALENT
CATION
ABSTR 8TH CONGR INT DIABETES FED, BRUSS, EXCERPTA MED,
AMST:78(1973)

BIRON P
PULMONARY FATE OF VASOACTIVE AGENTS IN ANIMALS AND HUMANS
CLIN RES 19: 124(1971)

BISHOP CL   FLACK JD
THE USE OF DRUGS TO ELUCIDATE THE MECHANISM OF LUTEOLYSIS IN
THE PSEUDOPREGNANT RAT
J REPROD FERTIL 35:597-598(1973)

BISHOP DW
SPERM TRANSPORT IN THE FALLOPIAN TUBE
IN: "PATHWAYS TO CONCEPTION. THE ROLE OF THE CERVIX AND THE
OVIDUCT IN REPRODUCTION," AI SHERMAN, CHAS C THOMAS, PUBL,
SPRINGFIELD, ILL:99-109(1971)

BITENSKY MW   GORMAN RE
CHEMICAL MEDIATION OF HORMONE ACTION
ANNU REV MED 23:263-284(1972)

BITENSKY MW   KEIRNS JJ   FREEMAN J
CYCLIC ADENOSINE MONOPHOSPHATE AND CLINICAL MEDICINE-PART I
AM J MED SCI 266:320-347(1973)

BITO LZ
ACCUMULATION AND APPARENT ACTIVE TRANSPORT OF PROSTAGLANDINS
PHARMACOLOGIST 13:293(1971)

BITO LZ
COMPARATIVE STUDY OF CONCENTRATIVE PROSTAGLANDIN
ACCUMULATION BY VARIOUS TISSUES OF MAMMALS AND MARINE
VERTEBRATES AND INVERTEBRATES
COMP BIOCHEM PHYSIOL (A)43:65-82(1972)

BITO LZ
ACCUMULATION AND APPARENT ACTIVE TRANSPORT OF
PROSTAGLANDINS BY SOME RABBIT TISSUES IN VITRO
J PHYSIOL (LOND) 221:371-387(1972)

BITO LZ
CONCENTRATIVE ACCUMULATION OF PROSTAGLANDINS BY THE
IRIS-CILIARY PROCESS IN VITRO AND FACILITATED REMOVAL OF
PROSTAGLANDIN E1 FROM INTRA-OCULAR FLUIDS IN VIVO
EXP EYE RES 14:175(1972)

BITO LZ
ABSORPTIVE TRANSPORT OF PROSTAGLANDINS FROM INTRAOCULAR
FLUIDS TO BLOOD: A REVIEW OF RECENT FINDINGS
EXP EYE RES 16:299-306(1973)

BITO LZ
INHIBITION OF UVEAL PROSTAGLANDIN TRANSPORT IN EXPERIMENTAL
UVEITIS
IN: "PROSTAGLANDINS AND CYCLIC AMP," RH KAHN, WEM LANDS,
ACADEMIC PRESS, NY : 213-214(1973)

BITO LZ   SALVADOR EV
INTRAOCULAR FLUID DYNAMICS. III. THE SITE AND MECHANISM OF
PROSTAGLANDIN TRANSFER ACROSS THE BLOOD INTRAOCULAR FLUID
BARRIERS
EXP EYE RES 14:233-241(1972)

BITO LZ   TURANSKY D   VAN VORIS A
CONCENTRATIVE ACCUMULATION OF PROSTAGLANDINS BY SOME TISSUES
OF MARINE INVERTEBRATES.
BIOL BULL 141:372(1971)

BLACKHAM A   FARMER JB   RADZIWONIK H
WESTWICK J
RABBIT MONOARTICULAR ARTHRITIS AND SYNOVIAL PROSTAGLANDINS
BR J PHARMACOL 48:343P-344P(1973)

BLAIR-WEST JR   COGHLAN JP   DENTON DA
FUNDER JW   SCOGGINS BA   WRIGHT RD
EFFECTS OF PROSTAGLANDIN E1 UPON THE STEROID SECRETION OF
THE ADRENAL OF THE SODIUM DEFICIENT SHEEP.
ENDOCRINOLOGY 88:367-371(1971)

BLIGH J   MILTON AS
THE THERMOREGULATOR EFFECTS OF PROSTAGLANDIN E1 WHEN INFUSED
INTO A LATERAL CEREBRAL VENTRICLE OF THE WELSH MOUNTAIN
SHEEP AT DIFFERENT AMBIENT TEMPERATURES
J PHYSIOL (LOND) 229:30P-31P(1973)

BLOCK AJ   VANE JR
RELEASE OF PROSTAGLANDINS (PGS) FROM RABBIT ISOLATED HEARTS
NAUNYN SCHMIEDEBERGS ARCH PHARMACOL 279:R19 (1973)

BLOOR CM   WHITE FC   SOBEL BE
CORONARY AND SYSTEMIC HAEMODYNAMIC EFFECTS OF PROSTAGLANDINS
IN THE UNANAESTHETIZED DOG
CARDIOVASC RES 7:156-166(1973)

BLUMENKRANTZ N   SONDERGAARD J
EFFECT OF PROSTAGLANDINS E1 AND F1A ON BIOSYNTHESIS OF
COLLAGEN
NATURE (NEW BIOL) 239:246(1972)

BOCEK RM   BEATTY CH
EFFECT OF PROSTAGLANDINS ON CYCLIC-AMP LEVELS OF FETAL
MUSCLE FROM RHESUS MONKEYS
IN: "PROSTAGLANDINS AND CYCLIC AMP," RH KAHN, WEM LANDS,
ACADEMIC PRESS, NY : 253-256(1973)

BODE HH   MEARA PA   JONES HS
CRAWFORD JD
SULFONYLUREAS INHIBIT PROSTAGLANDIN E1
PEDIATR RES 7:385(1973)

BOHLE E
EXPERIMENTAL FATTY LIVER AND PROSTAGLANDINS
Z GASTROENTEROL 11:233-242(1973)(GERMAN)

BOHMAN SO
SUBCELLULAR LOCALIZATION OF PROSTAGLANDIN SYNTHESIS IN
RABBIT RENAL PAPILLA
J ULTRASTRUCT RES 38:191(1972)

BOHMAN SO   MAUNSBACH AB
ULTRASTRUCTURE AND BIOCHEMICAL PROPERTIES OF SUBCELLULAR
FRACTIONS FROM RAT RENAL MEDULLA
J ULTRASTRUCT RES 38:225-245(1972)

BOJESEN E
INTRARENAL CONTROLLING SYSTEMS, THEORIES AND FACTS
ACTA PHYSIOL SCAND (SUPPL 396) : 28(1973)

BOJESEN E   BUCHAVE K
AN ISOTOPE DERIVATIVE METHOD FOR THE ANALYSIS OF
PROSTAGLANDINS E1 AND E2 IN SERUM USING 2-AMINO-(35S)
THIAZOLE AS THE REAGENT
BIOCHIM BIOPHYS ACTA 280:614-625(1972)

BOJESEN E   BUCHHAVE K
THE EFFECTS OF FASTING, BLEEDING AND SODIUM DEPLETION ON
PGE2 LEVELS OF RAT SERA
ABSTR, 9TH INT CONG BIOCHEM, STOCKH:398(1973)

BONDANI A   OLMEDO M
ACTION OF LOCAL ANESTHETICS IN THE ISOLATED UTERUS
CONTRACTION PRODUCED BY OXYTOCINE, PROSTAGLANDIN E1 AND E2
ABSTR 5TH INT CONGR PHARMACOL, SAN FRANC:25(1972)

BONDANI A   OLMEDO M
INHIBITION BY TETRAHYDROCANNIBINOL OF RAT UTERINE
CONTRACTIONS AND ITS DEPENDENCE ON CALCIUM
ACTA PHYSIOL LAT AM 23:ABSTR 316(1973):(SPANISH)

BONDANI A   OLMEDO-ZORILLA M
ACTION OF TETRODOTOXIN AND TETRACAINE IN THE IN VITRO
CONTRACTIONS OF THE RAT UTERUS PRODUCED BY OXYTOCIN,
PROSTAGLANDINS E1 AND E2
ARCH INVEST MED (MEX) 3(SUPPL 1):213-228(1972):(SPANISH)

BOOYSE FM   GUILIANI D   MARR JJ
RAFELSON ME JR
CYCLIC ADENOSINE 3',5'-MONOPHOSPHATE DEPENDENT PROTEIN
KINASE OF HUMAN PLATELETS: MEMBRANE PHOSPHORYLATION AND
REGULATION OF PLATELET FUNCTION
SER HAEMATOL 6:351-366(1973)

BOOYSE FM   RAFELSON ME JR
REGULATION AND MECHANISM OF PLATELET AGGREGATION
ANN N Y ACAD SCI 201:37-60(1972)

BORDA E   DE VARGAS IG   DE GIMENO MF
GIMENO AL
SPONTANEOUS AND PROSTAGLANDIN E1-INDUCED MOTILITY IN THE
ISOLATED MYOMETRIUM OF RATS IN HEAT. EFFECT OF METABOLISM
INHIBITORS
MEDICINA (B AIRES) 31:530(1971):(SPANISH)

BORKOWSKI A   DELCROIX C   LEVIN S
METABOLISM OF ADRENAL CHOLESTEROL IN MAN. II. IN VITRO
STUDIES INCLUDING A COMPARISON OF ADRENAL CHOLESTEROL
SYNTHESIS WITH THE SYNTHESIS OF THE GLUCOCORTICOSTEROID
HORMONES
J CLIN INVEST 51:1679-1687(1972)

BORN GVR
AGGREGATION AND RELATED PHENOMENA. APPLICABILITY OF IN VITRO
OBSERVATIONS ON THE AGGREGATION OF PLATELETS TO THEIR
FUNCTION IN VIVO
IN: "ERYTHROCYTES, THROMBOCYTES, LEUKOCYTES," E GERLACH,
K MOSER, E DEUTSCH, W WILMANNS, G THIEME, STUTTG
2:253-257(1973)

BORN GVR   FOULKS J   MICHAL F
SHARP DE
REVERSAL OF THE RAPID MORPHOLOGICAL REACTION OF PLATELETS
J PHYSIOL (LOND) 225:27P-28P(1972)

BORN GVR   MILLS DCB   SMITH JB
PHARMACOLOGY OF THE INHIBITION OF PLATELET AGGREGATION
BULL SCHWEIZ AKAD MED WISS 29:215-222(1973)

BOTELLA LLUISA J
THE PROSTAGLANDINS, WITH SPECIAL CONSIDERATION TO THEIR
GENITAL ACTIONS
ACTA GINECOL (MADR) 23:287-314(1972):(SPANISH)

BOTTECCHIA D
INFLUENCE OF TEMPERATURE AND PLASMA PRESERVATION ON THE
EFFECT OF PGE1 ON THE PLATELET AGGREGATION INDUCED BY ADP
BOLL SOC ITAL BIOL SPER 48:19-22(1972):(ITALIAN)

BOTTECCHIA D
INFLUENCE OF TEMPERATURE AND PRESERVATION OF PLASMA ON
PLATELET AGGREGATION INDUCED WITH ADP IN THE PRESENCE OF
PROSTAGLANDIN E1.
ARCH FISIOL 69:160-170(1972):(ITALIAN)

BOTTECCHIA D   SACCAROLA L
OBSERVATIONS OF THE EFFECT OF A PROSTAGLANDIN (PGE1) ON THE
PLATELETS OF THE RAT
BOLL SOC ITAL BIOL SPER 47:657-661(1971):(ITALIAN)

BOUCEK RJ   NOBLE NL
HISTAMINE, NOREPINEPHRINE, AND BRADYKININ STIMULATION OF
FIBROBLAST GROWTH AND MODIFICATION OF SEROTONIN RESPONSE
PROC SOC EXP BIOL MED 144:929-933(1973)

BOULLIN DJ   GREEN AR   PRICE KS
THE MECHANISM OF ADENOSINE DIPHOSPHATE INDUCED PLATELET
AGGREGATION: BINDING TO PLATELET RECEPTORS AND INHIBITION
OF BINDING AND AGGREGATION BY PROSTAGLANDIN E1
J PHYSIOL (LOND) 221:415-426(1972)

BOURNE H   MELMON K
BETA-ADRENERGIC RECEPTORS CONTROL SYNTHESIS OF CYCLIC AMP IN
HUMAN LEUKOCYTES
CLIN RES 19: 178(1971)

BOURNE HR
LEUKOCYTE CYCLIC AMP: PHARMACOLOGICAL REGULATION AND
POSSIBLE PHYSIOLOGICAL IMPLICATIONS
IN: "PROSTAGLANDINS IN CELLULAR BIOLOGY," PW RAMWELL AND B
B PHARRISS, PLENUM PRESS, NY/LOND:111-149(1972)

BOURNE HR
CHOLERA ENTEROTOXIN: FAILURE OF ANTI-INFLAMMATORY AGENTS TO
PREVENT CYCLIC AMP ACCUMULATION
NATURE (LOND) 241:399(1973)

BOURNE HR   EPSTEIN LB   MELMON KL
LYMPHOCYTE CYCLIC ADENOSINE MONOPHOSPHATE (AMP) SYNTHESIS
AND INHIBITION OF PHYTOHEMAGGLUTININ-INDUCED TRANSFORMATION.
J CLIN INVEST 50:10A(1971)

BOURNE HR   LEHRER RI   CLINE MJ
MELMON KL
CYCLIC 3',5'-ADENOSINE MONOPHOSPHATE IN THE HUMAN LEUCOCYTE:
SYNTHESIS, DEGRADATION, AND EFFECTS ON NEUTROPHIL
CANDIDACIDAL ACTIVITY.
J CLIN INVEST 50:920-929(1971)

BOURNE HR   LICHTENSTEIN LM   MELMON KL
LEUKOCYTE CYCLIC ADENOSINE MONOPHOSPHATE (AMP) INHIBITS
ANTIGENIC HISTAMINE RELEASE.
J CLIN INVEST 50:10A(1971)

BOURNE HR   LICHTENSTEIN LM   MELMON KL
PHARMACOLOGIC CONTROL OF ALLERGIC HISTAMINE RELEASE IN
VITRO: EVIDENCE FOR AN INHIBITORY ROLE OF 3',5'-ADENOSINE
MONOPHOSPHATE IN HUMAN LEUKOCYTES
J IMMUNOL 108:695-705(1972)

BOURNE HR   MELMON KL
ADENYL CYCLASE IN HUMAN LEUKOCYTES: EVIDENCE FOR
ACTIVATION BY SEPARATE BETA ADRENERGIC AND PROSTAGLANDIN
RECEPTORS
J PHARMACOL EXP THER 178:1-7(1971)

BOURNE HR   TOMKINS GM   DION S
REGULATION OF PHOSPHODIESTERASE SYNTHESIS: REQUIREMENT FOR
CYCLIC ADENOSINE MONOPHOSPHATE-DEPENDENT PROTEIN KINASE
SCIENCE 181:952-954(1973)

BOUSSER MG
EFFECTS OF COMBINED PROSTAGLANDIN E1 AND ASPIRIN ON
EXPERIMENTAL ARTERIAL THROMBOSIS IN RABBITS
BIOMEDICINE 19:90-93(1973)

BOUSSER MG
PROSTAGLANDINS, PLATELETS AND THROMBOSIS
ANN MED INTERNE (PARIS) 124:351-354(1973):(FRENCH)

BOUSSER MG
PROSTAGLANDIN-E1 AND PLATELETS
BIOMED 18:95-102(1973)

BOUSSER MG   LECRUBIER C
EFFECT OF PROSTAGLANDIN E1 ON EXPERIMENTAL THROMBOSIS AND
PLATELET AGGREGATION IN RABBITS
HAEMOSTASIS 1:294-303(1973)

BOWERY B   LEWIS GP
INHIBITION OF PROSTAGLANDIN SYNTHESIS AND FUNCTIONAL
HYPERAEMIA IN RABBIT ADIPOSE TISSUE
BR J PHARMACOL 45:146P-147P(1972)

BOWERY B   LEWIS GP
INHIBITION OF FUNCTIONAL VASODILATATION AND PROSTAGLANDIN
FORMATION IN RABBIT ADIPOSE TISSUE BY INDOMETHACIN AND
ASPIRIN
BR J PHARMACOL 47:305-314(1973)

BOYD GS   TRZECIAK WH
CHOLESTEROL METABOLISM IN THE ADRENAL CORTEX: STUDIES ON THE
MODE OF ACTION OF ACTH
ANN N Y ACAD SCI 212:361-377(1973)

BOZOVIC L   CASTENFORS J   DELIN A
EKESTROM S   GRANBERG PO
PRE-AND PEROPERATIVE EVALUATION AND FOLLOW UP OF
HYPERTENSIVE PATIENTS WITH RENAL ARTERY STENOSIS
SCAND J UROL NEPHROL 5:162-170(1971)

BRADLEY PB   SAMUELS GMR
PROSTAGLANDINS AND CHLORPROMAZINE
LANCET 2:612(1971)

BRADLEY PB   SAMUELS GMR
PROSTAGLANDINS AND NEURONAL MECHANISMS
HORMONES 3:262(1972)

BRAT T   DECOSTER JM
INTERRUPTION OF PREGNANCY WITH PROSTAGLANDIN F2A
J GYNECOL OBSTET BIOL REPROD 1:385-387(1972);(FRENCH)

BRAUN W
ROLE OF CYCLIC NUCLEOTIDES IN THE REGULATION OF IMMUNE
RESPONSES
IN PROSTAGLANDINS AND CYCLIC AMP RH KAHN WEM LANDS NEW YORK
ACADEMIC PRESS: 227-228(1973)

BRAUN W
IMMUNOLOGIC AND ANTINEOPLASTIC EFFECTS OF ENDOTOXIN: ROLE OF
MEMBRANES AND MEDIATION BY CYCLIC ADENOSINE-3',5'-
MONOPHOSPHATE
J INFECT DIS 128(SUPPL):S188-S197 (1973)

BRENNER WE   FISHBURNE JI   MCMILLAN CW
JOHNSON AM   HENDRICKS CH
PROSTAGLANDIN F2A INDUCED ABORTION'S EFFECT UPON COAGULATION
ABSTR SOC GYNECOL INVEST, 20TH ANNU MEET, ATLANTA:21(1973)

BRENNER WE   HENDRICKS CH   BRAAKSMA JT
FISHBURNE JI   STAUROVSKY LG
INTRA-AMNIOTIC ADMINISTRATION OF PROSTAGLANDIN F2A FOR
INDUCTION OF THERAPEUTIC ABORTION. A COMPARISON OF FOUR
DOSAGE SCHEDULES
J REPROD MED 9:456-463(1972)

BREWSTER D   MYERS M   ORMEROD J
SPINNER ME   TURNER S   SMITH ACB
DESIGN OF PROSTAGLANDIN SYNTHESIS
CHEM COMMUN 1235-1326(1972)

BREWSTER D   MYERS M   ORMEROD J
OTTER P   SMITH ACB   SPINNER ME
TURNER S
PROSTAGLANDIN SYNTHESIS: DESIGN AND EXECUTION
J CHEM SOC (PERKIN I) 22:2796-2804(1973)

BROCKLEHURST WE
ROLE OF KININS AND PROSTAGLANDINS IN INFLAMMATION
PROC R SOC MED 64:4-6(1971)

BROCKLEHURST WE
THE ASSAYS OF MEDIATORS IN HYPERSENSITIVITY REACTIONS
IN: "HANDBOOK OF EXPERIMENTAL IMMUNOLOGY. APPLICATION OF
IMMUNOLOGICAL METHODS," DM WEIR, BLACKWELL, OXFORD
3: 43.1-43.12(1973)

BRODIE GN   BAENZIGER NL   CHASE LR
MAJERUS PW
THE EFFECTS OF THROMBIN ON ADENYL CYCLASE ACTIVITY AND A
MEMBRANE PROTEIN FROM HUMAN PLATELETS
J CLIN INVEST 51:81-88(1972)

BRODY MJ   RYAN MJ   CLARK KE
EFFECTS OF PROSTAGLANDINS E1 AND F2A ON UTERINE
HEMODYNAMICS AND MOTILITY
ADV BIOSCI (SUPPL) 9:130(1972)

BROLIN SE   BORGLUND E   TEGNER L
WETTERMARK G
PHOTOKINETIC MICRO ASSAY BASED ON DEHYDROGENASE REACTIONS
AND BACTERIAL LUCIFERASE
ANAL BIOCHEM 42:124-135(1971)

BROTANEK V   HENDRICKS CH   BRENNER W
EKBLADH L
CHANGES IN UTERINE BLOOD FLOW DURING INFUSION OF
PROSTAGLANDINS
ABSTR 18TH ANNU MEET SOC GYNECOL INVEST, PHOENIX:15(1971)

BROWN RD
PROSTAGLANDIN E1 (PGE1): EFFECTS ON CONTRALATERAL
OLIVOCOCHLEAR BUNDLE (COCB) STIMULATION IN CATS
PHARMACOLOGIST 13:278(1971)

BROWN RD
EFFECTS OF PROSTAGLANDIN E1 ON CONTRALATERAL OLIVO-COCHLEAR
BUNDLE STIMULATION IN CATS
ARCH INT PHARMACODYN THER 198:372-376(1972)

BRUMMER HC
STORAGE LIFE OF PROSTAGLANDIN E2 IN ETHANOL AND SALINE
J PHARM PHARMACOL 23:804-805(1971)

BRUNNBERG FJ
PROSTAGLANDINS-HISTORY AND FUTURE OF A NEW SUBSTANCE
DTSCH APOTH 25:336-338(1973): (GERMAN)

BRYANT RE   SUTCLIFFE M
THE EFFECT OF CYCLIC AMP ON GRANULOCYTE ADHESIVENESS
CLIN RES 21:594(1973)

BRYDEN MM   PERRY C
INDUCTION OF EMBRYONIC DEATH IN SHEEP BY INTRAUTERINE
INJECTION OF A SMALL VOLUME OF NORMAL SALINE
J REPROD FERTIL 32:133-135(1973)

BULLOCK DW   KEYES PL
EFFECT OF PROSTAGLANDIN F2A ON ECTOPIC AND OVARIAN CORPORA
LUTEA OF THE RABBIT
ABSTR 53RD MEET ENDOCR SOC, SAN FRANC:A-119(1971)

BUNAG RD   WALASZEK EJ
IN VITRO INHIBITION BY SYNTHETIC PHOSPHOLIPIDS OF PRESSOR
RESPONSES TO RENIN
EUR J PHARMACOL 23:191-196(1973)

BUNDY G   LINCOLN F   NELSON N
PIKE J   SCHNEIDER W
NOVEL PROSTAGLANDIN SYNTHESES
ANN NY ACAD SCI 180:76-90(1971)

BUNDY GL
PROSTAGLANDINS AND RELATED COMPOUNDS
ANNU REP MED CHEM 7:157-168(1972)

BUNDY GL
RECENT ADVANCES IN PROSTAGLANDIN RESEARCH
ABSTR 7TH MIDDLE ATL REG MEET AM CHEM SOC PHILA:81(1972)

BUNDY GL   DANIELS EG   LINCOLN FH
PIKE JE
ISOLATION OF A NEW NATURALLY OCCURRING PROSTAGLANDIN,
5-TRANS-PGA2. SYNTHESIS OF 5-TRANS-PGE2 AND 5-TRANS-PGF2A
J AM CHEM SOC 94:2124(1972)

BUNDY GL   SCHNEIDER WP   LINCOLN FH
PIKE JE
THE SYNTHESIS OF PROSTAGLANDINS E2 AND F2A FROM (15R)- AND
(15S)-PGA2
J AM CHEM SOC 94:2123-2124(1972)

BUNDY GL   YANKEE EW   WEEKS JR
MILLER WL
SYNTHESIS AND BIOLOGICAL ACTIVITY OF A SERIES OF 15-METHYL
PROSTAGLANDINS
ADV BIOSCI (SUPPL) 9:17(1972)

BUNDY GL   YANKEE EW   WEEKS JR
MILLER WL
THE SYNTHESIS AND BIOLOGICAL ACTIVITY OF A SERIES OF
15-METHYL PROSTAGLANDINS
ADV BIOSCI 9:125-133(1973)

BURKE G
ASPIRIN AND INDOMETHACIN ABOLISH THYROTROPIN-INDUCED
INCREASE IN THYROID CELL PROSTAGLANDINS
PROSTAGLANDINS 2:413-415(1972)

BURKE G
EFFECTS OF THYROTROPIN AND N6, O2'-DIBUTYRYL CYCLIC
3',5'-ADENOSINE MONOPHOSPHATE ON PROSTAGLANDIN LEVELS IN
THYROID
PROSTAGLANDINS 3:291-297(1973)

BURKE G
COMPARATIVE EFFECTS OF PURINE NUCLEOTIDES OF THYROTROPIN-AND
PROSTAGLANDIN E1-RESPONSIVE ADENYLATE CYCLASE IN THYROID
PLASMA MEMBRANE
PROSTAGLANDINS 3:537-540(1973)

BURKE G   CHANG LL   SZABO M
THYROTROPIN AND CYCLIC NUCLEOTIDE EFFECTS ON PROSTAGLANDIN
LEVELS IN ISOLATED THYROID CELLS
SCIENCE 180:872-875(1973)

BURKE G   KOWALSKI K   BABIARZ D
EFFECTS OF THYROTROPIN, PROSTAGLANDIN E1 AND A PROSTAGLANDIN
ANTAGONIST ON IODIDE TRAPPING IN ISOLATED THYROID CELLS
LIFE SCI (II)10:513-521(1971)

BURKE G   KOWALSKI K   BABIARZ D
EFFECTS OF TSH, PROSTAGLANDIN E1 (PGE1), AND A
PROSTAGLANDIN ANTAGONIST ON IODIDE TRAPPING IN ISOLATED
THYROID CELLS
CLIN RES 19:368(1971)

BURKE G   SATO S
EFFECTS OF LONG-ACTING THYROID STIMULATOR AND PROSTAGLANDIN
ANTAGONISTS ON ADENYL CYCLASE ACTIVITY IN ISOLATED BOVINE
THYROID CELLS
LIFE SCI 10(2):969-981(1971)

BURKE G   SATO S   KOWALSKI K
BABIARZ D   SZABO M
THYROIDAL PROSTAGLANDIN RECEPTOR: EVIDENCE FOR AN ESSENTIAL
ROLE IN THE ACTION OF THYROTROPIN (TSH)
ABSTR 53RD MEET ENDOCR SOC, SAN FRANC:A-140(1971)

BURNS T   RADAWSKI D   UNDERWOOD R
DAUGHERTY R
EFFECTS OF PROSTAGLANDIN E1 (PGE1) ON VASCULAR RESISTANCES
AND WEIGHT OF THE JEJUNUM
ABSTR 5TH INT CONGR PHARMACOL, SAN FRANC:34(1972)

BURNS T   RADAWSKI D   UNDERWOOD R
DAUGHERTY R
EFFECTS OF PROSTAGLANDIN E (PGE1) ON VASCULAR RESISTANCES
AND WEIGHT OF THE JEJUNUM
CLIN RES 21:327(1973)

BURNSTOCK G   GANNON BJ   MALMFORS T
ROGERS DC
CHANGES IN THE PHYSIOLOGY AND FINE STRUCTURE OF THE TAENIA
OF THE GUINEA-PIG CAECUM FOLLOWING TRANSPLANTATION INTO THE
ANTERIOR EYE CHAMBER.
J PHYSIOL 219:139-154(1971)

BURSTEIN S
PROSTAGLANDINS AND CANNABIS: A POSSIBLE MODE OF ACTION FOR
THC
PSYCHOPHARMACOL BULL 9:25-26(1973)

BURSTEIN S   LEVIN E   VARANELLI C
PROSTAGLANDINS AND CANNABIS. II. INHIBITION OF BIOSYNTHESIS
BY THE NATURALLY OCCURRING CANNABINOIDS
BIOCHEM PHARMACOL 22:2905-2910(1973)

BURSTEIN S   RAZ A
INHIBITION OF PROSTAGLANDIN E2 BIOSYNTHESIS BY
DELTA1-TETRAHYDROCANNABINOL
PROSTAGLANDINS 2:369-374(1972)

BUTCHER RW
THE SECOND MESSENGER CONCEPT AND LIPID METABOLISM.
NAUNYN-SCHMIEDEBERGS ARCH PHARMAKOL 269:358-372(1971)

BUTCHER RW   HITTELMAN KJ
PROSTAGLANDINS AND THE CONTROL OF CYCLIC AMP LEVELS IN
ADIPOSE TISSUE
IN: "ENDOCRINOLOGY. PROC 4TH INT CONG ENDOCRINOL," RO SCOW,
FJG EBLING, IW HENDERSON, EXCERPTA MEDICA, AMST/NY:382-387
(1973)

BUZNIKOV GA   ZVEZDINA ND   PROKAZOVA NV
BERGEL'SON LD   TURPAEV TM
INFLUENCE OF GANGLIOSIDES ON THE SENSITIVITY OF EMBRYONIC
CELLS TO NEUROPHARMACOLOGICAL PREPARATIONS
DOKL AKAD NAUK SSSR(ENGL) 210:261-263(1973)

BYGDEMAN M
PROSTAGLANDINS AND THE UTERUS. STIMULATING EFFECT ON
CONTRACTILITY OF THE UTERUS
NORD MED 85:436(1971);(SWEDISH)

BYGDEMAN M   BEGUIN F   TOPPOZADA M
WIQVIST N
INTRA-UTERINE ADMINISTRATION OF PROSTAGLANDIN F2A FOR
INDUCTION OF ABORTION
ADV BIOSCI (SUPPL) 9:86(1972)

BYGDEMAN M   BEGUIN F   TOPPOZADA M
WIQVIST N
FURTHER EXPERIENCE WITH INTRAUTERINE PROSTAGLANDIN
ADMINISTRATION
J REPROD MED 9:392-396(1972)

BYGDEMAN M   TOPPOZADA M   WIQVIST N
PROSTAGLANDIN FOR INDUCTION OF ABORTION AND LABOUR
IN: "PROSTAGLANDINES 1973," INSERM, PARIS:279-295(1973)

BYGDEMAN M   TOPPOZADA M   WIQVIST N
PROSTAGLANDINS IN REPRODUCTION
MED TODAY 7:7-16(1973)

BYGDEMAN M   WIQVIST N
PROSTAGLANDINS AS ABORTION AGENTS
LAKARTIONINGEN 68:2732-2739(1971);(SWEDISH)

BYGDEMAN M   WIQVIST N
PROSTAGLANDINS IN REPRODUCTION
PROC WORLD CONGR FERTIL STERIL 7:142-144(1973)

BYLINSKY G
UPJOHN PUTS THE CELL'S OWN MESSENGERS TO WORK
FORTUNE 96-99,148-150,152(1972)

CABALLERO A   CORREDERA J   GARCIA-ALBERTOS F
ALONSO MAGAN JL
PROSTAGLANDIN PGF2A IN THE INDUCTION OF LABOR
TOKO-GINECOL PRACT 31:633-656(1972);(SPANISH)

CAEN JP   JENKINS CSP   MICHEL H
CHIVOT JJ   LEVY-TOLEDANO S   RENDU F
ADENOSINE METABOLISM IN PLATELETS AND PLASMA
SER HAEMATOL 6:317-332(1973)

CAIN MD   CERINI JC   CERINI MED
CHAMLEY WA   CUMMING IA   GODING JR
COMPETITIVE PROTEIN-BINDING ANALYSIS OF OVINE AND BOVINE
PLASMA PROGESTERONE
J REPROD FERTIL 28:148-150(1972)

CAIN MD   HALL RC   JACKSON RC
IRVINE RJ   GODING JR
RADIOIMMUNOASSAY OF PLASMA PROSTAGLANDIN F
ABSTR 6TH ANNU MEET, AUSTRALAS SOC CLIN PHARMACOL,
SYD:55(1972)

CALANDRA S   MONTAGUTI M
EFFECT OF PROSTAGLANDIN E1 ON CHOLESTEROL BIOSYNTHESIS IN
RAT LIVER
EXPERIENTIA 29:1361-1362(1973)

CALDWELL BV   BROCK WA   BURSTEIN S
SPEROFF L
RADIOIMMUNOASSAY OF PROSTAGLANDINS
ABSTR 18TH ANNU MEET SOC GYNECOL INVEST, PHOENIX:46(1971)

CALDWELL BV   BROCK WA   BURSTEIN S
SPEROFF L
RADIOIMMUNOASSAY OF PROSTAGLANDIN F2A
ABSTR 53RD MEET ENDOCR SOC, SAN FRANC:A-46(1971)

CALDWELL BV   BROCK WA   GORDON JW
SPEROFF L
SOME PRACTICAL CONSIDERATIONS IN THE RADIOIMMUNOASSAY OF THE
F PROSTAGLANDINS
IN: "PROSTAGLANDINS IN FERTILITY CONTROL 2," S BERGSTROM, K
GREEN AND B SAMUELSSON, W H O, KAROLINSKA INST, STOCKH:
83-91(1972)

CALDWELL BV   BURSTEIN S   BROCK WA
SPEROFF L
RADIOIMMUNOASSAY OF THE F PROSTAGLANDINS.
J CLIN ENDOCRINOL METAB 33:171-175(1971)

CALDWELL BV   SPEROFF L
EXPERIENCE WITH THE CLINICAL USE OF PROSTAGLANDINS
MED ACTUAL 9:403-422(1973);(SPANISH AND ENGLISH)

CALDWELL BV   SPEROFF L   BROCK WA
AULETTA FJ   GORDON JW   ANDERSON GG
HOBBINS JC
DEVELOPMENT AND APPLICATION OF A RADIOIMMUNOASSAY FOR F
PROSTAGLANDINS
J REPROD MED 9:361-371(1972)

CAMMOCK S
CONVERSION OF PGE1 TO A PGA1-LIKE COMPOUND BY RAT KIDNEY
HOMOGENATES
ADV BIOSCI (SUPPL) 9:10(1972)

CANALES PEREZ ES
THE PROSTAGLANDINS IN THE PHYSIOLOGY OF REPRODUCTION
GINECOL OBSTET MEX 30:315-329(1971);(SPANISH)

CANTOR EH   KELLIHER GJ
THE EFFECT OF PROSTAGLANDIN E1 ON DIGOXIN-INDUCED ARRHYTHMIA
CLIN RES 21:945(1973)

CAPRINO L   BORRELLI F   FALCHETTI R
EFFECT OF 4,5-DIPHENYL-2-BIS-(2-HYDROXYETHYL)AMINOXAZOL
(DITAZOL) ON PLATELET AGGREGATION, ADHESIVENESS AND BLEEDING
TIME
ARZNEIM FORSCH 23:1277-1283(1973)

CAPUTI AP
PLATELET AGGREGATION. PART II.
CLIN TER 67:449-501(1973);(ITALIAN)

CARAFOLI E   CROVETTI F
INTERACTIONS BETWEEN PROSTAGLANDIN E1 AND CALCIUM AT THE
LEVEL OF THE MITOCHONDRIAL MEMBRANE
ARCH BIOCHEM BIOPHYS 154:40-46(1973)

CARLSON HE   ROBBINS J
EFFECTS OF HORMONES AND NUCLEOTIDES ON CILIA
FED PROC 31:271(1972)

CARLSON JC   RUGG AE   GLEW ME
BARCIKOWSKI B   MCCRACKEN JA
LUTEOLYTIC PROPERTIES OF PROSTAGLANDIN F1A IN SHEEP
BIOL REPROD 7:108(1972)

CARLSON LA   BUTCHER RW
LEVELS OF CYCLIC AMP AND RATE OF FAT MOBILIZING LIPOLYSIS IN
HUMAN ADIPOSE TISSUE IN RESPONSE TO DIFFERENT ADRENERGIC
STIMULATORS
ADV CYCLIC NUCL RES 1:87-90(1972)

CARLSON LA   BUTCHER RW
CYCLIC AMP LEVELS IN HUMAN ADIPOSE TISSUE IN RESPONSE TO
CATECHOLAMINES, PROSTAGLANDIN E1 AND NICOTINIC ACID
ADV CYCLIC NUCL RES 1:569(1972)

CARLSON LA   ERICSSON M   ERIKSON U
PROSTAGLANDIN E1 (PGE1) IN PERIPHERAL ARTERIOGRAPHIES
ACTA RADIOL DIAGN STOCKH 14:583-587(1973)

CARLSON LA   ERIKSSON I
FEMORAL-ARTERY INFUSION OF PROSTAGLANDIN E1 IN SEVERE
PERIPHERAL VASCULAR DISEASE
LANCET 1:155-156(1973)

CARLSON LA   MICHELI H
STIMULATORY EFFECT OF PROSTAGLANDIN E1 ON FAT MOBILIZING
LIPOLYSIS IN ADIPOSE TISSUE OF RATS TREATED WITH NICOTINIC
ACID
IN:"METABOLIC EFFECTS OF NICOTINIC ACID AND ITS DERIVATIVES"
KF GEY AND LA CARLSON, HANS HUBER, BERN : 995-1001 (1971)

CARLSON LA   WALLDIUS G
THE NOBEL PRIZE IN PHYSIOLOGY OR MEDICINE. CYCLIC AMP IN
ADIPOSE TISSUE.
NORD MED 86:1261-1264(1971):(SWEDISH)

CARMINATI P   LERNER LJ
CHANGES IN ADENYL CYCLASE ACTIVITY IN SEVERAL AREAS OF
RABBIT OVIDUCT AND UTERUS AFTER INCUBATION WITH
PROSTAGLANDINS, THEOPHYLLINE, ISOPROTERENOL AND OXYTOCIN
ABSTR, 9TH INT CONG BIOCHEM, STOCKH:399(1973)

CARNEY JA   LEWIS A   WALKER BL
SLINGER SJ
EFFECT OF DIETARY RAPESEED OIL ON THE
ADRENOCORTICOTROPHIN-INDUCED PRODUCTION OF PROSTAGLANDINS IN
THE RAT ADRENAL
BIOCHIM BIOPHYS ACTA 280:211-214(1972)

CARPENTER CCJ JR
CHOLERA ENTEROTOXIN-RECENT INVESTIGATIONS YIELD INSIGHTS
INTO TRANSPORT PROCESSES
AM J MED 50:1-7(1971)

CARPENTER MP   MANNING L   WISEMAN B
PROSTAGLANDIN SYNTHESIS IN RAT TESTIS.
FED PROC 30:1081 ABS (1971)

CARR A
PROSTAGLANDIN (PGA1) BLOOD PRESSURE, PLASMA RENIN ACTIVITY
ANN INT MED 74:830(1971)

CARR AA
EFFECT OF PGA1 ON RENIN AND ALDOSTERONE IN MAN
PROSTAGLANDINS 3:621-628(1973)

CARRETERO OA   BUJAK B   HODARI AA
HODGKINSON CP   BUMPUS FM
IDENTIFICATION OF A PRESSOR POLYPEPTIDE IN HUMAN AMNIOTIC
FLUID
AM J OBSTET GYNECOL 111:1075-1082(1971)

CARRICK MJ
INTERACTIONS BETWEEN STEROIDS AND SPASMOGENS ON THE
MYOMETRIUM OF THE EWE
DISS, UNIV CALIF, DAVIS(1973)

CARRIERE S   FRIBURG J   GUAY JP
VASODILATORS, INTRARENAL BLOOD FLOW, AND NATRIURESIS IN THE
DOG
AM J PHYSIOL 221:92-98(1971)

CARRIERE S   FRIBORG J   GUAY JP
VASODILATORS, INTRARENAL BLOOD FLOW AND NATRIURESIS IN DOGS
PROC INT UNION PHYSIOL SCI, 25TH INT CONGR, MUNICH,
9:102(1971)

CARSTEN ME
CALCIUM BINDING IN UTERINE SARCOPLASMIC RETICULUM
ABSTR 18TH ANNU MEET SOC GYNECOL INVEST, PHOENIX:40(1971)

CARSTEN ME
PROSTAGLANDINS AND MYOMETRIAL CALCIUM TRANSPORT
ABSTR SOC GYNECOL INVEST 19TH ANNU MEET, PHOENIX:21(1972)

CARSTEN ME
EFFECTS OF PROSTAGLANDINS ON CALCIUM TRANSPORT IN UTERINE
SARCOPLASMIC RETICULUM
FED PROC 31:399(1972)

CARSTEN ME
PROSTAGLANDIN'S PART IN REGULATING UTERINE CONTRACTION BY
TRANSPORT OF CALCIUM
J REPROD MED 9:277-281(1972)

CARSTEN ME
CELLULAR MECHANISM OF PROSTAGLANDIN ACTION IN HUMAN
MYOMETRIUM
ABSTR SOC GYNECOL INVEST, 20TH ANNU MEET, ATLANTA:33(1973)

CARTER DC   KARIM SMM   BHANA D
GANESAN PA
THE EFFECT OF LOCALLY ADMINISTERED 15(R) 15 METHYL-E2
PROSTAGLANDIN ON BASAL AND PENTAGASTRIN-INDUCED ACID
SECRETION IN MAN
BR J SURG 60:320(1973)

CARTER DC   KARIM SMM   BHANA D
GANESAN PA
INHIBITION OF HUMAN GASTRIC SECRETION BY PROSTAGLANDIN
BR J SURG 60:828-831(1973)

CARTER DC   KARIM SMM   BHANA D
GANESAN PA
INHIBITION OF HUMAN GASTRIC SECRETION BY ORALLY OR
INTRAVENOUSLY ADMINISTERED PROSTAGLANDIN 16,16 DIMETHYL-E2
METHYL ESTER
BR J SURG 60:912(1973)

CASE RM
REVIEW. CELLULAR MECHANISMS CONTROLLING PANCREATIC EXOCRINE
SECRETION
ACTA HEPATO-GASTROENTEROL 20:435-444(1973)

CASE RM   SCRATCHERD T
PROSTAGLANDIN ACTION ON PANCREATIC BLOOD FLOW AND ON
ELECTROLYTE AND ENZYME SECRETION BY EXOCRINE PANCREAS IN
VIVO AND IN VITRO
J PHYSIOL (LOND) 226:93-405(1972)

CASE RM   SCRATCHERD T
THE ACTIONS OF DIBUTYRYL CYCLIC ADENOSINE
3',5'-MONOPHOSPHATE AND METHYL XANTHINES ON PANCREATIC
EXOCRINE SECRETION
J PHYSIOL (LOND) 223:649-667(1972)

CASEY PA   CASEY G   FLEISCH H
RUSSEL RGG
THE EFFECT OF POLYPHLORETIN PHOSPHATE, POLYESTRADIOL
PHOSPHATE, A DIPHOSPHONATE AND A POLYPHOSPHATE ON
CALCIFICATION INDUCED BY DIHYDROTACHYSTEROL IN SKIN, AORTA,
AND KIDNEY OF RATS.
EXPERIENTIA 28:137-138(1972)

CASSIDY F
THE PROSTAGLANDINS
REP PROG APPL CHEM 56:695-710(1972)

CASTENFORS J
EFFECTS OF PROSTAGLANDINS (PGA1 AND PGA2) ON THE
HEMODYNAMICS OF THE KIDNEY, SODIUM EXCRETION AND PLASMA
RENIN ACTIVITY
NORD MED 85:703(1971):(SWEDISH)

CATALONA WJ   ENGELMAN K   KETCHAM AS
HAMMOND WG
FAMILIAL MEDULLARY THYROID CARCINOMA, PHEOCHROMOCYTOMA, AND
PARATHYROID ADENOMA (SIPPLE'S SYNDROME)
CANCER 28:1245-1254(1971)

CATON MPL
THE PROSTAGLANDINS
PROG MED CHEM 8:317-376(1972)

CATON MPL
CHEMISTRY STRUCTURE AND AVAILABILITY
IN: "THE PROSTAGLANDINS. PHARMACOLOGICAL AND THERAPEUTIC
ADVANCES," MF CUTHBERT, HEINEMANN, LOND:1-22(1973)

CATON MPL   COFFEE ECJ   WATKINS GL
PROSTAGLANDINS - A NEW TOTAL SYNTHESIS OF
(+-)-11-DEOXYPROSTAGLANDIN
TETRAHEDRON LETT 773-774(1972)

CATON MPL   PARKER T   WATKINS GL
PROSTAGLANDINS II. SYNTHESIS OF
(+-)-9-DEOXY-13,14-DIHYDROPROSTAGLANDINS
TETRAHEDRON LETT 3341-3344(1972)

CHACHATY C   WOLKOWSKI Z   PIRIOU F
LUKACS G
SUBSTITUENT EFFECTS ON THE 13C RELAXATION TIME ALONG
ALIPHATIC CHAINS. APPLICATION TO A PROSTAGLANDIN
J C S CHEM COMM:951-952(1973)

CHANDLER JT   STRONG CG
THE ACTIONS OF PROSTAGLANDIN E1 ON ISOLATED RABBIT AORTA
ARCH INT PHARMACODYN THER 197:123-131(1972)

CHANG MC   HUNT DM
EFFECTS OF PROSTAGLANDINS ON THE TRANSPORTATION OF RABBIT
EGGS AND SPERMATOZOA
ADV BIOSCI (SUPPL) 9:134(1972)

CHANH PH   JUNSTAD M   WENNMALM A
AUGMENTED NORADRENALINE RELEASE FOLLOWING NERVE STIMULATION
AFTER INHIBITION OF PROSTAGLANDIN SYNTHESIS WITH
INDOMETHACIN
ACTA PHYSIOL SCAND 86:563-567(1972)

CHANNING CP
STIMULATORY EFFECTS OF PROSTAGLANDINS UPON LUTEINIZATION OF
RHESUS MONKEY GRANULOSA CELL CULTURES
PROSTAGLANDINS 2:331-349(1972)

CHARBONNEL B   SOUBRIER A   DRAY F
RADIOIMMUNOASSAY OF THE F PROSTAGLANDINS
ANN ENDOCRINOL (PARIS) 34:722-724(1973):(FRENCH)

CHARD T
THE ROLE OF OXYTOCIN IN THE INDUCTION OF LABOUR
IN: "ENDOCRINOLOGY. PROC 4TH INT CONG ENDOCRINOL," RO SCOW,
FJG EBLING, IW HENDERSON, EXCERPTA MEDICA, AMST/NY:1066-1070
(1973)

CHAYOTH R   EPSTEIN S   FIELD JB
ALTERATIONS IN THE ADENYLATE CYCLASE-CYCLIC 3'5'-ADENOSINE
MONOPHOSPHATE (CAMP) SYSTEM IN RAT HEPATOMA AND HUMAN
CARCINOMA OF THE COLON
J CLIN INVEST 51:19A(1972)

CHAYOTH R   FIELD JB
GLUCAGON AND PROSTAGLANDIN E1 (PGE1) EFFECTS ON ADENYLATE
CYCLASE (A.C.) AND CYCLIC AMP (CAMP) IN RAT HEPATOMA
CLIN RES 21:643(1973)

CHEDD G
ASPIRIN: AN ANSWER AT LAST?
NEW SCIENTIST AND SCI J 50:744-745(1971)

CHEN LC   RHODE JE   SHARP GWG
PROPERTIES OF ADENYL CYCLASE FROM HUMAN JEJUNAL MUCOSA
DURING NATURALLY ACQUIRED CHOLERA AND CONVALESCENCE
J CLIN INVEST 51:731-740(1972)

CHEN WM   SUNAHARA FA
INTERACTION OF PROSTAGLANDIN E1 AND NOREPINEPHRINE ON
MECHANICAL AND RB86 EFFLUX ACTIVITIES IN ISOLATED
VASCULAR TISSUE
FED PROC 30:625(1971)

CHENG CPK   WANG JCC
THE EFFECTS OF PROSTAGLANDIN E1 ON O2 CONSUMPTION IN
COLD-ADAPTED RATS
PROC INT UNION PHYSIOL SCI, 25TH INT CONGR, MUNICH 9:107
(1971)

CHI JY
INTERACTION AMONG PROSTAGLANDINS, NERVES AND HORMONES
CHIN MED J 9:558-562(1973);(CHINESE)

CHIANG TS
EFFECTS OF INTRAVENOUS INFUSION OF AUTACOIDS ON INTRAOCULAR
PRESSURE AND AQUEOUS HUMOR PROTEIN CONCENTRATION
PHARMACOLOGIST 15:209(1973)

CHIANG TS
EFFECTS OF EPINEPHRINE AND PROGESTERONE ON THE OCULAR
HYPERTENSIVE RESPONSE TO INTRAVENOUS INFUSION OF
PROSTAGLANDIN A2
PROSTAGLANDINS 4:415-419(1973)

CHIANG TS   THOMAS RP
CONSENSUAL OCULAR HYPERTENSIVE RESPONSE TO PROSTAGLANDIN E1
IN RABBITS
FED PROC 31:546(1972)

CHIANG TS   THOMAS RP
CONSENSUAL OCULAR HYPERTENSIVE RESPONSE TO PROSTAGLANDIN
INVEST OPHTHALMOL 11:169-176(1972)

CHIANG TS   THOMAS RP
EFFECTS OF PROGESTERONE AND EPINEPHRINE ON THE OCULAR
HYPERTENSIVE RESPONSE TO INTRAVENOUS INFUSION OF
PROSTAGLANDIN E1
ABSTR 5TH INT CONGR PHARMACOL, SAN FRANC:41(1972)

CHIANG TS   THOMAS RP
OCULAR HYPERTENSION FOLLOWING INTRAVENOUS INFUSION OF
PROSTAGLANDIN E1
ARCH OPHTHALMOL 88:418-420(1972)

CHIANG TS   THOMAS RP
CONSENSUAL OCULAR HYPERTENSIVE RESPONSE TO PROSTAGLANDIN E2
INVEST OPHTHALOL 11:845-849(1972)

CHIANG TS   THOMAS RP
EFFECTS OF PROGESTERONE ON THE OCULAR HYPERTENSIVE RESPONSE
TO PROSTAGLANDIN
EUR J PHARMACOL 22:304-310(1973)

CHIBA S   NAKAJIMA T   NAKANO J
EFFECT OF PROSTAGLANDINS E1 AND F2A ON HEART RATE BY DIRECT
INJECTION INTO THE CANINE SINUS NODE ARTERY
JAP J PHARMACOL 22:734-736(1972)

CHIMURA T   HIROI M   YAMUKI T
UTERINE CONTRACTILITY DURING LABOR INDUCED BY PROSTAGLANDIN
F2A
OBSTET GYNECOL (TOKYO) 46:787-792(1971);(JAPANESE)

CHIMURA T   OBATA N   SATO Y
TAKEUCHI S   INAGAWA T   SAWADA M
OTSUKA K
PROSTAGLANDINS LEVELS IN HUMAN BLOOD DURING LABOR
ACTA MED BIOL (NIIGATA) 20:163-169(1973)

CHRIST EJ
PROSTAGLANDIN BIOSYNTHESIS WITH SPECIAL REFERENCE TO ADIPOSE
TISSUE.
J AM OIL CHEM SOC 48:94A(1971)

CHRIST EJ
COMPARATIVE ASPECTS OF PROSTAGLANDIN BIOSYNTHESIS IN ANIMAL
TISSUES
ADV BIOSCI (SUPPL) 9:4(1972)

CHRIST EJ   VAN DORP DA
COMPARATIVE ASPECTS OF PROSTAGLANDIN BIOSYNTHESIS IN ANIMAL
TISSUES
BIOCHIM BIOPHYS ACTA 270:537-545(1972)

CHRIST EJ   VAN DORP DA
COMPARATIVE ASPECTS OF PROSTAGLANDIN BIOSYNTHESIS IN ANIMAL
TISSUES
ADV BIOSCI 9:35-38(1973)

CHRISTOPHE J   ROBBERECHT P   DESCHODT-LANCKM/ M
VANDERMEERS A   VANDERMEERS-PI/ MC CAMUS J
RATHE J
CELLULAR BIOLOGY OF THE EXOCRINE PANCREAS
BULL ACAD R MED BELG 12:323-359(1972);(FRENCH)

CIACERI G   ATTAGUILE G   MARINI PP
INFLUENCE OF SOME BIOFLAVONOIDS ON SPASMOGENIC ACTIVITY OF
PROSTAGLANDINS
G ITAL PATOL SCI AFFINI 20:41-49(1973);(ITALIAN)

CIERNIEWSKI C
THE PROSTAGLANDINS
POSTEPY BIOCHEM 19:5-18(1973);(POLISH)

CIOFALO FR
PROSTAGLANDINS AND SYNAPTOSOMAL TRANSPORT OF
3H-NOREPINEPHRINE AND 3H-5-HYDROXYTRYPTAMINE
RES COMMUN CHEM PATHOL PHARMACOL 5:551-554(1973)

CIZELJ T   ANDOLSEK L   PRETNAR A
USE OF PROSTAGLANDIN F2A FOR THERAPEUTIC ABORTION
ACTA EUR FERTIL 3:131-134(1972)

CLARK KE   RYAN MJ   BRODY MJ
EFFECTS OF PROSTAGLANDINS E1 AND F2A ON UTERINE ADRENERGIC
VASOCONSTRICTION AND CONTRACTILE ACTIVITY
ABSTR 5TH INT CONGR PHARMACOL, SAN FRANC:43(1972)

CLARKSON EM   DE WARDENER HE
INHIBITION OF SODIUM AND POTASSIUM TRANSPORT IN SEPARATED
RENAL TUBULE FRAGMENTS INCUBATED IN EXTRACTS OF URINE
OBTAINED FROM SALT-LOADED INDIVIDUALS
CLIN SCI 42:607-617(1972)

CLARKSON R
THE SYNTHESIS OF PROSTAGLANDINS
PROG ORG CHEM 8:1-28(1973)

CLASSEN M   KOCH H   BICKHARDT J
TOPF G   DEMLING L
THE EFFECT OF PROSTAGLANDIN E1 ON THE
PENTAGASTRIN-STIMULATED GASTRIC SECRETION IN MAN
DIGESTION 4:333-344(1971)

CLASSEN M   RUPPIN H
EFFECTS OF PROSTAGLANDINS ON THE GASTROINTESTINAL TRACT
Z GASTROENTEROL 11:217-222(1973);(GERMAN)

CLASSEN MP   STURZENHOFECKER P   KOCH H
DEMLING L
THE EFFECT OF PROSTAGLANDIN E1 ON THE SECRETION AND THE
MOTILITY OF THE HUMAN STOMACH
ACTA HEPATOGASTROENTEROL 20:159-162(1973)

CLAUSEN J   SRIVASTAVA KC
THE BIOSYNTHESIS OF PROSTAGLANDINS IN THROMBOCYTES
BIOCHEM J 128:4P(1972)

CLAUSEN J   SRIVASTAVA KC
THE SYNTHESIS OF PROSTAGLANDINS IN HUMAN PLATELETS
LIPIDS 7:246-250(1972)

CLAUSEN J   SRIVASTAVA KC
COMPARISON OF DIALYSIS, THIN LAYER AND SILICIC ACID COLUMN
CHROMATOGRAPHY FOR PROSTAGLANDIN ISOLATION FROM BIOLOGICAL
MATERIAL
LIPIDS 7:415-419(1972)

CLOWES GHA
THE ACTION OF POST-TRAUMATIC AND SEPSIS-LIBERATED
CIRCULATING PEPTIDES ON THE LUNG
MED WELT 24:1154-1155(1973); (GERMAN)

COCEANI F   PUGLISI L   LAVERS B
PROSTAGLANDINS AND NEURONAL ACTIVITY IN SPINAL CORD AND
CUNEATE NUCLEUS
ANN NY ACAD SCI 180:289-301(1971)

COCEANI F   VITI A
THE RELEASE OF PROSTAGLANDIN E1 FROM MICROPIPETTES IN VITRO
PROC CAN FED BIOL SOC 15:247(1972)

COCEANI F   VITI A
ACTIONS OF PROSTAGLANDIN E1 ON SPINAL NEURONS IN THE FROG
ADV BIOSCI (SUPPL) 9:77(1972)

COCEANI F   VITI A
THE RELEASE OF PROSTAGLANDIN E1 FROM MICROPIPETTES IN VITRO
BRAIN RES 45:469-477(1972)

COCEANI F   VITI A
ACTIONS OF PROSTAGLANDIN E1 ON SPINAL NEURONS IN THE FROG
ADV BIOSCI 9:481-487(1973)

COGHLAN JP   BLAIR-WEST JR   DENTON DA
SCOGGINS BA   WRIGHT RD
PERSPECTIVES IN ALDOSTERONE AND RENIN CONTROL
AUST N Z J MED 2:178-197(1971)

COHEN F   JAFFE BM
PRODUCTION OF PROSTAGLANDINS BY CELLS IN VITRO:
RADIOIMMUNOASSAY MEASUREMENT OF THE CONVERSION OF
ARACHIDONIC ACID TO PGE2 AND PGF2A
BIOCHEM BIOPHYS RES COMMUN 55:724-729(1973)

COHEN M   SZTOKALO J   HINSCH E
THE ANTIHYPERTENSIVE ACTION OF ARACHIDONIC ACID IN THE
SPONTANEOUS HYPERTENSIVE RAT AND ITS ANTAGONISM BY ANTI-
INFLAMMATORY AGENTS
LIFE SCI 13: 317-325(1973)

COHEN SL   MACINTYRE I   GRAHAME-SMITH D
WALKER JG
ALCOHOL-STIMULATED CALCITONIN RELEASE IN MEDULLARY
CARCINOMA OF THE THYROID
LANCET 2:1172-1174(1973)

COHN CK   BALL GG   HIRSCH J
HISTAMINE: EFFECT ON SELF-STIMULATION
SCIENCE 18:757-758 (1973)

COHN ML
THE INFLUENCE ON AMOBARBITAL-INDUCED SLEEPING TIME IN RATS
BY DRUGS AFFECTING CYCLIC AMP
IN: "PROSTAGLANDINS AND CYCLIC AMP," RH KAHN, WEM LANDS,
ACADEMIC PRESS, NY : 73-74(1973)

COLBERT JC
PROSTAGLANDINS ISOLATION AND SYNTHESIS
NOYES DATA CORP,PARK RIDGE NJ/(LOND) : (1973)

COLE B   ROBISON GA   HARTMANN RC
STUDIES ON THE ROLE OF CYCLIC AMP IN PLATELET FUNCTION.
ANN N Y ACAD SCI 185: 477-487 (1971)

COLE DF   NAGASUBRAMANIAN S
THE EFFECT OF NATURAL AND SYNTHETIC VASOPRESSINS AND OTHER
SUBSTANCES ON ACTIVE TRANSPORT IN CILIARY EPITHELIUM OF THE
RABBIT
EXP EYE RES 13:45-57(1972)

COLE DF   NAGASUBRAMANIAN S
SUBSTANCES AFFECTING ACTIVE TRANSPORT ACROSS THE CILIARY
EPITHELIUM AND THEIR POSSIBLE ROLE IN DETERMINING
INTRAOCULAR PRESSURE
EXP EYE RES 16:251-264(1973)

COLE DF   UNGER WG
THE INVOLVEMENT OF PROSTAGLANDIN IN OCULAR TRAUMA
EXP EYE RES 17:395(1973)

COLE DF   UNGER WG
PROSTAGLANDINS AS MEDIATORS FOR THE RESPONSES OF THE EYE TO
TRAUMA
EXP EYE RES 17:357-368(1973)

COLEMAN RA
EVIDENCE FOR A NON-ADRENERGIC INHIBITORY NERVOUS PATHWAY IN
GUINEA-PIG TRACHEA
BR J PHARMACOL 48:360P-361P(1973)

COLLIER HOJ
PROSTAGLANDINS AND ASPIRIN
NATURE(LOND)232:17-19(1971)

COLLIER HOJ
INTRODUCTION TO THE ACTIONS OF KININS AND PROSTAGLANDINS
PROC R SOC MED 64:1-4(1971)

COLLIER HOJ
DRUG DEPENDENCE: A PHARMACOLOGICAL ANALYSIS
BR J ADDICT 67:277-286(1972)

COLLIER HOJ
THE EXPERIMENTAL ANALYSIS OF DRUG-DEPENDENCE
ENDEAVOUR 31:123-129(1972)

COLLIER HOJ   FRANCIS DL   SCHNEIDER C
MODIFICATION OF MORPHINE WITHDRAWAL BY DRUGS INTERACTING
WITH HUMORAL MECHANISMS: SOME CONTRADICTIONS AND THEIR
INTERPRETATION
NATURE (LOND) 237:220-223(1972)

COLLIER HOJ   SAEED SA   SCHNEIDER C
WARREN BT
PROSTAGLANDINS AND THE ANALGESIC ACTION OF ASPIRIN
ADV BIOSCI (SUPPL) 9:64(1972)

COLLIER HOJ   SAEED SA   SCHNEIDER C
WARREN BT
ARACHIDONIC ACID AND THE ANALGESIC ACTION OF ASPIRIN-LIKE
DRUGS
ADV BIOSCI 9:413-418(1973)

COLLIER HOJ   SCHNEIDER C
NOCICEPTIVE RESPONSE TO PROSTAGLANDINS AND ANALGESIC ACTIONS
OF ASPIRIN AND MORPHINE
NATURE (NEW BIOL) 234:141-143(1972)

COLLIER JG
NEW DIALYSIS TECHNIQUE FOR THE CONTINUOUS MEASUREMENT OF THE
CONCENTRATION OF VASOACTIVE HORMONES
BR J PHARMACOL 44:383P(1972)

COLLIER JG   HERMAN AG   VANE JR
APPEARANCE OF PROSTAGLANDINS IN THE RENAL VENOUS BLOOD OF
DOGS IN RESPONSE TO ACUTE SYSTEMIC HYPOTENSION PRODUCED BY
BLEEDING OR ENDOTOXIN
J PHYSIOL (LOND) 230:19P-20P(1973)

COLLIER JG   KARIM SMM   ROBINSON B
SOMERS K
ACTION OF PROSTAGLANDINS A2, B1, E2 AND F2A ON SUPERFICIAL
HAND VEINS OF MAN
BR J PHARMACOL 44:374P-375P(1972)

COLLIER JG   KARIM SMM   ROBINSON B
SOMERS K
EFFECT OF PROSTAGLANDINS A1, A2, B1, E2 AND F2A ON THE
FOREARM ARTERIAL BED OF MAN
BR J PHARMACOL 46:551P-552P(1972)

COLLINS JA   JAFFE BM
EFFECTS OF PROSTAGLANDINS ON THE AFFINITY OF HEMOGLOBIN FOR
OXYGEN IN HUMAN WHOLE BLOOD IN VITRO
PROSTAGLANDINS 3:59-66(1973)

COLLINS M   PALMER GC   BACA G
SCOTT HR
STIMULATION OF CYCLIC AMP IN THE ISOLATED PERFUSED RAT LUNG
RES COMMUN CHEM PATHOL PHARMACOL 6:805-812(1973)

COMLINE RS   SILVER M   NATHANIELSZ PW
HALL LW
PARTURITION IN THE LARGER HERBIVORES
IN: FOETAL AND NEONATAL PHYSIOLOGY, LONDON, CAMBRIDGE UNIV
PRESS:606-612(1973)

CONCANNON PW
IN VITRO PROGESTERONE SYNTHESIS IN BOVINE CORPORA LUTEA AS
INFLUENCED BY ESTRADIOL AND PROSTAGLANDINS
DISS, CORNELL UNIV, ITHACA (1972)

CONSTANTOPOULOS A   NAJJAR VA
THE ACTIVATION OF ADENYLATE CYCLASE: II. THE POSTULATED
PRESENCE OF (A) ADENYLATE CYCLASE IN A PHOSPHO (INHIBITED)
FORM (B) A DEPHOSPHO (ACTIVATED) FORM WITH A CYCLIC
ADENYLATE STIMULATED MEMBRANE PROTEIN KINASE
BIOCHEM BIOPHYS RES COMMUN 53:794-799(1973)

CONWAY J   HATTON R
EFFECTS OF PROSTAGLANDINS E1, E2, A1 AND A2 ON RESISTANCE
AND CAPACITANCE VESSELS IN THE HIND LIMB OF THE DOG
J PHYSIOL (LOND) 230: 56P-57P (1973)

COOPER GF   FRIED J
CARBON-13 NUCLEAR MAGNETIC RESONANCE SPECTRA OF
PROSTAGLANDINS AND SOME PROSTAGLANDIN ANALOGS
PROC NATL ACAD SCI USA 70:1579-1584(1973)

COOPER KE   VEALE WL
EXCHANGE BETWEEN THE BLOOD-BRAIN AND CEREBROSPINAL FLUID OF
SUBSTANCES WHICH CAN INDUCE OR MODIFY FEBRILE RESPONSES
IN PHARMACOL THERMOREGUL PROC SATELL SYMP 27A-288(1972)
(KARGER BASEL) (1973)

COOPER RH   MCPHERSON M   SCHOFIELD JG
THE EFFECT OF PROSTAGLANDINS ON OX PITUITARY CONTENT OF
ADENOSINE 3':5'-CYCLIC MONOPHOSPHATE AND THE RELEASE OF
GROWTH HORMONE
BIOCHEM J 127:143-154(1972)

COORE HG   DENTON RM   MARTIN BR
RANDLE PJ
REGULATION OF ADIPOSE TISSUE PYRUVATE DEHYDROGENASE BY
INSULIN AND OTHER HORMONES
BIOCHEM J 125:115-127(1971)

COPER H
CLINICAL PHARMACOLOGY OF ORDINARY PAIN, SLEEP, AND SEDATIVE
DRUGS. (TRANQUILIZER)
INTERNIST (BERLIN) 13:169-178(1972);(GERMAN)

COREY EJ
STUDIES ON THE TOTAL SYNTHESIS OF PROSTAGLANDINS
ANN NY ACAD SCI 180:24-37(1971)

COREY EJ   ALBONICO SM   KOELLIKER U
SCHAAF TK   VARMA RK
NEW REAGENTS FOR STEREOSELECTIVE CARBONYL REDUCTION. AN
IMPROVED SYNTHETIC ROUTE TO THE PRIMARY PROSTAGLANDINS.
J AM CHEM SOC 93:1491-1493(1971)

COREY EJ   BECKER KB   VARMA RK
EFFICIENT GENERATION OF 15S CONFIGURATION IN PROSTAGLANDIN
SYNTHESIS. ATTRACTIVE INTERACTIONS IN STEREOCHEMICAL CONTROL
OF CARBONYL REDUCTION
J AM CHEM SOC 94:8616-8618(1972)

COREY EJ   ENSLEY HE
HIGHLY STEREOSELECTIVE CONVERSION OF PROSTAGLANDIN A2 TO THE
10,11ALPHA-OXIDO DERIVATIVE USING A REMOTELY PLACED
EXOGENOUS DIRECTING GROUP
J ORG CHEM 38:3187-3189(1973)

COREY EJ   ERICKSON BW   NOYORI R
A NEW SYNTHESIS OF ALPHA, BETA-UNSATURATED ALDHYDES
USING 1,3-BIS (METHYL-THIO) ALLYLLITHIUM
J AM CHEM SOC 93:1724-1729(1971)

COREY EJ   FUCHS PL
HOMOCONJUGATE ADDITION OF ORGANOCUPPER REAGENTS TO
CYCLOPROPANES AND ITS APPLICATION TO THE SYNTHESIS OF
PROSTANOIDS
J AM CHEM SOC 94:4014-4015(1972)

COREY EJ   GRIECO PA
HIGHLY EFFICIENT ROUTE TO A KEY INTERMEDIATE FOR THE
SYNTHESIS OF A PROSTAGLANDINS
TETRAHEDRON LETT 107-109(1972)

COREY EJ   KIM CU
IMPROVED SYNTHETIC ROUTES TO PROSTAGLANDINS UTILIZING
SULFIDE-MEDIATED OXIDATION OF PRIMARY AND SECONDARY ALCOHOLS
J ORG CHEM 38:1233-1234(1973)

COREY EJ   KOELLIKER U   NEUFFER J
METHOXYMETHYLATION OF THALLOUS CYCLOPENTADIENIDE. A
SIMPLIFIED PREPARATION OF A KEY INTERMEDIATE FOR THE
SYNTHESIS OF PROSTAGLANDINS
J AM CHEM SOC 93:1489-1490(1971)

COREY EJ   MANN J
A NEW STEREOCONTROLLED SYNTHESIS OF PROSTAGLANDINS VIA
PROSTAGLANDIN A2
J AM CHEM SOC 95: 6832-6833(1973)

COREY EJ   MOINET G
DIRECT, STEREOCONTROLLED SYNTHESIS OF A PROSTAGLANDINS USING
THE BICYCLO(2,2,1)HEPTENE APPROACH
J AM CHEM SOC 95: 6831-6832 (1973)

COREY EJ   MOINET G
SYNTHETIC ENTRY INTO THE PROSTAGLANDIN C SERIES
J AM CHEM SOC 95:7185-7186 (1973)

COREY EJ   RAVINDRANATHAN T
A SIMPLE ROUTE TO A KEY INTERMEDIATE FOR THE SYNTHESIS OF
11-DESOXYPROSTAGLANDINS
TETRAHEDRON LETT 4753-4755(1971)

COREY EJ   RAVINDRANATHAN T
A NEW SYNTHETIC APPROACH TO PROSTANOIDS VIA CYCLOPENTENE
VINYLATION
J AM CHEM SOC 94:4013-4014(1972)

COREY EJ   RAVINDRANATHAN T   TERASHIMA S
A NEW METHOD FOR THE 1,4 ADDITION OF THE METHYLENECARBONYL
UNIT (-CH2CO-) TO DIENES
J AM CHEM SOC 93:4326-4327(1971)

COREY EJ   SACHDEV HS
A SIMPLE SYNTHESIS OF 8-METHYLPROSTAGLANDIN C2
J AM CHEM SOC 95:8483-8484(1973)

COREY EJ   SHIRAHAMA H   YAMAMOTO H
TERASHIMA S   VENKATESWARLU A   SHAAF TK
STEREOSPECIFIC TOTAL SYNTHESIS OF PROSTAGLANDINS E3 AND F3A
J AM CHEM SOC 93:1490-1491(1971)

COREY EJ   SNIDER BB
A NEW SYNTHETIC ROUTE TO PROSTAGLANDINS
TETRAHEDRON LETT 3091-3094(1973)

COREY EJ   TERASHIMA S
A DRAMATIC CHANGE IN THE BALANCE BETWEEN SN2 AND E2
PATHWAYS WITH FORMATE AND OXALATE AS NUCLEOPHILE
TETRAHEDRON LETT 111-113(1972)

COREY EJ   TERASHIMA S   RAMWELL PW
JESSUP R   WEINSHENKER NM   FLOYD DM
CROSBY GA
11,15-EPIPROSTAGLANDIN E2 AND ITS ENANTIOMER. BIOLOGICAL
ACTIVITY AND SYNTHESIS
J ORG CHEM 37:3043-3044(1972)

COREY EJ   VARMA RK
SPECIFIC REDUCTION OF E PROSTAGLANDINS TO FA PROSTAGLANDINS
AND PROSTAGLANDIN E2 TO PROSTAGLANDIN E1.
J AM CHEM SOC 93:7319-7320(1971)

COREY EJ   VARMA RK
SOME RECENT DEVELOPMENTS IN THE SYNTHESIS OF PRIMARY
PROSTAGLANDINS
ABSTR 7TH MIDDLE ATL REG MEET AM CHEM SOC PHILA:81(1972)

COREY EJ   VENKATESWARLU A
PROTECTION OF HYDROXYL GROUPS AS TERT-BUTYLDIMETHYLSILYL
DERIVATIVES
J AM CHEM SOC 94:6190-6191(1972)

COREY EJ   WASHBURN WN   CHEN JC
STUDIES ON THE PROSTAGLANDIN A2 SYNTHETASE COMPLEX FROM
PLEXAURA HOMOMALLA
J AM CHEM SOC 95:2054-2055(1973)

CORLETT R   THORNEYCROFT IH   NAKAMURA RM
MISHELL DR JR
ABORTIFACIENT ACTIVITY OF VAGINALLY ADMINISTERED PGF2A
TABLETS
ADV BIOSCI (SUPPL) 9:95(1972)

CORNETTE JC   KIRTON KT   BARR KL
FORBES AD
RADIOIMMUNOASSAY OF PROSTAGLANDINS
J REPROD MED 9:355-360(1972)

CORTES-GALLEGOS V
LUTEOLYTIC AGENTS AND REPRODUCTION
GAC MED MEX 106:259-286(1973):(SPANISH)

CORTES-GALLEGOS V ORTEGA E   HENZL MR
SOJO-ARANDA I
EFFECT OF PROSTAGLANDIN E2 (PGE2) ON THE FUNCTION OF THE
HUMAN CORPUS LUTEUM
ARCH INVEST MED (MEX) 3(SUPPL 1): 173-186(1972):(SPANISH)

CORTES-GALLEGOS V ORTEGA EH   HENZL MR
SOJO-ARANDA I
THE EFFECT OF PROSTAGLANDIN E2 (PGE2) ON THE FUNCTION OF THE
HUMAN CORPUS LUTEUM
IN: "SIMPOSIO SOBRE PROSTAGLANDINS," IMPRESOS OFFSALI, MEX
CITY:42-61(1972):(SPANISH)

COUPAR IM   MCCOLL I
INHIBITION OF GLUCOSE ABSORPTION BY PROSTAGLANDINS E1, E2
AND F2A
J PHARM PHARMACOL 24:254-255(1972)

COURT JM   DUNLOP ME   LEONARD RF
HIGH-FREQUENCY OSCILLATION OF BLOOD FREE FATTY ACID LEVELS
IN MAN.
J APPL PHYSIOL 31(3):345-347(1971)

COUSINEAU D   GAGNON DJ
IN VITRO RELEASE OF VASOPRESSIN FROM THE POSTERIOR PITUITARY
BY ANGIOTENSIN II AND E2 PROSTAGLANDINS
UNION MED CAN 102:1577(1973):(FRENCH)

COUTINHO EM   MAIA H
THE MOTILITY OF THE HUMAN OVARY
PROC WORLD CONGR FERTIL STERIL 7:204(1973)

COVELLI I   FIMIANI V   FRATI L
ROLE OF PROSTAGLANDIN IN ACUTE INFLAMMATION
ACTA VITAMINOL ENZYMOL (MILANO) 27:265-266(1973):(ITALIAN)

COVIELLO A
HYDROSMOTIC EFFECT OF ANGIOTENSIN II IN THE TOAD BLADDER:
ROLE OF CYCLIC AMP
ACTA PHYSIOL LATINOAM XXIII:350-357(1973)

COX JP   KARNOVSKY ML
THE DEPRESSION OF PHAGOCYTOSIS BY EXOGENOUS CYCLIC
NUCLEOTIDES, PROSTAGLANDINS, AND THEOPHYLLINE
J CELL BIOL 59:480-490 (1973)

COX RI   THORBURN GD   CURRIE WB
RESTALL BJ   SCHNEIDER W
PROSTAGLANDIN F GROUP (PGF), PROGESTERONE AND OESTROGEN
CONCENTRATIONS IN THE UTERO-OVARIAN VENOUS PLASMA OF THE
CONSCIOUS EWE DURING THE OESTROUS CYCLE
ADV BIOSCI (SUPPL) 9:102-103(1972)

CRABBE P
RECENT PROGRESS IN THE SYNTHESIS OF NEW PROSTAGLANDINS
IN: "SIMPOSIO SOBRE PROSTAGLANDINS," IMPRESOS OFFSALI, MEX
CITY:2-41(1972):(SPANISH)

CRABBE P
RECENT DEVELOPMENTS IN THE SYNTHESIS OF NEW PROSTAGLANDINS
ARCH INVEST MED (MEX) 3(SUPPL 1): 151-172(1972):(SPANISH)

CRABBE P   CARPIO H
SYNTHESIS OF ALLENYL PROSTAGLANDINS
CHEM COMMUN 904-905(1972)

CRABBE P   CARPIO H   GUZMAN A
SYNTHESIS OF MODIFIED PROSTAGLANDINS
INTRA-SCI CHEM REP 6:55-63(1972)

CRABBE P   CERVANTES A
SYNTHESIS OF DIFLUOROMETHYLENE-PROSTAGLANDINS
TETRAHEDRON LETT 1319-1321(1973)

CRABBE P   CERVANTES A   GUZMAN A
SYNTHESIS OF A 9,11-BIS-DESOXY-PROSTAGLANDIN
TETRAHEDRON LETT 1123-1125(1972)

CRABBE P   CERVANTES A   MEANA MC
SYNTHESIS OF 11-DEOXY-10A-HYDROXYPROSTAGLANDINS
J CHEM SOC CHEM COMMUN 119-120(1973)

CRABBE P   GARCIA GA   RIUS C
SYNTHESIS OF NOVEL BICYCLIC PROSTAGLANDINS BY PHOTOCHEMICAL
CYCLO-ADDITION REACTIONS
J CHEM SOC PERKIN I : 810-816 (1973)

CRABBE P   GARCIA GA   VELARDE E
REGIOSELECTIVE INTRAMOLECULAR PHOTOCYCLISATION IN THE
PROSTAGLANDIN SERIES
J CHEM SOC CHEM COMMUN : 480-481 (1973)

CRABBE P   GUZMAN A
SYNTHESIS OF 11-DESOXY-PROSTAGLANDINS
TETRAHEDRON LETT 115-117(1972)

CRABBE P   GUZMAN A   VELARDE E
SYNTHESIS OF 10A-HYDROXY-PROSTAGLANDINS
CHEM COMMUN 1126(1972)

CRABBE P   GUZMAN A   VERA M
SYNTHESIS OF PROSTAGLANDIN C2
TETRAHEDRON LETT 3021-3022(1973):(FRENCH)

CRABBE P.   GUZMAN A   VERA M
ERRATUM. SYNTHESIS OF PROSTAGLANDIN C2. NO. 32, 3021(1973)
TETRAHEDRON LETT 4730(1973):(FRENCH)

CRABBE PG   GARCIA A   RIUS C
PHOTOCHEMICAL CYCLOADDITIONS IN THE PROSTAGLANDIN SERIES
TETRAHEDRON LETT 2951-2954(1972)

CRAFT I
ORAL PROSTAGLANDIN E2 AND AMNIOTOMY FOR INDUCTION OF LABOR
IN: "THE USE OF PROSTAGLANDINS E2 AND F2 ALPHA IN OBSTETRICS
AND GYNAECOLOGY," RG JACOMB AND RE HARDY, SYMPOSIA
SPECIALISTS, MIAMI:25-34(1973)

CRAFT IL
AMNIOTOMY AND ORAL PROSTAGLANDIN E2 TITRATION FOR THE
INDUCTION OF LABOUR
ADV BIOSCI (SUPPL) 9:99(1972)

CRISTALLI S   VILLANI F   CHIARRA A
PICCININI F
FURTHER INVESTIGATIONS ON THE CALCIUM TRANSPORT BY PGE2 IN
RAT UTERUS
ABSTR 5TH INT CONGR PHARMACOL, SAN FRANC:48(1972)

CROSSLAND J
MODERN VIEWS ON PHARMACOLOGY. 10. ANGIOTENSIN, KININS AND
PROSTAGLANDINS
PRACTITIONER 207:567-574(1971)

CROSSLEY NS
CYCLOHEXANE ANALOGUES OF THE PROSTAGLANDINS
TETRAHEDRON LETT 3327-3330(1971)

CROWSHAW K
THE DISTRIBUTION, BIOSYNTHESIS AND RELEASE OF RENAL
PROSTAGLANDINS
J AM OIL CHEM SOC 48:93A-94A(1971)

CROWSHAW K
PROSTAGLANDIN BIOSYNTHESIS FROM ENDOGENOUS PRECURSORS IN
RABBIT KIDNEY
NATURE (NEW BIOL) 231:240-242(1971)

CROWSHAW K
RENAL PROSTAGLANDINS: LOCALIZATION OF BIOSYNTHESIS IN THE
MEDULLA
ABSTR 5TH INT CONGR PHARMACOL, SAN FRANC:48(1972)

CROWSHAW K
THE INCORPORATION OF (1-14C) ARACHIDONIC ACID INTO THE
LIPIDS OF RABBIT RENAL SLICES AND CONVERSION TO
PROSTAGLANDINS E2 AND F2A
PROSTAGLANDINS 3:607-620(1973)

CRUNDWELL E   CRIPPS AL
SYNTHESIS OF 12-HYDROXYHEPTADECA-TRANS-8-TRANS-10-DIENOIC
ACID: A BY-PRODUCT OF PROSTAGLANDIN BIOGENESIS
CHEM IND (LOND) 3 JULY:767-768(1971)

CRUNKHORN P   WILLIS AL
CUTANEOUS REACTIONS TO INTRADERMAL PROSTAGLANDINS
BR J PHARMACOL CHEMOTHER 41:49-56(1971)

CRUNKHORN P   WILLIS AL
INTERACTION BETWEEN PROSTAGLANDINS E AND F GIVEN
INTRADERMALLY IN THE RAT.
BR J PHARMACOL 41:507-512(1971)

CRUTCHLEY DJ   PIPER PJ
INHIBITION OF THE INACTIVATION OF PROSTAGLANDINS IN GUINEA
PIG LUNGS
NAUNYN SCHMIEDEBERGS ARCH PHARMACOL 279:920 (1973)

CSAPO AI
ON THE MECHANISM OF THE ABORTIFACIENT ACTION OF
PROSTAGLANDIN F2A
J REPROD MED 9:400-412(1972)

CSAPO AI   CSAPO EF
OVARIECTOMY INDUCED PLACENTAL HYPERTROPHY
PROSTAGLANDINS 4:189-200(1973)

CSEPLI J   ERDELYI A
CIRCULATORY ACTIONS OF PROSTAGLANDIN F2A IN THE RAT
BIBL ANAT (12) : 449-452(1973)

CUATRECASAS P
CHOLERA TOXIN-FAT CELL INTERACTION AND THE MECHANISM OF
ACTIVATION OF THE LIPOLYTIC RESPONSE
BIOCHEMISTRY 12:3567-1377(1973)

CUMMINGS JH   MILTON-THOMPSON GJ BILLINGS JA
NEWMAN AN   MISIEWICZ JJ
THE EFFECT OF INTRAVENOUS PROSTAGLANDIN F2A ON SMALL
INTESTINAL FUNCTION
GUT 13:854(1972)

CUMMINGS JH   NEWMAN A   MISIEWICZ JJ
MILTON-THOMPSON GJ BILLINGS JA
EFFECT OF INTRAVENOUS PROSTAGLANDIN F2A ON SMALL INTESTINAL
FUNCTION IN MAN
NATURE (LOND) 243:169-171(1973)

CUNLIFFE WJ   COTTERILL JA   WILLIAMSON B
FORSTER RA
THE RELEVANCE OF SKIN SURFACE LIPIDS TO ACNE VULGARIS
BR J DERMATOL 86(SUPPL 8):10-15(1972)

CURNOW RT   NUTTALL FQ
EFFECT OF PROSTAGLANDIN E1 (PGE1) ON GLYCOGEN METABOLISM IN
THE RAT LIVER AND HEART IN VIVO
FED PROC 30:625(1971)

CURNOW RT   NUTTALL FQ
EFFECT OF PROSTAGLANDIN E1 ADMINISTRATION ON THE LIVER
GLYCOGEN SYNTHETASE AND PHOSPHORYLASE SYSTEMS
J BIOL CHEM 247:1892-1898(1972)

CURTIS GL   ELLIOTT JA   WILSON RB
RYAN WL
CYCLIC ADENOSINE 3',5'-MONOPHOSPHATE AND CELL VOLUME
CANCER RES 33:3273-3277(1973)

CUTHBERT MF
BRONCHODILATOR ACTIVITY OF AEROSOLS OF PROSTAGLANDINS E1
AND E2 IN ASTHMATIC SUBJECTS
PROC R SOC MED 64:15-16(1971)

CUTHBERT MF
PROSTAGLANDINS AND BRONCHIAL MUSCLE
THORAX 27:263(1972)

CUTHBERT MF
PROSTAGLANDINS AND THE RESPIRATORY SYSTEM
IN: "PROSTAGLANDINES 1973," INSERM, PARIS:317-329(1973)

CUTHBERT MF
PROSTAGLANDINS AND RESPIRATORY SMOOTH MUSCLE
IN: "THE PROSTAGLANDINS. PHARMACOLOGICAL AND THERAPEUTIC
ADVANCES," MF CUTHBERT, HEINEMANN, LOND:253-285(1973)

CUTHBERT MF   SMITH AP
EFFECTS OF INHALED PROSTAGLANDINS ON AIRWAY RESISTANCE IN
MAN
ABSTR 5TH INT CONGR PHARMACOL, SAN FRANC:50(1972)

CUTHBERT MF   SMITH AP
EFFECTS OF INHALED PROSTAGLANDINS ON AIRWAYS RESISTANCE IN
MAN
ADV BIOSCI (SUPPL) 9:37(1972)

CZECH MP   FAIN JN
ANTAGONISM OF INSULIN ACTION ON GLUCOSE METABOLISM IN WHITE
FAT CELLS BY DEXAMETHASONE
ENDOCRINOLOGY 91:518-522(1972)

CZEKANOWSKI R   MOSLER KH   SCHWALM H
THE INFLUENCE OF PROSTAGLANDIN F2A ON THE CONTRACTILITY OF
THE NONPREGNANT HUMAN UTERUS IN VITRO
Z GEBURTSHILFE PERINATOL 177:202-209(1973);(GERMAN)

CZERVIONKE RL   HOAK JC   FRY GL
EFFECT OF FREE FATTY ACIDS UPON CYCLIC AMP FORMATION IN
HUMAN BLOOD PLATELETS
CIRCULATION (SUPPL II) 46:142(1972)

CZERVIONKE RL   HOAK JC   FRY GL
EFFECT OF FREE FATTY ACIDS UPON CYCLIC AMP FORMATION IN
HUMAN PLATELETS
FED PROC 32:219(1973)

D'ARMIENTO M   JOHNSON GS   PASTAN I
REGULATION OF ADENOSINE 3':5'-CYCLIC MONOPHOSPHATE
PHOSPHODIESTERASE ACTIVITY IN FIBROBLASTS BY INTRACELLULAR
CONCENTRATIONS OF CYCLIC ADENOSINE MONOPHOSPHATE
PROC NATL ACAD SCI USA 69:459-462(1972)

D'ATRI G   GALIMBERTI E   MASCARETTI L
EFFECT OF A BIOLOGICAL PREPARATION AND ITS POLYPEPTIC
FRACTIONS ON THE CONTRACTILE ACTION OF PROSTAGLANDINS
BOLL CHIM FARM 111:616-625(1972);(ITALIAN)

D'ONOFRIO F   TORELLA R   SACCA L
EFFECTS OF PROSTAGLANDIN A1 ON GLUCOSE UTILIZATION BY RAT
DIAPHRAGM IN VITRO
FARMACO (SCI) 28:992-995(1973)

DAHLSTROM A
AMINERGIC TRANSMISSION-INTRODUCTION AND SHORT REVIEW
BRAIN RES 62:441-460(1973)

DALTON C   HOPE H   SHEPPARD H
MODULATION OF ADIPOSE TISSUE LIPOLYSIS BY GUANOSINE
TRIPHOSPHATE
FED PROC 32:801(1973)

DALTON C   HOPE HR
CYCLIC AMP GENERATION AND ITS DIRECT RELATIONSHIP TO RATE
OF LIPOLYSIS IN ISOLATED FAT CELLS
CIRCULATION (SUPPL IV) 8:IV-242 (1973)

DALTON C   HOPE HR
INABILITY OF PROSTAGLANDIN SYNTHESIS INHIBITORS TO AFFECT
ADIPOSE TISSUE LIPOLYSIS
PROSTAGLANDINS 4:641-651(1973)

DAMAS J
THE SLOW-REACTING SUBSTANCES (SRS)-BIBLIOGRAPHIC REVIEW
REV MED LIEGE 28:538-544(1973);(FRENCH)

DAMAS J   BOURDON V   NEURAY J
DEBY C
BRADYKININ AND THE BIOSYNTHESIS OF PROSTAGLANDINS IN VITRO
C R SOC BIOL (PARIS) 167: 787-790 (1973) (FRENCH)

DAMAS J   GEIGER R
NATURE OF THE CARDIOVASCULAR ACTION OF BRADYKININ IN THE RAT
C R SOC BIOL (PARIS) 167:1065-1068(1973);(FRENCH)

DANFORTH DN
GYNECOLOGY AND OBSTETRICS
SURG GYNECOL OBSTET 132:221-225(1971)

DANIEL V   BOURNE HR   TOMKINS GM
ALTERED METABOLISM AND ENDOGENOUS CYCLIC AMP IN CULTURED
CELLS DEFICIENT IN CYCLIC AMP-BINDING PROTEINS
NATURE (NEW BIOL) 244:167-169 (1973)

DANIELS EG
EXTRACTION OF RENOMEDULLARY PROSTAGLANDINS
IN: "KIDNEY HORMONES," JW FISHER, ACADEMIC PRESS, LOND:
507-524(1971)

DANON A   CHANG LCT
RELEASE OF PROSTAGLANDINS FROM RAT RENAL PAPILLA IN VITRO:
EFFECTS OF ARACHIDONIC ACID AND ANGIOTENSIN II
FED PROC 32:788(1973)

DANON A   CHANG LCT   NIES AS
OATES JA
RENAL PROSTAGLANDINS: EFFECTS OF VASOPRESSIN AND DIBUTYRYL
CYCLIC AMP IN VITRO
CLIN RES 21:683(1973)

DAO HAI N   VARGAFTIG BB
PHARMACOLOGY, MODE OF ACTION AND ANTAGONISM BY
ANTIINFLAMMATORY AGENTS OF "SLOW REACTING SUBSTANCE C"
RELEASED FROM EGG YOLK BY PHOSPHOLIPASE A
J PHARMACOL 2:220-221(1971)

DAO HAI N   VARGAFTIG BB
RELEASE OF VASOACTIVE SUBSTANCES FROM GUINEA-PIG LUNGS BY
SLOW REACTING SUBSTANCE C AND ARACHIDONIC ACID: BLOCKADE BY
NONSTEROID ANTIINFLAMMATORY AGENTS
J PHARMACOL 2:248-249(1971)

DAS A
PROSTAGLANDINS "THE WONDER DRUGS"
EAST PHARM 16(184):17-21(1973)

DAUGHERTY RM JR
EFFECTS OF IV AND IA PROSTAGLANDIN E1 ON DOG FORELIMB SKIN
AND MUSCLE BLOOD FLOW
AM J PHYSIOL 220:392-396(1971)

DAVIES BN   WITHRINGTON PG
ACTIONS OF PROSTAGLANDIN F2A ON THE SPLENIC VASCULAR
CAPSULAR AND SMOOTH MUSCLE IN THE DOG
BR J PHARMACOL CHEMOTHER 41:1-7(1971)

DAVIES BN   WITHRINGTON PG
THE ACTIONS OF DRUGS ON THE SMOOTH MUSCLE OF THE CAPSULE
AND BLOOD VESSELS OF THE SPLEEN
PHARMACOL REV 25:373-413(1973)

DAVIS HA
OUTPUT OF PROSTAGLANDINS FROM THE RABBIT KIDNEY
ADV BIOSCI (SUPPL) 9:55(1972)

DAVIS HA   HORTON EW
OUTPUT OF PROSTAGLANDINS FROM THE RABBIT KIDNEY, ITS
INCREASE ON RENAL NERVE STIMULATION AND ITS INHIBITION BY
INDOMETHACIN
BR J PHARMACOL 46:658-675(1972)

DAVIS HA   HORTON EW   JONES KB
QUILLIAM JP
IDENTIFICATION OF PROSTAGLANDINS IN PREVERTEBRAL VENOUS
BLOOD AFTER PREGANGLIONIC STIMULATION OF THE CAT SUPERIOR
CERVICAL GANGLION
BR J PHARMACOL 42:569-583(1971)

DAVISON P   RAMWELL PW   WILLIS AL
INHIBITION OF INTESTINAL TONE AND PROSTAGLANDIN SYNTHESIS
BY 5, 8, 11, 14 TETRAYNOIC ACID
BR J PHARMACOL 46:547P-548P(1972)

DAWSON MJ
THE ROLE OF CYCLIC AMP IN THE RESPIRATORY RESPONSE OF
SKELETAL MUSCLE TO CONTRACTURE-INDUCING AGENTS
IN: "PROSTAGLANDINS AND CYCLIC AMP," RH KAHN, WEM LANDS,
ACADEMIC PRESS, NY : 257-260(1973)

DAWSON MJ
THE ROLE OF CYCLIC AMP IN THE RESPIRATORY RESPONSE OF
SKELETAL MUSCLE TO CONTRACTURE-INDUCING AGENTS
DISS, UNIV PA(1973)

DAWSON WN JR   WHITE RP   ROBERTSON JT
PHARMACOLOGIC DETERMINATION OF THE ROLE OF AUTONOMIC
RECEPTOR SITES IN CEREBRAL VASOSPASM
SURG FORUM 24:455-457(1973)

DE ASUA LJ   SURIAN ES   FLAWIA MM
TORRES HN
EFFECT OF INSULIN ON THE GROWTH PATTERN AND ADENYLATE
CYCLASE ACTIVITY OF BHK FIBROBLASTS
PROC NATL ACAD SCI USA 70:1388-1392 (1973)

DE ATENOR MSB   BRAUCKMAN ES   COVIELLO A
EFFECT OF INHIBITORS OF ADENYL CYCLASE ON THE HYDROOSMOTIC
PRESSURE OF TOAD SKIN BY ANGIOTENSIN II AND VASOPRESSIN
ACTA PHYSIOL LAT AM 23:ABSTR 258(1973);(SPANISH)

DE GAETANO G   VERMYLEN J   VERSTRAETE M
INHIBITION OF PLATELET AGGREGATION: EXPERIMENTAL FINDINGS
AND CLINICAL PERSPECTIVES
NOUV REVUE FR HEMATOL 11:339-364(1971);(FRENCH)

DE GIMENO MF   LIMA F   BORDA E
GIMENO AL   BEDNERS AS
SPONTANEOUS AND PROSTAGLANDIN E1-INDUCED MOTILITY IN THE
ISOLATED MYOMETRIUM OF SPAYED RATS AND RATS IN HEAT. EFFECTS
OF POLYPHLORETIN PHOSPHATE, A PROSTAGLANDIN INHIBITOR
MEDICINA (B AIRES) 31:529-530(1971);(SPANISH)

DE GUIA D   MENDLOWITZ M   STRICKER J
RUSSO C
EFFECTS OF THREE DIRECT VASCULAR SMOOTH MUSCLE INHIBITORS IN
ESSENTIAL HYPERTENSION
CLIN PHARMACOL THER 14:133(1973)

DE GUIA D   STRICKER J   KRAKOFF LR
MODIFICATION OF NATRIURETIC EFFECT OF PROSTAGLANDIN A1
(PGA1) BY SODIUM DEPLETION
CLIN RES 20:591(1972)

DE HEMPTINNE D   SCHUDDINCK L   THIERY M
MARTENS G
NEONATAL BILIRUBINAEMIA-EFFECT OF OXYTOCIC COMPOUNDS
(OXYTOCIN AND PROSTAGLANDINS) AND VACUUM EXTRACTION
INT RES COMMUN SYST (73-12)10-14-3(1973)

DE MOOR P   BOUILLON R   VERHOEVEN G
BASIC DATA ON CALCIUM METABOLISM
ACTA CLIN BELG 28:323-357(1973);(DUTCH)

DE PURY GG   COLLINS FD
VERY LOW DENSITY LIPOPROTEINS AND LIPOPROTEIN LIPASE IN
SERUM OF RATS DEFICIENT IN ESSENTIAL FATTY ACIDS
J LIPID RES 13:268-275(1972)

DE-CICCO A
PHYSIOLOGICAL ACTION OF POLYUNSATURATED FATTY ACIDS
ACTA VITAMINOL ENZYMOL 25:13-23(1971);(ITALIAN)

DEBOER J   HOUTSMULLER UMT   VERGROESEN AJ
INOTROPIC EFFECTS OF PROSTAGLANDINS, FATTY ACIDS AND
ADENOSINE PHOSPHATES ON HYPODYNAMIC FROG HEARTS
PROSTAGLANDINS 3:805-825(1973)

DEBUCH H
LIPIDO-COMPOSITION OF LIPOPROTEINS
ANN BIOL CLIN(PARIS) 31:65-67(1973)

DEBY C   BACQ ZM
ACTION OF INDOMETHACIN, A PROSTAGLANDIN BIOSYNTHESIS
INHIBITOR, ON THE LIPIDS OF THE RAT
C R SOC BIOL (PARIS) 166:750-753(1972);(FRENCH)

DEBY C   BACQ ZM   SIMON D
IN VITRO INHIBITION OF THE BIOSYNTHESIS OF A PROSTAGLANDIN
BY GOLD AND SILVER
BIOCHEM PHARMACOL 22:3141-3143(1973)

DEBY C   MAGOTTEAUX G
MEASUREMENT OF PROSTAGLANDINS BY GAS CHROMATOGRAPHY AND
DETECTION OF ELECTRON CAPTURE
ARCH INT PHYSIOL BIOCHIM 79:824-825(1971);(FRENCH)

DEBY C   MAGOTTEAUX G   BARAC G
RENAL BLOOD OUTPUT OF ENDOGENOUS SMOOTH MUSCLE ACTING
PROSTAGLANDINS IN THE DOG
ARCH INT PHYSIOL BIOCHIM 79: 798-799(1971)

DEMERS LM   BEHRMAN HR   GREEP RO
EFFECTS OF PROSTAGLANDINS AND GONADOTROPHINE ON LUTEAL
PROSTAGLANDIN AND STEROID BIOSYNTHESIS
ADV BIOSCI (SUPPL) 9:114(1972)

DEMOULIN A   THIEBLOT P   FRANCHIMONT P
EFFECT OF VARIOUS STEROIDS AND PROSTAGLANDIN E1 ON SERUM
LEVELS OF GONADOTROPHINS IN THE MALE CASTRATE RAT
C R SOC BIOL (PARIS) 167:1684-1687(1973);(FRENCH)

DENKO CW
PROSTAGLANDINS, ESSENTIAL FATTY ACIDS, AND 35S INCORPORATION
INTO CARTILAGE
ARTHRITIS RHEUM 14(3):379(1971)

DENTON IC JR   WHITE RP   ROBERTSON JT
THE EFFECTS OF PROSTAGLANDINS E1, A1, AND F2A ON THE
CEREBRAL CIRCULATION OF DOGS AND MONKEYS
J NEUROSURG 36:34-42(1972)

DERUBERTIS FR   ADLER WH   ZENSER TV
HUDSON T
EFFECTS OF MITOGENS LYMPHOCYTE ADENYL CYCLASE ACTIVITY (AC),
CYCLIC-AMP CONTENT (CA) AND DNA SYNTHESIS
CLIN RES 21:57(1973)

DERUBERTIS FR   ZENSER TV   CURNOW RT
FERRARIS VA
PROSTAGLANDIN INHIBITION OF GLUCAGON MEDIATED INCREASES IN
HEPATIC CYCLIC-AMP IN VIVO
CLIN RES 21:620(1973)

DESIDERIO DM
CHEMICAL IONIZATION MASS SPECTROMETRY OF PROSTAGLANDINS
INTRA-SCI CHEM REP 6:97-98(1972)

DESIDERIO DM   HAGELE K
CHEMICAL IONIZATION MASS SPECTROMETRY OF PROSTAGLANDINS
CHEM COMMUN:1074-1075(1971)

DESIRAJU T
OBSERVATION OF MODIFICATION OF BRAIN FUNCTION BY INCREASING
PROSTAGLANDINS IN THE CSF OF CEREBRAL VENTRICLES
ADV BIOSCI (SUPPL) 9:81(1972)

DESIRAJU T
EFFECT OF INTRAVENTRICULARLY ADMINISTERED PROSTAGLANDIN E1
ON THE ELECTRICAL ACTIVITY OF CEREBRAL CORTEX AND BEHAVIOR
IN THE UNANESTHETIZED MONKEY
PROSTAGLANDINS 3:859-870(1973)

DESOLE E   TOSOLINI GC
PROSTAGLANDINS AND REPRODUCTION
FRIULI MED 27:463-473(1972):(ITALIAN)

DESSY F   MALEUX MR   COGNIOUL A
BRONCHOSPASMOLYTIC ACTIVITY OF SOME PROSTAGLANDINS IN THE
GUINEA-PIG
ARCH INT PHARMACODYN THER 206: 368-370 (1973)

DHALLA NS   BALASUBRAMANIAN V
EFFECT OF PROSTAGLANDINS AND CYCLIC AMP ON 3H-NOREPINEPHRINE
TRANSPORT ACROSS ADRENERGIC NEURONS IN HEART
ABSTR 5TH INT CONGR PHARMACOL, SAN FRANC:57(1972)

DHALLA NS   SULAKHE PV   MCNAMARA DB
STUDIES ON THE RELATIONSHIP BETWEEN ADENYLATE CYCLASE
ACTIVITY AND CALCIUM TRANSPORT BY CARDIAC SARCOTUBULAR
MEMBRANES
BIOCHIM BIOPHYS ACTA 323:276-284(1973)

DI ROSA M
BIOLOGICAL PROPERTIES OF CARRAGEENAN
J PHARM PHARMACOL 24:89-102(1972)

DI ROSA M   GIROUD JP   WILLOUGHBY DA
STUDIES OF THE MEDIATORS OF THE ACUTE INFLAMMATORY RESPONSE
INDUCED IN RATS IN DIFFERENT SITES BY CARRAGEENIN AND
TURPENTINE
J PATHOL 104:15-29(1971)

DI ROSA M   PAPADIMITRIOU JM   WILLOUGHBY DA
A HISTOPATHOLOGICAL AND PHARMACOLOGICAL ANALYSIS OF THE
MODE OF ACTION OF NON-STEROIDAL ANTI-INFLAMMATORY DRUGS
J PATHOL 105:239-256(1971)

DI ROSA M   WILLOUGHBY DA
MEDIATORS OF INCREASED VASCULAR PERMEABILITY AND CELLULAR
EMIGRATION IN CARRAGEENIN FOOT OEDEMA
NAUNYN-SCHMIEDEBERGS ARCH PHARMAKOL 269:482(1971)

DI ROSA M   WILLOUGHBY DA
SCREENS FOR ANTI-INFLAMMATORY DRUGS.
J PHARM PHARMACOL 23:297-298(1971)

DIAMOND EJ   MARTIN CR   MONDER C
EFFECT OF PROSTAGLANDIN AND INDOMETHACIN ON
CORTICOSTEROIDOGENESIS IN THE ISOLATED PERFUSED RAT ADRENAL
PHYSIOLOGIST 15:119(1972)

DICKSON RC   DOERY JCG   LEWIS AF
ULTRAVIOLET LIGHT: A NEW STIMULUS FOR THE INDUCTION OF
PLATELET AGGREGATION
SCIENCE 172:1140-1142(1971)

DIKSTEIN S
METABOLIC REQUIREMENTS FOR FLUID TRANSFER THROUGH THE
CORNEAL ENDOTHELIUM
EXP EYE RES 12:371-372(1971)

DILAWARI JB   NEWMAN A   POLEO J
MISIEWICZ JJ
THE EFFECT OF PROSTAGLANDINS AND OF ANTIINFLAMMATORY DRUGS
ON THE OESOPHAGUS AND THE CARDIAC SPHINCTER IN MAN
GUT 14:822(1973)

DILAWARI JB   POLEO JR   NEWMAN A
MISIEWICZ JJ
RESPONSE OF THE HUMAN OESOPHAGUS TO INTRAVENOUS
PROSTAGLANDIN F2A AND E2
REND ROM GASTROENTEROL 5:139-140(1973)

DIMOV V   GREEN K
THE METABOLISM OF PROSTAGLANDIN F3A IN THE RAT
BIOCHIM BIOPHYS ACTA 306:257-269(1973)

DINNENDAHL V   PETERS HD   SCHONHOFER PS
EFFECTS OF ANTILIPOLYTIC DRUGS ON CYCLIC 3',5'-AMP
DEPENDENT PROTEIN KINASE
NAUNYN SCHMIEDEBERGS ARCH PHARMACOL 278:293-300(1973)

DIPASQUALE G   RASSAERT C   RICHTER R
WELAJ P   TRIPP L
INFLUENCE OF PROSTAGLANDINS (PG) E2 AND F2A ON THE
INFLAMMATORY PROCESS
PROSTAGLANDINS 3:741-757(1973)

DIRKS JH   LENNHOFF M   WONG NLM
PROSTAGLANDINS - A POSSIBLE ROLE IN THE REDUCED PROXIMAL
REABSORPTION OF CONTRALATERAL KIDNEY CLAMPING
CLIN RES 20:592(1972)

DISTLER A   GROBECKER H   KREYE VAW
LAZAR J
PHYSIOLOGY AND PATHOLOGY OF VASCULAR RESPONSE
ARZNEIM FORSCH 23:18-30(1973)

DJERASSI C
STEROID CONTRACEPTIVES IN THE PEOPLE'S REPUBLIC OF CHINA
N ENGL J MED 289: 533-535 (1973)

DOBBS JW
STUDIES ON THE RELATIONSHIP OF CYCLIC AMP TO UTERINE
ADRENERGIC RESPONSES
DISS, VANDERBILT UNIV, NASHV (1971)

DOBRIN EI   BLOSS JL   POTTS WJ
NEUROPHARMACOLOGICAL AND BEHAVIORAL TOXICITY STUDIES WITH
PROSTAGLANDIN A2
TOXICOL APPL PHARMACOL 25:460-461(1973)

DONALD I
ABORTION AND THE OBSTETRICIAN
LANCET 1:1233(1971)

DONNELLY CH   MURPHY DL   MOSKOWITZ J
INHIBITION BY LITHIUM OF THE EFFECTS OF PROSTAGLANDIN E1
(PGE1) AND NOREPINEPHRINE (NE) ON CYCLIC AMP FORMATION IN
HUMAN PLATELETS
FED PROC 32:744(1973)

DORIA C   GAIO P   GANDOLFI C
PROSTAGLANDINS III: A MODIFIED ROUTE TO DL-PG1-SERIES FROM A
COREY'S INTERMEDIATE
TETRAHEDRON LETT 4307-4310(1972)

DORIGO P   MARAGNO I   BRESSA A
FASSINA G
REDUCED LIPOLYTIC RESPONSE IN VITRO TO CATECHOLAMINES, ACTH
AND CYCLIC ADENOSINE MONOPHOSPHATE IN BROWN FAT OF
COLD-ACCLIMATIZED RATS
BIOCHEM PHARMACOL 20:1201-1211(1971)

DOUGLAS JR JR   JOHNSON EM JR   MARSHALL GR
JAFFE BM   NEEDLEMAN P
STIMULATION OF SPLENIC PROSTAGLANDIN RELEASE BY ANGIOTENSIN
AND SPECIFIC INHIBITION BY CYSTEINE8-AII
PROSTAGLANDINS 3:67-74(1973)

DOUGLAS JR JR   MARSHALL GR   NEEDLEMAN P
BURTON RM
DISSOCIATION OF LIPOLYSIS FROM PROSTAGLANDIN-LIKE SUBSTANCE
RELEASE IN RABBIT SPLENIC FAT PAD
FED PROC 32:804(1973)

DOUGLAS JR JR   MINKES MS   NEEDLEMAN P
ORGAN PROFILE OF PROSTAGLANDIN RELEASE INDUCED BY ADENINE
NUCLEOTIDES AND ISCHEMIA
PHARMACOLOGIST 15:250(1973)

DOUSA TP
ROLE OF CYCLIC AMP IN THE ACTION OF ANTIDIURETIC HORMONE ON
KIDNEY
LIFE SCI 13:1033-1040(1973)

DOWNING DT
DIFFERENTIAL INHIBITION OF PROSTAGLANDIN SYNTHETASE AND
SOYBEAN LIPOXYGENASE
PROSTAGLANDINS 1:437-441(1972)

DOWNING DT   BARVE JA   GUNSTONE FD
JACOBSBERG FR   JIE MLK
STRUCTURAL REQUIREMENTS OF ACETYLENIC FATTY ACIDS FOR
INHIBITION OF SOYBEAN LIPOXYGENASE AND PROSTAGLANDIN
SYNTHETASE
BIOCHIM BIOPHYS ACTA 280:343-347(1972)

DOZI-VASSILIADES J KOVATSIS A
ACTION OF PROSTAGLANDIN E1 AND F1ALPHA ON RABBIT'S ISOLATED
INTESTINE IN THE PRESENCE OF ADP
J PHARM BELG 28:202-208(1973)

DRAY F   CHARBONNEL B
RADIOIMMUNOLOGIC DOSAGE OF PROSTAGLANDINS FA AND E1 IN THE
PERIPHERAL PLASMA OF NORMAL HUMANS
IN: "PROSTAGLANDINES 1973," INSERM, PARIS:133-158(1973):
(FRENCH)

DRAY F   MARON E   TILLSON SA
SELA M
IMMUNOCHEMICAL DETECTION OF PROSTAGLANDINS WITH
PROSTAGLANDIN-COATED BACTERIOPHAGE T4 AND BY
RADIOIMMUNOASSAY
ANAL BIOCHEM 50:399-408(1972)

DROLLER MJ
ULTRASTRUCTURE OF THE PLATELET RELEASE REACTION IN RESPONSE
TO VARIOUS AGGREGATING AGENTS AND THEIR INHIBITORS
LAB INVEST 29:595-606(1973)

DROLLER MJ
ULTRASTRUCTURAL VISUALIZATION OF THE THROMBIN-INDUCED
PLATELET RELEASE REACTION
SCAND J HAEMATOL 11:35-49(1973)

DROLLER MJ   WOLFE SM
THROMBIN-STIMULATED INCREASE OF CYCLIC AMP IN HUMAN
PLATELETS
BLOOD 38:791(1971)

DROLLER MJ   WOLFE SM
UNCERTAIN ROLE OF CAMP IN PLATELET FUNCTION
N ENGL J MED 286:948-949(1972)

DROSZCZ W
ADVANCES IN PULMONOLOGY. LUNG--AN ORGAN OF IMPORTANT
METABOLIC PROCESSES
POL ARCH MED WEWN 50:1245-1248(1973):(POLISH)

DUAX WL   EDMONDS JW
MOLECULAR CONFORMATION OF PROSTAGLANDIN A1
PROSTAGLANDINS 3:201-208(1973)

DUFFUS CM   DUFFUS JH   ROSIE B
ALPHA-AMYLASE AND ITS RELEASE BY PROSTAGLANDIN F2A IN
BARLEY ENDOSPERM SLICES
EXPERIENTIA 29:952(1973)

DUKES PP
POTENTIATION OF ERYTHROPOIETIN EFFECTS IN MARROW CELL-
CULTURES BY PROSTAGLANDIN-E1 OR CYCLIC 3', 5'-AMP.
BLOOD 38:822(1971)

DUKES PP
MODULATING EFFECTS OF ERYTHROPOIETIN, PROSTAGLANDINS AND
CYCLIC 3',5'-NUCLEOTIDES ON ERYTHROID DIFFERENTIATION
IN MARROW CELL CULTURES
FED PROC 31:487(1972)

DUKES PP
ERYTHROPOIETIC EFFECTS OF PROSTAGLANDINS
ADV BIOSCI (SUPPL) 9:29(1972)

DUKES PP
POTENTIATION OF ERYTHROPOIETIN EFFECTS IN MARROW CELL
CULTURES BY PROSTAGLANDIN E1 OR CYCLIC 3',5'-AMP
INTRA-SCI CHEM REP 6:73-75(1972)

DUKES PP
ERYTHROPOIETIC EFFECTS OF PROSTAGLANDINS
ADV BIOSCI 9:183-188(1973)

DUKES PP   SHORE NA   HAMMOND D
ORTEGA JA   DATTA MC
ENHANCEMENT OF ERYTHROPOIESIS BY PROSTAGLANDINS
J LAB CLIN MED 82:704-712(1973)

DUMONT JE   WILLEMS C   VAN SANDE J
NEVE P
REGULATION OF THE RELEASE OF THYROID HORMONES: ROLE OF
CYCLIC AMP
ANN N Y ACAD SCI 185:291-316(1971)

DUNHAM E   ZIMMERMAN BG
EFFECT OF THEOPHYLLINE ON PROSTAGLANDIN INDUCED
VASODILATATION
PHARMACOLOGIST 13:292(1971)

DUNHAM EW
STUDIES ON THE RELEASE AND VASCULAR EFFECTS OF
PROSTAGLANDINS
DISS, UNIV MINN, MINNEAP (1971)

DUNHAM EW   ANDERS MW
HIGH-SPEED LIQUID CHROMATOGRAPHIC ANALYSIS OF PROSTAGLANDINS
IN RAT KIDNEY
PROSTAGLANDINS 4:85-92(1973)

DUNHAM EW   HADDOX MK   GOLDBERG ND
ELEVATION OF CYCLIC GMP LEVELS IN ISOLATED VEINS DURING
PROSTAGLANDIN F2A INDUCED VENOCONSTRICTION
PHARMACOLOGIST 15:158(1973)

DUPONT A   CHAVANCY G
PROSTAGLANDINS AND CYCLIC AMP AS MEDIATORS OF
THYROTROPIN-RELEASING HORMONE ACTION
ABSTR 4TH INT CONGR ENDOCRINOL, INT CONGR SER 256, EXCERPTA
MEDICA, AMST:84(1972)

DUPONT A   CHAVANCY G
PROSTAGLANDINS AND CYCLIC AMP AS MEDIATORS OF
THYROTROPIN-RELEASING HORMONE
PROC CAN FED BIOL SOC 15:721(1972)

DUPONT A   CHAVANCY G   BORGEAT P
LABRIE F
ROLE OF CYCLIC AMP AND OF PROSTAGLANDINS AS MEDIATORS OF THE
ACTION OF TRH
UNION MED CAN 102:1580(1973)::(FRENCH)

DUPONT A   CHAVANCY G   LABRIE F
PROSTAGLANDINS AS MEDIATORS OF OF THYROTROPIN-RELEASING
HORMONE ACTION
ADV BIOSCI (SUPPL) 9:34(1972)

DUSLEAG LD   VASILIU V
PROSTAGLANDINS
OBSTET GINECOL (BUCHAR) 20:1-34(1972)::(RUMANIAN)

EAGLING EM   LOVELL J   PICKLES VR
PROSTAGLANDINS, MYOMETRIAL "ENHANCEMENT" AND CALCIUM.
J PHYSIOL (LOND) 213 : 53P-54P (1971)

EAKINS K
PROSTAGLANDIN ANTAGONISM BY POLYMERIC PHOSPHATES OF
PHLORETIN AND RELATED COMPOUNDS.
ANN NY ACAD SCI 180:386-395(1971)

EAKINS KE
OCULAR EFFECTS
IN: "THE PROSTAGLANDINS," PW RAMWELL, PLENUM PRESS, NY
1:219-237(1973)

EAKINS KE
RELEASE OF PROSTAGLANDIN-LIKE ACTIVITY IN OCULAR
INFLAMMATION
IN: "PROSTAGLANDINS AND CYCLIC AMP," RH KAHN, WEM LANDS,
ACADEMIC PRESS, NY : 211-212(1973)

EAKINS KE   FEX H   FREDHOLM B
HOGBERG B   VEIGE S
ON THE PROSTAGLANDIN INHIBITORY ACTION OF POLYPHLORETIN
ADV BIOSCI (SUPPL) 9:19(1972)

EAKINS KE   FEX H   FREDHOLM B
HOGBERG B   VEIGE S
ON THE PROSTAGLANDIN INHIBITORY ACTION OF POLYPHLORETIN
PHOSPHATE
ADV BIOSCI 9:135-138(1973)

EAKINS KE   MILLER JD   KARIM SMM
THE NATURE OF THE PROSTAGLANDIN-BLOCKING ACTIVITY OF
POLYPHLORETIN PHOSPHATE
J PHARMACOL EXP THER 176:441-447(1971)

EAKINS KE   SANNER JH
PROSTAGLANDIN ANTAGONISTS
IN: "THE PROSTAGLANDINS: PROGRESS IN RESEARCH," SMM KARIM,
MTP MED TECH PUBL CO, OXFORD:263-292(1972)

EAKINS KE   WHITELOCKE RA   PERKINS ES
BENNETT A   UNGER WG
RELEASE OF PROSTAGLANDINS INTO THE AQUEOUS HUMOUR IN
EXPERIMENTAL IMMUNOGENIC UVEITIS
EXP EYE RES 14:174-175(1972)

EAKINS KE   WHITELOCKE RAF   BENNETT A
MARTENET AC
PROSTAGLANDIN-LIKE ACTIVITY IN OCULAR INFLAMMATION
BR MED J 3:452-453(1972)

EAKINS KE   WHITELOCKE RAF   PERKINS ES
BENNETT A   UNGER WG
RELEASE OF PROSTAGLANDINS (PGS) INTO THE AQUEOUS HUMOUR IN
EXPERIMENTAL IMMUNOGENIC UVEITIS
ADV BIOSCI (SUPPL) 9:64(1972)

EAKINS KE   WHITELOCKE RAF   PERKINS ES
BENNETT A   UNGER WG
RELEASE OF PROSTAGLANDINS IN OCULAR INFLAMMATION IN THE
RABBIT
NATURE (NEW BIOL) 239:248-249(1972)

EAKINS KE   WHITELOCKE RAF   PERKINS ES
BENNETT A   UNGER WG
PROSTAGLANDIN RELEASE IN OCULAR INFLAMMATION IN RABBITS
AND. MAN
ADV BIOSCI 9:427-433(1973)

EAST PF   POTTS WJ
EFFECT OF PROSTAGLANDIN E2 ON A THREE CHOICE ODDITY
DISCRIMINATION TASK IN THE MONKEY
FED PROC 31:546(1972)

ECHT M   GAUER OH   KAPTEINA F
LANGE L
EFFECTS OF PROSTAGLANDIN E2 AND F2A ON CUTANEOUS ARTERIES
AND VEINS IN VIVO IN THE ISOLATED RABBIT EAR
ADV BIOSCI (SUPPL)9:158(1972)

ECHT M   GAUER OH   KAPTEINA F
LANGE L
EFFECTS OF PROSTAGLANDIN E2 AND F2A ON CUTANEOUS ARTERIES
AND VEINS IN VIVO IN THE ISOLATED RABBIT EAR
ADV BIOSCI 9:353-358(1973)

ECKENFELS A   VANE JR
PROSTAGLANDINS, OXYGEN TENSION AND SMOOTH MUSCLE TONE
BR J PHARMACOL 45:451-462(1972)

EHINGER B
LOCALIZATION OF THE UPTAKE OF PROSTAGLANDIN E1 IN THE EYE
EXP EYE RES 17:43-47(1973)

EHRENPREIS S   GREENBERG J
BLOCK OF CONTRACTIONS OF GUINEA PIG ILEUM BY MORPHINE,
INDOMETHACIN AND ACETYLSALICYLIC ACID INVOLVES INHIBITION
OF PROSTAGLANDIN CONTROLLED ACETYLCHOLINE RELEASE
FED PROC 32:788(1973)

EHRENPREIS S   GREENBERG J   BELMAN S
PROSTAGLANDINS REVERSE INHIBITION OF ELECTRICALLY-INDUCED
CONTRACTIONS OF GUINEA PIG ILEUM BY MORPHINE, INDOMETHACIN
AND ACETYLSALICYLIC ACID
NATURE(NEW BIOL) 245:280-282(1973)

EICHBERG J   SHEIN HM   SCHWARTZ M
HAUSER G
STIMULATION OF 32P-I INCORPORATION INTO PHOSPHATIDYLINOSITOL
AND PHOSPHATIDYLGLYCEROL BY CATECHOLAMINES AND BETA
ADRENERGIC RECEPTOR BLOCKING AGENTS IN RAT PINEAL ORGAN
CULTURES
J BIOL CHEM 248: 3615-3622 (1973)

EIK-NES K
PRODUCTION AND SECRETION OF TESTICULAR STEROIDS
RECENT PROG HORM RES 27:517-535(1971)

ELIAS JA
EXPERIENCE WITH PROSTAGLANDINS E2 AND F2A FOR INDUCTION OF
LABOR
J REPROD MED 9:307-310(1972)

ELIAS JA
PROSTAGLANDIN F2A VAGINAL PESSARIES FOR MID-TRIMESTER
ABORTION
ADV BIOSCI (SUPPL)9:96(1972)

ELIASSON R
PROSTAGLANDINS AND REPRODUCTION: A GENERAL SURVEY
J REPROD FERTIL (SUPPL) 18:127-132 (1973)

ELING TE   PARKES DG
STUDIES ON BIOSYNTHESIS OF PROSTAGLANDINS BY GUINEA PIG LUNG
PHARMACOLOGIST 15:209(1973)

ELING TE   PARKES DG
STUDIES ON THE BIOSYNTHESIS OF PROSTAGLANDINS BY GUINEA PIG
LUNG
ABSTR, 9TH INT CONG BIOCHEM, STOCKH:398(1973)

ELLIOTT DW
THE NORMAL PANCREAS-PHYSIOLOGY
IN:THE PANCREAS,LC CAREY,MOSBY,ST LOUIS:32-57(1973)

ELLIOTT RB
PROSTAGLANDINS AND FIBROCYSTIC DISEASE
NZ MED J 74:394(1971)

ELLIOTT RB   STARLING MB
THE EFFECT OF PROSTAGLANDIN F2A IN THE CLOSURE OF THE
DUCTUS ARTERIOSUS
PROSTAGLANDINS 2:399-403(1972)

ELLIS LC   JOHNSON JM   HARGROVE JL
CELLULAR ASPECTS OF PROSTAGLANDIN SYNTHESIS AND TESTICULAR
FUNCTION
IN: "PROSTAGLANDINS IN CELLULAR BIOLOGY," PW RAMWELL AND B
B PHARRISS, PLENUM PRESS, NY/LOND:385-398(1972)

ELSLAGER EF   MCLEAN JR   PERRICONE SC
POTOCZAK D   VELOSO H   WORTH DF
WHEELOCK RH
INHIBITORS OF PLATELET AGGREGATION. 1.
5,10-DIHYDRO-3-(PHENYL,THIENYL, AND FURYL)
THIAZOLO(3,2-B)(2,4) BENZODIAZEPINES AND RELATED COMPOUNDS
J MED CHEM 14(5):397-401(1971)

EMBREY MP
PROSTAGLANDINS.
PROC R SOC MED 64:1018-1020(1971)

EMBREY MP
THE USE OF PROSTAGLANDINS FOR THERAPEUTIC TERMINATION OF
PREGNANCY
IN:"THE USE OF PROSTAGLANDINS E2 AND F2ALPHA IN OBSTETRICS
AND GYNAECOLOGY," RG JACOMB AND RE HARDY, SYMPOSIA
SPECIALISTS, MIAMI:69-79(1973)

EMBREY MP   HILLIER K
EXTRA-AMNIOTIC PROSTAGLANDIN ADMINISTRATION FOR THE
INDUCTION OF ABORTION
ADV BIOSCI (SUPPL) 9:83(1972)

EMBREY MP   HILLIER K
THERAPEUTIC ABORTION BY EXTRA-AMNIOTIC ADMINISTRATION OF
PROSTAGLANDINS
J REPROD MED 9:420-424(1972)

EMBREY MP   HILLIER K   MAHENDRAN P
TERMINATION OF PREGNANCY BY EXTRAAMNIOTIC PROSTAGLANDINS AND
THE SYNERGISTIC ACTION OF OXYTOCIN
ADV BIOSCI 9:507-513(1973)

EMERSON TE JR   JELKS GW   DAUGHERTY RM JR
HODGMAN RE
EFFECTS OF PROSTAGLANDIN E1 AND F2A ON VENOUS RETURN AND
OTHER PARAMETERS IN THE DOG
AM J PHYSIOL 220:243-249(1971)

EMMELOT P
BIOCHEMICAL PROPERTIES OF NORMAL AND NEOPLASTIC CELL
SURFACES: A REVIEW
EUR J CANCER 9:319-333(1973)

EMORI KH
AGGREGATION OF NON-MAMMALIAN HEMOSTATIC CELLS AND ADENOSINE
3',5'-MONOPHOSPHATE
DISS, BOSTON UNIV(1973)

ENGEL JJ   SCRUGGS W   WILSON DE
FAILURE OF SC-19220 TO AFFECT PROSTAGLANDIN E1 (PGE1)
GASTRIC ANTISECRETORY ACTIONS
PROSTAGLANDINS 4:65-70(1973)

ENGEL JJ   WILSON DE
FAILURE OF A PROSTAGLANDIN INHIBITOR (SC-19220) TO AFFECT
PROSTAGLANDIN E1 (PGE1) GASTRIC ANTISECRETORY ACTIONS
CLIN RES 21:511(1973)

ENGELBRECHT JA   GREENBERG S   WILSON WR
MARCHAND G   WILLIAMSON HE
CARDIOVASCULAR AND RENAL PHARMACOLOGY OF PROSTAGLANDIN B2
(PGB2)
FED PROC 32:787(1973)

EPSTEIN SE   LEVEY GS   SKELTON CL
ADENYL CYCLASE AND CYCLIC AMP. BIOCHEMICAL LINKS IN THE
REGULATION OF MYOCARDIAL CONTRACTILITY
CIRCULATION 43:437-450(1971)

ERICSSON RJ
PROSTAGLANDINS (E1 AND E2) AND REPRODUCTION IN THE MALE RAT
ADV BIOSCI (SUPPL) 9:120(1972)

ERSPAMER GF   NEGRI L   PICCINELLI D
THE USE OF PREPARATIONS OF URINARY BLADDER SMOOTH MUSCLE FOR
BIOASSAY OF AND DISCRIMINATION BETWEEN POLYPEPTIDES
NAUNYN SCHMIEDEBERGS ARCH PHARMACOL 279:61-74(1973)

ERSPAMER V   ERSPAMER GF   INSELVINI M
NEGRI L
OCCURRENCE OF BOMBESIN AND ALYTESIN IN EXTRACTS OF THE SKIN
OF THREE EUROPEAN DISCOGLOSSID FROGS AND PHARMACOLOGICAL
ACTIONS OF BOMBESIN ON EXTRAVASCULAR SMOOTH MUSCLE
BR J PHARMACOL 45:333-348(1972)

EXTON JH   ROBISON GA   SUTHERLAND EW
PARK CR
STUDIES ON THE ROLE OF ADENOSINE 3',5'-MONOPHOSPHATE IN THE
HEPATIC ACTIONS OF GLUCAGON AND CATECHOLAMINES.
J BIOL CHEM 246:6166-6177(1971)

FABIYI JA   OKPAKO DT
ON THE RESPONSES OF THE ISOLATED RECTUM OF THE RAINBOW
LIZARD TO DRUGS AND ELECTRICAL FIELD STIMULATION
COMP GEN PHARMACOL 4:297-303(1973)

FAIN JN
INHIBITION OF ADENOSINE CYCLIC 3',5'-MONOPHOSPHATE
ACCUMULATION IN FAT CELLS BY ADENOSINE, N6-(PHENYLISOPROPYL)
ADENOSINE, AND RELATED COMPOUNDS
MOL PHARMACOL 9:595-604(1973)

FAIN JN   PSYCHOYOS S   CZERNIK AJ
FROST S   CASH WD
INDOMETHACIN, LIPOLYSIS AND CYCLIC AMP ACCUMULATION IN
WHITE FAT CELLS
ENDOCRINOLOGY 93:632-639(1973)

FANBURG BL
PROSTAGLANDINS AND THE LUNG
AM REV RESP DIS 108:482-489(1973)

FARGES JP
PROSTAGLANDINS TODAY: CARDIOVASCULAR AND RENAL EFFECTS
LYON MED 230(13):7-17(1973):(FRENCH)

FARMER JB   FARRAR DG   WILSON J
EFFECTS OF INDOMETHACIN ON RESPONSES OF THE ISOLATED TRACHEA
TO ARACHIDONIC ACID AND PROSTAGLANDIN F2A
NAUNYN SCHMIEDEBERGS ARCH PHARMACOL 279:R21 (1973)

FAVIER J   MULDER RH   RIETVELD WJ
GEVERS RH
INDUCTION OF LABOR WITH PROSTAGLANDIN E1
NED TIJDSCHR GENEESKD 116:1964-1965(1972):(DUTCH)

FELDBERG W   GUPTA KP
SAMPLING FOR BIOLOGICAL ASSAY OF CEREBROSPINAL FLUID FROM
THE THIRD VENTRICLE IN THE UNANESTHETIZED CAT
J PHYSIOL (LOND) 222:126P-129P(1972)

FELDBERG W   GUPTA KP
PYROGEN FEVER AND PROSTAGLANDIN-LIKE ACTIVITY IN
CEREBROSPINAL FLUID
J PHYSIOL (LOND) 228:41-53(1973)

FELDBERG W   GUPTA KP   MILTON AS
WENDLANDT S
EFFECT OF BACTERIAL PYROGEN AND ANTIPYRETICS ON
PROSTAGLANDIN ACTIVITY IN CEREBROSPINAL FLUID OF
UNANESTHETIZED CATS
BR J PHARMACOL 46:550P-551P(1972)

FELDBERG W   GUPTA KP   MILTON AS
WENDLANDT S
EFFECT OF PYROGEN AND ANTIPYRETICS ON PROSTAGLANDIN ACTIVITY
IN CISTERNAL CSF OF UNANAESTHETIZED CATS
J PHYSIOL (LOND) 234:279-303 (1973)

FELDBERG W   MILTON AS
PROSTAGLANDIN FEVER
IN: PHARMACOL THERMOREGUL PROC SATELL SYMP, 1972,
KARGER, BASEL : 302-310(1973)

FELDBERG W   MILTON AS
PROSTAGLANDIN-LIKE ACTIVITY IN THE CEREBROSPINAL FLUID
DURING PYREXIA
IN: PHARMACOL THERMOREGUL PROC SATELL SYMP, 1972,
KARGER, BASEL : 472(1973)

FELDBERG W   SAXENA PN
FEVER PRODUCED IN RABBITS AND CATS BY PROSTAGLANDIN E1
INJECTED INTO THE CEREBRAL VENTRICLES
J PHYSIOL (LOND) 215:23P-24P(1971)

FELDBERG W   SAXENA PN
FEVER PRODUCED BY PROSTAGLANDIN E1.
J PHYSIOL (LOND) 217 : 547-556 (1971)

FELDBERG W   SAXENA PN
FURTHER STUDIES ON PROSTAGLANDIN E1 FEVER IN CATS.
J PHYSIOL (LOND) 219 : 739-745(1971)

FENIUK W   LARGE BJ
SOME ACTIONS OF BRADYKININ ON MOUSE ISOLATED ILEUM
BR J PHARMACOL 49:144P-145P(1973)

FERDINANDI ES   JUST G
A SYNTHESIS OF D, 1-PROSTAGLANDIN E2 METHYL ESTER AND
RELATED COMPOUNDS
CAN J CHEM 49:1070-1084(1971)

FERGUSON WW   EDMONDS AW   STARLING JR
WANGENSTEEN SL
PROTECTIVE EFFECT OF PROSTAGLANDIN E1 (PGE1) ON LYSOSOMAL
ENZYME RELEASE IN SEROTONIN-INDUCED GASTRIC ULCERATION
ANN SURG 177:648-654(1973)

FERRARIS VA   HUDSON TH   DERUBERTIS FR
MITOGEN-STIMULATED RELEASE OF PROSTAGLANDIN FROM SPLEEN
CELLS
CLIN RES 21:578(1973)

FERREIRA SH
PROSTAGLANDINS, ASPIRIN-LIKE DRUGS AND ANALGESIA
NATURE (NEW BIOL) 240:200-203(1972)

FERREIRA SH
FATTY ACID HYDROPEROXIDES ARE POSSIBLE MEDIATORS OF PAIN IN
THE INFLAMMATORY PROCESS
ABSTR 5TH INT CONGR PHARMACOL SAN FRANC 67(1972)

FERREIRA SH
PROSTAGLANDINS, INFLAMMATION AND ASPIRIN-LIKE DRUGS
MED TODAY 7:29-40(1973)

FERREIRA SH   HERMAN A   VANE JR
PROSTAGLANDIN GENERATION MAINTAINS THE SMOOTH MUSCLE
TONE OF THE RABBIT ISOLATED JEJUNUM
BR J PHARMACOL 44:328P-329P(1972)

FERREIRA SH   MONCADA S
INHIBITION OF PROSTAGLANDIN SYNTHESIS AUGMENTS THE EFFECTS
OF SYMPATHETIC NERVE STIMULATION ON THE CAT SPLEEN
BR J PHARMACOL 43:419P-420P(1971)

FERREIRA SH   MONCADA S   VANE JR
INDOMETHACIN AND ASPIRIN ABOLISH PROSTAGLANDIN RELEASE FROM
THE SPLEEN
NATURE (NEW BIOL) 231:237-239(1971)

FERREIRA SH   MONCADA S   VANE JR
SOME EFFECTS OF INHIBITING ENDOGENOUS PROSTAGLANDIN
FORMATION ON THE RESPONSES OF THE CAT SPLEEN
BR J PHARMACOL 47:48-58(1973)

FERREIRA SH   MONCADA S   VANE JR
FURTHER EXPERIMENTS TO ESTABLISH THAT THE ANALGESIC ACTION
OF ASPIRIN-LIKE DRUGS DEPENDS ON THE INHIBITION OF
PROSTAGLANDIN BIOSYNTHESIS
BR J PHARMACOL 47:629P(1973)

FERREIRA SH   MONCADA S   VANE JR
PROSTAGLANDINS AND THE MECHANISM OF ANALGESIA PRODUCED BY
ASPIRIN-LIKE DRUGS
BR J PHARMACOL 49:86-97(1973)

FERREIRA SH   MONCADA S   VANE JR
THE BLOCKADE OF THE LOCAL GENERATION OF PROSTAGLANDINS
EXPLAINS THE ANALGESIC ACTION OF ASPIRIN
AGENTS ACTIONS 3:385-386(1973)

FERREIRA SH   NG KK   VANE JR
THE CONTINUOUS BIOASSAY OF THE RELEASE AND DISAPPEARANCE OF
HISTAMINE IN THE CIRCULATION
BR J PHARMACOL 49:543-553(1973)

FERREIRA SH   VANE JR
INHIBITION OF PROSTAGLANDIN BIOSYNTHESIS: AN EXPLANATION OF
THE THERAPEUTIC EFFECTS OF NONSTEROID ANTI-INFLAMMATORY
AGENTS
IN: "PROSTAGLANDINES 1973," INSERM, PARIS:345-357(1973)

FERREIRA SH   VARGAFTIG BB
GENERATION OF RABBIT AORTA CONTRACTING SUBSTANCE IN BLOOD BY
SLOW REACTING SUBSTANCE C(SRS-C) AND ITS BLOCKAGE BY
NON-STEROID ANTIINFLAMMATORY DRUGS
BIORHEOLOGY 10:288-289(1973)

FIANDINI G   TOTH E   FASSINA G
ACTION OF PAPAVERINE AND CALCIUM ON THE EFFECT OF
PROSTAGLANDIN E1 IN ISOLATED SHORT-CIRCUITED FROG SKIN
ABSTR 5TH INT CONGR PHARMACOL SAN FRANC 68(1972)

FICHMAN M   HORTON R
SIGNIFICANCE OF THE EFFECTS OF PROSTAGLANDINS ON RENAL AND
ADRENAL FUNCTION IN MAN
PROSTAGLANDINS 3:629-646(1973)

FICHMAN M   LITTENBERG G   WOO J
HORTON R
THE EFFECT OF PROSTAGLANDIN (PGA1) ON ADRENAL FUNCTION IN
MAN
ADV BIOSCI (SUPPL) 9:50(1972)

FICHMAN M   LITTENBERG G   WOO J
HORTON R
THE EFFECT OF PROSTAGLANDIN (PGA1) ON ADRENAL FUNCTION IN
MAN
ABSTR 5TH INT CONGR NEPHROL, MEX CITY:113(1972)

FICHMAN M   LITTENBERG G   WOO J
HORTON R
THE EFFECT OF PROSTAGLANDIN (PGA1) ON ADRENAL FUNCTION IN
MAN
ADV BIOSCI 9:313-320(1973)

FICHMAN MP
AUGMENTATION OF URINARY DIVALENT ION EXCRETION IN MAN BY
PROSTAGLANDIN A1 (PGA1)
CLIN RES 19:150+529(1971)

FICHMAN MP   LITTENBERG G   BROOKER G
HORTON R
EFFECT OF PROSTAGLANDIN A1 ON RENAL AND ADRENAL FUNCTION IN
MAN
CIRC RES 31(SUPPL II):II-19-II-35(1972)

FICHMAN MP   LITTENBERG G   WOO J
HORTON R
THE EFFECT OF PROSTAGLANDIN A1 (PGA1) ON ADRENAL FUNCTION IN
MAN
J CLIN INVEST 51:30A(1972)

FIELD FP   FREGLY MJ   KIM KJ
STUDIES ON A THYROID-DEPRESSING FACTOR OF RENAL ORIGIN IN
THE RAT
TOXICOL APPL PHARMACOL 21:556-568(1972)

FIELD J   DEKKER A   ZOR U
KANEKO T
IN VITRO EFFECTS OF PROSTAGLANDINS ON THYROID GLAND
METABOLISM
ANN NY ACAD SCI 180:278-288(1971)

FIELD JB   ZOR U   KANEKO T
YAMASHITA K   DEKKER A
COMPARISON OF EFFECTS OF TSH, LATS AND PROSTAGLANDINS ON
DOG THYROID SLICE ADENYL CYCLASE, CYCLIC AMP, COLLOID
DROPLET FORMATION AND INTERMEDIARY METABOLISM
IN: FURTHER ADV THYROID RES TRANS 6TH INT THYROID CONF
VIENNA 1970 2:817-824(1971)

FILHO HM   COUTINHO EM
CONTRACTILITY OF THE GUINEA PIG OVARY
ACTA PHYSIOL LAT AM 23:ABSTR 326(1973);(SPANISH)

FINCH N   DELLAVECCHIA L   FITT JJ
STEPHANI R   VLATTAS I
TOTAL SYNTHESIS OF DL-PROSTAGLANDIN E1
J ORG CHEM 38:4412-4424(1973)

FINCH N   FITT JJ   HSU IHC
CYCLOPENTENONE SYNTHESIS BY DIRECTED CYCLIZATION
J ORG CHEM 36:3191-3196(1971)

FINCK AD   KATZ RL
PREVENTION OF CHOLERA-INDUCED INTESTINAL SECRETION IN THE
CAT BY ASPIRIN
NATURE (LOND) 238:273-274(1972)

FINDLAY JK   CERINI MED   CERINI JC
CHAMLEY WA   HOOLEY RD   WILLIAMS DW
CUMMING IA   LEE CS   O'SHEA JD
PROSTAGLANDINS AND LUTEOLYSIS
IN: "PROSTAGLANDINES 1973," INSERM, PARIS:235-258(1973)

FISHBURNE JI   BRENNER WE   BRAAKSMA JT
STAUROVSKY LG   MUELLER RA   HENDRICKS CH
CARDIOPULMONARY RESPONSES TO INTRAVENOUS INFUSION OF
PROSTAGLANDIN F2A IN THE PREGNANT HUMAN FEMALE
ABSTR SOC GYNECOL INVEST 19TH ANNU MEET, PHOENIX:14(1972)

FJALLAND B
INHIBITION BY ANTI-INFLAMMATORY AGENTS OF THE RELEASE OF
RABBIT AORTA CONTRACTING SUBSTANCE (RCS) AND PROSTAGLANDINS
(PGS) FROM CHOPPED GUINEA-PIG LUNGS
ACTA PHYSIOL SCAND (SUPPL 396) : 105(1973)

FLACK JD
THE HYPOTHALAMUS-PITUITARY-ENDOCRINE SYSTEM
IN:"THE PROSTAGLANDINS," PW RAMWELL, PLENUM PRESS, NY
1:327-345(1973)

FLACK JD   RAMWELL PW
A COMPARISON OF THE EFFECTS OF ACTH, CYCLIC AMP, DIBUTYRYL
CYCLIC AMP, AND PGE2 ON CORTICOSTEROIDOGENESIS IN VITRO
ENDOCRINOLOGY 90:371-377(1972)

FLACK JD   RAMWELL PW   SHAW JE
ENDOCRINOLOGICAL IMPLICATIONS OF PROSTAGLANDINS
CURR TOP EXP ENDOCRINOL 1:199-228(1971)

FLEISCH JH   KENT KM   COOPER T
DRUG RECEPTORS IN SMOOTH MUSCLE
IN: "ASTHMA," KF AUSTEN, LM LICHTENSTEIN, ACADEMIC PRESS,NY:
139-167(1973)

FLORES AGA   SHARP GWG
ENDOGENOUS PROSTAGLANDINS AND OSMOTIC WATER FLOW IN THE TOAD
BLADDER
AM J PHYSIOL 223:1392-1397(1972)

FLOWER R   GRYGLEWSKI R   HERBACZYNSKA-CE/ K
VANE JR
EFFECTS OF ANTI-INFLAMMATORY DRUGS ON PROSTAGLANDIN
BIOSYNTHESIS
NATURE (NEW BIOL) 238:104-106(1972)

FLOWER RJ
ASPIRIN-LIKE DRUGS AND PROSTAGLANDINS
AM HEART J 86:844-846(1973)

FLOWER RJ   VANE JR
INHIBITION OF PROSTAGLANDIN SYNTHETASE IN BRAIN EXPLAINS
THE ANTI-PYRETIC ACTIVITY OF PARACETAMOL (4-ACETAMIDOPHENOL)
NATURE (LOND) 240:410-411(1972)

FLOYD DM   CROSBY GA   WEINSHENKER NM
NUCLEOPHILIC INVERSION AT C11 OF PROSTAGLANDIN INTERMEDIATES
TETRAHEDRON LETT 3265-3268(1972)

FLOYD DM   CROSBY GA   WEINSHENKER NM
THE SYNTHESIS OF (+)-11-EPIPGF2A AND (-)-11-EPI-PGE2
TETRAHEDRON LETT 3269-3272(1972)

FLOYD MB   WEISS MJ
SYNTHESIS OF PROSTAGLANDINS BY ALANATE ADDITIONS TO
CYCLOPENTENONES
PROSTAGLANDINS 3:921-924(1973)

FLYER RH   FINCH SC
THE EFFECTS OF PROSTAGLANDIN E1 ON HUMAN GRANULOCYTES DURING
PHAGOCYTOSIS
CLIN RES 19:726(1971)

FLYER RH   FINCH SC
THE EFFECTS OF PROSTAGLANDIN E1 ON HUMAN GRANULOCYTES DURING
PHAGOCYTOSIS
CLIN RES 20:486(1972)

FLYER RH   FINCH SC
THE EFFECTS OF PROSTAGLANDIN E1 ON HUMAN GRANULOCYTE
METABOLISM DURING PHAGOCYTOSIS
J RETICULOENDOTHEL SOC 14:325-331(1973)

FOA PP
THE ORALLY ACTIVE HYPOGLYCAEMIC AGENTS
IN: "FUNDAMENTALS OF BIOCHEMICAL PHARMACOLOGY," ZM BACQ,
R CAPEK, R PAOLETTI, J RENSON, PERGAMON PRESS,
OXF/NY/TORONTO:548-551(1971)

FONTAINE J   REUSE JJ
EFFECT OF METOCLOPRAMIDE AND PROCAINE ON ISOLATED FUNDIC
STRIPS OF RAT STOMACH
ARCH INT PHARMACODYN THER 204: 191-195 (1973); (FRENCH)

FORSTER W
PROSTAGLANDINS AND CIRCULATORY DISEASES
DTSCH GESUNDHEITSW 27:2264-2271(1972);(GERMAN)

FORSTER W
PROSTAGLANDINS AND NEW DIRECTIONS IN MEDICINE
DTSCH GESUNDHEITSW 28:385-390(1973);(GERMAN)

FORSTER W   MENTZ P
EFFECTS OF PGE1, PGE2, AND PGF2A ON ISOLATED NORMAL AND
DAMAGED HEART PREPARATIONS
ADV BIOSCI 9:379-384(1973)

FORSTER W   MEST HJ   MENTZ P
THE INFLUENCE OF PGF2A ON EXPERIMENTAL ARRHYTHMIAS
PROSTAGLANDINS 3:895-904(1973)

FORTE JG   SOLBERG LA JR
PHARMACOLOGY OF ISOLATED AMPHIBIAN GASTRIC MUCOSA
IN:INT ENCYCL PHARMACOL THER,G PETERS,PERGAMON PRESS,NY
2:195-260(1973)

FOSS P   TAKEGUCHI C   TAI H
SIH C
MECHANISM OF PROSTAGLANDIN BIOSYNTHESIS. 1802 STUDIES ON
PGF1A
ANN NY ACAD SCI 180:126-137(1971)

FOSS PS   SIH CJ   TAKEGUCHI C
SCHNOES H
BIOSYNTHESIS AND CHEMISTRY OF 9A,
15(S)-DIHYDROXY-11-OXO-13-TRANS-PROSTENOIC ACID
BIOCHEMISTRY 11:2271-2277(1972)

FRADL D   REEVE EB
PLASMA FIBRINOGEN RESPONSE TO PROSTAGLANDIN INFUSION
FED PROC 32:259(1973)

FRANK H   BRAUN T
STIMULATION OF RABBIT FAT CELL PLASMA MEMBRANE ADENYL
CYCLASE (AC) BY PROSTAGLANDINS
FED PROC 30:625(1971)

FRANKLIN TJ   FOSTER SJ
HORMONE-INDUCED DESENSITISATION OF HORMONAL CONTROL OF
CYCLIC AMP LEVELS IN HUMAN DIPLOID FIBROBLASTS
NATURE(NEW BIOL)246:146-148(1973)

FRANKLIN TJ   FOSTER SJ
LEAKAGE OF CYCLIC AMP FROM HUMAN DIPLOID FIBROBLASTS IN
TISSUE CULTURE
NATURE(NEW BIOL) 246 : 119-120 (1973)

FRANKS DJ   MACMANUS JP   WHITFIELD JF
THE EFFECT OF PROSTAGLANDINS ON CYCLIC AMP PRODUCTION AND
CELL PROLIFERATION IN THYMIC LYMPHOCYTES
BIOCHEM BIOPHYS RES COMMUN 44:(5)1177-1183(1971)

FRAZER A   WANG YC   PANDY G
MENDELS J
ADENYL CYCLASE ACTIVITY IN PLATELETS OF DEPRESSED PATIENTS
CLIN RES 21:265(1973)

FREDHOLM B   STRANDBERG K
ANTI-ANAPHYLACTIC EFFECT OF A PROSTAGLANDIN ANTAGONIST,
POLYPHLORETIN PHOSPHATE
ADV BIOSCI (SUPPL) 9:69(1972)

FREDHOLM B   STRANDBERG K
ANTIANAPHYLACTIC EFFECTS OF A PROSTAGLANDIN ANTAGONIST,
POLYPHLORETIN PHOSPHATE
ADV BIOSCI 9:447-451(1973)

FREDHOLM B
CATECHOLAMINES AND CYCLIC AMP
LAKARTIDNINGEN, 68(SUPP. II):43-48(1971);(SWEDISH)

FREDHOLM BB   HEDQVIST P
ROLE OF PRE- AND POSTJUNCTIONAL INHIBITION BY PROSTAGLANDIN
E2 OF LIPOLYSIS INDUCED BY SYMPATHETIC NERVE STIMULATION
IN DOG SUBCUTANEOUS ADIPOSE TISSUE IN SITU
BR J PHARMACOL 47:711-718(1973)

FREE MJ   JAFFE RA
FACTORS AFFECTING BLOOD FLOW AND PRESSURES IN TESTES OF
CONSCIOUS RATS
BIOL REPROD 7:119(1972)

FREE MJ   JAFFE RA
EFFECT OF PROSTAGLANDINS ON BLOOD FLOW AND PRESSURE IN THE
CONSCIOUS RAT
PROSTAGLANDINS 1:483-498(1972)

FREE MJ   KIEN ND
VENOUS ARTERIAL INTERACTIONS INVOLVING SEROTONIN IN THE
PAMPINIFORM PLEXUS OF THE RAT
PROC SOC EXP BIOL MED 143:284-288(1973)

FREEMAN PC   WEST GB
RESISTANCE OF RATS TO CARRAGEENAN AND TO ADJUVANT-INDUCED
ARTHRITIS.
BR J PHARMACOL 44:327P-328P(1972)

FREEMAN PC   WEST GB
SKIN REACTIONS TO PROSTAGLANDINS
J PHARM PHARMACOL 24:407-408(1972)

FREWIN DB   EAKINS KE   DOWNEY JA
BHATTACHERJEE P
PROSTAGLANDIN-LIKE ACTIVITY IN HUMAN ECCRINE SWEAT
AUST J EXP BIOL MED SCI 51:701-702(1973)

FRIED J
PROGRESS IN PROSTAGLANDIN SYNTHESIS
ABSTR 7TH MIDDLE ATL REG MEET AM CHEM SOC PHILA:82(1972)

FRIED J   LIN C   MEHRA M
KAO W   DALVEN P
SYNTHESIS AND BIOLOGICAL ACTIVITY OF PROSTAGLANDINS AND
PROSTAGLANDIN ANTAGONISTS
ANN NY ACAD SCI 180:38-63(1971)

FRIED J   LIN CH
SYNTHESIS AND BIOLOGICAL EFFECTS OF 13-DEHYDRO DERIVATIVES
OF NATURAL PROSTAGLANDIN F2A AND E2 AND THEIR 15-EPI
ENANTIOMERS
J MED CHEM 16:429-430(1973)

FRIED J   LIN CH   SIH JC
DALVEN P   COOPER GF
STEREOSPECIFIC TOTAL SYNTHESIS OF THE NATURAL AND RACEMIC
PROSTAGLANDINS OF THE E AND F SERIES
J AM CHEM SOC 94:4342-4343(1972)

FRIED J   MEHRA MM   GAEDE BJ
NOVEL SELECTIVE INHIBITORS OF HUMAN PLACENTAL
PG-15-DEHYDROGENASE
ADV BIOSCI (SUPPL) 9:18(1972)

FRIED J   MEHRA MM   KAO WL
SYNTHESIS OF (+)-AND (-)-7-OXAPROSTAGLANDIN F1A AND THEIR
15-EPIMERS
J AM CHEM SOC 93:5594-5595(1971)

FRIED J   SIH JC
TOTAL SYNTHESIS OF PROSTAGLANDINS. CONTROL OF
REGIOSPECIFICITY IN THE ALANE-EPOXIDE REACTION AND
SELECTIVE CATALYTIC OXIDATION OF ALKYNYLATION PRODUCTS
TETRAHEDRON LETT:3899-3902(1973)

FRIED J   SIH JC   LIN CH
DALVEN P
REGIOSPECIFIC EPOXIDE OPENING WITH ACETYLENIC ALANES. AN
IMPROVED TOTAL SYNTHESIS OF E AND F PROSTAGLANDINS
J AM CHEM SOC 94:4343-4345(1972)

FRIED JH   HARRISON IT   LEWIS B
RIEGL J   ROOKS W   TOMOLONIS A
STRUCTURE ACTIVITY RELATIONSHIP AMONG
6-SUBSTITUTED-2-NAPHTHYLACETIC ACIDS
SCAND J RHEUMATOL (SUPPL) 2 : 7-11(1973)

FRIIS-HANSEN B   CLAUSEN J
STUDIES ON SERUM LIPOPROTEIN IN THE NEONATAL PERIOD.
Z ERNAEHRUNGSWISS 10:(3)253-263(1971)

FRITH DA   SNART RS
INHIBITION OF VASOPRESSIN-STIMULATED WATER TRANSPORT ACROSS
THE ISOLATED TOAD BLADDER
COMP BIOCHEM PHYSIOL (A) 45:313-325(1973)

FROLICH JC   SWEETMAN BJ   CARR K
SPLAWINSKI J   WATSON JT   ANGGARD E
OATES JA
OCCURRENCE OF PROSTAGLANDINS IN HUMAN URINE
ADV BIOSCI 9:321-330(1973)

FROLICH JC   SWEETMAN BJ   SPLAWINSKI J
WATSON JT   OATES JA
OCCURRENCE OF PROSTAGLANDINS IN HUMAN URINE
ADV BIOSCI (SUPPL) 9:51(1972)

FROLICH JC   WILSON TW   CARR K
NIES AS   OATES JA
URINARY PROSTAGLANDINS: RELEASE FROM THE KIDNEY BY
ANGIOTENSIN IN DOG AND MAN
CLIN RES 21:687(1973)

FROLICH JC   WILSON TW   SWEETMAN BJ
SMIGEL M   CARR K   WATSON JT
OATES JA
MEASUREMENT OF PROSTAGLANDINS (PG) IN URINE
FED PROC 32:788(1973)

FUCHS AR
PROSTAGLANDIN EFFECTS ON PARTURITION AND UTERINE ACTIVITY
IN THE RAT
ABSTR SOC GYNECOL INVEST 19TH ANNU MEET, PHOENIX:21(1972)

FUCHS AR   COUTINHO EM
SUPPRESSION OF UTERINE ACTIVITY DURING MENSTRUATION BY
EXPANSION OF THE PLASMA VOLUME
ACTA ENDOCRINOL (KBH) 66:183-192(1971)

FUCHS AR   MOK E
INFLUENCE OF ESTROGEN AND PROGESTERONE ON
PROSTAGLANDIN-INDUCED UTERINE ACTIVITY IN THE RAT
ABSTR SOC GYNECOL INVEST 19TH ANNU MEET, PHOENIX:20(1972)

FUCHS F
INITIATION OF LABOUR-FACTS AND FANCIES
MEM SOC ENDOCRINOL 20:1-24(1973)

FUKUBA H   NAKATANI Y
PROSTAGLANDIN
YUKAGAKU 22:291-297(1973);(JAPANESE)

FULGRAFF G
EFFECTS OF PROSTAGLANDIN E2, ON RENAL TUBULAR REABSORPTION
OF SODIUM AND FLUID IN RATS
PROC INT UNION PHYSIOL SCI, 25TH INT CONGR, MUNICH,
9:191(1971)

FULGRAFF G
EVALUATION AND LOCALIZATION OF THE RENAL TUBULAR EFFECTS OF
PROSTAGLANDINS IN DIFFERENT STATES OF DIURESIS
(MICROPUNCTURE STUDIES)
ABSTR 5TH INT CONGR PHARMACOL SAN FRANC 74(1972)

FULGRAFF G
PROSTAGLANDINS. A NEW CLASS OF BIOGENIC ACTIVE SUBSTANCES
FORTSCHR MED 91:410-416(1973);(GERMAN)

FULGRAFF G
PROSTAGLANDINS
MED KLIN 68:195-201(1973);(GERMAN)

FULGRAFF G
PROSTAGLANDINS
IN: "CLINICAL PHARMACOLOGY AND PHARMACOTHERAPY, 2ND ED,"
HP KUMMERLE, ER GARRETT, KH SPITZY, URBAN & SCHWARZENBERG,
MUNICH: 313-330(1973);(GERMAN)

FULGRAFF G   BRANDENBUSCH G   MEIFORTH A
EFFECTS OF PGE2 ON RENAL TUBULAR FUNCTION
ADV BIOSCI (SUPPL) 9:48(1972)

FULGRAFF G   BRANDENBUSCH G   MEIFORTH A
EFFECTS OF PGE2 ON RENAL TUBULAR FUNCTION
ADV BIOSCI 9:301-305(1973)

FULGRAFF G   MEIFORTH A
EFFECTS OF PROSTAGLANDIN E2 ON EXCRETION AND REABSORPTION
OF SODIUM AND FLUID IN RAT KIDNEYS (MICROPUNCTURE STUDIES)
PFLUEGERS ARCH 330:243-256(1971)

FULGRAFF G   MEIFORTH A   SUDHOFF D
EFFECTS OF PROSTAGLANDIN E2 ON THE RENAL EXCRETION OF
WATER AND ELECTROLYTES IN RATS
NAUNYN-SCHMIEDEBERGS ARCH PHARMACOL 269:489-490(1971)

FURTADO MRF
OCCURRENCE OF A KININ-LIKE PEPTIDE IN THE URINARY BLADDER OF
THE TOAD BUFO MARINUS PARACNEMIS LUTZ
BIOCHEM PHARMACOL 21:118-124(1972)

FUTA R   TANIGUCHI C
OCULAR HYPERTENSION FOLLOWING LOCAL PROSTAGLANDIN AND
CHANGES OF ENZYMATIC ACTIVITY IN TRABECULAR MESHWORK AND
CILIARY EPITHELIUM
ACTA SOC OPHTHALMOL JAP 77:1607-1619(1973);(JAPANESE)

GAGNON DJ   COSINEAU D   BOUCHER PJ
RELEASE OF VASOPRESSIN BY ANGIOTENSIN II AND PROSTAGLANDIN
E2 FROM THE RAT NEURO-HYPOPHYSIS IN VITRO
LIFE SCI (I) 12:487-497(1973)

GAGNON DJ   SIROIS P
THE RAT ISOLATED COLON AS A SPECIFIC ASSAY ORGAN FOR
ANGIOTENSIN
BR J PHARMACOL 46:89-93(1972)

GALTON L
THE NEW MYSTERY-MAYBE MIRACLE-DRUG
NY TIMES MAG SECT 6:46-47,62,65,67-68,70,72(12/5/1971)

GANDINI A   LUALDI P   DELLA BELLA D
INFLUENCE OF RESERPINE ON THE RELEASE AND EFFECTS OF
PROSTAGLANDINS
ARCH INT PHARMACODYN THER 196:179-181(1972)

GANDINI A   LUALDI P   DELLA-BELLA D
RELEASE OF PROSTAGLANDINS AND ITS EFFECTS ON THE COLON
RESPONSES TO TRANSMURAL STIMULATION
NAUNYN-SCHMIEDEBERGS ARCH PHARMACOL 269:388-389(1971)

GANDOLFI C   DORIA G   GAIO P
PROSTAGLANDINS: BY-PRODUCTS IN BBR3-CLEAVAGE OF COREY'S
METHYLETHER INTERMEDIATE
TETRAHEDRON LETT 2063-2065(1972)

GANDOLFI C   DORIA G   GAIO P
PROSTAGLANDINS IV: 5-CIS-13-TRANS-14-CHLOROPROSTADIENOIC
AND 5-CIS-PROSTEN-13-YNOIC ACIDS
FARMACO (SCI) 27:1125-1129(1972)

GANDOLFI C   DORIA G   GAIO P
PROSTAGLANDINS II: 8,12-DIISO-PGE2 (ENT-11,15-EPI-PGE2)
TETRAHEDRON LETT 4303-4306(1972)

GANESAN PA   KARIM SMM
POLYPHLORETIN PHOSPHATE TEMPORARILY POTENTIATES
PROSTAGLANDIN E2 ON THE RAT FUNDUS, PROBABLY BY INHIBITING
PG15-HYDROXYDEHYDROGENASE
J PHARM PHARMACOL 25:229-233(1973)

GANS P
EFFECTS OF PROSTAGLANDINS AND OXYTOCIN ON THE UTERINE
MOTILITY OF THE LIVING RAT UNDER DIFFERENT HORMONAL
CONDITIONS
ADV BIOSCI (SUPPL) 9:131(1972)

GANS P
A SIDE EFFECT OF INDOMETHACIN?
NED TIJDSCHR GENEESK 117:1326-1327(1973);(DUTCH)

GARBERS DL   LUST WD   FIRST NL
LARDY HA
EFFECTS OF PHOSPHODIESTERASE INHIBITORS AND CYCLIC
NUCLEOTIDES ON SPERM RESPIRATION AND MOTILITY
BIOCHEMISTRY 10:1825-1831(1971)

GARCIA GA   DIAZ E   CRABBE P
LANTHANIDE-INDUCED SHIFTS IN PROTON NUCLEAR MAGNETIC
RESONANCE SPECTRA OF PROSTAGLANDINS
CHEM IND (LOND): 585-586 (1973)

GARDNER JD   GINZLER ER
SODIUM TRANSPORT IN HUMAN ERYTHROCYTES. ABSENCE OF AN EFFECT
OF PROSTAGLANDIN E1.
BIOCHEM BIOPHYS RES COMMUN 42:1063-1067(1971)

GASCON S
EFFECT OF PGE1 ON THE HORMONAL CONTROL OF LIPOLYSIS IN BIRDS
REV ESP FISIOL 27:69-72(1971);(SPANISH)

GAUTIER JC   SAMAMA M   BOUSSER MG
EXPERIMENTAL ARTERIAL THROMBOSES IN VIVO AND IN VITRO WITH
ASPIRIN AND PROSTAGLANDIN E1
SEM HOP PARIS 48:2699(1972);(FRENCH)

GAUTIER JC   SAMAMA M   BOUSSER MG
EXPERIMENTAL ARTERIAL THROMBOSES. IN VIVO AND IN VITRO
EFFECTS OF ASPIRIN AND PROSTAGLANDIN E1
ANN MED INTERNE (PARIS) 123:849-852(1972);(FRENCH)

GAVIN MA   MAESO EV   OLMEDO JAO
INHIBITION OF THE ANTIFERTILITY EFFECT OF PROSTAGLANDIN
F2A IN THE RAT BY PROGESTERONE
AN FAC VET LEON 18:591-602(1972);(SPANISH)

GAVIN MA   MAESO EV   OLMEDO JAO
PGF2A ACTION ON THE RAT DURING THE EMBRYONIC IMPLANTATION
PERIOD
AN FAC VET LEON, 18:571-583(1972);(SPANISH)

GAVIN MA   MAESO EV   OLMEDO JO
PRELIMINARY STUDY OF THE ACTION OF PGE2A ON THE IMPLANTATION
PHASE IN THE RAT
ACTA GINECOL 24:343-354(1973);(SPANISH)

GENTON E   WEILY HS   STEELE PP
PLATELETS, THROMBOSIS AND CORONARY ARTERY DISEASE
ADV CARDIOL 9:29-39(1973)

GEORGE JM
EFFECT OF MERCURY ON RESPONSE OF ISOLATED FAT CELLS TO
INSULIN AND LIPOLYTIC HORMONES
ENDOCRINOLOGY 89:1489-1498(1971)

GEORGE JM   KIER LB   HOYLAND JR
THEORETICAL CONSIDERATIONS OF ALPHA AND BETA ADRENERGIC
ACTIVITY.
MOL PHARMACOL 7:328-336(1971)

GEORGE WJ   PADDOCK RJ   KADOWITZ PJ
INFLUENCE OF PROSTAGLANDIN E1 ON CONTRACTILE FORCE AND
CYCLIC AMP LEVELS IN THE PERFUSED RAT HEART
ABSTR 5TH INT CONGR PHARMACOL SAN FRANC 80(1972)

GERRITSEN J   SCHREUDER-VANGE/ R TEN CATE JW
PLATELET KINETICS AND MORPHOLOGY IN THROMBOCYTHAEMIA
ABSTR 4TH INT CONGR THROMB HAEMOSTASIS, VIENNA:340(1973)

GERSHMAN H   POWERS E   LEVINE L
VAN VUNAKIS H
RADIOIMMUNOASSAY OF PROSTAGLANDINS, ANGIOTENSIN, DIGOXIN,
MORPHINE AND ADENOSINE-3, 5-CYCLIC-MONOPHOSPHATE WITH
NITROCELLULOSE MEMBRANES
PROSTAGLANDINS 1:407-423(1972)

GEUMEI A   BASHOUR F   SWAMY B
NAFRAWI A
EFFECTS OF PROSTAGLANDIN E1 ON HEPATIC ARTERIAL AND PORTAL
VENOUS VASCULATURE IN THE DOG
ABSTR 5TH INT CONGR PHARMACOL SAN FRANC 81(1972)

GEUMEI A   BASHOUR FA   SWAMY BV
NAFRAWI AG
PROSTAGLANDIN E1: ITS EFFECTS ON HEPATIC CIRCULATION IN DOGS
PHARMACOLOGY 9:336-347(1973)

GEVERS RH   RIETFELD WJ   FAVIER J
MULDER RH
LABOR INDUCTION WITH PROSTAGLANDIN E1
IN: "PERINATALE MEDIZIN ," E SALING AND JW DUDENHAUSEN,
GEORG THIEME, STUTTG 3:301-308(1972):(GERMAN)

GIELEN JE   NEBERT DW
ARYL HYDROCARBON HYDROXYLASE INDUCTION IN MAMMALIAN LIVER
CELL CULTURE
J BIOL CHEM 247:7591-7602(1972)

GIERTZ H   SCHUSTER J
THE ROLE OF MEDIATING SUBSTANCES IN ASTHMA IN ANIMAL TESTS
AND IN CLINICAL PRACTICE
MED KLIN 67:634-640(1972):(GERMAN)

GILDER SSB
LONDON LETTER
CAN MED ASSOC J 104:674(1971)

GILDER SSB
PROSTAGLANDIN E1 AND FEVER
CAN MED ASSOC J 109:971-972(1973)

GILLESPIE A
INTERRELATIONSHIP BETWEEN OXYTOCIN (EXOGENOUS AND
ENDOGENOUS) AND PROSTAGLANDINS
ADV BIOSCI (SUPPL) 9:123-125(1972)

GILLESPIE A
PROSTAGLANDINS AND HUMAN LABOUR
MEM SOC ENDOCRINOL 20:77-93(1973)

GILLESPIE E   LICHTENSTEIN LM
PHARMACOLOGIC CONTROL OF IGE-MEDIATED HISTAMINE RELEASE
FROM HUMAN LEUKOCYTES
INT ARCH ALLERGY APPL IMMUNOL 45:95-97(1973)

GILLESPIE E   LICHTENSTEIN LM
HISTAMINE RELEASE FROM HUMAN LEUKOCYTES: ROLE OF CYCLIC AMP
AND MICROTUBULES
ABSTR, 9TH INT CONG BIOCHEM, STOCKH:306(1973)

GILLETT PG   KINCH RAH   WOLFE LS
PACE-ASCIAK C
THERAPEUTIC ABORTION IN THE SECOND TRIMESTER BY
INTRA-AMNIOTIC ADMINISTRATION OF PROSTAGLANDIN F2A
ANN R COLL PHYS SURG CAN 5:52(1972)

GILLETT PG   KINCH RAH   WOLFE LS
PACE-ASCIAK C
THERAPEUTIC ABORTION IN THE SECOND TRIMESTER BY
INTRA-AMNIOTIC PROSTAGLANDIN F2A
J REPROD MED 9:416-419(1972)

GILLIS CN
METABOLISM OF VASOACTIVE HORMONES BY LUNG
ANESTHESIOLOGY 39:626-632(1973)

GILMAN AG
REGULATION OF CYCLIC AMP METABOLISM IN CULTURED CELLS OF
THE NERVOUS SYSTEM
ADV CYCLIC NUCL RES 1:389-410(1972)

GILMAN AG
INTERACTIONS OF PROSTAGLANDINS WITH NEURONS AND GLIA IN
TISSUE CULTURE
ABSTR 7TH MIDDLE ATL REG MEET AM CHEM SOC PHILA:72(1972)

GILMAN AG   MINNA J
ADENOSINE 3',5'-CYCLIC MONOPHOSPHATE (CAMP) REGULATION IN
SOMATIC CELL HYBRIDS
J CELL BIOL 55:86A(1972)

GILMAN AG   MINNA JD
EXPRESSION OF GENES FOR METABOLISM OF CYCLIC ADENOSINE
3':5'-MONOPHOSPHATE IN SOMATIC CELLS.I. RESPONSES TO
CATECHOLAMINES IN PARENTAL AND HYBRID CELLS
J BIOL CHEM 248:6610-6617(1973)

GILMAN AG   NIRENBERG M
REGULATION OF ADENOSINE 3',5'-CYCLIC MONOPHOSPHATE
METABOLISM IN CULTURED NEUROBLASTOMA CELLS
NATURE (LOND) 234 : 356-358 (1971)

GILMAN AG   NIRENBERG M
EFFECT OF CATECHOLAMINES ON THE ADENOSINE 3':5'-CYCLIC
MONOPHOSPHATE CONCENTRATIONS OF CLONAL SATELLITE CELLS OF
NEURONS
PROC NAT ACAD SCI 68:2165-2168(1971)

GILMAN AG   SCHRIER BK
ADENOSINE CYCLIC 3',5'-MONOPHOSPHATE IN FETAL RAT BRAIN CELL
CULTURES. 1. EFFECT OF CATECHOLAMINES
MOL PHARMACOL 8:410-416(1972)

GILMORE DP   SHAIKH AA
THE EFFECT OF PROSTAGLANDIN E2 IN INDUCING SEDATION ON THE
RAT
PROSTAGLANDINS 2:143-151(1972)

GILMORE NJ   VANE JR
HORMONES RELEASED INTO THE CIRCULATION WHEN THE URINARY
BLADDER OF THE ANAESTHETIZED DOG IS DISTENDED
CLIN SCI 41:69-83(1971)

GIMENO M   RETTORI V   COUTINHO EM
EFFECTS OF CHORIONIC GONADTROPINS AND HUMAN PLACENTAL
LACTOGEN ON THE MOTILITY OF THE ISOLATED RAT UTERUS
ACTA PHYSIOL LAT AM 23:ABSTR 319(1973):(SPANISH)

GIMENO MFDE   GIMENO AL   LIMA F
BORDA E
EFFECT OF POLYPHLORETIN-PHOSPHATE, AN INHIBITOR OF
PROSTAGLANDINS, ON THE SPONTANEOUS AND INDUCED MOTILITY OF
ISOLATED RAT UTERUS
ACTA PHYSIOL LATINOAM 23:105-112(1973)

GINSBORG BL   HIRST GDS
PROSTAGLANDIN E1 AND NORADRENALINE AT THE NEUROMUSCULAR
JUNCTION
BR J PHARMACOL 42:153-154(1971)

GIRDWOOD RH
ADVANCES IN MEDICINE
PRACTITIONER 211:407-418(1973)

GIROUD JP   TIMSIT J   WILLOUGHBY DA
PROSTAGLANDINS AND INFLAMMATION
RHUMATOLOGIE 3:55-61(1973):(FRENCH)

GLAVIANO VV   MASTERS T
INHIBITORY ACTION OF INTRACORONARY PROSTAGLANDIN E1 ON
MYOCARDIAL LIPOLYSIS
AM J PHYSIOL 220:1187-1193(1971)

GLAVIANO VV   MASTERS TN
DRUGS AFFECTING UPTAKE OF FREE FATTY ACIDS IN THE IN VIVO
CANINE MYOCARDIUM
ADV EXP MED BIOL 26:294-295(1972)

GLENN EM   BOWMAN BJ   LYSTER SC
ROHLOFF NA
AGGREGATION OF RED CELLS AND EFFECTS OF NONSTEROIDAL
ANTI-INFLAMMATORY DRUGS
PROC SOC EXP BIOL MED 138:235-240(1971)

GLENN EM   BOWMAN BJ   ROHLOFF NA
PRO-INFLAMMATORY EFFECTS OF CERTAIN PROSTAGLANDINS
IN: "PROSTAGLANDINS IN CELLULAR BIOLOGY," PW RAMWELL AND
BB PHARRISS, PLENUM PRESS, NY/LOND:239-343(1972)

GLENN EM   ROHLOFF N
ANTIARTHRITIC AND ANTIINFLAMMATORY EFFECTS OF CERTAIN
PROSTAGLANDINS
PROC SOC EXP BIOL MED 139:290-294(1972)

GLENN EM   ROHLOFF N   BOWMAN B
LYSTER S
ANTI-INFLAMMATORY AND PROINFLAMMATORY EFFECTS OF CERTAIN
PROSTAGLANDINS
ARTHRITIS RHEUM 15:110(1972)

GLENN EM   WILKS J   BOWMAN B
PLATELETS, PROSTAGLANDINS, RED CELLS, SEDIMENTATION RATES,
SERUM AND TISSUE PROTEINS AND NON-STEROIDAL
ANTI-INFLAMMATORY DRUGS
PROC SOC EXP BIOL MED 141:879-886(1972)

GLENN TM
ALTERATION OF THE COURSE OF FELINE POSTOLIGEMIC SHOCK BY
PROSTAGLANDIN INFUSION
FED PROC 31:545(1972)

GLENN TM
STUDIES ON THE MECHANISM OF THE PROTECTIVE ACTION OF
PROSTAGLANDINS IN CIRCULATORY SHOCK
ABSTR 5TH INT CONGR PHARMACOL, SAN FRANC:83(1972)

GLENN TM   RAFLOW GT   WANGENSTEEN SL
LEFER AM
BENEFICIAL EFFECT OF PROSTAGLANDINS E1 AND F2A IN ENDOTOXIN
SHOCK
PHARMACOLOGIST 13:293(1971)

GO VLW   SUMMERSKILL WHJ
DIGESTION, MALDIGESTION, AND THE GASTROINTESTINAL HORMONES
AM J CLIN NUTR 24:160-167(1971)

GODING JR   BAIRD DT   CUMMING IA
MCCRACKEN JA
FUNCTIONAL ASSESSMENT OF AUTOTRANSPLANTED ENDOCRINE ORGANS
ACTA ENDOCRINOL (SUPPL) (KBH) 158:169-199(1972)

GOETTER WE   GOLDBERG LI
DOPAMINE AND PROSTAGLANDIN A: THEIR SEPARATE AND COMBINED
EFFECTS ON RENAL HEMODYNAMICS
CIRCULATION 44: II65(1971)

GOLDFINE ID   ROTH J   PERLMAN RL
MUENZER J
TOLBUTAMIDE: AN INHIBITOR OF CYCLIC AMP PHOSPHODIESTERASE
IN ISLET CELLS AND OTHER TISSUES
CLIN RES 19:476(1971)

GOLDFINE ID   SHERLINE P
INSULIN ACTION IN ISOLATED RAT THYMOCYTES. II. INDEPENDENCE
OF INSULIN AND CYCLIC ADENOSINE MONOPHOSPHATE
J BIOL CHEM 247:6927-6931(1972)

GOLDHABER G   SULMAN FG
ASPIRIN AND PROSTAGLANDINS
HAREFUAH 83:40(1972);(HEBREW)

GOLDHABER P   RABADJIJA L   BEYER WR
KORNHAUSER A
BONE RESORPTION IN TISSUE CULTURE AND ITS RELEVANCE TO
PERIODONTAL DISEASE
J AM DENT ASSOC 87:1027-1033(1973)

GOLDSTEIN I   HOFFSTEIN S   GALLIN J
WEISSMANN G
MECHANISMS OF LYSOSOMAL ENZYME RELEASE FROM HUMAN
LEUKOCYTES: MICROTUBULE ASSEMBLY AND MEMBRANE FUSION
INDUCED BY A COMPONENT OF COMPLEMENT
PROC NATL ACAD SCI USA 70: 2916-2920 (1973)

GOLDSTEIN IM   BRAI M   OSLER AG
WEISSMANN G
LYSOSOMAL ENZYME RELEASE FROM HUMAN LEUKOCYTES: MEDIATION
BY THE ALTERNATE PATHWAY OF COMPLEMENT ACTIVATION
J IMMUNOL 111:33-37(1973)

GOLDYNE M   WINKELMANN RK
PROSTAGLANDINS AS MEDIATORS OF CUTANEOUS VASCULAR
REACTIVITY IN DOG AND MAN
CLIN RES 19:579(1971)

GOLDYNE M   WINKELMANN RK
PROSTAGLANDINS AS MEDIATORS OF CUTANEOUS VASCULAR
REACTIVITY IN DOG AND MAN
J CLIN INVEST 50:38A(1971)

GOLDYNE ME   WINKELMANN RK
IN VITRO EFFECTS OF PROSTAGLANDIN E2 ON CUTANEOUS VASCULAR
SMOOTH MUSCLE IN THE DOG AND IN MAN
J INVEST DERMATOL 60:258-262(1973)

GOLDYNE ME   WINKELMANN RK   RYAN RJ
PROSTAGLANDIN ACTIVITY IN HUMAN CUTANEOUS INFLAMMATION:
DETECTION BY RADIOIMMUNOASSAY
PROSTAGLANDINS 4:737-749(1973)

GOLDYNE ME   WINKELMANN RK   RYAN RJ
RADIOIMMUNOASSAY OF PROSTAGLANDIN IN CUTANEOUS DISEASE
CLIN RES 21:477(1973)

GOMER SK   ZIMMERMAN BG
DETERMINATION OF VASODILATOR INNERVATION IN NORMAL CANINE
KIDNEY
PHARMACOLOGIST 13:300(1971)

GOMER SK   ZIMMERMAN BG
EFFECT OF INDOMETHACIN AND OTHER AGENTS ON SUSTAINED
SYMPATHETIC VASODILATION IN THE DOG'S PAW
ABSTR 5TH INT CONGR PHARMACOL, SAN FRANC:84(1972)

GOMER SK   ZIMMERMAN BG
DETERMINATION OF SYMPATHETIC VASODILATOR RESPONSES DURING
RENAL NERVE STIMULATION
J PHARMACOL EXP THER 181:75-82(1972)

GOODSON JM
A POTENTIAL ROLE OF PROSTAGLANDINS IN THE ETIOLOGY OF
PERIODONTAL DISEASE
IN: "PROSTAGLANDINS AND CYCLIC AMP," RH KAHN, WEM LANDS,
ACADEMIC PRESS, NY : 215-216(1973)

GOODSON JM   DEWHIRST F   BRUNETTI A
PROSTAGLANDIN E2 LEVELS IN HUMAN GINGIVAL TISSUE
J DENT RES 52(SPEC):182(1973)

GORA S
PGE1 INFLUENCE ON FROG'S CALF MUSCLE CONTRACTILITY IN
PERFUSED PREPARATION
IN: "FARMAKOLOGIA AMIN KATECHOLOWYCH ORAZ LEKOW DZIALAJACYCH
NA UKLAD NERWOWY," I ZJAZDU, PANSTWOWY ZAKLAD WYDAWNICTW
LEKARSKICH, WARSAW : 261-264(1971);(POLISH)

GORDON AS
ERYTHROPOIETIN
VITAM HORM 31:105-174(1973)

GORMAN RR   MILLER OV
SPECIFIC PGE AND PGA RECEPTORS IN RAT ADIPOCYTE PLASMA
MEMBRANES
PHARMACOLOGIST 15:209(1973)

GORMAN RR   MILLER OV
SPECIFIC PROSTAGLANDIN E1 AND A1 BINDING SITES IN RAT
ADIPOCYTE PLASMA MEMBRANES
BIOCHIM BIOPHYS ACTA 323:560-572(1973)

GOTS R   FORMAL S   GIANNELLA R
INDOMETHACIN INHIBITION OF SALMONELLA, SHIGELLA, AND CHOLERA
TOXIN MEDIATED RABBIT ILEAL FLUID SECRETION
CLIN RES 5:975(1973)

GOUGH HG
A FACTOR ANALYSIS OF CONTRACEPTIVE PREFERENCES
J PSYCHOL 84:199-210(1973)

GOYAL RK   RATTAN S
MECHANISM OF THE LOWER ESOPHAGEAL SPHINCTER RELAXATION
ACTION OF PROSTAGLANDIN E1 AND THEOPHYLLINE
J CLIN INVEST 52:337-341(1973)

GOYAL RK   RATTAN S   HERSH T
DOSE-RESPONSE CURVES OF THE EFFECTS OF THE DIFFERENT
PROSTAGLANDINS ON THE LOWER ESOPHAGEAL SPHINCTER
CLIN RES 20:454(1972)

GOYAL RK   RATTAN S   HERSH T
COMPARISON OF THE EFFECTS OF PROSTAGLANDINS E1, E2, AND A2,
AND OF HYPOVOLUMIC HYPOTENSION ON THE LOWER ESOPHAGEAL
SPHINCTER
GASTROENTEROLOGY 65:608-612 (1973)

GRAHAME-SMITH DG
ENDOCRINE TUMORS PRODUCING GASTROINTESTINAL SYMPTOMS
IN:INT ENCYCL PHARMACOL THER,G PETERS,PERGAMON PRESS,NY
2:639-665(1973)

GRANDE F   PRIGGE WF
INFLUENCE OF PROSTAGLANDIN E1 ON THE ADIPOKINETIC EFFECT OF
GLUCAGON IN BIRDS
PROC SOC EXP BIOL MED 140:999-1004(1972)

GRANSTROM E
METABOLISM OF PROSTAGLANDIN F2A IN SWINE KIDNEY.
BIOCHIM BIOPHYS ACTA 239:120-125(1971)

GRANSTROM E
METABOLISM OF PROSTAGLANDIN F2A IN GUINEA PIG LUNG
EUR J BIOCHEM 20:451-458(1971)

GRANSTROM E
METABOLISM OF PROSTAGLANDIN F2A IN FEMALE SUBJECTS
IN:"PROSTAGLANDINS IN FERTILITY CONTROL 2," S BERGSTROM, K
GREEN AND B SAMUELSSON, W H O, KAROLINSKA INST, STOCKH:
18-37(1972)

GRANSTROM E
STRUCTURES OF C14 METABOLITES OF PROSTAGLANDIN F2A
ADV BIOSCI (SUPPL) 9:7(1972)

GRANSTROM E
STUDIES ON THE METABOLISM OF PROSTAGLANDIN F2A
DISS, KAROLINSKA INST ROY VET COLL, STOCKH (1972)

GRANSTROM E
STRUCTURES OF C14 METABOLITES OF PROSTAGLANDIN F2A
ADV BIOSCI 9:49-60(1973)

GRANSTROM E   SAMUELSSON B
ON THE METABOLISM OF PROSTAGLANDIN F2A IN FEMALE SUBJECTS.
II. STRUCTURES OF SIX METABOLITES
J BIOL CHEM 246:7470-7485(1971)

GRANSTROM E   SAMUELSSON B
STRUCTURE OF A DEOXYPROSTAGLANDIN IN MAN
J AM CHEM SOC 94:4380-4381(1972)

GRANSTROM E   SAMUELSSON B
DEVELOPMENT AND MASS SPECTROMETRIC EVALUATION OF A
RADIOIMMUNOASSAY FOR 9A, 11A-DIHYDROXY-15-KETOPROST-5-ENOIC
ACID
FEBS LETT 26:211-214(1972)

GRANT JA   LICHTENSTEIN L   ISHIZAKA K
MARSH D
HISTAMINE RELEASE FROM HUMAN LEUKOCYTES BY ANTI-IGG
FED PROC 31:747(1972)

GRANT JA   LICHTENSTEIN LM
REVERSED IN VITRO ANAPHYLAXIS INDUCED BY ANTI-IGG:
SPECIFICITY OF THE REACTION AND COMPARISON WITH
ANTIGEN-INDUCED HISTAMINE RELEASE
J IMMUNOL 109:20-25(1972)

GRANT NH   ALBURN HE
DUAL EFFECTS OF PROSTAGLANDINS ON HEAT DENATURATION OF
SERUM ALBUMIN
BIOCHEM PHARMACOL 20:429-436(1971)

GRAZIANI Y   LIVNE A
PROSTAGLANDIN, VASOPRESSIN AND WATER PERMEABILITY OF
ARTIFICIAL LIPID MEMBRANES
ISR J MED SCI 8:1006(1972)

GRAZIANI Y   LIVNE A
BILAYER LIPID MEMBRANE AS A MODEL FOR VASOPRESSIN,
PROSTAGLANDIN, AND CA2+ EFFECTS ON WATER PERMEABILITY
BIOCHIM BIOPHYS ACTA 291:612-620(1973)

GREAVES MW   MCDONALD-GIBSON W
INHIBITION OF PROSTAGLANDIN BIOSYNTHESIS BY CORTICOSTEROIDS
BR MED J 2:83-84(1972)

GREAVES MW   MCDONALD-GIBSON W
EXTRACTION OF PROSTAGLANDIN-LIKE ACTIVITY FROM WHOLE HUMAN
BLOOD
LIFE SCI(I) 11 : 73-81(1972)

GREAVES MW   MCDONALD-GIBSON W
PROSTAGLANDIN BIOSYNTHESIS BY SKIN AND EFFECTS OF
ANTI-INFLAMMATORY CORTICOSTEROIDS
ABSTR 5TH INT CONGR PHARMACOL SAN FRANC 87(1972)

GREAVES MW   MCDONALD-GIBSON W
PROSTAGLANDIN BIOSYNTHESIS BY SKIN AND EFFECTS OF
ANTI-INFLAMMATORY DRUGS
J INVEST DERMATOL 58:258(1972)

GREAVES MW   MCDONALD-GIBSON W
PROSTAGLANDIN BIOSYNTHESIS BY HUMAN SKIN AND ITS INHIBITION
BY CORTICOSTEROIDS
BR J PHARMACOL 44:172-175(1972)

GREAVES MW   MCDONALD-GIBSON W
EFFECT OF NON-STEROID ANTI-INFLAMMATORY DRUGS ON
PROSTAGLANDIN BIOSYNTHESIS BY SKIN
BR J DERMATOL 88:47-50(1973)

GREAVES MW   MCDONALD-GIBSON W
EFFECT OF NONSTEROID ANTI-INFLAMMATORY AND ANTIPYRETIC
DRUGS ON PROSTAGLANDIN BIOSYNTHESIS BY HUMAN SKIN
J INVEST DERMATOL 61:127-129(1973)

GREAVES MW   MCDONALD-GIBSON W
ITCH: ROLE OF PROSTAGLANDINS
BR MED J 3:608-609(1973)

GREAVES MW   MCDONALD-GIBSON W   SONDERGAARD JW
RECOVERY OF PROSTAGLANDIN-LIKE FATTY ACIDS FROM HUMAN
ALLERGIC CONTACT ECZEMA USING A SKIN PERFUSION METHOD
BR J PHARMACOL 41:416P(1971)

GREAVES MW   MCDONALD-GIBSON WJ
ANTI-INFLAMMATORY AGENTS AND PROSTAGLANDIN BIOSYNTHESIS
BR MED J 3:527(1972)

GREAVES MW   MCDONALD-GIBSON WJ MCDONALD-GIBSON RG
THE EFFECT OF VENOUS OCCLUSION, STARVATION AND EXERCISE ON
PROSTAGLANDIN ACTIVITY IN WHOLE HUMAN BLOOD
LIFE SCI (II) 11:919-924(1972)

GREAVES MW   SONDERGAARD J   MCDONALD-GIBSON W
RECOVERY OF PROSTAGLANDINS IN HUMAN CUTANEOUS INFLAMMATION
BR MED J 2:258-260(1971)

GREAVES MW   SONDERGAARD JS   HOLT G
MCDONALD-GIBSON W
RECOVERY OF PROSTAGLANDIN-LIKE FATTY ACIDS FROM HUMAN
ALLERGIC CONTACT ECZEMA BY THE USE OF A SKIN PERFUSION
METHOD
J PATHOL 103:PVIII-PIX(1971)

GREEN K
METABOLISM OF PROSTAGLANDIN E2 IN THE RAT
BIOCHEMISTRY 10:1072-1086(1971)

GREEN K
METABOLISM OF PROSTAGLANDIN F2A IN THE RAT
BIOCHIM BIOPHYS ACTA 231:419-444(1971)

GREEN K
QUANTITATIVE MASS SPECTROMETRIC ANALYSIS OF PROSTAGLANDINS
ADV BIOSCI 9:91-108(1973)

GREEN K
METHODS FOR QUANTITATIVE DETERMINATION OF PROSTAGLANDINS
USING GAS CHROMATOGRAPHY-MASS SPECTROMETRY
IN: "PROSTAGLANDINES 1973," INSERM, PARIS:13-132(1973)

GREEN K
PERMEABILITY PROPERTIES OF THE CILIARY EPITHELIUM IN
RESPONSE TO PROSTAGLANDINS
INVEST OPHTALMOL 12:752-758(1973)

GREEN K
QUANTITATIVE MEASUREMENT OF PROSTAGLANDINS USING GAS
CHROMATOGRAPHY-MASS SPECTROMETRY
ABSTR, 9TH INT CONG BIOCHEM, STOCKH:398(1973)

GREEN K   GRANSTROM E   SAMUELSSON B
QUANTITATIVE GAS CHROMATOGRAPHY-MASS SPECTROMETRY OF
PROSTAGLANDINS
IN: "PROSTAGLANDINS IN FERTILITY CONTROL 2," S BERGSTROM, K
GREEN AND B SAMUELSSON, W H O, KAROLINSKA INST, STOCKH:
92-106(1972)

GREEN K   SAMUELSSON B
QUANTITATIVE STUDIES ON THE SYNTHESIS IN-VIVO OF
PROSTAGLANDINS IN THE RAT--COLD STRESS INDUCED STIMULATION
OF SYNTHESIS
EUR J BIOCHEM 22:391-395(1971)

GREENBERG R   BEAULIEU G
THE BRONCHODILATOR ACTIVITY OF AY-22,093, A PROSTANOIC ACID
DERIVATIVE
PROC CAN FED BIOL SOC 15:466(1972)

GREENBERG S
DIFFERENTIATION OF THE CALCIUM STORES UTILIZED IN THE
CONTRACTILE RESPONSE TO NOREPINEPHRINE AND PROSTAGLANDINS
DISSERTATION UNIVERSITY OF IOWA (1972)

GREENBERG S   DIECKE FPJ   LONG P
STUDIES ON THE MECHANISM OF PROSTAGLANDIN A2 AND
B2-INDUCED VASOCONSTRICTION
CLIN RES 21:421(1973)

GREENBERG S   ENGELBRECHT JA   WILSON WR
DEPENDENCE OF PROSTAGLANDIN B-INDUCED VASOCONSTRICTION ON
DEPOLARIZATION AND GRANULAR NOREPINEPHRINE STORES
PHARMACOLOGIST 15:208(1973)

GREENBERG S   ENGELBRECHT JA   WILSON WR
PROSTAGLANDIN B2-INDUCED CUTANEOUS VASOCONSTRICTION - A
MODEL FOR RAYNAUD'S PHENOMENON
CIRCULATION (SUPPL IV) 8:IV-28 (1973)

GREENBERG S   ENGLEBRECHT JA   WILSON WR
CARDIOVASCULAR PHARMACOLOGY OF PROSTAGLANDIN B1 AND B2 IN
THE INTACT DOG
PROC SOC EXP BIOL MED 143: 1008-1013(1973)

GREENBERG S   KADOWITZ PJ   DIECKE FPJ
LONG JP
EFFECT OF PROSTAGLANDIN F2A (PGF2A) ON ARTERIAL AND VENOUS
CONTRACTILITY AND 45CA UPTAKE
ARCH INT PHARMACODYN THER 205 : 381-398 (1973)

GREENBERG S   KADOWITZ PJ   DIECKE FPJ
LONG JP
EFFECT OF PROSTAGLANDIN F2A ON RESPONSES OF VASCULAR SMOOTH
MUSCLE TO SEROTONIN, ANGIOTENSIN AND EPINEPHRINE
ARCH INT PHARMACODYN THER 206: 5-18 (1973)

GREENBERG S   LONG JP
ENHANCEMENT OF VASCULAR SMOOTH MUSCLE RESPONSES TO
VASOACTIVE STIMULI BY PROSTAGLANDIN E1 AND E2
ARCH INT PHARMACODYN THER 206: 94-104 (1973)

GREENBERG S   LONG JP   DIECKE FPJ
HEITZ DC   KADOWITZ PJ
A POSSIBLE MECHANISM OF PROSTAGLANDIN F2A (PGF2A) INDUCED
FACILITATION OF VASCULAR REACTIVITY
ABSTR 5TH INT CONGR PHARMACOL SAN FRANC 88(1972)

GREENBERG S   LONG JP   DIECKE FPJ
EFFECT OF PROSTAGLANDINS ON ARTERIAL AND VENOUS TONE AND
CALCIUM TRANSPORT
ARCH INT PHARMACODYN THER 204:373-389(1973)

GREENBERG S   WILSON WR   LONG JP
EFFECT OF METABOLIC INHIBITORS ON MESENTERIC ARTERIAL
SMOOTH MUSCLE
ARCH INT PHARMACODYN THER 206:213-228(1973)

GREENWAY CV   STARK RD
HEPATIC VASCULAR BED
PHYSIOL REV 51:23-65(1971)

GREEP RO
NATURE, ROLE AND CONTROL OF THE GONADOTROPHINS
J REPROD FERTIL (SUPPL) 18:1-13(1973)

GREEP RO
FACTORS INFLUENCING LUTEAL FUNCTION
PROC WORLD CONGR FERTIL STERIL 7:70-72(1973)

GRIECO PA   REAP JJ
PROSTAGLANDINS-A TOTAL SYNTHESIS OF (+1-)-11,15-DIDEOXY-PGE2
AND (+1-)-11-DEOXY-PGE2 METHYL ESTER
J ORG CHEM 38: 3413-3415(1973)

GRIEVES S   LIGGINS GC
ASSAY OF PROSTAGLANDINS BY GLC
ABSTR 4TH ASIA OCEANIA CONGR ENDOCR:99(1971)

GROLLMAN A
A SURVEY OF EXPERIMENTAL STUDIES IN RENAL HYPERTENSION
IN: "HYPERTENSION: MECHANISMS AND MANAGEMENT" G ONESTI KE
KIM AND JH MOYER,GRUNE & STRATTON,NY/LOND:581-589(1973)

GROSS F
NEW DRUG DEVELOPMENT-EXPECTATIONS AND SPECULATIONS
ACTA PHARMACOL SUEC 10:401-424(1973)

GROSS JB   BARTTER FC
EFFECTS OF PROSTAGLANDINS E1, A1, AND F2A ON RENAL HANDLING
OF SALT AND WATER
AM J PHYSIOL 225:218-224(1973)

GROSSMANN V   DOLEZAL S
THE EFFECT OF SOME PROSTAGLANDIN DERIVATIVES ON GASTRIC
SECRETION AND THEIR THERAPEUTIC VALUES AS ANTIULCERS
ABSTR 5TH INT CONGR PHARMACOL SAN FRANC 89(1972)

GROSTIC MF
DRUG METABOLISM
IN: "BIOCHEMICAL APPLICATIONS OF MASS SPECTROMETRY," GR
WALLER, WILEY-INTERSCIENCE, NY:573-590(1972)

GRUBY LA   ROWLANDS C   VARLEY BO
WYLLIE JH
THE FATE OF 5-HYDROXYTRYPTAMINE IN THE LUNGS
BR J SURG 58: 525-532(1971)

GRUDZINSKAS CV   WEISS MJ
PROSTAGLANDINS AND CONGENERS IV. THE SYNTHESIS OF CERTAIN
11-SUBSTITUTED DERIVATIVES OF 11-DEOXYPROSTAGLANDIN E2
AND F2A FROM 15-0-ACETYLPROSTAGLANDIN A2 METHYL ESTER
TETRAHEDRON LETT 141-144(1973)

GRUND VR   HUNNINGHAKE DB
HISTAMINE STIMULATED LIPOLYSIS IN ISOLATED FAT CELLS OF THE
DOG
ABSTR 5TH INT CONGR PHARMACOL SAN FRANC 90(1972)

GRYGLEWSKI R   FLOWER RJ   HERBACZYNSKA-CE/ K
VANE JR
INHIBITION OF PROSTAGLANDIN SYNTHETASE BY ANTI-INFLAMMATORY
DRUGS
ABSTR 5TH INT CONGR PHARMACOL SAN FRANC 90(1972)

GRYGLEWSKI R   VANE JR
RABBIT-AORTA CONTRACTING SUBSTANCE (RCS) MAY BE A
PROSTAGLANDIN PRECURSOR
BR J PHARMACOL 43:420P-421P(1971)

GRYGLEWSKI R   VANE JR
THE RELEASE OF PROSTAGLANDINS AND RABBIT AORTA CONTRACTING
SUBSTANCE (RCS) FROM RABBIT SPLEEN AND ITS ANTAGONISM BY
ANTI-INFLAMMATORY DRUGS
BR J PHARMACOL 45:37-47(1972)

GRYGLEWSKI R   VANE JR
THE GENERATION FROM ARACHIDONIC ACID OF RABBIT AORTA
CONTRACTING SUBSTANCE (RCS) BY A MICROSOMAL ENZYME
PREPARATION WHICH ALSO GENERATES PROSTAGLANDINS
BR J PHARMACOL 46:449-457(1972)

GUMULKA W   REWERSKI W   STRZALKOWSKA M
EFFECTS OF PROSTAGLANDINS OF THE F GROUP ON THE NERVOUS
SYSTEM
POL TYG LEK 28:1952-1954(1973);(POLISH)

GUMULKA W   REWERSKI W   STRZALKOWSKA M
SOME BIOLOGICAL AND PHARMACOLOGICAL PROPERTIES OF
PROSTAGLANDINS
POL TYG LEK 28:1904-1907(1973);(POLISH)

GUSTAFSSON L   HEDQVIST P   LAGERCRANTZ H
PROSTAGLANDIN MEDIATED ENHANCEMENT OF EFFECTOR RESPONSE TO
CHOLINERGIC NERVE STIMULATION
ACTA PHYSIOL SCAND (SUPPL 396) : 106(1973)

GUTIERREZ-CERN/ RM MORRILL LM   LEVINE L
PROSTAGLANDIN F2A LEVELS IN PERIPHERAL SERA OF MAN.
PROSTAGLANDINS 1:71-80(1972)

GUTKNECHT GO   JOHNSTON JO
DIRECT LUTEOLYTIC EFFECT OF LOCALLY ADMINISTERED
PROSTAGLANDIN F2A IN THE PREGNANT HAMSTER
BIOL REPROD 7:108-109(1972)

GUTTMACHER A
MEDICAL ASPECTS OF THE ABORTION EXPERIENCE
IN:THE ABORTION EXPERIENCE:PSYCHOLOGICAL & MEDICAL IMPACT,
HJ OSOFSKY AND JD OSOFSKY,HARPER & ROW,HAGERSTOWN MD:535-541
(1973)

GUZMAN A   CRABBE P
SYNTHESIS OF METHYLATED PROSTAGLANDINS
CHEM IND (LOND): 635-636 (1973)

GUZMAN A   CRABBE P
TOTAL SYNTHESIS OF A 9-DESOXY-PROSTAGLANDIN
CHEM LETT 1073-1075 (1973)

GYANG EA   DEUBEN RR   BUCKLEY JP
INTERACTION OF PROSTAGLANDIN E1 AND ANGIOTENSIN II ON
CENTRALLY MEDIATED PRESSOR ACTIVITIES IN THE CAT
PROC SOC EXP BIOL MED 142:532-537(1973)

HAAS E   GOLDBLATT H   LEWIS L
IDENTIFICATION OF THE ANGIOTENSIN PROTECTIVE COFACTOR AS
BRADYKININ
FED PROC 31:485(1972)

HAAS E   GOLDBLATT H   LEWIS L
GIPSON EC
INTERPLAY OF VASODILATOR AND VASOCONSTRICTOR SUBSTANCES
IN THE FEMORAL ARTERIAL BED. IDENTIFICATION OF THE
ANGIOTENSIN COFACTOR
LAB INVEST 28:1-7(1973)

HADHAZY P   ILLES P   KNOLL J
THE EFFECTS OF PGE1 ON RESPONSES TO CARDIAC VAGUS NERVE
STIMULATION AND ACETYLCHOLINE RELEASE
EUR J PHARMACOL 23:251-255(1973)

HADHAZY P   KNOLL J
EFFECT OF PROSTAGLANDIN E1 ON NEUROCHEMICAL MYOCARDIAL
TRANSMISSION
ORVOSTUDOMANY 24:81-88(1973);(HUNGARIAN)

HAEFFNER EW   PRIVETT OS
DEVELOPMENT OF DERMAL SYMPTOMS RESEMBLING THOSE OF AN
ESSENTIAL FATTY ACID DEFICIENCY IN IMMATURE
HYPOPHYSECTOMIZED RATS
J NUTR 103:74-79(1973)

HAGGENDAL J
REGULATION OF CATECHOLAMINE RELEASE
LIFE SCI 13:LXV-LXVI(1973)

HAGGITT RC   PITCOCK JA   MUIRHEAD EE
RENAL MEDULLARY FIBROSIS IN HYPERTENSION
HUM PATHOL 2:587-597(1971)

HAHN RA
AN INVESTIGATION INTO THE INTERACTION OF PROSTAGLANDIN F2A
WITH CHOLINERGIC MECHANISMS IN CANINE SALIVARY GLANDS
DISSERTATION OHIO STATE UNIVERSITY (1972)

HAHN RA   PATIL PN
SALIVATION INDUCED BY PROSTAGLANDIN F2A AND MODIFICATION OF
THE RESPONSE BY ATROPINE AND PHYSOSTIGMINE
BR J PHARMACOL 44:527-533(1972)

HAKANSON R   LIEDBERG G   OSCARSON J
EFFECTS OF PROSTAGLANDIN E1 ON ACID SECRETION, MUCOSAL
HISTAMINE CONTENT AND HISTIDINE DECARBOXYLASE ACTIVITY IN
RAT STOMACH
BR J PHARMACOL 47:498-503(1973)

HAKIM SAE
ASPIRIN INHIBITS OCULO-TENSIN AND PREVENTS RISE OF EYE
TENSION
INDIAN J OPHTHALMOL 19:145-154(1971)

HALES JRS   BAIRD JA
EFFECTS OF 5-HYDROXYTRYPTAMINE, NORADRENALINE,
CHOLINOMIMETIC SUBSTANCES & PROSTAGLANDINS E1 & E2 ON
THERMOREGULATION IN THE ECHIDNA
PROC AUST PHYSIOL PHARMACOL SOC 3:172(1972)

HALES JRS   BENNETT JW   BAIRD JA
FAWCETT AA
THERMOREGULATORY EFFECTS OF PROSTAGLANDINS E1, E2, F1A AND
F2A IN THE SHEEP
PFLUEGERS ARCH 339:125-133(1973)

HALL RC   CAIN MD   HODGE RL
IRVINE RI   JACKSON HR
PROSTAGLANDINS AND ENDOTOXIN SHOCK
ABSTR 6TH ANNU MEET, AUSTRALAS SOC CLIN PHARMACOL, SYD:
56(1972)

HALL WJ
SEASONAL CHANGE IN THE RESPONSE OF FROG SKIN TO
PROSTAGLANDIN
IR J MED SCI 141:96(1972)

HALL WJ
SEASONAL CHANGES IN THE SENSITIVITY OF FROG SKIN TO
PROSTAGLANDIN AND THE EFFECT OF EXTERNAL SODIUM AND CHLORIDE
ON THE RESPONSE
IR J MED SCI 142:230-243(1973)

HALL WJ   MARTIN JDG
THE INDEPENDENCE OF CALCIUM AND PROSTAGLANDIN E1 ACTIONS ON
SODIUM TRANSPORT IN FROG SKIN
IR J MED SCI 140:412(1971)

HALL WJ   MARTIN JDG
PROSTAGLANDIN E1 AND WATER MOVEMENT IN FROG SKIN
ADV BIOSCI (SUPPL) 9:33(1972)

HALL WJ   MARTIN JDG
PROSTAGLANDIN E1 AND NET WATER FLOW ACROSS ISOLATED FROG
SKIN
J PHYSIOL (LOND) 232:85P-86P(1973)

HALLINAN EA
SEPARATION OF PROSTAGLANDIN DIASTEREOMERS BY COUNTERCURRENT
DISTRIBUTION
J CHROMATOGR 79:368-369(1973)

HALPERN BD   AKKAPEDDI MK
HYDROGEN CERVICAL DILATOR: A NEW DELIVERY SYSTEM FOR
PROSTAGLANDINS
ABSTR SYMP ON CONTROLLED RELEASE OF BIOLOGICALLY ACTIVE
AGENTS, SOUTHERN RES INST, BIRM(1973)

HAM EA   CIRILLO VJ   ZANETTI M
SHEN TY   KUEHL FA JR
STUDIES ON THE MODE OF ACTION OF NON-STEROIDAL
ANTI-INFLAMMATORY AGENTS
IN: "PROSTAGLANDINS IN CELLULAR BIOLOGY," PW RAMWELL AND
BB PHARRISS, PLENUM PRESS, NY/LOND:345-352(1972)

HAMAMURA L   LEME JG
NEUROGENIC INFLAMMATION IN THE RAT. EVIDENCE FOR THE RELEASE
OF A PERMEABILITY FACTOR FROM SENSORY NERVES FOLLOWING
ELECTRICAL STIMULATION
AGENTS ACTIONS 3:381-382(1973)

HAMBERG M
STERIC ANALYSIS OF HYDROPEROXIDES FORMED BY LIPOXYGENASE
OXYGENATION OF LINOLEIC ACID
ANAL BIOCHEM 43:515-526(1971)

HAMBERG M
INHIBITION OF PROSTAGLANDIN SYNTHESIS IN MAN
BIOCHEM BIOPHYS RES COMMUN 49:720-726(1972)

HAMBERG M
A NOTE ON NOMENCLATURE
ADV BIOSCI 9:847-850(1973)

HAMBERG M
QUANTITATIVE STUDIES ON PROSTAGLANDIN SYNTHESIS IN MAN-II-
DETERMINATION OF THE MAJOR URINARY METABOLITE OF
PROSTAGLANDINS F1A AND F2A
ANAL BIOCHEM 55:368-378(1973)

HAMBERG M   ISRAELSSON U   SAMUELSSON B
METABOLISM OF PROSTAGLANDIN E2 IN GUINEA PIG LIVER
ANN NY ACAD SCI 180:164-180(1971)

HAMBERG M   JONSSON CE
INCREASED SYNTHESIS OF PROSTAGLANDINS IN THE GUINEA PIG
FOLLOWING SCALDING INJURY
ACTA PHYSIOL SCAND 87:240-245(1973)

HAMBERG M   SAMUELSSON B
METABOLISM OF PROSTAGLANDIN E2 IN GUINEA PIG LIVER
J BIOL CHEM 246:1073-1077(1971)

HAMBERG M   SAMUELSSON B
ON THE METABOLISM OF PROSTAGLANDINS E1 AND E2 IN THE GUINEA
PIG
J BIOL CHEM 247:3495-3502(1972)

HAMBERG M   SAMUELSSON B
DETECTION AND ISOLATION OF AN ENDOPEROXIDE INTERMEDIATE IN
PROSTAGLANDIN BIOSYNTHESIS
PROC NATL ACAD SCI USA 70: 899-903 (1973)

HAMBERG M   WILSON M
STRUCTURES OF NEW METABOLITES OF PGE2 IN MAN
ADV BIOSCI (SUPPL) 9:16(1972)

HAMBERG M   WILSON M
STRUCTURES OF NEW METABOLITES OF PROSTAGLANDIN E2 IN MAN
ADV BIOSCI 9:39-48(1973)

HAMMARSTROM S   SAMUELSSON B   BJURSELL G
PROSTAGLANDIN LEVELS IN NORMAL AND TRANSFORMED
BABY-HAMSTER-KIDNEY FIBROBLASTS
NATURE 243:50-51(1973)

HAMMARSTROM S   SAMUELSSON B   BJURSELL G
PROSTAGLANDIN LEVELS IN NORMAL AND TRANSFORMED
BABY-HAMSTER-KIDNEY FIBROBLASTS
ABSTR, 9TH INT CONG BIOCHEM, STOCKH:398(1973)

HAMPRECHT B   JAFFE BM   PHILPOTT GW
PROSTAGLANDIN PRODUCTION BY NEUROBLASTOMA, GLIOMA AND
FIBROBLAST CELL LINES: STIMULATION BY N6-02'-DIBUTYRYL
ADENOSINE 3':5'-CYCLIC MONOPHOSPHATE
FEBS LETT 36:193-198(1973)

HAMPRECHT B   SCHULTZ J
STIMULATION BY PROSTAGLANDIN E1 OF ADENOSINE 3':5'-CYCLIC
MONOPHOSPHATE FORMATION IN NEUROBLASTOMA CELLS IN THE
PRESENCE OF PHOSPHODIESTERASE INHIBITORS
FEBS LETT 34:85-89(1973)

HAMPRECHT B   SCHULTZ J
INFLUENCE OF NORADRENALIN, PROSTAGLANDIN E1 AND INHIBITORS
OF PHOSPHODIESTERASE ACTIVITY ON LEVELS OF THE CYCLIC
ADENOSINE 3':5'-MONOPHOSPHATE IN SOMATIC CELL HYBRIDS
HOPPE SEYLERS Z PHYSIOL CHEM 354:1633-1641(1973)

HANSON JM   MORLEY J   SORIA C
ANTI-INFLAMMATORY PROPERTY OF 401, A PEPTIDE FROM THE VENOM
OF THE BEE (APIS MELLIFICA L)
BR J PHARMACOL 46:537P-538P(1972)

HAPKE HJ
PROSTAGLANDINS: SOMATIC SUBSTANCES AND POSSIBLE DRUGS
DTSCH TIERAERZTL WOCHENSCHR 80:237-240(1973):(GERMAN)

HARBON S   VESIN MF   CLAUSER H
ROLE OF PROSTAGLANDINS IN THE REGULATION OF MOTILITY AND
ADENYLATE CYCLASE ACTIVITY IN UTERINE MUSCLE
IN: "PROSTAGLANDINES 1973," INSERM, PARIS:81-93(1973):
(FRENCH)

HARBON S   VESIN MF   CLAUSER H
INDEPENDENT EFFECTS OF PROSTAGLANDINS ON CYCLIC AMP LEVELS
AND RAT UTERUS MOTILITY. A COMPARATIVE STUDY WITH
EPINEPHRINE AND OXYTOCIN
ABSTR, 9TH INT CONG BIOCHEM, STOCKH:355(1973)

HARMS PG   OJEDA SR   MCCANN SM
PROSTAGLANDIN INVOLVEMENT IN HYPOTHALAMIC CONTROL OF
GONADOTROPIN AND PROLACTIN RELEASE
SCIENCE 181:760-761(1973)

HARPER MJK   SKARNES RC
THE ROLES OF PROSTAGLANDIN F AND SEROTONIN IN
ENDOTOXIN-INDUCED ABORTION AND FETAL DEATH
ADV BIOSCI (SUPPL) 9:132(1973)

HARRIS DN   SEMENUK NS   HESS SM
AGENTS AFFECTING CYCLIC AMP LEVELS
ANNU REP MED CHEM 8:224-233(1973)

HARRIS M   JENKINS MV   BENNETT A
WILLS MR
PROSTAGLANDIN PRODUCTION AND BONE RESORPTION BY DENTAL CYSTS
NATURE (LOND) 245:213-215(1973)

HARRIS M   JENKINS MV   BENNETT A
WILLS MR
PROSTAGLANDINS, BONE RESORPTION AND HYPERCALCEMIA
N ENGL J MED 289:865 (1973)

HARRISON IT   GRAYSHAN R   WILLIAMS T
SEMENOVSKI A   FRIED JH
SYNTHESIS OF BIS-HOMO-PROSTAGLANDINS
TETRAHEDRON LETT 5151-5154(1972)

HARRISON RG
RESPONSES OF THE MICROVESSELS TO TOPICALLY APPLIED
PROSTAGLANDINS
BIORHEOLOGY 10:288(1973)

HARRISON RG   WOLF J
LOCAL ACTIONS OF PROSTAGLANDINS ON THE MICROVASCULATURE
OF THE HAMSTER CHEEK POUCH
BIBL ANAT (12) : 453-458(1973)

HARWOOD JP   MOSKOWITZ J   KRISHNA G
DYNAMIC INTERACTION OF PROSTAGLANDIN AND NOREPINEPHRINE IN
THE FORMATION OF ADENOSINE 3',5'-MONOPHOSPHATE IN HUMAN AND
RABBIT PLATELETS
BIOCHIM BIOPHYS ACTA 261:444-456(1972)

HASHIKAWA T   SAKURAI K   SUZUKI T
ON THE PROSTAGLANDIN EXTRACTED FROM SKIN AND INTESTINE OF
FROG
PROC 44TH GEN MEET JAP PHARMACOL SOC, TOKYO:
45-46(1971)::(JAPANESE)

HASLAM RJ
INTERACTIONS OF THE PHARMACOLOGICAL RECEPTORS OF BLOOD
PLATELETS WITH ADENYLATE CYCLASE
SER HAEMATOL 6:333-350(1973)

HASLAM RJ   ROSSON GM
AGGREGATION OF HUMAN BLOOD PLATELETS BY VASOPRESSIN
AM J PHYSIOL 223:958-967(1972)

HASLAM RJ   TAYLOR A
ROLE OF CYCLIC 3',5'-ADENOSINE MONOPHOSPHATE IN PLATELET
AGGREGATION
IN: PLATELET AGGREGATION, J CAEN, MASSON, PARIS: 85-93(1971)

HASPELS AA   LUIJGIES JHH
INDUCTION OF ABORTION BY INTRAVENOUS AND INTRA-UTERINE
ADMINISTRATION OF PROSTAGLANDIN F2A
J REPROD MED 9:442-447(1972)

HASSIM AM
PROSTAGLANDINS IN OBSTETRICS AND GYNAECOLOGY
MED J ZAMBIA 5:24-29(1971)

HAUBRICH DR   PEREZ-CRUET J   REID WD
PROSTAGLANDIN E1 CAUSES SEDATION AND INCREASES
5-HYDROXYTRYPTAMINE TURNOVER IN RAT BRAIN
BR J PHARMACOL 48:80-87(1973)

HAYAISHI O   YAMAMOTO S
NEW BIOACTIVE SUBSTANCES, PROSTAGLANDINS
NIPPON YAKUZAISHIKAI ZASSHI 24(10):9-10(1972)::(JAPANESE)

HAYAISHI S
RECENT PROGRESS IN THE STUDY OF PROSTAGLANDINS
PROC SOC GERONTOL STUDY, OKAYAMA CITY, ONO PHARMACEUTICAL
CO, OSAKA:56-64(1972)::(JAPANESE)

HAYASHI M   MIYAKE H   TANOUCHI T
IGUCHI S   IGUCHI Y   TANOUCHI F
THE SYNTHESIS OF 16(R)- OR 16(S)- METHYLPROSTAGLANDINS
J ORG CHEM 38:1250-1251(1973)

HAYE B   CHAMPION S   JACQUEMIN C
CONTROL BY TSH OF A PHOSPHOLIPASE A2 ACTIVITY, A LIMITING
FACTOR IN THE BIOSYNTHESIS OF PROSTAGLANDINS IN THE THYROID
FEBS LETT 30:253-260(1973)

HAYE B   CHAMPION S   JACQUEMIN C
CONTROL BY THE THYROID STIMULATING HORMONE OF A
PHOSPHOLIPASE A2, LIMITING FACTOR OF THE BIOSYNTHESIS OF
PROSTAGLANDINS IN THE THYROID
ABSTR, 9TH INT CONG BIOCHEM, STOCKH:404(1973)

HEATHER JB   SOOD R   PRICE P
PERUZZOTTI GP   LEE SS   LEE LFH
SIH CJ
TOTAL SYNTHESIS OF PROSTAGLANDINS. V. A SYNTHESIS OF
(-)-PROSTAGLANDIN E2 VIA A TOTALLY ASYMMETRIC PROCESS
TETRAHEDRON LETT 2313-2316(1973)

HEDGE GA
EFFECTS OF CYCLIC AMP AND PROSTAGLANDINS ON ACTH SECRETION
PROC INT UNION PHYSIOL SCI, 25TH INT CONGR, MUNICH,
9:241(1971)

HEDGE GA
INCREASED ACTH SECRETION DUE TO HYPOTHALAMIC MICROINJECTION
OF PROSTAGLANDINS
ABSTR 4TH INT CONGR ENDOCRINOL, INT CONGR SER 256, EXCERPTA
MEDICA, AMST:22-23(1972)

HEDGE GA
EFFECTS OF CYCLIC AMP AND PROSTAGLANDINS ON ACTH SECRETION
PROC INT UNION PHYSIOL SCI, 25TH INT CONGR, MUNICH 9 : 241
(1972)

HEDGE GA   HANSON SD
THE EFFECTS OF PROSTAGLANDINS ON ACTH SECRETION
ENDOCRINOLOGY 91:925-933(1972)

HEDQVIST P
PROSTAGLANDIN E COMPOUNDS AND SYMPATHETIC NEUROMUSCULAR
TRANSMISSION.
ANN NY ACAD SCI 180:410-415(1971)

HEDQVIST P
PROSTAGLANDIN MEDIATED CONTROL OF NEUROTRANSMISSION IN
GUINEA PIG SEMINAL VESICLE
ABSTR 5TH INT CONGR PHARMACOL SAN FRANC 98(1972)

HEDQVIST P
PROSTAGLANDIN-INDUCED INHIBITION OF VASCULAR TONE AND
REACTIVITY IN THE CAT'S HINDLEG IN VIVO
EUR J PHARMACOL 17:157-162(1972)

HEDQVIST P
PROSTAGLANDIN MEDIATED CONTROL OF SYMPATHETIC NEUROMUSCULAR
TRANSMISSION
ADV BIOSCI (SUPPL) 9:75(1972)

HEDQVIST P
DISSOCIATION OF PROSTAGLANDIN AND ALPHA-RECEPTOR MEDIATED
CONTROL OF ADRENERGIC TRANSMITTER RELEASE
ACTA PHYSIOL SCAND 87:42A-43A(1973)

HEDQVIST P
PROSTAGLANDIN MEDIATED CONTROL OF SYMPATHETIC
NEUROEFFECTOR TRANSMISSION
ADV BIOSCI 9:461-473(1973)

HEDQVIST P
AUTONOMIC NEUROTRANSMISSION
IN:"THE PROSTAGLANDINS," PW RAMWELL, PLENUM PRESS, NY
1:101-131(1973)

HEDQVIST P
PROSTAGLANDIN ACTION AT ADRENERGIC NEUROEFFECTOR JUNCTIONS
IN: "PROSTAGLANDINES 1973," INSERM, PARIS:225-238(1973)

HEDQVIST P
PROSTAGLANDIN AND A-RECEPTOR MEDIATED CONTROL OF TRANSMITTER
RELEASE FROM ADRENERGIC NERVES
LIFE SCI 13:LXX-LXXII (1973)

HEDQVIST P
PROSTAGLANDIN AS A TOOL FOR LOCAL CONTROL OF TRANSMITTER
RELEASE FROM SYMPATHETIC NERVES
BRAIN RES 62:483-488(1973)

HEDQVIST P
ASPECTS ON PROSTAGLANDIN MEDIATED CONTROL OF ADRENERGIC
TRANSMITTER RELEASE
ACTA PHYSIOL SCAND (SUPPL 396) : 19(1973)

HEDQVIST P   HOLMGREN A   MATHE A
SVENBORG N
BRONCHIAL HYPERREACTIVITY TO PROSTAGLANDIN F2A AND
HISTAMINE IN PATIENTS WITH ASTHMA
ADV BIOSCI (SUPPL) 9:41(1972)

HEDQVIST P   HOLMGREN A   MATHE AA
EFFECT OF PROSTAGLANDIN F2A ON AIRWAY RESISTANCE IN MAN
ACTA PHYSIOL SCAND 82:29A(1971)

HEDQVIST P   STJARNE L   WENNMALM A
FACILITATION OF SYMPATHETIC NEUROTRANSMISSION IN THE CAT
SPLEEN AFTER INHIBITION OF PROSTAGLANDIN SYNTHESIS.
ACTA PHYSIOL SCAND 83:430-432(1971)

HEDQVIST P   WENNMALM A
COMPARISON OF THE EFFECTS OF PROSTAGLANDINS E1, E2 AND F2A
ON THE SYMPATHETICALLY STIMULATED RABBIT HEART.
ACTA PHYSIOL SCAND 83:156-162(1971)

HEDWALL PR   ABDEL-SAYED WA   MARK AL
ABBOUD FM
VASCULAR RESPONSES TO PROSTAGLANDIN E1 IN GRACILIS MUSCLE
AND HINDPAW OF THE DOG
AM J PHYSIOL 221:42-47(1971)

HEINEMANN HO
THE LUNG AS A METABOLIC ORGAN: AN OVERVIEW
FED PROC 32:1955-1956(1973)

HELLEM AJ
METABOLIC DISORDERS OF PLATELETS
ADV INT MED 17:171-187(1971)

HELMSTAEDT D
EFFECT OF PRENYLAMINE ON THROMBOCYTES IN VITRO AND IN VIVO
DISS, JOHANN WOLFGANG GOETHE-UNIVERSITAT, FRANKF AM (1971);
(GERMAN)

HELMSTAEDT D   JEUCK K   MAY B
BOHLE E
EFFECT OF PROSTAGLANDINS ON FATTY LIVER
FETTE SEIFEN ANSTRICHM 74:629(1972);(GERMAN)

HENNEY CS
ON THE MECHANISM OF T-CELL MEDIATED CYTOLYSIS
TRANSPLANT REV 17:37-70(1973)

HENNEY CS   BOURNE HR   LICHTENSTEIN LM
THE ROLE OF CYCLIC AMP IN THE SPECIFIC CYTOLYTIC ACTIVITY OF
LYMPHOCYTES
FED PROC 31:797(1972)

HENNEY CS   BOURNE HR   LICHTENSTEIN LM
THE ROLE OF CYCLIC 3',5'-ADENOSINE MONOPHOSPHATE IN THE
SPECIFIC CYTOLYTIC ACTIVITY OF LYMPHOCYTES
J IMMUNOL 108:1526-1534(1972)

HENNEY S
STUDIES ON THE MECHANISM OF T LYMPHOCYTE-MEDIATED CYTOLYSIS
INT ARCH ALLERGY APPL IMMUNOL 45:281-284(1973)

HENRICH H   LUTZ J
VASCULAR ESCAPE-PHENOMENON IN THE INTESTINAL CIRCULATION
AND ITS INDUCTION BY DIFFERENT VASOCONSTRICTOR AGENTS.
PFLUEGERS ARCH 329:82-94(1971);(GERMAN)

HENSON PM   OADES ZG   GOULD D
CYCLIC AMP AND ESTERASE ACTIVATION IN THE RELEASE OF
SEROTONIN FROM RABBIT PLATELETS
FED PROC 32:1010(1973)

HENZL MR   NORIEGA L   AZNAR R
ORTEGA E   SEGRE E
QUANTITATION OF UTERINE ACTIVITY AFTER VAGINAL
ADMINISTRATION OF PROSTAGLANDINS
ADV BIOSCI (SUPPL) 9:126(1972)

HERBACZYNSKA-CE/ K STASZEWSKA-BARC/ J JANCZEWSKA H
THE RELEASE OF A PROSTAGLANDIN-LIKE SUBSTANCE ACCOMPANYING
MUSCULAR EXERCISE IN THE HIND-LIMB OF THE DOG
INT RES COMMUN SYS (73-6)8-9-2(1973)

HERBACZYNSKA-CE/ K VANE JR
AN INTRA-RENAL ROLE FOR PROSTAGLANDIN PRODUCTION
ABSTR 5TH INT CONGR PHARMACOL SAN FRANC 100(1972)

HERBACZYNSKA-CE/ K VANE JR
LOCAL PROSTAGLANDIN PRODUCTION CONTRIBUTES TO BLOODFLOW
AUTOREGULATION IN THE DOG KIDNEY
ADV BIOSCI (SUPPL) 9:45(1972)

HERBACZYNSKA-CE/ K VANE JR
AN INTRA-RENAL ROLE FOR PROSTAGLANDIN PRODUCTION
EUR J CLIN INVEST 3:237(1973)

HERBACZYNSKA-CE/ K VANE JR
CONTRIBUTION OF INTRARENAL GENERATION OF PROSTAGLANDIN
TO AUTOREGULATION OF RENAL BLOOD FLOW IN THE DOG
CIRC RES 33:428-436 (1973)

HERMAN AG   ECKENFELS A   FERREIRA SH
VANE JR
RELATIONSHIP BETWEEN TONE OF ISOLATED SMOOTH MUSCLE
PREPARATIONS AND PRODUCTION OF PROSTAGLANDINS
ABSTR 5TH INT CONGR PHARMACOL SAN FRANC 100(1972)

HERRERA M
THE PROSTAGLANDINS. FACTS AND FANTASIES
REV CHIL OBSTET GINECOL 37:240-242(1972);(SPANISH)

HERTELENDY F
STUDIES ON GROWTH HORMONE SECRETION. II. STIMULATION BY
PROSTAGLANDINS IN VITRO.
ACTA ENDOCRINOL 68:355-362(1971)

HERTELENDY F   PEAKE G   TODD H
STUDIES ON GROWTH HORMONE SECRETION: III. INHIBITION OF
PROSTAGLANDIN, THEOPHYLLINE AND CYCLIC AMP STIMULATED
GROWTH HORMONE RELEASE BY VALINOMYCIN IN VITRO.
BIOCHEM BIOPHYS RES COMMUN 44:253-260(1971)

HERTELENDY F   TODD H   EHRHART K
BLUTE R
STUDIES ON GROWTH HORMONE SECRETION: IV. IN VIVO EFFECTS OF
PROSTAGLANDIN E1
PROSTAGLANDINS 2:79-91(1972)

HERXHEIMER H   ROETSCHER I
EFFECTS OF PROSTAGLANDIN E1 ON LUNG FUNCTION IN BRONCHIAL
ASTHMA
EUR J CLIN PHARMACOL 3:123-125(1971)

HICKLER RB
THE PROSTAGLANDINS
IN: "HYPERTENSION: MECHANISMS AND MANAGEMENT" G ONESTI KE
KIM AND JH MOYER,GRUNE & STRATTON,NY/LOND:671-679(1973)

HIGASHI GI   KREINER PW   KEIRNS JJ
BITENSKY MW
ADENOSINE 3',5'-CYCLIC MONOPHOSPHATE IN SCHISTOSOMA MANSONI
LIFE SCI 13:1211-1220(1973)

HIGGINS CB   BRAUNWALD E
THE PROSTAGLANDINS. BIOCHEMICAL, PHYSIOLOGIC AND CLINICAL
CONSIDERATIONS
AM J MED 53:92-112(1972)

HIGGINS CB   VATNER CF   FRANKLIN D
BRAUNWALD E
EFFECTS OF PROSTAGLANDIN A1 ON LEFT VENTRICULAR DYNAMIC IN
THE CONSCIOUS DOG
AM J PHYSIOL 222:1534-1538(1972)

HIGGINS CB   VATNER SF   BRAUNWALD E
REGIONAL HEMODYNAMIC EFFECTS F PROSTAGLANDIN A1 IN THE
CONSCIOUS DOG
AM HEART J 85:349-357(1973)

HIGGINS CB   VATNER SF   FRANKLIN D
BRAUNWALD E
EFFECTS OF PROSTAGLANDIN A1 (PGA1) ON LEFT VENTRICULAR
DYNAMICS IN CONSCIOUS DOGS
CLIN RES 19: 350(1971)

HIGGINS CB   VATNER SF   FRANKLIN D
PATRICK T   BRAUNWALD E
EFFECTS OF PROSTAGLANDIN A1 ON THE SYSTEMIC AND
CORONARY CIRCULATIONS IN THE CONSCIOUS DOG.
CIRC RES 28:638-648(1971)

HIGGS GA   VANE JR
AN IMPROVED METHOD FOR THE EXTRACTION OF PROSTAGLANDINS
PROSTAGLANDINS 4:695-699(1973)

HIGGS GA   YOULTEN LJF
PROSTAGLANDIN PRODUCTION BY RABBIT PERITONEAL
POLYMORPHONUCLEAR LEUKOCYTES IN VITRO
BR J PHARMACOL 44:330P(1972)

HILLIER K   DUTTON A   CORKER CS
THE EFFECT OF PROSTAGLANDIN F2A ON STEROID AND LH LEVELS IN
THE HUMAN FEMALE
ADV BIOSCI (SUPPL) 9:111(1972)

HILLIER K   DUTTON A   CORKER CS
SINGER A   EMBREY MP
PLASMA STEROID AND LUTEINIZING HORMONE LEVELS DURING
PROSTAGLANDIN F2A ADMINISTRATION IN LUTEAL PHASE OF
MENSTRUAL CYCLE
BR MED J 4:333-336(1972)

HILLIER K   HUNTER D
F-PROSTAGLANDINS--OCCURRENCE IN HUMAN OVARIAN, UTERINE AND
PERIPHERAL VENOUS PLASMA
INT RES COMMUN SYST (73-4)15-18-3(1973)

HINES JD   KAMEN B   CASTON D
ABNORMAL FOLATE BINDING PROTEIN(S) IN AZOTEMIC PATIENTS
BLOOD 42:997(1973)

HINMAN JW
PROSTAGLANDINS
ANNU REV BIOCHEM 41:161-178(1972)

HINMAN JW
ROUND TABLE DISCUSSION ON INFLAMMATION
IN: "PROSTAGLANDINS AND CYCLIC AMP," RH KAHN, WEM LANDS,
ACADEMIC PRESS, NY : 207-210(1973)

HINMAN JW
STUDIES ON EXPERIMENTAL HARVESTING AND REGROWTH OF PLEXAURA
HOMOMALLA IN GRAND CAYMAN WATERS
IN: FOOD-DRUGS FROM THE SEA PROCEEDINGS 1972, LR WORTHEN,
MARINE TECH SOC, WASH : 165-176(1973)

HIROSE T   SAID SI
RESPIRATORY EFFECTS OF PROSTAGLANDINS A1, E1 & F2A
CLIN RES 19:512(1971)

HIRSH J   DOERY JCG
PLATELET FUNCTION IN HEALTH AND DISEASE
PROG HEMATOL 7:185-234(1971)

HITTELMAN KJ   BUTCHER RW
CYCLIC AMP AND THE MECHANISM OF ACTION OF THE PROSTAGLANDINS
IN:"THE PROSTAGLANDINS. PHARMACOLOGICAL AND THERAPEUTIC
ADVANCES," MF CUTHBERT, HEINEMANN, LOND:151-165(1973)

HITTELMAN KJ   BUTCHER RW
EFFECTS OF ANTILIPOLYTIC AGENTS AND A-ADRENERGIC ANTAGONISTS
ON CYCLIC AMP METABOLISM IN HAMSTER WHITE ADIPOCYTES
BIOCHIM BIOPHYS ACTA 316:403-410(1973)

HO RJ   BOMBOY JD   SUTHERLAND EW
LOCAL FEEDBACK REGULATION OF CYCLIC NUCLEOTIDES
IN: "ENDOCRINOLOGY. PROC 4TH INT CONG ENDOCRINOL," RO SCOW,
FJG EBLING, IW HENDERSON, EXCERPTA MEDICA, AMST/NY:
352-358(1973)

HO RJ   SUTHERLAND EW
FORMATION AND RELEASE OF A HORMONE ANTAGONIST BY RAT
ADIPOCYTES
J BIOL CHEM 246:6822-6827(1971)

HO RJ   SUTHERLAND EW
OCCURENCE OF A HORMONE ANTAGONIST IN RAT EPIDIDYMAL FAT
CELLS
PHYSIOLOGIST 14:163(1971)

HOENSCH H
C-CELL CARCINOMA (PARAFOLLICULAR CELLS) OF THE THYROID
DTSCH MED WOCHENSCHR 96:126-129(1971);(GERMAN)

HOFFER BJ
INTERACTIONS OF PROSTAGLANDINS WITH SPECIFIC CENTRAL
NEURONS
ABSTR 7TH MIDDLE ATL REG MEET AM CHEM SOC PHILA:72-73(1972)

HOFFER BJ   SIGGINS GR   OLIVER AP
BLOOM FE
CYCLIC AMP MEDIATION OF NOREPINEPHRINE INHIBITION IN RAT
CEREBELLAR CORTEX: A UNIQUE CLASS OF SYNAPTIC RESPONSES
ANN N Y ACAD SCI 185:531-549(1971)

HOFFER BJ   SIGGINS GR   OLIVER AP
BLOOM FE
CYCLIC AMP-MEDIATED ADRENERGIC SYNAPSES TO CEREBELLAR
PURKINJE CELLS
ADV CYCLIC NUCL RES 1:411-423(1972)

HOFFER BJ   SIGGINS GR   OLIVER AP
BLOOM FE
ACTIVATION OF THE PATHWAY FROM LOCUS COERULEUS TO RAT
CEREBELLAR PURKINJE NEURONS: PHARMACOLOGICAL EVIDENCE OF
NORADRENERGIC CENTRAL INHIBITION
J PHARMACOL EXP THER 184:553-569(1973)

HOFFSOMMER RD   TAUB D   WENDLER NL
BY-PRODUCTS FORMED IN TRANSFORMATIONS OF PROSTAGLANDIN E1
TETRAHEDRON LETT 43:4085-4086(1971)

HOGAKI M
EFFECTS OF PROSTAGLANDIN F2A ON INDUCTION OF LABOR
IGAKU NO AYUMI 80:848-857(1972);(JAPANESE)

HOGAKI M   KINUSHITA K   SAKAMOTO S
WAGATSUMA T
COMPARATIVE STUDY ON ACTIONS OF PROSTAGLANDIN F2A AND
OXYTOCIN UPON PREGNANT RAT UTERUS
ACTA OBSTET GYNAECOL JAP 19:118-124(1972)

HOGG JA
DRUGS FROM NATURAL PRODUCTS -- ANIMAL SOURCES
ADV CHEM SER 108:14-32(1971)

HOLLAND JF   SWEELEY CC   THRUSH RE
TEETS RE   BIEBER MA
ON-LINE COMPUTER CONTROLLED MULTIPLE ION DETECTION IN
COMBINED GAS CHROMATOGRAPHY-MASS SPECTROMETRY
ANAL CHEM 45:308-314(1973)

HOLMES SW
THE OCCURRENCE, DISTRIBUTION AND ACTIONS OF PROSTAGLANDINS
IN THE CENTRAL NERVOUS SYSTEM
HORMONES 3:291(1972)

HOLZHUTER H   ANGELKORT B   WENZEL E
EFFECT OF PGE-1 ON PLATELET-FUNCTION, ITS REVERSIBILITY AND
INFLUENCE ON NA-51-CHROMATE UPTAKE
ADV BIOSCI (SUPPL) 9:36(1972)

HOLZHUTER H   WENZEL E   ANGELKORT B
THE UPTAKE OF DIFFERENT PERMEABLE SUBSTANCES BY HUMAN
THROMBOCYTES AND THE INFLUENCE OF THESE SUBSTANCES ON
PLATELET FUNCTION
IN: "ERYTHROCYTES, THROMBOCYTES, LEUKOCYTES," E GERLACH,
K MOSER, E DEUTSCH, W WILMANNS, G THIEME, STUTTG
2:297-300(1973)

HOLZMANN K
HYDATIDIFORM MOLE
GEBURTSHILFE FRAUENHEILK 33:338-360(1973)

HONG KJ   SHIM BS   KIM SK
THE EFFECTS OF PROSTAGLANDIN E1 UPON SERUM HAPTOGLOBIN
LEVEL IN RABBIT
ABSTR 20TH CONF KOREAN BIOCHEM SOC:5-6(1972)

HOPE WC   VANTRABERT TC   DALTON C
NOREPINEPHRINE-STIMULATED PROSTAGLANDIN PRODUCTION IN RAT
FAT CELLS
ABSTR 166TH NATL MEET AM CHEM SOC:BIOL-45(1973)

HOPKINS SJ
PROSTAGLANDINS: A PARADOX WITH REAL PROMISE
CHEM DRUG 196:23(JULY 3, 1971)

HORI T   HARADA Y
HYPERTHERMIC RESPONSES PRODUCED BY INTRACEREBRAL INJECTIONS
OF PROSTAGLANDINS E1 AND E2 AND NORADRENALINE
J PHYSIOL SOC JAP 35:434-435(1973)

HORNSTRA G
DEGREE AND DURATION OF PROSTAGLANDIN E1-INDUCED INHIBITION
OF PLATELET AGGREGATION IN THE RAT
EUR J PHARMACOL 15:343-349(1971)

HORNSTRA G
THE EFFECT OF PROSTAGLANDIN E1 AND LINOLEIC ACID ON
EXPERIMENTAL ARTERIAL THROMBOSIS IN RATS
J AM OIL CHEM SOC 48:94A(1971)

HORNSTRA G
THE INFLUENCE OF DIETARY SUNFLOWERSEED OIL AND HARDENED
COCONUT OIL ON INTRA-ARTERIAL OCCLUSIVE THROMBOSIS IN RATS
NUTR METAB 13:140-149(1971)

HORNYCH A   SAFAR M   PAPANICOLAOU N
MEYER P   MILLIEZ P
RENAL AND CARDIOVASCULAR EFFECTS OF PROSTAGLANDIN A2 IN
ESSENTIAL HYPERTENSION
PROC INT UNION PHYSIOL SCI, 25TH INT CONGR, MUNICH,
9:628(1971)

HORNYCH A   SAFAR M   PAPANICOLAOU N
MEYER P   MILLIEZ P
RENAL AND CARDIOVASCULAR EFFECTS OF PROSTAGLANDIN A2 IN
ESSENTIAL HYPERTENSION
EUR J CLIN INVEST 2:289(1972)

HORNYCH A   SAFAR M   PAPANICOLAOU N
MEYER P   MILLIEZ P
RENAL AND CARDIOVASCULAR EFFECTS OF PROSTAGLANDIN A2 IN
ESSENTIAL HYPERTENSION
ADV BIOSCI (SUPPL) 9:57(1972)

HORNYCH A   SAFAR M   PAPANICOLAOU N
MEYER P   MILLIEZ P
RENAL AND CARDIOVASCULAR EFFECTS OF PROSTAGLANDIN A2 IN
ESSENTIAL HYPERTENSION
ABSTR 5TH INT CONGR NEPHROL, MEX CITY:113(1972)

HORNYCH A   SAFAR M   PAPANICOLAOU N
MEYER P   MILLIEZ P
RENAL AND CARDIOVASCULAR EFFECT OF PROSTAGLANDIN A2 IN
HYPERTENSIVE PATIENTS
EUR J CLIN INVEST 3:391-398(1973)

HORNYCH A   SAFAR M   PAPANICOLAOU N
MEYER P   MILLIEZ P
HEMODYNAMIC EFFECTS OF PROSTAGLANDIN A2 IN ESSENTIAL
HYPERTENSION
ANN MED INTERNE (PARIS) 124:355-358(1973): (FRENCH)

HORTON E   THOMPSON C   JONES R
POYSER N
RELEASE OF PROSTAGLANDINS
ANN NY ACAD SCI 180:351-362(1971)

HORTON EW
PROSTAGLANDINS
IN: "PHYSIOLOGY AND BIOCHEMISTRY OF THE DOMESTIC FOWL," DJ
BELL, ACADEMIC PRESS, LOND 1:589-601(1971)

HORTON EW
PROSTAGLANDINS
IN: "INTERNATIONAL ENCYCLOPEDIA OF PHARMACOLOGY AND
THERAPEUTICS," PERGAMON, NY:1-28(1971)

HORTON EW
PROSTAGLANDINS
MONOGR ENDOCRINOL 7:1-197(1972)

HORTON EW
THE PROSTAGLANDINS
PROC R SOC LOND (BIOL) 182:411-426(1972)

HORTON EW
PROSTAGLANDINS AT ADRENERGIC NERVE-ENDINGS
BR MED BULL 29:148-151(1973)

HOSSLER FE   GORDON M   BARRNETT RJ
PHOSPHATASES ON THE MEMBRANES OF RABBIT SPERM HEADS
J CELL BIOL 55:117A(1972)

HOWELL SL   MONTAGUE W
ADENYLATE CYCLASE ACTIVITY IN ISOLATED RAT ISLETS OF
LANGERHANS -- EFFECTS OF AGENTS WHICH ALTER RATES OF INSULIN
SECRETION
BIOCHIM BIOPHYS ACTA 320:44-52(1973)

HOWIE PW   CALDER AA   FORBES CD
PRENTICE CRM
EFFECT OF INTRAVENOUS PROSTAGLANDIN E2 ON PLATELET FUNCTION,
COAGULATION, AND FIBRINOLYSIS
J CLIN PATHOL 26:354-358(1973)

HOWIE PW   PRENTICE CRM
THE EFFECT OF INTRAVENOUS PROSTAGLANDIN E2 ON PLATELET
FUNCTION, COAGULATION AND FIBRINOLYSIS
SCOTT MED J 17:229(1972)

HOYLAND JR   KIER LB
PREFERRED CONFORMATION OF PROSTAGLANDIN E1
J MED CHEM 15:84-86(1972)

HSIE AW   JONES C   PUCK TT
FURTHER CHANGES IN DIFFERENTIATION STATE ACCOMPANYING THE
CONVERSION OF CHINESE HAMSTER CELLS TO FIBROLASTIC FORM BY
DIBUTYRYL ADENOSINE CYCLIC 3':5'-MONOPHOSPHATE AND HORMONES
PROC NAT ACAD SCI 68:1648-1652(1971)

HUDSON DG
THE EFFECT OF PROSTAGLANDIN E1 AND E2 ON ISOLATED FROG
INTESTINE
J PHYSIOL (LOND) 221:3P-4P(1972)

HUGUES J
DISCUSSION
ACTA MED SCAND (SUPPL) 525:181(1971)

HUGUES J
SOME EVIDENCE THAT ADP IS NOT THE ONLY AGENT RESPONSIBLE FOR
COLLAGEN-INDUCED AGGREGATION
ACTA MED SCAND (SUPPL) 525:181-182(1971)

HUGUES J
CYCLIC AMP AND PLATELET AGGREGATION
ACTA UNIV CAROL (MED) (PRAHA) 53:179-184(1972)

HUIDOBRO HV   SMILG I   TONNELIER LS
RESPIRATORY EFFECTS OF THE PROSTAGLANDINS E1, E2 AND F2A
ACTA PHYSIOL LAT AM 23: ABSTR 423 (1973): (SPANISH)

HUIDOBRO HV   DE SMILG IJ
MODIFICATION BY PROSTAGLANDIN F2A OF THE BRONCHOCONSTRICTOR
EFFECTS OF SOME CHEMICAL MEDIATORS
ABSTR 5TH INT CONGR PHARMACOL SAN FRANC 109(1972)

HUKOVIC S
THE EFFECT OF PROSTAGLANDINS, POLYPEPTIDES AND AMINES
APPLIED ON SEROSE OR MUCOSE SIDE ON INDUCED CONTRACTIONS OF
ISOLATED STOMACH
ABSTR 5TH INT CONGR PHARMACOL SAN FRANC 110(1972)

HUMBERTSON AO JR   HALL JL
THE AUTONOMIC NERVOUS SYSTEM
PROG NEUROL PSYCHIATR 26:311-328(1971)

HUNTINGFORD PJ
CONTROL OF FERTILITY
PROC R SOC MED 64:952-955(1971)

HUTTON I   PARRATT JR   LAWRIE TDV
CARDIOVASCULAR EFFECTS OF PROSTAGLANDIN E1 IN EXPERIMENTAL
MYOCARDIAL INFARCTION
CARDIOVASC RES 7:149-155(1973)

HYMAN AL   GEORGE WJ   KADOWITZ PJ
SPECTRUM OF EFFECTS OF PROSTAGLANDINS E, A, AND F IN THE
PULMONARY CIRCULATION IN THE INTACT DOG
CIRCULATION (SUPPL II) 46:56(1972)

HYMAN AL   JOINER PD   KADOWITZ PJ
DIFFERENTIAL EFFECTS OF PROSTAGLANDINS E, A, AND F ON THE
PULMONARY CIRCULATION
FED PROC 32:787(1973)

HYMAN AL   PENNINGTON DG   JAQUES WE
PULMONARY VASCULAR RESPONSES TO ALLOXAN
J PHARMACOL EXP THER 181:92-97(1972)

HYMAN AL   WOOLVERTON WC   PENNINGTON DG
PULMONARY VASCULAR RESPONSES (PVR) TO ADENOSINE DIPHOSPHATE
(ADP) IN INTACT DOGS
CLIN RES 19:63(1971)

HYMAN AL   WOOLVERTON WC   PENNINGTON DG
JAQUES WE
PULMONARY VASCULAR RESPONSES TO ADENOSINE DIPHOSPHATE
J PHARMACOL EXP THER 178:549-561(1971)

HYNIE S   SHARP GWG
THE EFFECT OF CHOLERA TOXIN ON INTESTINAL ADENYL CYCLASE
ADV CYCLIC NUCL RES 1:163-174(1972)

ICHIKAWA A   MATSUMOTO H   SAKATO N
TOMITA K
EFFECT OF THYROID HORMONES ON EPINEPHRINE-INDUCED LIPOLYSIS
IN ADIPOSE TISSUE OF RATS
J BIOCHEM 69:1055-1064(1971)

IGNARRO LJ   ORONSKY AL   PERPER RJ
EFFECTS OF PROSTAGLANDINS ON RELEASE OF ENZYMES FROM
LYSOSOMES OF PANCREAS, SPLEEN AND KIDNEY CORTEX
LIFE SCI (I)12:193-201(1973)

IGUCHI S   TANOUCHI F   KIMURA K
HAYASHI M
THE SYNTHESIS OF 15-METHYL OR 15-16-DIMETHYL PROSTAGLANDINS
PROSTAGLANDINS 4:535-538(1973)

ILLES P   HADHAZY P   TORMA Z
VIZI ES   KNOLL J
THE EFFECT OF NUMBER OF STIMULI AND RATE OF STIMULATION ON
THE INHIBITION BY PGE1 OF ADRENERGIC TRANSMISSION
EUR J PHARMACOL 24:29-36 (1973)

ILLES P   TORMA Z   VIZI ES
KNOLL J
INHIBITION OF SYMPATHETIC TRANSMISSION BY PGE1, DEPENDENCE
ON THE DURATION AND INTENSITY OF STIMULATION
ORVOSTUDOMANY 24:89-98(1973);(HUNGARIAN)

ILLIANO G   CUATRECASAS P
ENDOGENOUS PROSTAGLANDINS MODULATE LIPOLYTIC PROCESSES IN
ADIPOSE TISSUE.
NATURE (NEW BIOL) 234:72-74(1971)

IMAI Y   KATOAKA K   SHENKMAN L
WAN L   HOLLANDER CS
IN VIVO EFFECTS OF PROSTAGLANDIN F2A ON THYROID HORMONE
RELEASE IN MAN
CLIN RES 21:494(1973)

IMPICCIATORE M   BERTACCINI G
THE BRONCHOCONSTRICTOR ACTION OF THE TETRADECAPEPTIDE
BOMBESIN IN THE GUINEA-PIG
J PHARM PHARMACOL 25:872-875(1973)

INAGAKI M   SAKAMOTO S
UTERINE CONTRACTILE EFFECT OF PGF2A IN CHRONIC RAT
PREPARATION
OBSTET GYNECOL (TOKYO) 39:631-639(1972);(JAPANESE)

INAGAWA T   OHKI S   SAWADA M
HIRATA F
STUDIES ON EXTRACTION, SEPARATION AND ESTIMATION OF
PROSTAGLANDINS BY RADIOIMMUNOASSAY
YAKUGAKU ZASSHI 92:1187-1194(1972);(JAPANESE)

INAGAWA T   OHKI S   SAWADA M
OHTSUKA K   HIRATA F
STUDIES ON EXTRACTION, SEPARATION AND ESTIMATION OF
PROSTAGLANDINS BY RADIOIMMUNOASSAY. II. SEPARATION OF
INDIVIDUAL PROSTAGLANDIN BY THIN-LAYER CHROMATOGRAPHY WITH
STEPWISE DEVELOPMENT, AND DETERMINATION OF PROSTAGLANDIN E2
YAKUGAKU ZASSHI 93:471-475(1973);(JAPANESE)

INGELMAN-SUNDBE/ IA SANDBERG F   RYDEN G
LINDGREN L   MOLFESE V
IN VITRO STUDIES OF THE MOTILITY OF THE HUMAN FALLOPIAN
TUBE
PROC WORLD CONGR FERTIL STERIL 7:163-166(1973)

INGERMAN C   SMITH JB   KOCSIS JJ
SILVER MJ
ARACHIDONIC ACID INDUCES PLATELET AGGREGATION AND PLATELET
PROSTAGLANDIN FORMATION
FED PROC 32:219(1973)

INOKI R
MECHANICS OF ACTIONS OF ANALGESICS - ESPECIALLY ON
PERIPHERAL ACTIONS
FOLIA PHARMACOL JAP 69:645-659(1973);(JAPANESE)

INSKEEP EK
POTENTIAL USES OF PROSTAGLANDINS IN CONTROL OF REPRODUCTIVE
CYCLES OF DOMESTIC ANIMALS
J ANIM SCI 36:1149-1157(1973)

ISAACS P   WHITTAKER S   TURNBERG LA
THE MECHANISM FOR THE DIARRHOEA ASSOCIATED WITH MEDULLARY
THYROID CARCINOMA
EUR J CLIN INVEST 3:240-241(1973)

ISHIMORI S
THE SUPPRESSIVE EFFECT OF PROSTAGLANDINS ON THE SECRETION OF
GASTRIC JUICE AND ITS CLINICAL SIGNIFICANCE
PROC SOC GERONTOL STUDY, OKAYAMA CITY, ONO PHARMACEUTICAL
CO, OSAKA:39-41(1972);(JAPANESE)

ISHIZAWA M   SAKABE K   MIYAZAKI E
CA++ AND NA+ ON THE INHIBITORY ACTION OF PROSTAGLANDIN (PG)
E IN CIRCULAR MUSCLES FROM GUINEA PIG STOMACH AND DOG
INTESTINE
J PHYSIOL SOC JAP 35:464-465(1973);(JAPANESE)

ITO H
MALE INFERTILITY AND PROSTAGLANDIN
YAKUBUTSU RYOHO 4:982-984(1971);(JAPANESE)

ITO H   KATAYAMA T
MALE INFERTILITY AND PROSTAGLANDIN
PROC WORLD CONGR FERTIL STERIL 7:483-485(1973)

ITO H   MITSUHASHI S   MOMOSE G
PROSTAGLANDIN AND RENAL FUNCTION - THE EFFECT OF
PROSTAGLANDIN A2 ON THE P.S.P. EXCRETION
PROSTAGLANDINS 3:359-365(1973)

ITO H   MOMOSE G   KATAYAMA T
EFFECT OF PROSTAGLANDIN ON THE SECRETION OF HUMAN GROWTH
HORMONE
FOLIA ENDOCRINOL JAP 47:359-360(1971);(JAPANESE)

ITO H   MOMOSE G   KATAYAMA T
TAKAGISHI H   ITO L   NAKAJIMA H
TAKEI Y
EFFECT OF PROSTAGLANDIN ON THE SECRETION OF HUMAN GROWTH
HORMONE
J CLIN ENDOCRINOL METAB 32:857-859(1971)

ITO H   MOMOSE G   KATAYAMA T
TAKAGISHI H   ITO R   NAKAJIMA H
TAKEI Y
EFFECT OF PROSTAGLANDIN E1 ON THE SECRETION OF HUMAN GROWTH
HORMONE
IGAKU NO AYUMI 76:733-735(1971);(JAPANESE)

ITO H   TAKAGISHI H   KATAYAMA H
MOMOSE K   ITO R   TAKEI T
NAKAJIMA H
SECRETION OF HUMAN GROWTH HORMONE BY PROSTAGLANDIN E1
FOLIA ENDOCRINOL JAP 46:35(1971)

ITO T   HIDAKA H   KATO T
YOSHITOSHI Y
RESPONSE OF PLASMA ALDOSTERONE TO POSTURAL CHANGE,
DIURETICS, ANGIOTENSIN II, SODIUM RESTRICTION OR LOADING AND
PROSTAGLANDIN
JAP HEART J 14:518-530(1973)

ITSKOVITZ HO   STEMPER J   PACHOLCZYK D
MCGIFF JC
RENAL PROSTAGLANDINS: DETERMINANTS OF INTRARENAL
DISTRIBUTION OF BLOOD FLOW IN THE DOG
CLIN SCI MOL MED 45: 321S-324S (1973)

IWATSUKI K   FURUTA Y   HASHIMOTO K
EFFECT OF PROSTAGLANDIN F2A ON THE SECRETION OF PANCREATIC
JUICE INDUCED BY SECRETIN AND BY DOPAMINE
EXPERIENTIA 29:319-320(1973)

JACKSON HR   HALL RC   HODGE RL
GIBSON E   KATIC F   STEVENS M
THE EFFECT OF ASPIRIN ON THE PULMONARY EXTRACTION OF PGF2A
AND CARDIOVASCULAR RESPONSE TO PGF2A
AUST J EXP BIOL MED SCI 51:837-845(1973)

JACKSON RT
PHARMACOLOGICAL MECHANISMS IN THE EUSTACHIAN TUBE
ANN OTOL RHINOL LARYNGOL 80:313-318(1971)

JACKSON RT
EFFECTS OF DRUGS ON THE NOSE AND MIDDLE EAR-LESSONS FROM AN
ANIMAL MODEL
CLIN PEDIATR 12:559-562(1973)

JACOB J   GRIMMER G
NORMALIZATION OF ESSENTIAL FATTY ACID DEFICIENCY INDUCED
FAT LIVER BY FEEDING OF CIS-12-OCTADECENOIC ACID
TRIGLYCERIDE
HOPPE SEYLERS Z PHYSIOL CHEM 352S:1445-1454(1971);(GERMAN)

JACOBS PM
SYNTHESIS OF 3,5-DIALKYL-1,2-DIOXOLANES: MODEL COMPOUNDS
FOR THE PROSTAGLANDIN BIOSYNTHESIS.
DISS, NORTHEAST UNIV (1973)

JACOBSON ED   SHEPHERD AP   MAO CC
SHANBOUR LL
MEDIATION OF MESENTERIC VASODILATION BY CYCLIC AMP
CLIN RES 21:90(1973)

JACOBSON HI   KEYES PL   BULLOCK DW
REGULATION OF UTERINE ESTROGEN RECEPTOR BY LUTEAL
PROGESTERONE IN THE PSEUDOPREGNANT RABBIT
BIOL REPROD 7:108(1972)

JACOBY HI   MARSHALL CH
ANTAGONISM OF CHOLERA ENTEROTOXIN BY ANTI-INFLAMMATORY
AGENTS IN THE RAT
NATURE (LOND) 235:163-165(1972)

JAFFE BM   PARKER CW
EXTRACTION OF PGE FROM HUMAN SERUM FOR RADIOIMMUNOASSAY
IN: "PROSTAGLANDINS IN FERTILITY CONTROL 2", S BERGSTROM K
GREEN AND B SAMUELSSON, W H O, KAROLINSKA INST, STOCKH:
69-82(1972)

JAFFE BM   PARKER CW
PROSTAGLANDIN RELEASE BY HUMAN CELLS IN VITRO
IN: "PROSTAGLANDINS IN CELLULAR BIOLOGY", PW RAMWELL AND
BB PHARRISS, PLENUM PRESS, NY/LOND:207-226(1972)

JAFFE BM   PARKER CW   MARSHALL GR
NEEDLEMAN P
RENAL CONCENTRATIONS OF PROSTAGLANDIN E IN ACUTE AND
CHRONIC RENAL ISCHEMIA
BIOCHEM BIOPHYS RES COMMUN 49:799-805(1972)

JAFFE BM   PARKER CW   PHILPOTT GW
IMMUNOCHEMICAL MEASUREMENT OF PROSTAGLANDIN OR
PROSTAGLANDIN-LIKE ACTIVITY FROM NORMAL AND NEOPLASTIC
CULTURED TISSUE
SURG FORUM 22:90-92(1971)

JAFFE BM   PHILPOTT GW   HAMPRECHT B
PARKER CW
PROSTAGLANDIN PRODUCTION BY CELLS IN VITRO
ADV BIOSCI 9:179-182(1973)

JAFFE BM   PHILPOTT GW   PARKER CW
PROSTAGLANDIN PRODUCTION BY CELLS IN VITRO
ADV BIOSCI (SUPPL) 9:28(1972)

JAFFE BM   PODOS SM   BECKER B
INDOMETHACIN BLOCKS ARACHIDONIC ACID-ASSOCIATED ELEVATION OF
AQUEOUS HUMOR PROSTAGLANDIN E
INVEST OPHTHALMOL 12:621-622(1973)

JAFFE BM   SMITH JW   NEWTON WT
PARKER CW
RADIOIMMUNOASSAY FOR PROSTAGLANDINS
SCIENCE 171:494-496(1971)

JAFFE BM   SMITH JW   PARKER CW
RADIOIMMUNOASSAY FOR PROSTAGLANDINS
J CLIN INVEST 50:48A-49A(1971)

JAISLE F
THE SIGNIFICANCE OF LIPASES IN PREGNANCY, DURING BIRTH AND
CONFINEMENT AND IN THE NEWBORN
IN LIPIDS DURING PREGNANCY AND DELIVERY FORTSCHR
GEBURTSHILFE GYNAEKOL 48:124-126:146-182(1972):(GERMAN)

JAISLE F
PROSTAGLANDINS, LIPID METABOLISM AND LABOR INDUCTION
IN:LIPIDS DURING PREGNANCY AND DELIVERY,FORTSCHR
GEBURTSHILFE GYNAEKOL 48:127-137:146-182(1972):(GERMAN)

JANSZEN FHA
HISTOCHEMICAL DETECTION OF PROSTAGLANDIN BIOSYNTHESIS
ADV BIOSCI (SUPPL) 9:46(1972)

JANSZEN FHA   NUGTEREN DH
A HISTOCHEMICAL STUDY OF THE PROSTAGLANDIN BIOSYNTHESIS
IN THE URINARY SYSTEM OF RABBIT, GUINEA PIG, GOLDHAMSTER,
AND RAT
ADV BIOSCI 9:287-292(1973)

JANUARY CT   SCHOTTELIUS BA
EFFECTS OF PROSTAGLANDIN F2A (PGF2A) ON MECHANICAL AND
ELECTRICAL PARAMETERS OF RAT PAPILLARY MUSCLE
PHYSIOLOGIST 14:167(1971)

JAWAHARLAL K   BERTI F
EFFECTS OF DIBUTYRYL CYCLIC AMP AND A NEW CYCLIC NUCLEOTIDE
ON GASTRIC ACID SECRETION IN THE RAT
PHARMACOL RES COMMUN 4:143-149(1972)

JAY AWL   ROWLANDS S   SKIBO L
RED BLOOD CELL DEFORMABILITY AND THE PROSTAGLANDINS
PROSTAGLANDINS 3:871-877(1973)

JENKINS CSP   PACKHAM MA   KINLOUGH-RATHB/ RL
MUSTARD JF
MODIFICATION OF PLATELET ADHERENCE TO PROTEIN COATED
SURFACES BY THE USE OF DRUGS
PROC CAN FED BIOL SCI 14:24(1971)

JENKINS CSP   PACKHAM MA   KINLOUGH-RATHBO/ RL
MUSTARD JF
INTERACTIONS OF POLYLYSINE WITH PLATELETS.
BLOOD 37:395-412(1971)

JESKE W
ROLE OF PROSTAGLANDINS IN THE ORGANISM
FARM POL 29:965-972(1973);(POLISH)

JEUCK K
PRIMARY AND SECONDARY EFFECTS OF PROSTAGLANDIN E1 (PGE1) ON
CARBOHYDRATE AND FAT METABOLISM
DISS, JOHANN WOLFGANG GOETHE-UNIVERSITAT, FRANKF AM (1971):
(GERMAN)

JEWELEWICZ R   CANTOR B   DYRENFURTH I
WARREN MP   PATTNER A   MURRAY T
BOWE E   VANDE WIELE RL
PLASMA GONADOTROPIN, GONADAL STEROIDS, AND RENIN CHANGES IN
PROSTAGLANDIN F2A INDUCED ABORTION
ABSTR 18TH ANNU MEET SOC GYNECOL INVEST, PHOENIX:16(1971)

JEWELEWICZ R   DYRENFURTH I   WARREN M
CANTOR B   VANDE WIELE RL
INTRAVENOUS INFUSION OF PROSTAGLANDIN F2A IN THE MID-LUTEAL
PHASE OF THE NORMAL HUMAN MENSTRUAL CYCLE
ABSTR SOC GYNECOL INVEST, 19TH ANNU MEET, PHOENIX:28(1972)

JIRAKULSOMCHOK D   MOORE WW
EFFECT OF PGE1 ON RENAL FUNCTION AND THE DISTRIBUTION OF
RENAL BLOOD FLOW.
PHYSIOLOGIST 14:168(1971)

JIRAKULSOMCHOK D   MOORE WW
THE EFFECTS OF PROSTAGLANDIN E1 ON RENAL FUNCTION
J MED ASSOC THAILAND 54:614-624(1971)

JOBKE A   PESKAR BA   PESKAR BM
ON THE SPECIFICITY OF ANTISERA AGAINST PROSTAGLANDINS A2
AND E2
FEBS LETT 37:192-196(1973)

JOHNS A   PICKLES VR   WOOSTER MJ
THE EFFECT OF PROSTAGLANDIN E1 ON THE K+-INDUCED
CONTRACTURE OF THE GUINEA-PIG URETER
J ENDOCRINOL 52:XII-XIII(1972)

JOHNSON AR   MORAN NC   MAYER SE
RELATIONSHIP OF CYCLIC AMP TO HISTAMINE RELEASE IN RAT
MAST CELLS
FED PROC 32:744(1973)

JOHNSON DG   FUJIMOTO WY   WILLIAMS RH
ENHANCED RELEASE OF INSULIN BY PROSTAGLANDINS IN ISOLATED
PANCREATIC ISLETS
DIABETES 22:658-663(1973)

JOHNSON DG   STROMBERG P   FUJIMOTO WY
WILLIAMS RH
ENHANCED RELEASE OF INSULIN BY PROSTAGLANDINS IN ISOLATED
PANCREATIC ISLETS
DIABETES 21:329(1972)

JOHNSON DG   THOA NB   WEINSHILBOUM R
AXELROD J   KOPIN IJ
ENHANCED RELEASE OF DOPAMINE-B-HYDROXYLASE FROM SYMPATHETIC
NERVES BY CALCIUM AND PHENOXYBENZAMINE AND ITS REVERSAL BY
PROSTAGLANDINS
PROC NAT ACAD SCI (USA) 68: 2227-2230 (1971)

JOHNSON DG   THOMPSON WJ   WILLIAMS RH
STIMULATION OF ADENYL CYCLASE FROM ISOLATED PANCREATIC
ISLETS BY PROSTAGLANDINS
FED PROC 32:801(1973)

JOHNSON GS   MORGAN WD   PASTAN I
REGULATION OF CELL MOTILITY BY CYCLIC AMP
NATURE (LOND) 235:54-56(1972)

JOHNSON GS   PASTAN I
CHANGE IN GROWTH AND MORPHOLOGY FIBROBLASTS BY
PROSTAGLANDINS.
J NAT CANCER INST 47:1357-1364(1971)

JOHNSON GS   PASTAN I
N6, O2-D BUTYRYL ADENOSINE 3',5'-MONOPHOSPHATE INDUCES
PIGMENT PRODUCTION IN MELANOMA CELLS
NATURE (NEW BIOL) 237:267-268(1972)

JOHNSON GS   PASTAN I   OYER DS
D'ARMIENTO M
PROSTAGLANDINS, CYCLIC AMP, AND TRANSFORMED FIBROBLASTS
ADV BIOSCI 9:173-178(1973)

JOHNSON GS   PEERY CV   OTTEN J
MORGAN WD   PASTAN I
GROWTH AND MORPHOLOGY OF TRANSFORMED FIBROBLASTS ARE
REGULATED BY CYCLIC AMP AND PROSTAGLANDINS
J CLIN INVEST 50:50A(1971)

JOHNSON H
DISCUSSION
ACTA MED SCAND (SUPPL) 525:181(1971)

JOHNSON M   DAVISON P   HOLLAND PC
RAMWELL PW
PREPARATION AND CHARACTERIZATION OF PROSTANOYL CARNITINE
LIPIDS 7:752-754(1972)

JOHNSON M   DAVISON P   RAMWELL PW
CARNITINE-DEPENDENT BETA-OXIDATION OF PROSTAGLANDINS
J BIOL CHEM 247:5656-5658(1972)

JOHNSON M   RABINOWITZ I   WILLIS A
WOLF P
PROSTAGLANDINS AND SICKLE CELL ANAEMIA
ADV BIOSCI (SUPPL) 9:30(1972)

JOHNSON M   RABINOWITZ I   WILLIS AL
WOLF PL
PROSTAGLANDINS AND SICKLE CELL ANEMIA
ADV BIOSCI 9:189-194(1973)

JOHNSON M   RABINOWITZ I   WILLIS AL
WOLF PL
DETECTION OF PROSTAGLANDIN INDUCTION OF ERYTHROCYTE SICKLING
CLIN CHEM 19:23-26(1973)

JOHNSON M   RAMWELL P
MODIFICATION OF MEMBRANE ENZYME ACTIVITY BY PROSTAGLANDINS
AND PROSTAGLANDIN ANTAGONISTS
ADV BIOSCI (SUPPL) 9:35(1972)

JOHNSON M   RAMWELL PW
PROSTAGLANDIN MODIFICATION OF MEMBRANE-BOUND ENZYME ACTIVITY
ADV BIOSCI 9:205-212(1973)

JOHNSON M   RAMWELL PW
PROSTAGLANDIN MODIFICATION OF MEMBRANE-BOUND ENZYME
ACTIVITY : A POSSIBLE MECHANISM OF ACTION?
PROSTAGLANDINS 3:703-719(1973)

JOHNSON M   RAMWELL PW
IMPLICATIONS OF PROSTAGLANDINS IN HEMATOLOGY
IN: "PROSTAGLANDINS AND CYCLIC AMP," RH KAHN, WEM LANDS,
ACADEMIC PRESS, NY : 275-304(1973)

JOHNSTON HH   HERZOG JP   LAULER DP
LOCAL RENAL TUBULAR ACTION OF LIPIDS FROM THE KIDNEY
IN: KIDNEY HORMONES, JW FISHER, ACADEMIC PRESS, LOND:
563-584 (1971)

JOINER PD   DAVIS LB   KADOWITZ PJ
HYMAN AL
EFFECTS OF PROSTAGLANDINS ON CANINE ISOLATED PULMONARY
LOBAR SMALL ARTERIES AND VEINS
PHARMACOLOGIST 15:208(1973)

JOINER PD   KADOWITZ PJ   DAVIS LB
HYMAN AL
RESPONSES OF CANINE ISOLATED PULMONARY LOBAR ARTERIES AND
VEINS TO PHARMACOLOGICAL AGENTS
CLIN RES 21:72(1973)

JONES G   RAPHAEL RA   WRIGHT S
STEREOCONTROLLED SYNTHESIS OF A PROSTANOID SYNTHON BY
OXIDATIVE CLEAVAGE OF SUBSTITUTED NORBORNENE DERIVATIVES
CHEM COMMUN 609-610(1972)

JONES JK   EDELMAN IS
EFFECTS OF PROSTAGLANDIN E1 AND ARGININE VASOTOCIN ON
SODIUM TRANSPORT AND ADENYL CYCLASE ACTIVITY IN DEVELOPING
TADPOLE SKIN
ADV BIOSCI (SUPPL) 9:32(1972)

JONES JK   EDELMAN IS
EFFECTS OF PROSTAGLANDIN E1 AND ARGININE VASOTOCIN ON
SODIUM TRANSPORT AND ADENYL CYCLASE ACTIVITY IN DEVELOPING
TADPOLE SKIN
ADV BIOSCI 9:201-204(1973)

JONES MF
THE USE OF RADIOISOTOPICALLY LABELLED METHYL ESTERS IN THE
DETERMINATION OF THE 15-EPIMER CONTENT OF PROSTAGLANDINS
J PHARM PHARMACOL 25:900-904(1973)

JONES RI   CAMMOCK S   HORTON EW
PARTIAL PURIFICATION AND PROPERTIES OF CAT PLASMA
PROSTAGLANDIN A ISOMERASE
BIOCHIM BIOPHYS ACTA 280:588-601(1972)

JONES RL
PROPERTIES OF A NEW PROSTAGLANDIN
BR J PHARMACOL 45:144P-145P(1972)

JONES RL
15-HYDROXY-9-OXOPROSTA-11,13-DIENOIC ACID AS THE PRODUCT
OF A PROSTAGLANDIN ISOMERASE
J LIPID RES 13:511-518(1972)

JONES RL
ACTIONS OF PGA2 AND ITS ISOMERS ON THE CARDIOVASCULAR SYSTEM
OF THE CAT
ADV BIOSCI (SUPPL) 9:8(1972)

JONES RL   CAMMOCK S
PARTIAL PURIFICATION AND PROPERTIES OF A PROSTAGLANDIN A
ISOMERASE
ADV BIOSCI (SUPPL) 9:9(1972)

JONES RL   CAMMOCK S
PURIFICATION, PROPERTIES, AND BIOLOGICAL SIGNIFICANCE OF
PROSTAGLANDIN A ISOMERASE
ADV BIOSCI 9:61-70(1973)

JONSSON CE
SMOOTH MUSCLE STIMULATING LIPIDS IN PERIPHERAL LYMPH AFTER
EXPERIMENTAL BURN INJURY
SCAND J PLAST RECONSTR SURG 5:1-5(1971)

JONSSON CE
PROSTAGLANDINS IN BURN INJURY, EXPERIMENTAL AND CLINICAL
STUDIES
DISS, UPPSALA UNIV (1972)

JONSSON CE   ANGGARD E
BIOSYNTHESIS AND METABOLISM OF PROSTAGLANDIN E2 IN HUMAN
SKIN
SCAND J CLIN INVEST 29:289-296(1972)

JONSSON CE   ARTURSON G   ANGGARD E
PROSTAGLANDINS IN THE LYMPH FROM BURNED TISSUES
NORD MED 86:908(1971);(SWEDISH)

JONSSON CE   ARTURSON G   ANGGARD E
APPEARANCE OF PROSTAGLANDINS IN LYMPH FROM BURNED TISSUE
IN: "RESEARCH IN BURNS," P MATTER, TL BARCLAY AND Z
KONICKOVA, HANS HUBER PUBL, STUTTG:515-519(1971)

JONSSON CE   HAMBERG M
PROSTAGLANDIS IN BURN INJURY
ACTA PHARMAKOL TOXICOL (KBH) 31:109(1972)

JONSSON CE   HAMBERG M
PROSTAGLANDINS IN BURN INJURY
ADV BIOSCI (SUPPL) 9:68(1972)

JONSSON CE·   HAMBERG M
PROSTAGLANDINS IN BURN INJURY
ADV BIOSCI 9:441-445(1973)

JORGENSEN HP   SONDERGAARD J
VASCULAR RESPONSES TO PROSTAGLANDIN E1
ACTA DERM VENEREOL (STOCKH) 53:203-206(1973)

JOSEFSSON JO   JOHANSSON P   HANSSON SE
DRUGS AND HORMONES AS MODIFIERS OF PINOCYTOSIS INDUCED BY
SODIUM IONS IN AMOEBA PROTEUS
ACTA PHARMACOL TOXICOL (SUPPL 1)31:82(1972)

JOUVENAZ GH   NUGTEREN DH   VAN DORP DA
GAS CHROMATOGRAPHIC DETERMINATION OF NANOGRAM AMOUNTS OF
PROSTAGLANDINS E AND F
PROSTAGLANDINS 3:175-187(1973)

JUBIZ W   FRAILEY J
GENERATION OF ANTIBODIES TO PROSTAGLANDINS E1 AND F2A
CLIN RES 19:127(1971)

JUBIZ W   FRAILEY J   CHILD C
PHYSIOLOGIC SIGNIFICANCE OF PROSTAGLANDINS OF THE E GROUP
(PGE)
ABSTR 4TH INT CONGR ENDOCRINOL, INT CONGR SER 256, EXCERPTA
MEDICA, AMST:191(1972)

JUBIZ W   FRAILEY J   CHILD C
BARTHOLOMEW K
PHYSIOLOGIC ROLE OF PROSTAGLANDINS OF THE E(PGE), F(PGF)
AND AB(PGAB) GROUPS. ESTIMATION BY RADIOIMMUNOASSAY IN
UNEXTRACTED HUMAN PLASMA
PROSTAGLANDINS 2:471-489(1972)

JUNGMANNOVA C   HAVRANEK F   HODR J
EFFECT OF PROSTAGLANDIN F2A ON THE PLACENTAL VESSELS IN
VITRO
CESK GYNEKOL 37:435-436(1972);(CZECH)

JUNGMANNOVA C   HAVRANEK F   HODR J
RELATIONSHIP BETWEEN F2A -- PROSTAGLANDIN CONCENTRATION
AND THE ACTIVITY OF PLACENTAL VESSELS IN VITRO
CESK GYNEKOL 38:141-142(1973);(CZECH)

JUNSTAD M   WENNMALM A
INCREASED RENAL EXCRETION OF NORADRENALINE IN RATS AFTER
TREATMENT WITH PROSTAGLANDIN SYNTHETASE INHIBITOR
INDOMETHACIN
ACTA PHYSIOL SCAND 85:573-576(1972)

JUNSTAD M   WENNMALM A
ON THE RELEASE OF PROSTAGLANDIN E2 FROM THE RABBIT HEART
FOLLOWING INFUSION OF NORADRENALINE
ACTA PHYSIOL SCAND 87:573-574(1973)

JUNSTAD M   WENNMALM A
PROSTAGLANDIN MEDIATED INHIBITION OF NORADRENALINE RELEASE
AT DIFFERENT NERVE IMPULSE FREQUENCIES
ACTA PHYSIOL SCAND 89:544-549(1973)

KACEW S   SINGHAL RL
STIMULATION OF KIDNEY CORTEX CYCLIC 3'-5'-ADENOSINE
MONOPHOSPHATE-H3 FORMATION BY DDT AND RELATED INSECTICIDES
FED PROC 32:802(1973)

KADOWITZ P
EFFECT OF PROSTAGLANDINS E1, E2 AND A2 ON VASCULAR
RESISTANCE AND RESPONSES TO ADRENERGIC STIMULI AND
ANGIOTENSIN IN THE DOG HINDLIMB
FED PROC 31:545(1972)

KADOWITZ PJ
EFFECT OF PROSTAGLANDINS E1, E2 AND A2 ON VASCULAR
RESISTANCE AND RESPONSES TO NORADRENALINE, NERVE
STIMULATION AND ANGIOTENSIN IN THE DOG HINDLIMB
BR J PHARMACOL 46:395-400(1972)

KADOWITZ PJ   BRODY MJ
DIFFERENTIAL EFFECTS OF PROSTAGLANDINS E1, E2, A1, F1A AND
F2A ON ADRENERGIC VASOCONSTRICTION IN THE DOG HINDPAW
FED PROC 30:625(1971)

KADOWITZ PJ   GEORGE WJ   HYMAN AL
EFFECT OF PROSTAGLANDINS E1 AND F2A ON ADRENERGIC RESPONSES
IN THE PULMONARY CIRCULATION
ADV BIOSCI (SUPPL) 9:80(1972)

KADOWITZ PJ   GEORGE WJ   JOINER PD
HYMAN AL
EFFECT OF PROSTAGLANDINS E1 AND F2A ON ADRENERGIC RESPONSES
IN THE PULMONARY CIRCULATION
ADV BIOSCI 9:501-506(1973)

KADOWITZ PJ   SWEET CS   BRODY MJ
POTENTIATION OF ADRENERGIC VENOMOTOR RESPONSES BY
ANGIOTENSIN, PROSTAGLANDIN F2A AND COCAINE
J PHARMACOL EXP THER 176:167-173(1971)

KADOWITZ PJ   SWEET CS   BRODY MJ
DIFFERENTIAL EFFECTS OF PROSTAGLANDINS E1, E2, F1A AND F2A
ON ADRENERGIC VASOCONSTRICTION IN THE DOG HINDPAW
J PHARMACOL EXP THER 177:641-649(1971)

KADOWITZ PJ   SWEET CS   BRODY MJ
BLOCKADE OF ADRENERGIC VASOCONSTRICTOR RESPONSES IN THE DOG
BY PROSTAGLANDINS E1 AND A1
J PHARMACOL EXP THER   179:563-572(1971)

KADOWITZ PJ   SWEET CS   BRODY MJ
EFFECT OF PROSTAGLANDINS ON ADRENERGIC NEUROTRANSMISSION TO
VASCULAR SMOOTH MUSCLE
IN: "PROSTAGLANDINS IN CELLULAR BIOLOGY," PW RAMWELL AND
BB PHARRISS, PLENUM PRESS, NY/LOND:479-511(1972)

KADOWITZ PJ   SWEET CS   BRODY MJ
ENHANCEMENT OF SYMPATHETIC NEUROTRANSMISSION BY
PROSTAGLANDIN F2A IN THE CUTANEOUS VASCULAR BED OF THE DOG
EUR J PHARMACOL 18:189-194(1972)

KADOWITZ PJ   SWEET CS   BRODY MJ
INFLUENCE OF PROSTAGLANDINS ON ADRENERGIC TRANSMISSION TO
VASCULAR SMOOTH MUSCLE
CIRC RES (SUPPL II) 31:36-50(1972)

KAFKA MS   PAK CHC
PEPTIDE HORMONES AT CELL MEMBRANES
FED PROC 31:255(1972)

KAFKA MS   PAK CYC
THE EFFECT OF POLYPEPTIDE HORMONES ON LIPID MONOLAYERS. III
THE EFFECT OF INSULIN, VASOPRESSIN, OXYTOCIN, ALBUMIN AND
PROSTAGLANDIN E1 ON THE SPECIFIC RESISTANCE TO THE
EVAPORATION OF WATER THROUGH MONOMOLECULAR FILMS OF
MONOOCTADECYL PHOSPHATE, STEARIC ACID, AND STEARYL ALCOHOL
J COLLOID INTERFACE SCI 41:148-155(1972)

KAFKA MS   PAK CYC
PEPTIDE HORMONES AT CELL MEMBRANES: STUDIES IN A MODEL
SYSTEM
J COLLOID INTERFACE SCI 41:388-390(1972)

KAJITA Y   HAYAISHI O
EFFECT OF PROSTAGLANDIN E1 ON THE INDUCTION OF RAT LIVER
TYROSINE AMINOTRANSFERASE
BIOCHIM BIOPHYS ACTA 261:281-283(1972)

KALEY G   MESSINA EJ   WEINER R
THE ROLE OF PROSTAGLANDINS IN MICROCIRCULATORY REGULATION
AND INFLAMMATION
IN: "PROSTAGLANDINS IN CELLULAR BIOLOGY," PW RAMWELL AND
BB PHARRISS, PLENUM PRESS, NY/LOND:309-327(1972)

KALEY G   WEINER R
PROSTAGLANDIN E1: A POTENTIAL MEDIATOR OF THE INFLAMMATORY
RESPONSE
ANN NY ACAD SCI 180:338-350(1971)

KALEY G   WEINER R
EFFECT OF PROSTAGLANDIN E1 ON LEUKOCYTE MIGRATION
NATURE (NEW BIOL) 234:114-115(1971)

KALINER M   TAUBER AI   AUSTEN KF
THE EFFECT OF PROSTAGLANDINS ON THE IMMUNOLOGIC RELEASE OF
HISTAMINE FROM HUMAN LUNG TISSUE
AM REV RESP DIS 107:1094-1085(1973)

KALISKER A
STUDIES ON THE INTERACTION OF VASOPRESSIN AND PROSTAGLANDIN
IN THE RABBIT RENAL MEDULLA
DISS, UNIV WASH, SEATTLE (1971)

KALISKER A
STUDIES ON THE INTERACTION OF VASOPRESSIN AND PROSTAGLANDIN
IN THE RABBIT RENAL MEDULLA
DISS,DEPT PHARMACOL,UNIV WASH (1971)

KALISKER A   DYER DC
PROSTAGLANDIN-VASOPRESSIN INTERACTION IN THE RENAL MEDULLA
PHARMACOLOGIST 13:293(1971)

KALISKER A   DYER DC
INHIBITION OF THE VASOPRESSIN-ACTIVATED ADENYL CYCLASE FROM
RENAL MEDULLA BY PROSTAGLANDINS
EUR J PHARMACOL 20:143-146(1972)

KALISKER A   DYER DC
IN VITRO RELEASE OF PROSTAGLANDIN FROM THE RENAL MEDULLA
EUR J PHARMACOL 19:305-309(1972)

KALMAN TI
A PLAUSIBLE ENZYMIC MECHANISM FOR THE FORMATION OF THE
CYCLOPENTANE RING OF PROSTAGLANDINS
INTRA-SCI CHEM REP 6:65-66(1972)

KALOYANIDES GJ   BASTRON RD   DIBONA GF
ROLE OF INCREASED RENAL BLOOD FLOW (RBF) AND BLOOD FLOW
DISTRIBUTION IN THE NATRIURESIS OF RENAL VASODILATATION AND
ELEVATED PERFUSION PRESSURE (PRA)
CLIN RES 21:691(1973)

KAMIKAWA Y   SAITO T   YOSHIDA H
HIRAI T
INTERACTIONS OF CATECHOLAMINES, PROSTAGLANDINS, STEROIDS
AND ADRENERGIC BLOCKING AGENTS ON THE ISOLATED UTERI FROM
CATS
JAP J PHARMACOL (SUPPL) 23:102(1973)

KANEKO T
PROSTAGLANDINS AND CYCLIC AMP
TAISHA 8:201-208(1971);:(JAPANESE)

KANEKO T   OKA H   SAITO S
MUNEMURA M   MUSA K   ODA T
YANAIHARA N·   YANAIHARA C
IN VITRO EFFECTS OF SYNTHETIC SOMATOTROPIN-RELEASE
INHIBITING FACTOR ON CYCLIC AMP LEVEL AND GH RELEASE IN RAT
ANTERIOR PITUITARY GLAND
ENDOCRINOL JAP 20 : 535-538(1973)

KANEKO T   SAITO S   OKA H
YANAIHARA N
SYNTHETIC THYROID STIMULATING HORMONE, RELEASING FACTOR AND
PITUITARY CYCLIC AMP
IGAKU NO AYUMI 76:785-786(1971);:(JAPANESE)

KANJE M   WALUM E   EDSTROM A
DIBUTYRYL CYCLIC AMP AND PROSTAGLANDIN E1 INDUCE
MORPHOLOGICAL ALTERATIONS IN CULTURED HUMAN GLIOMA CELLS
ACTA PHYSIOL SCAND (SUPPL 396) : 107(1973)

KANNEGIESSER H   LEE JB
DIFFERENCE IN HAEMODYNAMIC RESPONSE TO PROSTAGLANDINS A AND
E
NATURE(LOND)229:498-500(1971)

KAPLAN EL   JAFFE BM   PESKIN GW
A NEW PROVOCATIVE TEST FOR THE DIAGNOSIS OF THE CARCINOID
SYNDROME
AM J SURG 123:173-179(1972)

KAPLAN EL   PESKIN GW
PHYSIOLOGIC IMPLICATIONS OF MEDULLARY CARCINOMA OF THE
THYROID GLAND
SURG CLIN NORTH AM 51:125-137(1971)

KAPLAN EL   SAXENA N   PESKIN GW
PROSTAGLANDINS: A POSSIBLE MEDIATOR OF DIARRHEA IN ENDOCRINE
SYNDROMES
SURG FORUM 21:94-95(1971)

KAPLAN EL   SAXENA N   PESKIN GW
PROSTAGLANDINS: A POSSIBLE MEDIATOR OF DIARRHEA IN
ENDOCRINE SYNDROMES.
SURG FORUM 21:94-95(1971)

KAPLAN EL   SIZEMORE GW   PESKIN GW
JAFFE BM
HUMORAL SIMILARITIES OF CARCINOID TUMORS AND MEDULLARY
CARCINOMAS OF THE THYROID
SURGERY 74:21-29(1973)

KAPOOR K   LEVINE RA
EFFECT OF ADRENERGIC RECEPTOR BLOCKADE ON THE RESPONSE TO
PROSTAGLANDIN E1 IN CANINE ILEUM
ARCH INT PHARMACODYN THER 203:243-250(1973)

KARIM SMM
PROSTAGLANDINS
BR J HOSP MED 5:555-563(1971)

KARIM SMM
PROSTAGLANDINS AND HUMAN REPRODUCTION: PHYSIOLOGICAL ROLES
AND CLINICAL USES OF PROSTAGLANDINS IN RELATION TO HUMAN
REPRODUCTION
IN: "THE PROSTAGLANDINS: PROGRESS IN RESEARCH," SMM KARIM,
MTP MED TECHN PUBL CO, OXFORD:71-164(1972)

KARIM SMM
PROSTAGLANDINS
MED TODAY 7:3-6(1973)

KARIM SMM   AMY JJ
PROSTAGLANDINS - LIPIDS IN AMNIOTIC FLUID
IN: AMNIOTIC FLUID: RESEARCH AND CLINICAL APPLICATIONS, DVI
FAIRWEATHER AND TKAB ESKES, AMST,EXCERPTA MEDICA: 277-295
(1973)

KARIM SMM   CARTER DC   BHANA D
GANESAN PA
EFFECT OF ORALLY AND INTRAVENOUSLY ADMINISTERED
PROSTAGLANDIN 15(R)15-METHYL E2 ON GASTRIC SECRETION IN MAN
ADV BIOSCI 9:255-264(1973)

KARIM SMM   CARTER DC   BHANA D
GANESAN PA
EFFECT OF ORALLY ADMINISTERED PROSTAGLANDIN E2 AND ITS
15-METHYL ANALOGUES ON GASTRIC SECRETION
BR MED J 1:143-146(1973)

KARIM SMM   CARTER DC   BHANA D
GANESAN PA
INHIBITION OF BASAL AND PENTAGASTRIN INDUCED GASTRIC ACID
SECRETION IN MAN WITH PROSTAGLANDIN 16:16 DIMETHYL E2 METHYL
ESTER
INT RES COMMUN SYST (73-3)8-3-2(1973)

KARIM SMM   CARTER DC   BHANA D
GANESAN PA
THE EFFECT OF ORALLY AND INTRAVENOUSLY ADMINISTERED
PROSTAGLANDIN 16:16 DIMETHYL E2 METHYL ESTER ON HUMAN
GASTRIC ACID SECRETION
PROSTAGLANDINS 4:71-83(1973)

KARIM SMM   HILLIER K
GENERAL INTRODUCTION AND SOME PHARMACOLOGICAL ACTIONS OF
PROSTAGLANDINS
IN: "THE PROSTAGLANDINS: PROGRESS IN RESEARCH", SMM KARIM,
MTP MED TECH PUBL CO, OXFORD:1-46(1972)

KARIM SMM   SHARMA SD
ORAL ADMINISTRATION OF PROSTAGLANDIN E2 FOR THE INDUCTION
AND ACCELERATION OF LABOR
J REPROD MED 9:346-352(1972)

KARIM SMM   SHARMA SD   FILSHIE GM
TERMINATION OF PREGNANCY WITH 15 METHYL ANALOGUES OF
PROSTAGLANDINS E2 AND F2A
J REPROD MED 9:383-391(1972)

KARIM SMM   SHARMA SD   FILSHIE GM
TERMINATION OF SECOND TRIMESTER PREGNANCY WITH
INTRA-AMNIOTIC ADMINISTRATION OF PROSTAGLANDINS E2 AND F2A
J REPROD MED 9:427-433(1972)

KARIM SMM   SOMERS K
CARDIOVASCULAR AND RENAL ACTIONS OF PROSTAGLANDINS
IN: "THE PROSTAGLANDINS: PROGRESS IN RESEARCH", SMM KARIM,
MTP MED TECH PUBL CO, OXFORD:165-203(1972)

KARIM SMM   SOMERS K   HILLIER K
CARDIOVASCULAR AND OTHER EFFECTS OF PROSTAGLANDINS E2 AND
F2A IN MAN
CARDIOVASC RES 5:255-259(1971)

KASS MA   PODOS SM   MOSES RA
BECKER B
PROSTAGLANDIN E1 AND AQUEOUS HUMOR DYNAMICS
INVEST OPHTHALMOL 11:1022-1027(1972)

KATAGIRI N   TAKAHASHI N   NOGAMI K
IWAMOTO T   TAKAHASHI K
THE EFFECT OF PROSTAGLANDIN F2A ON PROMOTION AND INDUCTION
OF DELIVERY AND ON CHANGES OF UTERINE CONTRACTILITY
ABSTR JAP SOC OBSTET GYNECOL, KANTO PROVINCES MEET:24(1971);
(JAPANESE)

KATAGIRI N   TAKAHASHI N   NOGAMI K
IWAMOTO T   TAKAHASHI K
THE EFFECT OF PROSTAGLANDIN F2A ON PROMOTION AND INDUCTION
OF DELIVERY AND ON CHANGES OF UTERINE CONTRACTILITY
ABSTR JAP SOC OBSTET GYNECOL, KANTO PROV MEET:24(1971);
(JAPANESE)

KATO K
A METHOD OF INDUCING LABOR ACCORDING TO THE MECHANISM OF
UTERINE CONTRACTION.
SANFUJINKA NO JISSAI 21:645-651(1972)

KATO K   KAWASHIMA T   ISHIDA M
OSHIO Y   SAKAMOTO S   SHIDA T
KURODA K   SHU B
ORAL USE OF PROSTAGLANDIN E2 IN INDUCTION OF LABOR
SAN-FUJIN-KA NO SEKAI (WORLD OBSTET GYNECOL) 25:79-83(1973);
(JAPANESE)

KATO K   NEGISHI E   KURODA K
PLANNED CHILDBIRTH USING PROSTAGLANDIN F2A IN A GRAVIDA
WITH A COMPLICATION OF IDIOPATHIC THROMBOCYTOPENIC
PURPURA
OBSTET GYNECOL (TOKYO) 39:1222-1226(1972);(JAPANESE)

KATO Y
RECENT STUDY ON THE MECHANISM OF ACTION OF ASPIRIN
J JAP STOMATOL SOC 40:278(1973);(JAPANESE)

KATO Y   DUPRE J   BECK JC
PLASMA GROWTH HORMONE RESPONSES
ABSTR 4TH INT CONGR ENDOCRINOL, INT CONGR SER 256, EXCERPTA
MEDICA, AMST:37(1972)

KATO Y   DUPRE J   BECK JC
PLASMA GROWTH HORMONE IN THE ANESTHETIZED RAT: EFFECTS OF
DIBUTYRYL CYCLIC AMP, PROSTAGLANDIN E1, ADRENERGIC AGENTS,
VASOPRESSIN, CHLORPROMAZINE, AMPHETAMINE AND L-DOPA
ENDOCRINOLOGY 93:135-146(1973)

KATORI M
EFFECTS OF PROSTAGLANDINS ON THE CARDIOVASCULAR SYSTEM
PROC SOC GERONTOL STUDY, OKAYAMA CITY, ONO PHARMACEUTICAL
CO, OSAKA:1-4(1972);(JAPANESE)

KATSUBE J   SHIMOMURA H   MATSUI M
A NOVEL APPROACH TO THE SYNTHESIS OF PROSTAGLANDIN-F1
SKELETON
AGR BIOL CHEM 35:1828-1829(1971)

KATSUBE J   SHIMOMURA H   MATSUI M
SYNTHESIS OF PROSTAGLANDIN F1-RELATED COMPOUNDS
AGR BIOL CHEM 36:1997-2004(1972)

KATSUBE J   SHIMOMURA H   MURAYAMA E
TOKI K   MATSUI M
SYNTHETIC STUDIES ON CYCLOPENTANE DERIVATIVES. III. A NEW
SYNTHESIS OF PROSTAGLANDIN-F1 SKELETON
AGR BIOL CHEM 35:1768-1774(1971)

KATTLOVE HE
THE EFFECT OF COLD ON PLATELETS. III. ADENINE NUCLEOTIDE
METABOLISM AFTER BRIEF STORAGE AT COLD TEMPERATURE
BLOOD 42:557-564(1973)

KAUFMAN RG   FREEMAN RK   MISHELL DR JR
ABORTIFACIENT ACTIVITY OF INTRAVENOUSLY ADMINISTERED
PROSTAGLANDINS.
CONTRACEPTION 3:121-132(1971)

KAUKER ML
MICROPUNCTURE STUDY OF RENAL ACTION OF PROSTAGLANDIN A2
(PGA2) IN RATS
J PHARMACOL EXP THER 187:632-640(1973)

KAUMANN AJ
POSITIVE CHRONOTROPIC EFFECT OF THE PROSTAGLANDIN PGE1
ACTA PHYSIOL LAT AM 23:ABSTR 245(1973);(SPANISH)

KAWAGISHI Y   IDE K   MURAO M
EFFECTS OF PROSTAGLANDIN ON THE CARBOHYDRATE AND LIPID
METABOLISMS IN DIABETES
J JAP DIABETIC SOC 14:97(1971);(JAPANESE)

KAWAKAMI Y   UCHIYAMA K   IRIE T
MURAO M
EVALUATION OF AEROSOLS OF PROSTAGLANDINS E1 AND E2 AS
BRONCHODILATORS
EUR J CLIN PHARMACOL 6:127-132(1973)

KAWASAKI A   SHIBATA K   SAKANO M
INAGAWA J
PHARMACOLOGICAL STUDIES ON PROSTAGLANDIN F2A (PGF2A)
PHARMACOMETRICS 5:957-971(1971)

KAY AB   AUSTEN KF
THE IGE-MEDIATED RELEASE OF AN EOSINOPHIL LEUKOCYTE
CHEMOTACTIC FACTOR FROM HUMAN LUNG
J IMMUNOL 107:899-902(1971)

KAY AB   STECHSCHULTE DJ   AUSTEN KF
AN EOSINOPHIL LEUKOCYTE CHEMOTACTIC FACTOR OF ANAPHYLAXIS.
J EXP MED 133:602-619(1971)

KAY AB   STECHSCHULTE DJ   AUSTEN KF
AN EOSINOPHIL LEUKOCYTE CHEMOTACTIC FACTOR OF ANAPHYLAXIS
J EXP MED 133:602-619(1972)

KAZAL LA   MILLER OP   ERSLEV AJ
OBSERVATIONS ON PROSTAGLANDINS IN ERYTHROPOIETIN (E)
INHIBITOR CONCENTRATES
FED PROC 31:872(1972)

KECSKEMETI V   KELEMEN K   KNOLL J
EFFECT OF PROSTAGLANDIN E1 ON THE TRANSMEMBRANE POTENTIALS
OF THE MAMMALIAN HEART
ADV BIOSCI (SUPPL) 9:61(1972)

KECSKEMETI V   KELEMEN K   KNOLL J
EFFECT OF PROSTAGLANDIN E1 ON THE TRANSMEMBRANE POTENTIALS
OF THE MAMMALIAN HEART
ADV BIOSCI 9:373-377(1973)

KECSKEMETI V   KELEMEN K   KNOLL J
EFFECT OF PROSTAGLANDIN E1 ON ISOLATED MAMMALIAN HEART
PREPARATIONS
KISERL ORVOSTUD 25:184-192(1973);(HUNGARIAN)

KEIRNS JJ   CARRITT B   FREEMAN J
EISENSTADT JM   BITENSKY MW
ADENOSINE 3',5' CYCLIC MONOPHOSPHATE IN EUGLENA GRACILIS
LIFE SCI 13: 287-302(1973)

KEIRNS JJ   KREINER PW   BITENSKY MW
AN ABRUPT TEMPERATURE-DEPENDENT CHANGE IN THE ENERGY OF
ACTIVATION OF HORMONE-STIMULATED HEPATIC ADENYLYL CYCLASE
J SUPRAMOL STRUCT 1:368-379(1973)

KEIRSE MJNC   PATTEN PT   ANDERSON ABM
TURNBULL AC   JOHNS A   WOOSTER MJ
PICKLES VR
PREGNANT SHEEP MYOMETRIUM RESPONDS TO PROSTAGLANDINS IN
VITRO BUT NOT IN VIVO
INT RES COMMUN SYST (73-4)8-5-1(1973)

KEIRSE MJNC   TURNBULL AC
EXTRACTION OF PROSTAGLANDINS FROM HUMAN BLOOD
PROSTAGLANDINS 4:607-617(1973)

KELEMEN K   KECSKEMETI V   KNOLL J
EFFECT OF PROSTAGLANDIN E1 AND CELLULIN-A ON CARDIAC
TRANSMEMBRANE POTENTIALS
ABSTR 5TH INT CONGR PHARMACOL SAN FRANC 122(1972)

KELLIHER GJ   GLENN TM
EFFECT OF PROSTAGLANDIN E1 ON OUABAIN-INDUCED ARRHYTHMIA
EUR J PHARMACOL 24:410-414(1973)

KELLY LA   BUTCHER RW
THE EFFECTS OF DELTA-1-TETRAHYDROCANNABINOL ON CYCLIC AMP
LEVELS IN WI-38 FIBROBLASTS
BIOCHIM BIOPHYS ACTA 320:540-544(1973)

KELLY MT   WHITE A
A FEEDBACK MECHANISM OF HISTAMINE RELEASE INVOLVING
INTRACELLULAR CYCLIC AMP AND EXTRACELLULAR HISTAMINE
CLIN RES 20:531(1972)

KELLY MT   WHITE A
REABSORPTION OF LEUKOCYTE HISTAMINE AFTER EXPOSURE TO CYCLIC
AMP-ACTIVE AGENTS
CLIN RES 21:63(1973)

KELLY MT   WHITE A
HISTAMINE RELEASE INDUCED BY HUMAN LEUKOCYTE LYSATES-
REABSORPTION OF PREVIOUSLY RELEASED HISTAMINE AFTER EXPOSURE
TO CYCLIC AMP-ACTIVE AGENTS
J CLIN INVEST 52:1834-1840(1973)

KELLY MT   WHITE A
REABSORPTION OF LEUKOCYTE HISTAMINE AFTER EXPOSURE TO CYCLIC
AMP-ACTIVE AGENTS
CLIN RES 21:604(1973)

KELLY R   VANRHEENEN V
PROSTAGLANDIN SYNTHESIS. II. A NOVEL RESOLUTION OF ALDEHYDE
AND KETONE INTERMEDIATES
TETRAHEDRON LETT 1709-1712(1973)

KELLY RC   SCHLETTER I   JONES RL
TOTAL SYNTHESIS OF PGC2 METHYL ESTER-COMPARISON WITH
ENZYMATICALLY PRODUCED MATERIAL
PROSTAGLANDINS 4:653-660(1973)

KELLY RC   VANRHEENEN V   SCHLETTER I
PILLAI MD
PROSTAGLANDIN SYNTHESIS. I. AN IMPROVED SYNTHESIS THROUGH
BICYCLO(3.1.0)HEXANE INTERMEDIATES
J AM CHEM SOC 95:2746-2747(1973)

KELLY RGM   STARR MS
EFFECTS OF PROSTAGLANDINS AND A PROSTAGLANDIN ANTAGONIST ON
INTRAOCULAR PRESSURE AND PROTEIN IN THE MONKEY EYE
CAN J OPHTHALMOL 6:205-211(1971)

KELLY RW
MEASUREMENTS OF PROSTAGLANDINS IN BIOLOGICAL FLUIDS BY GAS
CHROMATOGRAPHY-MASS SPECTROMETRY COUPLED WITH A SMALL
ANALOGUE COMPUTER
ACTA ENDOCRINOL (KBH) SUPPL 155:221(1971)

KELLY RW
A SIMPLE DEVICE FOR ANALOGUE RECORDING OF THE ABUNDANCE OF
SELECTED IONS FROM A COMBINED GAS CHROMATOGRAPH-MASS
SPECTROMETER
J CHROMATOGR 71:337-339(1972)

KELLY RW
QUANTITATIVE MEASUREMENT OF PROSTAGLANDINS BY GAS
CHROMATOGRAPHY-MASS SPECTROMETRY
PROC INT SYMP GAS CHROMATOGR MASS SPECTROM :19-32(1972)

KELLY RW
METHOD FOR THE MEASUREMENT OF PROSTAGLANDIN F2A IN
BIOLOGICAL FLUIDS BY GAS CHROMATOGRAPHY-MASS
SPECTROMETRY
ANAL CHEM 45:2079-2082(1973)

KELLY RW
PHYSICAL METHODS OF MEASUREMENT OF PROSTAGLANDINS
CLIN ENDOCRINOL METAB 2:375-392(1973)

KENDALL-TAYLOR P
COMPARISON OF THE EFFECTS OF VARIOUS AGENTS ON THYROIDAL
ADENYL CYCLASE ACTIVITY WITH THEIR EFFECTS ON THYROID
HORMONE RELEASE
J ENDOCRINOL 54:137-145(1972)

KENDALL-TAYLOR P   MUNRO DS
THE LIPOLYTIC ACTIVITY OF LONG-ACTING THYROID STIMULATOR
BIOCHIM BIOPHYS ACTA 231:314-319(1971)

KESSEL E   OMRAN AR   BERNARD RP
RAVENHOLT RT   SPEIDEL JJ
INTERNATIONAL FERTILITY RESEARCH PROGRAM
PROC WORLD CONGR FERTIL STERIL 7:994-1002(1973)

KESSLER E   HUGHES RC   BENNETT EN
NADELA SM
EVIDENCE FOR THE PRESENCE OF PROSTAGLANDIN-LIKE MATERIAL IN
THE PLASMA OF DOGS WITH ENDOTOXIN SHOCK
J LAB CLIN MED 81:85-94(1973)

KEUSCH GT   KAPLAN MM   SMITH D
RAVANESI P
PERSISTENT, FULMINANT WATERY DIARRHEA COMPLICATING CHRONIC
ACTIVE HEPATITIS
GASTROENTEROLOGY 62 :   307-313(1972)

KEYNES WM   TILL AS
MEDULLARY CARCINOMA OF THE THYROID GLAND
Q J MED 40:443-456(1971)

KHOMMEL K   FISHER V   SHLIMOVICH S
PROSTAGLANDINS
SOV MED 36:77-82(1973);(RUSSIAN)

KIENZLE F   HOLLAND GW   JERNOW JL
KWOH S   ROSEN P
A VERSATILE PROSTAGLANDIN SYNTHESIS-USE OF A CARBOXY-
INVERSION REACTION
J ORG CHEM 38:3440-3442(1973)

KIER LB
MOLECULAR ORBITAL STUDIES OF CONFORMATION AND PHYSIOLOGICAL
ACTIVITY IN THE PROSTAGLANDINS
PHARMACOLOGIST 13:182(1971)

KILCOYNE MM   CANNON PJ
INFLUENCE OF THORACIC CAVAL OCCLUSION ON INTRARENAL BLOOD
FLOW DISTRIBUTION AND SODIUM EXCRETION.
AM J PHYSIOL 220:1220-1230(1971)

KIM SK   SHIM BS
A POTENTIAL ROLE OF PROSTAGLANDIN E1 IN ELEVATING SERUM
HAPTOGLOBIN LEVEL IN RABBIT
J CATHOLIC MED COL (SEOUL) 25: 1-13 (1973)

KIMBERG DV   FIELD M   JOHNSON J
HENDERSON A   GERSHON E
STIMULATION OF INTESTINAL MUCOSAL ADENYL CYCLASE BY CHOLERA
ENTEROTOXIN AND PROSTAGLANDINS
J CLIN INVEST 50:1218-1230(1971)

KINDAHL H   GRANSTROM E
METABOLISM OF PROSTAGLANDIN F1A IN THE GUINEA PIG
BIOCHIM BIOPHYS ACTA 280:466-471(1972)

KINOSHITA K   WAGATSUMA T   HOGAKI M
SAKAMOTO S
THE INDUCTION OF LABOR WITH PROSTAGLANDIN F2A
NISSANPU SHI 23:1027(1971);(JAPANESE)

KINZIE JL   ALPERS DH
BETA-ADRENERGIC STIMULATORS MEDIATE CAMP AUGMENTED AMINO
ACID TRANSPORT IN JEJUNAL MUCOSA
CLIN RES 21:516(1973)

KIPROV D   JURUKOVA Z   SOMOVA L
CHANGES IN THE INDEX OF GRANULATION OF THE JUXTAGLOMERULAR
CELLS IN EXPERIMENTAL HYPERTENSION
COR VASA 14:115-125(1972)

KIRTLAND SJ   BAUM H
PROSTAGLANDIN E1 MAY ACT AS A "CALCIUM IONOPHORE"
NATURE (NEW BIOL) 236:47-49(1972)

KIRTON KT   CORNETTE JC   BARR KL
CHARACTERIZATION OF ANTIBODY TO PROSTAGLANDIN F2A
BIOCHEM BIOPHYS RES COMMUN 47:903-909(1972)

KIRTON KT   CORNETTE JC   BARR KL
CHARACTERIZATION OF ANTIBODY TO PROSTAGLANDIN F2A
IN: "PROSTAGLANDINS IN FERTILITY CONTROL 2", S BERGSTROM
K GREEN AND B SAMUELSSON, W H O, KAROLINSKA INST, STOCKH:
60-68(1972)

KIRTON KT   FORBES AD
UTEROTROPIC ACTIVITY OF SEVERAL PROSTAGLANDINS IN RHESUS
MONKEYS
PHYSIOLOGIST 14:172(1971)

KIRTON KT   GUTKNECHT GD   BERGSTROM KK
WYNGARDEN LJ   FORBES AD
PROSTAGLANDINS AND REPRODUCTION
J REPROD MED 9:266-270(1972)

KISCHER CW
A REVALUATION OF A MORPHOGENETIC BLOCK PRODUCED BY
PROSTAGLANDIN-B1
ANAT REC 169:358(1971)

KISCHER CW
IN VITRO CONTRACTION OF EMBRYONIC SKIN PRODUCED BY
PROSTAGLANDINS
EXPERIENTIA 29:30-31(1973)

KISCHER CW
THE EPIDERMAL RESPONSE OF DEVELOPING SKIN TO
PROSTAGLANDIN-B1-MITOCHONDRIAL ALTERATIONS OBTAINED IN VITRO
EXP CELL RES 81:393-400 (1973)

KISCHER CW
DERMAL EVENTS DURING ORGANOGENESIS OF A SKIN DERIVATIVE:
ANALYSES OF NORMAL CHICK EMBRYO SKIN AND EXPLANTS TREATED
WITH PROSTAGLANDIN-B1 (PGB1)
TEX REP BIOL MED 31:489-505(1973)

KISCHER CW   KEETER JS
ANCHOR FILAMENT BUNDLES IN EMBRYONIC SKIN: ORIGIN AND
TERMINATION.
AM J ANAT 130:179-194(1971)

KITAMURA S
VASOACTIVE SUBSTANCE AND THE LUNG
RESPIR CIRC (TOKYO) 21: 688-700 (1973); (JAPANESE)

KITAMURA S   HOLDEN LD   MERRILL RW
FORD WT JR   SAID SI
NEWLY EXTRACTED POLYPEPTIDES AS BRONCHODILATORS AND
PULMONARY VASODILATORS
CLIN RES 21:665(1973)

KITAMURA S   PRESKITT J   YOSHIDA T
SAID SI
PROSTAGLANDIN RELEASE, RESPIRATORY ALKALOSIS, AND SYSTEMIC
HYPOTENSION DURING MECHANICAL VENTILATION
FED PROC 32:341(1973)

KITAMURA S   SAID SI   HOLDEN LD
MERRILL RW   FORD WT JR
NEWLY EXTRACTED POLYPEPTIDES AS BRONCHODILATORS AND
PULMONARY VASODILATORS
CLIN RES 21:100(1973)

KITZMILLER JL   LUCAS WE
STUDIES ON A MODEL OF AMNIOTIC FLUID EMBOLISM
OBSTET GYNECOL 39:626-627(1972)

KLAUSCH B   KYANK H
PROSTAGLANDINS IN GYNECOLOGY AND OBSTETRICS
ZENTRALBL GYNAKOL 94:705-719(1972)

KLEIN I   LEVEY GS
EFFECT OF PROSTAGLANDINS ON GUINEA PIG MYOCARDIAL ADENYL
CYCLASE
CLIN RES 19:323(1971)

KLEIN I   LEVEY GS
EFFECT OF PROSTAGLANDINS ON GUINEA PIG MYOCARDIAL ADENYL
CYCLASE
METABOL 20:890-896(1971)

KLOEZE J   HORNSTRA G
EFFECTS OF PROSTAGLANDIN ON PLATELET AGGREGATION AND
EXPERIMENTAL THROMBOSIS
IN: "PLATELET AGGREGATION, TABLE RONDE ROUSSEL," MASSON,
PARIS:165-171(1971)

KLUGE AF   UNTCH KG   FRIED JH
SYNTHESIS OF PROSTAGLANDIN MODELS AND PROSTAGLANDINS BY
CONJUGATE ADDITION OF A FUNCTIONALIZED ORGANOCOPPER REAGENT
J AM CHEM SOC 94:7827-7832(1972)

KLUGE AF   UNTCH KG   FRIED JH
STUDIES IN PROSTAGLANDINS. II. SYNTHESIS OF
13-CIS-PROSTAGLANDINS VIA A HIGHLY STEREOSELECTIVE CONJUGATE
ADDITION WITH A FUNCTIONALIZED ORGANOCOPPER REAGENT
J AM CHEM SOC 94:9256-9258(1972)

KNAPP DR   GAFFNEY TE
USE OF STABLE ISOTOPES IN PHARMACOLOGY-CLINICAL PHARMACOLOGY
CLIN PHARMACOL THER 13:307-316(1972)

KNOX FG   SCHNEIDER EG   WILLIS LR
STRANDHOY JW   OTT CE
SITE AND CONTROL OF PHOSPHATE REABSORPTION BY THE KIDNEY
KIDNEY INT 3:347-353(1973)

KOBAYASHI N
STUDY ON EXPERIMENTAL THROMBOSIS, WITH SPECIAL REFERENCE TO
THE PREVENTION OF THROMBUS FORMATION
JAP J CLIN PATHOL 21:879-885(1973);(JAPANESE)

KOBAYASHI T   NAKAYAMA R   KIMURA K
EFFECTS OF GLUCAGON, PROSTAGLANDIN E1 AND DIBUTYRYL CYCLIC
3',5'-AMP UPON THE TRANSMEMBRANE ACTION POTENTIAL OF GUINEA
PIG VENTRICULAR FIBER AND MYOCARDIAL CONTRACTILE FORCE
JAP CIRC J 35:807-819(1971)

KOBAYASHI Y   RYU T   MAEKAWA T
THE EFFECTS OF PROSTAGLANDIN E1 ON THE AGGLUTINATION OF
PLATELETS
PROC SOC GERONTOL STUDY, OKAYAMA CITY, ONO PHARMACEUTICAL
CO, OSAKA:9-11(1972);(JAPANESE)

KOCH H   DEMLING L   CLASSEN M
THE INFLUENCE OF PROSTAGLANDIN E2 ON THE BLOOD FLOW AND
SECRETION OF THE STOMACH STIMULATED WITH PENTAGASTRIN IN THE
ANAESTHETIZED CAT
ARCH FR MAL APP DIG 61:268C(1972)

KOCSIS JJ   HERNANDOVICH J   SILVER MJ
SMITH JB   INGERMAN C
DURATION OF INHIBITION OF PLATELET PROSTAGLANDIN FORMATION
AND AGGREGATION BY INGESTED ASPIRIN OR INDOMETHACIN
PROSTAGLANDINS 3:141-144(1973)

KOJIMA K   SAKAI K
SYNTHETIC STUDIES ON PROSTANOIDS. II. STEREOSPECIFIC TOTAL
SYNTHESIS OF PROSTAGLANDIN F1A
TETRAHEDRON LETT 3333-3336(1972)

KOJIMA K   SAKAI K
SYNTHETIC STUDIES ON PROSTANOIDS. III. STEREOCHEMISTRY OF
THE KEY INTERMEDIATE (IV) IN THE TOTAL SYNTHESIS OF
PROSTAGLANDIN F1A
TETRAHEDRON LETT 3337-3340(1972)

KOKOT F   KUSKA J   PAKULA E
KOZIAK H   PIETREK J   KOWALIK M
EFFECTS OF FAT LOAD ON SERUM-LIPIDS IN PATIENTS WITH
RENAL-DISEASE
NEPHRON 8:549-558(1971)

KOLENA J   CHANNING CP
STIMULATORY ACTION OF LH, FSH AND PROSTAGLANDINS (PG) UPON
CYCLIC 3',5'-AMP (CAMP) LEVELS IN PORCINE GRANULOSA CELLS
(GC).
ABSTR 53RD MEET ENDOCR SOC, SAN FRANC:A-118(1971)

KOLENA J   CHANNING CP
STIMULATORY EFFECTS OF LH, FSH AND PROSTAGLANDINS UPON
CYCLIC 3',5'-AMP LEVELS IN PORCINE GRANULOSA CELLS
ENDOCRINOLOGY 90:1543-1550(1972)

KOLENA J   CHANNING CP
GONADOTROPIN AND PROSTAGLANDIN STIMULI OF C-AMP LEVELS IN
PORCINE GRANULOSA CELLS
PHYSIOL BOHEMOSLOV 22:74(1973)

KOOMEN JM
THE ACTION OF PROSTAGLANDINS ON THE UTERUS
PHARM WEEKBL 108:357-368;409-418(1973);(DUTCH)

KOOPMAN WJ   DAVID JR
PHARMACOLOGIC MODULATION OF THE BIOLOGIC ACTIVITY OF MIF
IN:"INFLAMMATION MECHANISMS AND CONTROL," IH LEPOW AND
PA WARD, ACADEMIC PRESS, N Y:151-161(1972)

KOOPMAN WJ   GILLIS MH   DAVID JR
PHARMACOLOGIC MODULATION OF THE BIOLOGIC ACTIVITY OF MIF
IN: "PROC 7TH LEUCOCYTE CULTURE CONF," F DAGUILLARD,
ACADEMIC PRESS, NY/LOND 7:303-314(1973)

KOOPMAN WJ   ORANGE RP   AUSTEN KF
PROSTAGLANDIN INHIBITION OF THE IMMUNOLOGIC RELEASE OF SLOW
REACTING SUBSTANCE OF ANAPHYLAXIS IN THE RAT
PROC SOC EXP BIOL MED 137:64-67(1971)

KOPPANYI T   MALING HM
THE EFFECTS OF PRETREATMENT WITH RESERPINE,
ALPHA-METHYL-P-TYROSINE, OR PROSTAGLANDIN E1 ON ADRENERGIC
SALIVATION
PROC SOC EXP BIOL MED 140:787-793(1972)

KORDA A   SHEARMAN RP   SMITH ID
SOME OBSERVATIONS ON THE USE OF PGF2A FOR TERMINATION OF
PREGNANCY
AUST N Z J OBSTET GYNAECOL 13:44(1973)

KORMAN MG   HANSKY J   RITCHIE BC
WATTS JM   MALONEY JE
DISAPPEARANCE OF GASTRIN ACROSS THE LUNG
AUST J EXP BIOL MED SCI 51:679-687(1973)

KOROLKIEWICZ Z   POCWIARDOWSKA E
PGE1 INFLUENCE ON LIPOLYTIC PROPERTIES OF SOME FEVER
PRODUCING AGENTS AND ITS ACTION ON OXIDATIVE PHOSPHORYLATION
IN: "FARMAKOLOGIA AMIN KATECHOLOWYCH ORAZ LEKOW DZIALAJACYCH
NA UKLAD NERWOWY," I ZJAZDU, PANSTWOWY ZAKLAD WYDAWNICTW
LEKARSKICH, WARSAW : 265-271(1971);(POLISH)

KOROLKIEWICZ Z   POCWIARDOWSKA E
PGE1 AND IAA AS TOOLS FOR PRELIMINARY ANALYSIS OF METABOLIC
EFFECT OF LIPOPOLYSACCHARIDES FROM E. COLI
IN: PHOSPHOLIPIDS PROC INT SYMP, L SAMOCHOWIEC, J WOJCICKI,
INT SOC BIOCHEM PHARMACOL: 67-74(1973)

KOSS MC   GRAY JW   DAVISON M
NAKANO J
CARDIOVASCULAR ACTIONS OF PROSTAGLANDINS E1 AND F2A IN THE
CAT
EUR J PHARMACOL 24:151-157(1973)

KOSS MC   GRAY JW   NAKANO J
CARDIOVASCULAR ACTIONS OF PROSTAGLANDINS E1 AND F2A IN THE
CAT
FED PROC 31:545(1972)

KOSS MC   NAKANO J
EFFECTS OF PROSTAGLANDINS E1 AND F2A ON THE PERIPHERAL
CIRCULATION IN THE CAT
PROC SOC EXP BIOL MED 142:383-386(1973)

KOSS MC   RIEGER JA   NAKANO J
HEMODYNAMIC RESPONSES TO PROSTAGLANDIN F2A IN THE CAT:
SELECTIVE BLOCKADE BY MECLOFENAMIC ACID
FED PROC 32:787(1973)

KOTCHEN TA   MILLER MC
EFFECT OF PROSTAGLANDINS ON RENIN REACTIVITY
CLIN RES 21:694(1973)

KOWALSKI K   BABIARZ D   BURKE G
PHAGOCYTOSIS OF LATEX BEADS BY ISOLATED THYROID CELLS:
EFFECTS OF THYROTROPIN, PROSTAGLANDIN E1, AND DIBUTYRYL
CYCLIC AMP
J LAB CLIN MED 79:258-266(1972)

KOWALSKI K   SATO S   BURKE G
THYROTROPIN- AND PROSTAGLANDIN E2-RESPONSIVE ADENYL CYCLASE
IN THYROID PLASMA MEMBRANES
PROSTAGLANDINS 2:441-452(1972)

KRAEMER RJ   FOLTS JD
RELEASE OF PROSTAGLANDIN FOLLOWING TEMPORARY OCCLUSION OF
THE CORONARY ARTERY
FED PROC 32:454(1973)

KRAKOFF LR   DE GUIA D   VLACHAKIS N
STRICKER J   GOLDSTEIN M
EFFECT OF SODIUM BALANCE ON ARTERIAL BLOOD PRESSURE AND
RENAL RESPONSES TO PROSTAGLANDIN A1 IN MAN
CIRC RES 33:539-546(1973)

KRAM R   MAMONT P   TOMKINS GM
PLEIOTYPIC CONTROL BY ADENOSINE 3':5'-CYCLIC MONOPHOSPHATE:
A MODEL FOR GROWTH CONTROL IN ANIMAL CELLS
PROC NATL ACAD SCI U S A 70:1432-1436(1973)

KRANE SM
ACTION OF SALICYLATES
N ENGL J MED 286:317-318(1972)

KRAUSE W
THE PROSTAGLANDINS
DTSCH GESUNDHEITSW 28:1009-1018(1973);(GERMAN)

KREINER PW   GOLD CJ   KEIRNS JJ
BROCK WA   BITENSKY MW
HORMONAL CONTROL OF MELANOCYTES. MSH-SENSITIVE ADENYL
CYCLASE IN THE CLOUDMAN MELANOMA
YALE J BIOL MED 46:583-591(1973)

KREINER PW   GORMAN RE   BITENSKY MW
EFFECTS OF MSH AND PROSTAGLANDINS ON MELANOMA ADENYL CYCLASE
ABSTR 4TH INT CONGR ENDOCRINOL, INT CONGR SER 256, EXCERPTA
MEDICA, AMST:229(1972)

KREINER PW   KEIRNS JJ   BITENSKY MW
A TEMPERATURE-SENSITIVE CHANGE IN THE ENERGY OF ACTIVATION
OF HORMONE-STIMULATED HEPATIC ADENYLYL CYCLASE
PROC NATL ACAD SCI USA 70:1785-1789(1973)

400

KRISHNA G   GORSKI J   KRISHNAN N
THE INVOLVEMENT OF LACTIC ACID AND OTHER CONTROL MECHANISMS
IN THE HORMONE RECEPTOR INTERACTIONS PRECEDING THE FORMATION
OF CYCLIC AMP BY VARIOUS ADENYLATE CYCLASE SYSTEMS
ABSTR, 9TH INT CONG BIOCHEM, STOCKH:348(1973)

KRISHNA G   HARWOOD JP   BARBER AJ
JAMIESON GA
REQUIREMENT FOR GUANOSINE TRIPHOSPHATE IN THE
PROSTAGLANDIN ACTIVATION OF ADENYLATE CYCLASE OF PLATELET
MEMBRANES
J BIOL CHEM 247:2253-2254(1972)

KRISHNA G   HARWOOD JP   MOSKOWITZ J
DYNAMIC ASPECTS OF PROSTAGLANDIN INTERACTION WITH ADENYL
CYCLASE
ADV CYCLIC NUCL RES 1:578(1972)

KRISHNARAJ R   TALWAR GP
ROLE OF CYCLIC AMP IN MITOGEN INDUCED TRANSFORMATION OF
HUMAN PERIPHERAL LEUKOCYTES
J IMMUNOL 111:1010-1017(1973)

KU EC   WASVARY JM
INHIBITION OF PROSTAGLANDIN SYNTHETASE BY SU-21524
FED PROC 32:803(1973)

KUBISZ P   PETRICEK J
INFLUENCE OF PROSTAGLANDIN E1 FOR LIBERATING THE PLATELET
FACTOR 4(PF 4)
BRATISL LEK LISTY 58:408-412(1972);(SLOVAKIAN)

KUDRIN AN   PERSIANINOV LS   KOROZA GS
MECHANISM OF STIMULATING ACTION OF PROSTAGLANDIN F2A ON THE
CONTRACTILE ACTIVITY OF THE UTERUS
AKUSH GINEKOL (MOSK) 49(11):1-7(1973);(RUSSIAN)

KUEHL AF JR
PROSTAGLANDINS, CYCLIC NUCLEOTIDES AND CELL FUNCTION
IN: "PROSTAGLANDINES 1973," INSERM, PARIS:55-80(1973)

KUEHL FA JR
THE REGULATORY ROLE OF THE PROSTAGLANDINS ON THE
CYCLIC-3'-5'-AMP SYSTEM
ADV BIOSCI (SUPPL) 9:25(1972)

KUEHL FA JR
THE REGULATORY ROLE OF PROSTAGLANDINS ON CYCLIC NUCLEOTIDES
IN: "PROSTAGLANDINS AND CYCLIC AMP," RH KAHN, WEM LANDS,
ACADEMIC PRESS, NY : 223-225(1973)

KUEHL FA JR
THE REGULATORY ROLE OF PROSTAGLANDINS ON CYCLIC NUCLEOTIDES
ABSTR, 9TH INT CONG BIOCHEM, STOCKH:390(1973)

KUEHL FA JR   CIRILLO VJ   HAM EA
HUMES JL
THE REGULATORY ROLE OF THE PROSTAGLANDINS ON THE CYCLIC
3',5'-AMP SYSTEM
ADV BIOSCI 9:155-172(1973)

KUEHL FA JR   HUMES JL
DIRECT EVIDENCE FOR A PROSTAGLANDIN RECEPTOR AND ITS
APPLICATION TO PROSTAGLANDIN MEASUREMENTS
PROC NATL ACAD SCI USA 69:480-484(1972)

KUEHL FA JR   HUMES JL   MANDEL LR
CIRILLO VJ   ZANETTI ME   HAM EA
PROSTAGLANDIN ANTAGONISTS: STUDIES ON THE MODE OF ACTION OF
POLYPHLORETIN PHOSPHATE.
BIOCHEM BIOPHYS RES COMMUN 44:(6):1464-1470(1971)

KUHN DC   WILLIS AL
PROSTAGLANDIN E2, INFLAMMATION AND PAIN THRESHOLD IN RAT
PAWS
BR J PHARMACOL 49:183P-184P(1973)

KUHNLE H
CLINICAL EXPERIENCES WITH PROSTAGLANDIN F2A.
ARCH GYNAEKOL 214:461(1973);(GERMAN)

KUMAR R   SOLOMON LM
PROSTAGLANDINS IN CUTANEOUS BIOLOGY
ARCH DERMATOL 106:101-107(1972)

KUNIMOTO K   MIYAUCHI S
PROSTAGLANDIN
GENDAI IRYO 4:1227-1238(1972);(JAPANESE)

KUNZE H   VOGT W
SIGNIFICANCE OF PHOSPHOLIPASE A FOR PROSTAGLANDIN FORMATION
ANN NY ACAD SCI 180:123-125(1971)

KUO CH   TAUB D   WENDLER NL
AN EFFICIENT NEW APPROACH TO THE SYNTHESIS OF THE
PROSTAGLANDINS. SYNTHESIS OF PGE1
TETRAHEDRON LETT 5317-5320(1972)

KUO WN   KUO JF
REGULATION OF CYCLIC GMP AND CYCLIC AMP LEVELS IN RAT LUNG
AND OTHER TISSUES BY VARIOUS AGENTS AS DETERMINED BY
DOUBLE-PRELABELING WITH RADIOACTIVE GUANINE AND ADENINE
FED PROC 32:773(1973)

KUPIECKI FP
PHARMACOLOGICAL CONTROL OF FREE FATTY ACIDS.
PROG BIOCHEM PHARMACOL 6:274-316(1971)

KURAMOTO A
INHIBITORY MECHANISM OF PLATELET ENERGY PRODUCTION AND
PLATELET FUNCTIONS BY PGE1 AND UREMIC TOXINS
ACTA HAEMATOL JAP 36:477-487(1973);(JAPANESE)

KURAMOTO A   TAKETOMI Y   UCHINO H
RADIATION EFFECT ON PLATELET FUNCTION AND METABOLISM
ABSTR 4TH INT CONGR THROMB HAEMOSTASIS, VIENNA:295(1973)

KURODA M
RENIN ACTIVITY. RENAL ANTIHYPERTENSIVE SUBSTANCES
TAISHA 9:1137-1145(1972);(JAPANESE)

KUROZUMI S   TORU T   ISHIMOTO S
PREPARATION OF 2-(4-HYDROXY-1-OXOCYCLOPENT-2-ENE) HEPTANOIC
ACID AN IMPORTANT PROSTAGLANDIN SYNTHON
TETRAHEDRON LETT 4959-4960(1973)

LA BRECQUE M
PALLIATIVES AND PROSTAGLANDINS
THE SCIENCES 12:6-7(1972)

LABADIE P
THE PROSTAGLANDINS: UBIQUITOUS SUBSTANCES
REV PRAT 21:5004-5024(1971);(FRENCH)

LABHSETWAR AP
STUDIES ON THE ENDOCRINE PROFILE OF PROSTAGLANDIN E2
INTRA-SCI CHEM REP 6:31-42(1972)

LABHSETWAR AP
PITUITARY GONADOTROPHIC FUNCTION (FSH AND LH) IN VARIOUS
REPRODUCTIVE STATES
ADV REPROD PHYSIOL 6:97-183(1973)

LABHSETWAR AP   PERSER N
UTERINE UPTAKE OF (6,7-3H) OESTRADIOL IN THE PRESENCE OF
INTRAUTERINE CONTRACEPTIVE DEVICE IN CYCLIC AND PREGNANT
RATS
ACTA ENDOCRINOL (KBH) 69:583-588(1972)

LABORIT G
MECHANISMS AND THERAPEUTICS OF PAIN
REV BELG MED DENT 27:157-170(1972); (FRENCH)

LABORIT G   BARON C   LABORIT H
ASPIRIN, CATECHOLAMINES, & LACTACIDEMIA
AGRESSOLOGIE 14 : 25-30 (1973); (FRENCH)

LACUARA MC   LACUARA JL
CONTRACTILE RESPONSE OF THE ISOLATED PORTAL VEIN OF THE RAT
TO PROSTAGLANDINS E1 AND F2
MEDICINA (B AIRES) 31:519-520(1971);(SPANISH)

LAFFERTY JJ   KANNEGIESSER H   LEE JB
METABOLIC MECHANISMS OF ACTION OF THE RENAL PROSTAGLANDINS
ADV BIOSCI (SUPPL) 9:47(1972)

LAFFERTY JJ   KANNEGIESSER H   LEE JB
PARKER CW
METABOLIC MECHANISMS OF ACTION OF THE RENAL PROSTAGLANDINS
ADV BIOSCI 9:293-299(1973)

LAHIRI K
THE PROSTAGLANDINS
EAST PHARM 16(190):17-19(1973)

LAKE N   JORDAN LM   PHILLIS JW
MECHANISM OF NORADRENALINE ACTION IN CAT CEREBRAL CORTEX
NATURE (NEW BIOL) 240:249-250(1972)

LAKE N   JORDAN LM   PHILLIS JW
EVIDENCE AGAINST CYCLIC ADENOSINE 3'-5'-MONOPHOSPHATE (AMP)
MEDIATION OF NORADRENALINE DEPRESSION OF CEREBRAL CORTICAL
NEURONES
BRAIN RES 60:411-421 (1973)

LAMOND DR
THE ROLE OF THE BOVINE PRACTITIONER IN SYNCHRONIZATION AND
TWINNING IN CATTLE
BOVINE PRACT (8):2-8(1973)

LAMPRECHT SA   ZOR U   BAUMINGER S
LINDNER HR
MECHANISM OF ACTION OF LH ON CYCLIC AMP PRODUCTION BY THE
RAT: INDEPENDENCE OF PROSTAGLANDIN MEDIATION
ABSTR, 9TH INT CONG BIOCHEM, STOCKH:405(1973)

LANDON EJ   FORTE LR
CELLULAR MECHANISMS IN RENAL PHARMACOLOGY
ANNU REV OF PHARMACOL:171-188(1971)

LANDS W   LEE R   SMITH W
FACTORS REGULATING THE BIOSYNTHESIS OF VARIOUS
PROSTAGLANDINS
ANN NY ACAD SCI 180:107-122(1971)

LANDS WEM
ROUND TABLE DISCUSSION OF METABOLISM OF PROSTAGLANDIN AND
CYCLIC AMP
IN: "PROSTAGLANDINS AND CYCLIC AMP," RH KAHN, WEM LANDS,
ACADEMIC PRESS, NY : 15-16(1973)

LANDS WEM   LETELLIER PR   ROME LH
VANDERHOEK JY
INHIBITION OF PROSTAGLANDIN BIOSYNTHESIS
ADV BIOSCI 9:15-28(1973)

LANDS WEM   LETELLIER PR   VANDERHOEK JY
INHIBITION OF PROSTAGLANDIN BIOSYNTHESIS
ADV BIOSCI (SUPPL) 9:2(1972)

LANGE L   KAPTEINA F   ECHT M
KIRSCH K   SCHULTZE G
EFFECT OF PROSTAGLANDINS E2 AND F2A ON THE RESISTANCE AND
CAPACITY VESSELS OF THE ISOLATED RABBIT EAR
PFLUEGERS ARCH (SUPPL 332):R54(1972);(GERMAN)

LANGE RF   SHULMAN NR   TOMASULO PA
COLEMAN CN
METABOLIC EFFECTS OF ANTIBODIES ON PLATELETS
TRANS ASSOC AM PHYSICIANS 86:131-142(1973)

LANGSLOW DR
THE ANTI-LIPOLYTIC ACTION OF PROSTAGLANDIN E1 ON ISOLATED
CHICKEN FAT CELLS
BIOCHIM BIOPHYS ACTA 239:33-37(1971)

LAROS RK JR   WITTING WC   WORK BA JR
UTERINE ACTIVITY RESPONSE TO CONSTANT INFUSION OF
PROSTAGLANDIN F2A IN TERM HUMAN PREGNANCY
ABSTR SOC GYNECOL INVEST 20TH ANNU MEET, ATLANTA:46(1973)

LARSEN JW   HANSON T   CALDWELL BV
SPEROFF L
THE EFFECT OF ESTRADIOL INFUSION ON UTERINE ACTIVITY AND
PERIPHERAL LEVELS OF F PROSTAGLANDINS AND PROGESTERONE
ABSTR SOC GYNECOL INVEST 20TH ANNU MEET, ATLANTA:34(1973)

LARSSON C   ANGGARD E
ON THE FORMATION OF PROSTAGLANDIN E2 IN THE RABBIT RENAL
PAPILLA
ACTA PHARMACOL TOXICOL (KBH) 29 (SUPPL 4) : 29(1971)

LARSSON C   ANGGARD E
THE DISTRIBUTION OF ENZYMES INVOLVED IN PROSTAGLANDIN
SYNTHESIS AND METABOLISM IN THE RABBIT KIDNEY
ABSTR 5TH INT CONGR PHARMACOL, SAN FRANC:135(1972)

LARSSON C   ANGGARD E
FORMATION AND METABOLISM OF PROSTAGLANDINS IN THE RABBIT
KIDNEY, REGIONAL DIFFERENCES OF THE ENZYMES INVOLVED
ACTA PHARMACOL TOXICOL (KBH) 31:107(1972)

LARSSON C   ANGGARD E
FORMATION AND METABOLISM OF PROSTAGLANDINS IN THE RABBIT
KIDNEY, REGIONAL DIFFERENCES OF THE ENZYMES INVOLVED
ADV BIOSCI (SUPPL) 9:56(1972)

LARSSON C   ANGGARD E
REGIONAL DIFFERENCES IN THE FORMATION AND METABOLISM OF
PROSTAGLANDINS IN THE RABBIT KIDNEY
EUR J PHARMACOL 21:30-36(1973)

LARSSON C   ANGGARD E
ARACHIDONIC ACID LOWERS AND INDOMETHACIN INCREASES THE
BLOOD PRESSURE OF THE RABBIT
J PHARM PHARMACOL 25:653-655(1973)

LARSSON C   ANGGARD E
INCREASED JUXTAMEDULLARY BLOOD FLOW FOLLOWING STIMULATION
OF INTRARENAL PROSTAGLANDIN BIOSYNTHESIS
ACTA PHYSIOL SCAND (SUPPL 396) : 20(1974)

LAU IF   SAKSENA SK   CHANG MC
PREGNANCY BLOCKADE BY INDOMETHACIN, AN INHIBITOR OF
PROSTAGLANDIN SYNTHESIS: ITS REVERSAL BY PROSTAGLANDINS AND
PROGESTERONE IN MICE
PROSTAGLANDINS 4:795-803(1973)

LAUERSEN NH   FUCHS AR   FUCHS F
EFFECT OF ETHANOL ON OXYTOCIN AND PROSTAGLANDIN INDUCED
UTERINE ACTIVITY IN PREGNANT BABOONS
ABSTR SOC GYNECOL INVEST 20TH ANNU MEET, ATLANTA:34(1973)

LAURITZEN C
ENDOCRINOLOGY OF NORMAL PREGNANCY
MED KLIN 68:897-905(1973);(GERMAN)

LAVERY HA   LOWE RD   SCROOP GC
CENTRAL AUTONOMIC EFFECTS OF PROSTAGLANDIN F2A ON THE
CARDIOVASCULAR SYSTEM OF THE DOG
BR J PHARMACOL 41:454-461(1971)

LAYNE P   CONSTANTOPOULOS A   JUDGE JFX
RAUNER R   NAJJAR VA
THE OCCURRENCE OF FLUORIDE STIMULATED MEMBRANE
PHOSPHOPROTEIN PHOSPHATASE
BIOCHEM BIOPHYS RES COMMUN 53:800-805(1973)

LEACH BE   ARMSTRONG FB   GERMAIN GS
MUIRHEAD EE
VASODEPRESSOR ACTION OF PROSTAGLANDINS A2 AND E2 IN THE
SPONTANEOUSLY HYPERTENSIVE RAT (SH RAT): EVIDENCE FOR AN
ACTION MEDIATED BY THE VAGUS
J PHARMACOL EXP THER 185:479-485(1973)

LEACH BE   ARMSTRONG FB JR   GERMAIN GS
MUIRHEAD EE
ANTIHYPERTENSIVE ACTION OF RENOMEDULLARY PROSTAGLANDINS
(PGE2 AND PGA2) IN THE SPONTANEOUSLY HYPERTENSIVE RAT
(SH RAT): EVIDENCE FOR AN INDIRECT BLOOD-PRESSURE-LOWERING
EFFECT
J LAB CLIN MED 78:803-804(1971)

LEACH BE   MUIRHEAD EE   ARMSTRONG FB JR
GERMAIN GS
VASODEPRESSOR ACTION OF RENOMEDULLARY PROSTAGLANDINS (PGE2
AND PG2) IN THE SPONTANEOUSLY HYPERTENSIVE RAT (SH RAT):
EVIDENCE FOR A CENTRAL ACTION MEDIATED BY THE VAGUS
ADV BIOSCI 9:369-372(1973)

LEAF A
SURFACE COATING AND TRANSPORT
IN BIOMEMBRANES F KREUZER AND FG SLEGERS NEW YORK PLENUM
3:349-355(1972)

LEARY WP   VANE JR   LEDINGHAM JG
PROSTAGLANDIN RELEASE BY KIDNEYS IN HYPERTENSION
S AFR MED J 47:1761(1973)

LEE J   KANNEGIESSER H   O'TOOLE J
WESTURA E
HYPERTENSION AND THE RENOMEDULLARY PROSTAGLANDINS: A HUMAN
STUDY OF THE ANTIHYPERTENSIVE EFFECTS OF PGA1
ANN NY ACAD SCI 180:218-240(1971)

LEE JB
THE INTER-RELATIONSHIPS BETWEEN RENAL PROSTAGLANDINS AND
BLOOD PRESSURE REGULATION
AM J MED SCI 263:334-346(1972)

LEE JB
THE ANTIHYPERTENSIVE AND NATRIURETIC ENDOCRINE FUNCTION OF
THE KIDNEY: VASCULAR AND METABOLIC MECHANISMS OF THE RENAL
PROSTAGLANDINS
IN: "PROSTAGLANDINS IN CELLULAR BIOLOGY," PW RAMWELL AND BB
PHARRISS, PLENUM PRESS, NY/LOND:399-449(1972)

LEE JB
NATRIURETIC "HORMONE" AND THE RENAL PROSTAGLANDINS
PROSTAGLANDINS 1:55-70(1972)

LEE JB
RENAL HORMONES: WHITHER THE ANTIHYPERTENSIVE PROSTAGLANDINS?
IN: "MONOGRAPHS ON HYPERTENSION," MERCK SHARP & DOHME, WEST
POINT, PA:1-20(1972)

LEE JB
RENAL HOMEOSTASIS AND THE HYPERTENSIVE STATE: A UNIFYING
HYPOTHESIS
IN: "THE PROSTAGLANDINS," PW RAMWELL, PLENUM PRESS, NY
1:133-187(1973)

LEE JB
THE PROSTAGLANDINS AND THE REGULATION OF SYSTEMIC BLOOD
PRESSURE AND SODIUM AND WATER HOMEOSTASIS
IN: "PROSTAGLANDINES 1973," INSERM, PARIS:161-205(1973)

LEE JB
HYPERTENSION, NATRIURESIS AND THE RENOMEDULLARY
PROSTAGLANDINS: AN OVERVIEW
PROSTAGLANDINS 3:551-579(1973)

LEE JB
HOMEOSTASIS AND THE PROSTAGLANDINS
TRANS STUD COLL PHYSICIANS PHILA 41:6-27(1973)

LEE JB   MCGIFF JC   KANNEGIESSER H
AYKENT YY   MUDD JG   FRAWLEY TF
PROSTAGLANDIN A1: ANTIHYPERTENSIVE AND RENAL EFFECTS
ANN INT MED 74:703-710(1971)

LEE RA   LANDS WEM
COFACTORS IN THE BIOSYNTHESIS OF PROSTAGLANDINS F1A AND F2A
BIOCHIM BIOPHYS ACTA 260:203-211(1972)

LEE SJ   JOHNSON JG   SMITH CJ
HATCH FE
RENAL EFFECTS OF PROSTAGLANDIN A1 IN PATIENTS WITH ESSENTIAL
HYPERTENSION
KIDNEY INT 1:254-262(1972)

LEE SS
PROSTAGLANDINS
HWAHAK KWA KONGOP UI CHINBO 12:205-206(1972);(KOREAN)

LEE Y   SANNER JH   DOBRIN EI
THE ANTIDIARRHEAL ACTIVITY OF SC-26100 IN EXPERIMENTAL
ANIMALS
FED PROC 31:507(1972)

LEE YH   BIANCHI RG
THE ANTISECRETORY AND ANTIULCER ACTIVITY OF A PROSTAGLANDIN
ANALOG, SC-24665, IN EXPERIMENTAL ANIMALS
ABSTR 5TH INT CONGR PHARMACOL SAN FRANC 136(1972)

LEE YH   CHENG WD   BIANCHI RG
MOLLISON K   HANSEN J
EFFECTS OF ORAL ADMINISTRATION OF PGE2 ON GASTRIC SECRETION
AND EXPERIMENTAL PEPTIC ULCERATIONS
PROSTAGLANDINS 3:29-45(1973)

LEFEBVRE P
GLUCAGON AND THE METABOLISM OF ADIPOSE TISSUE-HORMONAL
INTERRELATIONS: INSULIN, PROSTAGLANDINS AND CATECHOLAMINES
ACTA DIABETOL LAT 10:372(1973);(ITALIAN)

LEFEBVRE PJ   LUYCKX AS
EFFECT OF PROSTAGLANDIN PGE1 ON BLOOD FLOW AND INSULIN
OUTPUT OF DOG PANCREAS IN SITU
DIABETES 21(SUPPL 1):369(1972)

LEFEBVRE PJ   LUYCKX AS
STIMULATION OF INSULIN SECRETION AFTER PROSTAGLANDIN PGE1
IN THE ANESTHETIZED DOG
BIOCHEM PHARMACOL 22:1773-1779(1973)

LEFEBVRE PJ   LUYCKX AS
EFFECT OF PROSTAGLANDIN PGE1 ON BLOOD FLOW AND INSULIN
OUTPUT OF DOG PANCREAS IN SITU
DIABETOLOGIA 9:77(1973)

LEHMANN F   PETERS F   BRECKWOLDT M
BETTENDORF G
PLASMA PROGESTINS AND PLASMA ESTROGENS LEVELS DURING
INFUSION OF PGF2A IN THE HUMAN
ADV BIOSCI (SUPPL) 9:112(1972)

LEHMANN FF   BRECKWOLDT M   BETTENDORF G
INDUCTION OF ABORTION WITH PROSTAGLANDIN F2A
GEBURTSHILFE FRAUENHEILKD 32:477-483(1972);(GERMAN)

LEMBERG A   DE WIKINSKI RLW   IZURIETA EM
PAGLIONE AM   HALPERIN H   DE NEUMAN PP
KRASNOKUKI D   JAMARDO N
HEPATIC LIPOLYTIC ACTIVITY IN RESPONSE TO HYPERLIPEMIC
SUBSTRATES. EFFECT OF NOREPINEPHRINE, PROSTAGLANDIN E1,
ADENOSINE CYCLIC MONOPHOSPHATE, AND A BLOCKING AGENT
(ALDPRENALOL).
REV ASOC BIOQUIM ARGENT 38(205-206):36-48(1973);(SPANISH)

LEMBERG A   WIKINSKI R   IZURIETA EM
HALPERIN H   PAGLIONE AM   DE NEUMAN P
JAUREGUI H
EFFECTS OF PROSTAGLANDIN E1 AND NOREPINEPHRINE ON GLUCOSE
AND LIPID METABOLISM IN ISOLATED PERFUSED RAT LIVER
BIOCHIM BIOPHYS ACTA 248:198-204(1971)

LEMBERG A   WIKINSKI R   IZURIETA EM
HALPERIN H   PAGLIONE AM   DE NEUMAN P
EFFECT OF PROSTAGLANDIN E1 AND NOREPINEPHRINE ON LIPID AND
GLUCOSE METABOLISM IN ISOLATED PERFUSED RAT LIVER OVERLOADED
WITH A LIPID SUBSTRATE
BIOCHIM BIOPHYS ACTA 280:458-465(1972)

LENZ R   BIERENBAUM ML   FLEISCHMAN AI
INHIBITION IN VIVO OF ADP INDUCED PLATELET AGGREGATION BY
PROSTAGLANDIN E1 IN THE RAT
CIRCULATION (SUPPL IV) 8:IV-190 (1973)

LEONARDI RG   ALEXANDER B   WHITE F
PREVENTION OF THE INHIBITORY EFFECT OF ASPIRIN ON PLATELET
AGGREGATION
FED PROC 31:248(1972)

LEONARDI RG   ALEXANDER B   WHITE F
LAWLOR D
PROTECTIVE EFFECT OF ARACHIDONIC ACID ON PLATELET TOXIC
AGENTS
ABSTR 4TH INT CONGR THROMB HAEMOSTASIS, VIENNA:80(1973)

LEONARDI RG   ALEXANDER B   WHITE F
LAWLOR D
PROTECTIVE EFFECT OF ARACHIDONIC ACID ON PLATELET TOXIC
AGENTS
CLIN RES 21:728(1973)

LERMAN RJ   PITCOCK JA   STEPHENSON P
MUIRHEAD EE
RENOMEDULLARY INTERSTITIAL CELL TUMOR (FORMERLY FIBROMA OF
RENAL MEDULLA)
HUMAN PATHOL 3:559-568(1972)

LERNER LJ   CARMINATI P
EFFECTS OF PGE2 AND 7-OXA-13-PROSTYNOIC ACID ON CAMP OF
RABBIT UTEROTUBAL JUNCTION AND OVIDUCTAL AMPULLA
ABSTR, 9TH INT CONG BIOCHEM, STOCKH:399(1973)

LESLIE CA   LEVINE L
EVIDENCE FOR THE PRESENCE OF A PROSTAGLANDIN E2-9-KETO
REDUCTASE IN RAT ORGANS
BIOCHEM BIOPHYS RES COMMUN 52:717-724(1973)

LEVEY GS
THE ROLE OF PHOSPHOLIPIDS IN HORMONE ACTIVATION OF ADENYLATE
CYCLASE
RECENT PROG HORM RES 29:383-384;386(1973)

LEVEY GS   KLEIN I
EFFECT OF PROSTAGLANDINS ON GUINEA PIG MYOCARDIAL ADENYL
CYCLASE
CLIN RES 19:82(1971)

LEVEY GS   KLEIN I
SOLUBILIZED MYOCARDIAL ADENYLATE CYCLASE: ACTIVATION BY
PROSTAGLANDINS
LIFE SCI 13:41-46 (1973)

LEVIN DL   PERLIA C   TASHJIAN AH JR
MEDULLARY CARCINOMA OF THE THYROID GLAND: THE COMPLETE
SYNDROME IN A CHILD
PEDIATRICS 52:192-196(1973)

LEVINE L
ANTIBODIES TO PHARMACOLOGICALLY ACTIVE MOLECULES:
SPECIFICITIES AND SOME APPLICATIONS OF ANTIPROSTAGLANDINS
PHARMACOL REV 25:293-307(1973)

LEVINE L   GUTIERREZ-CERN/ RM
PREPARATION AND SPECIFICITY OF ANTIBODIES TO
15-KETO-PROSTAGLANDIN F2A
PROSTAGLANDINS 2:281-294(1972)

LEVINE L   GUTIERREZ-CERN/ RM
LEVELS OF 13,14-DIHYDRO-15-KETO-PGF2A IN BIOLOGICAL FLUIDS
AS MEASURED BY RADIOIMMUNOASSAY
PROSTAGLANDINS 3:785-804(1973)

LEVINE L   GUTIERREZ-CERN/ RM   VAN VUNAKIS H
SPECIFICITIES OF PROSTAGLANDINS B1, F1A, AND F2A
ANTIGEN-ANTIBODY REACTIONS
J BIOL CHEM 246:6782-6785(1971)

LEVINE L   GUTIERREZ-CERN/ RM   VAN VUNAKIS H
SPECIFIC ANTIBODIES: REAGENTS FOR QUANTITATIVE ANALYSIS OF
PROSTAGLANDINS
ADV BIOSCI (SUPPL) 9:11-12(1972)

LEVINE L   GUTIERREZ-CERN/ RM   VAN VUNAKIS H
SPECIFIC ANTIBODIES: REAGENTS FOR QUANTITATIVE ANALYSIS OF
PROSTAGLANDINS
ADV BIOSCI 9:71-82(1973)

LEVINE L   HINKLE PM   VOELKEL EF
TASHJIAN AH JR
PROSTAGLANDIN PRODUCTION BY MOUSE FIBROSARCOMA CELLS IN
CULTURE: INHIBITION BY INDOMETHACIN AND ASPIRIN
BIOCHEM BIOPHYS RES COMMUN 47:888-896(1972)

LEVINE R
EFFECT OF PROSTAGLANDINS AND CYCLIC AMP ON GASTRIC SECRETION
ANN NY ACAD SCI 180:336-337(1971)

LEVINE RA
FAILURE OF PROSTAGLANDIN E1 TO STIMULATE CYCLIC AMP AND
GLYCOGENOLYSIS IN ISOLATED PERFUSED RAT LIVER
GASTROENTEROLOGY 64:186(1973)

LEVINE RA
THE ROLE OF CYCLIC AMP AND PROSTAGLANDINS IN HEPATIC AND
GASTROINTESTINAL FUNCTIONS
IN: "PROSTAGLANDINS AND CYCLIC AMP," RH KAHN, WEM LANDS,
ACADEMIC PRESS, NY : 75-117(1973)

LEVINE RA   WILSON DE
THE ROLE OF CYCLIC AMP IN GASTRIC SECRETION
ANN N Y ACAD SCI 185:363-375(1971)

LEVITT MJ
RAPID METHYLATION OF MICRO AMOUNTS OF NONVOLATILE ACIDS
ANAL CHEM 45:618-620(1973)

LEVITT MJ
ASSAY OF PROSTAGLANDINS
IN: "PROSTAGLANDINS AND CYCLIC AMP," RH KAHN, WEM LANDS,
ACADEMIC PRESS, NY : 17-18(1973)

LEVITT MJ   JOSIMOVICH JB
ANALYSIS OF PG BY GAS-CHROMATOGRAPHY WITH ELECTRON-CAPTURE
DETECTION
FEDERATION PROC 30:1081 ABS (1971)

LEVITT MJ   JOSIMOVICH JB   BROSKIN KD
ANALYSIS OF PROSTAGLANDINS BY ELECTRON-CAPTURE GAS
CHROMATOGRAPHY. I. THERMAL DECOMPOSITION OF
HEPTAFLUOROBUTYRATE METHYL ESTERS
PROSTAGLANDINS 1:121-131(1972)

LEVITT MJ   TOBON H   JOSIMOVICH JB
ENDOMETRIAL PGF2A CONTENT IN OVULATORY AND ANOVULATORY WOMEN
ABSTR SOC GYNECOL INVEST 20TH ANNU MEET, ATLANTA:52(1973)

LEVY JV
INHIBITORY EFFECTS OF ANTIINFLAMMATORY DRUGS ON PGE2-INDUCED
CONTRACTIONS OF HUMAN VEIN STRIPS IN VITRO
CLIN RES 20:211(1972)

LEVY JV
EFFECTS OF PAPAVERINE, AMINOPHYLLINE, PROSTAGLANDINS E1 AND
E2, AND CATHECHOLAMINES ON CONTRACTILE RESPONSES OF HUMAN
AND RABBIT MYOCARDIUM AND VASCULAR TISSUE
ADV CYCLIC NUCL RES 1:579-580(1972)

LEVY JV
EFFECT OF SYNTHETIC BRADYKININ ON CONTRACTILE TENSION OF
HUMAN SAPHENOUS VEIN STRIPS
BR J PHARMACOL 46:517-518(1972)

LEVY JV
EFFECTS OF NON-STEROID AND STEROID ANTIINFLAMMATORY DRUGS ON
CONTRACTILE RESPONSES OF ISOLATED HUMAN VEIN STRIPS
ABSTR 5TH INT CONGR PHARMACOL SAN FRANC 139(1972)

LEVY JV
DRUG STUDIES ON ARTERIAL TISSUE OF SPONTANEOUSLY
HYPERTENSIVE RATS (SHR)
FED PROC 32:749(1973)

LEVY JV
PAPAVERINE ANTAGONISM OF PROSTAGLANDIN E2 - INDUCED
CONTRACTION OF RABBIT AORTIC STRIPS
RES COMMUN CHEM PATHOL PHARMACOL 5:297-310(1973)

LEVY JV
STUDIES ON THE CONTRACTILE EFFECTS OF PROSTAGLANDINS ON
AORTIC STRIP PREPARATIONS FROM SPONTANEOUSLY HYPERTENSIVE
RATS
RES COMMUN CHEM PATHOL PHARMACOL 6:365-381(1973)

LEVY JV
CHRONOTROPIC AND INOTROPIC EFFECTS OF PGE2 ON ISOLATED RAT
ATRIA FROM NORMAL AND SPONTANEOUSLY HYPERTENSIVE RATS (SHR)
PROSTAGLANDINS 4:731-736(1973)

LEVY JV   KILLEBREW E
INOTROPIC EFFECT OF PROSTAGLANDIN E2 ON ISOLATED CARDIAC
TISSUE
PROC SOC EXP BIOL MED 136:1227-1231(1971)

LEWIS AJ
PROSTAGLANDINS
VET REV 23:33-39(1972)

LEWIS AJ
SOME OBSERVATIONS ON THE PULMONARY ARTERY OF THE
GUINEA-PIG
J PHARM PHARMACOL 25:156-167(1973)

LEWIS AJ   EYRE P
SOME CARDIOVASCULAR AND RESPIRATORY EFFECTS OF
PROSTAGLANDINS E1, E2 AND F2A IN THE CALF
PROSTAGLANDINS 2:55-64(1972)

LEWIS AJ   EYRE P   DOWNIE HG
RESPONSES TO AND RELEASE OF PUTATIVE MEDIATORS OF
ANAPHYLAXIS IN THE HORSE
FED PROC 31:747(1972)

LEWIS AJ   WELLS PW   EYRE P
CUTANEOUS REACTION TO PROSTAGLANDINS E1, E2 AND F2A IN THE
BOVINE
J PHARM PHARMACOL 24:326-328(1972)

LEWIS GP
ROLE OF KININS AND PROSTAGLANDINS AS MEDIATORS OF FUNCTIONAL
HYPERAEMIA
PROC R SOC MED 64:6-9(1971)

LIBERTI P   STANBURY JB
THE PHARMACOLOGY OF SUBSTANCES AFFECTING THE THYROID GLAND
IN: "ANNU REV PHARMACOL," HW ELLIOTT, R OKUN, RH DREISHBACH,
ANNUAL REVIEW INC, PALO ALTO, CALIF 11 : 113-142(1971)

LICHTENSTEIN LM
THE CONTROL OF IGE-MEDIATED HISTAMINE RELEASE: IMPLICATIONS
FOR THE STUDY OF ASTHMA
IN: "ASTHMA," KF AUSTEN, LM LICHTENSTEIN, ACADEMIC PRESS,NY:
91-107(1973)

LICHTENSTEIN LM
THE PHARMACOLOGICAL CONTROL OF ALLERGIC REACTIONS
IN: "MECHANISMS IN ALLERGY. REAGIN-MEDIATED
HYPERSENSITIVITY," L GOODFRIEND, AH SEHON, RP ORANGE,
MARCEL DEKKER, NY:395-412(1973)

LICHTENSTEIN LM   DEBERNARDO R
THE IMMEDIATE ALLERGIC RESPONSE: IN VITRO ACTION OF CYCLIC
AMP-ACTIVE AND OTHER DRUGS ON THE TWO STAGES OF HISTAMINE
RELEASE
J IMMUNOL 107:1131-1136(1971)

LICHTENSTEIN LM   GILLESPIE E   BOURNE HR
HENNEY CS
THE EFFECTS OF A SERIES OF PROSTAGLANDINS ON IN VITRO MODELS
OF THE ALLERGIC RESPONSE AND CELLULAR IMMUNITY
PROSTAGLANDINS 2:519-528(1972)

LICHTENSTEIN LM   HENNEY CS   BOURNE HR
PROSTAGLANDIN INHIBITION OF IMMEDIATE AND DELAYED
HYPERSENSITIVITY IN VITRO: MEDIATION BY CYCLIC ADENOSINE
MONOPHOSPHATE
J ALLERGY CLIN IMMUNOL 49:87(1972)

LIDDLE GW   HARDMAN JG
CYCLIC ADENOSINE MONOPHOSPHATE AS A MEDIATOR OF HORMONE
ACTION
N ENGL J MED 285:560-566(1971)

LIEDBERG G   HAKANSON R
EFFECT OF PROSTAGLANDIN E1 ON ACID SECRETION HISTIDINE
DECARBOXYLASE ACTIVITY AND HISTAMINE CONTENT IN RAT GASTRIC
MUCOSA
ACTA PHARMACOL TOXICOL (KBH) 31:108(1972)

LIGGINS GC
THE PHYSIOLOGICAL ROLE OF PROSTAGLANDINS IN PARTURITION
ABSTR 1ST INT PLANN PARENT FED SE ASIA OCEANIA REG MED SCI
CONGR, SYDNEY:15(1972)

LIGGINS GC
FOETAL PARTICIPATION IN THE PHYSIOLOGICAL CONTROLLING
MECHANISMS OF PARTURITION
IN: "FOETAL AND NEONATAL PHYSIOLOGY," CAMBRIDGE UNIV PRESS,
LOND : 562-678(1973)

LIGGINS GC   GRIEVES SA   KENDALL JZ
THE FETAL ROLE IN THE INITIATION OF PARTURITION
IN: "ENDOCRINOLOGY. PROC 4TH INT CONG ENDOCRINOL," RO SCOW,
FJG EBLING, IW HENDERSON, EXCERPTA MEDICA, AMST/NY:1061-1065
(1973)

LIGHT RJ
IDENTIFICATION AND ANALYSIS OF ARACHIDONIC ACID IN
PLEXAURA HOMOMALLA VAR R AND VAR S
BIOCHIM BIOPHYS ACTA 296:461-465(1973)

LIGHT RJ   SAMUELSSON B
IDENTIFICATION OF PROSTAGLANDINS IN THE GORGONIAN,
PLEXAURA HOMOMALLA
EUR J BIOCHEM 28:232-240(1972)

LILLIE JH
EFFECTS OF PROSTAGLANDIN E1 ON NORMAL AND STIMULATED RAT
PAROTID GLANDS: AN ELECTRON MICROSCOPIC AND BIOCHEMICAL
STUDY
DISS, UNIV MICH, ANN ARBOR(1972)

LIM R   MITSUNOBU K   LI WKP
MATURATION-STIMULATING EFFECT OF BRAIN ESTRACT AND DIBUTYRYL
CYCLIC AMP ON DISSOCIATED EMBRYONIC BRAIN CELLS IN CULTUR-
EXP CELL RES 79:243-246(1973)

LIMA JB   SMITH PD
SIPPLE'S SYNDROME (PHEOCHROMOCYTOMA AND THYROID CARCINOMA)
WITH BILATERAL BREAST CARCINOMA.
AM J SURG 121:732-735(1971)

LIMAS CJ   COHN JN
STIMULATION OF VASODILATORS OF THE (NA+K)-ATPASE OF
VASCULAR SMOOTH MUSCLE
FED PROC 32:406(1973)

LIMAS CJ   COHN JN
REGULATION OF MYOCARDIAL PROSTAGLANDIN DEHYDROGENASE
ACTIVITY. THE ROLE OF CYCLIC 3',5'-AMP AND CALCIUM IONS
PROC SOC EXP BIOL MED 142:1230-1234(1973)

LIMAS CJ   COHN JN
ISOLATION AND PROPERTIES OF MYOCARDIAL PROSTAGLANDIN
SYNTHETASE
CARDIOVASC RES 7:623-628(1973)

LINARELLI LG   BOBIK J   BOBIK C
THE EFFECT OF PARATHYROID HORMONE ON RABBIT RENAL CORTEX
ADENYL CYCLASE DURING DEVELOPMENT
PEDIATR RES 7:878-882(1973)

LINCOLN FH   SCHNEIDER WP   PIKE JE
PROSTANOIC ACID CHEMISTRY. II. HYDROGENATION STUDIES AND
PREPARATION OF 11-DEOXYPROSTAGLANDINS
J ORG CHEM 38:951-956(1973)

LINDMARK G   NILSSON BA   NYBERG R
ZADOR G
INDUCTION OF ABORTION WITH PROSTAGLANDIN F2A. INTRAVENOUS,
EXTRA-AMNIOTIC AND INTRA-AMNIOTIC ADMINISTRATION
LAKARTIDNINGEN 70:3939-3943(1973):(SWEDISH)

LIPPEL K   LLEWELLYN A   JARETT L
PALMITOYL-COA SYNTHETASE ACTIVITY IN ISOLATED RAT EPIDIDYMAL
FAT CELLS IN THE ABSENCE AND PRESENCE OF VARIOUS LIPOLYTIC
AND ANTI-LIPOLYTIC COMPOUNDS
BIOCHIM BIOPHYS ACTA 231:48-51(1971)

LIPPERT TH
PROSTAGLANDINS AND THE REPRODUCTIVE SYSTEM
DTSCH MED WOCHENSCHR 96:916-922(1971):(GERMAN)

LIPPERT TH
THERAPEUTIC ASPECTS OF PROSTAGLANDIN RESEARCH IN GYNECOLOGY
GEBURTSHILFE FRAUENHEILKD 32:13-19(1972):(GERMAN)

LIPPMANN W
INHIBITION OF GASTRIC ACID SECRETION IN THE RAT BY
SYNTHETIC PROSTAGLANDIN ANALOGUES
ANN NY ACAD SCI 180:332-335(1971)

LIPPMANN W
ORAL ANTIGASTRIC ACID SECRETORY ACTIVITY OF SYNTHETIC
PROSTAGLANDIN ANALOGUES (9-OXOPROSTANOIC ACIDS)
EXPERIENTIA 29:990-991(1973)

LIPPMANN W   SEETHALER K
ORAL ANTI-ULCER ACTIVITY OF A SYNTHETIC PROSTAGLANDIN
ANALOGUE (9-OXOPROSTANOIC ACID: AY-22,469)
EXPERIENTIA 29:993-995(1973)

LIPSON L   HYNIE S   SHARP G
EFFECT OF PROSTAGLANDIN E1 ON OSMOTIC WATER FLOW AND SODIUM
TRANSPORT IN THE TOAD BLADDER
ANN NY ACAD SCI 180:261-277(1971)

LIPSON LC   SHARP GWG
THE EFFECT OF PROSTAGLANDIN E1 ON SODIUM TRANSPORT AND
OSMOTIC WATER FLOW IN THE TOAD BLADDER
AM J PHYSIOL 220:1046-1052(1971)

LITTENBERG G
ADRENAL EFFECTS OF PROSTAGLANDIN A1 IN MAN
TEX REP BIOL MED 30:218(1972)

404

LITWACK G   FILLER R   ROSENFIELD S
LICHTASH N
MACROMOLECULAR BINDING OF(5,6-3H) PROSTAGLANDIN E1 IN LIVER
CYTOSOL IN VITRO
BIOCHEM BIOPHYS RES COMMUN 55:977-984(1973)

LIVOLSI VA
ANTI-METASTATIC EFFECT OF ASPIRIN
LANCET 2:263(1973)

LJUNGBERG O
ON MEDULLARY CARCINOMA OF THE THYROID
ACTA PATHOL MICROBIOL SCAND (SUPP 231):1-57(1972)

LLOYD JV   NISHIZAWA EE
PHOSPHOLIPID TURNOVER DURING ADP-INDUCED PLATELET
AGGREGATION
FED PROC 30:201(1971)

LLOYD JV   NISHIZAWA EE   MUSTARD JF
EFFECT OF ADP-INDUCED SHAPE CHANGE ON INCORPORATION OF
32P INTO PLATELET PHOSPHATIDIC ACID AND MONO-,DI-AND
TRIPHOSPHATIDYL INOSITOL
BR J HAEMATOL 25:77-99(1973)

LOBOTSKY J
PROSTAGLANDINS AT THE WORCESTER FOUNDATION
RES PROSTAGLANDINS 1(1):1,4(1971)

LOEFFLER LJ   LOVENBERG W   SJOERDSMA A
EFFECTS OF DIBUTYRYL-3',5'-CYCLIC ADENOSINE MONOPHOSPHATE,
PHOSPHODIESTERASE INHIBITORS AND PROSTAGLANDIN E1 ON
COMPOUND 48/80-INDUCED HISTAMINE RELEASE FROM RAT
PERITONEAL MAST CELLS IN VITRO.
BIOCHEM PHARMACOL 20:2287-2297(1971)

LOGAN ME   WIEDMEIER VT
CHANGES IN MYOCARDIAL ADENOSINE CONCENTRATIONS DURING
PROSTAGLANDIN-INDUCED CORONARY VASODILATION
FED PROC 32:787(1973)

LOMOVA MA   PERSIANINOV LS   LEONOV BV
RAVEKOVA RG   MASSAL'SKAIA LM
TOTAL PROSTAGLANDINS IN THE OVUM CELLS OF THE SEA URCHIN,
STRONGULOCENTROTUS INTERMEDIUS, IN THE PERIOD OF EARLY
EMBRYOGENESIS
AKUSH GINEKOL (MOSK) 49(11):71-73(1973):(RUSSIAN)

LONIGRO AJ   ITSKOVITZ HD   CROWSHAW K
MCGIFF JC
DEPENDENCY OF RENAL BLOOD FLOW ON PROSTAGLANDIN SYNTHESIS IN
THE DOG
CIRC RES 32:712-717(1973)

LONIGRO AJ   TERRAGNO NA   MALIK KU
MCGIFF JC
INTRARENAL RELEASE OF PROSTAGLANDINS AS DETERMINED BY THE
STATE OF SODIUM BALANCE
J CLIN INVEST 50:61A(1971)

LONIGRO AJ   TERRAGNO NA   MALIK KU
MCGIFF JC
A PROSTAGLANDIN MAY MEDIATE THE RENAL VASODILATOR ACTION OF
BRADYKININ.
J LAB CLIN MED 78:1016-1017(1971)

LONIGRO AJ   TERRAGNO NA   MALIK KU
MCGIFF JC
MODIFICATION OF THE RENAL VASODILATOR ACTION OF
BRADYKININ BY AN INHIBITOR OF PROSTAGLANDIN SYNTHESIS
FED PROC 31:545(1972)

LONIGRO AJ   TERRAGNO NA   MALIK KU
MCGIFF JC
DIFFERENTIAL INHIBITION BY PROSTAGLANDINS OF THE RENAL
ACTIONS OF PRESSOR STIMULI
PROSTAGLANDINS 3:595-606(1973)

LOOSE LD   DI LUZIO NR
THE EFFECTS OF PROSTAGLANDIN E1 ON CELLULAR AND HUMORAL
IMMUNE RESPONSES
J RETICULOENDOTHEL SOC 11:419-420(1972)

LOOSE LD   DILUZIO NR
EFFECT OF PROSTAGLANDIN E1 ON CELLULAR AND HUMORAL IMMUNE
RESPONSES
J RETICULOENDOTHEL SOC 13:70-77(1973)

LOTEN E
ON THE ANTILIPOLYTIC ACTION OF INSULIN AND PROSTAGLANDIN E1
DIABETOLOGIA 9:79(1973)

LOTEN E
THE MECHANISMS OF THE ANTILIPOLYTIC ACTIONS OF INSULIN AND
PROSTAGLANDIN E1
EUR J CLIN INVEST 3:251(1973)

LOTEN EG   LOVELL-SMITH CJ   SNEYD JGT
HORMONAL EFFECTS MEDIATED BY CYCLIC AMP IN THE REGULATION
OF LIPOLYSIS
IN: "ENDOCRINOLOGY. PROC 4TH INT CONG ENDOCRINOL," RO SCOW,
FJG EBLING, IW HENDERSON, EXCERPTA MEDICA, AMST/NY:365-369
(1973)

LOTEN EG   SNEYD JGT
EVIDENCE FOR SEPARATE SITES OF ACTION FOR THE ANTILIPOLYTIC
EFFECTS OF INSULIN AND PROSTAGLANDIN E1
ENDOCRINOLOGY 93:1315-1322(1973)

LOWRY PJ   MCMARTIN C   PETERS J
PROPERTIES OF A SIMPLIFIED BIOASSAY FOR ADRENOCORTICOTROPHIC
ACTIVITY USING THE STEROIDOGENIC RESPONSE OF ISOLATED
ADRENAL CELLS
J ENDOCRINOL 59:43-55(1973)

LOZADA ES   GOUAUX J   FRANKI N
APPEL GB   HAYS RM
STUDIES OF THE MODE OF ACTION OF THE SULFONYLUREAS AND
PHENYLACETAMIDES IN ENHANCING THE EFFECT OF VASOPRESSIN
J CLIN ENDOCRINOL METAB 34:704-712(1972)

LUDUENA FP   GRIGAS EO
EFFECT OF SOME BIOLOGICAL SUBSTANCES ON THE DOG RETRACTOR
PENIS IN VITRO
ARCH INT PHARMACODYN THER 196:269-274(1972)

LUIGIES JHH
CLINICAL REPORT ON THE USE OF PROSTAGLANDIN F2A FOR INDUCING
ABORTION IN THE FIRST TRIMESTER OF PREGNANCY
NED TIJDSCHR GENEESKD 116:2368-2369(1972):(DUTCH)

LUKACS G   PIRIOU F   GERO SD
VAN DORP DA   HAGAMAN EW   WENKERT E
CARBON-13 NUCLEAR MAGNETIC RESONANCE SPECTROSCOPY OF
NATURALLY OCCURRING SUBSTANCES. PROSTAGLANDINS.
TETRAHEDRON LETT 515-518(1973)

LUNDBORG G   SCHILDT B
MICROVASCULAR PERMEABILITY IN IRRADIATED RABBITS
ACTA RADIOL THER(STOCKH)10:311-320(1971)

LUNDHOLM L
CYCLIC AMP AS A SECOND MESSENGER
ACTA PHYSIOL SCAND (SUPPL 396) : 5(1973)

MACCHIA V   VARRONE S
THYROID CELL MEMBRANE AND MECHANISM OF TSH-ACTION
IN: "ENDOCRINOLOGY. PROC 4TH INT CONG ENDOCRINOL," RO SCOW,
FJG EBLING, IW HENDERSON, EXCERPTA MEDICA, AMST/NY:539-542
(1973)

MACGILLIVRAY I   DENNIS KJ
GYNAECOLOGICAL ASPECTS
IN: "EXPERIENCE WITH ABORTION. A CASE STUDY OF NORTH-EAST
SCOTLAND," G HOROBIN, CAMBRIDGE UNIV PRESS, CAMBRIDGE:
47-95(1973)

MACHIEDO GW   BROWN CS   LAVIGNE JE
RUSH BF JR
PROSTAGLANDIN E1 AS A THERAPEUTIC AGENT IN HEMORRHAGIC
SHOCK
SURG FORUM 24:12-14(1973)

MACKENZIE RD   HENDERSON JG   STEINBACH JM
THE EVALUATION OF A METHOD FOR ADENOSINE DIPHOSPHATE INDUCED
PLATELET AGGREGATION IN THE GUINEA PIG
THROMB DIATH HAEMORRH 25:30-40(1971)

MACLEOD RM   LEHMEYER JE
INDEPENDENCE OF PITUITARY TUMOR-MEDIATED SUPPRESSION OF
GROWTH HORMONE SECRETION AND CYCLIC 3',5'-ADENOSINE
MONOPHOSPHATE PRODUCTION
CANCER RES 33:843-848(1973)

MACLEOD RM   LEHMEYER LE   FONTHAM EH
STIMULATION OF GROWTH HORMONE RELEASE BY PROSTAGLANDIN AND
DIBUTYRYL CYCLIC AMP IN PITUITARY GLANDS OF TUMOR-BEARING
RATS
FED PROC 30:533(1971)

MACRI I   RIGON-MACRI P
EFFECTS OF HORMONES AND CHEMICAL MEDIATORS ON SOME
FUNCTIONAL PARAMETERS IN THE RABBIT HEART. IV. EFFECT OF
PROSTAGLANDIN E1.
BOLL SOC ITAL BIOL SPER 49:1217-1220(1973):(ITALIAN)

MADAN BR   GUPTA RS   MADAN V
SONI RK
EFFECTS OF PROSTAGLANDINS E1, E2, F1A AND F2A ON PULMONARY
CIRCULATION AND OTHER CARDIOCIRCULATORY PARAMETERS
INDIAN J PHARMACOL 5:384-389(1973)

MADAN BR   SONI RK   MADAN V
EFFECTS OF PROSTAGLANDINS E1, E2, F1A AND F2A ON
VENTRICULAR ARRHYTHMIA IN THE CORONARY-LIGATED DOGS
INDIAN J PHARMACOL 5:390-393(1973)

MADDOX IS
THE ROLE OF COPPER IN PROSTAGLANDIN SYNTHESIS
BIOCHIM BIOPHYS ACTA 306:74-81(1973)

MAEKAWA T
THE ORIGIN OF THROMBI AND PROSTAGLANDINS
PROC SOC GERONTOL STUDY, OKAYAMA CITY, ONO PHARMACEUTICAL
CO, OSAKA:5-8(1972):(JAPANESE)

MAGERLEIN BJ   DUCHARME DW   MAGEE WE
MILLER WL   ROBERT A   WEEKS JR
SYNTHESIS AND BIOLOGICAL PROPERTIES OF
16-ALKYLPROSTAGLANDINS
PROSTAGLANDINS 4:143-145(1973)

MAGOTTEAUX G   DEBY C   BARAC G
URINARY ELIMINATION OF SMOOTH MUSCLE ACTING PROSTAGLANDINS
IN THE DOG.
ARCH INT PHYSIOL BIOCHIM 79:799-801(1971)

MAILLARD JL   VOISIN GA
NON-HEMORRHAGIC ARTHUS REACTION AND ITS INDEPENDENCE FROM
CIRCULATING ANTIBODIES AND COMPLEMENT
INT ARCH ALLERGY APPL IMMUNOL 45 :190-192(1973)

MAIN IHM
PROSTAGLANDINS: ARE THEY LOCAL OR CIRCULATING HORMONES?
IN: "MODERN TRENDS IN ENDOCRINOLOGY," FTG PRUNTY AND
H GARDINER-HILL, BUTTERWORTHS, LOND 4:302-326(1972)

MAIN IHM
PROSTAGLANDINS AND THE GASTRO INTESTINAL TRACT
IN: "THE PROSTAGLANDINS. PHARMACOLOGICAL AND THERAPEUTIC
ADVANCES, "MF CUTHBERT, HEINEMANN, LOND:287-323(1973)

MAIN IHM   WHITTLE BJR
EFFECTS OF PROSTAGLANDINS OF THE E AND A SERIES ON RAT
GASTRIC MUCOSAL BLOOD FLOW AS DETERMINED BY 14C-ANILINE
CLEARANCE
ABSTR 5TH INT CONGR PHARMACOL SAN FRANC 145(1972)

MAIN IHM   WHITTLE BJR
EFFECTS OF PROSTAGLANDIN E2 ON RAT GASTRIC MUSCOSAL BLOOD
FLOW, AS DETERMINED BY 14C-ANILINE CLEARANCE
BR J PHARMACOL 44:331P-332P(1972)

MAIN IHM   WHITTLE BJR
THE RELATIONSHIP BETWEEN RAT GASTRIC MUCOSAL BLOOD FLOW AND
ACID SECRETION DURING ORAL OR INTRAVENOUS ADMINISTRATION OF
PROSTAGLANDINS AND CYCLIC AMP
ADV BIOSCI (SUPPL) 9:44(1972)

MAIN IHM   WHITTLE BJR
EFFECTS OF INDOMETHACIN ON RAT GASTRIC ACID SECRETION AND
MUCOSAL BLOOD FLOW
BR J PHARMACOL 47:666P(1973)

MAIN IHM   WHITTLE BJR
THE RELATIONSHIP BETWEEN RAT GASTRIC MUCOSAL BLOOD FLOW AND
ACID SECRETION DURING ORAL OR INTRAVENOUS ADMINISTRATION OF
PROSTAGLANDINS AND DIBUTYRYL CYCLIC AMP
ADV BIOSCI 9:271-275(1973)

MAIN IHM   WHITTLE BJR
POTENTIATION OF DIBUTYRYL CYCLIC 3'5'-AMP-INDUCED GASTRIC
ACID SECRETION IN RATS BY NON-STEROIDAL ANTI-INFLAMMATORY
DRUGS
BR J PHARMACOL 49:162P-163P(1973)

MAIN IHM   WHITTLE BJR
THE EFFECTS OF E AND A PROSTAGLANDINS ON GASTRIC MUCOSAL
BLOOD FLOW AND ACID SECRETION IN THE RAT
BR J PHARMACOL 49:428-436(1973)

MAJNO G   RYAN GB   GABBIANI G
HIRSCHEL BJ   IRLE C   JORIS I
CONTRACTILE EVENTS IN INFLAMMATION AND REPAIR
IN:"INFLAMMATION MECHANISMS AND CONTROL," IH LEPOW AND
PA WARD, ACADEMIC PRESS, N Y:13-27(1972)

MAJOR PW   KILPATRICK R
CYCLIC AMP AND HORMONE ACTION
J ENDOCRINOL 52:593-630(1972)

MAJUMDAR SK
PROSTAGLANDINS AND THE REPRODUCTIVE SYSTEM
J OBSTET GYNAECOL INDIA 23:10-14(1973)

MAJUMDAR SK
PROSTAGLANDINS IN THERAPY
ANTISEPTIC 70:705-712(1973)

MAKMAN MH
CONDITIONS LEADING TO ENHANCED RESPONSE TO GLUCAGON,
EPINEPHRINE, OR PROSTAGLANDINS BY ADENYLATE CYCLASE OF
NORMAL AND MALIGNANT CULTURED CELLS
PROC NAT ACAD SCI USA 68:2127-2130(1971)

MAKMAN MH   DVORKIN B   KEEHN E
HORMONAL REGULATION OF CYCLIC AMP IN AGING AND IN
VIRUS-TRANSFORMED HUMAN FIBROBLASTS AND COMPARISON WITH
OTHER CULTURED CELLS
ABSTR MEET CONTROL OF PROLIFERATION IN ANIMAL CELLS, COLD
SPRING HARBOR:63(1973)

MALOFIEJEW M
THE FUNCTIONAL DIFFERENTIATION OF RENOMEDULLAR
PROSTAGLANDINS IN RABBITS
ACTA PHYSIOL POL 22:53-63(1971):(POLISH)

MALOFIEJÉW M   LAUDANSKI T   KOSTRZEWSKA A
THE EFFECT OF PROSTAGLANDIN A2 UPON THE ELECTROBIOLOGICAL
PHENOMENA IN ISOLATED MYOMETRIUM OF RAT.
GINEKOL POL 44:225-229(1973):(POLISH)

MALOFIEJEW M   LAUDANSKI T   STRACZKOWSKI W
THE EFFECT OF A2 PROSTAGLANDIN (PGA2) UPON THE
CONTRACTILITY OF RAT'S MYOMETRIUM
GINEKOL POL 43:1289-1294(1972):(POLISH)

MANDL JP
THE EFFECT OF PROSTAGLANDIN E1 ON RABBIT SPERM TRANSPORT IN
VIVO
DISS, DEPT ZOOL, UNIV WIS, MILW (1971)

MANGANIELLO V   BRESLOW J   VAUGHAN M
AN EFFECT OF DEXAMETHASONE ON THE CYCLIC AMP CONTENT OF
HUMAN FIBROBLASTS STIMULATED BY CATECHOLAMINES AND
PROSTAGLANDIN E1
J CLIN INVEST 51:60A-61A(1972)

MANGANIELLO V   BRESLOW J   VAUGHAN M
INDEPENDENT MODIFICATION OF EFFECTS OF PROSTAGLANDIN E1
(PGE1) AND ISOPROTERENOL (IPT) ON CYCLIC AMP LEVELS IN HUMAN
FIBROBLASTS
PEDIATR RES 7:391(1973)

MANGANIELLO V   VAUGHAN M
PROSTAGLANDIN E1 EFFECTS ON ADENOSINE 3',5'-CYCLIC
MONOPHOSPHATE CONCENTRATION AND PHOSPHODIESTERASE
ACTIVITY IN FIBROBLASTS
PROC NATL ACAD SCI USA 69:269-273(1972)

MANGANIELLO V   VAUGHAN M
AN EFFECT OF INSULIN ON CYCLIC ADENOSINE 3':5'-MONOPHOSPHATE
PHOSPHODIESTERASE ACTIVITY IN FAT CELLS
J BIOL CHEM 248:7164-7170(1973)

MANN D   MEYER HG   FORSTER W
PRELIMINARY CLINICAL EXPERIENCE WITH THE ANTIARRHYTHMIC
EFFECT OF PGF2A
PROSTAGLANDINS 3:905-912(1973)

MANSEL-JONES D
PREPARATIONS OF PROSTAGLANDINS
LANCET ·1:708(1971)

MANSO C
PHYSIOLOGICAL IMPORTANCE OF CYCLIC AMP AND PROSTAGLANDINS
REV CIENC MED SERIES B 7:135-149,151-153(1971):(PORTUGUESE)

MANTHORPE T
THE EFFECT ON RENAL HYPERTENSION OF SUBCUTANEOUS
ISOTRANSPLANTATION OF RENAL MEDULLA FROM NORMAL OR
HYPERTENSIVE RATS
ACTA PATHOL MICROBIOL SCAND (A) 81:725-733(1973)

MANTOVANI P   BERTACCINI G
ACTION OF CAERULEIN AND RELATED SUBSTANCES ON
GASTRO-INTESTINAL MOTILITY OF THE ANAESTHETIZED DOG
ARCH INT PHARMACODY THER 193:362-371(1971)

MANTOVANI P   IMPICCIATORE M
ISOLATED RABBIT COLON PREPARATION: A SENSITIVE METHOD FOR
THE BIOASSAY OF CAERULEIN AND RELATED SUBSTANCES.
NAUNYN-SCHMIEDEBERGS ARCH PHARMACOL 271:330-334(1971)

MAOR D   NIVIY S   EYLAN E
EPINEPHRINE BLOCKADE OF RIBONUCLEASE ACTIVITY STIMULATION
IN LYMPHOID ORGANS BY CORTISONE AND PROSTAGLANDIN E1
BIOMEDICINE 19:374-378 (1973)

MARCHAND GR   GREENBERG S   WILSON WR
WILLIAMSON HE
EFFECTS OF PROSTAGLANDIN B2 ON RENAL HEMODYNAMICS AND
EXCRETION
PROC SOC EXP BIOL MED 143:93P-940(1973)

MARCO LA   COCEANI F
THE ACTION OF PROSTAGLANDIN E1 ON FROG SKELETAL MUSCLE
CAN J PHYSIOL PHARMACOL 51:627-634 (1973)

MARCUS AJ
OBSERVATIONS ON THE STRUCTURE AND FUNCTION OF HUMAN
PLATELET MEMBRANES
IN: "ERYTHROCYTES, THROMBOCYTES, LEUKOCYTES," E GERLACH,
K MOSER, E DEUTSCH, W WILMANNS, G THIEME, STUTTG
2:206-208(1973)

MARIONA FG   AIUTO CS
HUMAN AMNIOTIC FLUID PROSTAGLANDINS--THEIR ROLE IN THE
INITIATION OF LABOR
CLIN RES 5:966(1973)

MARK AL   SCHMID PG   ECKSTEIN JW
WENDLING MG
VENOUS RESPONSES TO PROSTAGLANDIN F2A
AM J PHYSIOL 220:222-226(1971)

MARKOV KM   PINELIS VG
THE INFLUENCE OF PROSTAGLANDIN A1 ON RENAL BLOOD FLOW,
SECRETION OF RENIN AND SODIUM EXCRETION IN DOGS
KARDIOLOGIIA 13:32-38(1973):(RUSSIAN)

MARKS RG
PROSTAGLANDINS COULD OUST THE PILL
PHARM J N Z 33:10(1971)

MARKWAROT F
INHIBITION OF PLATELET AGGREGATION BY ISOCHINOLINE
DERIVATIVES
ACTA UNIV CAROL (MED) (PRAHA) 52:79-82(1972)

MARLEY PB
ATTEMPTS TO INHIBIT UTERINE LUTEOLYSIN WITH NON-STEROIDAL
ANTI-INFLAMMATORY DRUGS IN THE GUINEA-PIG
ABSTR 5TH INT CONGR PHARMACOL SAN FRANC 149(1972)

MARLEY PB
INDOMETHACIN LENGTHENS THE OESTROUS CYCLE OF THE GUINEA-PIG
WHEN GIVEN ORALLY WITH OESTROGEN OR WHEN IMPLANTED WITHIN
THE UTERINE LUMEN
PROSTAGLANDINS 4:251-261(1973)

MARMO E   CAPUTI A   IMPERATORE A
VUOLO L
EXPERIMENTAL RESEARCH CARRIED OUT WITH SULPIRIDE IN RELATION
TO THE CARDIOVASCULAR SYSTEM AND THE INTESTINAL AND UTERINE
MUSCULATURE
GAZZ INT MED CHIR 76:758-795(1971);(ITALIAN)

MARRAZZI MA
PROSTAGLANDIN DEHYDROGENASE: PURIFICATION AND PROPERTIES
DISS, WASH UNIV, ST LOUIS(1972)

MARRAZZI MA   MATACHINSKY FM
REACTIONS AND BINDING SITES OF 15-OH-PROSTAGLANDIN
DEHYDROGENASE (PGDH)
PHARMACOLOGIST 13:292(1971)

MARRAZZI MA   MATSCHINSKY FM
PROPERTIES OF 15-HYDROXY PROSTAGLANDIN DEHYDROGENASE:
STRUCTURAL REQUIREMENTS FOR SUBSTRATE BINDING
PROSTAGLANDINS 1:373-388(1972)

MARRAZZI MA   SHAW JE   TAO FT
MATSCHINSKY FM
REVERSIBILITY OF 15-OH PROSTAGLANDIN DEHYDROGENASE (PGDH)
FROM PIG LUNG
ABSTR 5TH INT CONGR PHARMACOL SAN FRANC 150(1972)

MARRAZZI MA   SHAW JE   TAO FT
MATSCHINSKY FM
REVERSIBILITY OF 15-OH PROSTAGLANDIN DEHYDROGENASE FROM
SWINE LUNG
PROSTAGLANDINS 1:389-395(1972)

MARSHALL JM
THE PHYSIOLOGY OF THE MYOMETRIUM
IN: "THE UTERUS", HJ NORRIS, AT HERTIG, MR ABELL, WILLIAMS &
WILKINS, BALT:89-109(1973)

MARSHECK WJ   MIYANO M
MICROBIOLOGICAL OPTICAL RESOLUTION OF A RACEMIC SYNTHETIC
PROSTAGLANDIN INTERMEDIATE
BIOCHIM BIOPHY ACTA 316:363-365 (1973)

MARTEL J   TOROMANOFF E   MATHIEU J
NOMINE G
TOTAL STEREOSPECIFIC SYNTHESIS OF THE DL-PROSTAGLANDIN A2
TETRAHEDRON LETT 1491-1496(1972);(FRENCH)

MARTIN BR   DENTON RM   PASK HT
RANDLE PJ
MECHANISMS REGULATING ADIPOSE-TISSUE PYRUVATE DEHYDROGENASE
BIOCHEM J 129:763-773(1972)

MARTINEZ-MALDON/ M TSAPARAS N   EKNOYAN G
SUKI WN
RENAL ACTIONS OF PROSTAGLANDINS: COMPARISON WITH
ACETYLCHOLINE AND VOLUME EXPANSION
AM J PHYSIOL 222:1147-1152(1972)

MARTINEZ-MALDON/ M TSAPARAS N   SUKI WN
EKNOYAN G
COMPARISON OF THE EFFECT OF PROSTAGLANDINS (PGE) AND
ACETYLCHOLINE (ACH) ON RENAL SODIUM AND WATER EXCRETION
CLIN RES 19:87(1971)

MARTINEZ-MALDON/ M TSAPARAS N   SUKI WN
EKNOYAN G
COMPARISON OF THE EFFECT OF PROSTAGLANDINS (PGE) AND
ACETYLCHOLINE ON RENAL SODIUM AND WATER EXCRETION
PROC INT UNION PHYSIOL SCI, 25TH INT CONGR, MUNICH,
9:374(1971)

MARUMO F   ASANO Y
CHARACTERISTICS OF RENAL ADENYL CYCLASE
J PHYSIOL SOC JAP 33:534(1971);(JAPANESE)

MARUMO F   EDELMAN IS
EFFECTS OF CA AND PROSTAGLANDIN E1 ON VASOPRESSIN
ACTIVATION OF RENAL ADENYL CYCLASE
J CLIN INVEST 50:1613-1620(1971)

MASSION WH
PROTECTIVE EFFECTS OF ATP AND PROSTAGLANDIN E1 AGAINST
PULMONARY DAMAGE FOLLOWING SHOCK
ABSTR 5TH WORLD CONGR ANAESTHESIOL, KYOTO, EXCERPTA
MEDICA, AMST:63(1972)

MASSION WH
PROTECTIVE EFFECTS OF VARIOUS INHIBITOR AGENTS AGAINST
PULMONARY DAMAGE FOLLOWING SHOCK
FED PROC 32:371(1973)

MASSION WH   KUX M
MODIFICATION OF SHOCK-INDUCED LUNG CHANGES BY INHIBITORS
MED WELT 24:1156-1157(1973); (GERMAN)

MASUDA <
PROSTAGLANDINS THEIR EFFECTS ON INFLOW AND OUTFLOW OF
AQUEOUS HUMOR IN RABBITS
NIPPON GANKA GAKKAI ZASSHI 76:664-668(1972);(JAPANESE)

MASUDA K   MISHIMA S
EFFECTS OF PROSTAGLANDINS ON INFLOW AND OUTFLOW OF THE
AQUEOUS HUMOR IN RABBITS
JAP J OPHTHALMOL 17:300-309(1973)

MATHE AA   ASTROM A   PERSSON NA
SOME BRONCHOCONSTRICTING AND BRONCHODILATING RESPONSES OF
HUMAN ISOLATED BRONCHI: EVIDENCE FOR THE EXISTENCE OF
ALPHA-ADRENOCEPTORS
J PHARM PHARMACOL 23:905-910(1971)

MATHE AA   ASTROM A   PERSSON NA
SOME BRONCHOCONSTRICTING AND BRONCHODILATING RESPONSES OF
HUMAN ISOLATED BRONCHI: EVIDENCE FOR THE EXISTENCE OF
ALPHA-ADRENOCEPTORS
J PHARM PHARMACOL 23:905-910(1971)

MATHE AA   HEDQVIST P   HOLMGREN A
SVANBORG N
PROSTAGLANDIN F2A: EFFECT ON AIRWAY CONDUCTANCE IN HEALTHY
SUBJECTS AND PATIENTS WITH BRONCHIAL ASTHMA
ADV BIOSCI 9:241-245(1973)

MATHE AA   HEDQVIST P   HOLMGREN A
SVANBORG N
BRONCHIAL HYPERREACTIVITY TO PROSTAGLANDIN F2A AND
HISTAMINE IN PATIENTS WITH ASTHMA
BR MED J 1:193-196(1973)

MATHE AA   LEVINE L
RELEASE OF PROSTAGLANDINS AND METABOLITES FROM GUINEA-PIG
LUNG:INHIBITION BY CATECHOLAMINES
PROSTAGLANDINS 4:877-890(1973)

MATHE AA   STRANDBERG K
ANTAGONISM OF SLOW REACTING SUBSTANCE BY POLYPHLORETIN
PHOSPHATE ON ISOLATED HUMAN BRONCHI.
ACTA PHYSIOL SCAND 82:460-465(1971).

MATHE AA   STRANDBERG K   ASTROM A
BLOCKADE BY POLYPHLORETIN PHOSPHATE OF THE PROSTAGLANDIN
F2A ACTION ON ISOLATED HUMAN BRONCHI
NATURE (NEW BIOL) 230:215-216(1971)

MATHE AA   STRANDBERG K   FREDHOLM B
ANTAGONISM OF PROSTAGLANDIN F2A INDUCED BRONCHOCONSTRICTION
AND BLOOD PRESSURE CHANGES BY POLYPHLORETIN PHOSPHATE IN THE
GUINEA-PIG AND CAT
J PHARM PHARMACOL 24:378-382(1972)

MATHUR GP   GANDHI VM
PROSTAGLANDIN IN HUMAN AND ALBINO RAT SKIN
J INVEST DERMATOL 58:291-295(1972)

MATSUDA S   SHIMIZU T   KOGO T
PROSTAGLANDIN AND ABORTION IN EARLY PREGNANCY AND
CONTRACEPTION
OBSTET GYNECOL (TOKYO) 39:640-647(1972);(JAPANESE)

MATSUMOTO K   SASADA M   SAKAI T
EFFECT OF PROSTAGLANDIN F2A ON PREGNANT UTERUS AND INDUCTION
OF LABOUR AT TERM
PHARMACOMETRICS 5:941-956(1971)

MATSUOKA Y   FUJITA T   NOZATO T
YOKOHAMA H   ONISHI Y   OHTA K
TOXICITY AND TERATOGENICITY OF PROSTAGLANDIN F2A
RES PHARM (IYAKUHIN KENKYU) 2:403-413(1971);(JAPANESE)

MATUCHANSKY C   BERNIER JJ
EFFECTS OF PROSTAGLANDIN E1 ON NET AND UNIDIRECTIONAL
MOVEMENTS OF WATER AND ELECTROLYTES ACROSS JEJUNAL
MUCOSA IN MAN
GUT 12:854-855(1971)

MATUCHANSKY C   BERNIER JJ
EFFECT OF PROSTAGLANDIN E1 ON GLUCOSE, WATER, AND
ELECTROLYTE ABSORPTION IN THE HUMAN JEJUNUM
GASTROENTEROLOGY 64:1111-1118(1973)

MATUCHANSKY C   BERNIER JJ
EFFECTS OF PROSTAGLANDIN E1 ON JEJUNAL ABSORPTION IN MAN
DIGESTION 9:86-87(1973)

MATUCHANSKY C   BERNIER JJ
GENERAL REVIEW, PROSTAGLANDINS AND THE DIGESTIVE TRACT
BIOL GASTROENTEROL (PARIS) 6:251-268(1973);(FRENCH)

MATUCHANSKY C   MARY JY   BERNIER JJ
EFFECT OF PROSTAGLANDIN E1 UPON TRANSIT TIME, NET AND
UNIDIRECTIONAL MOVEMENTS OF WATER AND ELECTROLYTES IN THE
HUMAN JEJUNUM
BIOL GASTROENTEROL (PARIS) 5: 175-186(1972)

MATUCHANSKY C   MARY JY.   BERNIER JJ
EFFECT OF PROSTAGLANDIN E1 ON GLUCOSE ABSORPTION AND ON
TRANSINTESTINAL MOVEMENTS OF WATER AND ELECTROLYTES IN
HUMAN JEJUNUM
BIOL GASTROENTEROL (PARIS) 5:636C(1972);(FRENCH)

MATUCHANSKY C   MATUCHANSKY C
THE PROSTAGLANDINS
CAN MED 13:525-539(1972);(FRENCH)

MAULE WALKER FM   POYSER NL
PRODUCTION OF PROSTAGLANDINS BY THE EARLY PREGNANT GUINEA
PIG UTERUS.
ACTA ENDOCRINOL (SUPPL)(KBH)177:312(1973)

MAUVAIS-JARVIS P   GARDENAT JP
BIOLOGICAL ACTIVITY OF PROSTAGLANDINS
PATHOL BIOL (PARIS) 20:919-930(1972);(FRENCH)

MAY B
RELATIONS BETWEEN PROSTAGLANDINS AND CARBOHYDRATE
METABOLISM
J AM OIL CHEM SOC 48:94A(1971)

MAY B
ACTION OF PROSTAGLANDINS ON THE METABOLISM OF FATS AND
CARBOHYDRATES
Z GASTEROENTEROL 11:223-232(1973);(GERMAN)

MCAFEE DA   GREENGARD P
CYCLIC AMP: PHYSIOLOGICAL EVIDENCE FOR A ROLE AS THE
MEDIATOR OF DOPAMINERGIC TRANSMISSION IN RABBIT SUPERIOR
CERVICAL GANGLIA
FED PROC 31:269(1972)

MCCALL E   YOULTEN LJF
PROSTAGLANDIN E1 SYNTHESIS BY PHAGOCYTOSING RABBIT
POLYMORPHONUCLEAR LEUCOCYTES:ITS INHIBITION BY INDOMETHACIN
AND ITS ROLE IN CHEMOTAXIS
J PHYSIOL (LOND) 234: 98P-100P (1973)

MCCARTNEY JC   MULCARE DJ
ANURAN FORELIMB REGENERATION AFTER PROSTAGLANDIN TREATMENT
AM ZOOL 13:1351(1973)

MCCLATCHEY WM   CARR A
PROSTAGLANDIN (PGA1), ANGIOTENSIN AND RENAL FUNCTION
PROSTAGLANDINS 2:213-217(1972)

MCCLATCHEY WM   CARR AA
MODIFICATION OF RENAL EFFECTS OF ANGIOTENSIN BY A
PROSTAGLANDIN
CLIN RES 19:58(1971)

MCCLATCHEY WM   CARR AA
PROSTAGLANDIN A1 IN RENAL ARTERY STENOSIS
CLIN RES 20:64(1972)

MCCLATCHEY WM   CARR AA
PROSTAGLANDIN AND RENAL ARTERY STENOSIS
PROSTAGLANDINS 3:229-235(1973)

MCCLOY RB   PRANCAN AV   NAKANO J
CIRCULATORY AND RESPIRATORY EFFECTS OF DIFFERENT MIXTURES OF
PROSTAGLANDINS E2 AND F2A
CLIN RES 21:437(1973)

MCCLOY RB JR   NAKANO J
CARDIOVASCULAR EFFECTS OF THE INTRAVENOUS INFUSION OF
PROSTAGLANDINS E1, E2 AND A1
CIRCULATION (SUPPL II) 46:186(1972)

MCCLOY RB JR   NAKANO J
ALTERATIONS OF THE CARDIOVASCULAR RESPONSES TO CAROTID
ARTERIAL OCCLUSION BY PROSTAGLANDINS E1 AND E2
CLIN RES 21:950(1973)

MCCURDY JR   GREENFIELD LJ   PRANCAN AV
NAKANO J
CARDIOVASCULAR, HEMATOLOGICAL AND METABOLIC EFFECTS OF
PROSTAGLANDIN E1 IN ENDOTOXIN SHOCK
CIRCULATION (SUPPL II) 46:186(1972)

MCDONALD JWD
LYMPHOCYTE AND PLATELET ADENYL-CYCLASE.
CAN J BIOCHEM 49:316-319(1971)

MCDONALD JWD
CYCLIC AMP IN PLATELETS
CLIN RES 19:783(1971)

MCDONALD JWD   STUART RK
CYCLIC AMP MEDIATES A STABLE INHIBITORY EFFECT OF PGE-1 ON
PLATELET AGGREGATION
CLIN RES 20:934(1972)

MCDONALD JWD   STUART RK
REGULATION OF CYCLIC AMP LEVELS AND AGGREGATION IN HUMAN
PLATELETS BY PROSTAGLANDIN E1
J LAB CLIN MED 81:838-849(1973)

MCDONALD JWD   STUART RK
PROSTAGLANDIN E-2 IS A PARTIAL AGONIST OF PROSTAGLANDIN E-1
IN PLATELETS.
CLIN RES 21:1046(1973)

MCDONALD-GIBSON RG FLACK JD   RAMWELL PW
INHIBITION OF PROSTAGLANDIN BIOSYNTHESIS BY 7-OXA- AND
5-OXA-PROSTAGLANDIN ANALOGUES
BIOCHEM J 132:117-120(1973)

MCDONALD-GIBSON WJ MCDONALD-GIBSON RG GREAVES MW
METABOLISM OF PROSTAGLANDIN E1 BY HUMAN PLASMA
BIOCHEM J 127:40P-41P(1972)

MCDONALD-GIBSON WJ MCDONALD-GIBSON RG GREAVES MW
PROSTAGLANDIN E1 METABOLISM BY HUMAN PLASMA
PROSTAGLANDINS 2:251-263(1972)

MCELROY FA   KINLOUGH-RATHB/ RL ARDLIE NG
PACKHAM MA   MUSTARD JF
THE EFFECT OF AGGREGATING AGENTS ON OXIDATIVE METABOLISM OF
RABBIT PLATELETS
BIOCHIM BIOPHYS ACTA 253:64-77(1971)

MCGIFF JC   CROWSHAW K   TERRAGNO NA
LONIGRO AJ
RENAL PROSTAGLANDINS: THEIR BIOSYNTHESIS, RELEASE, EFFECTS
AND FATE
IN: "RENAL PHARMACOLOGY," JW FISHER, APPLETON, NY:211-240
(1971)

MCGIFF JC   CROWSHAW K   TERRAGNO NA
MALIK KU   LONIGRO AJ
DIFFERENTIAL EFFECT OF NORADRENALINE AND RENAL NERVE
STIMULATION ON VASCULAR RESISTANCE IN THE DOG KIDNEY AND THE
RELEASE OF A PROSTAGLANDIN E-LIKE SUBSTANCE
CLIN SCI 42:223-233(1972)

MCGIFF JC   CROWSHAW K   TERRAGNO NA
RENAL PROSTAGLANDINS: POSSIBLE PHYSIOLOGICAL ROLES
IN: "ENDOCRINOLOGY, PROC 4TH INT CONG ENDOCRINOL," RO SCOW,
FJG EBLING, IW HENDERSON, EXCERPTA MEDICA, AMST/NY:376-381
(1973)

MCGIFF JC   ITSKOVITZ HD
PROSTAGLANDINS AND THE KIDNEY
CIRC RES 33:479-488(1973)

MCGIFF JC   TERRAGNO NA   MALIK KU
LONIGRO AJ
RELEASE OF A PROSTAGLANDIN E-LIKE SUBSTANCE FROM CANINE
KIDNEY BY BRADYKININ
CIRC RES 31:36-43(1972)

MCGLAUGHLIN CL   BAILE CA   TRUEHEART P
HERRERA MG
GOLD THIOGLUCOSE (GTG) INDUCED VENTROMEDIAL HYPOTHALAMIC
(VMH) LESIONS AND GLUCOSE METABOLISM IN ALLOXANIZED-DIABETIC
AND HYPERGLYCEMIC OBESE (OBOB) MICE.
FED PROC 30:579(1971)

MCKIRDY HC
FUNCTIONAL RELATIONSHIP OF LONGITUDINAL AND CIRCULAR LAYERS
OF THE MUSCULARIS EXTERNA OF THE RABBIT LARGE INTESTINE
J PHYSIOL (LOND) 227:839-853(1972)

MCPHERSON MA   SCHOFIELD JG
CYTOCHALASIN B, PITUITARY METABOLISM, AND THE RELEASE OF
OF OX GROWTH HORMONE IN VITRO
FEBS LETT 24:45-48(1972)

MCQUEEN DS
THE EFFECT OF SOME PROSTAGLANDINS ON RESPIRATION IN RATS AND
CATS
BR J PHARMACOL 45:147P-148P(1972)

MCQUEEN DS
THE EFFECTS OF PROSTAGLANDIN E2, PROSTAGLANDIN F2A, AND
POLYPHLORETIN PHOSPHATE ON RESPIRATION AND BLOOD PRESSURE IN
ANAESTHETIZED GUINEA-PIGS
LIFE SCI (I) 12:163-172(1973)

MCQUEEN EG
ANTI-INFLAMMATORY DRUG MECHANISMS
DRUGS 6:104-117(1973)

MECS E   MARAZ A
THE EFFECT OF PROSTAGLANDIN E1 ON THE PRODUCTION OF
INTERFERON OF HUMAN LYMPHOCYTES AND TISSUE CULTURE CELLS
ADV BIOSCI (SUPPL) 9:70(1972)

MECS I   MARAZ A
THE EFFECT OF PROSTAGLANDIN E1 ON THE PRODUCTION OF
INTERFERON
ADV BIOSCI 9:453-455(1973)

MEERSON FE   BARBARASH NA
PROSTAGLANDINS AND ANTIHYPERTENSIVE FUNCTION OF THE KIDNEYS
KARDIOLOGIIA 11:138-154(1971);(RUSSIAN)

MELANDER A   SUNDLER F   INGBAR SH
INTERACTIONS BETWEEN STIMULATORS AND INHIBITORS OF THYROID
HORMONE SECRETION
ISR J MED SCI 8:1867-1868(1972)

MELANDER A   SUNDLER F   INGBAR SH
EFFECT OF POLYPHLORETIN PHOSPHATE ON THE INDUCTION OF
THYROID HORMONE SECRETION BY VARIOUS THYROID STIMULATORS
ENDOCRINOLOGY 92:1269-1273(1973)

MENDELSOHN J   BOONE R   MULTER MM
ADDITIVE EFFECTS OF CYCLIC-AMP AND CORTISOL ON PHA-INDUCED
LYMPHOCYTE TRANSFORMATION
BLOOD 38:799(1971)

MENDELSOHN J   MULTER MM   BOONE RF
ENHANCED EFFECTS OF PROSTAGLANDIN E1 AND DIBUTYRYL CYCLIC
AMP UPON HUMAN LYMPHOCYTES IN THE PRESENCE OF CORTISOL
J CLIN INVEST 52:2129-2137(1973)

MENDLOWITZ M
VASCULAR REACTIVITY IN SYSTEMIC ARTERIAL HYPERTENSION
IN: "HYPERTENSION: MECHANISMS AND MANAGEMENT" G ONESTI, KE
KIM AND JH MOYER,GRUNE & STRATTON,NY/LOND:51-58(1973)

MENENDEZ-CEPERO E   MIYARES-CAO C   GARCIA-GUTIERREZ A
ISOLATION OF PROSTAGLANDIN E2 (PGE2) FROM THE GASTRIC
MUCOSA AND GASTRIC JUICE OF ULCER PATIENTS
REV CUB FARM 7:53-57(1973):(SPANISH)

MENNIE AT   DALLEY V
ASPIRIN IN RADIATION-INDUCED DIARRHOEA
LANCET 1:1131(1973)

MESSINA EJ
PROSTAGLANDINS AND LOCAL REGULATION OF BLOOD FLOW
DISS, NY MED COLL (1973)

MESSINA EJ   WEINER R   KALEY G
EFFECTS OF PROSTAGLANDIN A1 ON THE RAT MICROCIRCULATION
PROC INT UNION PHYSIOL SCI, 25TH INT CONGR, MUNICH,
9:387(1971)

MESSINA EJ   WEINER R   KALEY G
EFFECTS OF INHIBITORS OF PROSTAGLANDIN SYNTHESIS ON
ARTERIOLAR RESPONSIVENESS
FED PROC 32:788(1973)

MEST HJ   FORSTER W
EVIDENCE FOR ANTIARRHYTHMIC EFFICIENCY OF ARACHIDONIC AND
LINOLEIC ACID-PRELIMINARY RESULTS
PROSTAGLANDINS 4:751-754(1973)

MEST HJ   SCHROR K   FORSTER W
ANTIARRHYTHMIC PROPERTIES OF PGE2: PRELIMINARY RESULTS
ADV BIOSCI 9:385-393(1973)

MICHAELSSON G   ROS AM
FAMILIAL LOCALIZED HEAT URTICARIA OF DELAYED TYPE.
ACTA DERM VENEROL(STOCKH)51:279-283(1971)

MICHELI H   CARLSON LA
A STUDY ON ISOLATED ADIPOSE TISSUE OF THE NICOTINIC
ACID-INDUCED FREE FATTY ACID REBOUND PHENOMENON WITH A
COMPARISON WITH PROSTAGLANDIN E1
IN:"METABOLIC EFFECTS OF NICOTINIC ACID AND ITS DERIVATIVES"
KF GEY AND LA CARLSON, HANS HUBER, BERN : 995-1001 (1971)

MICHOUD MC   HOGG JC
THE EFFECT OF BLOCKING VAGUS NERVES AND PROSTAGLANDIN
SYNTHESIS ON ALLERGIC BRONCHOCONSTRICTION IN UNANESTHETIZED
GUINEA PIGS
CLIN RES 21:1070(1973)

MICKEL HS   FOULDS EL   CLARK DA
LARKIN EC
EFFECTS OF PURE OXYGEN ATMOSPHERE IN VIVO ON PLASMA
LECITHIN-CHOLESTEROL ACYLTRANSFERASE REACTION
LIPIDS 6:740-744(1971)

MIDAK E
COMPARATIVE STUDIES ON THE EFFECT OF PROSTAGLANDIN E1 ON THE
CONTRACTILITY OF HUMAN AND RABBIT'S MYOMETRIUM AND ON
MODIFYING THE CONTRACTILE EFFECTS BY ADMINISTERING
CATECHOLAMINES AND ACETYLCHOLINE
GINEKOL POL 44:937-939(1973):(POLISH)

MIDAK E
MODIFICATION BY PGE1 OF CONTRACTILE EFFECTS OF
CATECHOLAMINES AND ACETYLCHOLINE ON ISOLATED HUMAN
NONPREGNANT MYOMETRIUM
ACTA PHYSIOL POL 24:841-846(1973)

MIDDLEDITCH BS   DESIDERIO DM
GAS-LIQUID CHROMATOGRAPHY OF TRIMETHYLSILYL AND ALKYL
OXIME-TRIMETHYLSILYL DERIVATIVES OF SOME PROSTAGLANDINS
PROSTAGLANDINS 2:115-121(1972)

MIDDLEDITCH BS   DESIDERIO DM
GAS CHROMATOGRAPHY OF PROSTAGLANDIN HEPTAFLUOROBUTYRATE
METHYL ESTERS
PROSTAGLANDINS 2:195-198(1972)

MIDDLEDITCH BS   DESIDERIO DM
MASS SPECTRA OF PROSTAGLANDINS. III. TRIMETHYLSILYL AND
ALKYL OXIME-TRIMETHYLSILYL DERIVATIVES OF PROSTAGLANDINS
OF THE E SERIES
J ORG CHEM 38:2204-2209 (1973)

MIDDLEDITCH BS   DESIDERIO DM
MASS SPECTRA OF PROSTAGLANDINS: II. TRIMETHYLSILYL AND
ALKYLOXIME-TRIMETHYLSILYL DERIVATIVES OF PROSTAGLANDINS
B1 AND B2
LIPIDS 8:267-270(1973)

MIDDLEDITCH BS   DESIDERIO DM
MASS SPECTRA OF PROSTAGLANDINS. I. TRIMETHYLSILYL AND
ALKYLOXIME-TRIMETHYLSILYL DERIVATIVES OF PROSTAGLANDIN A1
PROSTAGLANDINS 4:31-46(1973)

MIDDLEDITCH BS   DESIDERIO DM
MODIFIED PROSTAGLANDINS AS INTERNAL STANDARDS FOR
QUANTITATIVE ANALYSIS OF PROSTAGLANDINS OF THE E SERIES BY
COMBINED GAS CHROMATOGRAPHY-MASS SPECTROMETRY
PROSTAGLANDINS 4:459-462(1973)

MIDDLEDITCH BS   DESIDERIO DM
MASS SPECTRA OF PROSTAGLANDINS-IV-TRIMETHYLSILYL
DERIVATIVES OF PROSTAGLANDINS OF THE F SERIES
ANAL BIOCHEM 55:509-520(1973)

MIDDLEDITCH BS   KNIGHTS BA
THE MASS SPECTRA OF SOME O-METHYLOXIMES OF ALIPHATIC
ALDEHYDES AND KETONES
ORG MASS SPECTROM 6:179-188(1972)

MIKES F   SCHURIG V   GIL-AV E
COMPLEX-FORMING STATIONARY PHASES IN HIGH-SPEED LIQUID
CHROMATOGRAPHY
J CHROMATOGR 83:91-97(1973)

MIKESELL GT   YVER DR   WILLIAMS WF
PROSTAGLANDIN F2A-9-H3 TISSUE RESIDUES
J DAIRY SCI 55:1330(1972)

MILLER JD   EAKINS KE   ATWAL M
THE RELEASE OF PGE2-LIKE ACTIVITY INTO AQUEOUS HUMOR AFTER
PARACENTESIS AND ITS PREVENTION BY ASPIRIN.
INVEST OPHTHALMOL 12:939-942(1973)

MILLER OV
SPECIFICITY OF PROSTAGLANDIN BINDING SITES IN RAT
FORESTOMACH TISSUE AND THEIR POSSIBLE USE AS A QUANTITATIVE
ASSAY
ADV BIOSCI (SUPPL) 9:13(1972)

MILLER OV   MAGEE WE
SPECIFICITY OF PROSTAGLANDIN BINDING SITES IN RAT
FORESTOMACH TISSUE AND THEIR POSSIBLE USE AS A QUANTITATIVE
ASSAY
ADV BIOSCI 9:83-89(1973)

MILLS DCB
MECHANISMS AFFECTING CYCLIC AMP FORMATION IN HUMAN BLOOD
PLATELETS
CIRCULATION (SUPPL II) 44:81(1971)

MILLS DCB
EFFECTS OF AGGREGATING AGENTS ON PLATELET ADENYL CYCLASE
CIRCULATION (SUPPL II) 46:33(1972)

MILLS DCB   LIPSON C
PLATELET ENERGY CHARGE: EFFECTS OF AGGREGATING AGENTS AND
OF METABOLIC POISONS
FED PROC 32:732(1973)

MILLS DCB   SMITH JB
THE INFLUENCE ON PLATELET AGGREGATION OF DRUGS THAT AFFECT
THE ACCUMULATION OF ADENOSINE 3':5'-CYCLIC MONOPHOSPHATE IN
PLATELETS
BIOCHEM J 121:185-196(1971)

MILLS DCB   SMITH JB
THE CONTROL OF PLATELET RESPONSIVENESS BY AGENTS THAT
INFLUENCE CYCLIC AMP METABOLISM
ANN NY ACAD SCI 201:391-399(1972)

MILTON AS
PROSTAGLANDIN RELEASE IN THE CENTRAL NERVOUS SYSTEM DURING
ENDOTOXIN INDUCED FEVER
ADV BIOSCI (SUPPL) 9:79(1972)

MILTON AS
PROSTAGLANDIN E1 AND ENDOTOXIN FEVER, AND THE EFFECTS OF
ASPIRIN, INDOMETHACIN, AND 4-ACETAMIDOPHENOL
ADV BIOSCI 9:495-500(1973)

MILTON AS
THERMOREGULATORY EFFECTS OF PROSTAGLANDIN E1(PGE1)
IN: PHARMACOL THERMOREGUL PROC SATELL SYMP, 1972,
KARGER, BASEL : 498-499(1973)

MILTON AS   WENDLANDT S
THE EFFECTS OF 4-ACETAMIDOPHENOL (PARACETAMOL) ON THE
TEMPERATURE RESPONSE OF THE CONSCIOUS RAT TO THE
INTRACEREBRAL INJECTION OF PROSTAGLANDIN E1 ADRENALINE AND
PYROGEN
J PHYSIOL (LOND) 217:33-34P(1971)

MILTON AS   WENDLANDT S
EFFECTS ON BODY TEMPERATURE OF PROSTAGLANDINS OF THE A, E
AND F SERIES, ON INJECTION INTO THE THIRD VENTRICLE OF
UNANAESTHETIZED CATS AND RABBITS.
J PHYSIOL (LOND) 218 : 325-336(1971)

MILTON AS   WENDLANDT S
THE EFFECT OF DIFFERENT ENVIRONMENTAL TEMPERATURES ON THE
HYPERPYREXIA PRODUCED BY THE INTRAVENTRICULAR INJECTION OF
PYROGEN, 5-HYDROXYTRYPTAMINE AND PROSTAGLANDIN E1 IN THE
CONSCIOUS CAT
J PHYSIOL (PARIS) 63: 340-342 (1971)

MILTON-THOMPSON GJ   BILLINGS JA   CUMMINGS JH
NEWMAN A   MISIEWICZ JJ
THE EFFECT OF CIRCULATING PROSTAGLANDIN F2A ON THE
FUNCTION OF THE HUMAN SMALL INTESTINE
REND ROM GASTROENTEROL 5:139(1973)

MINNA JD   GILMAN AG
EXPRESSION OF GENES FOR METABOLISM OF CYCLIC ADENOSINE 3':5'
MONOPHOSPHATE IN SOMATIC CELLS-II-EFFECTS OF PROSTAGLANDIN
E1 AND THEOPHYLLINE ON PARENTAL AND HYBRID CELLS
J BIOL CHEM 248: 6618-6625 (1973)

MISIEWICZ JJ
SOME EFFECTS OF PROSTAGLANDINS IN MAN
PROC R SOC MED 64:14-15(1971)

MITSUHASHI S
PROSTAGLANDINS AND RENAL FUNCTION
JAP J UROL 64:705-706(1973):(JAPANESE)

MIYAKE H   HAYASHI M
THE SYNTHESIS OF 2-CARBOXY PROSTAGLANDINS
PROSTAGLANDINS 4:577-590(1973)

MIYANO M   DORN CR
PROSTAGLANDINS. V. SYNTHESIS OF DL-DIHYDROPROSTAGLANDIN E1
AND D8(12)-DEHYDROPROSTAGLANDIN E1
J ORG CHEM 37:1818-1823(1972)

MIYANO M   DORN CR
PROSTAGLANDINS. VI. CORRELATION OF THE ABSOLUTE
CONFIGURATION OF PYRETHROLONE WITH THAT OF THE
PROSTAGLANDINS
J AM CHEM SOC 95:2664-2669(1973)

MIYANO M   DORN CR   COLTON FB
MARSHECK WJ
THE MICROBIAL TRANSFORMATION OF PROSTAGLANDINS
J CHEM SOC CHEM COMMUN 9:425(1971)

MIYANO M   DORN CR   MUELLER RA
PROSTAGLANDINS. IV. A SYNTHESIS OF F-TYPE PROSTAGLANDINS.
A TOTAL SYNTHESIS OF PROSTAGLANDIN F2A
J ORG CHEM 37:1810-1818(1972)

MIYANO M   MUELLER RA   DORN CR
A TOTAL SYNTHESIS OF RACEMIC PGE1, PGF1A, DIHYDRO-PGE1,
AND THEIR STEREOISOMERS
INTRA-SCI CHEM REP 6:43-53(1972)

MIYANO M   STEALEY MA
STEREOSELECTIVE TOTAL SYNTHESIS OF PROSTAGLANDIN E1
J CHEM SOC CHEM COMMUN 180-181(1973)

MODIGLIANI R   RAMBAUD JC   BERNIER JJ
THE METHOD OF INTRALUMINAL PERFUSION OF THE HUMAN SMALL
INTESTINE. II. ABSORPTION STUDIES IN HEALTH
DIGESTION 9:264-290(1973)

MODY NJ
EFFECTS OF PROSTAGLANDINS ON PLATELET FUNCTION
IN: "THE PROSTAGLANDINS: PROGRESS IN RESEARCH," SMM KARIM,
MTP MED TECH PUBL CO, OXFORD:239-262(1972)

MOGHISSI KS   RANGARAJAN NS   LACROIX GE
INDUCTION OF LABOR WITH PROSTAGLANDIN F2A AND OXYTOCIN: A
MATCHED STUDY
J REPROD MED 9:304-306(1972)

MOLFESE A
ACTION OF THE PROSTAGLANDINS ON FEMALE GENITALIA
MINERVA GINECOL 23:205-229(1971):(ITALIAN)

MOLFESE A
SOME EFFECTS OF PROSTAGLANDINS ON THE FEMALE GENITAL SYSTEM
MINERVA GINECOL 25:586-593(1973):(ITALIAN)

MONCADA S   FERREIRA SH   VANE JR
DOES BRADYKININ CAUSE PAIN THROUGH PROSTAGLANDIN
PRODUCTION?
ABSTR 5TH INT CONGR PHARMACOL, SAN FRANC:160(1972)

MONCADA S   FERREIRA SH   VANE JR
PROSTAGLANDINS, ASPIRIN-LIKE DRUGS AND THE OEDEMA OF
INFLAMMATION
NATURE (LOND) 246:217-219(1973)

MONKHOUSE DC   VAN CAMPEN L   AGUIAR AJ
KINETICS OF DEHYDRATION AND ISOMERIZATION OF
PROSTAGLANDINS E1 AND E2
J PHARM SCI 62:576-580(1973)

MONTGOMERY RG   PATEL NC   LEE JG
A COMPARISON OF THE DIURETIC EFFECTS OF PROSTAGLANDIN A1,
SODIUM ETHACRYNATE, AND PLACEBO
PROSTAGLANDINS 4:381-394(1973)

MONTOREANO R   CANDIA O
THE ADENYL CYCLASE-CAMP SYSTEM IN THE ACTIVE TRANSPORT OF
CHLORIDE IN THE ISOLATED FROG CORNEA
ACTA PHYSIOL LAT AM 23:ABSTR 341(1973):(SPANISH)

MOODY FG
ROLE OF MUCOSAL BLOOD FLOW IN THE PATHOGENESIS OF GASTRIC
ULCERS
IN:INT ENCYCL PHARMACOL THER,G PETERS,PERGAMON PRESS,NY
2:339-360(1973)

MOORE WV   WOLFF J
BINDING OF PROSTAGLANDIN E1 TO BEEF THYROID MEMBRANES
J BIOL CHEM 248:5705-5711(1973)

MORGAN H   WHITE RP   PENNINK M
ROBERTSON JT
PROSTAGLANDINS AND EXPERIMENTAL CEREBRAL VASOSPASM
SURG FORUM 23:447-448(1972)

MORI K
EFFECTS OF SEX HORMONES ON PROSTAGLANDIN BIOSYNTHESIS
FOLIA ENDOCRINOL JAP 47:664(1972):(JAPANESE)

MORI T   KANEOKA T   SEKIBA K
INTRAVENOUS PROSTAGLANDIN F2A AND OXYTOCIN FOR THE INDUCTION
OF TERM LABOR
J JAP SOC OBSTET GYNAECOL (NIPPON SANKA FUJINKA GAKKAI
ZASSHI) 25:265-273(1973)

MORIWAKI K   OKADA F
THE HYPOTENSIVE EFFECT OF PGE1 ON HYPERTENSIVE CASES OF
VARIOUS TYPES
PROC SOC GERONTOL STUDY, OKAYAMA CITY, ONO PHARMACEUTICAL
CO, OSAKA:24-27(1972):(JAPANESE)

MORRIS JM
MECHANISMS INVOLVED IN PROGESTERONE CONTRACEPTION AND
ESTROGEN INTERCEPTION
AM J OBSTET GYNECOL 117:167-176(1973)

MOSKOWITZ J   HARWOOD JP   FORN J
KRISHNA G   ROGERS B   MORROW A
EFFECT OF NORADRENALINE AND PROSTAGLANDIN E1 ON ADENOSINE
3',5'-MONOPHOSPHATE FORMATION IN ISOLATED PERICARDIAL FAT
CELLS OF MAN
NATURE(NEW BIOL) 230:214-215(1971)

MOSKOWITZ J   HARWOOD JP   REID WD
KRISHNA G
THE INTERACTION OF NOREPINEPHRINE AND PROSTAGLANDIN E1 ON
THE ADENYL CYCLASE SYSTEM OF HUMAN AND RABBIT BLOOD
PLATELETS
BIOCHIM BIOPHYS ACTA 230:279-285(1971)

MOSKOWITZ J   KRISHNA G
THE EFFECT OF PROSTAGLANDIN E1 ON NOREPINEPHRINE-INDUCED
MEMBRANE DEPOLARIZATION AND CYCLIC AMP ACCUMULATION IN BROWN
FAT CELLS
PHARMACOLOGY 10:129-135(1973)

MOSLER KH   CZEKANOWSKI R   DORNHOFER W
COMPARATIVE STUDIES OF THE EFFECT OF OXYTOCIN AND
PROSTAGLANDIN F2A IN THE UTERUS
ADV BIOSCI (SUPPL) 9:122(1972)

MOSORA N
PROSTAGLANDINS
CLUJUL MED 45:585-591(1972):(ROUMANIAN)

MOURA AM   SIMPKINS H
CYCLIC AMP LEVELS IN CULTURED MYOCARDIAL CELLS UNDER THE
INFLUENCE OF AGENTS INDUCING CHRONOTROPIC AND INOTROPIC
RESPONSES
CLIN RES 21:1013(1973)

MOUTSCHEN J   MOUTSCHEN-DAHMEN M
HAZARDS OF PROSTAGLANDINS FOR CHROMOSOME DAMAGE
MUTAGEN RES 21:42-43(1973)

MOVAT HZ   MACMORINE DRL   TAKEUCHI Y
THE ROLE OF PMN-LEUKOCYTE LYSOSOMES IN TISSUE INJURY,
INFLAMMATION AND HYPERSENSITIVITY
VIII. MODE OF ACTION AND PROPERTIES OF VASCULAR
PERMEABILITY FACTORS RELEASED BY PMN-LEUKOCYTES DURING
IN VITRO PHAGOCYTOSIS
INT ARCH ALLERGY APPL IMMUNOL 40:218-235(1971)

MOVAT HZ   URIUHARA T   TAKEUCHI T
MACMORINE DRL
THE ROLE OF PMN-LEUKOCYTE LYSOSOMES IN TISSUE INJURY,
INFLAMMATION AND HYPERSENSITIVITY.
VII. LIBERATION OF VASCULAR PERMEABILITY FACTORS FROM
PMN-LEUKOCYTES DURING IN VITRO PHAGOCYTOSIS
INT ARCH ALLERGY APPL IMMUNOL 40:197-217(1971)

MOZES E   SHEARER GM   MELMON KL
BOURNE HR
IN VITRO CORRECTION OF ANTIGEN-INDUCED IMMUNE SUPPRESSION:
EFFECTS OF POLY(A)-POLY(U) AND PROSTAGLANDIN E1
CELL IMMUNOL 9:226-233(1973)

MUELLER RA
PROSTAGLANDINS AND RELATED COMPOUNDS
ANNU REP MED CHEM 8:172-182(1973)

MUELLER RA   FISHBURNE JI JR   BRENNER WE
BRAAKSMA JT   HOFFER JL   HENDRICKS CH
CHANGES IN HUMAN PLASMA CATECHOLAMINES AND
DOPAMINE-BETA-HYDROXYLASE PRODUCED BY PROSTAGLANDIN F2A
PROSTAGLANDINS 2:219-229(1972)

MUIRHEAD CR
THE FILTER LOOP TECHNIQUE AS A METHOD OF MEASURING PLATELET
AGGREGATION IN THE FLOWING BLOOD OF THE RAT: THE INHIBITORY
ACTIVITY OF 5-OXO-1-CYCLOPENTENE-1-HEPTANOIC ACID
(AY-16,804) ON PLATELET AGGREGATION
THROMB DIATH HAEMORRH 30:138-147(1973)

MUIRHEAD EE
VASOACTIVE AND ANTI-HYPERTENSIVE EFFECTS OF PROSTAGLANDINS
AND OTHER RENOMEDULLARY LIPIDS
IN: "THE PROSTAGLANDINS. PHARMACOLOGICAL AND THERAPEUTIC
ADVANCES," MF CUTHBERT, HEINEMANN, LOND:201-251(1973)

MUIRHEAD EE   BROOKS B   BROWN GB
LEACH BE   BYERS LW   PITCOCK JA
THE RENAL MEDULLA AND THE ANTIHYPERTENSIVE FUNCTION OF THE
KIDNEY
CLIN RES 19:558(1971)

MUIRHEAD EE   BROOKS B   PITCOCK JA
STEPHENSON P
RENOMEDULLARY ANTIHYPERTENSIVE FUNCTION IN ACCELERATED
(MALIGNANT) HYPERTENSION OBSERVATIONS ON RENOMEDULLARY
INTERSTITIAL CELLS
J CLIN INVEST 51:181-190(1972)

MUIRHEAD EE   GERMAIN G   LEACH BE
PITCOCK JA   STEPHENSON P   BROOKS B
BROSIUS WL   DANIELS EG   HINMAN JW
PRODUCTION OF RENOMEDULLARY PROSTAGLANDINS BY RENOMEDULLARY
INTERSTITIAL CELLS GROWN IN TISSUE CULTURE
CIRC RES 31(SUPPL II):II-161-II-172(1972)

MUIRHEAD EE   GERMAIN GS   LEACH BE
BROOKS B   PITCOCK JA   STEPHENSON P
BROSIUS WL JR   HINMAN JW   DANIELS EG
RENOMEDULLARY PROSTAGLANDINS (PG) DERIVED FROM RENOMEDULLARY
INTERSTITIAL CELLS (RIC) GROWN IN TISSUE CULTURE (TC)
CLIN RES 20:69(1972)

MUIRHEAD EE   GERMAIN GS   LEACH BE
BROOKS B   STEPHENSON P
RENOMEDULLARY INTERSTITIAL CELLS (RIC), PROSTAGLANDINS (PG)
AND THE ANTIHYPERTENSIVE FUNCTION OF THE KIDNEY
PROSTAGLANDINS 3:581-594(1973)

MUIRHEAD EE   GERMAIN GS   LEACH BE
PITCOCK JA   BYERS LW   ARMSTRONG FB
STEPHENSON P
RENOMEDULLARY INTERSTITIAL CELLS (RIC):ENDOCRINE-LIKE
ANTIHYPERTENSIVE STRUCTURES.
CLIN RES 21:83 (1973)

MUIRHEAD EE   GERMAIN GS   LEACH BE
PITCOCK JA   STEPHENSON P   BROOKS B
BROSIUS WL JR   DANIELS EG   HINMAN JW
SYNTHESIS OF PROSTAGLANDINS (PGS) BY RENOMEDULLARY
INTERSTITIAL CELLS (RIC) GROWN IN TISSUE CULTURE
ADV BIOSCI 9:341-347(1973)

MUIRHEAD EE   LEACH BE   ARMSTRONG FB
GERMAIN GS
CENTRALLY MEDIATED ANTIHYPERTENSIVE ACTION OF
PROSTAGLANDINS E2 AND A2 IN THE SPONTANEOUSLY HYPERTENSIVE
RAT (SH RAT)
FED PROC 31:393(1972)

MUIRHEAD EE   LEACH BE   ARMSTRONG FB JR
GERMAIN GS
VASODEPRESSOR ACTION OF RENOMEDULLARY PROSTAGLANDINS
(PGE2 AND PGA2) IN THE SPONTANEOUSLY HYPERTENSIVE RAT
(SH RAY): EVIDENCE FOR A CENTRAL ACTION MEDIATED BY THE
VAGUS
ADV BIOSCI (SUPPL) 9:60(1972)

MUIRHEAD EE   LEACH BE   BYERS LW
BROOKS B   DANIELS EG   HINMAN JW
ANTIHYPERTENSIVE NEUTRAL RENOMEDULLARY LIPIDS (ANRL)
IN: "KIDNEY HORMONES," JW FISHER, ACADEMIC PRESS, LOND:
485-506(1971)

MUIRHEAD EE   LEACH BE   BYERS LW
BROOKS B   PITCOCK JA
ANTIHYPERTENSIVE FUNCTION OF THE RENAL MEDULLA
IN: "HYPERTENSION: MECHANISMS AND MANAGEMENT" G ONESTI KE
KIM AND JH MOYER,GRUNE & STRATTON,NY/LOND:631-643(1973)

MUIRHEAD EE   LEACH BE   GERMAIN GS
BYERS LW   ARMSTRONG FB
THE RENOMEDULLARY ANTIHYPERTENSIVE ENDOCRINE SYSTEM
CLIN RES 21:717(1973)

MUKHOPADHYAY A   RATTAN S   GOYAL RK
ESOPHAGEAL RESPONSE TO PROSTAGLANDIN E1 (PGE1) INFUSION
CLIN RES 21:52(1973)

MULLER AF
PRESIDENTIAL DISCOURSE
HELV MED ACTA 36:85-109(1972);(FRENCH)

MULLER-SCHWEINI/ E
INVESTIGATIONS ON THE MODE OF ACTION OF DIHYDROERGOTAMINE
IN HUMAN SAPHENOUS AND DOG SAPHENOUS AND FEMORAL VEINS
NAUNYN SCHMIEDEBERGS ARCH PHARMACOL 279:R44 (1973)

MUNDY GR   LUBEN RA   RAISZ LG
OPPENHEIM J   BUELL DN
BONE RESORBING ACTIVITY IN SUPERNATANTS FROM HUMAN LYMPHOID
CELL LINES.
CLIN RES 21:980(1973)

MUNSON PL
EFFECTS OF MORPHINE AND RELATED DRUGS ON THE CORTICOTROPHIN
(ACTH)-STRESS REACTION
PROG BRAIN RES 39:361-372(1973)

MURAD F
CYCLIC AMP LEVELS IN TRACHEAL PREPARATIONS: EFFECTS OF
EPINEPHRINE, (EPI) THEOPHYLLINE AND PROSTAGLANDIN E1
CLIN RES 21:73(1973)

MURAI K   MIZUSHIMA N   GAKUGA K
THE EFFECTS OF PROSTAGLANDINS ON EXPERIMENTAL GASTRIC ULCER
PROC SOC GERONTOL STUDY, OKAYAMA CITY, ONO PHARMACEUTICAL
CO, OSAKA:33-38(1972);(JAPANESE)

MURAKAMI M   ODAKE K   TAKASE M
YOSHINO K
POTENTIATING EFFECT OF ADENOSINE ON OTHER INHIBITORS OF
PLATELET AGGREGATION
THROMB DIATH HAEMORRH 27:252-262(1972)

MURARI R   NATARAJAN S   SESHADRI TR
RAMASWAMI AS
SOME NEW PHARMACOLOGICAL PROPERTIES OF BLACK TEA POLYPHENOLS
CURR SCI 41:435-437(1972)

MURER EH
COMPOUNDS KNOWN TO AFFECT THE CYCLIC ADENOSINE MONOPHOSPHATE
LEVEL IN BLOOD PLATELETS: EFFECT ON THROMBIN-INDUCED CLOT
RETRACTION AND PLATELET RELEASE
BIOCHIM BIOPHYS ACTA 237:310-315(1971)

MURER EH
EFFECT OF PROSTAGLANDIN E1 ON CLOT RETRACTION
NATURE (LOND) 229:112-113(1971)

MURER EH
FACTORS INFLUENCING THE INITIATION AND THE EXTRUSION PHASE
OF THE PLATELET RELEASE REACTION
BIOCHIM BIOPHYS ACTA 261:435-443(1972)

MURER EH   DAY HJ
RELEASE FROM PLATELETS INDUCED BY SR++ OR CA++: EFFECT OF
ANTI-INFLAMMATORY DRUGS
CIRCULATION 48 (SUPPL 4) : IV-198 (1973)

MURPHY OL   DONNELLY C   MOSKOWITZ J
INHIBITION BY LITHIUM OF PROSTAGLANDIN E1 AND NOREPINEPHRINE
EFFECTS ON CYCLIC ADENOSINE MONOPHOSPHATE PRODUCTION IN
HUMAN PLATELETS
CLIN PHARMACOL THER 14:810-814(1973)

MURRAY-LYON IM   SANDLER M   CHEETHAM HD
WATTS JAE   WILLIAMS R
5-HYDROXYINDOLE-SECRETING RECTAL CARCINOID TUMOUR
GUT 13:385-386(1972)

MUSTARD JF   PACKHAM MA
COMMENTARY. DRUGS INHIBITING PLATELET FUNCTION
BIOCHEM PHARMACOL 22:3151-3156(1973)

NAFTOLIN F
PROSTAGLANDINS: BASIC REPRODUCTIVE CONSIDERATION AND NEW
CLINICAL APPLICATIONS
IN: "PROSTAGLANDINS AND CYCLIC AMP," RH KAHN, WEM LANDS,
ACADEMIC PRESS, NY : 157-182(1973)

NAFTOLIN F   KIRSHEN EJ   RYAN KJ
THERAPEUTIC ABORTION UTILIZING LOCAL APPLICATION OF
PROSTAGLANDIN F2A
J REPROD MED 9:437-441(1972)

NAJJAR VA   CONSTANTOPOULOS A
THE ACTIVATION OF ADENYLATE CYCLASE. I. A POSTULATED
MECHANISM FOR FLUORIDE AND HORMONE ACTIVATION OF ADENYLATE
CYCLASE
MOL CELL BIOCHEM 2:87-93(1973)

NAKAMURA H
LIVER LIPIDS AND PGE1
PROC SOC GERONTOL STUDY, OKAYAMA CITY, ONO PHARMACEUTICAL
CO, OSAKA:32(1972);(JAPANESE)

NAKANO J
EFFECTS OF THE METABOLITES OF PROSTAGLANDIN E1 ON THE
SYSTEMIC AND PERIPHERAL CIRCULATIONS IN DOGS
PROC SOC EXP BIOL MED 136:1265-1268(1971)

NAKANO J
PROSTAGLANDINS AND THE CIRCULATION
MOD CONCEPTS CARDIOVASC DIS 40:49-54(1971)

NAKANO J
RELATIONSHIP BETWEEN THE CHEMICAL STRUCTURE OF
PROSTAGLANDINS AND THEIR VASOACTIVITIES IN DOGS
BR J PHARMACOL 44:63-70(1972)

NAKANO J
GENERAL PHARMACOLOGY OF PROSTAGLANDINS
IN: "THE PROSTAGLANDINS. PHARMACOLOGICAL AND THERAPEUTIC
ADVANCES," MF CUTHBERT, HEINEMANN, LOND:23-124(1973)

NAKANO J
CARDIOVASCULAR ACTIONS
IN: "THE PROSTAGLANDINS," PW RAMWELL, PLENUM PRESS, NY
1:238-316(1973)

NAKANO J
THE PROSTAGLANDINS: THEIR EFFECT ON 14 CLINICAL CONDITIONS
RESIDENT STAFF PHYSICIAN 19:93-99,102-106(1973)

NAKANO J   CHANG ACK   FISHER RG
EFFECTS OF PROSTAGLANDINS E2 AND F2A ON THE CAROTID ARTERIAL
BLOOD FLOW, CEREBROSPINAL FLUID PRESSURE AND INTRAOCULAR
PRESSURE IN DOGS
PROC SOC EXP BIOL MED 140:866-869(1972)

NAKANO J   CHANG ACK   FISHER RG
EFFECTS OF PROSTAGLANDINS E1, E2, A1, A2, AND F2A ON CANINE
CAROTID ARTERIAL BLOOD FLOW, CEREBROSPINAL FLUID PRESSURE,
AND INTRAOCULAR PRESSURE
J NEUROSURG 38:32-39(1973)

NAKANO J   CHANG AK   FISHER RG
EFFECTS OF PROSTAGLANDINS E1, E2, A2 AND F2A ON THE CAROTID
ARTERIAL BLOOD FLOW AND CEREBROSPINAL FLUID PRESSURE
CLIN RES 20:389(1972)

NAKANO J   CHIBA S   NAKAJIMA T
EFFECT OF PROSTAGLANDINS E1 AND F2A ON HEART RATE BY DIRECT
INJECTION INTO THE CANINE SINUS NODE ARTERY
CLIN RES 19:645(1971)

NAKANO J   MCCLOY RB   PRANCAN AV
CIRCULATORY AND PULMONARY AIRWAY RESPONSES TO DIFFERENT
MIXTURES OF PROSTAGLANDINS E2 AND F2A IN DOGS
EUR J PHARMACOL 24:61-66(1973)

NAKANO J   MCCLOY RB JR
EFFECTS OF INDOMETHACIN ON THE PULMONARY VASCULAR AND AIR-
WAY RESISTANCE RESPONSES TO PULMONARY MICROEMBOLIZATION
PROC SOC EXP BIOL MED 143:218-221(1973)

NAKANO J   MORSY NH
BETA-OXIDATION OF PROSTAGLANDINS E1 AND E2 IN RAT LUNG AND
KIDNEY HOMOGENATE
CLIN RES 19:142(1971)

NAKANO J   PRANCAN AV
RELATIONSHIP BETWEEN STRUCTURE OF PROSTAGLANDINS AND THEIR
VASOACTIVITY IN DOGS
PHARMACOLOGIST 13:292(1971)

NAKANO J   PRANCAN AV
EFFECT OF ACUTE HYPOXIA AND RENAL ARTERIAL CONSTRICTION ON
THE METABOLISM OF PROSTAGLANDIN E1 IN THE DOG RENAL MEDULLA
CLIN RES 20:503(1972)

NAKANO J   PRANCAN AV
EFFECT OF PROSTAGLANDINS E1 AND A1 ON THE GASTRIC
CIRCULATION IN DOGS
PROC SOC EXP BIOL MED 139:1151-1154(1972)

NAKANO J   PRANCAN AV
METABOLIC DEGRADATION OF PROSTAGLANDIN E1 IN THE LUNG AND
KIDNEY OF RATS IN ENDOTOXIN SHOCK
PROC SOC EXP BIOL MED 144:506-508(1973)

NAKANO J   PRANCAN AV   KESSINGER CL
METABOLISM OF PROSTAGLANDIN E1 (PGE1) BY CEREBRAL AND
CEREBELLAR HOMOGENATES.
J LAB CLIN MED 78:998(1971)

NAKANO J   PRANCAN AV   KESSINGER JM
EFFECT OF PROSTAGLANDINS E1 AND A1 ON THE GASTRIC
CIRCULATION IN DOGS
CLIN RES 19:399(1971)

NAKANO J   PRANCAN AV   MCCURDY JR
GREENFIELD LJ
EFFECTS OF PROSTAGLANDIN E1 ON THE CARDIOVASCULAR,
HEMATOLOGICAL AND METABOLIC RESPONSES TO ENDOTOXIN
CLIN RES 21:239(1973)

NAKANO J   PRANCAN AV   MOORE S
EFFECT OF THE PROSTAGLANDIN ANTAGONISTS ON THE
VASOACTIVITIES OF PROSTAGLANDINS E1 (PGE1), E2 (PGE2), A1
(PGA1), AND F2A (PGF2A)
CLIN RES 19:712(1971)

NAKANO J   PRANCAN AV   MOORE SE
METABOLISM OF PROSTAGLANDIN E1 IN THE CEREBRAL CORTEX AND
CEREBELLUM OF THE DOG AND RAT
BRAIN RES 39:545-548(1972)

NAKANO J   PRANCAN AV   MORSY NH
METABOLISM OF PROSTAGLANDIN E1 IN STOMACH, JEJUNUM CHYLE AND
PLASMA OF THE DOG AND THE RAT
JAP J PHARMACOL 23:355-361(1973)

NANRA RS   CHIRAWONG P   KINCAID-SMITH P
MEDULLARY ISCHAEMIA IN EXPERIMENTAL ANALGESIC NEPHROPATHY-
THE PATHOGENESIS OF RENAL PAPILLARY NECROSIS
AUST N Z J MED 3:580-586(1973)

NASEEM SM   HOLLANDER VP
EFFECTS OF PROSTAGLANDINS, CYCLIC AMP AND ITS DERIVATIVES ON
THE GROWTH, MORPHOLOGY AND ONCOGENICITY OF CULTURED MOUSE
MYELOMA (MPC-11) CELLS
FED PROC 32:610(1973)

NASEEM SM   HOLLANDER VP
INSULIN REVERSAL OF GROWTH INHIBITION OF PLASMA CELL TUMOR
BY PROSTAGLANDIN OR ADENOSINE 3',5'-MONOPHOSPHATE
CANCER RES 33:2909-2912(1973)

NATHANIELSZ PW   ABEL M   SMITH GW
HORMONAL FACTORS IN PARTURITION IN THE RABBIT
IN: FOETAL AND NEONATAL PHYSIOLOGY, LONDON, CAMBRIDGE UNIV
PRESS:594-602(1973)

NEEDLEMAN P   DOUGLAS JR   JOHNSON EM
JAKSCHIK B   STOECKLEIN PB
CATECHOLAMINE INDUCED RENAL PROSTAGLANDIN BIOSYNTHESIS
PHARMACOLOGIST 15:208(1973)

NEEDLEMAN P   KAUFFMAN AH   DOUGLAS JR JR
JOHNSON EM JR   MARSHALL GR
SPECIFIC STIMULATION AND INHIBITION OF RENAL
PROSTAGLANDIN RELEASE BY ANGIOTENSIN ANALOGS
AM J PHYSIOL 224:1415-1419(1973)

NEEDLEMAN P   KAUFFMAN AH   STOECKLEIN PA
DOUGLAS JR JR   JOHNSON EM JR   MARSHALL GR
STIMULATION AND INHIBITION OF RENAL PROSTAGLANDIN RELEASE BY
ANGIOTENSIN ANALOGUES
FED PROC 32:788(1973)

NEEDLEMAN P   MARSHALL GR   DOUGLAS JR JR
PROSTAGLANDIN RELEASE FROM VASCULATURE BY ANGIOTENSIN II:
DISSOCIATION FROM LIPOLYSIS
EUR J PHARMACOL 23:316-319(1973)

NEKRASOVA A   LAUNTSBERG L   SOKOLOVA R
PROSTAGLANDIN-LIKE RENAL VASODEPRESSOR LIPIDS AND
ELECTROLYTIC EXCHANGE IN THE KIDNEY
ADV BIOSCI (SUPPL)9:49(1972)

NEKRASOVA AA   SEREBROVSKAYA YA   LANTSBERG LA
THE CONTENT OF PROSTAGLANDIN-LIKE SUBSTANCES, AND THE
ACTIVITY OF RENIN IN THE KIDNEYS OF RABBITS WITH
RENOVASCULAR HYPERTENSION, AND WITH UNILATERAL NEPHRECTOMY
PATOFIZIOL EKSPED TER 15:74-76(1971);(RUSSIAN)

NEKRASOVA AA   SOKULVA RI   LANTSBERG LA
PROSTAGLANDINLIKE RENAL VASODEPRESSOR LIPIDS AND ELECTROLYTE
EXCHANGE IN THE KIDNEY
ADV BIOSCI 9:307-312(1973)

NEMOTO T   AOKI H   IKE A
YAMADA K   KONDO I   KOBAYASHI S
INAGAWA T
SERUM PROSTAGLANDIN LEVEL IN BRONCHIAL ASTHMA
ARERUGI 22:728-734(1973);(JAPANESE)

NEUFELD AH   CHAVIS RM   SEARS ML
DEGENERATION RELEASE OF NOREPINEPHRINE CAUSES TRANSIENT
OCULAR HYPEREMIA MEDIATED BY PROSTAGLANDINS
INVEST OPHTHALMOL 12:167-175(1973)

NEUFELD AH   CHAVIS RM   SEARS ML
EVIDENCE FOR PROSTAGLANDIN SYNTHESIS AND RELEASE CAUSED BY
DEGENERATING SYMPATHETIC NERVE TERMINALS IN THE RABBIT EYE
IN: "PROSTAGLANDINS AND CYCLIC AMP," RH KAHN, WEM LANDS,
ACADEMIC PRESS, NY : 25-27(1973)

NEUFELD AH   JAMPOL LM   SEARS ML
ASPIRIN PREVENTS THE DISRUPTION OF THE BLOOD-AQUEOUS BARRIER
IN THE RABBIT EYE
NATURE (LOND) 238:158-159(1972)

NEUFELD AH   JAMPOL LM   SEARS ML
ASPIRIN PREVENTS THE DISRUPTION OF THE BLOOD-AQUEOUS
BARRIER IN THE RABBIT EYE
NATURE (LOND) 238:158-159(1972)

NEUFELD AH   SEARS ML
PROSTAGLANDIN AND EYE
PROSTAGLANDINS 4:157-175(1973)

NEUFELD AH   SEARS ML
THE SITE OF ACTION OF PROSTAGLANDIN E2 ON THE DISRUPTION OF
THE BLOOD-AQUEOUS BARRIER IN THE RABBIT EYE
EXP EYE RES 17:445-448(1973)

NEWTON JR   COLLINS WP
PLASMA HORMONE CHANGES FOLLOWING INTRAVENOUS F2A AND E2 FOR
THERAPEUTIC ABORTION
ADV BIOSCI (SUPPL)9:113(1972)

NEZAMIS JE   ROBERT A   STOWE DF
INHIBITION BY PROSTAGLANDIN E1 OF GASTRIC SECRETION IN THE
DOG.
J PHYSIOL (LOND) 218 : 369-383(1971)

NG AYH   CHENG MCE   RATNAM SS
TERMINATION OF MID-TRIMESTER PREGNANCY BY INTRA-AMNIOTIC
INJECTION OF PROSTAGLANDINS E2 AND F2A:
SINGAPORE MED J 14:490-493(1973)

NG KKF
FATE OF ANGIOTENSIN I IN THE TOAD BUFO MELANOSTICTUS
BR J PHARMACOL 48:456-463(1973)

NIELSEN JG
PROSTAGLANDINS. A SURVEY
DAN TIDSSKR FARM 46:149-164(1972);(DANISH)

NIES A   TANNENBAUM J   SPLAWINSKI J
OATES J
NATRIURETIC EFFECT OF PROSTAGLANDIN E2 AND ITS PRECURSOR
ARACHIDONIC ACID
CLIN RES 21:77(1973)

NIES A   TANNENBAUM J   SPLAWINSKI J
OATES J
PROSTAGLANDIN FORMED ENDOGENOUSLY CAN AFFECT RENAL FUNCTION
CLIN RES 21:701(1973)

NIEWIAROWSKI S   GOLDSTEIN S
INTERACTION OF CULTURED HUMAN FIBROBLASTS WITH FIBRIN:
MODIFICATION BY DRUGS AND AGING IN VITRO
J LAB CLIN MED 82:605-610(1973)

NIEWIAROWSKI S   REGOECZI E   MUSTARD JF
PLATELET INTERACTION WITH FIBRINOGEN AND FIBRIN: COMPARISON
OF THE INTERACTION OF PLATELETS WITH THAT OF FIBROBLASTS,
LEUKOCYTES AND ERYTHROCYTES
ANN N Y ACAD SCI 201:72-83(1972)

NIEWIAROWSKI S   REGOECZI E   STEWART GJ
SENYI AF   MUSTARD JF
PLATELET INTERACTION WITH POLYMERIZING FIBRIN
J CLIN INVEST 51:685-700(1972)

NISHIBORI T   MATSUOKA Y
STUDIES ON ABSORPTION, DISTRIBUTION, EXCRETION AND
METABOLISM OF 9 8-3H-PROSTAGLANDIN F2A (3H-PG F2A) IN RATS
RES PHARM (IYAKUHIN KENKYU) 2:397-402(1971);(JAPANESE)

NISTICO G   MARLEY E
CENTRAL EFFECTS OF PROSTAGLANDIN E1 IN ADULT FOWLS
NEUROPHARMACOLOGY 12:1009-1016(1973)

NISWENDER GD   MENON KMJ   JAFFE RB
REGULATION OF THE CORPUS LUTEUM DURING THE MENSTRUAL CYCLE
AND EARLY PREGNANCY
FERTIL STERIL 23:432-442(1972)

NITSCHKOFF ST
ABOUT THE EFFECT OF PROSTAGLANDINS (PGA1 AND PGE1) ON THE
CAROTID SINUS REFLEX IN RABBITS
DTSCH GESUNDHEITSW 51:2393-2396(1971);(GERMAN)

NOMURA T   OGATA H   ITO M
OCCURRENCE OF PROSTAGLANDINS IN FISH TESTIS
TOHOKU J AGRIC RES 24:138-144(1973)

NORIEGA GUERRA L   N AREVALO T   C GUERRERO B
E ORTEGA H   R AZNAR RAMOS
EFFECT OF PROSTAGLANDINS E2 AND F2A ON UTERINE CONTRACTILITY
IN: "SIMPOSIO SOBRE PROSTAGLANDINAS," IMPRESOS OFFSALI,
MEX CITY:89-107(1972);(SPANISH)

NORIEGA-GUERRA L   AREVALO N   GUERRERO C
ORTEGA E   AZNAR-RAMOS R
EFFECT OF PROSTAGLANDINS E2 AND F2 ON UTERINE CONTRACTIONS
ARCH INVEST MED (MEX) 3(SUPPL 1): 201-212(1972);(SPANISH)

NORINS AL
ATOPIC DERMATITIS
PEDIATR CLIN NORTH AM 18:801-838(1971)

NOVALES RR   NOVALES BJ
SODIUM-FREE AND CYTOCHALASIN B INHIBITION OF PROSTAGLANDIN
A2 ACTION ON AMPHIBIAN CHROMATOPHORES
AM ZOOL 13:1277(1973)

NOVY MJ
DISTRIBUTION OF OVARIAN BLOOD FLOW IN RABBITS AS MEASURED
BY RADIOACTIVE MICROSPHERES
BIOL REPROD 7:105-106(1972)

NOVY MJ
EFFECT OF UTERINE CONTRACTIONS ON MYOMETRIAL, PLACENTAL AND
UMBILICAL BLOOD FLOW IN THE RHESUS MONKEY
ABSTR SOC GYNECOL INVEST 20TH ANNU MEET, ATLANTA:25(1973)

NOVY MJ   PIASECKI G   JACKSON BT
FETAL HEMODYNAMIC RESPONSES TO GRADED REDUCTIONS IN
UTERINE BLOOD FLOW AND ADMINISTRATION OF PROSTAGLANDINS
ABSTR SOC GYNECOL INVEST 19TH ANNU MEET, PHOENIX:22(1972)

NUGTEREN DH
ESSENTIAL FATTY-ACIDS AND PROSTAGLANDINS
EXP EYE RES 12:368(1971)

NUGTEREN DH   HAZELHOF E
ISOLATION AND PROPERTIES OF INTERMEDIATES IN PROSTAGLANDIN
BIOSYNTHESIS
BIOCHIM BIOPHYS ACTA 326:448-461(1973)

NUKADA T
THE EFFECTS OF PROSTAGLANDIN E1 ON PHEOCHROMOCYTOMA
PROC SOC GERONTOL STUDY, OKAYAMA CITY, ONO PHARMACEUTICAL
CO, OSAKA:28-29(1972);(JAPANESE)

NUNEZ J   MEDAOUI S   RAPPAPORT L
PROSTAGLANDINS AND THYROID HORMONES
IN: "PROSTAGLANDINES 1973," INSERM, PARIS:95-100(1973);
(FRENCH)

NUTTER DO   CRUMLY HR JR
CANINE CORONARY VASCULAR AND CARDIAC RESPONSES TO THE
PROSTAGLANDINS
CARDIOVASC RES 6:217-225(1972)

NUTTER DO   RATTS T
MYOCARDIAL EFFECTS OF PROSTAGLANDINS.
CIRCULATION 44:II-206(1971)

NUTTER DO   RATTS T
INOTROPIC AND CHRONOTROPIC ACTIONS OF PROSTAGLANDINS E1, A1,
AND F2A
CLIN RES 20:29(1972)

NUTTER DO   RATTS T
DIRECT ACTIONS OF PROSTAGLANDINS E1, A1, AND F2A ON
MYOCARDIAL CONTRACTION
PROSTAGLANDINS 3:323-336(1973)

NYBERG R
THERAPEUTIC ABORTION BY INTRA-AMNIOTIC ADMINISTRATION OF
PROSTAGLANDIN F2A
ADV BIOSCI (SUPPL)9:87(1972)

NYSTROM E   SJOVALL J
SEPARATION OF PROSTAGLANDINS ON A HYDROPHOBIC SEPHADEX
DERIVATIVE
ANAL LETT 6:155-161(1973)

O'BRIEN JR
ASPIRIN, THROMBOSIS AND THE FUTURE
LANCET 2:486(1971)

O'CONNOR NE   MORGAN AP
POTENTIATION BY PROSTAGLANDIN E1 OF AMBIENT PRESSURE OXYGEN
TOXICITY
CLIN RES 19:517(1971)

OESTERLING TO   MOROZOWICH W   ROSEMAN TJ
PROSTAGLANDINS
J PHARM SCI 61:1861-1895(1972)

OHASHI S   OHMURO S   SUGAWARA I
KUWATA K   OKAMOTO E
EFFECTS OF PROSTAGLANDIN E1 ON CANINE GASTROINTESTINAL
MOTILITY IN VIVO AND HUMAN ISOLATED SMOOTH MUSCLE IN VITRO
JAP J SMOOTH MUSCLE RES 9:69-77(1973);(JAPANESE)

OHNO H   MIYAMOTO T   NISHIBORI S
HIRATA F
USE OF ENZYME ON THE FLUOROMETRY FOR PROSTAGLANDINS
YAKUGAKU ZASSHI 92:1111-1116(1972);(JAPANESE)

OIEN HG   HUMES JL   KUEHL FA JR
STRUCTURAL REQUIREMENTS FOR THE BINDING OF PROSTAGLANDINS
AND ANALOGS TO A LIPOCYTE RECEPTOR PREPARATION
FED PROC 32:803(1973)

OKA H   KANEKO T   YAMASHITA K
SUZUKI S   ODA T
THE GLUCAGON AND FLUORIDE SENSITIVE ADENYL CYCLASE IN
PLASMA MEMBRANE OF RAT LIVER
ENDOCRINOL JAP 20:263-270(1973)

OKADA F
STUDIES ON MECHANISM OF HYPOTENSIVE ACTION OF ENDOGENOUS
VASODEPRESSOR, PROSTAGLANDIN E1
OSAKA DAIGAKU IGAKU ZASSHI (OSAKA UNIV MED J) 23:21-34
(1971);(JAPANESE)

OKPAKO DT
THE ACTIONS OF HISTAMINE AND PROSTAGLANDINS F2A AND E2 ON
PULMONARY VASCULAR RESISTANCE OF THE LUNG OF THE GUINEA-PIG
J PHARM PHARMACOL 24:40-46(1972)

OLDHAM S
HORMONES
MANUF CHEM AEROSOL NEWS 42:21-27,58(1971)

OLLEY PM   COCEANI F
THE IN VITRO RESPONSE OF THE LAMB DUCTUS ARTERIOSUS TO
PROSTAGLANDINS
PEDIATR RES 6:341(1972)

OLLEY PM   COCEANI F
THE IN VITRO RESPONSE OF THE LAMB DUCTUS ARTERIOSUS TO
PROSTAGLANDINS
PROC CAN FED BIOL SOC 15:703(1972)

OLLEY PM   COCEANI F   KENT G
PULMONARY INACTIVATION OF PROSTAGLANDIN E1 IN THE FETAL AND
NEWBORN LAMB
PEDIATR RES 7:425(1973)

OLMEDO OJAO
CONTRIBUTION TO THE STUDY OF THE OVULATORY MECHANISMS AND
LUTEOLYSIS IN THE RABBIT
DISS, U OVIEDO, SPAIN (1973);(SPANISH)

ONAYA T   TAKEYA Y   KAJIWARA T
YAMADA R   SHICHIJO K
THE EFFECTS OF PROSTAGLANDINS ON THE PITUITARY AND THYROID
GLANDS
PROC SOC GERONTOL STUDY, OKAYAMA CITY, ONO PHARMACEUTICAL
CO, OSAKA:30-31(1972);(JAPANESE)

ONO S   OBARA Y   HATANO M
THE EFFECT OF PROSTAGLANDIN E1 ON THE PHOSPHOLIPID
METABOLISM IN THE LENS
OPHTHALMIC RES 4:281-283(1973)

OPLER LA   MAKMAN MH
MEDIATION BY CYCLIC AMP OF HORMONE-STIMULATED
GLYCOGENOLYSIS IN CULTURED RAT ASTROCYTOMA CELLS
BIOCHEM BIOPHYS RES COMMUN 46:1140-1145(1972)

ORANGE RP
THE IMMUNOLOGICAL RELEASE OF CHEMICAL MEDIATORS FROM HUMAN
LUNG
IN: "MECHANISMS IN ALLERGY. REAGIN-MEDIATED
HYPERSENSITIVITY," L GOODFRIEND, AH SEHON, RP ORANGE,
MARCEL DEKKER, NY:439-450(1973)

OREHEK J   DOUGLAS JS   LEWIS AJ
BOUHUYS A
PROSTAGLANDIN REGULATION OF AIRWAY SMOOTH MUSCLE TONE
NATURE (NEW BIOL) 245: 84-85 (1973)

OREHEK J   DOUGLAS JS   LEWIS AJ
BOUHUYS A
SYNTHESIS OF PROSTAGLANDINS BY THE ISOLATED GUINEA PIG
TRACHEA AND THE EFFECTS OF ITS INHIBITION
J PHYSIOL (PARIS) 67:297A-298A(1973);(FRENCH)

OREHEK J   LEWIS AJ   DOUGLAS JS
BOUHUYS A
LOCAL MODULATION OF INDUCED AIRWAY CONSTRICTION BY
PROSTAGLANDIN-LIKE SUBSTANCES (PLS)
FED PROC 32:389(1973)

ORLOFF J
INTRODUCTORY REMARKS: SECRETION AND MEMBRANE FUNCTION
ANN N Y ACAD SCI 185:342-344(1971)

ORTEGA E   RODRIGUEZ C   CORTES-GALLEGOS V
NORIEGA-GUERRA L   HENZL MR
CLINICAL PHARMACOLOGY OF PROSTAGLANDIN E2 IN NORMAL
NON-PREGNANT WOMEN
ARCH INVEST MED (MEX) 3(SUPPL 1):187-200(1972);(SPANISH)

ORTEGA HE   RODRIGUEZ C   CORTES-GALLEGOS V
NORIEGA L   HENZL MR
CLINICAL PHARMACOLOGY OF PROSTAGLANDIN E2 IN NORMAL
NON-PREGNANT WOMEN
IN: "SIMPOSIO SOBRE PROSTAGLANDINAS," IMPRESOS OFFSALI,
MEX CITY:62-88(1972);(SPANISH)

OSSWALD W
LACK OF INFLUENCE OF INDOMETHACIN ON NEUROGENIC
VASODILATATION IN THE HIND-LIMB AND SPLEEN OF THE DOG
J PHARM PHARMACOL 25:655-656(1973)

OTTEN J   DUMONT JE
GLUCOSE METABOLISM IN NORMAL HUMAN THYROID TISSUE IN VITRO
EUR J CLIN INVEST 2:213-219(1972)

OTTEN J   JOHNSON GS   PASTAN I
REGULATION OF CYCLIC AMP LEVELS IN FIBROBLASTS BY INSULIN,
TRYPSIN, SERUM AND PROSTAGLANDIN E1
FED PROC 31:440(1972)

OZAWA S   ASANO S
HYPOTENSIVE EFFECT OF PROSTAGLANDINS
PROC SOC GERONTOL STUDY, OKAYAMA CITY, ONO PHARMACEUTICAL
CO, OSAKA:16-21(1972);(JAPANESE)

OZAWA Y   HIGASHI F   KITAMOTO K
ARAI J   ASANO S   KOIDE S
INOUE T   NAKAMURA H   SHINOZAKI Y
OOTANI N   SHINAGAWA S
RENAL EFFECT OF PROSTAGLANDIN
JAP CIRC J 35:1425(1971)

OZEN S   YUKI Y   KONDO O
IMAMURA K   TOYAMA K   ABE M
INHIBITION OF GASTRIC JUICE SECRETION BY PROSTAGLANDIN E1
CLIN ENDOCRINOL (TOKYO) 19:741-745(1971);(JAPANESE)

OZER A   SHARP GWG
EFFECT OF PROSTAGLANDINS AND THEIR INHIBITORS ON OSMOTIC
WATER FLOW IN THE TOAD BLADDER
AM J PHYSIOL 222:674-680(1972)

OZER A   SHARP GWG
ANTAGONISM OF PROSTAGLANDIN BY CHLORPROPAMIDE
ADV BIOSCI (SUPPL)9:23(1972)

OZER A   SHARP GWG
MODULATION OF ADENYL CYCLASE ACTION IN TOAD BLADDER BY
CHLORPROPAMIDE: ANTAGONISM TO PROSTAGLANDIN E1
EUR J PHARMACOL 22:227-232(1973)

OZER H   HOLLENBERG NK   ABRAMS HL
PGE1 IN PHARMACOANGIOGRAPHY OF KIDNEYS
ADV BIOSCI (SUPPL)9:52(1972)

OZER H   HOLLENBERG NK   ABRAMS HL
EFFECT OF PGE1 ON CONTRAST VISUALIZATION OF THE RENAL
VASCULAR BED: COMPARISON WITH OTHER VASODILATORS
ADV BIOSCI 9:331-334(1973)

PACE-ASCIAK C
POLYHYDROXY CYCLIC ETHERS FORMED FROM TRITIATED
ARACHIDONIC ACID BY ACETONE POWDERS OF SHEEP SEMINAL
VESICLES
BIOCHEMISTRY 10:3664-3669(1971)

PACE-ASCIAK C
STIMULATION OF PROSTAGLANDIN BIOSYNTHESIS IN HOMOGENATES
OF RAT STOMACH FUNDUS BY CATECHOLAMINES
FED PROC 30:1081 ABS (1971)

PACE-ASCIAK C
PROSTAGLANDIN SYNTHETASE ACTIVITY IN THE RAT STOMACH FUNDUS.
ACTIVATION BY L-NOREPINEPHRINE AND RELATED COMPOUNDS
BIOCHIM BIOPHYS ACTA 280:161-171(1972)

PACE-ASCIAK C
CATECHOLAMINE INDUCED INCREASE IN PROSTAGLANDIN E IN RAT
STOMACH FUNDUS HOMOGENATES
ADV BIOSCI (SUPPL)9:3(1972)

PACE-ASCIAK C   MILLER D
PROSTAGLANDINS DURING DEVELOPMENT-I-AGE-DEPENDENT ACTIVITY
PROFILES OF PROSTAGLANDIN 15-HYDROXYDEHYDROGENASE AND
13,14-REDUCTASE IN LUNG TISSUE FROM LATE PRENATAL, EARLY
POSTNATAL AND ADULT RATS.
PROSTAGLANDINS 4:351-362(1973)

PACE-ASCIAK C   WOLFE LS
A NOVEL PROSTAGLANDIN DERIVATIVE FORMED FROM ARACHIDONIC
ACID BY RAT STOMACH HOMOGENATES
BIOCHEMISTRY 10:3657-3664(1971)

PACE-ASCIAK C   WOLFE LS
N-BUTYLBORONATE DERIVATIVES OF THE F PROSTAGLANDINS
RESOLUTION OF PROSTAGLANDINS OF THE E AND F SERIES BY
GAS-LIQUID CHROMATOGRAPHY
J CHROMATOGR 56:129-133(1971)

PACE-ASCIAK CR
CATECHOLAMINE INDUCED INCREASE IN PROSTAGLANDIN E
BIOSYNTHESIS IN HOMOGENATES OF THE RAT STOMACH FUNDUS
ADV BIOSCI 9:29-33(1973)

PACKHAM MA   GUCCIONE MA
REACTIONS OF POLYLYSINE WITH PLATELETS IN PLASMA AND IN
SUSPENSIONS OF WASHED PLATELETS
ABSTR 4TH INT CONGR THROMB HAEMOSTASIS, VIENNA:177(1973)

PACKHAM MA   JENKINS CSP   KINLUUGH-RATHB/ RL
MUSTARD JF
AGENTS INFLUENCING PLATELET ADHESION TO SURFACES AND THE
RELEASE REACTION.
CIRCULATION 44:II67(1971)

PACKHAM MA   MUSTARD JF
PLATELET REACTIONS
SEMIN HEMATOL 8(1):30-64(1971)

PACKHAM MA   RADOJEWSKI AM   PERRY DW
MUSTARD JF
LOSS OF PLATELET MUCOPOLYSACCHARIDES DURING ADP-INDUCED
AGGREGATION AND RELEASE REACTION
CIRCULATION (SUPPL II)46:32(1972)

PAGE LB
HYPERTENSION AND A SMALL KIDNEY
N ENGL J MED 289:736-743(1973)

PALMER GC
CYCLIC 3',5'-ADENOSINE MONOPHOSPHATE RESPONSE IN RABBIT
LUNG -- ADULT PROPERTIES AND DEVELOPMENT
BIOCHEM PHARMACOL 21:2907-2914(1972)

PALMER GC
CYCLIC AMP RESPONSE IN THE RABBIT LUNG: DEVELOPMENT AND
ADULT PROPERTIES
ABSTR 5TH INT CONGR PHARMACOL SAN FRANC 175(1972)

PALMER MA   PIPER PJ   VANE JR
RELEASE OF RABBIT AORTA CONTRACTING SUBSTANCE (RCS) AND
PROSTAGLANDINS INDUCED BY CHEMICAL OR MECHANICAL STIMULATION
OF GUINEA-PIG LUNGS
BR J PHARMACOL 49:226-242(1973)

PAOLETTI R
PROSTAGLANDINS AND LIPID MOBILIZATION
ACTA DIABETOL LAT 10:403(1973);(ITALIAN)

PAOLETTI R
ESSENTIAL FATTY ACIDS AND PROSTAGLANDINS
ACTA DIABETOL LAT 10:921-922(1973);(ITALIAN)

PAOLETTI R   PUGLISI L
PHARMACOLOGICAL CONTROL OF LIPID TRANSPORT
NAUNYN-SCHMIEDEBERGS ARCH PHARMAKOL 269:317-330(1971)

PAOLETTI R   PUGLISI L
LIPID METABOLISM
IN: "THE PROSTAGLANDINS," PW RAMWELL, PLENUM PRESS, NY
1:317-326(1973)

PAPANICOLAOU N
INVESTIGATION ON THE MECHANISM OF PROSTAGLANDINS RELEASE, ON
THEIR POSSIBLE HOMEOSTATIC ROLE, AND ON THEIR RELATIONSHIP
WITH THE THIRD FACTOR
ABSTR 5TH INT CONGR PHARMACOL, SAN FRANC:175(1972)

PAPANICOLAOU N
INVESTIGATION ON THE MECHANISM OF PROSTAGLANDINS RELEASE
EXPERIENTIA 28:275-276(1972)

PAPANICOLAOU N   MEYER P
INACTIVATION OF PROSTAGLANDINS E2 AND A2 ON THEIR SINGLE
PASSAGE THROUGH THE PULMONARY VASCULAR BED IN ANAESTHETIZED
RATS
REV CAN BIOL 31:313-316(1972)

PAPANICOLAOU N   MEYER P   MILLIEZ P
INTERACTIONS BETWEEN PROSTAGLANDIN E1 AND NORADRENALINE IN
ANAESTHETIZED RATS.
J PHARM PHARMACOL 23:454-455(1971)

PAPANICOLAOU N   TCHERNIA G   BOHOUN C
MEYER P   SCHWEISGUTH O   MILLIEZ P
DETECTION OF PROSTAGLANDINS IN A FUNCTIONING RENAL TUMOR
ADV BIOSCI (SUPPL)9:82(1972)

PAPANIKOLAOU N
PROSTAGLANDINS: THEIR RELEASE AND THEIR POSSIBLE HOMEOSTATIC
ROLE ON BLOOD PRESSURE, BLOOD VOLUME AND SODIUM AND WATER
BALANCE REGULATION
ABSTR 5TH INT CONGR NEPHROL, MEX CITY:113(1972)

414

PAPPO R   COLLINS P   JUNG C
NEW SYNTHETIC APPROACH IN THE PROSTAGLANDIN FIELD
ANN NY ACAD SCI 180:64-75(1971)

PAPPO R   COLLINS P   JUNG C
RESOLUTION AND CONFIGURATIONAL ASSIGNMENTS OF
METHYL-3-HYDROXY-5-OXO-CYCLOPENT-1-ENEHEPTANOATE, AN
IMPORTANT PROSTAGLANDIN INTERMEDIATE
TETRAHEDRON LETT 943-944(1973)

PAPPO R   COLLINS PW
SYNTHESIS OF PROSTAGLANDIN ANALOGS BY NOVEL 1,4 ADDITION
REACTIONS
TETRAHEDRON LETT 2627-2630(1972)

PARISI M   PICCINNI ZF
ASPIRIN POTENTIATES THE HYDRO-OSMOTIC EFFECT OF
ANTIDIURETIC HORMONE IN TOAD URINARY BLADDER
BIOCHIM BIOPHYS ACTA 279:209-212(1972)

PARISI M   PICCINNI ZF   RIPOCHE P
BOURGUET J
SYNERGISM BETWEEN CELLULAR TONICITY AND ANTIDIURETIC HORMONE
IN THE REGULATION OF THE ABSORPTION OF WATER: MEDICINA
EFFECTS OF NORADRENALINE, PROSTAGLANDIN E1 AND ASPIRIN
MEDICIN (B AIRES) 32:782(1972):(SPANISH)

PARK IJ   WENTZ AC   JONES HW JR
THE VIABILITY OF FETAL SKIN OF ABORTUSES INDUCED BY SALINE
OR PROSTAGLANDIN
AM J OBSTET GYNECOL 115:274-275(1973)

PARK MK   DYER DC
EFFECT OF POLYPHLORETIN PHOSPHATE AND 7-OXA-13-PROSTYNOIC
ACID ON THE VASOACTIVE ACTIONS OF PROSTAGLANDIN E2 AND
5-HYDROXYTRYPTAMINE ON ISOLATED HUMAN UMBILICAL ARTERIES.
PROSTAGLANDINS 3:913-920(1973)

PARK MK   DYER DC   VINCENZI FF
PROSTAGLANDIN E2 AND ITS ANTAGONISTS: EFFECTS ON AUTONOMIC
TRANSMISSION IN THE ISOLATED SINO-ATRIAL NODE
PROSTAGLANDINS 4:717-730(1973)

PARKER CP
CYCLIC AMP
IN:"INFLAMMATION MECHANISMS AND CONTROL," IH LEPOW AND
PA WARD, ACADEMIC PRESS, N Y:239-259(1972)

PARKER CW
THE ROLE OF PROSTAGLANDINS IN THE IMMUNE RESPONSE
IN: "PROSTAGLANDINS IN CELLULAR BIOLOGY" PW RAMWELL AND B
B PHARRISS, PLENUM PRESS, NY/LOND:173-194(1972)

PARKER CW
ADRENERGIC RESPONSIVENESS IN ASTHMA
IN: "ASTHMA," KF AUSTEN, LM LICHTENSTEIN, ACADEMIC PRESS,NY:
185-210(1973)

PARKER CW   BAUMANN ML   HUBER MG
ALTERATIONS IN CYCLIC AMP METABOLISM IN HUMAN BRONCHIAL
ASTHMA. II. LEUKOCYTE AND LYMPHOCYTE RESPONSES TO
PROSTAGLANDINS
J CLIN INVEST 52:1336-1341(1973)

PARKER CW   HUBER MG   BAUMANN ML
ALTERATIONS IN CYCLIC AMP METABOLISM IN HUMAN BRONCHIAL
ASTHMA III. LEUKOCYTE AND LYMPHOCYTE RESPONSES TO STEROIDS
J CLIN INVEST 52:1342-1348 (1973)

PARKER CW   JAFFE BM
PROSTAGLANDINS AND THE IMMUNE SYSTEM
ABSTR 7TH MIDDLE ATL REG MEET AM CHEM SOC PHILA:72(1972)

PARKER CW   MORSE SI
BORDETELLA PERTUSSIS-INDUCED ALTERATIONS IN LYMPHOCYTE
CYCLIC-AMP METABOLISM
J CLIN INVEST 51:72A-73A(1972)

PARKER CW   MORSE SI
THE EFFECT OF BORDETELLA PERTUSSIS ON LYMPHOCYTE CYCLIC AMP
METABOLISM
J EXP MED 137:1078-1090(1973)

PARKER J   JELKS G   EMERSON T JR
CARDIOVASCULAR EFFECTS OF PROSTAGLANDIN A1 IN THE DOGS
FED PROC 31:815(1972)

PARKER JL   EMERSON TE JR   DAUGHERTY RM JR
VASCULAR RESPONSES IN THE CANINE FORELIMB DURING
INTRAARTERIAL INFUSION OF PROSTAGLANDIN A1
PHYSIOLOGIST 14:206(1971)

PARRATT JR   STURGESS R
THE EFFECT OF INDOMETHACIN ON THE CARDIOVASCULAR RESPONSES
OF CATS TO E. COLI ENDOTOXIN
BR J PHARMACOL 49:163P-164P(1973)

PARTRIDGE JJ   CHADHA NK   USKOKOVIC MR
ASYMMETRIC SYNTHESIS OF PROSTAGLANDIN INTERMEDIATES
ABSTR 166TH NATL MEET AM CHEM SOC:ORGN-88(1973)

PARTRIDGE JJ   CHADHA NK   USKOKOVIC MR
ASYMMETRIC SYNTHESIS OF PROSTAGLANDIN INTERMEDIATES
J AM CHEM SOC 95:7171-7172(1973)

PASTAN I   JOHNSON GS   OTTEN J
PEERY CV   D'ARMIENTO M
ROLE OF CYCLIC AMP IN THE ABNORMAL GROWTH AND MORPHOLOGY OF
TRANSFORMED FIBROBLASTS
FED PROC 30:1047(1971)

PATERSON CA   ECK BA
PROSTAGLANDIN E1 AND THE OCULAR LENS
OPHTHALMOL RES 2:246-248(1971)

PATERSON CA   ECK BA
AN EFFECT OF PROSTAGLANDIN E1 ON THE LENS
OPHTHALMOL RES 3:16(1972)

PATERSON CA   ECK BA
INFLUENCE OF PROSTAGLANDINS ON CATION MOVEMENT IN THE LENS
EXP EYE RES 15:767-778(1973)

PATERSON CA   PFISTER RR
OCULAR HYPERTENSIVE RESPONSE TO ALKALI BURNS IN THE
MONKEY
EXP EYE RES 17:449-453(1973)

PATILLO RA   HUSSA RO   GARANCIS J
INHIBITION OF STEROID SYNTHESIS BY PROSTAGLANDIN INHIBITORS
IN A HUMAN TROPHOBLASTIC CELL LINE (JAR CELLS)
ABSTR 53RD MEET ENDOCR SOC, SAN FRANC:A-243(1971)

PATRONO C
ESTIMATION OF PICOGRAM QUANTITIES OF PROSTAGLANDIN F2A BY
RADIOIMMUNOASSAY
ABSTR 5TH INT CONGR PHARMACOL, SAN FRANC:177(1972)

PATTILLO RA   HUSSA RO   TERRAGNO NA
STORY MT   MATTINGLY RF
ABSENCE OF PROSTAGLANDIN SYNTHESIS IN THE MALIGNANT HUMAN
TROPHOBLAST IN CULTURE
AM J OBSTET GYNECOL 115:91-94(1973)

PATULEIA MC
THE PROSTAGLANDINS
REV PORT PEDIATR 4:99-117(1973):(PORTUGUESE)

PAULO LG   WILKERSON RD   ROH BL
GEORGE WL   FISHER JW
THE EFFECTS OF PROSTAGLANDIN E1 ON ERYTHROPOIETIN PRODUCTION
PROC SOC EXP BIOL MED 142:771-775(1973)

PAUST J   KONIG H
PROSTAGLANDINS
NACHR CHEM TECH 19:291-294(1971):(GERMAN)

PAWAR SS
STORAGE AND ACTIVITY OF PROSTAGLANDIN-LIKE SUBSTANCE IN BODY
TISSUES OF RATS FED HIGH FAT DIETS
INDIAN J BIOCHEM BIOPHYS 10:71-73(1973)

PEAKE RL   ARAI K   ANDERSON FG
WALLACE JM
MEDULLARY CARCINOMA OF THE THYROID -- EVIDENCE FOR
SECRETION OF VASOACTIVE PROSTAGLANDIN
ABSTR 53RD MEET ENDOCR SOC, SAN FRANC:A-243(1971)

PEARSON J
A MEDICAL MARVEL
DETROIT NEWS:1,16(JUNE 12,1972)

PEASLEE MH
FROG SKIN DARKENING EFFECT OF PROSTAGLANDIN-E1 AND
PROSTAGLANDIN-F2A
AM ZOOL 11:651(1971)

PEERY CV   JOHNSON GS   PASTAN I
ADENYL CYCLASE IN NORMAL AND TRANSFORMED FIBROBLASTS IN
TISSUE CULTURE. ACTIVATION BY PROSTAGLANDINS
J BIOL CHEM 246:5785-5790(1971)

PELLETIER G
ULTRASTRUCTURAL STUDY OF ANTERIOR PITUITARY SECRETION
UNION MED CAN 102:546-550(1973):(FRENCH)

PELOFSKY S   JACOBSON ED   FISHER RG
EFFECTS OF PROSTAGLANDIN E1 ON EXPERIMENTAL CEREBRAL
VASOSPASM
J NEUROSURG 36:634-639(1972)

PENNEY DP   AVERILL K   OLSON J
PROJECTIONS OF ADRENOCORTICAL CELLS INTO SINUSOIDAL LUMINA,
INFLUENCE OF PROSTAGLANDIN E1
AM J ANAT 135:135-140(1972)

PENNEY DP   AVERILL K   OLSON J
MARINETTI GV
EFFECTS OF PROSTAGLANDINS ON THE ADRENAL CORTEX
ANAT REC 175:409(1973)

PENNEY DP   OLSON J   AVERILL K
FINE STRUCTURAL STUDIES OF RAT ADRENAL CORTICES FOLLOWING
PROSTAGLANOIN ADMINISTRATION
Z ZELLFORSCH MIKROSK ANAT 146:297-307(1973)

PENNEY DP   OLSON J   MARINETTI GV
VAALA S   AVERILL K
LOCALIZATION OF TRITIATED PROSTAGLANDIN E1 IN RAT ADRENAL
CORTICES. AUTORADIOGRAPHIC AND BIOCHEMICAL STUDIES
Z ZELLFORSCH MIKROSK ANAT 146: 309-317(1973)

PENNINGTON DG   HYMAN AL   JAQUES WE
PULMONARY VASCULAR RESPONSE TO ENDOTOXIN IN INTACT DOGS
SURGERY 73:246-255(1973)

PENNINGTON DG   HYMAN AL   WOOLVERTON WC
PULMONARY VASCULAR RESPONSES TO SELECTIVE LUNG COOLING IN
INTACT DOGS
PROC SOC EXP BIOL MED 137:1375-1380(1971)

PENNINK M   WHITE RP   CROCKARELL JR
ROBERTSON JT
ROLE OF PROSTAGLANDIN F2A IN GENESIS OF EXPERIMENTAL
CEREBRAL VASOSPASM. ANGIOGRAPHIC STUDY IN DOGS
J NEUROSURG 37:398-406(1972)

PEREZ G   MCGUCKIN J
CELLULAR LOCALIZATION OF PROSTAGLANDIN A2 IN THE RAT KIDNEY
PROSTAGLANDINS 2:393-398(1972)

PEREZ-CRUET J   HAUBRICH D   REID WD
PROSTAGLANDIN E1 (PGE1) INDUCES "PARADOXICAL" SLEEP AND
INCREASES BRAIN ACETYLCHOLINE (ACH) LEVELS AND SEROTONIN
(5-HT) TURNOVER
PHARMACOLOGIST 13:278(1971)

PERKINS ES   UNGER WG   BASS M
THE ROLE OF PROSTAGLANDIN IN THE OCULAR RESPONSES TO LASER
IRRADIATION OF THE IRIS
EXP EYE RES 17:394-395(1973)

PERKLEV T
BLOCKADE OF PROSTAGLANDIN SYNTHESIS BY AN ANTITUMOR DRUG,
ESTRACYT
IN: "PROSTAGLANDINS AND CYCLIC AMP," RH KAHN, WEM LANDS,
ACADEMIC PRESS, NY : 219-221(1973)

PERSAUD TVN
PROSTAGLANDIN E2 EFFECTS ON PLACENTA AND FETUS FOLLOWING
INTRAAMNIOTIC ADMINISTRATION IN RATS
INT RES COMMUN SYST (73-9)8-5-4(1973)

PERSIANINOV LS
THEORY AND PRACTICE OF ANTENATAL CARE OF THE FETUS
VESTN AKAD MED NAUK SSSR 28:61-69(1973):(RUSSIAN)

PERSIANINOV LS   CHERNUKHA EA
THE USE OF PROSTAGLANDIN F2A TO PROVOKE AN ABORTION
AKUSH GINEKOL (MOSK) 49(10):37-44(1973):(RUSSIAN)

PERSIANINOV LS   LEONOV BV   MASSAL'SKAYA LM
LOMOVA MA
EFFECT OF PROSTAGLANDIN F2A ON SEA URCHIN EMBRYOS AT
DIFFERENT STAGES OF DEVELOPMENT AND ON THEIR SENSITIVITY TO
NEUROPHARMACOLOGICAL AGENTS
BULL EXP BIOL MED (USSR) 76:1229-1231(1973)

PERSIANINOV LS   LOMOVA MA   LEONOV BV
PROSTAGLANDINS AND THE CONTRACTILE ACTIVITY OF THE UTERUS
AKUSH GINEKOL (MOSK) 47:3-7(1971):(RUSSIAN)

PERSIANINOV LS   MANUILOVA IA   CHERNUKHA EA
USE OF PROSTAGLANDINS F2A FOR INDUCTION AND STIMULATION OF
LABOUR
ADV BIOSCI (SUPPL)9:98(1972)

PERSIANINOV LS   MANUILOVA IA   CHERNUKHA EA
THE USE OF PROSTAGLANDIN F2A TO INDUCE AND STIMULATE THE
LABOUR PAINS
AKUSH GINEKOL (MOSK) 48:3-8(1972):(RUSSIAN)

PERSSON NA   HEDQVIST P
REDUCED INTESTINAL MUSCULAR RESPONSE TO ADRENERGIC NERVE
STIMULATION AFTER THE ADMINISTRATION OF PROSTAGLANDINS
ACTA PHYSIOL SCAND (SUPPL 396) : 108(1973)

PESCETTO G
PROBLEMS ASSOCIATED WITH THE USE OF CONTRACEPTIVES
CLIN TER (SUPPL)65:63-85(1973):(ITALIAN)

PESCLE C   SASSO GF   MASTROBERARDINO G
CHIARIELLO M
EXPERIMENTAL OBSERVATION ON THE EFFECT OF PGE1 ON
ERYTHROPOIETIC ACTIVITY
BLOOD 42:481-482(1973)

PESKIN GW   BEKOE S   KAPLAN EL
THE RELATIONSHIP OF GASTROINTESTINAL TUMORS TO DIARRHEA
SURG CLIN NORTH AM 51:233-239(1971)

PETERS HD   WESTHOFEN P   KARZEL K
REGULATION OF MUCOPOLYSACCHARIDE SYNTHESIS IN FIBROBLAST
TISSUE CULTURES
NAUNYN SCHMIEDEBERGS ARCH PHARMACOL 277(SUPPL):R54(1973)

PFEIFFER CJ   LEWANDOWSKI LG
INFLUENCE OF PROSTAGLANDIN AND ASPIRIN ON GASTRIC SECRETION
IN HAMSTERS
LEBER MAGEN DARM 2:142-144(1972):(GERMAN)

PHAM-HUU-CHANH -   NICOT-CARTHERY MC
PROSTAGLANDIN-RELEASING EFFECT OF EXOGENOUS NORADRENALINE
IN ISOLATED RABBIT HEART
NATURWISSENSCHAFTEN 60:482(1973)

PHANG JM   DOWNING SJ
AMINO ACID TRANSPORT IN BONE: STIMULATION BY CYCLIC AMP
AM J PHYSIOL 224:191-196(1973)

PHARRISS BB
THE EFFECT OF PROSTAGLANDINS ON LUTEAL STEROIDOGENESIS
IN: "HORMONAL STEROIDS," PROC 3RD INT CONGR, HAMB, INT CONGR
SER 219, EXCERPTA MEDICA, AMST:680-684(1971)

PICKARD JD
THE MECHANISM OF ACTION OF PROSTAGLANDIN F2A ON CEREBRAL
FLOW IN THE BABOON
J PHYSIOL (LOND) 234:46P-47P(1973)

PICKARD JD   MACKENZIE ET
INHIBITION OF PROSTAGLANDIN SYNTHESIS AND THE RESPONSE OF
BABOON CEREBRAL CIRCULATION TO CARBON DIOXIDE
NATURE(NEW BIOL) V 245:187-188(1973)

PICKENS J   WEST GB   WHELAN CJ
RELEASE BY KININ OF A SUBSTANCE CONTRACTING RABBIT AORTA
(RCS) FROM GUINEA-PIG LUNG
BR J PHARMACOL 45:140P(1972)

PICKLES VR
PROSTAGLANDINS AND ASPIRIN
NATURE (LOND) 239:33-34(1972)

PICKLES VR
PROSTAGLANDINS AND CONTRACEPTION
RES REPROD 5:1-2(1973)

PIERCE NF   CARPENTER CCJ JR   ELLIOT HL
GREENOUGH WB III
EFFECTS OF PROSTAGLANDINS, THEOPHYLLINE, AND CHOLERA
EXOTOXIN UPON TRANSMUCOSAL WATER AND ELECTROLYTE MOVEMENT
IN THE CANINE JEJUNUM
GASTROENTEROLOGY 60:22-32(1971)

PIKE JE
PROSTAGLANDINS
SCI AM 225(5):84-92(1971)

PIKE JE
RECENT ADVANCES IN THE CHEMISTRY OF PROSTAGLANDINS
IN: "PROSTAGLANDINES 1973," INSERM, PARIS:103-111(1973)

PION RJ   HALE RW   REICH L
VAGINAL ADMINISTRATION OF PROSTAGLANDINS AND EARLY ABORTION
J REPROD MED 9:413-415(1972)

PIPER P   VANE J
THE RELEASE OF PROSTAGLANDIN FROM LUNG AND OTHER TISSUES
ANN NY ACAD SCI 180:363-385(1971)

PIPER PJ
DISTRIBUTION AND METABOLISM
IN: "THE PROSTAGLANDINS, PHARMACOLOGICAL AND THERAPEUTIC
ADVANCES," MF CUTHBERT, HEINEMANN, LOND:125-150(1973)

PIPER PJ
SUBSTANCES RELEASED FROM PASSIVELY SENSITIZED HUMAN LUNG
TISSUE DURING CHALLENGE
INT ARCH ALLERGY APPL IMMUNOL 45:87-90(1973)

PIPKIN FB   MOTT JC   ROBERTON NRC
ANGIOTENSIN II-LIKE ACTIVITY IN CIRCULATING ARTERIAL BLOOD
IN IMMATURE AND ADULT RABBITS
J PHYSIOL (LOND) 218: 385-403 (1971)

PITCOCK JA
SYNTHESIS OF RENOMEDULLARY PROSTAGLANDINS (PG'S) BY
RENOMEDULLARY INTERSTITIAL CELLS (RIC) GROWN IN TISSUE
CULTURE
ADV BIOSCI (SUPPL)9:54(1972)

PLAUT M   LICHTENSTEIN LM   GILLESPIE E
HENNEY CS
STUDIES ON THE MECHANISM OF LYMPHOCYTE-MEDIATED CYTOLYSIS-
IV-SPECIFICITY OF THE HISTAMINE RECEPTOR ON EFFECTOR T CELLS
J IMMUNOL 111:389-394(1973)

PLISHKER GA   GREEN JW
ENHANCEMENT OF CALCIUM STIMULATED ATPASE ACTIVITIES OF
HUMAN ERYTHROCYTE GHOSTS BY PROSTAGLANDIN E1
FED PROC 32:803(1973)

PODOS SM   BECKER B   KASS MA
PROSTAGLANDIN SYNTHESIS, INHIBITION, AND INTRAOCULAR
PRESSURE
INVEST OPHTHALMOL 12:426-433(1973)

PODOS SM   BECKER B   KASS MA
INDOMETHACIN BLOCKS ARACHIDONIC ACID-INDUCED ELEVATION OF
INTRAOCULAR PRESSURE
PROSTAGLANDINS 3:7-16(1973)

PODOS SM   JAFFE BM   BECKER B
PROSTAGLANDINS AND GLAUCOMA
BR MED J 4:232(1972)

POLET H   LEVINE L
SERUM PROSTAGLANDIN A1 ISOMERASE.
BIOCHEM BIOPHYS RES COMMUN 45:(5)1169-1176(1971)

POLETTI CE   WEPSIC JG   SWEET WH
MIDDLE CEREBRAL ARTERIAL SPASM FROM SUBARACHNOID BLOOD:
SPASMOLYSIS WITH TOPICAL USE OF NITROGLYCERIN
SURG FORUM 23:449-450(1972)

POLGAR P   EMERSON CP JR   RUTENBURG AM
A STUDY OF THE CYCLIC AMP SYSTEM IN LEUKEMIA
PROC AM ASSOC CANCER RES 13:119(1972)

POLGAR P   VERA JC   DESFORGES J
RUTENBURG AM
CYCLIC AMP RESPONSE TO PROSTAGLANDIN STIMULATION IN NORMAL
AND LEUKEMIC HUMAN LYMPHOCYTES
FED PROC 32:881(1973)

POLIS BD   GRANDIZIO AM   POLIS E
SOME IN VITRO AND IN VIVO EFFECTS OF A NEW PROSTAGLANDIN
DERIVATIVE
ADV EXP MED BIOL 33:213-220(1972)

POLIS BD   POLIS E   KWONG S
WYETH J
REACTIVATION OF OXIDATIVE PHOSPHORYLATION IN DEGENERATE
MITOCHONDRIA BY A STABLE FREE RADICAL PROSTAGLANDIN
DERIVATIVE
ABSTR, 9TH INT CONG BIOCHEM, STOCKH:405(1973)

POMENBURG R   BROSENS I   ROBERTSON WB
PROSTAGLANDIN (F2A) INDUCED HISTOLOGICAL CHANGES IN
CONCEPTUSES AND DECIDUOMATA OF PREGNANT GOLDEN HAMSTERS
INT RES COMMUN SYST (73-12)8-5-9(1973)

POSNER J
THE ACTION OF PROSTAGLANDINS ON THE BOVINE SPHINCTER
PUPILLAE
J PHYSIOL (LOND) 217:25-26P(1971)

POSNER J
PROSTAGLANDIN E2 AND THE BOVINE SPHINCTER PUPILLAE
BR J PHARMACOL 49:415-427(1973)

POSTNOV YV
ON THE ROLE OF SALT REGIME IN THE DEVELOPMENT OF ARTERIAL
HYPERTENSION
KARDIOLOGIIA 12(7):5-12(1972):(RUSSIAN)

POTTS WJ   EAST PF
THE EFFECT OF PROSTAGLANDIN E2 ON CONDITIONED AVOIDANCE
RESPONSE PERFORMANCE IN RATS.
ARCH INT PHARMACODYN THER 191:74-79(1971)

POTTS WJ   EAST PF
EFFECTS OF PROSTAGLANDINS AND PROSTAGLANDIN PRECURSORS ON
THE CONDITIONED AVOIDANCE RESPONSE (CAR) IN RATS.
PHARMACOLOGIST 13:292(1971)

POTTS WJ   EAST PF
EFFECTS OF PROSTAGLANDIN E2 ON THE BODY TEMPERATURE OF
CONSCIOUS RATS AND CATS
ARCH INT PHARMACODYN THER 197:31-36(1972)

POTTS WJ   EAST PF
THE EFFECTS OF PROSTAGLANDIN E2 ON THE CONDITIONED AVOIDANCE
RESPONSE IN THE RHESUS MONKEY
ABSTR 5TH INT CONGR PHARMACOL, SAN FRANC:185(1972)

POTTS WJ   EAST PF   LANDRY D
THE EFFECTS OF PROSTAGLANDIN E2 ON RAT CONDITIONED AVOIDANCE
RESPONSE (CAR) PERFORMANCE AND ELECTROENCEPHALOGRAM (EEG)
WHEN GIVEN BY THE INTRAVENTRICULAR ROUTE
FED PROC 31:546(1972)

POTTS WJ   EAST PF   LANDRY D
THE EFFECTS OF PROSTAGLANDIN E1 ON CONDITIONED AVOIDANCE
RESPONSE BEHAVIOR AND THE ELECTROENCEPHALOGRAM
ADV BIOSCI (SUPPL)9:78(1972)

POTTS WJ   EAST PF   LANDRY D
DIXON JP
THE EFFECTS OF PROSTAGLANDIN E2 ON CONDITIONED AVOIDANCE
RESPONSE BEHAVIOR AND THE ELECTROENCEPHALOGRAM
ADV BIOSCI 9:489-494(1973)

POWELL JA   DUELL EA   VOORHEES JJ
BETA-ADRENERGIC STIMULATION OF ENDOGENOUS EPIDERMAL CYCLIC
AMP FORMATION
ARCH DERMATOL 104:359-365(1971)

POWELL JR.   BRODY MJ
ENHANCEMENT OF REFLEX VASOCONSTRICTION BY PROSTAGLANDIN F2A
(PGF2A)
PHARMACOLOGIST 15:208(1973)

POWELL JR   BRODY MJ
PERIPHERAL FACILITATION OF REFLEX VASOCONSTRICTION BY
PROSTAGLANDIN F2A
J PHARMACOL EXP THER 187:495-500(1973)

POWELL W   HAMMARSTROM S   SAMUELSSON B
A PROSTAGLANDIN RECEPTOR IN OVINE CORPUS LUTEUM
ABSTR, 9TH INT CONG BIOCHEM, STOCKH:404(1973)

POWLES TJ   EASTY DM   EASTY GC
BONDY PK   MUNRO-NEVILLE A
ASPIRIN INHIBITION OF IN VITRO OSTEOLYSIS STIMULATED BY
PARATHYROID HORMONE AND PGE1
NATURE (NEW BIOL) 245: 83-84 (1973)

POWLES TJ   EASTY DM   EASTY GC
NEVILLE AM
TUMOUR-INDUCED OSTEOLYSIS
LANCET 2:504(1973)

POYSER NL
THE FORMATION OF PROSTAGLANDINS BY THE GUINEA-PIG UTERUS AND
THE EFFECT OF INDOMETHACIN
ADV BIOSCI (SUPPL)9:104(1972)

POYSER NL
THE ROLE OF PROSTAGLANDINS IN FERTILITY CONTROL
BIOCHEM SOC TRANS 1:535-539(1973)

POYSER NL
THE PHYSIOLOGY OF PROSTAGLANDINS
CLIN ENDOCRINOL METAB 2:393-410(1973)

PRASAD KN
CYCLIC AMP AND THE DIFFERENTIATION OF NEUROBLASTOMA CELL
CULTURE
PROC AM ASSOC CANCER RES 13:16(1972)

PRASAD KN
MORPHOLOGICAL DIFFERENTIATION INDUCED BY PROSTAGLANDIN IN
MOUSE NEUROBLASTOMA CELLS IN CULTURE
NATURE (NEW BIOL) 236:49-52(1972)

PRASAD KN
NEUROBLASTOMA CLONES: PROSTAGLANDIN VERSUS DIBUTYRYL CYCLIC
AMP, 8-BENZYLTHIO-CYCLIC AMP, PHOSPHODIESTERASE INHIBITORS
AND X-RAYS
PROC SOC EXP BIOL MED 140:126-129(1972)

PRASAD KN
RADIOPROTECTIVE EFFECT OF PROSTAGLANDIN AND AN INHIBITOR OF
CYCLIC NUCLEOTIDE PHOSPHODIESTERASE ON MAMMALIAN CELLS
IN CULTURE
INT J RADIAT BIOL 22:187-189(1972)

PRASAD KN
CYCLIC AMP-INDUCED DIFFERENTIATED MOUSE NEUROBLASTOMA CELLS
LOSE TUMOURGENIC CHARACTERISTICS
CYTOBIOS 6:163-166(1972)

PRASAD KN   GILMER K   KUMAR S
MORPHOLOGICALLY "DIFFERENTIATED" MOUSE NEUROBLASTOMA CELLS
INDUCED BY NONCYCLIC AMP AGENTS: LEVELS OF CYCLIC AMP,
NUCLEIC ACID AND PROTEIN
PROC SOC EXP BIOL MED 143:1168-1171(1973)

PRASAD KN   KUMAR S
CYCLIC AMP AND THE DIFFERENTIATION OF MOUSE NEUROBLASTOMA
CELLS IN CULTURE
ABSTR MEET CONTROL OF PROLIFERATION OF ANIMAL CELLS, COLD
SPRING HARBOR:39(1973)

PRASAD KN   KUMAR S
CYCLIC 3',5'-AMP PHOSPHODIESTERASE ACTIVITY DURING CYCLIC
AMP-INDUCED DIFFERENTIATION OF NEUROBLASTOMA CELLS IN
CULTURE
PROC SOC EXP BIOL MED 142:406-409(1973)

PRASAD KN   KUMAR S
LEVELS OF CYCLIC AMP AND CYCLIC AMP PHOSPHODIESTERASE DURING
DIFFERENTIATION OF MOUSE NEUROBLASTOMA CELLS IN CULTURE.
PROC AM ASSOC CANCER RES 14:6(1973)

PRASAD KN   MANDAL B
CHOLINE ACETYLTRANSFERASE LEVEL IN CYCLIC AMP-INDUCED
MORPHOLOGICALLY DIFFERENTIATED NEUROBLASTOMA CELLS IN
CULTURE
J CELL BIOL 55:206A(1972)

PRASAD KN   MANDAL B
CATECHOL-O-METHYL-TRANSFERASE ACTIVITY IN DIBUTYRYL
CYCLIC AMP, PROSTAGLANDIN AND X-RAY-INDUCED DIFFERENTIATED
NEUROBLASTOMA CELL CULTURE
EXP CELL RES 74:532-534(1972)

PRASAD KN   MANDAL B
CHOLINE ACETYLTRANSFERASE LEVEL IN CYCLIC AMP AND X-RAY
INDUCED MORPHOLOGICALLY DIFFERENTIATED NEUROBLASTOMA CELLS
IN CULTURE
CYTOBIOS 8:75-80(1973)

PRASAD KN   MANDAL B   WAYMIRE JC
LEES GJ   VERNADAKIS A   WEINER N
BASAL LEVEL OF NEUROTRANSMITTER SYNTHESIZING ENZYMES AND
EFFECT OF CYCLIC AMP AGENTS ON THE MORPHOLOGICAL
DIFFERENTIATION OF ISOLATED NEUROBLASTOMA CLONES
NATURE (NEW BIOL) 241:117-119(1973)

PRASAD KN   SHEPPARD JR
NEUROBLASTOMA CELL CULTURE: MEMBRANE CHANGES DURING CYCLIC
AMP-INDUCED MORPHOLOGICAL DIFFERENTIATION
PROC SOC EXP BIOL MED 141:240-243(1972)

PRASAD N
EFFECT OF CYTOCHALASIN B AND VINBLASTINE ON X-RAY, DIBUTYRYL
CYCLIC AMP AND PROSTAGLANDIN-INDUCED DIFFERENTIATION OF
MOUSE NEUROBLASTOMA CELL CULTURE
CYTOBIOLOGIE 5:265-271(1972)

PREZYNA A   ATTALLAH A   VANCE K
SCHOOLMAN M   LEE J
THE RENOMEDULLARY BODY A NEWLY RECOGNIZED STRUCTURE OF
RENOMEDULLARY INTERSTITIAL CELL ORIGIN ASSOCIATED WITH HIGH
PROSTAGLANDIN CONTENT
PROSTAGLANDINS 3:669-678(1973)

PRIANO LL   MILLER TH   WILSON RD
TRABER DL
TREATMENT OF HYPOVOLEMIC SHOCK WITH PROSTAGLANDIN E1
FED PROC 31:817(1972)

PRINCE A   ALVAREZ FS   YOUNG J
PREPARATION OF PROSTAGLANDINS A2, E2 AND 11-EPI-E2 FROM
THEIR 15 ACETATE METHYL ESTERS USING AN ENDOGENOUS ENZYME
SYSTEM PRESENT IN PLEXAURA HOMOMALLA (ESPER)
PROSTAGLANDINS 3:531-535(1973)

PRIVETT OS   PHILLIPS F   FUKAZAWA T
KALTENBACH CC   SPRECHER HW
STUDIES ON THE RELATIONSHIP OF THE SYNTHESIS OF
PROSTAGLANDINS TO THE BIOLOGICAL ACTIVITY OF ESSENTIAL FATTY
ACIDS
BIOCHIM BIOPHYS ACTA 280:348-355(1972)

PUGLISI L
OPPOSITE EFFECTS OF PROSTAGLANDINS E AND F ON TRACHEAL
SMOOTH MUSCLES AND THEIR INTERACTION WITH CALCIUM IONS
ADV BIOSCI (SUPPL)9:38(1972)

PUGLISI L
OPPOSITE EFFECTS OF PROSTAGLANDINS E AND F ON TRACHEAL
SMOOTH MUSCLES AND THEIR INTERACTION WITH CALCIUM IONS
ADV BIOSCI 9:219-227(1973)

PUGLISI L   BERTI F   FOLCO GG
CYCLIC-GMP INTERACTION WITH THE PARASYMPATHETIC SYSTEM OF
ISOLATED RAT STOMACH
PHARMACOL RES COMMUN 4:227-235(1972)

PURI CP   HINGORANI V   LAUMAS KR
EFFECT OF PROSTAGLANDIN F2A ON PROGESTIN SYNTHESIS IN THE
HUMAN AND THE RABBIT OVARY
ADV BIOSCI (SUPPL)9:109(1972)

QUAGLIATA F   LAWRENCE VJW   PHILLIPS-QUAGL/ JM
SELECTIVE AND SYNERGISTIC EFFECTS OF PROSTAGLANDIN (PG) E1
AND OF PROCARBAZINE ON VARIOUS PARAMETERS OF THE IMMUNE
RESPONSE IN THE MOUSE
FED PROC 31:799(1972)

QUAGLIATA F   LAWRENCE VJW   PHILLIPS-QUAGL/ JM
PROSTAGLANDIN E1 AS A REGULATOR OF LYMPHOCYTE FUNCTION.
SELECTIVE ACTION ON B LYMPHOCYTES AND SYNERGY WITH
PROCARBAZINE IN DEPRESSION OF IMMUNE RESPONSES
CELL IMMUNOL 6:457-465(1973)

QUILLIGAN EJ
MATERNAL FACTORS INFLUENCING THE ONSET OF LABOR
CLIN OBSTET GYNECOL 16:150-158(1973)

RABINOWITZ I   JOHNSON M   WOLF PL
PROSTAGLANDIN INTERACTION WITH THE SICKLE ERYTHROCYTE
MEMBRANE
FED PROC 32:803(1973)

RABINOWITZ I   RAMWELL P   DAVISON P
CONFORMATION OF PROSTAGLANDINS
NATURE (NEW BIOL) 233:88-90(1971)

RABITO CA   FASCIOLO JC
EFFECTS OF PROSTAGLANDINS PGE1 AND PGF2A ON OXYGEN
CONSUMPTION, SODIUM AND POTASSIUM CONTENT OF RENAL TISSUE
EXPERIENTIA 29:673-675(1973)

RACHOW TE
17ALPHA-ACETOXY-11BETA-METHYL-19-NORPREG-4-ENE-3,20-DIONE
(SC-21009) AND PROSTAGLANDIN F2A (PGF2A) ON THE VESICULAR
FOLLICLES AND CORPORA LUTEA IN BOVINE OVARIES.
DISS, SOUTH ILL UNIV (1973)

RADAWSKI D   VEENENDAAL M   DAUGHERTY R JR
EMERSON T JR
EFFECTS OF PROSTAGLANDIN E1 (PGE1) ON THE CANINE CEREBRAL
CIRCULATION
FED PROC 32:411(1973)

RADMANOVIC BZ
THE EFFECT OF PROSTAGLANDIN E1 ON THE PERISTALTIC REFLEX OF
THE ISOLATED GUINEA-PIG ILEUM
PROC INT UNION PHYSIOL SCI, 25TH INT CONGR, MUNICH,
9:464(1971)

RADMANOVIC BZ
EFFECT OF PROSTAGLANDIN E1 ON THE PERISTALTIC ACTIVITY OF
THE GUINEA-PIG ISOLATED ILEUM
ARCH INT PHARMACODYN THER 200:396-404(1972)

RADZIALOWSKI FM   NOVAK L
REVERSAL OF ANTILIPOLYTIC EFFECT OF PROSTAGLANDIN E2 BY AN
OXAZEPINE DERIVATIVE (SC-19220)
LIFE SCI (1)10:1261-1265(1971)

RADZIALOWSKI FM   ROSENBERG LN
EFFECT OF SC-19220, A PROSTAGLANDIN INHIBITOR ON THE
ANTILIPOLYTIC ACTION OF PROSTAGLANDIN E2, PROPRANALOL AND
INSULIN IN THE ISOLATED RAT ADIPOCYTE
LIFE SCI (II) 12:337-343(1973)

RAFLO GT   WANGENSTEEN SL   GLENN TM
LEFER AM
MECHANISM OF THE PROTECTIVE EFFECTS OF PROSTAGLANDINS E1
AND F2A IN CANINE ENDOTOXIN SHOCK
EUR J PHARMACOL 24:86-95 (1973)

RAGGATT PR   ENGEL LL   SYMINGTON T
FATTY ACID COMPOSITION OF THE STEROL ESTER FRACTION OF HUMAN
ADRENAL CORTEX IN CUSHING'S SYNDROME AND AFTER TREATMENT
WITH ANIMOGLUTETHIMIDE
LIPIDS 7:474-482(1972)

RAGNI MV   PREUSS HG
PGE1 STIMULATION OF 3H-THYMIDINE INCORPORATION INTO RENAL
SLICE DNA
CLIN RES 19:545(1971)

RAISZ LG   TRUMMEL CL   SIMMONS H
INDUCTION OF BONE RESORPTION IN TISSUE CULTURE: PROLONGED
RESPONSE AFTER BRIEF EXPOSURE TO PARATHYROID HORMONE OR
25-HYDROXYCHOLECALCIFEROL
ENDOCRINOLOGY 90:744-751(1972)

RAISZ LG   TRUMMEL CL   WENER JA
SIMMONS H
EFFECT OF GLUCOCORTICOIDS ON BONE RESORPTION IN TISSUE
ENDOCRINOLOGY 90:961-967(1972)

RALL HJS
PROSTAGLANDINS: A SYNOPTIC STUDY
S AFR MED J 46:468-471(1972)

RALL TW
THE METABOLISM AND FUNCTION OF CYCLIC AMP IN THE CENTRAL
NERVOUS SYSTEM
IN: "PROSTAGLANDINS AND CYCLIC AMP," RH KAHN, WEM LANDS,
ACADEMIC PRESS, NY : 57-71(1973)

RAMASWAMY AS
PROSTAGLANDIN ANTAGONISM BY QUERCETIN
ABSTR 7TH WORLD CONGR FERTIL STERIL, INT CONGR SER 234A,
EXCERPTA MEDICA, AMST:11(1971)

RAMASWAMY AS
PROSTAGLANDIN ANTAGONISM (PGE1) BY DEUTERIUM OXIDE (D2O)
ADV BIOSCI (SUPPL)9:24(1972)

RAMWELL P   SHAW J
THE BIOLOGICAL SIGNIFICANCE OF THE PROSTAGLANDINS
ANN NY ACAD SCI 180:10-13(1971)

RAMWELL PW
PROSTAGLANDIN RELEASE AND METABOLISM DURING DIFFERENT
PHYSIOLOGICAL STATES
OFFICE OF NAVAL RES AD REPORT 767385 FEBRUARY 28 (1973)

RAMWELL PW   KURY P
PROSTAGLANDINS & MEMBRANES
IN: "PROSTAGLANDINES 1973," INSERM, PARIS:43-54(1973)

RAMWELL PW   RABINOWITZ I
INTERACTION OF PROSTAGLANDINS AND CYCLIC AMP
IN: "EFFECTS OF DRUGS ON CELLULAR CONTROL MECHANISMS," BR
RABIN AND RB FREEDMAN, UNIV PARK PRESS, BALT:207-235(1972)

RANGACHARI PN
PROSTAGLANDINS
J SCI IND RES 31:26-35(1972)

RAO KS   RENO FE   MCCONNELL RG
SUBACUTE TOXICITY STUDIES WITH PROSTAGLANDIN E1 (PGE1 IN
LABORATORY ANIMAL SPECIES)
TOXICOL APPL PHARMACOL 25:492(1973)

RAO VSN   SHARMA PL
EFFECT OF (+)-INPEA ON CONTRACTIONS OF THE ISOLATED RAT
UTERUS EVOKED BY PROSTAGLANDINS
J REPROD FERTIL 34:523-525(1973)

RASHID S
THE RELEASE OF PROSTAGLANDIN FROM THE OESOPHAGUS AND THE
STOMACH OF THE FROG (RANA TEMPORARIA)
J PHARM PHARMACOL 23:456-457(1971)

RATNER A   WILSON MC   PEAKE GT
ANTAGONISM OF PROSTAGLANDIN-PROMOTED PITUITARY CYCLIC AMP
ACCUMULATION AND GROWTH HORMONE SECRETION IN VITRO BY
7-OXA-13-PROSTYNOIC ACID
PROSTAGLANDINS 3:413-419(1973)

RATTAN S   GOYAL RK
EVIDENCE FOR CYCLIC 3',5'-ADENOSINE MONOPHOSPHATE (CAMP)
PARTICIPATION IN THE LOWER ESOPHAGEAL SPHINCTER (LES)
RELAXATION
CLIN RES 21:90(1973)

RATTAN S   HERSH T   GOYAL RK
EFFECTS OF PROSTAGLANDINS ON THE LOWER ESOPHAGEAL SPHINCTER
CLIN RES 19:660(1971)

RATTAN S   HERSH T   GOYAL RK
EFFECT OF PROSTAGLANDIN F2A AND GASTRIN PENTAPEPTIDE ON THE
LOWER ESOPHAGEAL SPHINCTER
PROC SOC EXP BIOL MED 141:573-575(1972)

RAVENHOLT RT   SPEIDEL JJ
THE 'PENICILLIN' OF CONTRACEPTION MAY BE HERE
MED OPIN 7(5):66-71(1971)

RAVINA A   RAVINA JH
THE PROSTAGLANDINS. HAS THE YEAR 1970 WITNESSED THE BIRTH OF
A MIRACLE MEDICATION?
PRESSE MED 79(4):139-140(1971);(FRENCH)

RAWLINS MD   LUFF RH   CRANSTON WI
REGIONAL BRAIN SALICYLATE CONCENTRATIONS IN AFEBRILE AND
FEBRILE RABBITS
BIOCHEM PHARMACOL 22:2639-2642(1973)

RAZ A
METABOLITES FORMED AFTER INTRAVENOUS ADMINISTRATION OF FREE
OR ALBUMIN-BOUND PROSTAGLANDIN E2 IN THE RAT
FEBS LETT 27:245-247(1972)

RAZ A
INTERACTION OF PROSTAGLANDINS WITH BLOOD PLASMA PROTEINS
I. BINDING OF PROSTAGLANDIN E2 TO HUMAN PLASMA PROTEINS AND
ITS EFFECT ON THE PHYSIOLOGICAL ACTIVITY OF PROSTAGLANDIN E2
IN VITRO AND IN VIVO
BIOCHIM BIOPHYS ACTA 280:602-613(1972)

RAZ A
INTERACTION OF PROSTAGLANDINS WITH BLOOD PLASMA PROTEINS
COMPARATIVE BINDING OF PROSTAGLANDINS A2,F2A AND E2 TO
HUMAN PLASMA PROTEINS
BIOCHEM J 130:631-636(1972)

RAZ A
INTERACTION OF PROSTAGLANDINS WITH BLOOD PLASMA PROTEINS.
III. RATE OF DISAPPEARANCE AND METABOLITES FORMATION AFTER
INTRAVENOUS ADMINISTRATION OF FREE OR ALBUMIN-BOUND
PROSTAGLANDINS F2A AND A2.
LIFE SCI(II) 11:965-974(1972)

RAZ A
INHIBITION OF PROSTAGLANDIN SYNTHESIS BY ANTIINFLAMMATORY
DRUGS: DIFFERENTIAL EFFECTS OF INDOMETHACIN, ASPIRIN AND
FLUFENAMIC ACID ON SHEEP SEMINAL VESICLES MICROSOMAL
PROSTAGLANDIN SYNTHETASE AND ON SEMINAL VESICLES SLICES
ABSTR, 9TH INT CONG BIOCHEM, STOCKH:405(1973)

RAZ A   KENIG-WAKSHAL R
RADIOIMMUNOASSAY OF THE E PROSTAGLANDINS: SPECIFICITY OF
PGE1 AND PGE2 ANTISERA AGAINST PROSTAGLANDINS E, F, A AND B.
ISR J MED SCI 9:552-553(1973)

RAZ A   STYLOS WA
SPECIFICITY OF PROSTAGLANDIN A1 ANTISERUM AGAINST
PROSTAGLANDINS A1 AND B1
FEBS LET 30:21-24(1973)

REEDER DD   BECKER HD   THOMPSON JC
EFFECT OF PROSTAGLANDIN E1 ON FOOD STIMULATED GASTRIN AND
GASTRIC SECRETION IN DOGS
PHYSIOLOGIST 15:246(1972)

REEVES JT   GROVER RF
BLOCKADE OF ACUTE HYPOXIC PULMONARY HYPERTENSION BY
ENDOTOXIN
CIRCULATION (SUPPL IV) 8:IV-133(1973)

REGOLI D   PARK WK   RIOUX F
II. PHARMACOLOGY OF ANGIOTENSIN ANTAGONISTS
CAN J PHYSIOL PHARMACOL 51:114-121(1973)

REIMERS HJ   PACKHAM MA   KINLOUGH-RATHB/ RL
MUSTARD FJ
EFFECT OF REPEATED TREATMENT OF RABBIT PLATELETS WITH LOW
CONCENTRATIONS OF THROMBIN ON THEIR FUNCTION, METABOLISM AND
SURVIVAL
BR J HAEMATOL 25:675-689(1973)

RETTORI V   GIMENO M   COUTINHO EM
MOTILITY OF THE GUINEA PIG FALLOPIAN TUBE
ACTA PHYSIOL LAT AM 23:ABSTR 323(1973);(PORTUGESE)

REVAZ C
PROSTAGLANDINS AND THE PHYSIOLOGY OF REPRODUCTION
REV MED SUISSE ROMANDE 92:797-806(1972);(FRENCH)

REWERSKI W   GUMULKA W
PROSTAGLANDINS AND THE CENTRAL NERVOUS SYSTEM
POL TYG LEK 28:1286-1288(1973);(POLISH)

RICHELSON E
STIMULATION OF TYROSINE HYDROXYLASE ACTIVITY IN A CLONE OF
MOUSE NEUROBLASTOMA BY DIBUTYRYL CYCLIC AMP (DBCAMP)
ABSTR 5TH INT CONGR PHARMACOL, SAN FRANC:193(1972)

RICHENS A
DEVELOPMENTS IN DRUG THERAPY
NURS MIRROR 137:30-32(1973)

RICO JMGT   FERREIRA JMC   CRAVO AC
EFFECT OF INDOMETHACIN ON THE INCREASED CAPILLARY
PERMEABILITY INDUCED BY CERTAIN PHLOGISTIC SUBSTANCES
C R SOC BIOL 166:1569-1572(1972);(FRENCH)

RICO JMGT   RICO JT   FERREIRA JMC
CRAVO AC
THE EFFECT OF PROSTAGLANDINS PGE1, PGE2, AND PGF2A ON THE
VASCULAR PERMEABILITY OF THE RAT
C R SOC BIOL 167:1315-1318(1973);(FRENCH)

RIGLER GL   RIEDESEL ML
IN VIVO RESPONSES OF OVARIAN CYCLIC ADENOSINE MONOPHOSPHATE
TO PROSTAGLANDIN E, LUTEINIZING AND FOLLICLE STIMULATING
HORMONE IN RATS
FED PROC 32:214(1973)

RIGON-MACRI P   MACRI I
INHIBITORY EFFECT OF PROSTAGLANDIN E1 ON THE RESPIRATION OF
RAT CEREBRAL CORTICAL SLICES IN COMBINATION WITH THE ACTION
OF HEPARIN
BOLL SOC ITAL BIOL SPER 48:862-864(1972);(ITALIAN)

RING A
CLINICAL EXPERIENCE WITH PROSTAGLANDIN F2A FOR INDUCTION OF
LABOR
J REPROD MED 9:311-313(1972)

RIOUX F   PARK WK   REGOLI D
APPLICATION OF DRUG-RECEPTOR THEORIES TO ANGIOTENSIN
CAN J PHYSIOL PHARMACOL 51:665-672(1973)

RIPOCHE P   BOURGUET J   PARISI M
THE EFFECT OF HYPERTONIC MEDIA ON WATER PERMEABILITY OF FROG
URINARY BLADDER. INHIBITION BY CATECHOLAMINES AND
PROSTAGLANDIN E1
J GEN PHYSIOL 61:110-124(1973)

RIVKIN I   BECKER EL
POSSIBLE IMPLICATION OF CYCLIC 3'5'-ADENOSINE MONOPHOSPHATE
IN THE CHEMOTAXIS OF RABBIT PERITONEAL POLYMORPHONUCLEAR
LEUKOCYTES (PMN'S)
FED PROC 31:657(1972)

ROBAYE B   DEBY C   BARAC G
ULTRAFILTRATION OF PLASMA PROSTAGLANDINS ACTING ON THE
SMOOTH MUSCLE, AND PROSTAGLANDINURIA IN THE DOG
ARCH INT PHYSIOL BIOCHIM 80:616(1972);(FRENCH)

ROBERT A
DUODENAL ULCERS IN THE RAT: PRODUCTION AND PREVENTION
IN: "PEPTIC ULCER," CJ PFEIFFER, MUNKSGAARD, COPENH:21-33
(1971)

ROBERT A
PROSTAGLANDINS AND THE DIGESTIVE SYSTEM
IN: "PROSTAGLANDINES 1973," INSERM, PARIS:297-315(1973)

ROBERT A   LANCASTER C   NEZAMIS JE
BADALAMENTI JN
A GASTRIC ANTISECRETORY AND ANTIULCER PROSTAGLANDIN WITH
ORAL AND LONG-ACTING ACTIVITY
GASTROENTEROLOGY 64:790(1973)

ROBERT A   MAGERLEIN BJ
15-METHYL PGE2 AND 16,16-DIMETHYL PGE2: POTENT INHIBITORS
OF GASTRIC SECRETION
ADV BIOSCI 9:247-253(1973)

ROBERT A   NEZAMIS JE   LANCASTER C
15-METHYL-PROSTAGLANDIN E2: A POTENT INHIBITOR OF GASTRIC
SECRETION
ADV BIOSCI (SUPPL)9:42(1972)

ROBERT A   STANDISH WL
PRODUCTION OF DUODENAL ULCERS, IN RATS, WITH ONE INJECTION
OF HISTAMINE
FED PROC 32:322(1973)

ROBERT A   STOWE DF   NEZAMIS JE
PREVENTION OF DUODENAL ULCERS BY ADMINISTRATION OF
PROSTAGLANDIN E2 (PGE2)
SCAND J GASTROENTEROL 6:303-305(1971)

ROBERTS G   GOMERSALL R   ADAMS M
TURNBULL AC
THERAPEUTIC ABORTION BY TRANSAMNIOTIC INJECTION OF
PROSTAGLANDINS
ADV BIOSCI (SUPPL)9:92(1972)

ROBERTSON AL   KHAIRALLAH PA
EFFECTS OF ANGIOTENSIN II AND SOME ANALOGUES ON VASCULAR
PERMEABILITY IN THE RABBIT
CIRC RES 31:923-931(1972)

ROBERTSON RP
INHIBITION OF INSULIN SECRETION BY PROSTAGLANDIN (PG)E1
INDEPENDENT OF HYPOTENSIVE OR ALPHA ADRENERGIC EFFECTS
DIABETES (SUPPL I) 22:305(1973)

ROBERTSON RP   GAVARESKI DJ   PORTE D JR
BIERMAN EL
PROSTAGLANDIN E1: INHIBITION OF GLUCOSE-INDUCED INSULIN
SECRETION IN DOGS
CLIN RES 21:219(1973)

ROBERTSON RP   GAVARESKI DJ   PORTE D JR
BIERMAN EL
PROSTAGLANDIN (PG) E1: INHIBITION OF GLUCOSE-STIMULATED
INSULIN SECRETION IN THE INTACT DOG
CLIN RES 21:635(1973)

ROBERTSON RP   GAVARESKI DJ   PORTE D JR
BIERMAN EL
PROSTAGLANDIN E1 (PGE):INHIBITION OF GLUCOSE-INDUCED
INSULIN SECRETION IN DOGS
ABSTR 8TH CONGR INT DIABETES FED, BRUSS, EXCERPTA MED,
AMST:38(1973)

ROBINSON BF   COLLIER JG   KARIM SMM
SOMERS K
EFFECT OF PROSTAGLANDINS A1, A2, B1, E2 AND F2A ON FOREARM
ARTERIAL BED AND SUPERFICIAL HAND VEINS OF MAN
CLIN SCI 44:367-376(1973)

ROBINSON DR   SMITH H   LEVINE L
PROSTAGLANDIN (PG) SYNTHESIS BY HUMAN SYNOVIAL CULTURES AND
ITS STIMULATION BY COLCHICINE
ARTHRITIS RHEUM 16:129(1973)

ROBISON G   ARNOLD A   COLE B
HARTMANN R
EFFECTS OF PROSTAGLANDINS ON FUNCTION AND CYCLIC AMP
LEVELS OF HUMAN BLOOD PLATELETS
ANN NY ACAD SCI 180:324-331(1971)

ROBISON GA   SUTHERLAND EW
ON THE RELATION OF CYCLIC AMP TO ADRENERGIC RECEPTORS AND
SYMPATHIN
ADV CYTOPHARMACOL 1:263-272(1971)

RODBELL M
IN VITRO ASSAYS OF ADENYL CYCLASE
ACTA ENDOCRINOL (KBH) 153:337-347(1971)

RODGERS BM   STARUSCIK RN   REIS RL
EFFECTS OF AMNIOTIC FLUID ON CARDIAC CONTRACTILITY AND
VASCULAR RESISTANCE
AM J PHYSIOL 220:1979-1982(1971)

ROLAND M
NEW HORIZONS IN CONTRACEPTION
IN: " MAJOR PROBLEMS IN OBSTETRICS AND GYNECOLOGY",
EA FRIEDMAN, WB SAUNDERS, PHILA/LOND/TORONTO 5:141-155(1973)

ROMERO JC   KOZAK TJ   HOOBLER SW
THE EFFECT OF NEPHRECTOMY AND OF RENOMEDULLARY EXTRACTS ON
THE BLOOD PRESSURE OF EXPERIMENTALLY HYPERTENSIVE RABBITS
PROC SOC EXP BIOL MED 140:651-656(1972)

ROONEY PJ   LEE P   BROOKES P
DICK WC
REFLECTIONS ON POSSIBLE MECHANISMS OF ACTION OF
ANTI-INFLAMMATORY DRUGS
CURR MED RES OPIN 1:501-516(1973)

ROSCA V   NECHIFOR M
PHYSIOLOGICAL AND PHARMACOLOGICAL IMPLICATIONS OF
PROSTAGLANDINS
REV MED CHIR SOC MED NAT IASI 77:433-439(1973);(ROUMANIAN)

ROSELL S   CHISOLM G   INTAGLIETTA M
SYMPATHETIC NERVE ACTIVITY AND ISOVOLUMETRIC CAPILLARY
PRESSURE (PCI) IN SUBCUTANEOUS ADIPOSE TISSUE
FED PROC 32:323(1973)

ROSEMAN TJ   SIMS B   STEHLE RG
STABILITY OF PROSTAGLANDINS
AM J HOSP PHARM 30:236-239(1973)

ROSEMAN TJ   SIMS B   STEHLE RG
STABILITY OF PROSTAGLANDINS
ADV BIOSCI 9:851-852(1973)

ROSEMAN TJ   YALKOWSKY SH
PHYSICO-CHEMICAL PROPERTIES OF PROSTAGLANDINS I: SOLUBILITY
BEHAVIOR, IONIZATION CONSTANTS, AND SURFACE PROPERTIES OF
PGF2A (THAM)
ABSTR 13TH NATL MEET APHA, ACAD PHARM SCI 2(2):5(1972)

ROSEMAN TJ   YALKOWSKY SH
PHYSICOCHEMICAL PROPERTIES OF PROSTAGLANDIN F2A(TROMETHAMINE
SALT): SOLUBILITY BEHAVIOR, SURFACE PROPERTIES, AND
IONIZATION CONSTANTS
J PHARM SCI 62:1680-1685(1973)

ROSEN OM   HIRSCH AH   GOREN EN
FACTORS WHICH INFLUENCE CYCLIC AMP FORMATION AND DEGRADATION
IN AN ISLET CELL TUMOR OF THE SYRIAN HAMSTER
ARCH BIOCHEM BIOPHYS 146:660-663(1971)

ROSEN P   HOLLAND GW   JERNOW JL
KIENZLE F   KWOH S
A VERSATILE PROSTAGLANDIN SYNTHESIS. UTILIZATION OF A
CARBOXY-INVERSION REACTION
ABSTR 166TH NATL MEET AM CHEM SOC:MEDI-61(1973)

ROSENFELD MG   ABRASS IB   MENDELSOHN J
ROOS BA   BOONE RF   GARREN LD
CONTROL OF TRANSCRIPTION OF RNA RICH IN POLYADENYLIC ACID IN
HUMAN LYMPHOCYTES
PROC NATL ACAD SCI USA 69:2306-2311(1972)

ROSENQVIST U
ADRENERGIC RECEPTOR RESPONSE IN HYPOTHYROIDISM. AN IN VITRO
STUDY ON HUMAN ADIPOSE TISSUE AND RABBIT AORTA
ACTA MED SCAND (SUPPL 532) : 1-28 (1972)

ROSENQVIST U   EFENDIC S
SYNERGISTIC EFFECT OF PROSTAGLANDIN E1 ON NORADRENALINE
INDUCED LIPOLYSIS IN ADIPOSE TISSUE FROM HYPOTHYROID
SUBJECTS
ACTA ENDOCRINOL (KBH) SUPPL 155:18(1971)

ROSENQVIST U   EFENDIC S
STIMULATORY EFFECT IN VITRO OF PROSTAGLANDIN E1 ON
NORADRENALINE-INDUCED LIPOLYSIS IN SUBCUTANEOUS ADIPOSE
TISSUE FROM HYPOTHYROID SUBJECTS
ACTA MED SCAND 190:341-345(1971)

ROSENTHALE ME
BRONCHODILATOR ACTIVITY OF THE PROSTAGLANDINS
INTRA-SCI CHEM REP 6:11-23(1972)

ROSENTHALE ME   DERVINIS A   KASSARICH J
BRONCHODILATOR ACTIVITY OF THE PROSTAGLANDINS E1 AND E2
J PHARMACOL EXP THER 178:541-548(1971)

ROSENTHALE ME   DERVINIS A   KASSARICH J
SINGER S
PROSTAGLANDINS AND ANTI-INFLAMMATORY DRUGS IN THE DOG KNEE
JOINT
J PHARM PHARMACOL 24:149-150(1972)

ROSENTHALE ME   DERVINIS A   KASSARICH J
SINGER S   GLUCKMAN MI
FURTHER STUDIES ON THE BRONCHODILATING PROPERTIES OF THE
PROSTAGLANDINS
ADV BIOSCI (SUPPL)9:39(1972)

ROSENTHALE ME   DERVINIS A   KASSARICH J
BLUMENTHAL A   GLUCKMAN MI
BRONCHODILATING PROPERTIES OF THE PROSTAGLANDIN F2H IN THE
GUINEA PIG AND CAT
PROSTAGLANDINS 3:767-772(1973)

ROSENTHALE ME   DERVINIS A   KASSARICH J
SINGER S   GLUCKMAN MI
COMPARATIVE STUDIES ON THE BRONCHODILATING PROPERTIES OF
THE PROSTAGLANDIN F2H
ADV BIOSCI 9:229-233(1973)

ROSENTHALE ME   DERVINIS A   KASSARICH J
BLUMENTHAL A   GLUCKMAN MI
BRONCHODILATING PROPERTIES OF THE PROSTAGLANDIN PGF2B
IN: "PROSTAGLANDINS AND CYCLIC AMP," RH KAHN, WEM LANDS,
ACADEMIC PRESS, NY : 53-56(1973)

ROSS G
ESCAPE OF MESENTERIC VESSELS FROM ADRENERGIC AND
NONADRENERGIC VASOCONSTRICTION.
AM J PHYSIOL 221(5):1217-1222(1971)

ROSSI EC   LEVIN NW
INHIBITION OF PLATELET AGGREGATION BY FUROSEMIDE (LASIX)
CLIN RES 20:498(1972)

ROSSI EC   LEVIN NW
INHIBITION OF ADP-INDUCED PLATELET AGGREGATION BY FUROSEMIDE
J LAB CLIN MED 81:140-147(1973)

ROSSI EC   LEVIN NW
INHIBITION OF PRIMARY ADP-INDUCED PLATELET AGGREGATION IN
NORMAL SUBJECTS AFTER ADMINISTRATION OF NITROFURANTOIN
(FURADANTIN)
J CLIN INVEST 52:2457-2467(1973)

ROSSINI AA   LEE JB   FRAWLEY TF
AN UNPREDICTABLE LACK OF EFFECT OF PROSTAGLANDINS ON INSULIN
RELEASE IN ISOLATED RAT ISLETS.
DIABETES 20:374(1971)

ROTH HJ
PROSTAGLANDINS
PHARM ZTG 118:1562-1568(1973);(GERMAN)

ROZENGURT E   DE ASUA LJ
ROLE OF CYCLIC 3':5'-ADENOSINE MONOPHOSPHATE IN THE EARLY
TRANSPORT CHANGES INDUCED BY SERUM AND INSULIN IN QUIESCENT
FIBROBLASTS
PROC NATL ACAD SCI USA 70 : 3609-3612(1973)

RUCINSKA E
THE INFLUENCE OF PGF1 ON THE CONTRACTION OF THE STRIATED
MUSCLE
IN: "FARMAKOLOGIA AMIN KATECHOLOWYCH ORAZ LEKOW DZIALAJACYCH
NA UKLAD NERWOWY," I ZJAZDU, PANSTWOWY ZAKLAD WYDAWNICTW
LEKARSKICH, WARSAW : 278-281(1971);(POLISH)

RUDICK J   GONDA M   DREILING DA
JANOWITZ HD
EFFECTS OF PROSTAGLANDIN E1 ON PANCREATIC EXOCRINE FUNCTION
GASTROENTEROLOGY 60:272-278(1971)

RUEGG M   JACQUES R
A SIMPLE IN VITRO METHOD OF CHARACTERIZING NARCOTIC
ANTAGONISTS
EXPERIENTIA 28:1525-1526(1972)

RUEGG M   JAQUES R
A SIMPLE IN VITRO METHOD OF ASSESSING THE ANALGESIC ACTION
OF DRUGS
ADV BIOSCI (SUPPL)9:73(1972)

RUTENBURG AM   POLGAR P   VERA C
PROSTAGLANDIN EFFECT ON ADENYL CYCLASE IN LEUKEMIC
LEUKOCYTES
IN: "PROSTAGLANDINS AND CYCLIC AMP," RH KAHN, WEM LANDS,
ACADEMIC PRESS, NY : 217(1973)

RYAN JW   NIEMEYER RS   GOODWIN DW
METABOLIC FATES OF BRADYKININ, ANGIOTENSIN I, ADENINE
NUCLEOTIDES AND PROSTAGLANDINS E1 AND F1A IN THE PULMONARY
CIRCULATION
ADV EXP MED BIOL 21:259-265(1972)

RYAN JW   ROBLERO J   STEWART JM
LEARY WPP
METABOLISM OF VASOACTIVE POLYPEPTIDES IN THE PULMONARY
CIRCULATION
CHEST(SUPPL 5)59:8S-9S(1971)

RYAN JW   SMITH U
METABOLISM OF ADENOSINE 5'-MONOPHOSPHATE DURING CIRCULATION
THROUGH THE LUNGS
TRANS ASSOC AM PHYSICIANS 84:297-306(1971)

RYAN M   ZIMMERMAN B   KRAFT E
POTENTIATION OF VASOCONSTRICTOR RESPONSES (VCR) TO
NOREPINEPHRINE (NE) BY PROSTAGLANDIN SYNTHESIS INHIBITORS IN
THE KREBS PERFUSED DOG'S PAW
PHARMACOLOGIST 15:250(1973)

RYAN MJ
ANALYSIS OF FACTORS REGULATING UTERINE BLOOD FLOW
DISS, UNIV IOWA (1972)

RYAN MJ   CLARK KE   VAN ORDEN DE
FARLEY D   EDVINSSON L   SJOBERG NO
VAN ORDEN LS   BRODY MJ
ROLE OF PROSTAGLANDINS IN ESTROGEN-INDUCED UTERINE HYPEREMIA
PROSTAGLANDINS 4:629-640(1973)

RYAN MJ   ZIMMERMAN BG
EFFECT OF PROSTAGLANDIN PRECURSORS, DIHOMO-GAMMA-LINOLENIC
ACID (DLA) AND ARACHIDONIC ACID (AA), ON THE VASOCONSTRICTOR
RESPONSE (VCR) TO INTRAARTERIAL (IA) NOREPINEPHRINE (NE) IN
THE DOG PAW
FED PROC 32:803(1973)

SABATINI-SMITH S
THE EFFECTS OF PROSTAGLANDINS E1 AND F2A, NOREPINEPHRINE,
AND 3'-5' ADENOSINE MONOPHOSPHATE ON CALCIUM TRANSPORT IN
ELECTRICALLY STIMULATED CARDIAC SARCOPLASMIC RETICULUM
FED PROC 30:625(1971)

SABATINI-SMITH S
TRANSPORT OF CALCIUM IN ISOLATED SARCOPLASMIC RETICULUM AS
AFFECTED BY PROSTAGLANDINS, NOREPINEPHRINE AND 3'-5'
ADENOSINE MONOPHOSPHATE
PROC INT UNION PHYSIOL SCI, 25TH INT CONGR, MUNICH,
9:486(1971)

SABIR M   SINGH M   BHIDE NK
SPASMOLYTIC EFFECT OF POLYSORBATES (TWEENS) 80 AND 20
ISOLATED TISSUES
INDIAN J PHYSIOL PHARMACOL 16:193-199(1972)

SACCA L   RENGO F   CHIARIELLO M
CONDORELLI M
GLUCOSE INTOLERANCE AND IMPAIRED INSULIN SECRETION BY
PROSTAGLANDIN A1 IN FASTING ANESTHETIZED DOGS
ENDOCRINOLOGY 92:31-34(1973)

SACCA L   RENGO F   PEREZ G
CHIARIELLO M   CONDORELLI M
EFFECTS OF HYPOTENSIVE DOSES OF PROSTAGLANDIN A1 ON PLASMA
GLUCOSE, FREE FATTY ACIDS AND INSULIN IN RESERPINIZED DOGS
PHARMACOLOGY 9:159-163(1973)

SACKEYFIO AC
ANAPHYLATOXIN-INDUCED RELEASE OF A SUBSTANCE WITH
PROSTAGLANDIN-LIKE ACTIVITY IN ISOLATED PERFUSED
GUINEA-PIG LUNGS
BR J PHARMACOL 46:544P-545P(1972)

SACKS PV   KANAREK D
TREATMENT OF ACUTE PLEURITIC PAIN. COMPARISON BETWEEN
INDOMETHACIN AND A PLACEBO
AM REV RESP DIS 108:666-669(1973)

SADOW J   PENN R
THE EFFECT OF PROSTAGLANDIN E1 ON THE
CORTICOTROPHIN-RELEASING ACTIVITY OF VASOPRESSIN AND
HYPOTHALAMIC EXTRACT IN THE ISOLATED PITUITARY GLAND
J ENDOCRINOL 57:XXVII-XXVIII(1973)

SAEED SA   ROY AC
PURIFICATION OF 15-HYDROXY PROSTAGLANDIN DEHYDROGENASE FROM
BOVINE LUNG
BIOCHEM BIOPHYS RES COMMUN 47:96-102(1972)

SAID SI
METABOLIC ACTIVITY OF THE LUNG AND ITS ROLE IN PULMONARY
DISEASE
CLIN NOTES RESP DIS 10:3-9(1971)

SAID SI
THE LUNG IN RELATION TO VASOACTIVE HORMONES
FED PROC 32:1972-1976(1973)

SAID SI
THE LUNG: AN ENDOCRINE ORGAN?
JAP J THORACIC DIS 11(SUPPL): 1 (1973)

SAID SI
THE LUNG-AN ENDOCRINE ORGAN?
JAP J THORAC DIS 11:509-512(1973)

SAID SI   KITAMURA S   VREIM C
PROSTAGLANDINS: RELEASE FROM LUNG DURING MECHANICAL
VENTILATION AT LARGE TIDAL VOLUMES
J CLIN INVEST 51:83A-84A(1972)

SAID SI   KITAMURA S   VREIM C
RELEASE OF PROSTAGLANDINS FROM THE LUNG DURING MECHANICAL
VENTILATION AT LARGE TIDAL VOLUMES
CLIN RES 20:87(1972)

SAINI RK   ROSSI F   CAPUTI AP
MARMO E
EFFECTS OF BR 750 (2,6-DICHLOROBENZYLIDINEAMINOGUANIDINE
ACETATE) ON CARDIAC AND SOME SMOOTH MUSCLES. AN IN VITRO
EXPERIMENTAL STUDY.
FARMACO (PRAT) 28:642-658(1973)

SAKADA Y   TAKAGI S   SATO T
SAWAZAKI C
LUTEOLYTIC ACTION OF PROSTAGLANDIN
OBSTET GYNECOL (TOKYO) 39:653-657(1972);(JAPANESE)

SAKAI K
CHEMISTRY OF PROSTAGLANDIN
J SOC ORG SYNTH CHEM (TOKYO) 29:205-226(1971);(JAPANESE)

SAKAI K   IDE J   ODA O
NAKAMURA N
SYNTHETIC STUDIES ON PROSTANOIDS. 1. SYNTHESIS OF METHYL
9-OXO-PROSTANOATE
TETRAHEDRON LETT 1287-1290(1972)

SAKAMOTO S   AGATSUMA T   BOKU K
HOGAKI M   NOZUE E   OH M
KURAMOCHI K
INDUCTION OF LABOR WITH PROSTAGLANDIN F2A
OBSTET GYNECOL (TOKYO) 46:1116-1124(1971);(JAPANESE)

SAKAMOTO S   HOGAKI M   KINOSHITA K
WAGATSUMA T
EFFECT OF PROSTAGLANDIN F2A ON UTERINE CONTRACTION EXPRESSED
IN INTRAUTERINE PRESSURE CURVE
SANFUJINKA NO JISSAI 21:209-218(1972);(JAPANESE)

SAKAMOTO S   SATO K   KINOSHITA K
PHARMACODYNAMICS OF PROSTAGLANDINS.
FARUMASHIA 9:399-402(1973);(JAPANESE)

SAKATA J
PROSTAGLANDINS IN GYNECOLOGY
PROC SOC GERONTOL STUDY, OKAYAMA CITY, ONO PHARMACEUTICAL
CO, OSAKA:50-55(1972);(JAPANESE)

SAKSENA SK   STEELE R   HARPER MJK
EFFECTS OF EXOGENOUS OESTRADIOL AND PROGESTERONE ON SERUM
LEVELS OF PROSTAGLANDIN F AND LUTEINIZING HORMONE IN
CHRONICALLY OVARIECTOMIZED RATS
J REPROD FERTIL 32:495-499(1973)

SALOMY M   HALBRECHT I
IMMEDIATE AND LATE EFFECT OF PGE1 ON THE RABBIT OVIDUCT
(IN-VIVO-STUDIES)
ADV BIOSCI (SUPPL)9:133(1972)

SALZMAN EW
CYCLIC AMP AND PLATELET FUNCTION
N ENGL J MED 286:358-363(1972)

SALZMAN EW
PROSTAGLANDINS IN THE FUNCTION OF BLOOD PLATELETS
IN: "PROSTAGLANDINES 1973," INSERM, PARIS:331-343(1973)

SALZMAN EW   KENSLER PC   LEVINE L
CYCLIC 3',5'-ADENOSINE MONOPHOSPHATE IN HUMAN BLOOD
PLATELETS. IV. REGULATORY ROLE OF CYCLIC AMP IN PLATELET
FUNCTION
ANN N Y ACAD SCI 201:61-71(1972)

SALZMAN EW   LEVINE L
CYCLIC 3',5'-ADENOSINE MONOPHOSPHATE IN HUMAN BLOOD
PLATELETS. II. EFFECT OF N6-2'-O-DIBUTYRYL CYCLIC 3',5'-
ADENOSINE MONOPHOSPHATE ON PLATELET FUNCTION.
J CLIN INVEST 50:131-141(1971)

SALZMAN EW   STEAD N   DEYKIN D
INTERRELATIONS OF PLATELET PROSTAGLANDIN SYNTHESIS AND
CYCLIC AMP METABOLISM
ABSTR 4TH INT CONGR THROMB HAEMOSTASIS, VIENNA:78(1973)

SALZMAN EW   WEISENBERGER H
ROLE OF CYCLIC AMP IN PLATELET FUNCTION
ADV CYCLIC NUCL RES 1:231-247(1972)

SAMLOFF IM
PEPSINOGENS, PEPSINS, AND PEPSIN INHIBITORS
GASTROENTEROLOGY 60:586-604(1971)

SAMUELSSON B
BIOSYNTHESIS OF PROSTAGLANDINS
FED PROC 31:1442-1450(1972)

SAMUELSSON B
THE SYNTHESIS AND BIOLOGICAL ROLE OF PROSTAGLANDINS
BIOCHEM J 128:4P(1972)

SAMUELSSON B
QUANTITATIVE ASPECTS ON PROSTAGLANDIN SYNTHESIS IN MAN
ADV BIOSCI 9:7-14(1973)

SAMUELSSON B
ROUND-TABLE DISCUSSION ON ANALYTICAL METHODS
ADV BIOSCI 9:121-123(1973)

SAMUELSSON B
BIOSYNTHESIS AND METABOLISM OF PROSTAGLANDINS
IN: "PROSTAGLANDINES 1973," INSERM, PARIS:21-41(1973)

SAMUELSSON B
BIOSYNTHESIS AND METABOLISM OF PROSTAGLANDINS
IN: "ENDOCRINOLOGY. PROC 4TH INT CONG ENDOCRINOL," RO SCOW,
FJG EBLING, IW HENDERSON, EXCERPTA MEDICA. AMST/NY:370-375
(1973)

SAMUELSSON B
METABOLISM OF PROSTAGLANDINS
ABSTR, 9TH INT CONG BIOCHEM, STOCKH:390(1973)

SAMUELSSON B   GRANSTROM E   GREEN K
HAMBERG M
METABOLISM OF PROSTAGLANDINS
ANN NY ACAD SCI 180:139-163(1971)

SAMUELSSON B   WENNMALM A
INCREASED NERVE STIMULATION INDUCED RELEASE OF NORADRENALINE
FROM THE RABBIT HEART AFTER INHIBITION OF PROSTAGLANDIN
SYNTHESIS.
ACTA PHYSIOL SCAND 83:163-168(1971)

SANDLER M
MIGRAINE: A PULMONARY DISEASE?
LANCET 1:618-619(1972)

SANNER J
PROSTAGLANDIN INHIBITION WITH A DIBENZOXAZEPINE HYDRAZIDE
DERIVATIVE AND MORPHINE.
ANN NY ACAD SCI 180:396-409(1971)

SANNER JH
PROSTAGLANDIN ANTAGONISM.
J AM OIL CHEM SOC 48:94A(1971)

SANNER JH
DIBENZOXAZEPINE HYDRAZIDES AS PROSTAGLANDIN ANTAGONISTS
INTRA-SCI CHEM REP 6:1-9(1972)

SANNER JH   MUELLER RA   SCHULZE RH
STRUCTURE-ACTIVITY RELATIONSHIPS OF SOME DIBENZOXAZEPINE
DERIVATIVES AS PROSTAGLANDIN ANTAGONISTS
ADV BIOSCI (SUPPL)9:20(1972)

SANNER JH   MUELLER RA   SCHULZE RH
STRUCTURE-ACTIVITY RELATIONSHIPS OF SOME DIBENZOXAZEPINE
DERIVATIVES AS PROSTAGLANDIN ANTAGONISTS
ADV BIOSCI 9:139-148(1973)

SANZ F   MARTINEZ MR   ASTUDILLO MD
THE EFFECT OF CERTAIN DRUGS ON THE ACTION OF PROSTAGLANDINS
IN THE ACTIVE SODIUM TRANSPORT THROUGH BIOLOGICAL MEMBRANES
PROC 1ST EUR BIOPHYS CONGR:333-337(1971)

SARUTA T   KAPLAN NM
PROSTAGLANDINS AND OTHER STIMULI OF BEEF ADRENAL
STEROIDOGENESIS
ABSTR 53RD MEET ENDOCR SOC, SAN FRANC:A-89(1971)

SARUTA T   KAPLAN NM
THE EFFECTS OF PROSTAGLANDINS UPON BEEF ADRENAL
STEROIDOGENESIS.
CLIN RES 19:69(1971)

SARUTA T   KAPLAN NM
ADRENOCORTICAL STEROIDOGENESIS: THE EFFECTS OF
PROSTAGLANDINS
J CLIN INVEST 51:2246-2251(1972)

SATINOFF E
SALICYLATE: ACTION ON NORMAL BODY TEMPERATURE IN RATS
SCIENCE 176:532-533(1972)

SATO A   ONAYA T   KOTANI M
HARADA G   TSUKUI T   YAMADA T
CHANGES IN CYCLIC AMP CONTENT OF CEREBRAL CORTEX,
HYPOTHALAMUS, AND ANTERIOR PITUITARY LOBES BY ACTIVE AMINES
CLIN ENDOCRINOL (TOKYO) 21:1035-1038(1973);(JAPANESE)

SATO K   KINOSHITA M   SAKAMOTO M
PROSTAGLANDIN
CLIN ENDOCRINOL (TOKYO) 21: 245-269 (1973); (JAPANESE)

SATO K   KUWABARA Y   KINOSHITA K
SAKAMOTO S
METHODS FOR THE ANALYSIS OF PROSTAGLANDIN
OBSTET GYNECOL (TOKYO) 39:669-677(1972);(JAPANESE)

SATO S
STUDIES ON THE MECHANISM OF ACTION OF THYROID STIMULATING
SUBSTANCES
SHINSHU MED J 20:233-243(1972);(JAPANESE)

SATO S   KOWALSKI K   BURKE G
EFFECTS OF A PROSTAGLANDIN ANTAGONIST, POLYPHLORETIN
PHOSPHATE, ON BASAL AND STIMULATED THYROID FUNCTION
PROSTAGLANDINS 1:345-363(1972)

SATO S   KOWALSKI K   SZABO M
BURKE G
EFFECTS OF A PROSTAGLANDIN ANTAGONIST, POLYPHLORETIN
PHOSPHATE (PPP), ON BASAL AND STIMULATED THYROID FUNCTION.
J LAB CLIN MED 78:823-824(1971)

SATO S   SZABO M   KOWALSKI K
BURKE G
ROLE OF PROSTAGLANDIN IN THYROTROPIN ACTION ON THYROID
ENDOCRINOLOGY 90:343-356(1972)

SATO S   YAMADA T
ACTIVATION OF ADENYL CYCLASE PRESENT IN THE PLASMA MEMBRANE
OF BOVINE THYROID GLAND CELLS BY TSH, LONG ACTING THYROID
STIMULATOR, AND PROSTAGLANDIN E2.
CLIN ENDOCRINOL (TOKYO) 21:725-729(1973);(JAPANESE)

SATO T
INDUCTION OF LABOR AND PROSTAGLANDIN
CLIN ENDOCRINOL (TOKYO) 19:369-372(1971)

SATO T   ABI H
PROSTAGLANDIN AND CONTRACEPTION AND TERMINATION OF PREGNANCY
(EXPERIENCE WITH PROSTAGLANDIN VAGINAL SUPPOSITORY)
OBSTET GYNECOL (TOKYO) 39:648-652(1972);(JAPANESE)

SATO T   AMI K   MAYUZUMI R
THE EFFECT OF PROSTAGLANDIN F2A ON GONADOTROPIN SECRETION,
UTERUS AND OVARY OF THE FEMALE MATURE RAT
FOLIA ENDOCRINOL JAP 47:383(1971);(JAPANESE)

SATO T   AMI OH   MAYUZUMI R
EBINHARA H   HIRONO M   TAMURA S
MATSUMOTO S
EFFECT OF PROSTAGLANDIN F2A ON PITUITARY GONADOTROPIN, THE
UTERUS AND OVARY OF THE RAT
FOLIA ENDOCRINOL JAP 46:48(1971);(JAPANESE)

SATOH K   SAKAMOTO S
CYCLIC AMP AND PROSTAGLANDINS.
IGAKU NO AYUMI 81:269-278(1972);(JAPANESE)

SAUNDERS RN   MOSER C
INCREASED VASCULAR RESISTANCE BY PROSTAGLANDIN B2 (PGB2) IN
THE ISOLATED RAT PANCREAS.
PHARMACOLOGIST 13:292(1971)

SAUNDERS RN   MOSER CA
CHANGES IN VASCULAR RESISTANCE INDUCED BY PROSTAGLANDINS E1
AND E2 IN THE ISOLATED RAT PANCREAS
ARCH INT PHARMACODYN THER 197:86-92(1972)

SAUNDERS RN   MOSER CA
INCREASED VASCULAR RESISTANCE BY PROSTAGLANDINS B1 AND B2
IN THE ISOLATED RAT PANCREAS
NATURE (NEW BIOL) 237:285(1972)

SAUNDERS RN   MUELLER RA   MOSER CA
BECKER DV
THE EFFECT OF NATURAL PROSTAGLANDINS AND STRUCTURAL ANALOGS
ON VASCULAR RESISTANCE TO PERFUSION IN THE ISOLATED RAT
PANCREAS
ABSTR 5TH INT CONGR PHARMACOL SAN FRANC 201(1972)

SAUNDERS RN   MUELLER RA   MOSER CA
BECKER DV
RELATION OF PROSTAGLANDIN STRUCTURE TO VASCULAR RESISTANCE
ACTIVITY IN THE ISOLATED RAT PANCREAS
ADV BIOSCI (SUPPL)9:21(1972)

SAUNDERS RN   MUELLER RA   MOSER CA
BECKER DV
RELATION OF PROSTAGLANDIN STRUCTURE TO VASCULAR RESISTANCE
IN THE ISOLATED RAT PANCREAS
ADV BIOSCI 9:149-154(1973)

SAUNDERS RN   ROZEK LF   JIU J
MOSER C
INCREASED VASCULAR RESISTANCE DEMONSTRATED BY PROSTAGLANDIN
A2 IN THE ISOLATED RAT PANCREAS
FED PROC 31:545(1972)

SAWAZAKI C
PROSTAGLANDINS IN GYNECOLOGY
PROC SOC GERONTOL STUDY, OKAYAMA CITY, ONO PHARMACEUTICAL
CO, OSAKA:47-49(1972);(JAPANESE)

SAWAZAKI C
EFFECT OF PROSTAGLANDIN ON LABOR INDUCTION AND PROMOTION
OBSTET GYNECOL (TOKYO) 39:588-594(1972);(JAPANESE)

SAWAZAKI C
COMPARISON OF LABOR INDUCTION EFFECT BETWEEN PROSTAGLANDIN
F2A AND OXYTOCIN
OBSTET GYNECOL (TOKYO) 39:595-618(1972);(JAPANESE)

SAWAZAKI C
LABOR INDUCTION AND PROMOTION EFFECT OF ORAL PGE2
OBSTET GYNECOL (TOKYO) 39:619-630(1972);(JAPANESE)

SAWAZAKI C
PROSTAGLANDINS IN OBSTETRICS AND GYNECOLOGY
OBSTET GYNECOL (TOKYO) 40:701-715(1973);(JAPANESE)

SAWAZAKI C   TAKEUCHI K   YANAGISAWA Y
SAKADA Y
PROSTAGLANDIN AND INFERTILITY
OBSTET GYNECOL (TOKYO) 39:659-668(1972);(JAPANESE)

SAXENA PN
HYPERTHERMIA PRODUCED BY INTRACEREBROVENTRICULAR INJECTION
OF PROSTAGLANDIN E1 (PGE1) IN PIGEONS
ABSTR 5TH INT CONGR PHARMACOL SAN FRANC 202(1972)

SCARAMUZZI OE   BAILE CA   MAYER J
PROSTAGLANDINS AND FOOD INTAKE OF RATS
EXPERIENTIA 27:256-257(1971)

SCHAAF TK   COREY EJ
A TOTAL SYNTHESIS OF PROSTAGLANDINS F1A AND E1
J ORG CHEM 37:2921-2922(1972)

SCHAEFER AK
INVESTIGATIONS ON THE EFFECT OF PROSTAGLANDINS E2 AND F2A
ON THE BRONCHIAL TONE OF THE CAT
DISS, FREE UNIV BERL (1973);(GERMAN)

SCHAFER A   FREY HH
EFFECTS OF PROSTAGLANDINS E2 AND F2A ON BRONCHIAL TONE IN
CATS
ACTA PHYSIOL SCAND (SUPPL 396) : 105(1973)

SCHAFER DE   LUST WD   POLSON JB
HEDTKE J   SIRCAR B   THAKUR AK
GOLDBERG ND
STUDIES ON THE POSSIBLE ROLE OF CYCLIC AMP IN SOME ACTIONS
OF CHOLERA TOXIN
ANN NY ACAD SCI 185:376-385(1971)

SCHAUB RE   WEISS MJ
PROSTAGLANDINS AND CONGENERS II. THE CONJUGATE ADDITION OF
3-T-BUTOXYOCTYL MAGNESIUM BROMIDE TO CYCLOPENTENONES. A
SYNTHESIS OF RAC. 11-DEOXY-13-DIHYDROPROSTAGLANDIN E1
TETRAHEDRON LETT 129-130(1973)

SCHAUMBURG BP
BINDING OF PROSTAGLANDIN E1 TO RAT THYMOCYTES
BIOCHIM BIOPHY ACTA 326:127-139(1973)

SCHECHTER J   WEINER R
ULTRASTRUCTURAL CHANGES IN THE EPENDYMAL LINING OF THE
MEDIAN EMINENCE FOLLOWING THE INTRAVENTRICULAR
ADMINISTRATION OF CATECHOLAMINE
ANAT REC 172:643-650(1972)

SCHIMMER BP
EFFECTS OF CATECHOLAMINES AND MONOVALENT CATIONS ON
ADENYLATE CYCLASE ACTIVITY IN CULTURED GLIAL TUMOR CELLS
BIOCHIM BIOPHYS ACTA 252:567-573(1971)

SCHLENK H
ODD NUMBERED AND NEW ESSENTIAL FATTY ACIDS
FED PROC 31:1430-1435(1972)

SCHMUCKER M   ZIEVE PD
STUDIES OF GLYCOSIDASES IN FRESH AND STORED HUMAN PLATELETS
J LAB CLIN MED 80:635-643(1972)

SCHNEIDER EG   STRANDHOY JW   WILLIS LR
KNOX FG
RELATIONSHIP BETWEEN PROXIMAL SODIUM REABSORPTION AND
EXCRETION OF CALCIUM, MAGNESIUM AND PHOSPHATE
KIDNEY INT 4:369-376(1973)

SCHNEIDER V   BIELING C   SCHINDLER AE
SADOWSKI P   BURKLE G   GEISBE H
STRANGULATED HERNIA OF TREITZ AS A RARE CAUSE OF ILEUS OF
PREGNANCY
GEBURTSHILFE FRAUENHEILKD 33:877-881(1973);(GERMAN)

SCHNEIDER WP
THE CHEMISTRY OF THE PROSTAGLANDINS
IN: "THE PROSTAGLANDINS: PROGRESS IN RESEARCH," SMM KARIM,
MTP MED TECH PUBL CO, OXFORD:293-319(1972)

SCHNEIDER WP   HAMILTON RD   RHULAND LE
OCCURRENCE OF ESTERS OF (15S)-PROSTAGLANDIN A2 AND E2 IN
CORAL
J AM CHEM SOC 94:2122-2123(1972)

SCHNEIDER WP   MURRAY HC
MICROBIOLOGICAL REDUCTION AND RESOLUTION OF PROSTAGLANDINS.
SYNTHESIS OF NATURAL PGF2A AND ENT-PGF2B METHYL ESTERS
J ORG CHEM 38:397-398(1973)

SCHNEIDER WP   RHULAND LE   HAMILTON RD
BUNDY GL   DANIELS EG   LINCOLN FH
PIKE JE
PROSTAGLANDINS FROM MARINE SOURCES
IN FOOD-DRUGS FROM THE SEA PROCEEDINGS 1972 LR WORTHEN
WASHINGTON MARINE TECHNOLOGY SOCIETY:151-155(1973)

SCHOENE NW   DUTKY RC   IACONO JM
BIOSYNTHESIS OF PROSTAGLANDIN E2 IN HUMAN BLOOD PLATELETS
FROM 1-14C-ARACHIDONIC ACID
CIRCULATION (SUPPL II)46:32(1972)

SCHOENE NW   IACONO JM
METABOLISM OF LINOLEIC AND ARACHIDONIC ACIDS IN HUMAN BLOOD
PLATELETS
FED PROC 32:919(1973)

SCHOFIELD JG
CYTOCHALASIN B AND RELEASE OF GROWTH HORMONE
NATURE (NEW BIOL) 234:215-216(1971)

SCHOMBERG DW
PROSTAGLANDIN F2A DOES NOT INDUCE CYTOLYSIS OF PROGESTERONE
SECRETING MONOLAYER CULTURES
ABSTR SOC GYNECOL INVEST 19TH ANNU MEET, PHOENIX:29(1972)

SCHONHOFER PS   DINNENDAHL V   PADBERG P
INFLUENCE OF DRUGS ON MUCOPOLYSACCHARIDE SYNTHESIS IN
FIBROBLAST TISSUE CULTURES
NAUNYN SCHMIEDEBERGS ARCH PHARMACOL 277(SUPPL):R68(1973)

SCHONHOFER PS   SKIDMORE IF   PAUL MI
DITZION BR   PAUK GL   KRISHNA G
EFFECTS OF GLUCOCORTICOIDS ON ADENYL CYCLASE AND
PHOSPHODIESTERASE ACTIVITY IN FAT CELL HOMOGENATES AND THE
ACCUMULATION OF CYCLIC AMP IN INTACT FAT CELLS
NAUNYN SCHMIEDEBERGS ARCH PHARMACOL 273:267-282(1972)

SCHOOLEY JC   MAHLMANN LJ
STIMULATION OF ERYTHROPOIESIS IN PLETHORIC MICE BY
PROSTAGLANDINS AND ITS INHIBITION BY ANTIERYTHROPOIETIN
PROC SOC EXP BIOL MED 138:523-524(1971)

SCHRODER CH   HSIE AW
MORPHOLOGICAL TRANSFORMATION OF ENUCLEATED CHINESE HAMSTER
OVARY CELLS BY DIBUTYRYL ADENOSINE 3',5'-MONOPHOSPHATE AND
HORMONES
NATURE(NEW BIOL) 246:58-60(1973)

SCHUBART U   UDEM L   BAUM S
ROSEN OM
REGULATION OF CYCLIC NUCLEOTIDE PHOSPHODIESTERASE (PDE)
ACTIVITY IN AN ISLET CELL TUMOR OF THE SYRIAN HAMSTER
DIABETES (SUPPL I)22:306(1973)

SCHULTZ G   JAKOBS KH   BOHME E
SCHULTZ K
EFFECT OF VARIOUS HORMONES ON THE FORMATION OF
ADENOSINE-3',5'-MONOPHOSPHATE AND
GUANOSINE-3',5'-MONOPHOSPHATE BY SPECIAL PREPARATIONS FROM
RAT KIDNEYS
EUR J BIOCHEM 24:520-529(1972);(GERMAN)

SCHULTZ J   HAMPRECHT B
ADENOSINE 3',5'-MONOPHOSPHATE IN CULTURED NEUROBLASTOMA
CELLS: EFFECT OF ADENOSINE, PHOSPHODIESTERASE INHIBITORS AND
BENZAZEPTINES
NAUNYN SCHMIEDEBERGS ARCH PHARMACOL 278:215-225(1973)

SCHUSTER MM   VANASIN B
ALTERATION OF SMOOTH MUSCLE ELECTRICAL AND MOTOR ACTIVITY BY
PROSTAGLANDINS.
J CLIN INVEST 50:83A(1971)

SCHUSTER MM   VANASIN B
PROSTAGLANDIN EFFECTS ON ELECTRICAL AND MOTOR ACTIVITY OF
COLONIC MUSCLE.
REND ROM GASTROENTEROL 3:109-110(1971)

SCHWABE U   EBERT R   ERBLER HC
ADENOSINE RELEASE FROM ISOLATED FAT CELLS AND ITS
SIGNIFICANCE FOR THE EFFECTS OF HORMONES ON CYCLIC 3',5'-AMP
LEVELS AND LIPOLYSIS
NAUNYN SCHMIEDEBERGS ARCH PHARMACOL 276:133-148(1973)

SCOMMEGNA A   DMOWSKI WP
DYSFUNCTIONAL UTERINE BLEEDING
CLIN OBSTET GYNECOL 16:221-254(1973)

SCOTT TW
LIPID METABOLISM OF SPERMATOZOA
J REPROD FERTIL (SUPPL)18:65-76(1973)

SCRATCHERD T   CASE RM
THE ROLE OF ADENYL CYCLASE IN THE GASTROINTESTINAL TRACT
IN:INT ENCYCL PHARMACOL THER,G PETERS,PERGAMON PRESS,NY
21:547-612(1973)

SELIGMAN ML   DEMOPOULOS HB
SPIN-PROBE ANALYSIS OF MEMBRANE PERTURBATIONS PRODUCED BY
CHEMICAL AND PHYSICAL AGENTS
ANN N Y ACAD SCI 222:640-667(1973)

SELKURT EE   ABEL FL   EDWARDS JL
YUM MN
RENAL FUNCTION IN DOGS WITH HYPERTENSION INDUCED BY
IMMUNOLOGIC NEPHRITIS
PROC SOC EXP BIOL MED 144:295-303(1973)

SEN AK   SUNAHARA FA   TALESNIK J
CYCLIC AMP AND METABOLICALLY-INDUCED CORONARY
VASODILATION IN THE ISOLATED PERFUSED RAT HEART
ABSTR 5TH INT CONGR PHARMACOL, SAN FRANC:208(1972)

SERRANO RIOS M   HAWKINS FG   ESCOBAR F
PROSTAGLANDINS: THEIR MECHANISM OF ACTION. RELATION TO 3-5
AMP
REV CLIN ESP 125:285-294(1972);(SPANISH)

SETH SOS   MUKHOPADHYAY A   BAGCHI N
PRABHAKAR MC   ARORA RB
ANTIHISTAMINIC AND SPASMOLYTIC EFFECTS OF MUSK
JAP J PHARMACOL 23:673-679(1973)

SETTIPANE GA
NEW CONCEPTS OF PATHOGENIC MECHANISMS IN ALLERGY
R I MED J 56:325-328:339(1973)

SEYBERTH HW   SCHMIDT-GAYK H   JAKOBS KH
HACKENTHAL E
CYCLIC ADENOSINE MONOPHOSPHATE IN PHAGOCYTIZING GRANULOCYTES
AND ALVEOLAR MACROPHAGES
J CELL BIOL 57:567-571(1973)

SEZILLE G
THE PROSTAGLANDINS
IN: "PROBLEMES ACTUELS DE BIOCHIMIE GENERALE," P BOULANGER
AND J POLONOVSKI, MASSON, PARIS:23-54(1972);(FRENCH)

SEZILLE G
THE PROSTAGLANDINS
LILLE MED 18:1001-1003(1973);(FRENCH)

SHAIKH AA   SAKSENA S
PERIOVULATORY LEVELS OF F PROSTAGLANDINS IN THE UTERINE
VENOUS PLASMA OF THE CYCLIC HAMSTER
ADV BIOSCI (SUPPL)9:105(1972)

SHARE L   CLAYBAUGH JR
REGULATION OF BODY FLUIDS
ANNU REV PHYSIOL 34:235-260(1972)

SHARMA SC
SENSITIVE RADIOIMMUNE ASSAY FOR PROSTAGLANDINS F
J PHYSIOL (LOND) 226:74P-75P(1972)

SHARMA SC   FITZPATRICK RJ   LANHAM J
DEVELOPMENT OF A POTENT ANTISERUM TO PROSTAGLANDIN F AND
MEASUREMENT OF PROSTAGLANDIN F IN THE PERIPHERAL CIRCULATION
OF PARTURIENT SHEEP.
J REPROD FERTIL 35:598-599(1973)

SHARMA SK
INHIBITORS AND ACTIVATORS OF ENZYMES REGULATING THE CELLULAR
CONCENTRATIONS OF CYCLIC AMP
IN: "METABOLIC INHIBITORS," RM HOCHSTER, M KATES AND JH
QUASTEL, ACADEMIC PRESS, NY 4:389-433(1973)

SHARP GWG   FLORES J   WITKUM PA
STUDIES ON THE MECHANISM OF ACTION OF CHOLERA TOXIN:
INTERRELATIONSHIP WITH HORMONE RECEPTORS.
EUR J CLIN INVEST 3:268-269(1973)

SHARP GWG   HYNIE S
STIMULATION OF INTESTINAL ADENYL CYCLASE BY CHOLERA TOXIN
NATURE (LOND) 229:266-269(1971)

SHARP GWG   HYNIE S   LIPSON LC
PARKINSON D
ACTION OF CHOLERA TOXIN TO STIMULATE ADENYL CYCLASE.
CLIN RES 19:577(1971)

SHARP GWG   OZER A   FLORES AA
LIPSON LC
THE EFFECTS OF ANTIDIURETIC HORMONE, PROSTAGLANDIN E1, AND
PROSTAGLANDIN ANTAGONISTS ON OSMOTIC WATER FLOW ACROSS THE
TOAD BLADDER
INTRA-SCI CHEM REP 6:77-84(1972)

SHAW J   GIBSON W   JESSUP S
RAMWELL P
THE EFFECT OF PGE1 ON CYCLIC AMP AND ION MOVEMENTS IN TURKEY
ERYTHROCYTES.
ANN NY ACAD SCI 180:241-260(1971)

SHAW JE   JESSUP SJ   RAMWELL PW
PROSTAGLANDIN-ADENYL CYCLASE RELATIONSHIPS
ADV CYCLIC NUCL RES 1:479-491(1972)

SHAW JE   RAMWELL PW
PROSTAGLANDINS--A GENERAL REVIEW
RES PROSTAGLANDINS 1(1):1-4(1971)

SHAW JE   URQUHART J
PARAMETERS OF THE CONTROL OF ACID SECRETION IN THE ISOLATED
BLOOD-PERFUSED STOMACH
J PHYSIOL (LOND) 226:107P-108P(1972)

SHEA SM   CAULFIELD JB   BURKE JF
MICROVASCULAR ULTRASTRUCTURE IN THERMAL INJURY: A
RECONSIDERATION OF THE ROLE OF MEDIATORS
MICROVASC RES 5:87-96(1973)

SHEARMAN R   SMITH I   KORDA A
SECOND TRIMESTER TERMINATION BY INTRA-UTERINE PROSTAGLANDIN
F2A. CLINICAL AND HORMONAL RESULTS WITH OBSERVATIONS ON
INDUCED LACTATION AND CHRONOPERIODICITY
J REPROD MED 9:448-452(1972)

SHEARMAN RP
THE PROSTAGLANDINS
MED J AUST 2:86-90(1972)

SHEN TY
PERSPECTIVES IN NONSTEROIDAL ANTI-INFLAMMATORY AGENTS
ANGEW CHEM (ENGL) 11:460-472(1972)

SHENTON BK   FIELD EJ   ROGERS AF
CARNEY JA   SYKES JAC   MERTIN J
PROSTAGLANDINS AND CELLULAR IMMUNITY: A REGULATORY
MECHANISM?
INT RES COMMUN SYST (73-12)8-10-1(1973)

SHEPHERD AP   MAO CC   JACOBSON ED
SHANBOUR LL
THE ROLE OF CYCLIC AMP IN MESENTERIC VASODILATION
MICROVASC RES 6: 332-341 (1973)

SHEPHERD AP   SOLOMON N   SHANBOUR LL
JACOBSON ED
THE ROLE OF CYCLIC AMP IN MESENTERIC VASODILATION
GASTROENTEROLOGY 62:811(1972)

SHEPPARD JR
DIFFERENCE IN THE CYCLIC ADENOSINE 3',5'-MONOPHOSPHATE
LEVELS IN NORMAL AND TRANSFORMED CELLS
NATURE (NEW BIOL) 236:14-16(1972)

SHEPPARD JR   PRASAD KN
CYCLIC AMP LEVELS AND THE MORPHOLOGICAL DIFFERENTIATION OF
MOUSE NEUROBLASTOMA CELLS
LIFE SCI (II) 12:431-439(1973)

SHERMAN AI   VAKHARIYA VR
AN EVALUATION OF PROSTAGLANDIN F2A FOR THE INDUCTION OF
LABOR AT TERM
J REPROD MED 9:292-299(1972)

SHIBUYA E
CHEMOTACTIC ACTIVITY IN THE AQUEOUS HUMOR OF PATIENTS
WITH BEHCET'S DISEASE
ACTA SOC OPHTHALMOL JAP 77:1434-1442(1973);(JAPANESE)

SHIGENOBU K   SPERELAKIS N
CALCIUM CURRENT CHANNELS INDUCED BY CATECHOLAMINES IN CHICK
EMBRYONIC HEARTS WHOSE FAST SODIUM CHANNELS ARE BLOCKED BY
TETRODOTOXIN OR ELEVATED POTASSIUM
CIRC RES 31:932-952(1972)

SHIMIZU K   YAMAMOTO M   YOSHITOSHI Y
EFFECTS OF SALINE INFUSION ON PROSTAGLANDIN-LIKE MATERIALS
IN RENAL VENOUS BLOOD AND MEDULLA OF CANINE KIDNEY
JAP HEART J 14:140-145(1973)

SHIMIZU T   FUJIMOTO S   ENDO K
THE EFFECT OF PROSTAGLANDIN F2A ON LABOR INDUCTION
SANFUJINKA NO SEKAI 23:847-850(1971);(JAPANESE)

SHIO H
PLATELET AGGREGATION AND PROSTAGLANDINS
J PHYSIOL SOC JAP 35:408(1973);(JAPANESE)

SHIO H   RAMWELL PW
PROSTAGLANDIN E2 STIMULATION OF HUMAN PLATELET AGGREGATION.
PHYSIOLOGIST 14:230(1971)

SHIO H   RAMWELL PW
PROSTAGLANDIN E(1) AND E(2): QUALITATIVE DIFFERENCE IN
PLATELET AGGREGATION
IN: "PROSTAGLANDINS IN CELLULAR BIOLOGY," PW RAMWELL AND BB
PHARRISS, PLENUM PRESS, NY/LOND:77-92(1972)

SHIO H   RAMWELL PW
EFFECT OF PROSTAGLANDIN E2 AND ASPIRIN ON THE SECONDARY
AGGREGATION OF HUMAN PLATELETS
NATURE (NEW BIOL) 236:45-46(1972)

SHIO H   RAMWELL PW
PROSTAGLANDIN E1 IN PLATELET HARVESTING: AN IN VITRO STUDY
SCIENCE 175:536-538(1972)

SHIO H   RAMWELL PW
THE PLATELET AS A MODEL OF PROSTAGLANDIN ACTION
INTRA-SCI CHEM REP 6:25-29(1972)

SHIO H   RAMWELL PW   JESSUP SJ
PROSTAGLANDIN E2: EFFECTS ON AGGREGATION, SHAPE CHANGE AND
CYCLIC AMP OF RAT PLATELETS
PROSTAGLANDINS 1:29-36(1972)

SHIO H   SHAW J   RAMWELL P
RELATION OF CYCLIC AMP TO THE RELEASE AND ACTIONS OF
PROSTAGLANDINS
ANN NY ACAD SCI 185:327-335(1971)

SHIRAHAMA H
SYNTHESIS OF PROSTAGLANDIN
KAGAKU (KYOTO) 28:332-341(1973);(JAPANESE)

SHIRAKURA T   YOSHIMATSU H   MIYAO S
EFFECT OF PROSTAGLANDIN E1 ON THE ERYTHROPOIESIS PROMOTING
ACTION OF TESTOSTERONE
MED BIOL (IGAKU TO SEIBUTSUGAKU) 84:295-298(1972);(JAPANESE)

SHIRAKURA T   YOSHIMATSU H   MIYAO S
MAEKAWA T
EFFECT OF PROSTAGLANDIN E1 ON BLOOD MAKING CAPABILITY
KETSUEKI TO MYAKKAN 3:337-340(1972);(JAPANESE)

SHKVATSABAYA IK   NEKRASOVA AA   SEREBROVSKAYA YA
HUMORAL PRESSOR-DEPRESSOR FUNCTION OF THE KIDNEYS IN THE
PATHOGENESIS OF EXPERIMENTAL RENAL HYPERTENSIONS
KARDIOLOGIIA 11:25-32(1971):(RUSSIAN)

SHULMAN NR   LANGE RF   TOMASULO PA
METABOLIC EFFECTS OF ANTIBODIES ON PLATELETS
CLIN RES 21:729(1973)

SIGGINS G   HOFFER B   BLOOM F
PROSTAGLANDIN-NOREPINEPHRINE INTERACTIONS IN BRAIN:
MICROELECTRO-PHORETIC AND HISTOCHEMICAL CORRELATES
ANN NY ACAD SCI 180:302-323(1971)

SIGGINS GR
PROSTAGLANDINS AND THE MICROVASCULAR SYSTEM: PHYSIOLOGICAL
AND HISTOCHEMICAL CORRELATIONS
IN: "PROSTAGLANDINS IN CELLULAR BIOLOGY," PW RAMWELL AND
BB PHARRISS, PLENUM PRESS, NY/LOND:451-478(1972)

SIGGINS GR
PROSTAGLANDINS AND THE MICROVASCULAR SYSTEM
ABSTR 7TH MIDDLE ATL REG MEET AM CHEM SOC PHILA:72(1972)

SIGGINS GR   HOFFER BJ   BLOOM FE
STUDIES ON NOREPINEPHRINE-CONTAINING AFFERENTS TO PURKINJE
CELLS OF RAT CEREBELLUM. III. EVIDENCE FOR MEDIATION OF
NOREPINEPHRINE EFFECTS BY CYCLIC 3',5'-ADENOSINE
MONOPHOSPHATE.
BRAIN 25:535-553(1971)

SIH CJ   HEATHER JB   PERUZZOTTI GP
PRICE P   SOOD R   LEE LFH
TOTAL SYNTHESIS OF PROSTAGLANDINS. IV. A COMPLETELY
STEREOSPECIFIC SYNTHESIS OF PROSTAGLANDIN E1
J AM CHEM SOC 95:1676-1677(1973)

SIH CJ   PRICE P   SOOD R
SALOMON RG
TOTAL SYNTHESIS OF PROSTAGLANDINS. II. PROSTAGLANDIN E1
J AM CHEM SOC 94:3643-3644(1972)

SIH CJ   SALOMON RG   PRICE P
PERUZZOTTI G   SOOD R
TOTAL SYNTHESIS OF (+-)-15-DEOXY-PROSTAGLANDIN E1
CHEM COMMUN 240-242(1972)

SIH CJ   SALOMON RG   PRICE P
SOOD R   PERUZZOTTI G
TOTAL SYNTHESIS OF PROSTAGLANDINS. III.
11-DESOXYPROSTAGLANDINS
TETRAHEDRON LETT 2435-2437(1972)

SIH CJ   TAKEGUCHI CA
BIOSYNTHESIS
IN: "THE PROSTAGLANDINS," PW RAMWELL, PLENUM PRESS, NY
1:83-100(1973)

SILVER MJ   SMITH JB   INGERMAN C
KOCSIS JJ
HUMAN BLOOD PROSTAGLANDINS: FORMATION DURING CLOTTING
PROSTAGLANDINS 1:429-436(1972)

SILVER MJ   SMITH JB   INGERMAN C
KOCSIS JJ
ARACHIDONIC ACID-INDUCED HUMAN PLATELET AGGREGATION AND
PROSTAGLANDIN FORMATION
PROSTAGLANDINS 4:863-875(1973)

SILVER MJ   SMITH JB   INGERMAN CC
KOCSIS JJ
PROSTAGLANDINS IN BLOOD: MEASUREMENT, SOURCES, AND EFFECTS
PROG HEMATOL 8 : 235-257(1973)

SILVER MJ   SMITH JB   WEBSTER GR
PHOSPHOLIPASE ACTIVITY IN HUMAN PLATELETS
PHARMACOLOGIST 13:276(1971)

SIMARD-DUQUESNE N   DVORNIK D
THE EFFECT OF PGE1 ON ADP-INDUCED AGGREGATION IN REFRACTORY
PLATELETS
PROSTAGLANDINS 3:457-460(1973)

SIMON EJ   HILLER JM   EDELMAN I
STEREOSPECIFIC BINDING OF THE POTENT NARCOTIC ANALGESIC
(3H)ETORPHINE TO RAT-BRAIN HOMOGENATE
PROC NATL ACAD SCI USA 70:1947-1949 (1973)

SIMPKINS H   PANKO E
THE INTERACTION OF PROSTAGLANDIN E1 WITH THE HUMAN
ERYTHROCYTE.
CLIN RES 19:794(1971)

SIMPSON LL   RAPPORT MM
THE BINDING OF BOTULINUM TOXIN TO MEMBRANE LIPIDS:
SPHINGOLIPIDS, STEROIDS AND FATTY ACIDS
J NEUROCHEM 18:1751-1759(1971)

SINCLAIR RJ
THE EFFECTS OF HISTAMINE AND PROSTAGLANDIN E2 ON RENAL
FLUID DYNAMICS, RENAL LYMPH FLOW AND COMPOSITION IN THE DOG
DISS, UNIV OKLA (1973)

SINGH KP   BHANDARI DS
NEUROPHARMACOLOGICAL STUDY OF PROSTAGLANDIN (PG)F2A
J INDIAN MED ASSOC 61:423-427(1973)

SINGH U   SEBUWUFU PHS
PROSTAGLANDINS AND NEUROSECRETION: A HISTOCHEMICAL STUDY
EAST AFR MED J 50:199-206(1973)

SINGLEY JA   CHAVIN W
CIRCULATING CORTISOL AND ACTH TITERS OF NORMAL GOLDFISH,
CARASSIUS AURATUS L. IN RESPONSE TO PROSTAGLANDINS
AM ZOOL 13:1280-1281(1973)

SIPIDO V
SYNTHESIS OF PROSTAGLANDINS
IND CHIM BELG 36:965-966(1971):(DUTCH)

SIXMA JJ   HOLMSEN H   TRIESCHNIGG ACM
UPTAKE OF ADENINE IN HUMAN BLOOD PLATELETS
ERYTHROCYTES THROMBOCYTES LEUKOCYTES RECENT ADV MEMBRANE
AND METABOL RES INT SYMP 2ND:228-230(1973)

SLADEN A
DRUGS USED IN RESPIRATORY THERAPY
IN: "CLINICAL ANESTHESIA. PHARMACOLOGY OF ADJUVANT DRUGS,"
HL ZAUDER, FA DAVIS CO, PHILA 10:315-341(1973)

SLATES HL   ZELAWSKI ZS   TAUB D
WENDLER NL
A NEW STEREOSELECTIVE TOTAL SYNTHESIS OF PROSTAGLANDIN E1
AND ITS OPTICAL ANTIPODES
CHEM COMMUN 304-305(1972)

SMEBY RR   BUMPUS FM
RENIN INHIBITORS
IN: "KIDNEY HORMONES," JW FISHER, ACADEMIC PRESS,
LOND:207-216(1971)

SMEJKAL V
PROSTAGLANDINS, A GROUP OF SUBSTANCES OF BIOLOGICAL AND
PHARMACOLOGICAL INTEREST
FARMAKOTER ZPR 19:113-125(1973):(CZECH)

SMIGEL M   FLEISCHER S
LOCALIZATION OF PROSTAGLANDIN E RECEPTORS IN THE PLASMA
MEMBRANE OF RAT LIVER
FED PROC 32:454(1973)

SMITH AD
CELLULAR CONTROL OF THE UPTAKE, STORAGE AND RELEASE OF
NORADRENALINE IN SYMPATHETIC NERVES
BIOCHEM SOC SYMP 36:103-131(1972)

SMITH AD
MECHANISMS INVOLVED IN THE RELEASE OF NORADRENALINE FROM
SYMPATHETIC NERVES
BR MED BULL 29:123-129(1973)

SMITH AP
RESPONSE OF ASPIRIN-ALLERGIC PATIENTS TO CHALLENGE BY SOME
ANALGESICS IN COMMON USE.
BR MED J 2:494-496(1971)

SMITH AP
EFFECTS OF PROSTAGLANDINS ON THE RESPIRATORY SYSTEM
IN: "THE PROSTAGLANDINS: PROGRESS IN RESEARCH," SMM KARIM,
MTP MED TECH PUBL CO, OXFORD:223-238(1972)

SMITH AP
SIDE-EFFECTS OF PROSTAGLANDINS
LANCET 2:655(1972)

SMITH AP
LUNGS
IN: "THE PROSTAGLANDINS," PW RAMWELL, PLENUM PRESS, NY
1:203-218(1973)

SMITH AP
PROSTAGLANDINS AND THE LUNG
MED TODAY 7:17-20(1973)

SMITH AP
ROLE OF PROSTAGLANDINS IN THE PATHOGENESIS AND TREATMENT OF
ASTHMA
IN: "ASTHMA," KF AUSTEN, LM LICHTENSTEIN, ACADEMIC PRESS,NY:
267-277(1973)

SMITH AP   CUTHBERT MF
EFFECTS OF I.V. INFUSED PROSTAGLANDIN ON AIRWAY RESISTANCE
IN MAN
ABSTR 5TH INT CONGR PHARMACOL SAN FRANC 217(1972)

SMITH AP   CUTHBERT MF
PROSTAGLANDINS AND RESISTANCE TO BETA-ADRENOCEPTOR
STIMULANTS
BR MED J 2:166(1972)

SMITH AP   CUTHBERT MF
ANTAGONISTIC ACTION OF AEROSOLS OF PROSTAGLANDINS F2A AND E2
ON BRONCHIAL MUSCLE TONE IN MAN
BR MED J 3:212-213(1972)

SMITH AP   CUTHBERT MF
THE EFFECTS OF INHALED PROSTAGLANDINS ON BRONCHIAL TONE IN
MAN
ADV BIOSCI 9:213-217(1973)

SMITH GM
THE ASSAY OF SUBSTANCES AFFECTING GASTRIC ACID SECRETION
IN:INT ENCYCL PHARMACOL THER,G PETERS,PERGAMON PRESS,NY
2:123-171(1973)

SMITH GM   FREULER F
THE MEASUREMENT OF INTRAVASCULAR AGGREGATION BY
CONTINUOUS PLATELET COUNTING
BIBL ANAT 12:229-234(1973)

SMITH GW   ABEL M   NATHANIELSZ PW
UTERINE SENSITIVITY TO PROSTAGLANDINS IN THE PREGNANT
RABBIT: A METHOD FOR TESTING THE EFFECT OF PREVIOUS
ADMINISTRATION OF SMALL DOSES OF PGF2A FOR SHORT PERIODS
PROSTAGLANDINS 3:525-530(1973)

SMITH JB   INGERMAN C   KOCSIS JJ
SILVER MJ
FORMATION OF PROSTAGLANDINS DURING THE AGGREGATION OF HUMAN
BLOOD PLATELETS
J CLIN INVEST 52:965-969(1973)

SMITH JB   KOCSIS JJ   INGERMAN C
SILVER MJ
PLATELET PHOSPHOLIPASES - HYDROLYSIS OF PHOSPHATIDYLCHOLINE
(PC)
ABSTR 5TH INT CONGR PHARMACOL SAN FRANC 217(1972)

SMITH JB   KOCSIS JJ   INGERMAN C
SILVER MJ
INACTIVATION OF PROSTAGLANDIN A1 AND A2 BY HUMAN RED CELLS
PHARMACOLOGIST 15:208(1973)

SMITH JB   WILLIS AL
ASPIRIN SELECTIVELY INHIBITS PROSTAGLANDIN PRODUCTION IN
HUMAN PLATELETS
NATURE (NEW BIOL) 231:235-237(1971)

SMITH JW   STEINER AL   NEWBERRY WM JR
PARKER CW
CYCLIC ADENOSINE 3',5'-MONOPHOSPHATE IN HUMAN LYMPHOCYTES.
ALTERATIONS AFTER PHYTOHEMAGGLUTININ STIMULATION
J CLIN INVEST 50:432-441(1971)

SMITH JW   STEINER AL   PARKER CW
HUMAN LYMPHOCYTE METABOLISM. EFFECTS OF CYCLIC AND NONCYCLIC
NUCLEOTIDES ON STIMULATION BY PHYTOHEMAGGLUTININ.
J CLIN INVEST 50:442-448(1971)

SMITH MJH   DAWKINS PD
SALICYLATE AND ENZYMES
J PHARM PHARMACOL 23:729-744(1971)

SMITH U   RYAN JW
PINOCYTOTIC VESICLES OF THE PULMONARY ENDOTHELIAL CELL.
CHEST 59:SUPPL 12S-15S(1971)

SMITH WL   LANDS WEM
OXYGENATION OF POLYUNSATURATED FATTY ACIDS DURING
PROSTAGLANDIN BIOSYNTHESIS BY SHEEP VESICULAR GLAND
BIOCHEMISTRY 11:3276-3285(1972)

SMITH WL JR
OXYGENATION OF POLYUNSATURATED FATTY ACIDS BY SHEEP
VESICULAR AND SOYBEAN DIOXYGENASES
DISS, UNIV MICH, ANN ARBOR (1971)

SMYTHIES JR
A POSSIBLE ROLE FOR NEUCLEOSIDE-PROTEIN COMPLEXES IN
MEMBRANE
BR J PHARMACOL 41:419P-420P(1971)

SMYTHIES JR
THE MOLECULAR NATURE OF THE ACETYLCHOLINE RECEPTOR: A
STEREOCHEMICAL STUDY
EUR J PHARMACOL 14:268-279(1971)

SMYTHIES JR
MOLECULAR SPECIFICATION OF THE A-AND B-ADRENERGIC RECEPTORS
PHARMACOLOGIST 13:201(1971)

SMYTHIES JR
MOLECULAR MECHANISMS OF ADRENERGIC RECEPTORS
J THEOR BIOL 35:93-101(1972)

SMYTHIES JR
THE ADENYL CYCLASE COMPLEX: A MODEL OF ITS MECHANISM OF
ACTION IN RELATION TO THE ALPHA- AND BETA-ADRENERGIC
RECEPTORS
EUR J PHARMACOL 19:18-24(1972)

SMYTHIES JR   ANTUN F   YANK G
YORKE C
MOLECULAR MECHANISMS OF STORAGE OF TRANSMITTERS IN
SYNAPTIC TERMINALS
NATURE (NEW BIOL) 231 : 185-188(1971)

SNEDDON JD   SMYTHE A   SATCHELL D
BURNSTOCK G
AN INVESTIGATION OF THE IDENTITY OF THE TRANSMITTER
SUBSTANCE RELEASED BY NON-ADRENERGIC, NON-CHOLINERGIC
EXCITATORY NERVES SUPPLYING THE SMALL INTESTINE OF SOME
LOWER VERTEBRATES
COMP GEN PHARMACOL 4:53-60(1973)

SNIDER BB
THE TOTAL SYNTHESIS OF RACEMIC FUMAGILLIN; A NEW SYNTHESIS
OF PROSTAGLANDINS
DISS, HARVARD UNIV (1973)

SNYDER DS   EAGLSTEIN WH
PROSTAGLANDINS AND SUNBURN
J INVEST DERMATOL 60:110-111(1973)

SNYDER DS   EAGLSTEIN WH
PROSTAGLANDINS AND SUNBURN
CLIN RES 21:742(1973)

SOLEZ K   FOX JA   KRAMER EC
HEPTINSTALL RH
EFFECTS OF INDOMETHACIN ON RENAL MEDULLARY PLASMA FLOW AND
ON NEPHROTOXIC ACUTE RENAL FAILURE
CLIN RES 5:994(1973)

SOLIANI F
REVIEW OF PROSTAGLANDINS
POLYCLINICO (PRAT) 79:445-455(1972):(ITALIAN)

SOLIS RT   WRIGHT CB   GIBBS MB
A MODEL FOR QUANTITATING IN-VIVO PLATELET AGGREGATION
BIBL ANAT 12:223-228(1973)

SOMERS K   KARIM SMM
CARDIOVASCULAR PHARMACOLOGICAL ACTIONS OF PROSTAGLANDINS
IN MAN
MED TODAY 7:24-28(1973)

SOMOVA IL   DOCHEV DH
STUDIES ON THE MECHANISM OF DIURETIC ACTION OF
PROSTAGLANDINE-E1
DOKL BOLG AKAD NAUK 26:1277-1280(1973)

SOMOVA L
A STUDY ON THE VASODEPRESSOR LIPID ISOLATED FROM KIDNEYS OF
HYPERTENSIVE ANIMALS
NEPHRON 8:575-583(1971)

SOMOVA L
INHIBITION OF PROSTAGLANDIN SYNTHESIS IN THE KIDNEYS BY
ASPIRIN-LIKE DRUGS
ADV BIOSCI (SUPPL)9:53(1972)

SOMOVA L
EFFECT OF PROSTAGLANDIN E1 AND E2 ON THE VASCULAR
RESPONSIVENESS TO ADRENALINE, NORADRENALINE, ANGIOTENSIN AND
VASOPRESSIN
COR VASA 14:213-221(1972)

SOMOVA L
INHIBITION OF PROSTAGLANDIN SYNTHESIS IN THE KIDNEYS BY
ASPIRINLIKE DRUGS
ADV BIOSCI 9:335-339(1973)

SOMOVA L   ORBETZOVA V   KIROV CH
INVESTIGATION OF CERTAIN HAEMATOLOGIC PARAMETERS IN RATS
TREATED WITH PROSTAGLANDINS (E1,E2,A1,A2,RESP)
COR VASA 15:125-135(1973)

SOMOVA L   PETKOV O
PROSTAGLANDINS AND ASPIRIN.
SUVREM MED 24(2):41-45(1973):(BULGARIAN)

SOMOVA L   PETKOV O   ORBETSOVA V
QUANTITATIVE CHANGES OF PROSTAGLANDINS IN KIDNEYS OF NORMAL
AND HYPERTONIC ANIMALS TREATED WITH ASPIRIN-LIKE
PREPARATIONS
EKSP MED MORFOL 11:168-172(1972):(RUSSIAN)

SOMOVA LI
HYPOTENSIVE EFFECT OF SYNTHETIC PROSTAGLANDINS OF THE E AND
A SERIES (PGE1, PGE2, PGA1 AND PGA2)
DOKL BOLG AKAD NAUK 24:1719-1722(1971)

SOMOVA LI
ANTIHYPERTENSIVE EFFECT OF SYNTHETIC PROSTAGLANDINS OF E
AND A SERIES
DOKL BOLG AKAD NAUK 25:137-140(1972)

SOMOVA LI
HYPOTENSIVE AND DIURETIC EFFECTS OF A VASODEPRESSOR LIPID OF
RENAL ORIGIN
BYULL EKSP BIOL MED 75:36-40(1973)

SOMOVA LI
A STUDY OF THE ACTION MECHANISM OF A VASODEPRESSIVE LIPID OF
RENAL ORIGIN AND ITS ASSOCIATION WITH ANTIHYPERTENSIVE
FUNCTION OF THE KIDNEYS
PATOL FIZIOL EKSP TER 17-22(1973):(RUSSIAN)

SOMOVA LI   DOCHEV DH
DIURETIC AND NATRIURETIC EFFECT OF PROSTAGLANDINS OF THE E
AND A SERIES
C R ACAD BULG SCI 24:1125-1128(1971)

SOMOVA LI   DOCHEV DH
THE EFFECT OF PROSTAGLANDINS (PGE1, PGE2, PGA1 AND PGA2) ON
THE ELECTROLYTE BALANCE OF ANIMALS WITH EXPERIMENTAL
HYPERTENSION
C R ACAD BULG SCI 24:1275-1278(1971)

SONDERGAARD J
SKIN
IN: "THE PROSTAGLANDINS," PW RAMWELL, PLENUM PRESS, NY
1:189-202(1973)

426

SONDERGAARD J
PROSTAGLANDINS IN INFLAMMATION:NEW PERSPECTIVES
UGESKR LAEGER 135:1012-1017(1973);(DANISH)

SONDERGAARD J
CUTANEOUS INFLAMMATION INDUCED BY PROSTAGLANDIN E1
ACTA PHYSIOL SCAND (SUPPL 396) : 21(1973)

SONDERGAARD J   GREAVES MW
PGE1-EFFECT ON HUMAN CUTANEOUS VASCULATURE AND SKIN
HISTAMINE.
BR J DERMATOL 84:424-428(1971)

SONDERGAARD J   HELIN P   JORGENSEN HP
HUMAN CUTANEOUS INFLAMMATION INDUCED BY PROSTAGLANDIN E1
J PATHOL 109:239-243(1973)

SONDERGAARD J   JORGENSEN HP
BLOCKADE BY POLYPHLORETIN PHOSPHATE OF THE PROSTAGLANDIN
E1-INDUCED HUMAN CUTANEOUS REACTION
BR J DERM 88:51-54(1973)

SONDERGAARD J   WOLF-JURGENSEN P
THE CELLULAR EXUDATE OF HUMAN CUTANEOUS INFLAMMATION INDUCED
BY PROSTAGLANDINS E1 AND F1A
ACTA DERM VENEREOL (STOCKH) 52:361-364(1972)

SORRELLS K   ERDOS EG   MASSION WH
EFFECT OF PROSTAGLANDIN E1 ON THE PULMONARY VASCULAR
RESPONSE TO ENDOTOXIN
PROC SOC EXP BIOL MED 140:310-313(1972)

SORRENTINO L   CAPASSO F   DI ROSA M
INDOMETHACIN AND PROSTAGLANDINS
EUR J PHARMACOL 17:306-308(1972)

SPAT A   SARKADI B   INTODY Z
KORNER A   SZANTO J
EFFECT OF RENAL PAPILLARY LIPIDS AND PROSTAGLANDIN E2 ON
CORTICOSTEROID PRODUCTION IN THE RAT
ACTA PHYSIOL ACAD SCI HUNG 40:187-199(1971)

SPAT A   SARKADI B   INTODY Z
KORNER A   SZANTO J
INFLUENCE OF LIPIDS OF THE RENAL PAPILLA AND OF
PROSTAGLANDIN E2 ON THE PRODUCTION OF HORMONES BY THE
ADRENAL CORTEX
KISERL ORVOSTUD 24:12-25(1972);(HUNGARIAN)

SPECTOR D   ZUSMAN R   SCHNEIDER G
CALDWELL B   SPEROFF L   MULROW P
EFFECT OF DIETARY SODIUM INTAKE ON CIRCULATING
PROSTAGLANDINS IN HUMANS
CLIN RES 20:868(1972)

SPELLACY WN   GALL SA
PROSTAGLANDIN F2A AND OXYTOCIN FOR TERM LABOR INDUCTION
J REPROD MED 9:300-303(1972)

SPEROFF L
PROSTAGLANDINS AND ABORTION
IN:THE ABORTION EXPERIENCE:PSYCHOLOGICAL & MEDICAL IMPACT,HJ
OSOFSKY & JD OSOFSKY,HARPER & ROW,HAGERSTOWN MD:415-435
(1973)

SPEROFF L   AULETTA FJ   CALDWELL BV
BROCK WA
PROSTAGLANDIN F2A AND THE PRIMATE CORPUS LUTEUM
ABSTR SOC GYNECOL INVEST 19TH ANNU MEET, PHOENIX:28(1972)

SPEROFF L   BROCK WA   CALDWELL BV
ANDERSON GG   HOBBINS JC
PERIPHERAL PLASMA ESTROGEN AND PROGESTERONE LEVELS DURING
PROSTAGLANDIN INFUSION FOR THE TERMINATION OF PREGNANCY
ABSTR 18TH ANNU MEET SOC GYNECOL INVEST, PHOENIX:16(1971)

SPEROFF L   GLASS RH   KASE NG
HORMONE BIOSYNTHESIS, METABOLISM, AND MECHANISM OF ACTION
IN: "CLINICAL GYNECOLOGIC ENDOCRINOLOGY AND INFERTILITY,"
L SPEROFF, RH GLASS AND NG KASE, WILLIAMS & WILKINS, BALT:
1-19(1973)

SPEROFF L   GLASS RH   KASE NG
MALE INFERTILITY
IN: "CLINICAL GYNECOLOGIC ENDOCRINOLOGY AND INFERTILITY,"
L SPEROFF, RH GLASS & NG KASE, WILLIAMS & WILKINS, BALT:
204-213(1973)

SPIRA PJ   WELCH KMA   LANCE JW
THE EFFECT OF HUMORAL AGENTS ON THE CRANIAL CIRCULATION OF
THE MONKEY
PROC AUST ASSOC NEUROL 10:97-103(1973)

SPLAWINSKI JA   NIES AS   BIECK PR
OATES JA
MECHANISM OF THE CONTRACTION INDUCED BY ARACHIDONIC ACID
(AA) ON THE RAT STOMACH LONGITUDINAL MUSCLE STRIP.
PHARMACOLOGIST 13:291(1971)

SPLAWINSKI JA   NIES AS   OATES JA
COMPARISON OF THE EFFECTS OF PROSTAGLANDINS E2 AND F2A WITH
THEIR PRECURSOR , ARACHIDONIC ACID,ON RENAL FUNCTION IN DOGS
ADV BIOSCI (SUPPL)9:62(1972)

SPLAWINSKI JA   NIES AS   SWEETMAN B
OATES JA
THE EFFECTS OF ARACHIDONIC ACID, PROSTAGLANDIN E2 AND
PROSTAGLANDIN F2A ON THE LONGITUDINAL STOMACH STRIP OF THE
RAT
J PHARMACOL EXP THER 187:501-510(1973)

SPRAGGINS RL
PGA2 AND ISOMERS FROM CORAL PROSTAGLANDINS
TETRAHEDRON LETT 4343-4346(1972)

SPRAGGINS RL
TRANSITION METAL ION ASSISTED CHROMATOGRAPHY. SEPARATION OF
PROSTAGLANDINS PGA2 AND PGB2
J ORG CHEM 38:3661-3662(1973)

SPRAGGINS RL
PGA2 TYPE PROSTAGLANDIN INTERMEDIATES IN THE PRODUCTION OF
PGA2 FROM 15-EPI-PGA2
IN: FOOD-DRUGS FROM THE SEA PROCEEDINGS 1972, LR WORTHEN,
MARINE TECH SOC, WASH : 157-163(1973)

SRIVASTAVA KC   CLAUSEN J
SYNTHESIS OF PROSTAGLANDINS (PGE, PGF, PGA AND PGB) IN HUMAN
PLATELETS
LIPIDS 7:762-765(1972)

SRIVASTAVA KC   CLAUSEN J
STABILITY OF PROSTAGLANDIN E COMPOUNDS IN SOLUTION
LIPIDS 8:592-594(1973)

STAMFORD IF   UNGER WG
IMPROVED PURIFICATION AND CHROMATOGRAPHY OF EXTRACTS
CONTAINING PROSTAGLANDINS
J PHYSIOL (LOND) 225:4P-5P(1972)

STAMM T   WELLMANN W
EFFECTS OF CATECHOLAMINES AND PROSTAGLANDINS ON CYCLIC AMP
LEVELS IN BRAIN IN VIVO
NAUNYN SCHMIEDEBERGS ARCH PHARMACOL (SUPPL) 277:R74(1973)

STARKE K   MONTEL H
SYMPATHOMIMETIC INHIBITION OF NORADRENALINE RELEASE:
MEDIATED BY PROSTAGLANDINS?
NAUN SCHMIEDEBERG'S ARCH PHARMACOL 278:111-116(1973)

STARKE K   MONTEL H
INTERACTION BETWEEN INDOMETHACIN, OXYMETAZOLINE AND
PHENTOLAMINE ON THE RELEASE OF (3H)NORADRENALINE FROM BRAIN
SLICES
J PHARM PHARMACOL 25:758-759(1973)

STARKE K   MONTEL H
LOCAL FEED-BACK CONTROL OF NORADRENALIN RELEASE AS POSSIBLE
SITES OF ACTION OF ANTI-HYPERTENSIVE DRUGS
THER WOCHE 23:4263-4264(1973)

STARR MS
EFFECTS OF PROSTAGLANDINS ON BLOOD FLOW IN THE RABBIT EYE
EXP EYE RES 11:161-169(1971)

STARR MS
FURTHER STUDIES ON THE EFFECT OF PROSTAGLANDIN ON
INTRAOCULAR-PRESSURE IN RABBIT
EXP EYE RES 11:170-177(1971)

STARR MS
SOME EFFECTS OF PROSTAGLANDINS ON RABBIT AND MONKEY EYES
EXP EYE RES 12:368(1971)

STAUNTON MD   GREENING WP
CLINICAL DIAGNOSIS OF THYROID CANCER
BR MED J 4: 532-535 (1973)

STECHSCHULTE DJ   ORANGE RP ,   AUSTEN KF
DETECTION OF SLOW REACTING SUBSTANCE OF ANAPHYLAXIS (SRS-A)
IN PLASMA OF GUINEA PIGS DURING ANAPHYLAXIS
J IMMUNOL 111:1585-1589(1973)

STEFANINI P   BENEDETTI-VALEN/ F FIORANI P
FARAGLIA V   PISTOLESE GR
PROBLEMS IN SURGICAL MANAGEMENT OF RENOVASCULAR HYPERTENSION
J CARDIOVASC SURG (TORINO)13:55-73(1972)

STEINBERG D
HORMONAL CONTROL OF LIPOLYSIS IN ADIPOSE TISSUE
ADV EXP MED BIOL 26:77-88(1972)

STEINER L   FORSTER DMC   BERGVALL U
CARLSON LA
EFFECT OF PROSTAGLANDIN E1 ON CEREBRAL CIRCULATORY
DISTURBANCES FOLLOWING SUBARACHNOID HEMORRHAGE IN MAN
PANMINERVA MED 13:190(1971)

STEINER L   FORSTER DMC   BERGVALL U
CARLSON LA
EFFECT OF PROSTAGLANDIN E1 ON CEREBRAL CIRCULATORY
DISTURBANCES FOLLOWING SUBARACHNOID HEMORRHAGE IN MAN.
PANMINERVA MED 13:190(1971)

STEINER L   FORSTER DMC   BERGVALL U
CARLSON LA
EFFECT OF PROSTAGLANDIN E1 ON CEREBRAL CIRCULATORY
DISTURBANCES
EUR NEUROL (PART II)8:23-31(1972)

STEINER L   FORSTER DMC   BERGVALL U
CARLSON LA
EFFECT OF PROSTAGLANDIN E1 ON CEREBRAL CIRCULATORY
DISTURBANCES FOLLOWING SUBARACHNOID HEMORRHAGE IN MAN
NEURORADIOLOGY 4:20-24(1972)

STEWART JM
ROLE OF THE LUNGS IN THE METABOLISM OF CIRCULATING HORMONES
CHEST 59(SUPPL):7S-8S(1971)

STIBBE J   ONG GL   TEN HOOR F
NAUTA J
THE INFLUENCE OF PGE1 ON PLATELET DECREASE IN THE HEART-LUNG
MACHINE
ABSTR 4TH INT CONGR THROMB HAEMOSTASIS, VIENNA:412(1973)

STITT JT
PROSTAGLANDIN E1 FEVER INDUCED IN RABBITS
J PHYSIOL (LOND) 232 : 163-179(1973)

STITT JT   HARDY JD
EFFECT OF PROSTAGLANDIN E1 INJECTED INTO THE BRAINSTEM ON
THE BODY TEMPERATURE OF RABBITS
FED PROC 31:364(1972)

STITT JT   HARDY JD
EVIDENCE THAT PROSTAGLANDIN E1 MAY BE A MEDIATOR IN
PYROGENIC FEVER IN RABBITS
BIOMETEOROLOGY 5:112(1972)

STITT JT   PIERCE JB
THE EFFECTS OF AMBIENT AND HYPOTHALAMIC TEMPERATURE ON
PROSTAGLANDIN E1 FEVERS INDUCED IN RABBITS
FED PROC 32:407(1973)

STJARNE L
HYPEREXCRETION OF CATECHOLAMINES INDUCED BY INDOMETHACINE
ACTA PHYSIOL SCAND 83:574-576(1971)

STJARNE L
HYPEREXCRETION OF CATECHOLAMINES INDUCED BY INDOMETHACINE.
ACTA PHYSIOL SCAND 83:574-576(1971)

STJARNE L
PROSTAGLANDIN E RESTRICTING NORADRENALINE SECRETION --
NEURAL IN ORIGIN?
ACTA PHYSIOL SCAND 86:574-576(1972)

STJARNE L
ENHANCEMENT BY INDOMETHACIN OF COLD-INDUCED HYPERSECRETION
OF NORADRENALINE IN THE RAT IN VIVO--BY SUPPRESSION OF PG-
MEDIATED FEEDBACK CONTROL?
ACTA PHYSIOL SCAND 86:388-397(1972)

STJARNE L
KINETICS OF SECRETION OF SYMPATHETIC NEUROTRANSMITTER AS A
FUNCTION OF EXTERNAL CALCIUM: MECHANISM OF INHIBITORY EFFECT
OF PROSTAGLANDIN E
ACTA PHYSIOL SCAND 87:428-430(1973)

STJARNE L
INHIBITORY EFFECT OF PROSTAGLANDIN E2 ON NORADRENALINE
SECRETION FROM SYMPATHETIC NERVES AS A FUNCTION OF EXTERNAL
CALCIUM
PROSTAGLANDINS 3:105-109(1973)

STJARNE L
DUAL ALPHA-ADRENOCEPTOR MEDIATED CONTROL OF SECRETION OF
SYMPATHETIC NEUROTRANSMITTER: ONE MECHANISM DEPENDENT AND
ONE INDEPENDENT OF PROSTAGLANDIN E
PROSTAGLANDINS 3:111-116(1973)

STJARNE L
COMPARISON OF SECRETION OF SYMPATHETIC NEUROTRANSMITTER
INDUCED BY NERVE STIMULATION WITH THAT EVOKED BY HIGH
POTASSIUM, AS TRIGGERS OF DUAL ALPHA-ADRENOCEPTOR
MEDIATED NEGATIVE FEEDBACK CONTROL OF NORADRENALINE
SECRETION
PROSTAGLANDINS 3:421-426(1973)

STJARNE L
ROLE OF ALPHA-ADRENOCEPTORS IN PROSTAGLANDIN E MEDIATED
NEGATIVE FEEDBACK CONTROL OF THE SECRETION OF
NORADRENALINE FROM THE SYMPATHETIC NERVES OF ISOLATED
GUINEA-PIG VAS DEFERENS
PROSTAGLANDINS 4:845-851(1973)

STJARNE L
LACK OF CORRELATION BETWEEN PROFILES OF TRANSMITTER
EFFLUX AND OF MUSCULAR CONTRACTION IN RESPONSE TO NERVE
STIMULATION IN ISOLATED GUINEA-PIG VAS DEFERENS
ACTA PHYSIOL SCAND 88:137-144(1973)

STJARNE L
SYNAPTIC DYNAMICS I: MECHANISMS OF CATECHOLAMINE SECRETION
LIFE SCI 13:CLIII-CLVI(1973)

STJARNE L
FREQUENCY DEPENDENCE OF DUAL NEGATIVE FEEDBACK CONTROL OF
SECRETION OF SYMPATHETIC NEUROTRANSMITTER IN GUINEA-PIG VAS
DEFERENS
BR J PHARMACOL 49:358-360(1973)

STJARNE L
MICHAELIS-MENTEN KINETICS OF CALCIUM-DEPENDENCE OF
SYMPATHETIC NEUROTRANSMITTER SECRETION AND OF PROSTAGLANDIN
DEPENDENT AND-INDEPENDENT FEEDBACK CONTROL OF THIS FUNCTION
ACTA PHYSIOL SCAND (SUPPL 396) : 106(1973)

STJARNE L   GRIPE K
PROSTAGLANDIN-DEPENDENT AND -INDEPENDENT FEEDBACK CONTROL OF
NORADRENALINE SECRETION IN VASOCONSTRICTOR NERVES OF
NORMOTENSIVE HUMAN SUBJECTS
NAUNYN SCHMIEDEBERGS ARCH PHARMACOL 280:441-446(1973)

STOCK K   PRILOP M
EFFECTS OF PHENYLISOPROPYLADENOSINE AND PGE ON CAMP-CONTENT
AND GLYCEROL RELEASE OF RAT FAT CELLS
NAUNYN SCHMIEDEBERGS ARCH PHARMACOL 277(SUPPL):R77(1973)

STOLC V
REGULATION OF IODINE METABOLISM IN HUMAN LEUKOCYTES BY
ADENOSINE 3',5'-MONOPHOSPHATE
BIOCHIM BIOPHYS ACTA 264:285-288(1972)

STOLC V
REGULATION OF CYCLIC ADENOSINE MONOPHOSPHATE (CAMP)
FORMATION IN HUMAN POLYMORPHONUCLEAR NEUTROPHILS
FED PROC 32:536(1973)

STONER J   MANGANIELLO VC   VAUGHAN M
EFFECTS OF BRADYKININ AND INDOMETHACIN ON CYCLIC GMP AND
CYCLIC AMP IN LUNG SLICES
PROC NATL ACAD SCI USA PART II 70:3830-3833(1973)

STRAND JC   MILLER MP   MCGIFF JC
COMPARISON OF THE BIOLOGICAL ACTIVITIES OF PROSTAGLANDIN E2
METHYL ESTER (ME) AND 15(S)-15-METHYL PROSTAGLANDIN E2
METHYL ESTER (15S) WITH PROSTAGLANDIN E2 (PGE2)
FED PROC 32:787(1973)

STRANDBERG K
EFFLUX OF SLOW REACTING SUBSTANCE (SRS) AND PROSTAGLANDINS
FROM CAT PAWS PERFUSED WITH D-TUBOCURARINE AND N-OCTYLAMINE.
ACTA PHYSIOL SCAND 82:509-520(1971)

STRANDBERG K
SMOOTH MUSCLE STIMULATING LIPIDS APPEARING ON HISTAMINE
RELEASE IN THE CAT AND GUINEA-PIG
DISS, DEPT PHARMACOL, KAROLINSKA INST, STOCKH (1971)

STRANDBERG K
CA++ DEPENDENCE OF HISTAMINE RELEASE AND FORMATION OF SLOW
REACTING SUBSTANCE IN THE CAT PAW
ACTA PHYSIOL SCAND 82:500-508(1971)

STRANDBERG K
ANTI-ANAPHYLACTIC EFFECTS OF A PROSTAGLANDIN ANTAGONIST,
POLYPHLORETIN PHOSPHATE, IN THE GUINEA PIG
ACTA PHARMACOL TOXICOL (KBH) (SUPPL) 31:110(1972)

STRANDBERG K
BRONCHOCONSTRICTOR ACTIVITIES OF HISTAMINE, PROSTAGLANDIN
F2A AND SLOW REACTING SUBSTANCE
ACTA PHYSIOL SCAND (SUPPL 396) : 21(1973)

STRANDBERG K   MATHE AA   FREDHOLM B
PROTECTIVE EFFECT OF POLYPHLORETIN PHOSPHATE IN ANAPHYLAXIS
IN THE GUINEA PIG
LIFE SCI (I)11:701-712(1972)

STRANDBERG K   UVNAS B
PURIFICATION AND PROPERTIES OF THE SLOW REACTING SUBSTANCE
FORMED IN THE CAT PAW PERFUSED WITH COMPOUND 48/80
ACTA PHYSIOL SCAND 82:358-374(1971)

STRANDHOY JW   HAAS JA   KNOX FG
ALTERATION OF PERITUBULE STARLING FORCES AND PROXIMAL SODIUM
REABSORPTION BY PROSTAGLANDINS E1 AND E2
FED PROC 32:398(1973)

STREETEN DHP   DALAKOS TG   SOUMA M
THE PATHOGENIC ROLE OF THE UPRIGHT POSTURE AND THE
THERAPEUTIC ACTION OF DEXTROAMPHETAMINE IN IDIOPATHIC EDEMA
CLIN PHARMACOL THER 12:302(1971)

STRICKER J   MENDLOWITZ M   RUSSO C
GITLOW SE   BERTANI LM
THE EFFECT OF CONTINUOUS INTRAVENOUS PROSTAGLANDIN A1
INFUSION IN PATIENTS WITH ESSENTIAL HYPERTENSION
CLIN RES 19:713(1971)

STRIKE D   SMITH H
A NOVEL APPROACH TO THE TOTAL SYNTHESIS OF PROSTAGLANDINS.
PREPARATION OF A STEREOISOMERIC MIXTURE CONTAINING
(+-)-13,14-DIHYDROPROSTAGLANDIN E1
ANN NY ACAD SCI 180:91-100(1971)

STROM TB   CARPENTER CB   AUSTEN KF
MERRILL JP
MODULATION OF LYMPHOCYTE MEDIATED CYTOTOXICITY BY CYCLIC
NUCLEOTIDES
FED PROC 32:969(1973)

STROM TB   CARPENTER CB   GAROVOY MR
AUSTEN KF   MERRILL JP   KALINER M
THE MODULATING INFLUENCE OF CYCLIC NUCLEOTIDES UPON
LYMPHOCYTE-MEDIATED CYTO-TOXICITY.
J EXP MED 138:381-393(1973)

STROM TB   DEISSEROTH A   MORGANROTH J
CARPENTER CB   MERRILL JP
ALTERATION OF THE CYTOTOXIC ACTION OF SENSITIZED LYMPHOCYTES
BY CHOLINERGIC AGENTS AND ACTIVATORS OF ADENYLATE CYCLASE
PROC NATL ACAD SCI USA 69:2995-2999(1972)

STRONG CG   CHANDLER JT
INTERACTIONS OF PROSTAGLANDIN E(1) AND CATECHOLAMINES IN
ISOLATED VASCULAR SMOOTH MUSCLE
IN: "PROSTAGLANDINS IN CELLULAR BIOLOGY," PW RAMWELL AND
BB PHARRISS, PLENUM PRESS, NY/LOND:369-383(1972)

STUART RK   MCDONALD JWD
PROSTAGLANDIN E-1 (PGE-1), CYCLIC AMP, AND CALCIUM IN
PLATELET AGGREGATION
CLIN RES 20:502(1972)

STYLOS W   BURSTEIN S   RIVETZ B
GUNSALUS P   SKARNES R
THE PRODUCTION OF ANTI-F PROSTAGLANDIN SERUM AND ITS USE IN
RADIOIMMUNOASSAY
INTRA-SCI CHEM REP 6:67-71(1972)

STYLOS WA   RIVETZ B
PREPARATION OF SPECIFIC ANTISERUM TO PROSTAGLANDIN A
PROSTAGLANDINS 2:103-113(1972)

SU JY   HIGGINS CB   FRIEDMAN WF
CHRONOTROPIC AND INOTROPIC EFFECTS OF PROSTAGLANDINS E1, A1,
AND F2A ON ISOLATED MAMMALIAN CARDIAC TISSUE
PROC SOC EXP BIOL MED 143:1227-1230(1973)

SULLIVAN W
ASPIRIN AND THE PROSTAGLANDINS
J PHARM (AUSTRALAS) 52: 606(1971)

SUN FF   STAFFORD JE
DISPOSITION OF PROSTAGLANDIN F2A IN THE RHESUS MONKEY
ABSTR 13TH NATL MEET APHA, ACAD PHARM SCI 2(2):15(1972)

SUNAHARA FA   SEN AK   TALESNIK J
STUDIES ON REGULATION OF CORONARY CIRCULATION. I.
PROSTAGLANDINS AND CORONARY INSUFFICIENCY
CAN MED ASSOC J 107:634(1972)

SUNAHARA FA   TALESNIK J
INTERACTION OF PROSTAGLANDIN E1 AND NORADRENALINE ON
CORONARY CIRCULATION
PROC CAN FED BIOL SCI 14:100(1971)

SUNAHARA FA   TALESNIK J   SEN AK
EFFECTS OF PROSTAGLANDIN AND DIAZOXIDE ON CORONARY
CIRCULATION OF NOREPINEPHRINE AND CALCIUM STIMULATED HEART
HEART
ABSTR 5TH INT CONGR PHARMACOL SAN FRANC 225(1972)

SUNJIC V   SUNJIC V   KAIFEZ F
PROSTAGLANDINS
FARM GLAS 28:239-249(1972);(CROATIAN)

SVANBORG K   BYGDEMAN M
METABOLISM OF PROSTAGLANDIN F2A IN THE RABBIT
EUR J BIOCHEM 28:127-135(1972)

SVANBORG N   HAMBERG M   HEDQVIST P
ASPECTS ON PROSTAGLANDIN ACTION IN ASTHMA
ACTA PHYSIOL SCAND (SUPPL 396) : 22(1973)

SWEAT FW   WINCEK TJ
THE STIMULATION OF HEPATIC ADENYLATE CYCLASE BY
PROSTAGLANDIN E1
BIOCHEM BIOPHYS RES COMMUN 55:522-529 (1973)

SWEDIN G
STUDIES ON NEUROTRANSMISSION MECHANISMS IN THE RAT AND
GUINEA-PIG VAS DEFERENS
ACTA PHYSIOL SCAND (SUPPL) 369:1-34(1971)

SWEET CS
STUDIES ON NEUROGENIC AND HUMORAL MECHANISMS OF ACUTE RENAL
HYPERTENSION
DISS, UNIV IOWA, IOWA CITY (1971)

SWEET CS   KADOWITZ PJ   BRODY MJ
ANOTHER HUMORAL SUBSTANCE THAT ENHANCES ADRENERGIC
RESPONSIVENESS DURING ACUTE RENAL ISCHAEMIA
NATURE 231:263-265(1971)

SWEET CS   KADOWITZ PJ   BRODY MJ
A HYPERTENSIVE RESPONSE TO INFUSION OF PROSTAGLANDIN F2A
INTO THE VERTEBRAL ARTERY OF THE CONSCIOUS DOG
EUR J PHARMACOL 16:229-232(1971)

SWEET CS   KADOWITZ PJ   BRODY MJ
DEPRESSION OF ADRENERGIC TRANSMISSION BY A FACTOR IN RENAL
VENOUS BLOOD.
PHARMACOLOGIST 13:293(1971)

SWEET CS   KADOWITZ PJ   FORKER EL
BRODY MJ
DEPRESSION OF ADRENERGIC TRANSMISSTION BY A FACTOR IN RENAL
VENOUS BLOOD: NEW EVIDENCE FOR AND ANTIHYPERTENSIVE FUNCTION
OF THE KIDNEY
ARCH INT PHARMACODYN THER 198:229-237(1972)

SWEETMAN BJ   FROLICH JC   WATSON JT
CONVERSION OF PGE1 OR PGE2 TO PGA OR PGB AS MORE ·SUITABLE
DERIVATIVES FOR VAPOR PHASE ANALYSIS INTO THE SUBNANOGRAM
RANGE
ABSTR 5TH INT CONGR PHARMACOL SAN FRANC 227(1972)

SWEETMAN BJ   FROLICH JC   WATSON JT
QUANTITATIVE DETERMINATION OF PROSTAGLANDINS A, B AND E IN
THE SUB-NANOGRAM RANGE
PROSTAGLANDINS 3:75-87(1973)

SWERDLOFF RS   GROVER PK
SEARCH FOR INHIBIN-EFFECT OF STEROIDS AND PROSTAGLANDINS ON
SERUM FSH AND LH LEVELS
CLIN RES 21:504(1973)

SWIDERSKA-KULIK/ B
CLINICAL USE OF PROSTAGLANDINS TYPE E (PGE) AND TYPE A (PGA)
POL ARCH MED WEWN 47:61-67(1971);(POLISH)

SYKES JAC   MADDOX IS
PROSTAGLANDIN PRODUCTION BY EXPERIMENTAL TUMOURS AND EFFECTS
OF ANTI-INFLAMMATORY COMPOUNDS
NATURE (NEW BIOL) 237:59-61(1972)

SYLWESTROWICZ W   NIEDZWIECKI J   METLER S
EMERICH J
LABOR INDUCTION BY MEANS OF PROSTAGLANDIN F2A
GINEKOL POL 44:1043-1047(1973);(POLISH)

SZADUJKIS-SZADU/ L KOROLKIEWICZ Z
THE INFLUENCE OF METABOLIC INHIBITORS ON ACTIVE SODIUM
TRANSPORT STIMULATED BY CATECHOLAMINES AND PGE1
IN: PHOSPHOLIPIDS PROC INT SYMP, L SAMOCHOWIEC, J WOJCICKI,
INT SOC BIOCHEM PHARMACOL: 137-150(1973)

SZONTAGH F
PROSTAGLANDINS IN HUMAN REPRODUCTION
ORV HETIL 114:2337-2341(1973);(HUNGARIAN)

SZONTAGH F   RESCH B   BARTFAI G
HERCZEG J   TEKULICS A
EFFECT OF PROSTAGLANDIN F2A ON THE PREGNANT UTERUS
ORV HETIL 113:919-922(1972);(HUNGARIAN)

T'ANG MI
CLINICAL APPLICATION OF PROSTAGLANDIN IN OBSTETRICS
CHIN MED J 9:563-569(1973);(CHINESE)

TAI HH   HOLLANDER CS
PROSTAGLANDIN (PG) BIOSYNTHESIS: REGULATION BY HORMONES
POSSESSING SPECIFIC AROMATIC RING STRUCTURES
CLIN RES 20:558(1972)

TAI HH   HOLLANDER CS
REGULATION OF PROSTAGLANDIN SYNTHETASE ACTIVITY IN RABBIT
KIDNEY MEDULLA: A POSSIBLE MECHANISM OF HORMONAL AND DRUG
ACTION
ADV BIOSCI (SUPPL) 9:5(1972)

TAI HH   TAI CL   HOLLANDER CS
REGULATION OF 15-HYDROXYPROSTAGLANDIN DEHYDROGENASE (PGDH)
ACTIVITY BY POSITIVE AND NEGATIVE MODULATORS
ABSTR, 9TH INT CONG BIOCHEM, STOCKH:405(1973)

TAIRA N   SATOH S
PROSTAGLANDIN F2A AS A POTENT EXCITANT OF THE
PARASYMPATHETIC POSTGANGLIONIC NEURONS OF THE DOG SALIVARY
LIFE SCI 13: 501-506(1973)

TAKAKI S   YOSHIDA T   T'IEN K
SAKATA H   SUZUKI K   SAWASAKI C
THE BEHAVIOR OF PROSTAGLANDINS IN THE PERIPHERAL BLOOD
CIRCULATION DURING THE MENSTRUAL CYCLE
J JAP OBSTET GYNECOL SOC 25:541-542(1973)

TAKAMITSU Y   URAKABE S   SHIRAI D
ABE H
EFFECT OF PROSTAGLANDINS ON THE WATER PERMEABILITY AND
SODIUM TRANSPORT IN THE TOAD BLADDER
ABSTR 5TH INT CONGR NEPHROL, MEX CITY:112(1972)

TAKAMITSU Y   URAKABE S   SUGITA M
SHIRAI D
EFFECT OF PROSTAGLANDINS ON THE WATER PERMEABILITY AND
SODIUM TRANSPORT IN THE TOAD BLADDER
J PHYSIOL SOC JAP 35:394(1973);(JAPANESE)

TAKANO S
THE INHIBITORY EFFECT OF PROSTAGLANDIN E1 ON THE CONTRACTION
OF ISOLATED SPLEEN STRIPS PRODUCED BY NORADRENALINE, OR
ACETYLCHOLINE
FUKUSHIMA J MED SCI 18:35-41(1971)

TAKANO S
THE INHIBITORY EFFECT OF PROSTAGLANDIN E1 ON THE CONTRACTION
OF ISOLATED SPLEEN STRIPS PRODUCED BY NORADRENALINE, OR
ACETYLCHOLINE
FUKUSHIMA J MED SCI 18:35-41(1971)

TAKAO M
HUMAN IRIS EXTRACT
ACTA SOC OPHTHALMOL JAP 75:1131-1132(1971);(JAPANESE)

TAKEGUCHI C   KOHNO E   SIH CJ
MECHANISM OF PROSTAGLANDIN BIOSYNTHESIS. I. CHARACTERIZATION
AND ASSAY OF BOVINE PROSTAGLANDIN SYNTHETASE
BIOCHEMISTRY 10:2372-2376(1971)

TAKEGUCHI CA
THE BIOSYNTHESIS OF PROSTAGLANDINS
DISS, UNIV WIS (1973)

TAKEUCHI H   MORI A   KOHSAKA M
A PHARMACOLOGICAL STUDY ON A GIANT NEURONE, IDENTIFIED IN
THE SUB-ESOPHAGEAL GANGLION OF THE GIANT AFRICAN SNAIL
(ACHATINA FULICA FERRUSAC), SENSITIVE TO 5-HYDROXY
TRYPTAMINE AND DOPAMINE
C R SOC BIOL (PARIS) 167:602-610(1973)

TAKEYA N
ACTION OF PGF2A ON THE ISOLATED HEART OF THE GUINEA PIG
ABSTR 43RD KINKI REG MEET JAP PHARMACOL SOC 15:(1973):
(JAPANESE)

TALESNIK J   SUNAHARA FA
ENHANCEMENT OF METABOLIC CORONARY DILATATION BY ASPIRIN-LIKE
SUBSTANCES BY SUPPRESSION OF PROSTAGLANDIN FEEDBACK CONTROL?
NATURE (LOND) 244:351-353(1973)

TAN L   FALARDEAU P   WANG HM
LEHOUX JG
BINDING OF PROSTAGLANDINS WITH CYTOCHROME P-450
ABSTR, 9TH INT CONG BIOCHEM, STOCKH:398(1973)

TAN L   WANG HM   LEHOUX JG
BINDING OF PROSTAGLANDINS AND CYTOCHROME P-450
INT RES COMMUN SYST (73-3)3-5-1(1973)

TAN L   WANG HM   LEHOUX JG
BINDING OF PROSTAGLANDINS AND CYTOCHROME P-450
PROSTAGLANDINS 4:9-16(1973)

TAN WC   PRIVETT OS
STUDIES ON THE SYNTHESIS OF PROSTAGLANDINS IN THE VESICULAR
GLANDS OF ESSENTIAL FATTY ACID-DEFICIENT AND
HYPOPHYSECTOMIZED RATS
BIOCHIM BIOPHYS ACTA 296:586-592(1973)

TAN WC   PRIVETT OS
STUDIES ON DETECTION AND SYNTHESIS OF PROSTAGLANDINS IN TAIL
SKIN OF THE RAT
LIPIDS 8:166-169(1973)

TANABE K   YAMAZAKI M   MATSUMOTO Y
MISHIMA K
CENTRAL NERVOUS ACTION OF PROSTAGLANDIN F2A
ABSTR 47TH KANTO REG MEET, JAP PHARMACOL SOC, TOKYO:19
(1972):(JAPANESE)

TANDON AK   DWIVEDI S   SINGH RH
SOMANI PN   UDUPA KN
PROSTAGLANDIN-LIKE ACTIVITY IN ISCHAEMIC HEART DISEASE AND
ESSENTIAL HYPERTENSION
INT RES COMMUN SYS (73-11): 11-11-7 (1973)

TASAKA K   OHYAMA M
EFFECT OF CERTAIN AUTACOIDS ON THE REGIONAL BLOOD FLOW IN
THE RABBIT MIDBRAIN RETICULAR FORMATION
JAP J PHARMACOL (SUPPL) 23:72(1973)

TASHJIAN AH JR   VOELKEL EF   GOLDHABER P
LEVINE L
SUCCESSFUL TREATMENT OF HYPERCALCEMIA BY INDOMETHACIN IN
MICE BEARING A PROSTAGLANDIN-PRODUCING FIBROSARCOMA
PROSTAGLANDINS 3:515-524(1973)

TASHJIAN AH JR   VOELKEL F   LEVINE L
GOLDHABER P
EVIDENCE THAT THE BONE RESORPTION-STIMULATING FACTOR
PRODUCED BY MOUSE FIBROSARCOMA CELLS IS PROSTAGLANDIN E2. A
NEW MODEL FOR THE HYPERCALCEMIA OF CANCER
J EXP MED 136:1329-1343(1972)

TAUB D
DISCUSSION. (PART I. CHEMISTRY OF PROSTAGLANDINS).
ANN NY ACAD SCI 180:101-103(1971)

TAUB D
A NEW STEREOSELECTIVE SYNTHESIS OF RACEMIC AND NATURALLY
OCCURRING PROSTAGLANDIN E1
INTRA-SCI CHEM REP 6:99-104(1972)

TAUB D   HOFFSOMMER RD   KUO CH
SLATES HL   ZELAWSKI ZS   WENDLER NL
A STEREOSELECTIVE TOTAL SYNTHESIS OF (+-)
PROSTAGLANDIN E1
TETRAHEDRON 29:1447-1456(1973)

TAUBER AI   KALINER M   STECHSCHULTE DJ
AUSTEN KF
THE EFFECT OF PROSTAGLANDINS ON THE IMMUNOLOGIC RELEASE OF
HISTAMINE FROM HUMAN LUNG TISSUE
J ALLERGY CLIN IMMUNOL 51:106(1973)

TAUBER AI   KALINER M   STECHSCHULTE DJ
AUSTEN KF
IMMUNOLOGIC RELEASE OF HISTAMINE AND SLOW REACTING SUBSTANCE
OF ANAPHYLAXIS FROM HUMAN LUNG. V. EFFECTS OF PROSTAGLANDINS
ON RELEASE OF HISTAMINE
J IMMUNOL 111:27-32(1973)

TAUBER AI   KALINER MA   STECHSCHULTE DJ
AUSTEN KF
PROSTAGLANDINS AND THE IMMUNOLOGICAL RELEASE OF CHEMICAL
MEDIATORS FROM HUMAN LUNG
IN: "PROSTAGLANDINS AND CYCLIC AMP," RH KAHN, WEM LANDS,
ACADEMIC PRESS, NY : 29-48(1973)

TAYLOR RE   STITT ES   ROBISON GA
HARTMANN RC
CHARACTERIZATION OF PROSTAGLANDIN-CYCLIC AMP INTERACTIONS IN
HUMAN BLOOD PLATELETS
BLOOD 42:994(1973)

TAYLOR SH   MAJID PA
INSULIN AND THE HEART
J MOL CELL CARDIOL 2:293-317(1971)

TAYLOR SI   JUNGAS RL
EFFECTS OF ANTILIPOLYTIC AGENTS ON PYRUVATE DEHYDROGENASE
ACTIVITY OF RAT ADIPOSE TISSUE
FED PROC 31:244(1972)

TAYLOR SI   MUKHERJEE C   JUNGAS RL
STUDIES ON THE MECHANISM OF ACTIVATION OF ADIPOSE TISSUE
PYRUVATE DEHYDROGENASE BY INSULIN
J BIOL CHEM 248:73-81(1973)

TAYLOR WA
THE EFFECT OF DISODIUM CROMOGLYCATE AND PGF2A ON ACUTE
CUTANEOUS REACTIONS OF THE GUINEA PIG AND RAT
INT ARCH ALLERGY APPL IMMUNOL 45:82-83(1973)

TCHILINGUIRIAN NGO
COMPARISON OF PROSTAGLANDIN F2A AND PITOCIN IN THE
INDUCTION OF LABOR IN HIGH RISK PREGNANT WOMEN
J REPROD MED 9:331-338(1972)

TERAKI Y   MIYAZAKA S   KAKUO S
INFLUENCES OF PROSTAGLANDIN, 5-HYDROXYTRYPTAMINE, AND
POLYPEPTIDE ON THE BLOOD VESSELS OF MESOMETRIUM AND
MESENTERIUM
ABSTR 48TH KANTO REG MEET,JAP PHARMACOL SOC : 28(1973):
(JAPANESE)

TERASHIMA R   ANDERSON FL   JUBIZ W
INFLUENCE OF ADRENERGIC DRUGS AND SODIUM CHLORIDE ON THE
RELEASE OF PROSTAGLANDIN E FROM THE KIDNEY
CLIN RES 21:276(1973)

TERASHIMA S
SYNTHESIS OF PROSTAGLANDINS, ITS IMPROVEMENT AND
DEVELOPMENT
J SYNTH ORG CHEM JAP 31:353-374(1973):(JAPANESE)

TERRAGNO DA   STRAND JC   PACHOLCZYK DA
MCGIFF JC
PROSTAGLANDIN E2, AND INTRARENAL HORMONE
IN: "PROSTAGLANDINES 1973," INSERM, PARIS:207-223(1973)

TERRAGNO NA   CROWSHAW K   MALIK KU
LONIGRO AJ   MCGIFF JC
LOSS OF ADRENERGIC-INDUCED RENAL VASOCONSTRICTION AND THE
SELECTIVE RELEASE OF A PROSTAGLANDIN E-LIKE SUBSTANCE INTO
CANINE RENAL VENOUS BLOOD
FED PROC 30:486(1971)

TERRAGNO NA   LONIGRO AJ   MALIK KU
MCGIFF JC
BRADYKININ-INDUCED VASODILATION AND THE RELEASE OF
PROSTAGLANDINS.
CIRCULATION 44:II-118(1971)

TERRAGNO NA   LONIGRO AJ   MALIK KU
MCGIFF JC
THE RELATIONSHIP OF THE RENAL VASODILATOR ACTION OF
BRADYKININ TO THE RELEASE OF A PROSTAGLANDIN E-LIKE
SUBSTANCE
EXPERIENTIA 28:437-439(1972)

TERRAGNO NA   MCGIFF JC   ITSKOVITZ HD
PROSTAGLANDIN SYNTHESIS INHIBITION AND THE ACTIONS OF
BRADYKININ (BK) AND ELEDOISIN (EL) ON THE DISTRIBUTION OF
BLOOD FLOW IN THE ISOLATED CANINE KIDNEY
FED PROC 32:353(1973)

TERRANOVA T
SOME ASPECTS OF THE GENERAL PATHOLOGY OF INFLAMMATION
CLIN TER SUPPL 65:1-48(1973):(ITALIAN)

THIBAULT P
CURRENT INFORMATION ON THE PROSTAGLANDINS
PRESSE MED 79:400(1971):(FRENCH)

THIERY M
THE EFFECT OF PROSTAGLANDINS ON THE HUMAN REPRODUCTIVE CYCLE
VERH K VLAAM ACAD GENEESKD BELG 35:347-364(1973):(DUTCH)

THIERY M   DE HEMPTINNE D   VANDERHEYDEN K
VROMAN S   DEROM R   VAN KETS H
MARTENS G
INTRAVENOUS PROSTAGLANDIN F2A AND AMNIOTOMY FOR THE ELECTIVE
INDUCTION OF LABOR AT TERM
J PERINAT MED 1:268-282(1973)

THIERY M   VROMAN S   VANDERHEYDEN K
DE HEMPTINNE D   DEROM R   VAN KETS H
MARTENS G
THE FETAL EFFECT OF PROSTAGLANDIN F2A APPLIED IN THE
ELECTIVE INDUCTION OF LABOR AT TERM
J REPROD MED 9:314-326(1972)

THOMAS G   WEST GB
PROSTAGLANDINS AS REGULATORS OF BRADYKININ RESPONSES
J PHARM PHARMACOL 25:747-748(1973)

THOMPSON CJ   GOODE CN
THE ANALYSIS OF PROSTAGLANDINS USING A COMPUTER FOR PLOTTING
ABUNDANCE VERSUS TIME
ADV BIOSCI (SUPPL) 9:16(1972)

430

THOMPSON CJ   GOODE CN
THE ANALYSIS OF PROSTAGLANDINS USING A COMPUTER FOR PLOTTING
ION ABUNDANCE VERSUS TIME
ADV BIOSCI 9:117-120(1973)

THOMPSON GF   COLLINS JM   SCHMALZRIED LM
TOTAL RATE EQUATION FOR DECOMPOSITION OF PROSTAGLANDIN E2
J PHARM SCI 62: 1738-1739(1973)

THOMPSON RB   KAUFMAN CE   DISCALA VA
EFFECT OF RENAL VASODILATATION ON DIVALENT ION EXCRETION AND
TMPAH IN ANESTHETIZED DOGS
AM J PHYSIOL 221:1097-1104(1971)

THOMPSON RB   KAUFMAN CE   DISCALA VA
EFFECT OF RENAL VASODILATATION ON DIVALENT ION EXCRETION
CLIN RES 19:550(1971)

THORBURN GD   CURRIE WB
THE PHYSIOLOGY OF THE INITIATION OF LABOUR
IN: "PROSTAGLANDINES 1973," INSERM, PARIS:259-277(1973)

THORBURN GD   NICOL DH
REGRESSION OF THE OVINE CORPUS LUTEUM FOLLOWING INFUSION OF
PROSTAGLANDIN F2A INTO THE OVARIAN ARTERY AND UTERINE VEIN
J REPROD FERTIL 28:155(1972)

TILFORD CH   MACKENZIE RD   BLOHM TR
GRISAR JM
SUBSTITUTED 3,4-PENTADIENYLDIAMINES AS INHIBITORS OF
PLATELET AGGREGATION
J MED CHEM 16:688-693 (1973)

TIMSIT J   GIROUD JP   DECHEZLEPRETRE S
LECHAT P
RESEARCH ON THE ANTI-INFLAMMATORY ACTION OF ADRENALINE
ARCH INT PHARMACODYN THER 205:153-172(1973):(FRENCH)

TISMAN G   HERBERT V
STUDIES OF EFFECTS OF CYCLIC ADENOSINE 3',5'-MONOPHOSPHATE
IN REGULATION OF HUMAN HEMOPOIESIS IN VITRO
IN VITRO 9:86-91(1973)

TISMAN G   HERBERT V   GO LT
BRENNER L
EFFECTS ON HUMAN BONE MARROW OF CYCLIC AMP (CAMP),
DIBUTYRYL CAMP (DCAMP), THEOPHYLLINE (THEO), AND
PROSTAGLANDIN (PGE1)
CLIN RES 20:503(1972)

TOBIAN L
RENAL MEDULLARY INTERSTITIAL CELLS AND THE ANTIHYPERTENSIVE
ACTION OF NORMAL AND "HYPERTENSIVE" KIDNEYS
IN: "HYPERTENSION: MECHANISMS AND MANAGEMENT" G ONESTI KE
KIM AND JH MOYER,GRUNE & STRATTON,NY/LOND:645-652(1973)

TOBIAN L   AZAR S
FUNCTIONS OF RENAL PAPILLA.
CLIN RES 19:577(1971)

TOBIAN L   AZAR S
ANTIHYPERTENSIVE AND OTHER FUNCTIONS OF THE RENAL PAPILLA
TRANS ASSOC AM PHYSICIANS 84:281-288(1971)

TOFTE T
BRONCHODILATING AGENTS. PREVENTION AND THERAPY
TIDSSKR NOR LAEGEFOREN 93:777-778(1973):(DANISH)

TOJO S   KANAZAWA S   MOCHIZUKI M
NIIYA T   SAKAI T   NISHIDA Y
A BASIC AND CLINICAL STUDY ON THE LABOR INDUCING EFFECT OF
PROSTAGLANDIN F2A
J JAP OBSTET GYNAECOL SOC 23:441-449(1971)

TOJO S   MOCHIZUKI M   NIIYA T
KANAZAWA S   SAKAI T
STUDIES ON IN VITRO PRESERVATION OF THE HUMAN UTERUS
PROC WORLD CONGR FERTIL STERIL 7:421-423(1973)

TOKI S   TAKENOSHITA R   KAGEURA E
STUDIES ON 3-HYDROXYHEXOBARBITAL DEHYDROGENASE
IN: "ABSORPTION, METABOLISM AND EXCRETION OF DRUGS," K
KAKEMI, HIROKAWA PUBL CO, TOKYO:165-180(1971):(JAPANESE)

TOMKINS GM   KRAM R   MAMONT P
DANIEL V   LITWACK G   BOURNE HR
MECHANISMS OF GROWTH CONTROL IN CULTURED CELLS
ABSTR MEET CONTROL OF PROLIFERATION IN ANIMAL CELLS,
COLD SPRING HARBOR:69(1973)

TOMLINSON RV   RINGOLD HJ   QURESHI MC
FORCHIELLI E
RELATIONSHIP BETWEEN INHIBITION OF PROSTAGLANDIN SYNTHESIS
AND DRUG EFFICACY. SUPPORT FOR THE CURRENT THEORY ON MODE
OF ACTION OF ASPIRIN-LIKE DRUGS
BIOCHEM BIOPHYS RES COMMUN 46:552-559(1972)

TONNESEN MG   JUBIZ W   FRAILEY J
MOORE JG
STUDIES OF ENDOGENOUS GASTRIC PGE PRODUCTION IN HUMAN
SUBJECTS.
CLIN RES 21:276(1973)

TOWNLEY RG   ADOLPHSON RL
RELATIONSHIP OF PROSTAGLANDINS TO AIRWAY SMOOTH MUSCLE
IN: "PROSTAGLANDINS AND CYCLIC AMP," RH KAHN, WEM LANDS,
ACADEMIC PRESS, NY : 49-51(1973)

TOYODA T   KIKUCHI H   GOTO Y
EFFECT OF PROSTAGLANDIN E1 ON PANCREATIC SECRETION OF
INSULIN AND GLUCAGON
IGAKU NO AYUMI 87:20-21(1973):(JAPANESE)

TRAVERSO G   PIRILLO D   VILLA A
SYNTHESIS OF (+-)-9-DEOXY-11-DEHYDRO-13,14-DIHYDRO-15-R,
S-PROSTAGLANDIN
FARMACO (SCI) 28:1040-1042(1973)

TREAT E   ULANO HB   SHANBOUR LS
JACOBSON ED
SELECTIVE DILATION WITH PROSTAGLANDIN E1, GLUCAGON, AND
ISOPROTERENOL OF THE CONSTRICTED SUPERIOR MESENTERIC ARTERY
SURG FORUM 22:371-373(1971)

TREPTOW K   HECHT K   MORITZ V
POPPEI M   CHOINOWSKI S   NITSCHKOFF S
THE EFFECT OF PROSTAGLANDINS (PGE1 AND PGE2) UPON CENTRAL
NERVOUS PROCESSES OF LEARNING OF A CONDITIONED REFLEX AND
THE FORMATION OF A TIME-MEMORY
ACTA BIOL MED 31:111-120(1973); (GERMAN)

TRICOMI V
INDUCTION OF LABOR-A CONTEMPORARY VIEW
CLIN OBSTET GYNECOL 16:226-242(1973)

TSAFRIRI A   LINDNER HR   ZOR U
IN VITRO INDUCTION OF MEIOTIC DIVISION IN FOLLICLE-ENCLOSED
OVA BY LH AND PROSTAGLANDIN E2
PROC WORLD CONGR FERTIL STERIL 7:404-411(1973)

TSAI TH   PARMETER L   WHITE HL
MAXWELL RA
EFFECT OF INDOMETHACIN AND ASPIRIN ON THE RESPONSE OF
ISOLATED GUINEA-PIG ILEUM TO ARACHIDONIC ACID, A PRECURSOR
OF PROSTAGLANDIN E2
ABSTR 5TH INT CONGR PHARMACOL, SAN FRANC:237(1972)

TSONEV IT
PROSTAGLANDINS-NEW CLASS OF BIOLOGICALLY ACTIVE SUBSTANCES
SUVREM MED 24(10):3-7(1973):(RUSSIAN)

TUCHMANN-DUPLES/ H MERCIER-PAROT L
ACTION OF PROSTAGLANDIN F2A ON THE CORPUS LUTEUM OF THE
PREGNANT RAT
C R ACAD SCI (D) (PARIS) 275:2033-2035(1972):(FRENCH)

TURCK WPG   ZEITLIN IJ   SMITH AN
GRANT IWB
AIRWAYS OBSTRUCTION IN THE CARCINOID SYNDROME. A STUDY OF
THE EFFECT OF DRUGS AND OTHER FACTORS 1. CLINICAL AND
LABORATORY FINDINGS
SCOTT MED J 17:237-243(1972)

TURKER RK   KIRAN BK   VURAL H
DUAL EFFECTS OF PROSTAGLANDIN E1 ON THE CAT ISOLATED
PAPILLARY MUSCLE
ARZNEIM FORSCH 21:989-991(1971)

TURKER RK   ONUR R
EFFECT OF PROSTAGLANDIN E1 ON INTESTINAL MOTILITY OF THE CAT
ARCH INT PHYSIOL BIOCHIM 79:535-543(1971)

TURKER RK   YAMAMOTO M   BUMPUS FM
KHAIRALLAH PA
LUNG PERFUSION WITH ANGIOTENSINS I AND II. EVIDENCE OF
RELEASE OF MYOTROPIC AND INHIBITORY SUBSTANCES.
CIRC RES 28:559-567(1971)

TURNBERG LA
ABSORPTION AND SECRETION OF SALT AND WATER BY THE SMALL
INTESTINE
DIGESTION 9:357-381(1973)

TURNER S
CONTROL ELEMENTS IN ORGANIC SYNTHESIS
CHEM BR 7:191-195(1971)

TURPIE AGG   STEWART IO
THE EFFECT OF BACTERIA ON PLATELET AGGREGATION AND THE
PLATELET RELEASE REACTION
FED PROC 32:835(1973)

TYLER ET
HOW SOON WILL WE HAVE THE "IDEAL" CONTRACEPTIVE?
J AM MED ASSOC 219:1333(1972)

UBATUBA FB
THE USE OF THE HAMSTER STOMACH IN VITRO AS AN ASSAY
PREPARATION FOR PROSTAGLANDINS
BR J PHARMACOL 49:662-666(1973)

UEDA E   HATANAKA Y   ITO T
KOKUBU T   YAMAMURA Y
METABOLISM OF VASOACTIVE SUBSTANCES IN THE LUNG
JAP CIRC J 37:1255-1259(1973)

ULANO HB   TREAT E   SHANBOUR LL
JACOBSON ED
SELECTIVE DILATION OF THE CONSTRICTED SUPERIOR MESENTERIC
ARTERY
GASTROENTEROLOGY 62:39-47(1972)

UNBEHAUN V   CONRADT A   GUNTHER J
TREATMENT OF HYDATIDIFORM MOLE BY INDUCTION OF LABOR WITH
PROSTAGLANDIN F2A
GEBURTSHILFE FRAUENHEILK 33:436-439(1973):(GERMAN)

UNGER WG
BINDING OF PROSTAGLANDIN TO HUMAN SERUM ALBUMIN
J PHARM PHARMACOL 24:470-477(1972)

UNGER WG   STAMFORD IF   BENNETT A
EXTRACTION OF PROSTAGLANDINS FROM HUMAN BLOOD.
NATURE.(LOND) 233 : 336-337 (1971)

URAKABE S   SHIRAI O
EFFECT OF VASOPRESSIN, CYCLIC 3',5'-AMP AND CHLORPROPAMIDE
ON WATER PERMEABILITY OF THE URINARY BLADDER OF BUFO BUFO
JAPONICUS
SAISHIN IGAKU 26:689-696(1971):(JAPANESE)

URQUHART J   SHAW JE
EFFECT OF PROSTAGLANDIN E2 ON THE ISOLATED PERFUSED
STOMACH OF THE DOG
GASTROENTEROLOGY 64:872(1973)

VAINER H   LUKASIEWICZ H   JANNEAU C
CAEN J
BIOSYNTHETIC ACTIVITIES IN NORMAL AND PATHOLOGICAL
PLATELET POPULATIONS
ACTA UNIV CAROL (MED) (PRANA) 53:105-112(1972)

VAISRUB S
CHOLERA, PROSTAGLANDINS AND CAMP
J AM MED ASSOC 219:213(1972)

VALCAVI U
SYNTHESIS OF 19,20-DI-NOR-9-KETO-8(12)-PROSTENOIC ACID
FARMACO (SCI) 27:610-619(1972)

VALE W   RIVIER C   GUILLEMIN R
A "PROSTAGLANDIN RECEPTOR" IN THE MECHANISMS INVOLVED IN THE
SECRETION OF ANTERIOR PITUITARY HORMONES.
FED PROC 30:363ABS(1971)

VALERI CR
RESEARCH ON PROSTAGLANDINS IN LONDON
ONR LOND REP R-30-71:1-7(1971)

VALERI CR   ZAROULIS CG   ROGERS JC
HANDIN RI   MARCHIONNI LD
PROSTAGLANDINS IN THE PREPARATION OF BLOOD COMPONENTS
SCIENCE 175:539-542(1972)

VALERI CR   ZAROULIS CG   ROGERS JC
HANDIN RI   MARCHIONNI LD
USE OF PROSTAGLANDINS IN THE PREPARATION OF BLOOD COMPONENTS
IN: "PROSTAGLANDINS IN CELLULAR BIOLOGY," PW RAMWELL AND
BB PHARRISS, PLENUM PRESS, NY/LOND:5-25(1972)

VAN DE VEERDO/ FCG BROUWER E
ROLE OF CALCIUM AND PROSTAGLANDIN (PGE1) IN THE MSH-INDUCED
ACTIVATION OF ADENYLATE CYCLASE IN XENOPUS LAEVIS
BIOCHEM BIOPHYS RES COMMUN 52:130-136(1973)

VAN DORP D
RECENT DEVELOPMENTS IN THE BIOSYNTHESIS AND THE ANALYSES OF
PROSTAGLANDINS
ANN NY ACAD SCI 180:181-199(1971)

VAN DORP DA
ESSENTIAL FATTY ACIDS AND PROSTAGLANDINS
IN: "FETTSTOFFWECHSELSTORUNGEN, IHRE ERKENNUNG UND
BEHANDLUNG," G SCHETTLER, G THIEME, STUTTG:
152-177(1971):(GERMAN)

VAN DORP DA
CYCLIZATION REACTIONS IN LIPID BIOCHEMISTRY
ABSTR, 9TH INT CONG BIOCHEM, STOCKH:389(1973)

VAN HULLE F   SIPIDO V   VANDEWALLE M
CYCLOPENTANONES. VIII. A STEREOSELECTIVE SYNTHESIS OF (DL)
8-EPIPROSTAGLANDIN F1A.
TETRAHEDRON LETT 2213-2216(1973)

VAN ORDEN D6   FARLEY DB
PROSTAGLANDIN F2A RADIOIMMUNOASSAY UTILIZING POLYETHYLENE
GLYCOL SEPARATION TECHNIQUE
PROSTAGLANDINS 4:215-233(1973)

VAN PRAAG D   FARBER SJ   PROSE PH
THE EFFECT OF INCREASED OSMOLALITY ON THE METABOLISM OF PGE
IN RABBIT KIDNEY
ABSTR, 9TH INT CONG BIOCHEM, STOCKH:399(1973)

VAN WIJK R   WICKS WD   CLAY K
EFFECTS OF DERIVATIVES OF CYCLIC 3',5'-ADENOSINE
MONOPHOSPHATE ON THE GROWTH, MORPHOLOGY, AND GENE EXPRESSION
OF HEPATOMA CELLS IN CULTURE
CANCER RES 32:1905-1911(1972)

VANCE JE   BUCHANAN KD   WILLIAMS RH
GLUCAGON AND INSULIN RELEASE. INFLUENCE OF DRUGS AFFECTING
THE AUTONOMIC NERVOUS SYSTEM.
DIABETES 20:78-82(1971)

VANCE VK   ATTALLAH A   PREZYNA A
LEE JB
HUMAN RENAL PROSTAGLANDINS
PROSTAGLANDINS 3:647-667(1973)

VANCE VK   LEE JB
HYPERTENSION AND HUMAN RENAL PROSTAGLANDINS
MED ANN 42:419-422(1973)

VANDENBROUCKE-/ MF DE VISSCHER M
PHYSIOLOGICAL CONTROL OF THYROGLOBULIN PINOCYTOSIS AND
HYDROLYSIS BY IN VITRO STUDY OF PIG THYROID SLICES
ISR J MED SCI 8:1865(1972)

VANDERHOEK J   LANDS WEM
THE ANTIOXIDANT INHIBITION OF THE FATTY ACID OXYGENASE OF
SHEEP VESICULAR GLAND
IN: "PROSTAGLANDINS AND CYCLIC AMP," RH KAHN, WEM LANDS,
ACADEMIC PRESS, NY : 19-20(1973)

VANE JR
INHIBITION OF PROSTAGLANDIN SYNTHESIS AS A MECHANISM OF
ACTION FOR ASPIRIN-LIKE DRUGS
NATURE (NEW BIOL) 231:232-235(1971)

VANE JR
MEDIATORS OF THE ANAPHYLACTIC REACTION
CIBA FOUND STUDY GROUP 38:121-131(1971)

VANE JR
INACTIVATION OF PROSTAGLANDINS AND KININS
PROC R SOC MED 64:16(1971)

VANE JR
PROSTAGLANDINS AND THE ASPIRIN-LIKE DRUGS
HOSP PRACT 7(3):61-71(1972)

VANE JR
PROSTAGLANDINS IN THE INFLAMMATORY RESPONSE
IN: "INFLAMMATION: MECHANISMS AND CONTROL," IH LEPOW AND PA
WARD, ACADEMIC PRESS, NY:261-279(1972)

VANE JR
INHIBITION OF PROSTAGLANDIN BIOSYNTHESIS AS THE MECHANISM OF
ACTION OF ASPIRINLIKE DRUGS
ADV BIOSCI 9:395-411(1973)

VARGAFTIG BB
THE PHARMACOLOGY OF SLOW REACTING SUBSTANCE C AND OF
ARACHIDONIC ACID
AGENTS ACTIONS 3:357-365(1973)

VARGAFTIG BB   DAO HAI N
MECHANISM OF ACTION AND ANTAGONISM OF SRS-C RELEASED FROM
PHOSPHOLIPASE C, FROM EGG YOLK
J PHARMACOL 2:287-304(1971):(FRENCH)

VARGAFTIG BB   DAO HAI N
RELEASE OF VASOACTIVE SUBSTANCES FROM GUINEA-PIG LUNGS BY
SLOW-REACTING SUBSTANCE C AND ARACHIDONIC ACID
PHARMACOLOGY 6:99-108(1971)

VARGAFTIG BB   DAO HAI N
INTERFERENCE OF SOME THIOL DERIVATIVES WITH THE
PHARMACOLOGICAL EFFECTS OF ARACHIDONIC ACID AND SLOW
REACTING SUBSTANCE C AND WITH THE RELEASE OF RABBIT AORTA
CONTRACTING SUBSTANCES
EUR J PHARMACOL 18:43-55(1972)

VARGAFTIG BB   DAO HAI N
SELECTIVE INHIBITION BY MEPACRINE OF THE RELEASE OF "RABBIT
AORTA CONTRACTING SUBSTANCE" EVOKED BY THE ADMINISTRATION OF
BRADYKININ
J PHARM PHARMACOL 24:159-161(1972)

VARGAFTIG BB   DAO HAI N
PARADOXICAL INHIBITION OF THE EFFECTS OF BRADYKININ BY SOME
SULFHYDRYL REAGENTS
EXPERIENTIA 28:59-62(1972)

VARGAFTIG BB   DAO HAI N
INHIBITION OF THE RAT STOMACH CONTRACTIONS DUE TO
ARACHIDONIC ACID: A SIMPLE PROCEDURE FOR DETECTION OF
INHIBITORS OF PROSTAGLANDIN SYNTHESIS
ADV BIOSCI (SUPPL) 9:74(1972)

VARGAFTIG BB   DAO HAI N
INHIBITION BY SULFHYDRYL REAGENTS OF THE EFFECTS OF
BRADYKININ, ARACHIDONIC ACID AND "SLOW REACTING SUBSTANCE C"
IN: "VASOPEPTIDES CHEMISTRY PHARMACOLOGY AND
PATHOPHYSIOLOGY," N BACK, F SICUTERI, PLENUM PUBL, NY:
155-166(1972)

VARGAFTIG BB   DAO HAI N
INHIBITION BY ACETAMIDOPHENOL OF THE PRODUCTION OF
PROSTAGLANDIN-LIKE MATERIAL FROM BLOOD PLATELETS IN VITRO
IN RELATION TO SOME IN VIVO ACTIONS
EUR J PHARMACOL 24:283-288(1973)

VARGAFTIG BB   DE VOS CJ
ANTIINFLAMMATORY PROPERTIES OF SULFHYDRYL AND ANTIOXIDANT
REAGENTS
ABSTR 5TH INT CONGR PHARMACOL, SAN FRANC:240(1972)

VARGAFTIG BB   ZIRINIS P
PLATELET AGGREGATION INDUCED BY ARACHIDONIC ACID IS
ACCOMPANIED BY RELEASE OF POTENTIAL INFLAMMATORY MEDIATORS
DISTINCT FROM PGE2 AND PGF2A
NATURE (NEW BIOL) 244:114-116(1973)

VASTIK JF   GIMENO M   LIMA F
GIMENO AL
SPONTANEOUS MOTILITY AND THE PRESENCE OF PROSTAGLANDINS
IN DISTINCT SEGMENTS OF THE HUMAN FALLOPIAN TUBE
ACTA PHYSIOL LAT AM 23:ABSTR 324(1973);(SPANISH)

VEALE WL   COOPER KE
DOES THE CEREBROSPINAL FLUID OF THE VENTRICULAR SYSTEM
SERVE AS A ROUTE OF EGRESS FOR PROSTAGLANDIN FROM
HYPOTHALAMIC TISSUE?
PROC CAN FED BIOL SOC 15:330(1972)

VEALE WL   COOPER KE
SPECIES DIFFERENCES IN THE PHARMACOLOGY OF TEMPERATURE
REGULATION
IN: PHARMACOL THERMOREGUL PROC SATELL SYMP, 1972,
KARGER, BASEL : 289-301(1973)

VELASQUEZ MT   NOTARGIACOMO AV   COHN JN
INFLUENCE OF CORTICAL PLASMA TRANSIT-TIME ON
P-AMINOHIPPURATE EXTRACTION DURING INDUCED RENAL
VASODILATATION IN ANAESTHETIZED DOGS
CLIN SCI 43:401-411(1972)

VELO GP   DUNN CJ   GIROUD JP
TIMSIT J   WILLOUGHBY DA
DISTRIBUTION OF PROSTAGLANDINS IN INFLAMMATORY EXUDATE
J PATHOL 111:149-158(1973)

VENKIAH KR
PROSTAGLANDINS
INDIAN COUNC MED RES TECH SER 6:72-81(1971)

VENTON DL   COUNSELL RE
SYNTHESIS OF MODEL STEROID CARBOXYLIC ACIDS AS CARBOCYCLIC
PROSTAGLANDIN ANALOGS
ABSTR 166TH NATL MEET AM CHEM SOC:MEDI-17(1973)

VERDY M   BEAULIEU R   DEMERS L
STURTRIDGE WC   THOMAS P   ASHWINI KUMAR M
PLASMA CALCITONIN ACTIVITY IN A PATIENT WITH THYROID
MEDULLARY CARCINOMA AND HER CHILDREN WITH OSTEOPETROSIS
J CLIN ENDOCRINOL METAB 32:216-221(1971)

VERGROESEN AJ   DE BOER J
THE EFFECTS OF PROSTAGLANDINS ON THE HYPODYNAMIC FROG
HEART COMPARED WITH THOSE OF FATTY ACIDS, EPINEPHRINE AND
ADENOSINE PHOSPHATES
J AM OIL CHEM SOC 48:94A(1971)

VERGROESEN AJ   GANS P   GOTTENBOS JJ
TEN HOOR F
PROSTAGLANDINS IN THE CLINIC
KLIN WOCHENSCHR 49:889-895(1971);(GERMAN)

VERGROESEN AJ   GANS P   GOTTENBOS JJ
TEN HOOR F
THE CLINICAL USES OF PROSTAGLANDINS. PROBLEMS AND
PERSPECTIVES
NED TIJDSCHR GENEESKD 116:1839-1848(1972);(DUTCH)

VERNE J   HEBERT S   RICHSHOFFER N
THE ACTION OF PROSTAGLANDIN E1 ON HEPATOCYTES IN
HISTIOTYPIC CULTURE
C R SOC BIOL (PARIS) 167:825-827(1973);(FRENCH)

VERRY M   DECHAVANNE M   DESNOYERS P
STOLTZ JF   VAINER H
ANTIPLATELET PROPERTIES OF GLICLAZIDE, A NEW ORAL
ANTIDIABETIC AGENT
ABSTR 8TH CONGR INT DIABETES FED, BRUSS, EXCERPTA MED,
AMST:207-208(1973)

VIGDAHL RL   MONGIN J JR   MARQUIS NR
PLATELET AGGREGATION. IV. PLATELET PHOSPHODIESTERASE AND ITS
INHIBITION BY VASODILATORS.
BIOCHEM BIOPHYS RES COMMUN 42 : 1088-1094(1971)

VILLANUEVA R   HINDS L   KATZ RL
EAKINS KE
ANTAGONISM OF SOME ACTIONS OF PROSTAGLANDIN F2A (PGF2A) BY
POLYPHLORETIN PHOSPHATE (PPP) IN VIVO
FED PROC 30:626(1971)

VILLANUEVA R   HINDS L   KATZ RL
EAKINS KE
THE EFFECT OF POLYPHLORETIN PHOSPHATE ON SOME SMOOTH MUSCLE
ACTIONS OF PROSTAGLANDINS IN THE CAT
J PHARMACOL EXP THER 180:78-85(1972)

VILLEE CA
PHARMACOLOGIC PHYSIOLOGY OF THE ENDOMETRIUM AND STEROID
INTERACTIONS
IN: "THE UTERUS", HJ NORRIS, AT HERTIG, MR ABELL, WILLIAMS &
WILKINS, BALT:80-88(1973)

VIZI ES   TOROK T   KNOLL J
STUDY OF THE MECHANISM OF ACTION OF PROSTAGLANDINS E1 AND E2
BY THE SUCROSE GAP TECHNIQUE, ON TAENIA COLI PREPARATION
ORVOSTUDOMANY 24:99-110(1973);(HUNGARIAN)

VLACHOS K   KOVATSIS A   TSOUKALI E
KOTSAKI V
ISOLATION, IDENTIFICATION, AND BIOLOGICAL ACTIVITY OF
PROSTAGLANDINS IN STEER SPERM
DTSCH TIERAERZTL WOCHENSCHR 80:100-102(1973);(GERMAN)

VLATTAS I   DELLAVECCHIA L   FITT JJ
O-(1-ALKYL- OR -ARYLTHIOALKYL)HYDROXYLAMINES. A NEW CLASS OF
OXIME REAGENTS, THEIR PREPARATION AND SYNTHETIC UTILITY
J ORG CHEM 38:3749-3752(1973)

VOGEL P   CRABBE P
METHYLENE-10,11-AND METHYL-11-PROSTAGLANDINS
HELV CHIM ACTA 56:557-560(1973);(FRENCH)

VON EULER US
INTRODUCTORY REMARKS
ANN NY ACAD SCI 180:6-9(1971)

VON EULER US
PIECES IN THE PUZZLE
ANNU REV PHARMACOL:1-12(1971)

VON EULER US
SYNTHESIS, STORAGE, AND RELEASE OF THE ADRENERGIC
NEUROTRANSMITTER
KLIN WOCHENSCHR 49:524-529(1971);(GERMAN)

VON EULER US
REGULATION OF CATECHOLAMINE METABOLISM IN THE SYMPATHETIC
NERVOUS SYSTEM. CHAIRMAN'S REMARKS
PHARMACOL REV 24:365-369(1972)

VON EULER US
SOME ASPECTS OF THE ACTIONS OF PROSTAGLANDINS (THE FIRST
HEYMANS MEMORIAL LECTURE)
ARCH INT PHARMACODYN THER 202(SUPPL):295-307(1973)

VOORHEES JJ   DUELL EA
PSORIASIS AS A POSSIBLE DEFECT OF THE ADENYL CYCLASE-CYCLIC
AMP CASCADE. A DEFECTIVE CHALONE MECHANISM?
ARCH DERMATOL 104:352-358(1971)

VULLIEMOZ P
THE PROSTAGLANDINS
MED HYG 29:2001-2005(1971);(FRENCH)

VURAL H   BAYSAL F
THE EFFECTS OF SOME PROSTAGLANDINS ON THE ISOLATED
GALL-BLADDER OF GUINEA-PIG
TIP FAK MECM 25:6-12(1972);(TURKISH)

VURAL H   BAYSAL F   KOCAK N
PROSTAGLANDINS AND HUMAN APPENDIX
DIYARBAKIR TIP FAK DERG 1:149-154(1972);(TURKISH)

WADA T   ISHIZAWA M
FUNDAMENTAL AND CLINICAL STUDIES ON THE EFFECTS OF
PROSTAGLANDINS ON THE ALIMENTARY CANAL
PROC SOC GERONTOL STUDY, OKAYAMA CITY, ONO PHARMACEUTICAL
CO, OSAKA:42-46(1972);(JAPANESE)

WAGNER R   BITENSKY M   KREINER P
BARRNETT R
ADENYLATE CYCLASE IN ISOLATED CAPILLARY ENDOTHELIUM
J CELL BIOL 55:270A(1972)

WAGNER WD   PETERSON RA   CENEDELLA RJ
THE EFFECTS OF COLD AND PROSTAGLANDIN E1 ON LIPID
MOBILIZATION IN THE CHICKEN.
CAN J PHYSIOL PHARMACOL 49:394-398(1971)

WAITZMAN MB
HYPOTHALAMUS AND OCULAR PRESSURE
SURV OPHTHALMOL 16(1):1-23(1971)

WAITZMAN MB
PROSTAGLANDINS IN SPONTANEOUS AND STREPTOZOTOCIN-INDUCED
DIABETES
FED PROC 32:364(1973)

WAITZMAN MB
PROSTAGLANDINS AND DIABETIC RETINOPATHY
EXP EYE RES 16:307-313(1973)

WAITZMAN MB   KUCK JFR JR   WOODS WD
EFFECT OF PROSTAGLANDINS (PGS) ON GALACTOSE UPTAKE BY LENSES
AND ON ADENYLATE CYCLASE OF ISOLATED LENS CELLS
FED PROC 31:384(1972)

WAITZMAN MB   WOODS WD
SOME CHARACTERISTICS OF AN ADENYL CYCLASE PREPARATION FROM
RABBIT CILIARY PROCESS TISSUE
EXP EYE RES 12:99-111(1971)

WALKER E
RESEARCH ON PROSTAGLANDINS REVEALS POTENTIAL USES
MOD NURS HOME 30:4,9,14-15(1973)

WALKER JL
THE REGULATORY FUNCTION OF PROSTAGLANDINS IN THE RELEASE
OF HISTAMINE FROM PASSIVELY SENSITISED HUMAN LUNG TISSUE
ADV BIOSCI (SUPPL) 9:40(1972)

WALKER JL
THE REGULATORY FUNCTION OF PROSTAGLANDINS IN THE RELEASE OF
HISTAMINE AND SRS-A FROM PASSIVELY SENSITIZED HUMAN LUNG
TISSUE
ADV BIOSCI 9:235-240(1973)

WALLACH EE   VIRUTAMASEN P   WRIGHT KH
RABBIT OVARIAN CONTRACTIONS IN HCG-INDUCED OVULATION
PROC WORLD CONGR FERTIL STERIL 7:600-601(1973)

WALLACH S   CHAUSMER AB   SHERMAN BS
HORMONAL EFFECTS ON CALCIUM TRANSPORT IN LIVER
CLIN ORTHOP 78:40-46(1971)

WALLER SL
PROSTAGLANDINS AND THE GASTROINTESTINAL TRACT
GUT 14:402-417(1973)

WATERHOUSE RK
UROLOGY
SURG GYNECOL OBSTET 132:245-246(1971)

WATSON JT   PELSTER DR   SWEETMAN BJ
FROLICH JC   OATES JA
DISPLAY-ORIENTED DATA SYSTEM FOR MULTIPLE ION DETECTION WITH
GAS CHROMATOGRAPHY-MASS SPECTROMETRY IN QUANTIFYING
BIOMEDICALLY IMPORTANT COMPOUNDS
ANAL CHEM 45:2071-2078(1973)

WATSON JT   SWEETMAN BJ
PROSTAGLANDIN ANALYSIS: SURVEY OF DERIVATIVES USED FOR
IDENTIFICATION AND DETECTION BY GAS CHROMATOGRAPH-MASS
SPECTROMETER-COMPUTER SYSTEM
J AM OIL CHEM SOC 48:333A-334A(1971)

WATSON JT   SWEETMAN BJ   FROLICH FC
QUANTIFICATION OF PROSTAGLANDINS AT PHYSIOLOGICALLY
SIGNIFICANT LEVELS WITH A GC-MS-COMPUTER SYSTEM
ABSTR 5TH INT CONGR PHARMACOL, SAN FRANC:248(1972)

WEBB DR   STITES DP   PERLMAN JD
LUONG D   FUDENBERG HH
LYMPHOCYTE ACTIVATION: THE DUALISTIC EFFECT OF CAMP
BIOCHEM BIOPHYS RES COMMUN 53:1002-1008(1973)

WEBER U
PROSTAGLANDINS AND HORMONAL REGULATION
MED MONATSSCHR 27:290-295(1973):(GERMAN)

WEDEEN RP   GOLDSTEIN M   LEVITT MF
MECHANISMS OF EDEMA AND THE USE OF DIURETICS
PEDIATR CLIN NORTH AM 18:561-576(1971)

WEEKS JR
BIOLOGICAL SIGNIFICANCE OF PROSTAGLANDINS WITH SPECIAL
REFERENCE TO THEIR EFFECTS ON METABOLISM.
NAUNYN SCHMIEDEBERGS ARCH PHARMACOL 269:347-357(1971)

WEEKS JR
PROSTAGLANDINS
ANNU REV PHARMACOL 12:317-336(1972)

WEEKS JR
TACHYPHYLACTIC RESPONSE OF THE RAT UTERUS IN VIVO TO
PROSTAGLANDINS E2 AND F2A
ADV BIOSCI (SUPPL) 9:129(1972)

WEEKS JR
THE PROSTAGLANDINS: THEIR NATURE, FORMATION AND GENERAL
PHARMACOLOGY
IN: "PROSTAGLANDINS AND CYCLIC AMP," RH KAHN, WEM LANDS,
ACADEMIC PRESS, NY : 1-14(1973)

WEHLE B   BERGSTROM A   CASTENFORS J
SERNER I
EFFECT OF PROSTAGLANDIN ON KIDNEY FUNCTION AND BLOOD
PRESSURE IN PATIENTS WITH KIDNEY DISEASES
NORD MED 85:782-783(1971):(SWEDISH)

WEINER R   KALEY G
LYSOSOMAL FRAGILITY INDUCED BY PROSTAGLANDIN F2A
NATURE (NEW BIOL) 236:46-47(1972)

WEINRYB I   MICHEL IM
ALPHA-METHYLFLUORENE-2-ACETIC ACID: EFFECTS ON ADENYLATE
CYCLASES IN VITRO
ABSTR 164TH NATL MEET, AM CHEM SOC, NY:BIOL-203(1972)

WEINRYB I   MICHEL IM.   HESS SM
CHARACTERIZATION OF ADENYLATE CYCLASE FROM GUINEA PIG LUNG
AND MEASUREMENT OF ITS INHIBITION BY SUBSTRATE ANALOGS AND
CYCLIC NUCLEOTIDES
ABSTR 5TH INT CONGR PHARMACOL SAN FRANC 249(1972)

WEINRYB I   MICHEL IM   HESS SM
ADENYLATE CYCLASE FROM GUINEA PIG LUNG: FURTHER
CHARACTERIZATION AND INHIBITORY EFFECTS OF SUBSTRATE ANALOGS
AND CYCLIC NUCLEOTIDES
ARCH BIOCHEM BIOPHYS 154:240-249(1973)

WEINSCHENKER NM   STEPHENSON R
BASIC HYDROGEN PEROXIDE CLEAVAGE OF A BICYCLIC KETONE. A NEW
PROCEDURE FOR A PROSTAGLANDIN INTERMEDIATE
J ORG CHEM 37:3741(1972)

WEINSHENKER NM
IMPROVED PREPARATION OF A PROSTAGLANDIN INTERMEDIATE:
2-CHLORO-2-CYANO-DELTA5-7-SYN-METHOXYMETHYL BICYCLO (2.2.1)
HEPTENE
PROSTAGLANDINS 3:219-222(1973)

WEINSHENKER NM   ANDERSEN NH
CHEMISTRY
IN: "THE PROSTAGLANDINS," PW RAMWELL, PLENUM PRESS, NY
1:5-82(1973)

WEINSHENKER NM   LONGWELL A
QUANTITATIVE DETERMINATION OF 15-EPI-PGF2A IN PGF2A
PROSTAGLANDINS 2:207-211(1972)

WEINSHENKER NM   RAMWELL PW
REVIEW OF THE INTRA-SCIENCE RESEARCH FOUNDATION SYMPOSIUM ON
THE CHEMISTRY AND PHARMACOLOGY OF PROSTAGLANDINS
PROSTAGLANDINS 1:83-85(1972)

WEISBLAT DI
THE PROSTAGLANDIN STORY
DRUG COSMET IND 111:34-37,121-122(1972)

WEISENBERGER H
CYCLIC AMP AND PLATELET FUNCTION
IN: "ERYTHROCYTES, THROMBOCYTES, LEUKOCYTES," E GERLACH,
K MOSER, E DEUTSCH, W WILMANNS, G THIEME, STUTTG
2:327-333(1973)

WEISS L
STUDIES ON CELLULAR ADHESION IN TISSUE CULTURE XIIA. SOME
EFFECTS OF PROSTAGLANDINS AND CYCLIC NUCLEOTIDES
EXP CELL RES 81:57-62(1973)

WEISSMAN G
MECHANISM OF CYCLIC AMP-MEDIATED INHIBITION OF LYSOSOMAL
ENZYME RELEASE IN TISSUE INJURY
ABSTR 164TH NATL MEET AM CHEM SOC, NY:MEDI-2(1972)

WEISSMAN G   DUKOR P   ZURIER RB
EFFECT OF CYCLIC AMP ON RELEASE OF LYSOSOMAL ENZYMES FROM
PHAGOCYTES
NATURE (NEW BIOL) 231:131-135(1971)

WEISSMANN G
MECHANISM OF PROSTAGLANDIN-MEDIATED INHIBITION OF LYSOSOMAL
ENZYME RELEASE IN INFLAMMATION
ADV BIOSCI (SUPPL) 9:67(1972)

WEISSMANN G
LYSOSOMAL MECHANISMS OF TISSUE INJURY IN ARTHRITIS
N ENGL J MED 286:141-147(1972)

WEISSMANN G   ZURIER RB   HOFFSTEIN S
EFFECT OF PROSTAGLANDIN E1 ON LYSOSOMAL ENZYME RELEASE FROM
HUMAN LEUKOCYTES IN THE PRESENCE OF CYTOCHALASIN B
ADV BIOSCI 9:435-440(1973)

WEISSMANN G   ZURIER RB   HOFFSTEIN S
LEUKOCYTES AS SECRETORY ORGANS OF INFLAMMATION
AGENTS ACTIONS 3:370-379(1973)

WEISSMANN G   ZURIER RB   SPIELER PJ
GOLDSTEIN IM
MECHANISMS OF LYSOSOMAL ENZYME RELEASE FROM LEUKOCYTES
EXPOSED TO IMMUNE COMPLEXES AND OTHER PARTICLES
J EXP MED 134:#3,PT.2 SUPPL 149S-165S(1971)

WEISSMANN G   ZURIER RB   TSUNG PK
HOFFSTEIN S
MECHANISM OF CYCLIC AMP-MEDIATED INHIBITION OF LYSOSOMAL
ENZYME RELEASE IN TISSUE INJURY
J CLIN INVEST 51:102A(1972)

WELLMANN W   SCHWABE U
EFFECTS OF PROSTAGLANDINS E1, E2 AND F2A ON CYCLIC AMP
LEVELS IN BRAIN IN VIVO
BRAIN RES 9:371-378(1973)

WENDLER NL   TAUB D   KUO CH
SLATES HL   ZELAWSKI ZF
TOTAL SYNTHESIS OF PROSTAGLANDINS
ABSTR 7TH MIDDLE ATL REG MEET AM CHEM SOC PHILA:82(1972)

WENDLING MG   DUCHARME DW   GRAHAM BE
EFFECTS OF CHRONIC ADMINISTRATION OF PROSTAGLANDIN E1 (PGE1)
ON ARTERIAL BLOOD PRESSURE OF UNANESTHETIZED HYPERTENSIVE
RATS
PHYSIOLOGIST 15:301(1972)

WENDT RL   BAUM T
AEROSOL ADMINISTRATION OF PROSTAGLANDINS E1 AND E2 AND
ISOPROTERENOL: STUDIES ON THE CARDIOVASCULAR SYSTEM
EUR J PHARMACOL 17:141-151(1972)

WENER JA   GORTON SJ   RAISZ LG
ESCAPE FROM INHIBITION OR RESORPTION IN CULTURES OF FETAL
BONE TREATED WITH CALCITONIN AND PARATHYROID HORMONE
ENDOCRINOLOGY 90:752-759(1972)

WENNMALM A
STUDIES ON MECHANISMS CONTROLLING THE SECRETION OF
NEUROTRANSMITTERS IN THE RABBIT HEART
ACTA PHYSIOL SCAND SUPPL 365:1-36(1971)

WENNMALM A   HEDQVIST P
INHIBITION BY PROSTAGLANDIN E1 OF PARASYMPATHETIC
NEUROTRANSMISSION IN THE RABBIT HEART
LIFE SCI (I) 10:465-470(1971)

WENNMALM A   JUNSTAD M
ENDOGENOUS PROSTAGLANDIN MEDIATED INHIBITION OF
PARASYMPATHETIC NEUROTRANSMISSION IN THE RABBIT HEART?
ACTA PHYSIOL SCAND (SUPPL 396) : 22(1973)

WENNMALM A   STJARNE L
INHIBITION OF THE RELEASE OF ADRENERGIC TRANSMITTER BY A
FATTY ACID IN THE PERFUSATE FROM SYMPATHETICALLY STIMULATED
RABBIT HEART.
LIFE SCI(I) 10 : 471-479(1971)

WENTZ AC   JONES GS   ROCCO L
BLEDSOE T
STIMULATION OF CORTISOL SYNTHESIS IN FEMALES RECEIVING
PROSTAGLANDIN F2A (PGF2A)
ABSTR 4TH INT CONGR ENDOCRINOL, INT CONGR SER 256, EXCERPTA
MEDICA, AMST:190(1972)

WERNING C   VETTER W   WEIDMANN P
SCHWEIKERT HU   STIEL D   SIEGENTHALER W
EFFECT OF PROSTAGLANDIN E1 ON RENIN IN THE DOG
AM J PHYSIOL 220:852-856(1971)

WHITE JG   GOLDBERG ND   ESTENSEN RD
HADDOX MK   RAO GHR
AN ASSOCIATION BETWEEN INCREASED LEVELS OF PLATELET CYCLIC
3',5'-GUANOSINE MONOPHOSPHATE (C-GMP) AND THE RELEASE
REACTION
ABSTR 4TH INT CONGR THROMB HAEMOSTASIS, VIENNA:172(1973)

WHITE JG   KRUMWIEDE M
INFLUENCE OF CYTOCHALASIN B ON THE SHAPE CHANGE INDUCED IN
PLATELETS BY COLD
BLOOD 41:823-832(1973)

WHITE RP   DENTON IC   ROBERTSON JT
DIFFERENTIAL EFFECTS OF PROSTAGLANDINS A1, E1, AND F2A ON
CEREBROVASCULAR TONE IN DOGS AND RHESUS MONKEYS
FED PROC 30:625(1971)

WHITE RP   HEATON JA   DENTON IC
PHARMACOLOGICAL COMPARISON OF PROSTAGLANDIN F2A, SEROTONIN
AND NOREPINEPHRINE ON CEREBROVASCULAR TONE OF MONKEY
EUR J PHARMACOL 15:300-309(1971)

WHITE RP   PENNINK M
REVERSAL OF THE PRESSOR RESPONSE OF PROSTAGLANDIN F2A BY
POLYPHLORETIN PHOSPHATE IN DOGS
ARCH INT PHARMACODYN THER 197:274-281(1972)

WHITE RP   PENNINK M   ROBERTSON JT
PROSTAGLANDIN F2A AND EXPERIMENTAL CEREBRAL VASOSPASM IN
DOGS.
PHARMACOLOGIST 13:292(1971)

WHITELOCKE RAF   EAKINS KE
COMPARISON OF THE EFFECTS OF PROSTAGLANDINS E1, E2 AND F2A
ON THE RABBIT EYE USING FLUORESCEIN ANGIOGRAPHY
EXP EYE RES 14:175(1972)

WHITELOCKE RAF   EAKINS KE
VASCULAR CHANGES IN THE ANTERIOR UVEA OF THE RABBIT PRODUCED
BY PROSTAGLANDINS
ARCH OPHTHALMOL 89:495-499(1973)

WHITELOCKE RAF   EAKINS KE
A COMPARISON OF SOME DERIVATIVES OF PHLORETIN AS
PROSTAGLANDIN ANTAGONISTS
EXP EYE RES 17:395-396(1973)

WHITELOCKE RAF   EAKINS KE   BENNETT A
ACUTE ANTERIOR UVEITIS AND PROSTAGLANDINS
PROC R SOC MED 66:429-434(1973)

WHITFIELD JF   MACMANUS JP
CALCIUM-MEDIATED EFFECTS OF THYMOCYTE PROLIFERATION BY
PROSTAGLANDIN E1
PROC SOC EXP BIOL MED 139:818-824(1972)

WHITFIELD JF   MACMANUS JP   BRACELAND B
GILLAN DJ
INHIBITION BY CALCIUM OF THE CYCLIC AMP-MEDIATED STIMULATION
OF THYMIC LYMPHOBLAST PROLIFERATION BY PROSTAGLANDIN E1
HORM METAB RES 4:304-308(1972)

WHITFIELD JF   MACMANUS JP   BRACELAND BM
GILLAN DJ
THE INFLUENCE OF CALCIUM ON THE CYCLIC AMP-MEDIATED
STIMULATION OF DNA SYNTHESIS AND CELL PROLIFERATION BY
PROSTAGLANDIN E1
J CELL PHYSIOL 79:333-362(1972)

WHITFIELD JF   MACMANUS JP   FRANKS DJ
BRACELAND BM   GILLAN DJ
CALCIUM-MEDIATED EFFECTS OF CALCITONIN ON CYCLIC AMP
FORMATION AND LYMPHOBLAST PROLIFERATION IN THYMOCYTE
POPULATIONS EXPOSED TO PROSTAGLANDIN E1
J CELL PHYSIOL 80:315-328(1972)

WHITFIELD JF   MACMANUS JP   YOUDALE T
FRANKS DJ
THE ROLES OF CALCIUM AND CYCLIC AMP IN THE STIMULATOR ACTION
OF PARATHYROID HORMONE ON THYMIC LYMPHOCYTE PROLIFERATION
J CELL PHYSIOL 78:355-368(1971)

WHITFIELD JF   MACMANUS JP   YOUDALE T
FRANKS DJ
THE ROLES OF CALCIUM AND CYCLIC AMP IN THE STIMULATORY
ACTION OF PARATHYROID HORMONE ON THYMIC LYMPHOCYTE
PROLIFERATION
J CELL PHYSIOL 78:355-368(1971)

WHITFIELD JF   RIXON RH   MACMANUS JP
BALK SD
CALCIUM, CYCLIC ADENOSINE 3',5'-MONOPHOSPHATE, AND THE
CONTROL OF CELL PROLIFERATION: A REVIEW
IN VITRO 8:257-278(1973)

WHITTLE BJR
STUDIES ON THE MODE OF ACTION OF CYCLIC 3'5'-AMP AND
PROSTAGLANDIN E2 ON RAT GASTRIC ACID SECRETION AND MUCOSAL
BLOOD FLOW
BR J PHARMACOL 46:546P-547P(1972)

WICHA J
CHEMISTRY OF PROSTAGLANDINS
FARM POL 29:947-963(1973):(POLISH)

WICKRAMASINGHE JAF MOROZOWICH W   HAMLIN WE
SHAW SR
DETECTION OF PGF2A AS THE PENTAFLUOROBENZYL ESTER BY
ELECTRON CAPTURE GAS CHROMATOGRAPHY
ABSTR 13TH NATL MEET APHA, ACAD PHARM SCI 2(2):14(1972)

WICKRAMASINGHE JAF MOROZOWICH W   HAMLIN WE
SHAW SR
DETECTION OF PROSTAGLANDIN F2A AS PENTAFLUOROBENZYL ESTER BY
ELECTRON-CAPTURE GLC
J PHARM SCI 62:1428-1431(1973)

WICKRAMASINGHE JAF SHAW SR
QUANTITATIVE DETERMINATION OF PROSTAGLANDIN F2A IN
BIOLOGICAL FLUIDS BY AN ELECTRON-CAPTURE GAS CHROMATOGRAPHIC
METHOD
INT CONGR PHARM SCI ABSTR 33:42(1973)

WILBER JF   SEIBEL MJ
TRH INTERACTIONS WITH A SPECIFIC ANTERIOR PITUITARY MEMBRANE
RECEPTOR
ABSTR 4TH INT CONGR ENDOCRINOL, INT CONGR SER 256, EXCERPTA
MEDICA, AMST:85(1972)

WILKERSON RD   PAULO LG   GEORGE WJ
ROH BL   FISHER JW
THE EFFECTS OF PROSTAGLANDIN E1 (PGE-1) ON ERYTHROPOIETIN
ABSTR 5TH INT CONGR PHARMACOL SAN FRANC 252(1972)

WILKS JW   FORBES KK   NORLAND JF
SYNTHESIS OF PROSTAGLANDIN F2A BY THE OVARY AND UTERUS
J REPROD MED 9:271-276(1972)

WILLIAMS ED
ROLE OF KININS AND PROSTAGLANDINS IN AMINE-SECRETING TUMOR
SYNDROMES
PROC R SOC MED 64:16(1971)

WILLIAMS TJ   MORLEY J
PROSTAGLANDINS AS POTENTIATORS OF INCREASED VASCULAR
PERMEABILITY IN INFLAMMATION
NATURE (LOND) 246:215-217(1973)

WILLINGHAM MC   CARCHMAN RA   PASTAN IH
A MUTANT OF 3T3 CELLS WITH CYCLIC AMP METABOLISM SENSITIVE
TO TEMPERATURE CHANGE
PROC NATL ACAD SCI USA 70: 2906-2910 (1973)

WILLIS AL
BIOSYNTHESIS OF PROSTAGLANDINS E2 AND F2A GENERATES LABILE
MATERIAL WHICH INDUCES PLATELET AGGREGATION
ABSTR 4TH INT CONGR THROMB HAEMOSTASIS, VIENNA:79(1973)

WILLIS AL
PLATELET SYNTHESIS OF PRO-AGGREGATING MATERIAL FROM
ARACHIDONATE AND ITS BLOCKADE BY ASPIRIN
CIRCULATION (SUPPL IV) 8:IV-55 (1973)

WILLIS AL   CORNELSEN M
REPEATED INJECTION OF PROSTAGLANDIN E2 IN RAT PAWS INDUCED
CHRONIC SWELLING AND A MARKED DECREASE IN PAIN THRESHOLD
PROSTAGLANDINS 3:353-357(1973)

WILLIS AL   DAVISON P   RAMWELL PW
BROCKLEHURST WE   SMITH B
RELEASE AND ACTIONS OF PROSTAGLANDINS IN INFLAMMATION AND
FEVER: INHIBITION BY ANTIINFLAMMATORY AND ANTIPYRETIC DRUGS
IN: "PROSTAGLANDINS IN CELLULAR BIOLOGY," PW RAMWELL AND
BB PHARRISS, PLENUM PRESS, NY/LOND:227-268(1972)

WILLIS AL   JOHNSON M   RABINOWITZ I
WOLF P
PROSTAGLANDIN E2 MAY INDUCE AND ENHANCE SICKLE CELL CRISIS
CLIN CHEM 18:700(1972)

WILLIS AL   KUHN DC
A NEW POTENTIAL MEDIATOR OF ARTERIAL THROMBOSIS WHOSE
BIOSYNTHESIS IS INHIBITED BY ASPIRIN
PROSTAGLANDINS 4:127-130(1973)

WILLIS AL   WEISS HJ
A CONGENITAL DEFECT IN PLATELET PROSTAGLANDIN PRODUCTION
ASSOCIATED WITH IMPAIRED HEMOSTASIS IN STORAGE POOL DISEASE
PROSTAGLANDINS 4 : 783-794(1973)

WILLOUGHBY DA
ASPIRIN-LIKE DRUGS AND PROSTAGLANDINS
LANCET 2:545(1971)

WILLOUGHBY DA
STUDIES ON THE INFLAMMATORY RESPONSE
HELV ODONTOL ACTA 15:89(1971)

WILLOUGHBY DA   DI ROSA M
A UNIFYING CONCEPT FOR INFLAMMATION: A NEW APPRAISAL OF SOME
OLD MEDIATORS
IN: "IMMUNOPATHOLOGY OF INFLAMMATION," INT CONGR SER 229,
EXCERPTA MEDICA, AMST:28-38(1971)

WILLOUGHBY DA   DI ROSA M
STUDIES ON MODE OF ACTION OF NON-STEROIDAL ANTIINFLAMMATORY
DRUGS
ABSTR 5TH INT CONGR PHARMACOL SAN FRANC 253(1972)

WILLOUGHBY DA   GIROUD JP   DI ROSA M
VELO GP
THE CONTROL OF THE INFLAMMATORY RESPONSE WITH SPECIAL
REFERENCE TO THE PROSTAGLANDINS
IN: "PROSTAGLANDINS AND CYCLIC AMP," RH KAHN, WEM LANDS,
ACADEMIC PRESS, NY : 187-206(1973)

WILMUT I   ROWSON LEA
EXPERIMENTS ON THE LOW-TEMPERATURE PRESERVATION OF COW
EMBRYOS
VET REC 92:686-690 (1973)

WILSON DB   KITABCHI AE
STEROIDOGENESIS IN ISOLATED ADRENAL CELLS (IAC) AND ITS
RELATION TO ADENYL CYCLASE (AC) AND PHOSPHODIESTERASE (PDE).
CLIN RES 19:654(1971)

WILSON DE
PROSTAGLANDINS AND THE GASTROINTESTINAL TRACT
PROSTAGLANDINS 1:281-293(1972)

WILSON DE   HANKEWYCH M
PROSTAGLANDIN E1 EFFECTS ON HEPATIC GLYCOGENOLYSIS
GASTROENTEROLOGY 65:576(1973)

WILSON DE   LEVINE RA
THE EFFECT OF PROSTAGLANDIN E1 ON CANINE GASTRIC ACID
SECRETION AND GASTRIC MUCOSAL BLOOD FLOW
AM J DIG DIS 17:527-532(1972)

WILSON DE   PHILLIPS C   LEVINE RA
INHIBITION OF GASTRIC SECRETION IN MAN BY PROSTAGLANDIN A1
GASTROENTEROLOGY 61:201-206(1971)

WILSON L
THE LEVELS OF PROSTAGLANDINS IN THE UTERUS OF NONPREGNANT,
PREGNANT AND PROGESTERONE-TREATED EWES
DISS, WEST VA UNIV, MORGANTOWN (1971)

WILSON TW   FROLICH JC   CARR K
OATES JA
URINARY PROSTAGLANDIN E IN PATIENTS WITH RENAL ARTERY
STENOSIS
CLIN RES 21:460(1973)

WINKELMANN RK
MOLECULAR INFLAMMATION OF THE SKIN
J INVEST DERMATOL 57:197-208(1971)

WINKELMANN RK
THE PHARMACOLOGY OF ISOLATED CUTANEOUS VASCULAR SMOOTH
MUSCLE
G ITAL DERMATOL MINERVA DERMATOL 108:34-38(1973)

WINKELMANN RK   SAMS WM JR   GOLDYNE ME
CONTRACTION OF ISOLATED CUTANEOUS VASCULAR SMOOTH MUSCLE AND
ITS RESPONSE TO PROSTAGLANDINS
IN: "PROSTAGLANDINS IN CELLULAR BIOLOGY," PW RAMWELL AND
BB PHARRISS, PLENUM PRESS, NY/LOND:353-368(1972)

WIQVIST N
PROSTAGLANDINS AND THE UTERUS. CLINICAL USE
NORD MED 85:436(1971):(SWEDISH)

WIQVIST N   BEGUIN F   BYGDEMAN M
TOPPOZADA M
RECENT ASPECTS ON SYSTEMIC ADMINISTRATION OF PROSTAGLANDIN
J REPROD MED 9:378-382(1972)

WIQVIST N   BYGDEMAN M   TOPPOZADA M
ADMINISTRATION OF PROSTAGLANDIN FOR TERMINATION OF PREGNANCY
PROC WORLD CONGR FERTIL STERIL 7:137-141(1973)

WISKONT-BUCZKOW/ H
ANEMIZATION OF THE KIDNEY WITH IMPAIRMENT OF ITS EXCRETION
FUNCTION AS LABOR-PROVOKING FACTOR IN EXPERIMENTAL ANIMALS
GINEKOL POL 44:1353-1361(1973):(POLISH)

WITTING WC   WORK BA JR   LAROS RK JR
UTERINE ACTIVITY RESPONSE TO CONSTANT INFUSION OF
PROSTAGLANDIN F2 IN TERM HUMAN PREGNANCY
J REPROD MED 9:283-286(1972)

WLODAWER P   SAMUELSSON B
ON THE ORGANIZATION AND MECHANISM OF PROSTAGLANDIN
SYNTHETASE
J BIOL CHEM 248:5673-5678(1973)

WLODAWER P   SAMUELSSON B   ALBONICO SM
COREY EJ
SELECTIVE INHIBITION OF PROSTAGLANDIN SYNTHETASE BY A
BICYCLO (2.2.1) HEPTENE DERIVATIVE
J AM CHEM SOC 93:2815-2816(1971)

WOLF PL   RABINOWITZ I   JOHNSON M
A BIOCHEMICAL AND ELECTRON MICROSCOPIC STUDY OF
PROSTAGLANDIN-INDUCED ERYTHROCYTE SICKLING
AM J PATHOL 70:83A-84A(1973)

WOLFF J   COOK GH
ACTIVATION OF THYROID MEMBRANE ADENYLATE CYCLASE BY PURINE
NUCLEOTIDES
J BIOL CHEM 248:350-355(1973)

WOLFF J   JONES AB
THE PURIFICATION OF BOVINE THYROID PLASMA MEMBRANES AND
THE PROPERTIES OF MEMBRANE-BOUND ADENYL CYCLASE
J BIOL CHEM 246:3939-3947(1971)

WOLFF J   MOORE WV
THE EFFECT OF INDOMETHACIN ON THE RESPONSE OF THYROID
TISSUE TO THYROTROPIN
BIOCHEM BIOPHYS RES COMMUN 51:34-39(1973)

WONG PDY   BEDWANI JR   CUTHBERT AW
HORMONE ACTION AND THE LEVELS OF CYCLIC AMP AND
PROSTAGLANDINS IN THE TOAD BLADDER
NATURE (NEW BIOL) 238:27-31(1972)

WOOD C
PROSTAGLANDINS: THE UNIQUE MOLECULES
SPECTRUM 10:154-156(1972)

WOODWARD DJ   HOFFER BJ   SIGGINS GR
BLOOM FE
THE ONTOGENETIC DEVELOPMENT OF SYNAPTIC JUNCTIONS, SYNAPTIC
ACTIVATION AND RESPONSIVENESS TO NEUROTRANSMITTER SUBSTANCES
IN RAT CEREBELLAR PURKINJE CELLS
BRAIN RES 34:73-97(1971)

WOODWARD RB   GOSTELI J   ERNEST I
FRIARY RJ   NESTLER G   RAMAN H
SITRIN R   SUTER C   WHITESELL JK
A NOVEL SYNTHESIS OF PROSTAGLANDIN F2A
J AM CHEM SOC 95: 6853-6855(1973)

WOOSTER MJ
EFFECTS OF PROSTAGLANDIN E1 ON DOG URETER IN VITRO
J PHYSIOL 213:51P-53P(1971)

WRIGHT RK   HSIA SL
EFFECTS OF INSULIN, PROSTAGLANDIN E2 AND EPINEPHRINE ON THE
FORMATION OF CYCLIC AMP BY HUMAN SKIN.
FED PROC 30:1205ABS(1971)

WYLLIE AM   WYLLIE JH
PROSTAGLANDINS AND GLAUCOMA
BR MED J 3:615-617(1971)

YAMAMOTO G   SHIMIZU S   YOSHIRI K
DETECTION OF PROSTAGLANDINS IN THE RENAL TISSUES AND RENAL
VENOUS BLOOD OF DOGS
PROC SOC GERONTOL STUDY, OKAYAMA CITY, ONO PHARMACEUTICAL
CO, OSAKA:22-23(1972):(JAPANESE)

YAMAMOTO L   FEINDEL W   WOLFE L
HODGE C
REDUCTION OF CEREBRAL BLOOD FLOW BY PROSTAGLANDINS
ANN R COLL PHYS SURG CAN 5:64(1972)

YAMAMOTO YL   FEINDEL L   WOLFE LS
KATOH H   HODGE CP
EXPERIMENTAL VASOCONSTRICTION OF CEREBRAL ARTERIES BY
PROSTAGLANDINS
J NEUROSURG 37:385-397(1972)

YAMAMOTO YL   FEINDEL W   WOLFE LS
KATOH H   HODGE CP
EFFECTS OF PROSTAGLANDINS ON CEREBRAL BLOOD FLOW IN DOGS
PANMINERVA MED 13:171-172(1971)

YAMAMOTO YL   FEINDEL W   WOLFE LS
KATOH H   HODGE CP
EFFECTS OF PROSTAGLANDINS ON CEREBRAL BLOOD FLOW
EUR NEUROL 6:144-152(1971)

YAMAMOTO YL   FEINDEL W   WOLFE LS
HODGE CP
PROSTAGLANDIN INDUCED VASOCONSTRICTION OF CEREBRAL ARTERIES
AND ITS REVERSAL BY ETHANOL
ADV BIOSCI (SUPPL) 9:59(1972)

YAMAMOTO YL   FEINDEL W   WOLFE LS
HODGE CP
PROSTAGLANDIN INDUCED VASOCONSTRICTION OF CEREBRAL ARTERIES
AND ITS REVERSAL BY ETHANOL
ADV BIOSCI 9:359-367(1973)

YAMASHITA K   BLOOM G   FIELD JB
EFFECTS OF IONS ON THYROTROPIN AND PROSTAGLANDIN E1
STIMULATION OF GLUCOSE OXIDATION AND ADENYL CYCLASE-CYCLIC
AMP SYSTEM IN DOG THYROID SLICES
METABOLISM 20 : 943-953(1971)

YAMASHITA K   FIELD JB
STIMULATION OF CYCLIC 3',5'-GUANOSINE MONOPHOSPHATE (CGMP)
IN DOG THYROID SLICES BY ACETYLCHOLINE (ACH) AND SODIUM
FLUORIDE (NAF)
CLIN RES 20:445(1972)

YAMASHITA K   FIELD JB
ELEVATION OF CYCLIC GUANOSINE 3',5'-MONOPHOSPHATE LEVELS IN
DOG THYROID SLICES CAUSED BY ACETYLCHOLINE AND SODIUM
FLUORIDE
J BIOL CHEM 247:7062-7066(1972)

YAMATAKE Y   KATO J   TAKAGI K
ACTIONS OF PROSTAGLANDINS ON THE VESSELS OF DOG'S FORE-LEG
ABSTR 48TH KANTO REG MEET,JAP PHARMACOL SOC : 27(1973);
(JAPANESE)

YANG NST   MARSH JM   BEHRMAN HR
LEMAIRE WJ
ACCUMULATION OF PROSTAGLANDINS IN RABBIT GRAAFIAN FOLLICLES
ABSTR SOC GYNECOL INVEST 20TH ANNU MEET, ATLANTA:27(1973)

YANKEE EW   BUNDY GL
(15S)-15-METHYLPROSTAGLANDINS
J AM CHEM SOC 94:3651-3652(1972)

YANKEE EW   LIN CH   FRIED J
A GENERAL METHOD FOR THE CONVERSION OF F INTO E
PROSTAGLANDINS
CHEM COMMUN 1120-1121(1972)

YARGER WE   AYNEDJIAN HS   BANK N
A MICROPUNCTURE STUDY OF POSTOBSTRUCTIVE DIURESIS IN THE RAT
J CLIN INVEST 51:625-637(1972)

YATES FE   RUSSELL SM   MARAN JW
BRAIN-ADENOHYPOPHYSIAL COMMUNICATION IN MAMMALS
ANNU REV PHYSIOL 33:393-444(1971)

YOSHIDA Y   HAMADA F   KAINO H
SHA K   USUI T   KOSAKI T
NOAKI Y   UDAGAWA K   KOSUGE O
INOUE Y   IWASAKI H
THE EFFECT OF PROSTAGLANDIN F2A (PGF2A) ON INDUCTION OF
LABOR IN WOMEN PAST TERM
ABSTR JAP SOC OBSTET GYNECOL, KANTO PROVINCES MEET:24(1971);
(JAPANESE)

YOSHITOSHI Y   MASUYAMA Y
REVIEW OF RESEARCH ON ESSENTIAL HYPERTENSION
JAP J CLIN MED 31:3118-3123(1973);(JAPANESE)

YU SC   BURKE G
ANTIGENIC ACTIVITY OF PROSTAGLANDINS: SPECIFICITIES OF
PROSTAGLANDINS E1, A1 AND F2A ANTIGEN-ANTIBODY REACTIONS
PROSTAGLANDINS 2:11-22(1972)

YU SC   CHANG L   BURKE G
THYROTROPIN INCREASES PROSTAGLANDIN LEVELS IN ISOLATED
THYROID CELLS
ABSTR 4TH INT CONGR ENDOCRINOL, INT CONGR SER 256, EXCERPTA
MEDICA, AMST:192(1972)

YU SC   CHANG L   BURKE G
THYROTROPIN INCREASES PROSTAGLANDIN LEVELS IN ISOLATED
THYROID CELLS
J CLIN INVEST 51:1038-1042(1972)

YURUKOVA S   SOMOVA L
ON THE LIPID CYTOPLASMIC INCLUSIONS IN THE INTERSTITIAL
CELLS OF THE RENAL MEDULLA. I. A STUDY ON THE AMOUNT OF
LIPID INCLUSIONS IN ANIMALS WITH EXPERIMENTAL NEUROGENIC
HYPERTENSION
EKSP MED MORFOL 11:173-182(1972);(RUSSIAN)

ZADUNAISKY JA
INTERACTIONS OF EPINEPHRINE AND PROSTAGLANDINS ON THE
CORNEAL EPITHELIUM
AM J OPHTHALMOL 76:1020-1021(1973)

ZAKARIJA M   MCKENZIE JM   BASTOMSKY CH
STIMULATION OF THE ADENYL CYCLASE-CYCLIC AMP SYSTEM IN THE
THYROID OF THE RAT
ENDOCRINOLOGY 92:1349-1353(1973)

ZAMECNIK J   GRINWICH DL   ARMSTRONG DT
PREOVULATORY ELEVATION OF PROSTAGLANDIN F LEVELS IN RABBIT
OVARIAN FOLLICLES BY LUTEINIZING HORMONE
PROC CAN FED BIOL SOC 16:29(1973)

ZANELLA J JR   RALL TW
EVALUATION OF ELECTRICAL PULSES AND ELEVATED LEVELS OF
POTASSIUM IONS AS STIMULANTS OF ADENOSINE 3', 5'-
MONOPHOSPHATE (CYCLIC AMP) ACCUMULATION IN GUINEA-PIG BRAIN
J PHARMACOL EXP THER 186:241-252(1973)

ZARDAY Z   GOUAUX J   HAYS R
THE REGULATORY ROLE OF CALCIUM ON THE ACTION OF VASOPRESSIN
AND PROSTAGLANDIN E1 (PGE1) IN THE TOAD BLADDER
CLIN RES 19:744(1971)

ZARDAY Z   GOUAUX J   HAYS R
INHIBITION OF TOAD BLADDER ADENYL CYCLASE BY PROSTAGLANDIN
E1 (PGE1); THE CRITICAL ROLE OF MAGNESIUM AND THE
PGE1/VASOPRESSIN RATIO
ABSTR 5TH INT CONGR NEPHROL, MEX CITY:113(1972)

ZARDAY Z   GOUAUX J   HAYS RM
EFFECT OF PROSTAGLANDIN E1 (PGE1) ON TOAD BLADDER ADENYL
CYCLASE: THE CRITICAL ROLE OF MAGNESIUM
J CLIN INVEST 51:106A(1972)

ZAWILSKA K   IZRAEL V
PLATELETS AND INFLAMMATION
PATHOL BIOL (PARIS) 21:771-780(1973);(FRENCH)

ZAWILSKA K   TIMSIT J   GIROUD JP
CAEN JP
PLATELETS AND INFLAMMATION-11-INFLUENCE OF CARRAGEENAN ON
HUMAN AND RAT PLATELETS AGGREGATION-COMPARISON WITH THE
EFFECT OF PROSTAGLANDIN E2
PATHOL BIOL (PARIS) 21(SUPPL):57-59(1973);(FRENCH)

ZDICHYNEC B
POSSIBILITIES OF THE RELATIONSHIP OF PROSTAGLANDINS TO THE
ETIOLOGY OF ARTERIOSCLEROSIS
VNITRI LEK 19:506-507(1973);(CZECH)

ZEMAN A   SCHARMANN H
MASS SPECTROMETRY OF A COLLECTION OF LIPIDS. III.
FETTE SEIFEN ANSTRICHM 75:170-180(1973); (GERMAN)

ZENSER TV   CURNOW RT   DERUBERTIS FR
EFFECTS OF PROSTAGLANDIN E1 (PGE1) ON LIVER ADENYL CYCLASE
ACTIVITY (ACA)
CLIN RES 21:70(1973)

ZIA P   HORTON R
A CHROMATOGRAPHIC TECHNIQUE (SEPHADEX LH-20) FOR THE
SEPARATION OF MAJOR PROSTAGLANDINS
PROSTAGLANDINS 4:543-551(1973)

ZIBOH VA
DISCUSSION
IN: "PROSTAGLANDINS IN CELLULAR BIOLOGY," PW RAMWELL AND
BB PHARRISS, PLENUM PRESS, NY/LOND:285-291(1972)

ZIBOH VA
BIOSYNTHESIS OF PROSTAGLANDIN E2 IN HUMAN SKIN: SUBCELLULAR
LOCALIZATION AND INHIBITION BY UNSATURATED FATTY ACIDS AND
ANTI-INFLAMMATORY DRUGS
J LIPID RES 14:377-384(1973)

ZIBOH VA   BLANK H
THE ROLE OF PROSTAGLANDINS IN THE SKIN
HAUTARZT 24:519-522(1973); (GERMAN)

ZIBOH VA   HSIA SL
INHIBITION OF STEROL ESTER BIOSYNTHESIS BY PROSTAGLANDIN E2
(PGE2) IN SKIN OF ESSENTIAL FATTY ACID (EFA)-DEFICIENT RATS
ABSTR 162 NATL MEET AM CHEM SOC WASH BIOL:152(1971)

ZIBOH VA   HSIA SL
PROSTAGLANDIN E2: BIOSYNTHESIS AND EFFECTS ON GLUCOSE AND
LIPID METABOLISM IN RAT SKIN
ARCH BIOCHEM BIOPHYS 146:100-109(1971)

ZIBOH VA   HSIA SL
EFFECTS OF PROSTAGLANDIN E2 ON RAT SKIN: INHIBITION OF
STEROL BIOSYNTHESIS AND CLEARING OF SCALY LESIONS IN
ESSENTIAL FATTY ACID DEFICIENCY.
J LIPID RES 13:458-467(1972)

ZIBOH VA   MCELLIGOTT T   HSIA SL
PROSTAGLANDIN E2 BIOSYNTHESIS IN HUMAN SKIN: SUBCELLULAR
LOCALIZATION AND INHIBITION BY UNSATURATED FATTY ACIDS AND
ANTI-INFLAMMATORY AGENTS
ADV BIOSCI (SUPPL) 9:71(1972)

ZIBOH VA   MCELLIGOTT T   HSIA SL
PROSTAGLANDIN E2 BIOSYNTHESIS IN HUMAN SKIN: SUBCELLULAR
LOCALIZATION AND INHIBITION BY UNSATURATED FATTY ACIDS AND
ANTI-INFLAMMATORY AGENTS
ADV BIOSCI 9:457-460(1973)

ZIEVE PD   SCHMUKLER M
THE EFFECT OF CYCLIC AMP ON GLYCOGENOLYSIS AND GLYCOLYSIS IN
HUMAN PLATELETS
BIOCHIM BIOPHYS ACTA 252:280-284(1971)

ZIJLSTRA WG   BRUNSTING JR   TEN HOOR F
VERGROESEN AJ
PROSTAGLANDIN E1 AND CARDIAC ARRHYTHMIA
EUR J PHARMACOL 18:392-395(1972)

ZIMMERMAN BG   RYAN MJ   GOMER S
KRAFT E
EFFECT OF THE PROSTAGLANDIN SYNTHESIS INHIBITORS
INDOMETHACIN AND EICOSA-5,8,11,14-TETRAYNOIC ACID ON
ADRENERGIC RESPONSES IN DOG CUTANEOUS VASCULATURE.
J PHARMACOL EXP THER 187:315-323(1973)

ZINK HA   PODOS SM   BECKER B
INHIBITION BY IMIDAZOLE OF THE INCREASE IN INTRAOCULAR
PRESSURE INDUCED BY TOPICAL PROSTAGLANDIN E
NATURE (NEW BIOL) 245:21-23(1973)

ZOR U   BAUMINGER S   LAMPRECHT SA
KOCH Y   CHOBSIENG P   LINDNER HR
STIMULATION OF CYCLIC AMP PRODUCTION IN THE RAT OVARY BY
LUTEINIZING HORMONE: INDEPENDENCE OF PROSTAGLANDIN MEDIATION
PROSTAGLANDINS 4:499-507(1973)

ZOTOV YA   PINELIS VG   IVANOVA IA
MARKOVA KM
EFFECT OF DENERVATION OF THE AORTIC ARCH ON THE DEVELOPMENT
OF EXPERIMENTAL RENOVASCULAR HYPERTENSION
KARDIOLOGIIA 11:100-106(1971);(RUSSIAN)

ZUCKER MB
PROTEOLYTIC INHIBITORS, CONTRACT AND OTHER VARIABLES IN THE
RELEASE REACTION OF HUMAN PLATELETS
THROMB DIATH HAEMORRH 28:393-407(1972)

ZURIER RB   BALLAS M
PROSTAGLANDIN E1 (PGE1) SUPPRESSION OF ADJUVANT ARTHRITIS
HISTOPATHOLOGY
ARTHRITIS RHEUM 16:251-258(1973)

ZURIER RB   DUKOR P   WEISSMANN G
EFFECT OF CYCLIC AMP AND COLCHICINE ON HYDROLASE RELEASE
FROM PHAGOCYTES
CLIN RES 19:453(1971)

ZURIER RB   HOFFSTEIN S   WEISSMANN G
EFFECTS OF "INFLAMMATORY" AND "ANTI-INFLAMMATORY"
PROSTAGLANDINS ON LYSOSOMAL ENZYME DISCHARGE FROM HUMAN
LEUKOCYTES
ARTHRITIS RHEUM 15:133(1972)

ZURIER RB   HOFFSTEIN S   WEISSMANN G
MECHANISMS OF LYSOSOMAL ENZYME RELEASE FROM HUMAN
LEUKOCYTES- I-EFFECT OF CYCLIC NUCLEOTIDES AND COLCHICINE
J CELL BIOL 58:27-41(1973)

ZURIER RB   HOFFSTEIN S   WEISSMANN G
SUPPRESSION OF ACUTE AND CHRONIC INFLAMMATION IN
ADRENALECTOMIZED RATS BY PHARMACOLOGIC AMOUNTS OF
PROSTAGLANDINS
ARTHRITIS RHEUM 16:606-618 (1973)

ZURIER RB   HOFFSTEIN S   WEISSMANN G
PMN LEUCOCYTES AS SECRETORY ORGANS OF INFLAMMATION: EFFECTS
OF CYCLIC NUCLEOTIDES AND AUTONOMIC AGONISTS
CLIN RES 21:729(1973)

ZURIER RB   MITNICK H   BLOOMGARDEN D
WEISSMANN G
EFFECT OF BEE VENOM ON EXPERIMENTAL ARTHRITIS
ANN RHEUM DIS 32:466-470(1973)

ZURIER RB   QUAGLIATA F
SUPPRESSION OF ADJUVANT ARTHRITIS BY PROSTAGLANDIN E1
ARTHRITIS RHEUM 14:426-427(1971)

ZURIER RB   QUAGLIATA F
EFFECT OF PROSTAGLANDIN E1 ON ADJUVANT ARTHRITIS.
NATURE (NEW BIOL) 234 : 304-305(1971)

ZURIER RB   WEISSMAN G
EFFECTS OF PROSTAGLANDINS ON ENZYME RELEASE FROM HUMAN
LEUKOCYTES
BULL NY ACAD MED 48:1058(1972)

ZURIER RB   WEISSMANN G
INHIBITION BY PROSTAGLANDIN E1 OF LYSOSOMAL ENZYME DISCHARGE
FROM HUMAN POLYMORPHS
ARTHRITIS RHEUM 14:191-192(1971)

ZURIER RB   WEISSMANN G
INHIBITION BY PROSTAGLANDIN E1 (PGE1) OF LYSOSOMAL ENZYME
DISCHARGE FROM HUMAN POLYMORPHS
FED PROC 30:654(1971)

ZURIER RB   WEISSMANN G
EFFECT OF PROSTAGLANDINS UPON ENZYME RELEASE FROM
LYSOSOMES AND EXPERIMENTAL ARTHRITIS
IN: "PROSTAGLANDINS IN CELLULAR BIOLOGY," PW RAMWELL AND
BB PHARRISS, PLENUM PRESS, NY/LOND:151-172(1972)

ZURIER RB   WEISSMANN G   MITNICK H
EFFECTS OF PROSTAGLANDINS ON ENZYME RELEASE FROM HUMAN BLOOD
LEUCOCYTES ,
FED PROC 31:254(1972)

ZUSMAN RM
QUANTITATIVE CONVERSION OF PGA OR PGE TO PGB
PROSTAGLANDINS 1:167-168(1972)

ZUSMAN RM.   CALDWELL BV   MULROW PJ
SPEROFF L
THE ROLE OF PROSTAGLANDIN A IN THE CONTROL OF SODIUM
HOMEOSTASIS AND BLOOD PRESSURE
PROSTAGLANDINS 3:679-690(1973)

ZUSMAN RM   CALDWELL BV   SPEROFF L
RADIOIMMUNOASSAY OF THE A PROSTAGLANDINS
PROSTAGLANDINS 2:41-53(1972)

ZUSMAN RM   FORMAN BH   SCHNEIDER G
CALDWELL BV   SPEROFF L   MULROW PJ
THE EFFECT OF CHRONIC SODIUM LOADING AND SODIUM RESTRICTION
ON PLASMA AND RENAL CONCENTRATIONS OF PROSTAGLANDIN A IN
NORMAL WISTAR AND SPONTANEOUSLY HYPERTENSIVE AOKI RATS
CLIN SCI MOL MED 45(SUPPL 1):325S-329S(1973)

ZUSMAN RM   SCHNEIDER G   FORMAN B
MULROW PJ
STUDIES ON PROSTAGLANDIN A IN THE SPONTANEOUSLY HYPERTENSIVE
RAT
CIRCULATION (SUPPL II) 46:102(1972)

ZUSMAN RM   SPECTOR D   CALDWELL BV
SPEROFF L   SCHNEIDER G   MULROW PJ
THE EFFECT OF CHRONIC SODIUM LOADING AND SODIUM RESTRICTION
ON PLASMA PROSTAGLANDIN A, E AND F CONCENTRATIONS IN NORMAL
HUMANS
J CLIN INVEST 52:1093-1098(1973)

SUBJECT INDEX

AUTHOR INDEX